THE BIODIVERSITY CRISIS AND CRUSTACEA

CRUSTACEAN ISSUES

12

General editor
FREDERICK R. SCHRAM
Zoological Museum, University of Amsterdam

CRC Press
Taylor & Francis Group
Boca Raton London New York

CRC Press is an imprint of the
Taylor & Francis Group, an **informa** business
A BALKEMA BOOK

THE BIODIVERSITY CRISIS AND CRUSTACEA

PROCEEDINGS OF THE FOURTH INTERNATIONAL CRUSTACEAN CONGRESS, AMSTERDAM, NETHERLANDS, 20-24 JULY 1998, VOLUME 2

Edited by

J. CAREL VON VAUPEL KLEIN
Division of Systematic Zoology, Leiden University

FREDERICK R. SCHRAM
Zoological Museum, University of Amsterdam

CRC Press
Taylor & Francis Group
Boca Raton London New York

CRC Press is an imprint of the
Taylor & Francis Group, an **informa** business

A BALKEMA BOOK

Volume 12 of Crustacean Issues is Volume 2 of the Fourth International Crustacean Congress, Amsterdam, Netherlands, 20-24 July 1998

CRC Press
Taylor & Francis Group
6000 Broken Sound Parkway NW, Suite 300
Boca Raton, FL 33487-2742

Contents

2 Ecology and Behavior

3 Toxicity and Physiology

4 Reproduction

5 Larvae

6 Fisheries and Aquaculture

7 *Biodiversity*

Preface

'Crustaceans and the Biodiversity Crisis' was the theme of the Fourth International Crustacean Congress held in Amsterdam, The Netherlands, from 20 to 24 July, 1998. The congress also hosted the Third European Crustacean Conference, the 1998 Summer Meeting of The Crustacean Society, the Eighth Meeting of GEREC (Groupe d'Études et de Réflections sur l'Évolution des Crustacés), and the Third International Workshop on Sea Lice.

The congress was sponsored by a variety of bodies including: Stichting Crustacea (Foundation of Dutch Carcinologists), The Brazilian Society of Carcinology, The Carcinological Society of Japan, The Crustacean Society, The Crustacean Society of China, the Groupe d'Études et de Réflections sur l'Évolution des Crustacés (GEREC), the Division of Systematic Zoology of Leiden University, the Working Group on the Biology of Sea Lice, the Zoological Museum of the University of Amsterdam, the Board of Regents of the University of Amsterdam, the Rijksinstituut voor de Zuivering van Afvalwater (Institute for Inland Water Management and Waste Water Management, RIZA), the Royal Dutch Academy of Sciences (KNAW), the Dutch Ministries of Foreign Affairs and Developmental Cooperation, The European Union, and the Mayor and Aldermen of the City of Amsterdam.

In all, there were slightly over 500 registered participants from nations all around the world, together presenting 14 invited plenary lectures, some 250 contributed oral papers, and about 270 poster papers. The official Proceedings are being published in two volumes; this is the second, while number 1 constitutes a Special Volume published by Koninklijke Brill N.V. Academic Publishers of Leiden. In these two volumes, a total of some 156 contributions have been compiled. The papers have been grouped according to the main sub-themes of the congress, and in this volume the papers that primarily deal with the biodiversity issues of Invasive Crustacea, Ecology and Behavior, and Fisheries and Aquaculture are to be found. In addition, a few contributions from the other sub-themes are also included dealing with Diversity in Space and Time (including Systematics, Morphology and Anatomy, Phylogeny, and Palaeontology), General Biogeography, Larvae, and Physiology and Biochemistry (including Molecular Biology and Genetics).

Furthermore, the contributions from the Sea Lice Conference are being published in the journal *Contributions to Zoology* (Amsterdam), volume 69. Also, the papers from the Jan H. Stock Memorial Symposium will be published in a special Jan H. Stock memorial issue of the journal *Crustaceana* 72(8), November, 1999.

We hope the present volume may contribute to an intensification of research efforts on all aspects of crustacean biology and that such, in its turn, may serve to enhance the chances for crustacean biodiversity to be preserved well into the 21st century.

We want to take this opportunity to offer a special thanks to Dr Matthijs van Couwelaar, University of Amsterdam, who was particularly helpful in processing and editing the original congress abstracts, as well as with the editing and formatting of the proceedings manuscripts in this volume.

J.C. von Vaupel Klein, Leiden
Frederick R. Schram, Amsterdam
May, 1999

1 Invasive crustaceans and distribution

Ecological impact of crustacean invaders: General considerations and examples from the Rhine River

GERARD VAN DER VELDE, SANJEEVI RAJAGOPAL & BARRY KELLEHER
Laboratory of Aquatic Ecology, University of Nijmegen, Nijmegen, Netherlands
ILONA B. MUSKÓ
Balaton Limnological Research Institute, Hungarian Academy of Sciences, Tihany, Hungary
ABRAHAM BIJ DE VAATE
Institute for Inland Water Management and Waste Water Treatment, Lelystad, Netherlands

ABSTRACT

The paper reviews the impact of and success factors for crustacean invaders in aquatic systems. As an example, recent invasions of crustaceans in the Rhine River are considered in more detail. In this river, 19 crustacean invaders have been recorded since 1800, with the rate of invasion increasing in recent years. Two species (rule of ten) appear to be most successful: the amphipods *Gammarus tigrinus* and *Corophium curvispinum*. Their success is not due to one factor but to a combination of behavioural, physiological and environmental factors. The outcome of interspecies competition may also be the result of this combination of factors. *G. tigrinus* has been found to be very resistant to invasions by other gammarids, but *Dikerogammarus villosus* may be able to compete with it successfully under present day conditions. When in very high densities, *C. curvispinum* is able to change its habitat conditions, which appears to depend on water flow, chlorophyll-*a* concentrations and water quality. At these high densities, it accumulates so much mud on the stones that the macroinvertebrate community changes to a mud community with a reduced range of species compared to the previous situation. *Dreissena polymorpha* is outcompeted under such circumstances, a case of spatial competition by 'swamping'. Both successful amphipods exert extensive influence on the food chain, not only horizontally (competition, swamping), but also vertically, with predators switching diets at each successful invasion.

1 INTRODUCTION

Invasion by exotic species is considered to be one of the major threats to biodiversity worldwide, with preponderance on plants and vertebrates. Two aquatic macroinvertebrates currently receiving much attention with respect to invasions are the predatory cladoceran *Bythotrephes cederstroemi* and the Zebra mussel *Dreissena polymorpha*, both invading the Great Lakes in the USA (Morton 1997). Historically, many unintentional and intentional introductions or immigrations of such aquatic organisms have been reported in the literature. Most introductions are not very successful, in the sense that those species are caught only incidentally, are restricted to certain habitats, often occupying under-utilized niches like cooling water discharge areas, or showing populations remaining small with negligible effect on the ecosystem. Sometimes, however, such species can threaten endangered or protected indigenous species, e.g. by hybridization, predation or competition. Many invasions, however, are reflections of other changes rather than being agents of change themselves and in such cases, biological invasions can be regarded as indicators of environmental change (Vitousek et al. 1996). Biological invasions can also lead to interactions between species that often have never met before and to replacement of autochthonous species by exotic relatives (Conlan 1994). An explosive and catastrophic growth in the population size of invaders can cause drastic changes in an ecosystem as a whole due to depressed local populations, individual species extinction and influence on vertical and horizontal food-chain processes, even leading to a restructuring of the ecosystem. Even then, in most cases, there is a lack of evidence that such an invasion has a significant ecological impact. This is due to the unpredictable nature of invasions, which means that pre-invasion data are usually not available. The speed with which changes can occur is such that changes immediately after the start of the invasion are missed. The impact of a mass invader is usually large at its population peak, after which the population fluctuates and other factors influencing the communities may mask the effect of the invasion.

Mass invasions attract special attention when they cause economic problems, while concern is also aroused by related problems of ecological changes, possible genetic influences on native species and the introduction into native populations of non-endemic pathogens; even human health may be endangered (Sinderman 1992). Economic problems caused by crustaceans are mostly related to professional and recreational fishery (crayfish, Chinese mitten crab (*Eriocheir sinensis*), Spiny water flea (*Bythotrephes*), aquaculture (crabs), damage to crops, especially rice, by crayfish, damage to ditch walls, stream banks and shorelines (burrowing crayfish, the isopod *Sphaeroma*, Chinese mitten crab) and biofouling (barnacles, woodboring isopods). The Chinese mitten crab is also a second intermediate host to the Oriental liver fluke (*Paragonimus westermanii*); eating raw or inadequately cooked infected crabs can infect the human host. Some snail species are the first intermediate hosts; if these species are not present the cycle can not be completed (Cohen & Carlton 1995).

Among the invertebrates, crustacean species are often important and successful invaders; their success is only rivaled by some mollusk species (Morton 1997). Kinzelbach (1995) mentions 20 crustacean species (four of them classified as successful to extremely successful) as exotic invaders in Central and Western European inland waters, as well as 15 mollusk species (seven of them successful to extremely successful) and 8 species belonging to other groups of macroinvertebrates. Glassner-Shwayder (1996) mentions five crustaceans as belonging to the most important invaders with respect to ecological and economic impact in the United States. For British marine waters, Clare Eno et al. (1997) recorded 30 exotic macroinvertebrate species, of which 9 were Mollusca, 8 were Annelida and 7 were Crustacea.

Cohen & Carlton (1995) compared the numbers of introduced marine, estuarine and freshwater species in four regional studies and found Crustacea always to rank number one or two in the list of introduced macroinvertebrate species. In freshwater areas (Great Lakes, Hudson River), mollusk species outnumber crustacean species, while in estuarine and marine environments (Baltic and Swedish coast, San Francisco Bay) crustacean species outnumber mollusks.

2 SUCCESS FACTORS

Exotic crustacean invasions occur only when natural barriers to dispersal are circumvented due to human activities. Human impact can lift barriers for dispersal in several ways:
   – By the construction of canals, rivers and lakes, linking large catchment areas,
   – By the construction of a web of smaller waters like ditches and ponds which can facilitate further local dispersal,
   – By the construction of harbors at river mouths and the discharge of ballast water by ships, allowing the introduction of estuarine and other brackish water tolerant species,
   – By the increase in transcontinental as well as local shipping, especially over the last fifty years, that has increased the number of immigrant species (Cohen & Carlton 1998),
   – By intentional introduction of species, for example as fish food or for aquaculture purposes,
   – Through escapes from aquaculture, aquaria, garden ponds etc., and
   – By trade and transport, which may facilitate the dispersal of species, either unintentionally, for example the non-target species transported with plants and animals transplanted for aquaculture ('hitchhikers'), or intentionally, for example, species collected and discarded afterwards.

3 WHAT DETERMINES SUCCESS?

In spite of increased opportunities for the immigration of exotic species, not all are successful. To explain this phenomenon, the rule of ten was formulated (Williamson 1996). This states that only ten percent of the established immigrant species can develop populations dense enough to turn them into pests. Species that were successful during a previous invasion on one continent have also proved successful as they entered new areas on another continent. This means that such species become cosmopolitans, making ecosystems all over the world more and more similar with respect to species composition.

   Successful invasions by new species usually proceed via a number of stages. The initial introduction must occur across a natural barrier. Enormous spatial leaps may be made, e.g. from one continent to another. Subsequently, the invasive species must be able to settle or adapt to the habitat in a new area that is conducive to its survival. As a next stage, it must be able to reproduce and complete its life cycle. The species will then spread over the new range successfully, usually by gradual local dispersal but also by jumps aided by transport means. It will show exponential population growth, to densities sometimes higher than ever recorded, followed by somewhat lower steady-state fluctuations.

Attempts have been made to define the possible factors likely to be important for successful invasions. These include:

– Invader properties, adaptations, genetic characteristics, mode of reproduction, growth rate and dispersive capability,

  – Competition, predation, diseases and parasites, and

  – Climate match, habitat modification, vacant niche and ancestral habitat.

With respect to invader properties, the qualities of successful crustacean invaders are generally in accordance with those of so-called opportunistic species, often also termed r-strategists. Morton (1997) summed up the qualities for mollusk invaders. In a similar way, we can list the qualities for crustacean invaders:

  – Short life span and generation time,

  – Rapid growth with early sexual maturity,

  – High fecundity,

  – Female able to colonize alone,

  – Larger than most relatives,

  – Euryoecious (the ability to colonize a wide range of habitat types),

  – Eurytopic (a wide range of physiological tolerances),

  – Wide genetic variability and phylogenetic plasticity,

  – Gregarious behaviour,

  – Suspension feeding, omnivorous,

  – An ability to repopulate defaunated habitats, following population crashes caused by extreme disturbances,

  – Some form of association with human activities.

Success by exotic invaders can also be explained by a lack of native competitors, predators, diseases and/or parasites, allowing uncontrolled population expansion. Lack of predators and parasites can make the performance of crustacean invaders much better than that of their native conspecifics, as has been found for the European green crab (*Carcinus maenas*) on the east coast of the USA (Torchin et al. 1998). On the other hand, exotic invaders may import diseases to which they are more tolerant than their indigenous relatives. An example is the introduction into Europe of a fungus called crayfish pest (*Aphanomyces astaci)* via the American crayfish, *Orconectes limosus*. The indigenous crayfish *Astacus astacus* turned out to be very susceptible to this pest.

Furthermore, immigrant species appear to be successful especially in regions with a climate similar to where they originate. Range extension can also be aided by human modification of habitats, which may mean that new species are preadapted due to the resulting similarity to their ancestral habitat and so colonize very quickly, while indigenous species simultaneously are weakened by the modifications. Sometimes a new niche is exploited because no equivalent indigenous species are present. Pollution calamities in rivers, for example, can create open space for new colonizers, giving newcoming species a great opportunity to colonize. Persistent pollution can also create such empty niches.

Rivers provide interesting case studies with respect to invasions of exotic invaders because they are migration channels that have been modified in several ways. Exotic invaders can intrude in rivers from various sources: from estuaries or harbors in the upstream direction, or from one river catchment to another in the downstream direction. This process of internationalization of the actors in the food web is occurring in rivers and lakes all over the world, often involving the same or related invading species.

4 THE RHINE: A MODIFIED RIVER SYSTEM

The Rhine is a regulated ecosystem, having long been subject to anthropogenic influences on its physical, chemical and biotic characteristics (Van der Velde et al. 1991, Admiraal et al. 1993). The physical variety of the Rhine has been greatly reduced by changes to the river's geomorphology and hydrology. The river, especially the Lower Rhine, is now extremely normalized with high current velocities causing very low connectivity with its floodplain waters. The main river bed has been fixed by stone groynes, and the banks of the various branches have also often been protected against erosion by stones. Water quality has been degraded over the last century; the Rhine is highly eutrophic, due to nutrient-rich discharges from agricultural, municipal and other sources in the entire drainage area, leading to high phytoplankton productivity in the Lower Rhine (De Ruyter van Steveninck et al. 1992). Temperature, salinity and pollutant concentrations have also increased (Den Hartog et al. 1992). Recent improvements in water quality, such as higher oxygen levels and reduced concentrations of ammonium, heavy metals (Table 1) and cholinesterase inhibitors, have led to the establishment of more and more exotic invaders, which now dominate the Rhine, having replaced indigenous species (Kureck 1992, Bij de Vaate & Greijdanus-Klaas 1993, Van den Brink et al. 1996).

Furthermore, the estuary of the rivers Rhine and Meuse in the Netherlands has been closed by dams provided with sluice gates that caused the disappearance of the former brackish water gradient (Smit et al. 1997). The disappearance of estuarine conditions caused large changes in the communities, such as the amphipod fauna (Platvoet & Pinkster 1995). The former estuary has filled up with Rhine water, and only two amphipod species were able to profit from the new situation: *Gammarus tigrinus* and *Corophium curvispinum*, both exotic species.

Shipping traffic on the Rhine is very intensive; it is a major transport route from the harbors of Rotterdam and Amsterdam to Germany, France and Switzerland. Shipping provides great opportunities for the transport for immigrant estuarine species from the harbors upstream. In the interest of shipping, the Rhine has also been connected upstream to eastern rivers (Ems, Weser, Oder, Wista) by canals like the Mittelland Canal. However, one of the most important connections is the Main-Danube Canal, a canal completed and officially opened in September 1992, which links the Rhine catchment area via the Main tributary with the Danube River (Tittizer 1996a) (Fig. 1). This has created a migration route from the Black Sea to the North Sea, allowing Ponto-Caspian species to migrate westwards, while species from the Rhine can invade the Danube, such as the freshwater shrimp *Atyaephyra desmaresti* (Tittizer 1996b, Moog et al. 1999, Wittmann et al. 1999). The water transfer is from the Danube to the Main, which is towards the Rhine, but vessels can also transport exotic invaders in the opposite direction.

Sometimes a pollution calamity can free a niche, which may be of advantage for new colonizers. In the Rhine, a chemical spill known as the Sandoz accident occurred in November 1986 (Heil 1990). From Basel downstream, a stretch of 400 km was vir-

Table 1. Water quality data (annual mean values) for the Rhine at Lobith (Rijkswaterstaat, RIWA).

	1987			1988			1989		
	mean	min.	max.	mean	min.	max.	mean	min.	max.
General									
Discharge (m^3/sec)	2860	1370	7640	2820	1180	10270	1820	855	4530
Water temperature (°C)	12.3	0.8	22.1	13.7	5.9	23.1	14.1	5.7	23.8
Dissolved oxygen (mg/l)	9.3	6.2	13.0	9.3	6.2	12.3	9.2	7.1	11.4
Oxygen saturation (%)	85	70	126	88	50	111	89	68	126
pH	7.7	7.3	8.5	7.7	7.5	8.0	7.7	7.4	8.5
Suspended matter (mg/l)	40	19	130	45	21	112	34	23	57
Chlorophyll-a (µg/l)	12	BDL	63	22	BDL	92	30	BDL	150
Nutrients									
Ammonium (mg/l N)	0.56	0.07	2.33	0.34	BDL	0.76	0.34	0.02	1.11
Total nitrogen (mg/l N)	1.5	0.8	3.3	1.4	1.0	1.8	1.4	1.0	2.4
Nitrate (mg/l N)	4.0	2.7	5.1	3.7	2.6	4.9	4.5	3.1	5.9
Orthophosphate (mg/l P)	0.19	0.11	0.35	0.15	0.05	0.26	0.13	0.02	0.21
Total phosphate (mg/l P)	0.38	0.26	0.71	0.34	0.25	0.46	0.34	0.24	0.43
Inorganic substances									
Conductivity (20°C, mS/m)	NA			80	48	120	89	54	135
Chloride (mg/l)	140	50	212	150	67	214	182	95	321
Sulphate (mg/l)	66	51	84	63	47	89	NA		
Sodium (mg/l)	81	51	114	79	39	118	98	52	178
Potassium (mg/l)	5.7	3.6	7.5	6.0	4.1	7.8	7.0	5.3	10.7
Calcium (mg/l)	76	61	95	81	45	106	84	69	110
Magnesium (mg/l)	11	8.8	13	11	8.3	14	12	9.7	14
Total hardness (mmol/l)	NA			NA			NA		
Heavy metals									
Iron (mg/l)	1.41	0.72	4.65	1.36	0.79	4.13	1.07	0.77	2.10
Cadmium (µg/l)	0.10	0.04	0.44	0.10	0.04	0.38	0.11	0.05	0.27
Copper (µg/l)	5.3	3.4	10.9	5.1	0.7	10.1	6.0	4.0	8.3
Mercury (µg/l)	0.05	BDL	0.29	0.05	BDL	0.18	0.05	0.02	0.16
Lead (µg/l)	4.0	1.8	10.8	3.4	0.6	6.8	4.9	2.2	8.2
Nickel (µg/l)	3.9	1.7	8.2	3.7	2.2	5.9	4.6	2.5	8.0
Selenium (µg/l)	NA			NA			NA		
Zinc (µg/l)	34	4	73	34	13	57	31	19	46

NA = data not available, BDL = below detectable level.

tually cleared of macroinvertebrates (Güttinger & Stumm 1990). Measures have subsequently been taken to improve the water quality and to provide ecological rehabilitation (IRC 1987, 1992). This gradual improvement of the water quality has led to a greater biodiversity in the river.

5 CRUSTACEAN INVADERS IN THE RHINE

Remaining pollution levels and habitat changes have in recent years led to rising numbers of crustacean immigrants in the Rhine (Van den Brink et al. 1990). Major potential causes include the altered ionic composition of the river water as a result of mining activities, especially discharge from French potassium mines; thermal pollu-

Table 1. Continued.

	1990			1991			1992		
	mean	min.	max.	mean	min.	max.	mean	min.	max.
General									
Discharge (m³/sec)	1860	900	7030	1750	790	6710	2010	870	4920
Water temperature (°C)	14.3	2.7	23.0	13.9	3.4	23.9	13.7	5.3	23.8
Dissolved oxygen (mg/l)	9.8	7.5	11.6	9.4	6.7	12.7	9.6	6.9	11.7
Oxygen saturation (%)	94	70	118	91	73	122	89	79	102
pH	7.7	7.5	8.2	7.8	7.5	8.4	7.6	7.3	7.8
Suspended matter (mg/l)	34	BDL	130	32	10	53	29	19	60
Chlorophyll-a (µg/l)	23	BDL	105	27	BDL	97	13	BDL	52
Nutrients									
Ammonium (mg/l N)	0.39	0.04	1.16	0.42	0.07	1.18	0.30	0.09	1.11
Total nitrogen (mg/l N)	1.3	0.7	2.1	1.5	0.1	2.1	1.2	0.6	2.2
Nitrate (mg/l N)	4.1	2.6	6.1	3.8	2.2	5.0	3.9	2.6	5.3
Orthophosphate (mg/l P)	0.11	BDL	0.20	0.08	0.02	0.17	0.09	0.02	0.17
Total phosphate (mg/l P)	0.30	0.24	0.43	0.27	0.18	0.46	0.24	0.17	0.38
Inorganic substances									
Conductivity (20°C, mS/m)	89	50	130	90	48	125	82	54	115
Chloride (mg/l)	187	85	266	199	89	322	163	74	266
Sulphate (mg/l)	NA			NA			NA		
Sodium (mg/l)	99	43	154	105	41	164	87	42	139
Potassium (mg/l)	7.0	4.6	9.9	7.1	4.3	9.6	6.4	4.4	9.0
Calcium (mg/l)	88	66	98	89	65	100	80	58	100
Magnesium (mg/l)	11	8.4	13	12	8.3	14	10	9.0	13
Total hardness (mmol/l)	NA			NA			2.4	1.8	3.1
Heavy metals									
Iron (mg/l)	1.13	0.58	2.70	0.99	0.57	1.85	0.96	0.44	1.73
Cadmium (µg/l)	0.10	0.05	0.21	0.10	BDL	0.24	0.06	0.04	0.09
Copper (µg/l)	5.7	0.7	11.0	6.5	4.5	8.7	5.8	4.3	9.1
Mercury (µg/l)	0.04	0.02	0.11	0.05	BDL	0.14	0.05	0.02	0.09
Lead (µg/l)	5.1	2.0	12.7	4.6	BDL	6.7	4.0	2.6	6.1
Nickel (µg/l)	3.8	1.9	6.2	4.6	3.3	6.4	4.0	3.0	7.4
Selenium (µg/l)	NA			< 2.0	< 2.0	< 2.0	< 2.0	< 2.0	< 2.0
Zinc (µg/l)	46	14	184	30	6	98	23	15	38

NA = data not available, BDL = below detectable level.

tion canceling out the effects of severe winters; and the fact that crustaceans show greater tolerance to all sorts of pollution than insects and, to a lesser degree, mollusks, as has often been demonstrated for originally estuarine and brackish water species.

In the near future, many species can be expected to migrate to the Rhine via the Main-Danube Canal, especially those Ponto-Caspian species which are already observed in the Upper Danube, like the amphipods *Obesogammarus obesus*, *Echinogammarus trichiatus* and *Corophium sowinskyi* (Jazdzewski & Konopacka 1993, Weinzierl et al. 1996, 1997). This route to the Rhine has already been successfully traversed by some amphipod species such as *Dikerogammarus haemobaphes* and *D. villosus*, as well as the isopod *Jaera istri* and the mysid *Limnomysis benedeni*

Table 1. Continued.

	1993			1994			1995		
	mean	min.	max.	mean	min.	max.	mean	min.	max.
General									
Discharge (m^3/sec)	2010	1200	10940	2530	1100	8100	2790	1080	11890
Water temperature (°C)	13.8	6.0	22.8	15.2	4.7	26.0	13.4	5.3	24.4
Dissolved oxygen (mg/l)	10.0	8.2	11.4	9.5	6.3	11.9	10.0	8.1	11.9
Oxygen saturation (%)	93	77	149	97	75	145	98	86	120
pH	7.8	7.6	7.9	7.7	7.3	8.6	7.7	7.5	7.9
Suspended matter (mg/l)	33	18	130	31	5	98	35	14	140
Chlorophyll-a (µg/l)	19	BDL	135	6.8	BDL	41	8.0	BDL	57
Nutrients									
Ammonium (mg/l N)	0.25	0.09	0.68	0.17	0.02	0.51	0.15	BDL	0.56
Total nitrogen (mg/l N)	1.2	0.3	2.7	1.1	0.4	4.9	0.9	0.2	2.5
Nitrate (mg/l N)	3.7	2.4	5.2	3.4	2.4	4.3	3.3	2.2	4.5
Orthophosphate (mg/l P)	0.08	BDL	0.12	0.08	BDL	0.14	0.08	BDL	0.17
Total phosphate (mg/l P)	0.22	0.15	0.58	0.21	0.10	0.44	0.20	0.10	0.46
Inorganic substances									
Conductivity (20°C, mS/m)	80	58	100	74	47	90	71	38	99
Chloride (mg/l)	144	85	194	120	64	170	116	49	190
Sulphate (mg/l)	65	45	93	63	36	80	59	42	83
Sodium (mg/l)	81	47	106	70	33	90	67	28	100
Potassium (mg/l)	6.0	4.3	7.5	5.4	3.9	6.7	5.4	3.9	7.2
Calcium (mg/l)	80	60	97	75	59	91	76	51	95
Magnesium (mg/l)	11	8.8	14	11	8.5	18	11	7.3	15
Total hardness (mmol/l)	2.5	1.9	3.0	2.3	1.8	2.9	2.3	1.6	2.9
Heavy metals									
Iron (mg/l)	1.05	0.55	4.70	1.10	0.56	2.95	1.33	0.60	6.10
Cadmium (µg/l)	0.05	BDL	0.11	0.07	BDL	0.27	0.07	0.04	0.23
Copper (µg/l)	5.2	3.6	8.8	4.9	2.5	8.2	4.6	3.2	10.2
Mercury (µg/l)	0.03	0.02	0.08	0.02	BDL	0.06	0.02	BDL	0.09
Lead (µg/l)	4.6	2.3	17.8	4.3	1.7	9.6	3.7	0.9	13.5
Nickel (µg/l)	3.8	2.3	10.0	3.8	1.6	8.7	2.9	1.7	8.0
Selenium (µg/l)	< 2	< 2.0	< 2.0	BDL	BDL	2.0	BDL	BDL	BDL
Zinc (µg/l)	23	10	38	25	5	105	21	10	58

NA = data not available, BDL = below detectable level.

(Schleuter et al. 1994, Bij de Vaate & Klink 1995, Schleuter & Schleuter 1995, Tittizer 1995, Reinhold & Tittizer 1998).

The following exotic crustaceans, grouped by origin, have been found in the Rhine.

5.1 Northern-Europe

5.1.1 *Bythotrephes longimanus (Cladocera)*

This species was recorded for the first time in 1987 in the untreated water storage reservoirs of the drinking water companies de 'Brabantse Biesbosch' in the Netherlands. Since then it has been found in numbers reaching 10,000 per m^3. From 1988, it

Table 1. Continued.

	1996			1997		
	mean	min.	max.	mean	min.	max.
General						
Discharge (m^3/sec)	1760	1010	4360	1910	930	6930
Water temperature (°C)	13.1	2.1	23.2	14.1	1.1	24.5
Dissolved oxygen (mg/l)	10.5	7.1	13.8	10.1	7.0	14.2
Oxygen saturation (%)	98	84	120	99	82	127
pH	7.8	7.4	8.1	7.8	7.6	8.5
Suspended matter (mg/l)	29	3	120	31	1	310
Chlorophyll-a (µg/l)	8.6	BDL	57	9.6	BDL	63
Nutrients						
Ammonium (mg/l N)	0.22	0.03	0.98	0.10	0.01	0.39
Total nitrogen (mg/l N)	0.8	0.1	1.6	0.8	0.1	2.7
Nitrate (mg/l N)	3.6	2.3	4.8	3.2	0.9	4.5
Orthophosphate (mg/l P)	0.10	0.03	0.14	0.10	0.01	0.20
Total phosphate (mg/l P)	0.22	0.13	0.51	0.21	0.11	0.46
Inorganic substances						
Conductivity (20°C, mS/m)	NA			NA		
Chloride (mg/l)	155	71	249	147	58	246
Sulphate (mg/l)	65	43	88	60	45	86
Sodium (mg/l)	85	45	127	80	44	112
Potassium (mg/l)	6.1	4.2	7.6	6.3	4.5	8.0
Calcium (mg/l)	82	69	98	85	69	105
Magnesium (mg/l)	11	8.9	14	12	9.1	13
Total hardness (mmol/l)	2.5	2.1	3.0	2.6	2.1	3.2
Heavy metals						
Iron (mg/l)	1.23	0.51	3.08	1.16	0.80	2.07
Cadmium (µg/l)	0.07	0.01	0.16	0.06	0.01	0.17
Copper (µg/l)	5.2	2.3	7.0	5.0	3.4	8.3
Mercury (µg/l)	0.04	0.01	0.22	0.04	0.02	0.07
Lead (µg/l)	4.5	1.9	10.8	3.9	2.1	8.3
Nickel (µg/l)	3.1	1.6	6.2	3.5	1.6	5.9
Selenium (µg/l)	0.4	BDL	1.1	0.2	BDL	0.4
Zinc (µg/l)	41	2	100	24	10	60

NA = data not available, BDL = below detectable level.

has been found in Lake Volkerak-Zoommeer. Peak abundances of *Bythotrephes longimanus* and *Leptodora kindti*, which is a native predatory cladoceran, seem to lead to drastic decreases in the abundance of daphnids, especially in July. Copepod densities were also slightly lower after the invasion. *Leptodora* abundance is not affected by *Bythotrephes*. Perhaps both species are controlled by fish. The impact of *B. longimanus* has not been as profound as that of *B. cederstroemi* in the Great Lakes of the USA.

B. *longimanus* must have reached the Netherlands via long distance dispersal, because the species is present in the catchment area of the Rhine in the Alps (though not that of the Meuse), but also in north-eastern Germany, Poland, Belarus, the Baltic states, Scandinavia (Norway and Sweden), Great Britain, and Ireland. In the Rhine

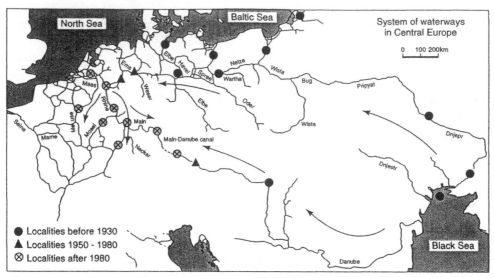

Figure 1. Major rivers systems to the east of the Rhine, with canals linking the rivers and the main dispersal routes of *Corophium curvispinum* (from Tittizer et al. 1994).

catchment area, *B. longimanus* is known to be present in Lake Constance (Bodensee), which means that specimens drifting in the river could arrive in the Netherlands within 5-10 days. A few specimens have been found in the Dutch part of the Maas River (Meuse). Although it cannot be ruled out that water birds have introduced the species into the basins, it is likely that resting eggs were transported downstream and hatched in standing waters like the storage basins and the slow flowing parts of the river. *Bythotrephes* specimens caught in the river had only one lateral spine on the caudal spine, indicating that they had not yet molted. *B. longimanus* can reproduce parthenogenetically.

Large populations of *B. longimanus* occur in the storage basins and in Lake Volkerak-Zoommeer, both large and deep artificial lakes (5-27 m) containing river water. These man-made lakes house few predatory fish, so that zooplankton is well developed. In spite of their eutrophic character, sedimentation processes make the water clear, which makes it easier for *B. longimanus* to detect its prey. This type of water did not occur in the Netherlands in the past, so *B. longimanus* has made use of a totally new, man-made opportunity to settle (Ketelaars & Van Breemen 1993, Ketelaars et al. 1993, Ketelaars & Gille 1994).

5.2 *Southern Europe*

5.2.1 *Atyaephyra desmaresti (Decapoda)*

This freshwater shrimp originates from the Mediterranean, from where it has spread actively over Europe by using the canals that connect European river basins, including that of the Rhine. The species invaded the Main-Danube Canal in 1994 and the Danube River basin in 1998 (Moog et al. 1999, Wittmann et al. 1999). In the Netherlands, it was first noted in 1915; it is now widespread in the rivers, canals and lakes. In harsh winters, the numbers of records are low, reflecting its southern origin

(Van den Brink & Van der Velde 1986b). The species is phytophilous, eurythermous and brackish water tolerant (Tittizer 1996b).

5.2.2 *Echinogammarus berilloni (Amphipoda)*

This amphipod species originates from the Mediterranean. According to Pinkster (1973), it is a typical species of middle courses of streams and rivers. It is an active migrant, which has made use of canals to reach the Rhine. It is found in the Rhine from Düsseldorf to Basel, where it occurs in low densities, and in the Mosel (Moselle), a tributary of the Rhine. It is salt tolerant and eurythermous (Tittizer 1996b).

5.2.3 *Orchestia cavimana (Amphipoda)*

This species from Mediterranean-Ponto-Caspian origin was already recorded in the Rhine (Waal) near Tiel in 1898, from where it spread over the Netherlands, where it is still a common species. Via the Scheldt (1906) in Belgium, it reached the Seine (1913) from where it migrated via the Rhine-Marne Canal to the Mosel and hence to the Upper Rhine (1937). From there, it spread downstream to the Middle and Lower Rhine in 1962-1971 (Kinzelbach 1972). Its mode of dispersal is unknown. The species inhabits the littoral margins of the river and lakes ('spray zone') and flood marks where organic matter accumulates between stones or macrophytes that have washed ashore, etc. Its food consists of organic matter like macrophyte tissue, sponge tissue, diatoms, bluegreen and green algae. Organic water pollution seems to be a factor favoring this species, although it behaves as a terrestrial species and the animals only seek the water to escape from enemies. In fact, it invaded an empty space, where only some insect species can also occur (Kinzelbach 1972). According to Den Hartog (1963), the occurrence of *O. gammarella* may limit *O. cavimana* at the isohaline of 1.8‰ chloride at high tide in the estuaries.

5.2.4 *Proasellus meridianus and Proasellus coxalis (Isopoda)*

Both species immigrated from the western Mediterranean and were perhaps introduced by ships in the 20th century. *P. meridianus* was first found in the Lower Rhine in 1948. It spread along the river and inhabits slow flowing waters as well as still waters. Both species are known from the entire Rhine. They feed on detritus and dead plant material. Both species are salt tolerant (Tittizer 1996b).

5.3 *Ponto-Caspian*

5.3.1 *Corophium curvispinum (Amphipoda)*

This tube-building species has been extending its range to the west since the start of 20th century via the rivers Pripyat and Wista and their interlinking canal systems. In 1987 it was found in the German and Dutch parts of the Rhine (Van den Brink et al. 1989; Schöll 1990a, Tittizer et al. 1990). The Rhine was reached via the Havel-Spree area, Mittelland Canal and Dortmund-Ems Canal (Tittizer 1996b), from where it spread along the whole river and its tributaries (Bobbe 1994). The Main-Danube Canal has been colonized by this species from two directions, viz. from the rivers Rhine (1987) and Main (1988) and from the Upper Danube (1959); it was found in this canal in 1993 (Tittizer 1996b).

Dispersal routes for *C. curvispinum* include the ballast water in ships. Great Britain was probably reached in 1918 via the ballast water in ships trading with ports on the Elbe River in northern Germany (Harris 1991) and via rivers connected. Dispersal modes include ballast water, ships' hulls and drift or active migration (Harris 1991). With respect to dispersal speed, its spread over Europe was similar to that of the zebra mussel (*D. polymorpha*), but at a much slower rate (Fig. 1).

Reproduction occurs via a brood pouch with a maximum clutch of 42 eggs. The breeding season is strongly correlated with high water temperatures. Reproduction begins in March and continues until October (Rajagopal et al. 1999b).

In the Lower Rhine, there are three generations per year (Van den Brink et al. 1993, Rajagopal et al. 1999a). The life span of the animals is less than eight months. Populations may reach densities of hundreds of thousands of individuals occurring on one m^2 of hard substrate. It is now the dominant macroinvertebrate on stones in the Rhine (Van den Brink et al. 1993, Rajagopal et al. 1999a, b).

The food of this filter feeder consists of phytoplankton and detritus (Van den Brink et al. 1991, 1993). Eutrophication favors high population densities of the species (Waterstraat & Köhn 1989), which requires a constant supply of food and silt particles (Muskó et al. 1998). *C. curvispinum* builds its mud tubes on hard substrates such as stones and wooden piles, as well as on aquatic macrophytes (Crawford 1937, Entz 1949, Muskó 1990, 1992, 1994, Sebestyén 1938, Van den Brink et al. 1991). *C. curvispinum* can also be found in the sediment (juveniles) or swimming in the water at lower densities.

According to Crawford (1937), the species is variable, adaptable and widely distributed and found in salt, brackish and fresh water. Several varieties have been described, including one freshwater adapted variety (var. *devium* Wundsch). The species occurs up to a salinity of 6‰ (Romanova 1975). With respect to salinity, *C. curvispinum* var. *devium* needs ion-rich, hard freshwater with a minimum $[Na^+]$ of 0.5 mMol L^{-1} (Harris & Aladin 1997). Taylor & Harris (1986a, b), Harris & Bayliss (1990) and Harris (1991) demonstrated that *C. curvispinum* is leakier to Na^+ and Cl^- and water than freshwater gammarid species. There is considerable physiological variability within populations of *C. curvispinum* from various localities with respect to the retaining and replacing of Na^+ and Cl^-. Its adaptation to freshwater by means of a lower permeability than the other varieties must have evolved very recently. At higher salinities, the 'freshwater form' will possibly compete with brackish water *Corophium* species, like *C. lacustre* and *C. multisetosum* (Herbst & Bäthe 1993).

C. curvispinum in tubes consume twice the amount of oxygen consumed by specimens outside the tubes (Muskó et al. 1995, 1998, Harris & Muskó 1999). Much oxygen seems to be required to irrigate the tubes. At low oxygen levels, the animals leave the tubes and begin to swim. In order to obtain enough oxygen while in their tubes, they need current or wave action.

Rather high water temperatures seem to be required. Jazdzewski & Konopacka (1990) mentioned the co-existence of *E. ischnus*, *C. curvispinum* and *D. polymorpha* in a heated lake in Poland with a temperature amplitude of 7.0-31.8°C. *C. curvispinum* is absent at heavy organic pollution and low oxygen levels (Jazdzewski 1980, Harris & Muskó 1999).

Parasites are unknown, while predators include birds, fish, crayfish and other predatory macoinvertebrates (Biró 1974, Van den Brink et al. 1993, Marguillier et al. 1998, Kelleher et al. 1998, 1999).

5.3.2 *Dikerogammarus haemobaphes (Amphipoda)*

This gammarid species migrated from the Upper Danube (1976), via the Main-Danube Canal (1993) and the Main (1993) to the Rhine (1994). *D. haemobaphes* is an euryoecious species occurring on stones, macrophytes and filamentous algae in large rivers and lakes (Kititsyna 1980, Muskó 1993, Tittizer 1996b). This species is brackish water tolerant up to 8‰ (Ponomareva 1975) and is able to endure a wide range of temperature variations (6-30°C) (Kititsyna 1980). It normally reproduces from April to October, but does so all the year round in the warm water of the cooling water discharge of an electric power plant (Kititsyna 1980). In Lake Balaton, this species reproduces from May to the end of August on submerged macrophytes (Muskó 1993). Its food consists of various plant material (diatoms, filamentous algae and detritus) (Ponyi 1961, Romanova 1963). Because of its high reproduction capacity, suggestions have been made to cultivate *D. haemobaphes* as fish food (Vorob'eva & Nikonova 1987).

5.3.3 *Dikerogammarus villosus (Amphipoda)*

D. villosus took the same immigration route as *D. haemobaphes,* viz., from the Upper Danube (1992), along the Main-Danube canal (1993) and the Main (1994) to the Rhine (1995). This actively migratory and euryecious species inhabits large rivers and lakes (Tittzer 1996b). It is more salt-tolerant than *D. haemobaphes* (Bij de Vaate & Klink 1995 and literature therein). Since its arrival in the Upper Danube, *D. villosus* has outcompeted *D. haemobaphes,* which was successful and numerous before (Weinzierl et al. 1996). *D. villosus* has an enormous reproductive capacity, as was shown in the lower reaches of the Don River. In a frost-free brook, it appeared to reproduce all the year round (Mordukhai-Boltovskoi 1949). Its food includes both plant and animal food and sometimes it even eats the conspecific newborns and weak adults (Mordukhai-Boltovskoi 1949). In the Lower Rhine, it reaches its highest densities on hard substrates, where it is already an important contributor to dietary biomass of fish (Kelleher et al. 1998).

5.3.4 *Echinogammarus ischnus (Amphipoda)*

E. ischnus was first recorded from the Rhine in 1989, reaching densities of 100 ind. m^{-2} of hard substrate. The species was already present in northern German canals from 1900. At the end of the 1980s, there were reports from the Rhine-Herne Canal and from the Wesel-Dattel Canal (Schöll 1990b). In 1991, it was found in the Dutch part of the Rhine (Van den Brink et al. 1993). *E. ischnus* migrated to the Main-Danube Canal (1995) via the Upper Danube (1989). The species is salt tolerant and eurythermous (Tittizer 1996b). Present-day densities in the Lower Rhine are below 50 individuals per m^2 of hard substrate.

5.3.5 *Hemimysis anomala (Mysidacea)*

This euryhaline species was recorded from stony banks of the Neckar River and the Middle Rhine in 1997 (Schleuter et al. 1998). The way the species entered the Rhine is unclear. It may have entered either from the Baltic Sea or from the Danube. It seems possible, however, that it came from the Danube via the Main-Danube Canal (Wittmann et al. 1999). Perhaps it arrived via the ballast water in a ship from the Baltic Sea. In the Netherlands, the species was recently found in a small artificial lake not far from the Noordzeekanaal near Amsterdam (Faasse 1998). Specimens were also found in fish taken from the Lower Rhine in 1997, so it is possible that the species was overlooked for some years because of its habit to hide in and near cavities in stony banks of lakes and rivers and because of its nocturnal activity pattern. Its impact on the river communities is unknown, but in the Biesbosch reservoirs it became very numerous as a voracious predator on zooplankton and also an omnivorous feeder (Ketelaars et al. 1999).

5.3.6 *Limnomysis benedeni (Mysidacea)*

This species could have reached the Rhine via the Danube, where it has spread upstream to reach the Upper Danube in 1994, via the Main-Danube Canal (Wittmann 1995). *Limnomysis* was discovered in 1997 in a side channel of the middle Rhine with some *Myriophyllum spicatum* (Geissen 1997). Recently it was also found in a river-connected sandpit (1997) and a floodplain pond (1998) along the Waal River, as well as in the Lek River (1998). The Waal and Lek are both side branches of the Rhine, and in the Zuiderdiep, which is connected with the Haringvliet, the regulated estuary of the Waal and Meuse (Kelleher et al. 1999b). In 1998, it was discovered in the Main-Danube Canal (Reinhold & Tittizer 1998). Its discovery in the Netherlands preceded that in the Middle Rhine by some months (Kelleher et al. 1999b).

Vessels seem to be the most likely dispersal mode, in view of the widespread localities in the Rhine. In contrast to *H. anomala*, *L. benedeni* is associated to aquatic vegetation and can be considered to be potamophilous, phytophilous and salt tolerant.

5.3.7 *Jaera istri (Isopoda)*

This species originates from the Black and Caspian Seas, where it inhabits hard substrates in estuaries and rivers in the littoral zone. This isopod came from the Danube (1958), and migrated upstream, via the Main-Danube Canal (1993) to the Main (1995), then to the Rhine (1995) and subsequently further downstream (Tittizer 1996b). In 1997, the species was also recorded from the Lower Rhine branches in the Netherlands (Kelleher et al. 1998). Population densities have been increasing on hard substrates. The animals may spread by means of vessels. Its food consists of algae, plant remains and detritus (Tittizer 1996). The species is salt tolerant.

5.4 *East Asia*

5.4.1 *Eriocheir sinensis (Decapoda)*

The Chinese mitten crab lives most of its life in freshwater but reproduces in the sea. It migrates actively into the river via the estuaries. In the Yang-tse-Kiang, it migrates 1300 km upstream. It is dispersed via the ballast water in ships and is widespread in the USA (Cohen & Carlton 1995).

In the Netherlands, it was first recorded in 1929 and spread quickly over the country, even over land, also invading the Rhine-Meuse estuaries. From the estuaries, the animals migrate far upstream into Germany; it has even been recorded north of Basel (Tittizer 1996b). The crabs are omnivorous. Sometimes they develop very dense populations, perhaps competing for food with fish. The species is salt tolerant and eurythermous (Tittizer 1996b).

5.5 North America

5.5.1 Callinectes sapidus (Decapoda)

The adult Blue crab is an omnivorous inhabitant of estuaries on the east coast of North America, where it can live in nearly fresh to hyposaline water. Just as the Chinese mitten crab, females migrate to the sea for reproduction, while juvenile animals migrate into the estuaries. The species tolerates a temperature range of 2-35°C, but the animals are only active above 15°C. It seems that this species was introduced in the Netherlands by ballast water in ships in 1932 (Adema 1991). Its present distribution includes the Noordzeekanaal, which is connected with the Rhine via the Amsterdam-Rhine Canal. However, it has not yet been observed in either the Rhine estuaries (mostly blocked by dams) or the river itself.

5.5.2 Crangonyx pseudogracilis (Amphipoda)

This amphipod species has inhabited central and southern England and Wales since 1975, from where it spread to the north, and Ireland. In 1979, it was discovered in the north of the Netherlands (Province of Groningen) (Pinkster et al. 1992). *C. pseudogracilis* reproduces mainly in the warm summer months; its reproductive cycle is slower than that of *G. tigrinus* and *G. pulex* (Pinkster & Platvoet 1983). The species is euryoecious inhabiting all types of water ranging from fresh or brackish, clean or polluted. In the Dutch part of the Rhine, the species has been discovered once, in a temporary backwater near Millingen aan de Rijn, where three specimens were collected (Bij de Vaate & Klink 1995). The species is stagnophilous (Garland 1981) and thermophilous (Sutcliffe & Carrick 1981).

5.5.3 Gammarus tigrinus (Amphipoda)

Sexton first described this species in Britain in 1939, in brackish water and ion-rich polluted water. It was later found to have originated from the USA. This species was introduced into Germany in the Weser and Werra Rivers as fish food, because of its tolerance to salinated and ion-rich polluted water. Specimens from Britain were released in these rivers in 1957 and a mass development began. The species cleans the water of detritus and dead animals, but it is also very rapacious and voracious, attacking wounded or weakened fish (Schmitz 1960).

G. tigrinus was first recorded in the Netherlands in lake IJsselmeer in 1960, where it was also introduced in low numbers by a fisheries institute (Pinkster & Stock 1967). It may also have been released into the IJsselmeer by ballast water in ships from Germany (Nijssen & Stock 1966). From the IJsselmeer it spread quickly over the Netherlands, establishing populations in still as well as flowing water. In 1973, it was recorded from the Haringvliet, the estuary of the Meuse and Rhine (Dieleman & Pinkster 1977). Since 1982/83, it has been found in the Lower Rhine itself (Berndt 1984, Pinkster et al. 1992).

Dispersal takes place via an existing network of canals and rivers by active migration (e.g. Northern Ireland, Germany and The Netherlands) and by human interference (Pinkster et al. 1980) as well as ballast water in ships (Hynes 1955, Jazdzewski

1980). Its dispersal speed has been calculated as up to 40 km per year (Pinkster et al. 1980). Allelic frequencies (PGI, MPI and PGM) have demonstrated that *G. tigrinus* populations from Germany and the Netherlands show considerable genetic differences reflecting different origins (Bulnheim 1985). There has, however, been a second invasion into the Netherlands from Germany, as was proven by electrophoretic tests (Pinkster et al. 1992).

With respect to reproduction, 15-16 generations are theoretically possible during one year (Pinkster 1975). Females attain sexual maturity beyond 4 mm body length, and only 27-29 days are required to reach this maturity at 20°C. Reproduction stops below 5°C (Pinkster et al. 1992). The highest density of *G. tigrinus* ever found was 24,000 ind. m^{-2} of bottom in the Frisian lake Tjeukemeer (Chambers 1977). In the Rhine, *G. tigrinus* has been found in densities of thousands per m^2 of hard substrate. In recent years, populations have declined, probably due to the invasion of *D. villosus*. *G. tigrinus* is omnivorous, feeding on animals, algae, plants and detritus, and functions both as a primary and secondary consumer of the first and second order. *G. tigrinus* tolerates a salinity of 1-16‰ (Dorgelo 1974) and a temperature of up to 30°C (Savage 1982). It prefers well aerated water. Principally a brackish water species, it has successfully invaded freshwater habitats. Furthermore, it is pollution tolerant (Streit & Kuhn 1994).

As regards pathogens, one fungus, *Saprolegnia* sp., affects *G. tigrinus* but not *G. pulex* (Dieleman & Pinkster 1977). Fish are its main predators, and the species is often introduced as a feeding supplement.

5.5.4 *Orconectes limosus (Decapoda)*
O. limosus originates from the eastern United States. In 1890, 100 specimens were released into fish ponds near Berlin (Germany), from where the species spread via the Mietzel River to the Oder, Warthe and Netze Rivers. It was also introduced in France. Until 1940, the western border of its geographical area was the Elbe. In the Netherlands, the species actively migrated via the Meuse from Belgium, as well as from Germany via the Twenthe Canal (Geelen 1975). Since the first half of the 20th century, it has spread over the whole of Europe (Tittizer 1996b), including the Rhine. It feeds on macroinvertebrates, detritus, algae and macrophytes.

This crayfish species is well adapted to pollution and low oxygen concentrations, and is immune to the crayfish pest that wiped out the indigenous *Astacus astacus*. Although it is also found in lakes, it prefers canals and rivers. It is euryecious and salt tolerant (Tittizer 1996b).

5.5.5 *Rhithropanopeus harrisii (Decapoda)*
This species originates from brackish water and estuaries on the North-American east coast. It was present in The Netherlands in the second half of the 19th century, and was even regarded as an endemic species from the Zuiderzee, being described as a new species in 1874. It must have been introduced via ballast water in ships. It is a common species of brackish water (Adema 1991), migrating into the river from the estuary. It has been recorded several times in the estuary as well as upstream in the Lower Rhine itself, until just across the German border near Rees, although in very low numbers (Den Hartog et al. 1989, Van der Velde et al. 1990, Fontes & Schöll 1994, Bij de Vaate & Klink 1995).

6 IMPACT

Of the 19 crustacean species recorded from the Rhine, two species (10%; rule of ten) can be regarded as particularly successful, in the sense of having established high density populations over long stretches of the river, viz., *G. tigrinus* and *C. curvispinum*. Why have other species been less successful? The reasons for the limited success of an invader may be suboptimal conditions, poor adaptation to present conditions leading to reduced fitness, severe interspecific competition, lack of food, heavy predation or combinations of all of these factors. Most crustacean invaders of the Rhine show brackish to hard water adaptations, are euryecious, tolerant to pollution in general and thermophilous. With respect to feeding habits they are omnivores, deposit or filter feeders. They may also be successful invaders because they possess a brood pouch, which means that one female can establish a population. The brood pouch protects the eggs at an early stage against environmental changes and perhaps also against the influence of pollution. Dick et al. (in press) have shown that *Crangonyx pseudogracilis* females alter their brood care activities in response to dissolved oxygen and temperature fluctuations, and this species is a successful invader of such waters.

Invasions of gammarids in the Rhine have led to the replacement of one gammarid species by another, viz. *G. pulex* by *G. tigrinus*, probably as a result of competitive displacement and intraguild predation. Pinkster (1975) and Pinkster et al. (1977) recorded the displacement of several native *Gammarus* species by the invasion of *G. tigrinus*. One factor in this process is the enormous reproductive capacity of *G. tigrinus* in oligohaline water as compared to that of *G. zaddachi*, *G. duebeni* and *G. pulex*, as was demonstrated by Pinkster et al. (1977). However, severe and prolonged winters can reduce the populations of *G. tigrinus* (Pinkster et al. 1992). Another factor explaining its success compared to *G. pulex* is a better tolerance of salt pollution, which means a lower energy demand due to ionic regulation and consequently a better competitive strength at high ionic concentrations (Koop 1997). Dick & Platvoet (1996) found that *G. tigrinus* excluded *G. pulex* from oligohaline water and freshwater with a high conductivity, whereas *G. pulex* excluded *G. tigrinus* in freshwater with a low conductivity. Between these two species there is frequent mutual predation of moulting individuals. In fresh water, *G. pulex* showed significantly higher predation frequencies on *G. tigrinus* than vice versa. In oligohaline water, however, there was no clear difference in predation frequencies between *G. tigrinus* and *G. pulex*. Dick & Platvoet (1996) developed a model for interactions with relevant population parameters, which could identify the complex switches in species dominance with transient periods of co-existence. The intensity of gammarid species interactions thus seems mediated by behavioural, physiological and environmental factors (Dick & Platvoet 1996).

Streit & Kuhn (1994) demonstrated that *G. tigrinus* is much more tolerant to organophosphorus insecticides than *G. pulex* and *G. fossarum*. This means that the latter species could be competively weakened by pesticides, another factor that could explain the dominance of *G. tigrinus* in the Rhine and the disappearance of *G. pulex*.

The next gammarid species invading the Rhine was *Echinogammarus ischnus*, which was, however, unable to develop very dense populations. It may be hindered by *C. curvispinum,* which could prevent its success as a stone dweller by covering the stone surfaces in mud and by reducing the supply of the Zebra mussel (*D. poly-morpha*). That *E. ischnus* is capable of outcompeting other gammarid species has recently been demonstrated by its invasion of the Great Lakes. In 1995, *E. ischnus* reached the Great Lakes in the US, probably via ballast water (Witt et al. 1997). It outcompeted *G. fasciatus*, that had previously increased in abundance by hiding and feeding in the mussel beds built up after the invasion of the Zebra mussel (*D. poly-morpha*) (Dermott et al. 1998 and literature therein). Waterstraat & Köhn (1989) and Köhn & Waterstraat (1990) mentioned that *E. ischnus* is closely associated with *Dreissena* clumps in Lake Kummerow (Germany). It seems that *E. ischnus* dominates in spite of its somewhat lower fecundity and life span than *G. fasciatus,* being more of a specialist on rocky habitats than *G. fasciatus*. In dense submerged vegetation or quiet, turbid water, *E. ischnus* shows low densities and here *G. fasciatus,* well equipped to cling to vegetation, is more numerous; *Cladophora* growing on rocks favours *G. fasciatus* (Dermott 1998, Dermott et al. 1998). This means that successful invasions clearly depend on the properties of the invader, which must have an advantage in some habitats compared to the properties of indigenous species. This example is also a good demonstration of a habitat being conditioned for *E. ischnus* by an earlier Ponto-Caspian invader, the Zebra mussel.

E. ischnus is able to outcompete other gammarid species especially in stony habitats. In addition to the above-mentioned factors, it is possible that *G. tigrinus* and the later invasion of *D. villosus* have prevented its further success in the Rhine or that *D. villosus* outcompeted *G. tigrinus*, which allowed a third subdominant gammarid, in this case *E. ischnus*, to settle in the deeper parts of the river.

All *Gammarus* species can prevent successful invasion by *Crangonyx pseudo-gracilis*; Dick (1996) found that this species is preyed upon heavily by *G. pulex* as well as *G. tigrinus,* regardless of the molt state of the victims; frequencies of predation upon reproductive females were high, reaching 70%. Until now, this species has not been successful in the Rhine.

The most promising species with respect to recent success is *D. villosus*, which is a much larger amphipod than the other species. It is currently developing very dense populations. *D. villosus* is suspected to be more predatory than the other gammarids. Isotope analysis ($\delta^{15}N$) has indicated that they are active at the same trophic level as fish species (Marguillier 1998). Dick (1996) mentioned that differences in predatory/survival abilities of gammarid species are influenced by body size, with larger species and sexes preying upon the smaller. So *D. villosus* may have considerable impact on the *G. tigrinus* population (Fig. 2). *D. villosus* may dominate the stone communities and *G. tigrinus* the sandy communities, as *G. tigrinus* is still dominant in the river-connected sandpits in the Rhine (Kelleher et al. 1998). As more gammarid species enter the Rhine system, the outcome of subsequent invasions will be unpredictable in the long term. However, the history of invasions over the last decade has generated data sets that will allow some assessment of their initial impact; such data were not available for the initial colonizers like *C. curvispinum* and *E. ischnus*. The co-existence of various gammarid species in the Rhine is made possible by differences in habitat selection and differential predation and competitive interactions.

By far the most successful invasion is that of the tube-building corophiid *C. curvi-spinum*. This species shows many of the properties mentioned in Section 3. Recorded for the first time in the Rhine in 1987, it developed very dense populations within a

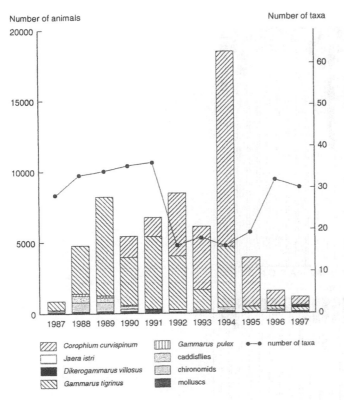

Figure 2. Development of the macroinvertebrate fauna on standardized artificial substrates (for a description of the method see De Pauw et al. 1994, Pashkevich et al. 1996) over the years in the Rhine near Lobith (the Netherlands) starting in 1987 (after the Sandoz accident and the first year that *Corophium curvispinum* was recorded). Numbers of animals from two artificial substrate samplers are averaged; numbers of taxa are total numbers from both samplers. Observations have been carried out in September.

few years (Fig. 2), so that it became the most numerous macroinvertebrate on the stones along the river (Fig. 3). Due to its tube building, it fixes much mud on the stones, altering the stone communities (Fig. 4). Monitoring data on artificial substrates over the years have shown that the macroinvertebrate species richness was reduced at the highest densities of *C. curvispinum* (Fig. 2). According to Kinzelbach (1997), *C. curvispinum* also outcompetes the freshwater isopod *Asellus aquaticus* and several species of chironomid larvae. As a result, the numbers of their predators, like leeches, have decreased. What was clear from the beginning was the strongly negative influence of *C. curvispinum* on another Ponto-Caspian invader, the Zebra mussel (*D. polymorpha*), which was already present in high numbers in the Rhine (Van den Brink et al. 1991, 1993, Van der Velde et al. 1994, 1998).

At other localities, for example, in Lake Balaton, the Zebra mussel and *C. curvispinum* may have arrived about the same time during their expansion from the Ponto-Caspian region (1932 and 1935, respectively). Sebestyén (1938) noted that where *D. polymorpha* was present, *C. curvispinum* was also present, so these species seem to be associated in some way. It is therefore interesting that in the Rhine, *C. curvi-*

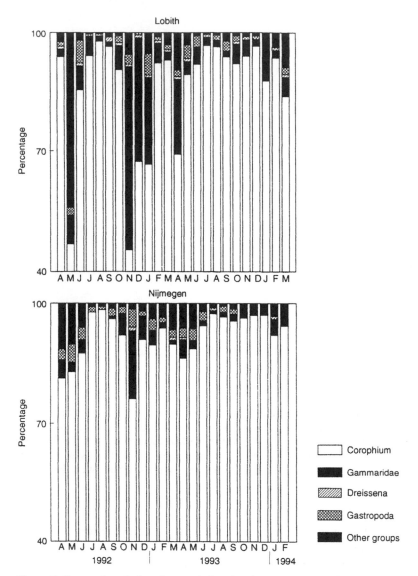

Figure 3. Seasonal variations in population density percentages of various macroinvertebrates from April 1992 to March 1994 on stones (*n* = 5) of groynes in the Lower Rhine at Lobith and Waal at Nijmegen, the Netherlands.

spinum 'swamps' *D. polymorpha* and reduces its populations. *C. curvispinum* simply changes the habitat so that the Zebra mussel disappears. Amounts of mud fixed on the stones are clearly correlated with the densities of *C. curvispinum* (Fig. 5). Sometimes in summer, the mud layer can be 4 cm thick (Paffen et al. 1994). By a thick layer of mud especially in the summer months, when densities of *C. curvispinum* were at a peak, Zebra mussel larvae lack bare stone surfaces to settle, and adult Zebra mussels were smothered to death. Jantz (1996) found that at high densities of *C. curvispinum*, there was a shift in the Zebra mussels' size classes, from high frequency of specimens

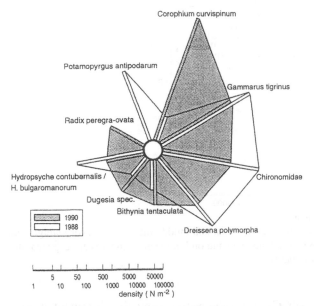

Figure 4. Changes in the densities of the major macroinvertebrates on stones in the Rhine at the start of the population explosion of *Corophium curvispinum* and afterwards (data from Van den Brink et al. 1993).

with a shell length of 5-15 mm to a predominance of specimens smaller than 5 mm, while many overgrown dead specimens were found. The reduction of the Zebra mussel populations by *C. curvispinum* can be regarded as a case of spatial competition.

With respect to food competition between these two species in the Rhine, stable isotope analysis showed very similar values for $\delta^{13}C$ (carbon source) and $\delta^{15}N$ (trophic level) for the stone-dwelling *C. curvispinum*, *D. polymorpha* and the sand dwelling Asiatic clams (*Corbicula fluminea*, *Corbicula fluminalis*), indicating a common diet of phytoplankton and particulate organic matter (POM) among these filter feeding animals (Marguillier 1998). This means that competition between the Zebra mussel and *C. curvispinum* can also be regarded as a case of interspecific exploitative competition, although competition for space seems to be the main aspect. Van den Brink et al. (1991, 1993) found detritus as well as planktonic algae in the digestive tract of *C. curvispinum*. The mean number of eggs per brood of *C. curvispinum* during the season showed a positive correlation with chlorophyll-*a* in the Rhine (Rajagopal et al. 1999b). The growth rate of *C. curvispinum* was positively correlated with the temperature, but increased growth rate correlated with an increase in chlorophyll-*a* concentrations (Rajagopal et al. 1999a). This means that the high densities of *C. curvispinum* in the Rhine can arise due to high chlorophyll-*a* concentrations in the eutrophicated river. In 1994, *C. curvispinum* reached a peak density (Fig. 2) but chlorophyll-*a* concentrations dropped to a low level and remained low in 1995 and 1996 (Table 1). This drop in chlorophyll-*a* concentrations may have been caused by the reduced levels of nutrients, especially phosphate (in 1986 annual mean total phosphate was 0.45 mg l^{-1}, range 0.12-0.78 mg l^{-1}) leading to decreased plank-

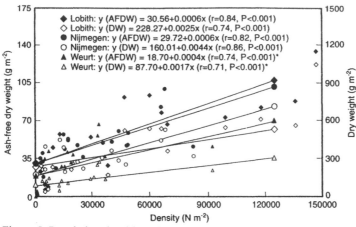

Figure 5. Population densities of *Corophium curvispinum* and muddy tube material (dry weight and ash-free dry weight) on stones at Lobith and Nijmegen and on long-term cumulatively exposed tiles at Weurt (Maas-Waal Canal), the Netherlands.

tonic algal development. Although *C. curvispinum* remained the dominant macro-invertebrate on the stones, its densities decreased in these years. At the same time macroinvertebrate species richness increased after the decline of *C. curvispinum* (Fig. 2), but this was mainly due to the occurrence of new hard substrate exotics.

Another factor that appeared to be important in the development of a high density population able to outcompete the Zebra mussel is stream velocity (Van der Velde et al. 1998). A high stream velocity favors *C. curvispinum* by transporting silt and planktonic algae, as well as causing high oxygen levels, which are necessary for metabolism. This is clearly illustrated by investigations along the longitudinal axis of the Rhine and Meuse (Figs 6-8). From these results, it can be concluded that a low stream velocity, such as that in the Meuse, prevents the building up of highly dense populations of *C. curvispinum* (Fig. 7). However, the water quality of the Meuse could also be a limiting factor because Na^+ concentrations are low and near the tolerance limit of *C. curvispinum* (Van der Velde et al. 1998). In the Meuse, *D. polymorpha* is one of the dominant macroinvertebrates on the stones. In the Rhine stream velocity drops downstream, allowing *D. polymorpha* to maintain its dominant position at downstream sites (Fig. 8).

As pointed out above, mass invaders influence both horizontal (competition, swamping) and vertical food chain processes. *G. tigrinus* and *C. curvispinum* both play an important role in the food web of the Rhine and are consumed by fish. Kelleher et al. (1998) found that Perch (*Perca fluviatilis*) and Eel (*Anguilla anguilla*) consume both amphipod species. Eels foraging at groynes appear to have switched their diet completely from *G. tigrinus* to *C. curvispinum*. Perch, being more active hunters on mobile prey, maintain a preference for *G. tigrinus*. In river-connected sandpits, eels consume more *G. tigrinus* and *O. limosus* than *C. curvispinum* in terms of biomass. In another study in the Rhine, Ruffe (*Gymnocephalus cernuus*), Perch and juvenile Pike-perch (*Stizostedion lucioperca*) were found to feed on *G. tigrinus* and *D. villosus* as well as on *C. curvispinum*. Besides these percids, Eel and Roach (*Rutilus rutilus*) were also found to feed on *C. curvispinum* (Kelleher et al. 1999a).

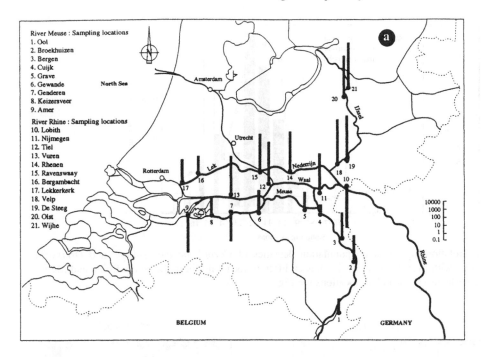

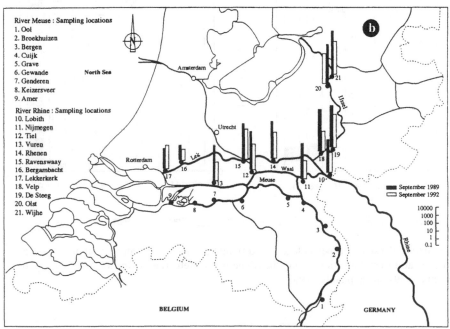

Figure 6. Map showing the sampling localities in the Rhine and Meuse in the Netherlands. A) Mean population densities of *Corophium curvispinum* (individuals per m^2 stone surface at 1 m depth) at different sampling locations in the Rhine and Meuse during September 1993. B) Mean population densities of *Dreissena polymorpha* (individuals per m^2 stone surface at 1 m depth) at different sampling localities in the Rhine, before (September 1989) and after (September 1992) the population explosion of *C. curvispinum*.

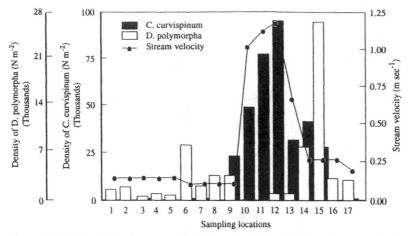

Figure 7. Relationships between population densities of *Corophium curvispinum* and *Dreissena polymorpha*, with average stream velocities at different locations along the rivers Rhine and Meuse in the Netherlands (for sampling locations see Fig. 6).

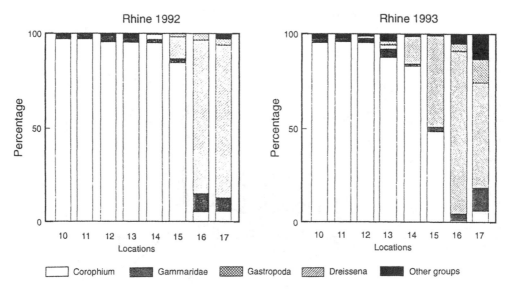

Figure 8. Population densities (in percentages) of different macroinvertebrates on stones ($n = 3$-5) from the groynes at different locations along the Rhine in the Netherlands (for location see Fig. 6) during September 1992 and September 1993.

This means that there is a rapid flux in exotic species composition in the food base of these predatory species with differential preferences. *D. villosus* seems set to become the next important food for many of these species, because of its increasing population and because it is bigger than the other amphipod species (Kelleher et al. 1998). Diet switching can be important as a controlling factor of the amphipod populations, since predation pressure on these exotic amphipods can change rapidly. If a predator switches to another additional prey, the other prey species may suffer less from predation and their population could increase again.

In summary, it can be stated that gammarids and tube-building corophiids have different strategies for coping with their competitors. Gammarids show aggressive behaviour against their relatives, while tubicolous corophiids like *C. curvispinum* introduce mud to their habitat and so prevent the development of other competitors. The outcome is in both cases not only dependent on behavior but also on physiological as well as environmental factors, as was illustrated above. This means that the success of crustacean invaders can be explained by linking detailed information on behavior and physiology to monitoring of the macroinvertebrate communities and environmental factors.

Continuous biological monitoring in the Rhine can provide data with respect to successive invasions, their interactions and their impact in relation to water quality changes. Further improvement of the water quality may change the outcome of the competition between the crustacean invaders. A decrease in salt concentration when the French potassium mines become exhausted, probably in 2004, may have a strong impact especially on the crustaceans in the river. Thus, the changes in the Rhine and other rivers can be regarded as large-scale ecological experiments.

ACKNOWLEDGEMENTS

The authors would like to thank the Ministry of Transport and Public Works, the Ministry of Housing, Physical Planning and the Environment, the Netherlands Organisation of the Advancement of Pure Scientific Research (NWO/BION) and the Beijerinck-Popping Foundation for financial support.

REFERENCES

Adema, J.P.H.M. 1991. *De krabben van Nederland en België*. Leiden: Nationaal Natuurhistorisch Museum.

Admiraal, W., Van der Velde, G., Smit, H. & Cazemier, W.G. 1993. The rivers Rhine and Meuse in the Netherlands: present state and signs of ecological recovery. *Hydrobiologia* 265: 97-128.

Berndt, J. 1984. Nachweis von *Gammarus tigrinus* Sexton für den unteren Niederrhein. *Decheniana* 137: 168-169.

bij de Vaate, A. & Greijdanus-Klaas, M. 1993. Monitoring macroinvertebrates in the River Rhine. Results of a study executed in the Dutch part in 1990. Institute for inland Water Management. Report nr. 52-1993 of the project 'Ecological Rehabilitation of the River Rhine': 1-22.

bij de Vaate, A. & Klink, A.G. 1995. *Dikerogammarus villosus* Sowinsky (Crustacea: Gammaridae) a new immigrant in the Dutch part of the Lower Rhine. *Lauterbornia* 20: 51-54.

Biró, P. 1974. Observations on the food of eel (*Anguilla anguilla* L.) in Lake Balaton. *Annal. Biol. Tihany* 41: 133-151.

Bobbe, Th. 1994. Erstnachweis von *Corophium curvispinum* Sars in der Lahn. *Lauterbornia* 19: 69-70.

Bulnheim, H.-P. 1985. Genetic differentiation between natural populations of *Gammarus tigrinus* (Crustacea, Amphipoda) with reference to its range extension in European continental waters. *Arch. Hydrobiol.* 102: 273-290.

Chambers, M.R. 1977. The population ecology of *Gammarus tigrinus* (Sexton) in the reed beds of the Tjeukemeer. *Hydrobiologia* 53: 155-164.

Clare Eno, N., Clark, R.A. & Sanderson, W.G. (eds) 1997. *Non-native marine species in British waters: a review and directory.* Peterborough: Joint Nature Conservation Committee.

Cohen, A.N. & Carlton, J.T. 1995. *Biological study. Nonindigenous aquatic species in a United States estuary: a case study of the biological invasions of the San Francisco bay and delta.* Washington D.C.: United States Fish and Wildlife Service.

Cohen, A.N. & Carlton, J.T. 1998. Accelerating invasion rate in a highly invaded estuary. *Science* 279: 555-558.

Conlan, K.E. 1994. Amphipod crustaceans and environmental disturbance: a review. *J. Nat. Hist.* 28: 519-554.

Crawford, G.I. 1937. A review of the amphipod genus *Corophium*, with notes on the British species. *J. mar. biol. Ass. UK* 21: 589-630.

De Pauw, N., Lambert, V., Van Kenhove, A. & Bij de Vaate, A. 1994. Comparison of two artificial substrate samplers for macroinvertebrates in biological monitoring of large and deep rivers and canals in Belgium and The Netherlands. *J. Env. Mon. & Ass.* 30: 25-47.

De Ruyter van Steveninck, E.D., Admiraal, W., Breebaart, L., Tubbing, G.M.J. & van Zanten, B. 1992. Plankton in the River Rhine: structural and functional changes observed during downstream transport. *J. Plankton Res.* 14: 1351-1368.

Den Hartog, C. 1963. The amphipods of the Deltaic region of the rivers Rhine, Meuse and Scheldt in relation to the hydrography of the area. Part II. The Talitridae. *Neth. J. Sea Res.* 2: 40-67.

Den Hartog, C., Van den Brink, F.W.B. & Van der Velde, G. 1989. Brackish-water invaders in the River Rhine. A bioindication for increased salinity level over the years. *Naturwiss.* 76: 80-81.

Den Hartog, C., Van den Brink, F.W.B. & Van der Velde, G. 1992. Why was the invasion of the Rhine by *Corophium curvispinum* and *Corbicula* species so successful? *J. Nat. Hist.* 26: 1121-1129.

Dermott, R. 1998. Competitive displacement of *Gammarus fasciatus* by the European amphipod *Echinogammarus ischnus* in the Great Lakes. Eighth International Zebra Mussel and aquatic nuisance species conference, 16 to 19 March, 1998. Abstract book: 145.

Dermott, R., Witt, J., Um, Y.M. & González, M. 1998. Distribution of the Ponto-Caspian Amphipod *Echinogammarus ischnus* in the Great Lakes and replacement of native *Gammarus fasciatus. J. Great Lakes Res.* 24: 442-452.

Dick, J.T.A., 1996. Post-invasion amphipod communities of Lough Neagh, Northern Ireland: influences of habitat selection and mutual predation. *J. Anim. Ecol.* 65: 756-767.

Dick, J.T.A. & Platvoet, D. 1996. Intraguild predation and species exclusions in amphipods: the interaction of behaviour, physiology and environment. *Freshwat. Biol.* 36: 375-383.

Dick, J.T.A., Faloon, S.E. & Elwood, R.W. (1998). Active brood care in an amphipod: Influences of embryonic development, temperature and oxygen. *Anim. Behav.* 56: 663-672.

Dieleman, J. & Pinkster, S. 1977. Further observations on the range extension of the alien amphipod *Gammarus tigrinus* Sexton 1939 in the Netherlands. *Bull. zool. Mus. Amsterdam* 6(3): 21-29.

Dorgelo, J. 1974. Comparative ecophysiology of gammarids (Crustacea: Amphipoda) from marine, brackish and fresh-water habitats, exposed to the influence of salinity-temperature combinations. 1. Effect on survival. *Hydrobiol. Bull.* 8 (1-2): 90-108.

Entz, B. 1949. Beiträge zur Kenntnis der Morphologie und Biologie des *Corophium curvispinum* G.O. Sars forma *devium* Wundsch in Ungarn. *Arch. Hydrobiol.* 42: 423-469.

Faasse, M. 1998. The ponto-caspian mysid, *Hemimysis anomala* Sars, 1907, new to the fauna of the Netherlands. *Bull. zool. Mus. Amsterdam* 16(10): 73-76.

Fontes, R.-J. & Schöll, F. 1994. *Rhithropanopeus harrisii* (Gould 1841) – eine neue Brackwasserart im deutschen Rheinabschnitt (Crustacea, Decapoda, Brachyura). *Lauterbornia* 15: 111-113.

Garland, E.M. 1981. The colonization of Windermere by *Crangonyx pseudogracilis* 1961 to 1964. *Freshwat. Biol. Assoc. Occasion. Publ.* 12: 1-13.

Geelen, J.F.M. 1975. *Orconectes limosus* (Raf.) and *Astacus astacus* L. (Crustacea, Decapoda) in the Netherlands. *Hydrobiol. Bull.* 9(3): 109-113.

Geissen, H.-P. 1997. Nachweis von *Limnomysis benedeni* Czerniavski (Crustacea: Mysidacea) im Mittelrhein. *Lauterbornia* 31: 125-127.

Glassner-Shwayder, K. 1996. Biological invasions. How aquatic nuisance species are entering North American waters, the harm they cause and what can be done to solve the problem. Great Lakes Panel on Aquatic Nuisance Species.

Güttinger, H. & Stumm, W. 1990. Ökotoxikologie am Beispiel der Rheinverschmutzung durch den Chemie-Unfall bei Sandoz in Basel. *Naturwiss.* 77: 253-261.

Harris, R.R. 1991. Amphipod also invades Britain. *Nature* 354: 194.

Harris, R.R. & Aladin, N.V. 1997. The ecophysiology of osmoregulation in Crustacea. In N. Hazon, F. B. Eddy & G. Flik (eds), *Ionic regulation in animals: a tribute to professor W.T.W. Potts:* 1-25, Berlin: Springer Verlag.

Harris, R.R. & Bayliss, D. 1990. Osmoregulation in *Corophium curvispinum* (Crustacea: Amphipoda), a recent coloniser of freshwater. III. Evidence for adaptive changes in sodium regulation. *J. Comp. Physiol. B* 160: 85-92.

Harris, R.R. & Muskó, I.B. 1999. Oxygen consumption, hypoxia and tube-dwelling in the invasive amphipod *Corophium curvispinum. J. Crustacean Biol.* 19: 224-234.

Heil, K.H. 1990. Die Auswirkung des Sandoz-Unfalles auf die Biozönose des Rheins. In G. Friedrich & R. Kinzelbach (eds), *Die Biologie des Rheins. Limnologie aktuell* 1: 11-26.

Herbst, V. & Bäthe, J. 1993. Die aktuelle Verbreitung der Gattung *Corophium* (Crustacea: Amphipoda) in der Weser. *Lauterbornia* 13: 27-35.

Hynes, H.B.N. 1955. Distribution of some freshwater Amphipoda in Britain. *Proc. Int. Assoc. theor. appl. Limnol.* 12: 620-628.

IRC 1987. Rhine action Programme. Report of the International Commission for Protection of the River Rhine against Pollution. Koblenz: 1-24.

IRC 1992. Ökologisches Gesamtkonzept für den Rhein. Report of the International Commission for Protection of the River Rhine against Pollution. Koblenz: 1-22.

Jazdzewski, K. 1980. Range extensions of some gammaridean species in european inland waters caused by human activity. *Crustaceana*, Suppl. 6: 84-107.

Jazdzewski, K. & Konopacka, A. 1990. New, interesting locality of Ponto-caspian gammarid *Echinogammarus ischnus* (Stebbing, 1898) (Crustacea, Amphipoda) in Poland. *Przeglad Zoologiczny* 34(1): 101-111 (in Polish).

Jazdzewski, K. & Konopacka, A. 1993. Remarks on the morphology, taxonomy and distribution of *Corophium curvispinum* G.O. Sars, 1895 and *Corophium sowinskyi* Martynov, 1924 (Crustacea, Amphipoda, Corophiidae). *Boll. Mus. civ. St. nat. Verona* 20: 487-501.

Jantz, B. 1996. Wachstum, Reproduktion, Populationsentwicklung und Beeinträchtigung der Zebramuschel (*Dreissena polymorpha*) in einem grossen Fliessgewässer, dem Rhein. Thesis University of Köln, Germany.

Kelleher, B., Bergers, P.J.M., Van den Brink, F.W.B., Giller, P.S., Van der Velde, G. & Bij de Vaate, A. 1998. Effects of exotic amphipod invasions on fish diet in the Lower Rhine. *Arch. Hydrobiol.* 143: 363-382.

Kelleher, B., Van der Velde, G., Giller, P.S. & Bij de Vaate, A. 1999a. Dominant role of exotic mass invaders in the diet of important fish species of the lower Rhine River. *Crustacean Issues* 12: 35-46.

Kelleher, B., Van der Velde, G., Wittmann, K.J., Faasse, M.A. & Bij de Vaate, A. (1999b). Current status of the freshwater Mysidae in the Netherlands with records of *Limnomysis benedeni*: Czerniavsky, 1882, a pontocaspian species in Dutch Rhine branches. *Bull. zool. Mus. Amsterdam.* (16/13): 89-96.

Ketelaars, H.A.M. & Van Breemen, L.W.C.A. 1993. The invasion of the predatory cladoceran *Bythotrephes longimanus* Leydig and its influence on the plankton communities in the Biesbosch reservoirs. *Verh. Internat. Verein. Limnol.* 25: 1168-1175.

Ketelaars, H.A.M. & Gille, L. 1994. Range extensions of the predatory cladoceran *Bythotrephes longimanus* Leydig 1860 (Crustacea, Onychopoda) in Western Europe. *Neth. J. Aquat. Ecol.* 28(2): 175-180.

Ketelaars, H.A.M., Van der Velden, J.A., Schutten, H.J. & Bijkerk, R. 1993. Invasie van rovende Staartwatervlo in Nederland. *BioNieuws* 3(17): 5.

Ketelaars, H.A.M., Lambregts-van de Clundert, Carpentier, C.J., Wagenvoort, A.J. & Hoogenboezem, W. 1999. Ecological effects of the mass occurrence of the Ponto-Caspian invader, *Hemimysis anomala*, G.O. Sars, 1907 (Crustacea: Mysidacea), in a freshwater reservoir in the Netherlands, with notes on its autecology and new records. *Hydrobiologia* 394: 233-248.

Kinzelbach, R. 1972. Zur Verbreitung und Ökologie des Süsswasser-Strandflohs *Orchestia cavimana* Heller, 1865 (Crustacea: Amphipoda: Talitridae). *Bonn. zool. Beitr.* 23: 267-282.

Kinzelbach, R. 1995. Neozoans in European waters. Exemplifying the worldwide process of invasion and species mixing. *Experientia* 51: 526-538.

Kinzelbach, R. 1997. Aquatische Neozoen in Europa. *Neozoen. Newsletter der Arbeitsgruppe Neozoen, Allgemeine und Spezielle Zoologie*; Universität Rostock. Nr. 1/1997: 7-8.

Kititsyna, L.A. 1980. Ecological and physiological pecularities of *Dikerogammarus haemobaphes* (Eichw.) in the region of the Tripolye State Supercentral Electric Station heated water discharge. *Gidrobiol. Zh.* 16(4): 77-85 (in Russian).

Köhn, J. & Waterstraat, A. 1990. The amphipod fauna of Lake Kummerow (Mecklenburg, German Democratic Republic) with reference to *Echinogammarus ischnus* Stebbing, 1899. *Crustaceana* 58: 74-82.

Koop, J.H.E. 1997. Survival strategies of amphipods in saltpolluted rivers: information from microcalorimetry and ^{31}P NMR spectroscopy. *Limnologica* 27: 307-309.

Kureck, A. 1992. Neue Tiere im Rhein. Die Wiederbesiedlung des Stroms und die Ausbreitung der Neozoen. *Naturwiss.* 79: 533-540.

Marguillier, S. 1998. Stable isotopes ratios and food web structure of aquatic ecosystems. Thesis Vrije Universiteit Brussel, Belgium.

Marguillier, S., Dehairs, F., Van der Velde, G., Kelleher, B. & Rajagopal, S. 1998. Initial results on the trophic relationships based on *Corophium curvispinum* in the Rhine traced by stable isotopes. In P.H. Nienhuis, R.S.E.W. Leuven & A.M.J. Ragas (eds), *New concepts for sustainable management of river basins*: 171-177, Leiden: Backhuys Publishers.

Meijering, M.P.D., Jazdzewski, K. & Köhn, J. 1995. Ecotypes of Amphipoda in central Europe inland waters. *Pol. Arch. Hydrobiol.* 42: 527-536.

Moog, O., Nesemann, H., Zitek, A. & Melcher, A. 1999. Erstnachweis der Süsswassergarnele *Atyaephyra desmaresti* (Millet 1831) (Decapoda) in Österreich. *Lauterbornia* 35: 67-70.

Mordukhai-Boltovskoi, F.D. 1949. Life cycle of some Caspian gammarids. *Dokl. Akad. Nauk SSSR* 46(5): 997-999 (in Russian).

Morton, B. 1997. The aquatic nuisance species problem: a global perspective and review. In F.M. D'Itri (ed.), *Zebra mussels and aquatic nuisance species*: 1-54, Chelsea, Michigan: Ann Arbor Press.

Muskó, I.B. 1990. Qualitative and quantitative relationships of Amphipoda (Crustacea) living on macrophytes in Lake Balaton. *Hydrobiologia* 191: 269-274.

Muskó, I.B. 1992. Life history of *Corophium curvispinum* G.O. Sars (Crustacea: Amphipoda) living on macrophytes in Lake Balaton. *Hydrobiologia* 243/244: 197-202.

Muskó, I.B. 1993. The life history of *Dikerogammarus haemobaphes* (Eichw.) (Crustacea: Amphipoda) living on macrophytes in Lake Balaton (Hungary). *Arch. Hydrobiol.* 127: 227-238.

Muskó, I.B. 1994. Occurrence of Amphipoda in Hungary since 1853. *Crustaceana* 66: 144-152.

Muskó, I.B., Tóth, L.G.& Szabó, E. 1995. Respiration and respiratory electron transport system (ETS) activity of two amphipods: *Corophium curvispinum* G.O. Sars and *Gammarus fossarum* Koch. *Pol. Arch. Hydrobiol.* 42: 547-558.

Muskó, I.B., Tóth, L.G. & Szabó, E. 1998. Respiratory energy loss of *Corophium curvispinum* (Crustacea: Amphipoda) in Lake Balaton (Hungary) during the vegetation period. *Verh. Internat. Verein. Limnol.* 26: 2107-2114.

Nijssen, H. & Stock, J.H. 1966. The amphipod, *Gammarus tigrinus* Sexton, 1939, introduced in the Netherlands (Crustacea). *Beaufortia* 13 (160): 197-206.

Paffen, B.G.P, Van den Brink, F.W.B., Van der Velde, G. & Bij de Vaate, A. 1994. The population explosion of the amphipod *Corophium curvispinum* in the Dutch Lower Rhine. *Wat. Sci. Tech.* 29: 53-55.

Pashkevich, A., Pavluk, T. & Bij de Vaate, A. 1996. Efficiency of a standardized artificial substrate for biological monitoring of river water quality. *J. Env. Mon. & Ass.* 40: 143-156.

Pinkster, S. 1973. The *Echinogammarus berilloni*-group, a number of predominantly Iberian amphipod species (Crustacea). *Bijdr. Dierk.* 43(1): 1-38.

Pinkster, S. 1975. The introduction of the alien amphipod *Gammarus tigrinus* Sexton, 1939 (Crustacea, Amphipoda) in the Netherlands and its competition with indigenous species. *Hydrobiol. Bull.* 9: 131-138.

Pinkster, S. & Platvoet, D. 1983. Further observations on the distribution and biology of two alien amphipods, *Gammarus tigrinus* Sexton, 1939, and *Crangonyx pseudogracilis* Bousfield, 1958, in the Netherlands (Crustacea, Amphipoda). *Bull. zool. Mus. Amsterdam* 9(17): 153-164.

Pinkster, S., Smit, H. & Brandse-de Jong, N. 1977. The introduction of the alien amphipod *Gammarus tigrinus* Sexton, 1939, in the Netherlands and its competition with indigenous species. *Crustaceana* Suppl. 4: 91-105.

Pinkster, S., Scheepmaker, M., Platvoet, D. & Broodbakker, N. 1992. Drastic changes in the amphipod fauna (Crustacea) of Dutch inland waters during the last 25 years. *Bijdr. Dierk.* 61: 193-204.

Platvoet, D. & Pinkster, S. 1995. Changes in the amphipod fauna (Crustacea) of the Rhine, Meuse and Scheldt estuary due to the 'Delta plan' coastal engineering works. *Neth. J. Aquat. Ecol.* 29: 5-30.

Ponomareva, Z.A. 1976. Distribution of some amphipods of the Caspian relict complex under different temperature conditions. *Izv. VNIORKH* 110: 36-40 (in Russian).

Ponyi, J. 1961. Über die Ernährung einiger Amphipoden (Crustacea) in Ungarn. *Annal. Biol. Tihany* 28: 117-123.

Rajagopal, S., Van der Velde, G., Paffen, B.G.P. & Bij de Vaate, A. (1999a). Growth and production of *Corophium curvispinum* G.O. Sars, 1895 (Amphipoda), an invader in the Lower Rhine. In F.R. Schram & J.C. von Vaupel Klein (eds), Crustaceans and the biodiversity crisis. Proc. 4th Internat. Crust. Congr., Amsterdam, The Netherlands, July 20-24, 1998. Vol I: 457-472. Leiden: Brill.

Rajagopal, S., Van der Velde, G., Paffen, B.G.P., Van den Brink, F.W.B & Bij de Vaate, A. (1999b). Life history and reproductive biology of the invasive amphipod *Corophium curvispinum* (Crustacea: Amphipoda) in the Lower Rhine. *Arch. Hydrobiol.* 144: 305-325.

Reinhold, M. & Tittizer,T. 1998. *Limnomysis benedeni* Czerniavsky 1882 (Crustacea: Mysidacea), ein weiteres pontokaspisches Neozoon im Main-Donau-Kanal. *Lauterbornia* 33: 37-40.

Romanova, N.N. 1963. Sposoby pitaniya u pushchevye gruppirovki donnykh bespozvonochnykh Severnogo Kaspiya. *Trudy Vsesosyuznogo gidrobiologicheskogo obshchestva*, T. 13. M. izd. AN SSSR, 1963.

Romanova, N.N. 1975. Quantitative distribution and ecology of Corophiids (Crustacea, Amphipoda, *Corophium*) in the Caspian Sea. *Byul. M. O-va Isp. Prirodi, Otd. biol.* 80(3): 51-63 (in Russian).

Savage, A.A. 1982. The survival and growth of *Gammarus tigrinus* Sexton (Crustacea: Amphipoda) in relation to salinity and temperature. *Hydrobiologia* 94: 201-212.

Schleuter, M. & Schleuter, A. 1995. *Jaera istri* (Veuille) (Janiridae, Isopoda) aus der Donau erreicht über den Main-Donau-Kanal den Main. *Lauterbornia* 21: 177-178.

Schleuter, A., Geissen, H.-P.& Wittmann, K.J. 1998. *Hemimysis anomala* G.O. Sars 1907 (Crustacea: Mysidacea), eine euryhaline pontokaspische Schwebgarnele in Rhein und Neckar – Erstnachweis für Deutschland. *Lauterbornia* 32: 67-71.

Schleuter, M., Schleuter, A., Potel, S. & Banning, M. 1994. *Dikerogammarus haemobaphes* (Eichwald 1841) (Gammaridae) aus der Donau erreicht über den Main-Donau-Kanal den Main. *Lauterbornia* 19: 155-159.

Schmitz, W. 1960. Die Einbürgerung von *Gammarus tigrinus* Sexton auf dem europäischen Kontinent. *Arch. Hydrobiol.* 57(1/2): 223-225.

Schöll, F. 1990a. Zur Bestandssituation von *Corophium curvispinum* Sars im Rheingebiet. *Lauterbornia* 5: 67-70.

Schöll, F. 1990b. Erstnachweis von *Chaetogammarus ischnus* Stebbing im Rhein. *Lauterbornia* 5: 71-74.

Sebestyèn, O. 1938. Colonization of two new fauna-elements of Pontus-origin (*Dreissensia poly-morpha* Pall. and *Corophium curvispinum* G.O. Sars forma *devium* Wundsch) in Lake Balaton. *Verh. Int. Verein. Limnol.* 8: 169-182.

Sinderman, C.J. 1992. Disease risks associated with importation of non-indigenous marine animals. *Mar. Fish. Rev.* 54: 1-10.

Smit, H., Van der Velde, G., Smits, R.G. & Coops, H. 1997. Ecosystem responses in the Rhine-Meuse Delta during two decades after enclosure and steps toward estuary restoration. *Estuaries* 20: 504-520.

Streit, B. & Kuhn, K. 1994. Effects of organophosphorus insecticides on autochthonous and introduced *Gammarus* species. *Wat. Sci. Tech.* 29: 233-240.

Sutcliffe, D.W. & Carrick, T.R. 1981. Effect of temperature on the duration of egg development, and moulting and growth in juveniles of *Crangonyx pseudogracilis* (Crustacea: Amphipoda) in the laboratory. *Freshwat. Biol.* 11: 511-522.

Taylor, P.M. & Harris, R.R. 1986a. Osmoregulation in *Corophium curvispinum* (Crustacea: Amphipoda), a recent coloniser of freshwater. I. Sodium regulation. *J. Comp. Physiol. B.* 156: 323-329.

Taylor, P.M. & Harris, R.R. 1986b. Osmoregulation in *Corophium curvispinum* (Crustacea: Amphipoda), a recent coloniser of freshwater II. Water balance and the functional anatomy of the antennary organ. *J. Comp. Physiol. B* 156: 331-337.

Tittizer, T. 1995. Faunenaustausch zwischen Main und Donau über den Main-Donau-Kanal mit besonderer Berücksichtigung der Neozoen. Fachgespräch Faunen- und Florenveränderung durch Gewässerausbau – Neozoen und Neophyten. Berlin, 16. November 1995. Umweltbundesamt, Fachgebiet II 1.3. Umweltforschungsplan des Bundesministeriums für Umwelt, Naturschutz und Reaktorsicherheit-Bodenschutz. Forschungsbericht 107 99 999/03. UBA-FB 96-096. Texte 74/96: 67-75.

Tittizer, T. 1996a. Main-Danube canal now a short cut for fauna. *Danube Watch* 2(3): 7-8.

Tittizer, T. 1996b. Vorkommen und Ausbreitung aquatischer Neozoen (Makrozoobenthos) in den Bundeswasserstrassen. In H. Gebhardt, R. Kinzelbach & S. Schmidt-Fischer (eds), *Gebietsfremde Tierarten. Auswirkungen auf einheimischen Arten, Lebensgemeinschaften und Biotope. Situationsanalyse.* Umweltministerium/Baden Württemberg. Landsberg, Ecomed Verlagsgesellschaft AG & Co. KG: 49-86.

Tittizer, T., Schöll, F. & Dommermuth, M. 1994. The development of the macrozoobenthos in the River Rhine in Germany during the 20th century. *Wat. Sci. Tech.* 29: 21-28.

Tittizer, T., Schöll, F. & Schleuter, M. 1990. Beitrag zur Struktur und Entwicklungsdynamik der Benthalfauna des Rheins von Basel bis Düsseldorf in den Jahren 1986 und 1987. In R. Kinzelbach & G. Friedrich (eds), *Biologie des Rheins. Limnologie aktuell* 1: 293-323.

Torchin, M.E., Lafferty, K.D. & Kuris, A.M. 1998. The European Green crab: its escape from natural enemies, and the safety and effiacacy of biological control. Eighth International Zebra Mussel and other aquatic nuisance species conference, March 16 to 19, 1998. Sacramento. Abstract book: 84.

van den Brink, F.W.B. & Van der Velde, G. 1986. Observations on the seasonal and yearly occurrence and the distribution of *Atyaephyra desmaresti* (Millet, 1831) (Crustacea, Decapoda, Natantia) in the Netherlands. *Hydrobiol. Bull.* 19: 193-198.

van den Brink, F.W.B., Van der Velde, G. & Bij de Vaate, A. 1989. A note on the immigration of *Corophium curvispinum* Sars, 1895 (Crustacea: Amphipoda) into the Netherlands via the river Rhine. *Bull. zool. Mus. Amsterdam* 11: 211-213.

van den Brink, F.W.B., Van der Velde, G. & Bij de Vaate, A. 1991. Amphipod invasion on the Rhine. *Nature* 352: 576.

van den Brink, F.W.B., Van der Velde, G. & Bij de Vaate, A. 1993. Ecological aspects, explosive range extension and impact of a mass invader, *Corophium curvispinum* Sars, 1895 (Crustacea: Amphipoda), in the Lower Rhine (Netherlands). *Oecologia* 93: 224-232.

van den Brink, F.W.B., Van der Velde, G. & Cazemier, W.G. 1990. The faunistic composition of the freshwater section of the river Rhine in The Netherlands: present state and changes since 1900. In R. Kinzelbach & G. Friedrich (eds), *Die Biologie des Rheins. Limnologie aktuell* 1: 191-216.

van den Brink, F.W.B., Paffen, B.G.P., Oosterbroek, F.M.J. & Van der Velde, G. 1993. Immigration of *Echinogammarus* (Stebbing, 1899) (Crustacea: Amphipoda) into the Netherlands via the Lower Rhine. *Bull. zool. Mus. Amsterdam* 13: 167-169.

van den Brink, F.W.B., Van der Velde, G., Buijse, A.D. & Klink, A.G. 1996. Biodiversity in the lower Rhine and Meuse river-floodplains: its significance for ecological river management. *Neth. J. Aquat. Ecol.* 30: 129-149.

van der Velde, G., Van den Brink, F.W.B., Van der Gaag, M. & Bergers, P.J.M. 1990. Changes in numbers of mobile macroinvertebrates and fish in the river Waal in 1987, studied by sampling the cooling-water intakes of a power plant: first results of a Rhine biomonitoring project. In R. Kinzelbach & G. Friedrich (eds), *Die Biologie des Rheins. Limnologie aktuell* 1: 325-343.

van der Velde, G., Van Urk, G., Van den Brink, F.W.B., Colijn, F., Bruggeman, W.A. & Leuven, R.S.E.W. 1991. Rein Rijnwater, een sleutelfactor in chemisch oecosysteemherstel. In G.P. Hekstra & F.J.M. van Linden (ed.), *Fauna en flora chemisch onder druk*: 231-266, Wageningen: Pudoc.

van der Velde, G., Paffen, B.G.P., Van den Brink, F.W.B., Bij de Vaate, A. & Jenner, H.A. 1994. Decline of Zebra mussel populations in the Rhine: competition between two mass invaders (*Dreissena polymorpha* and *Corophium curvispinum*). *Naturwiss.* 81: 32-34.

van der Velde, G., Rajagopal, S., van den Brink, F.W.B., Kelleher, B., Paffen, B.G.P., Kempers, A.J. & Bij de Vaate, A. 1998. Ecological impact of an exotic amphipod invasion in the River Rhine. In P.H. Nienhuis, R.S.E.W. Leuven & A.M.J. Ragas (eds), *New concepts for sustainable management of river basins*: 159-169, Leiden: Backhuys Publishers.

Vitousek, P.M., D'Antonio, C.M., Loope, L.L. & Westbrooks, R. 1996. Biological invasions as global environmental change. *Amer. Scientist* 84: 468-478.

Vorob'eva, A.A. & Nikonova, R.S. 1987. Rearing of the gammarids *Dikerogammarus haemobaphes* and *Niphargoides maeoticus*. *Gidrobiol. Zh.* 23(6): 52-56 (in Russian).

Waterstraat, A. & Köhn, J. 1989. Ein Beitrag zur Fauna des Kummerower Sees, Erstnachweis des Amphipoden *Echinogammarus ischnus* Stebbing, 1899, in der DDR. *Arch. Freunde Naturg. Mecklb.* 29: 93-106.

Weinzierl, A., Potel, S. & Banning, M. 1996. *Obesogammarus obesus* (Sars 1894) in der oberen Donau (Amphipoda, Gammaridae). *Lauterbornia* 26: 87-89.

Weinzierl, A., Seitz, G. & Thannemann 1997. *Echinogammarus trichiatus* (Amphipoda) und *Atyaephyra desmaresti* (Decapoda) in der bayerischen Donau. *Lauterbornia* 31: 31-32.

Williamson, M., 1996. *Biological invasions*. London: Chapman & Hall.

Witt, J.D.S., Hebert, P.D.N. & Morton, W.B. 1997. *Echinogammarus ischnus*: another crustacean invader in the Laurentian Great Lakes basin. *Can. J. Fish. Aquat. Sci.* 54: 264-268.

Wittmann, K.J. 1995. Zur Einwanderung potamophiler Malacostraca in die oberen Donau: *Limnomysis benedeni* (Mysidacea), *Corophium curvispinum* (Amphipoda) und *Atyaephyra desmaresti* (Decapoda). *Lauterbornia* 20: 77-85.

Wittman, K.J., Theiss, J. & Banning, M. 1999. Die Drift von Mysidacea und Decapoda und ihre Bedeutung für die Ausbreitung von Neozoen im Main-Donau-System. *Lauterbornia* 35: 53-66.

Dominant role of exotic invertebrates, mainly Crustacea, in diets of fish in the lower Rhine River

BARRY KELLEHER & GERARD VAN DER VELDE
Laboratory of Aquatic Ecology, Department of Ecology, University of Nijmegen, Nijmegen, Netherlands

PAUL S. GILLER
Department of Zoology & Animal Ecology, National University of Ireland, Lee Maltings, Cork, Ireland

ABRAHAM BIJ DE VAATE
Institute for Inland Water Management and Waste Water Treatment (RIZA), Lelystad, Netherlands

ABSTRACT

Recent water quality improvements to the stressed Rhine ecosystem have led to an increase in macroinvertebrate species richness. However, this enrichment is not only due to the return of some native macroinvertebrates, but mainly to an even greater number and rate of entry by exotic species. A qualitative assessment of the diet of dominant fish species taken from the filter screens of a power station situated on the lower Rhine indicates the high importance of exotic amphipods as food items for macrozoobenthivorous fish. *Corophium curvispinum* dominated the diet of eel (*Anguilla anguilla*), while exotic gammarids, *Gammarus tigrinus* and *Dikerogammarus villosus*, together with *C. curvispinum*, were important for ruffe (*Gymnocephalus cernuus*), perch (*Perca fluviatilis*) and juvenile stages of pikeperch (*Stizostedion lucioperca*). To a lesser extent, the exotic bivalves *Dreissena polymorpha* and *Corbicula* spp. were consumed by the cyprinids *Abramis brama*, *Abramis bjoerkna* and *Rutilus rutilus*. Comparison of the data from the present study with earlier research indicates that the dietary importance of exotic species to lower Rhine fish has been constant over the last decade; the generalist feeding of these fish has allowed them to cope with the relatively rapid flux in exotic species composition in the food base over the same period. The importance of exotic species to the Rhine's ecological processes and structure is discussed, as is the urgent need for their consideration in future plans for the Rhine's restoration.

1 INTRODUCTION

Recent improvements in the water quality of the lower Rhine, as part of the Rhine restoration plan initiated after the Sandoz chemical spill in 1986 (Capel et al. 1988), have not only encouraged the desired recolonisation by a small number of native macroinvertebrates, e.g. *Ephoron virgo* (Bij de Vaate et al. 1992), but also an even greater colonization by exotic species originating from different areas of the world (Den Hartog et al. 1992, Bij de Vaate 1993). The majority of these new species are molluscs and crustaceans. Their high rate of entry is thought to be related to the Rhine's high degree of connectivity with European and Eurasian systems via trade canals, and the intercontinental carriage of some brackish water species in the ballast water tanks of international shipping (Bij de Vaate & Klink 1995 and references therein). The Rhine is ion-rich, salinated and heavily polluted via eutrophication and pesticides (Den Hartog et al. 1989, 1992, Admiraal et al. 1993). This, in combination with the suitably euryhaline tolerance of these exotics (Den Hartog et al. 1989, Harris & Bayliss 1990, Van den Brink et al. 1990), leads to the high rate of colonisation. After invasion, it appears that many of these species occupy empty niche spaces and experience subsequent population explosions, facilitated by an r-strategy and abundant food supply (Van den Brink et al. 1993). While some species stabilize at sub-dominant densities in the Rhine after this initial phase, e.g. *Echinogammarus ischnus*, there are others that continue to maintain high population densities and these dominate the present macroinvertebrate community: the amphipods *Corophium curvispinum* and *Gammarus tigrinus*, and the bivalves *Corbicula fluminea* and *Corbicula fluminalis*.

Biological invasions can cause a number of disruptions to existing ecosystems (Drake et al. 1989, Fahnenstiel et al. 1995). Due to this, and with the relative densities of native lower Rhine macroinvertebrates being low, many questions regarding the importance of these exotic species in the ecological functioning of this unbalanced system have been posed (Den Hartog et al. 1992, Pinkster et al. 1992, Van den Brink et al. 1993). However, until now, only a few answers have been given. One of the questions that should be addressed deals with the trophic relationship between these exotes and the fish fauna. While much literature deals with the deliberate introduction of exotic amphipods for fishery improvement (see Conlan 1994 for review), considerably less work has been directed towards the study of the relationships between native predators and invasive exotic amhipods.

The lower Rhine fish fauna is largely dominated by euryecious species like roach (*Rutilus rutilus*), common bream (*Abramis brama*) and silver bream (*Abramis bjoerkna*), ruffe (*Gymnocephalus cernuus*), perch (*Perca fluviatilis*), eel (*Anguilla anguilla*) and the introduced pikeperch (*Stizostedion lucioperca*) (Van der Velde et al. 1990). While a study on the feeding ecology of these fish species was carried out shortly after the Sandoz chemical spill in 1986 (Bergers 1991), little can be said of the present status of trophic relationships between the macroinvertebrate and fish assemblage given the recent changes in macroinvertebrate species composition. This gap in knowledge has recently been indicated by a study on diet changes in eel and perch following exotic amphipod invasions, in common biotopes of the lower Rhine branch, the IJssel River (Kelleher et al. 1998). A significant change in the diet composition of both species, compared with the data of Bergers (1991), was noted, mainly as a consequence of the recent population explosion of *C. curvispinum*. Its abundance and the effect of its tubiculous mode of settlement (mud fixation) (Van der Velde et al. 1994, 1998) greatly influence the available food supply of fish. However, current dietary data are lacking for these fishes from other areas of the lower Rhine and more importantly, no data exist for the present feeding habits of other dominant fish species.

In order to fill this gap in knowledge, and thereby assess the food role of exotic invertebrates, the diet of eel, pikeperch, perch, ruffe and bream, sampled from the cooling water filters of a power station situated on the main branch of the lower Rhine, the river Waal, were investigated between 1994 and 1996. Sampling at power stations for the purpose of diet analysis has been used previously with success (Van den Broek 1978). This paper provides an indication of the proportion of exotic macroinvertebrates in the diets of the fishes studied, based on qualitative and semi-quantitative methods. The response of predators to new additions in the food base over the last decade was examined by hypothesizing that the recent dynamics of exotic species invasion and colonization of the lower Rhine could be demonstrated through analysis of temporal changes in their occurrence in the diets of the dominant fish species.

2 MATERIALS AND METHODS

Fish were collected from May to September in 1994, 1995 and 1996, at monthly intervals. Samples were taken from the filter screens of a cooling water intake of the Gelderland power station that has an open connection with the river Waal (Fig. 1) via the northern end of the Maas-Waal Canal, near Nijmegen. These filters prevent the entry of fishes and large debris taken in with the cooling water that would damage the intake machinery. For further details on the filtering process and machinery, refer to Hadderingh et al. (1983) and Van der Velde et al. (1990).

Only fish entering the screens at the time of sampling (9.00-11.00 am) were collected for the purpose of diet assessment, as feeding is likely to be highest in the early morning for most of the species sampled. This bias was intentionally introduced to sampling in order to exclude fish drawn in through the cooling water inlets, which

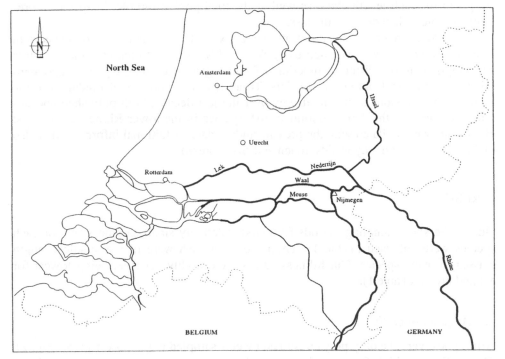

Figure 1. Map of the lower Rhine in the Netherlands, showing the sampling site at Nijmegen on the river Waal.

may remain on the sieve trays for a maximum of 24 hours. During this period the further digestion, or loss through regurgitation, of stomach contents may occur. Such specimens were thought to serve no useful purpose in the study as they may add an unreliable indication of the feeding habits of fish species. Chosen specimens were placed in plastic bags and stored in an ice filled cooler box while being transported back to the laboratory where the standard length and fresh weight of all individuals of each species were recorded (in order to divide each species into different length classes so as to relate fish size to diet).

Individual gut contents were identified using a stereomicroscope and counted. The number of fish with empty digestive tracts was recorded and expressed as a percentage of the total number collected for each species. An index of gut fullness, ranging from 1 (< 5 prey items) to 4 (full gut), was applied to each fish containing food. As the focus of the study was on the importance of macroinvertebrates in the food base of predators, the relative importance of prey to each fish species was assessed by graphical depiction of percentage frequency (% F) against percentage occurrence (% O) of prey in fish containing food (Costello 1990). This method was chosen as prey sizes were considered large enough not be overrated by a numerical method (see Hyslop 1980 for review of analysis methods). Furthermore, the model of Costello (1990) is a very convenient method to counter the disadvantages of these parameters and provides many indications of feeding strategies and food spectrum use by predators. Percentage biomass was not calculated due to the negative effects of power station intake on gut fullness of fish (Hadderingh et al. 1983). To examine variation in the relative importance of prey types to different life stages, plots of % F against % O are given for each length class, as well as the total pooled sample, in species represented by more than one length class.

The current importance of different exotic taxa with regards to the food spectrum of each fish species was assessed using % F and % O so as to indicate their relative consumption by the different functional feeding groups of fish, e.g. macrozoobenthivorous and zooplanktivorous feeders. To assess the responses of predators to the high rate of invasions of potential prey over the last decade, both published and unpublished data on the diet of dominant fish species in the lower Rhine over the last eight years were collated with the present study's data. Additional information is also included for two more cyprinids, roach and silver bream.

3   RESULTS

Due to the short sampling periods (2 hours), relatively small sample sizes for each species were collected (Table 1). The majority of fish were of the smallest length classes for each species. Gut fullness values were highly variable, being lowest for eel and bream (Table 1).

3.1   *Anguilla anguilla*

Arthropods were the only prey consumed by eels sampled at the power plant (Fig. 2). The exotic corophiid, *C. curvispinum*, was the only macrocrustacean present and was the predominant prey type. This exotic amphipod formed virtually the entire

Table 1. Total number, percentage of empty digestive tracts of different length classes (cm) and the median gut fullness value (M.G.F.) of fish containing food collected at a power station on the Dutch lower Rhine between 1994 and 1996.

Species	*n*	% Empty	M.G.F.
A. anguilla	25	44.0	2
24-33 cm	9	55.6	2
< 24 cm	16	37.5	2
*G. cernuus**	28	10.7	2
S. lucioperca	30	33.3	3
> 40 cm	3	0.0	4
20-40 cm	15	33.3	3.5
< 20 cm	12	41.7	1
P. fluviatilis	45	46.7	3.5
12-23 cm	2	0.0	4
< 12 cm	43	48.8	2
*A. brama***	23	52.2	2

*All < 10 cm, ** All > 28 cm.

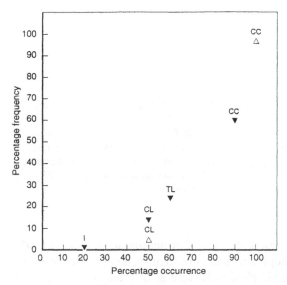

Figure 2. Relative importance of prey types to eel (*A. anguilla*) of < 24 cm (▼) and 24-33 cm (Δ) length. CC = *Corophium curvispinum*, TL = Trichopteran larvae, CL = Chironomid larvae, I = other Insecta.

diet composition of medium sized (24-33 cm) eels. In addition to *C. curvispinum*, small eels (< 24 cm) also consumed relatively small amounts of Chironomidae and Hydropsychidae. Other benthic taxa, *Corbicula* spp. and *Dreissena polymorpha* were also included in its diet in other parts of the lower Rhine (Table 2).

3.2 *Gymnocephalus cernuus*

All fish collected were less than 10 cm in length. Sediment was found in all of the fish, indicating a benthic foraging strategy that was reflected in the prey taxa found

Table 2. Relative importance of exotic macroinvertebrates in the diet of dominant fish species sampled at a power station and other locations (Kelleher unpubl. data, Kelleher et al. 1998) on the Dutch lower Rhine from 1994 to 1996.

Species	Bivalvia	Corophiidae	Gammaridae
A. anguilla	*	***	*
G. cernuus	–	**	***
S. lucioperca	–	**	*
P. fluviatilis	–	**	***
A. brama	–	–	*
A. bjoerkna	*	*	–
R. rutilus	**	**	–

* = Low importance, ** = Moderate importance, *** = High importance, – = Absent.

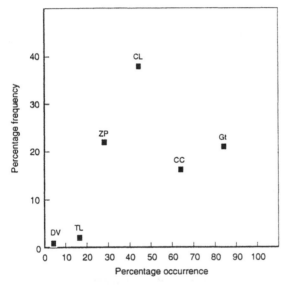

Figure 3. Relative importance of prey types to ruffe (*G. cernuus*) of < 10 cm length. CC = *Corophium curvispinum*, TL = Trichopteran larvae, CL = Chironomid larvae, Gt = *Gammarus tigrinus*, DV = *Dikerogammarus villosus*, ZP = Zooplankton.

(Fig. 3). Important prey types for most fish were Chironomidae and amphipods, *C. curvispinum* and *G. tigrinus*. Trichopteran larvae (*Hydropsyche contubernalis* and *Ecnomus tenellus*), daphnids and benthic chydorids were also found in some fish. A specimen of a recently invading exotic gammarid, *Dikerogammarus villosus*, was found in one fish sampled in May 1996.

3.3 *Stizostedion lucioperca*

Corophium curvispinum was the most important macroinvertebrate prey in the overall sample, and this importance was even greater when the predominantly piscivorous habit of larger pikeperch was depicted separately from the macrozoobenthivorous juveniles (Fig. 4). The corophiid was the only prey type consumed by fish of lengths less than 20 cm. The largest adult fish fed exclusively upon juvenile percids. Medium

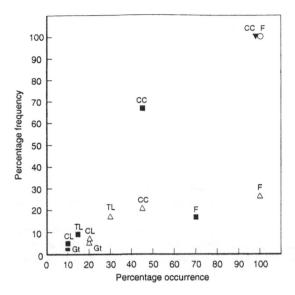

Figure 4. Relative importance of prey types to pikeperch (*S. lucioperca*) of < 20 cm (▼) and 20-40 cm (Δ), > 40 cm (O) length, and the total sample (■). CC = *Corophium curvispinum*, TL = Trichopteran larvae, CL = Chironomid larvae, Gt = *Gammarus tigrinus*, F = Teleostei.

sized fish had a more general diet, but the most important prey species were small percids. Trichoptera larvae (*E. tenellus* and Hydropsychidae) and *C. curvispinum* were the most important macroinvertebrate prey types for this size class, but were only preyed upon by a small number of specimens, which also contained juvenile fish.

3.4 *Perca fluviatilis*

The majority of perch sampled were juveniles. Only two mature specimens were collected. *C. curvispinum* and exotic gammarids were the most important prey species (Fig. 5). The recent flux of exotic gammarid species in the lower Rhine is particularly well reflected in the diet of perch (see Table 2). *G. tigrinus* was always the dominant gammarid taken over the sampling period, while *E. ischnus* was found only in 1994 in juvenile fish, *D. villosus* was first noted in 1995 in juveniles and later, as well as in juveniles, in a mature perch in 1996. Both of these less abundant gammarid species contributed 7% to diet and occurred in the same percentage of fish (23%). Some young perch fed heavily on microcrustaceans and Chironomidae.

3.5 *Abramis brama*

Exotic macroinvertebrates did not predominate in the diet of bream. Dietary composition was extremely variable between individuals (Fig. 6) but, numerically, Chironomidae and the native molluscs, *Bithynia tentaculata* and *Pisidium* sp., were the most important macroinvertebrate prey taxa overall. *G. tigrinus,* microcrustaceans and *E. tenellus* were present in minor quantities. Tables 2 and 3 show that exotics, particularly the bivalves *C. fluminea* and *C. fluminalis*, occurred to a greater extent in the other abundant cyprinids, roach and silver bream.

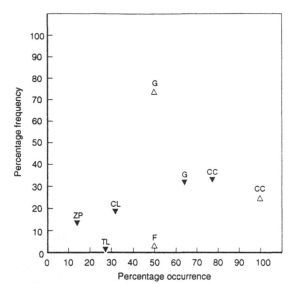

Figure 5. Relative importance of prey types to perch (*P. fluviatilis*) of < 12 cm (▼) and 12-23 cm (Δ) length. CC = *Corophium curvispinum*, TL = Trichopteran larvae, CL = Chironomid larvae, G = Gammaridae, F = Teleostei, ZP = Zooplankton.

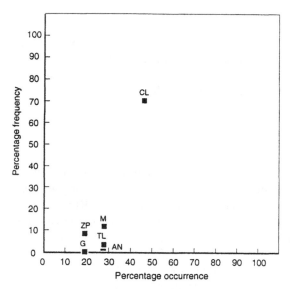

Figure 6. Relative importance of prey types to bream (*A. brama*) of > 28 cm length. AN = Annelida, M = Mollusca, TL = Trichopteran larvae, CL = Chironomid larvae, ZP = Zooplankton, G = Gammaridae.

4 DISCUSSION

The sampling of fish from a power station, while unconventional in diet studies, provided interesting data in a logistically convenient manner. In large, regulated rivers such as the Rhine, where river utilities inhibit more conventional ways of data collection (Petts et al. 1989), this method, where possible, should be seriously considered for the regular provision of data in fisheries research and management. Large numbers of small fish were sampled (Table 1), probably because small fish are more easily sucked into the cooling water intakes. Large fish collected were probably those in a weakened condition following spawning. The low gut fullness values and high percentage of empty guts observed were probably due to the physical effect of intake on the fishes, as the time of sampling coincided with early morning feeding of fish. These two factors seem intrinsically linked to power station sampling as suggested by results of Hadderingh et al. (1983). Nevertheless, while the general trend of gut fullness was low, all scores on the gut fullness index were well represented. The selection of fish being restricted to those entering during the sampling period was therefore a useful measure as the effect of intake on gut fullness affected the quantitative aspect of diet only. The exclusion of percentage biomass as a dietary parameter was therefore justified. However, Kelleher et al. (1998) reported that exotic amphipod species are important contributors to dietary biomass of eel and perch in groyne and sandpit biotopes of the IJssel River (Fig. 1).

Qualitatively, the data are indicative of the dietary components and feeding habits of these fish in other areas (Popova & Sytina 1977, Eie & Borgstrøm 1981, Bergman 1988, Muskó 1992 and references therein). In general, exotic amphipods are the most important prey taxa for dominant macrozoobenthivorous fish in the lower Rhine, with eel displaying a preference for more sessile animals (*C. curvispinum*) in comparison to the small length classes of the Percidae which consumed gammarids to a greater extent (Table 3). These amphipods are also included in the diet of these fish in other systems (references cited in Johnsen 1965, Biró 1974, Van den Brink et al. 1993, Muskó & Russo 1999). In contrast, the importance of exotic species is much less evident in the other feeding group made up of the dominant cyprinids (Table 3), where chironomids were most important in addition to exotic bivalves. Despite this apparent unimportance, trophic interaction between planktivorous/filter-feeding assemblages may be significant through indirect pathways (Mills & Forney 1988, Bronmark & Klosiewski 1992). The dietary indication on the dynamics of exotes entering, expanding and either declining or stabilizing (Table 2) compares favourably with information on sampled densities in the lower Rhine main channel over the same period. It indicates that, despite this quite rapid turnover of species, the high consumption of exotics by the fish fauna has been constant. This is probably due to these consumers employing a generalist feeding strategy in relation to the high abundances of the various exotes during their respective population expansion period.

Bergers (1991) and Den Hartog et al. (1992) described the lower Rhine in the period between the Sandoz chemical spill in 1986 and the population explosion of *C. curvispinum* around 1990, as a simplified ecosystem lacking complex trophic linkages. The present study has provided a qualitative overview of the developing food base in the lower Rhine following the breakdown in trophic structure in 1986, and shows that invasions of exotic amphipods over the last decade have restored predator-prey relationships. However, as deliberate introductions of exotic amphipods can incur negative effects on ecosystems (see Conlan 1994), it is possible that the invasive amphipods forming the lower Rhine food base may also negatively influence predator populations. Further study should be directed towards the influence of the changes in the available diet of fish, brought about by the high number of exotic amphipod invasions, over the structuring of predator populations and the fish community.

The dietary data and densities of exotes in the lower Rhine indicate that *C. curvispinum* and invasive gammarids play a significant role in energy mediation and tro-

Table 3. Population dynamics of exotic macroinvertebrate species in the lower Rhine as deduced by their occurrence and frequency in the diet of dominant fish species between 1988 (Bergers 1991) and 1994-1996.

Species	88	94	95	96	88	94	95	96	88	94	95	96
	D. polymorpha				*Corbicula* spp.				*C. curvispinum*			
A. anguilla	***	*	*	*	–	*	*	*	*	***	***	***
P. fluviatilis	*	–	–	–	–	–	–	–	–	*	**	**
S. lucioperca	–	–	–	–	–	–	–	–	*	***	***	***
G. cernuus	–	–	–	–	–	–	–	–	*	**	***	***
A. brama	**	–	–	–	–	–	–	–	–	–	–	–
A. bjoerkna	**	–	–	–	–	–	**	*	–	–	*	–
R. rutilus	**	**	–	–	–	*	–	–	–	**	–	–
	G. tigrinus				*E. ischnus*				*D. villosus*			
A. anguilla	***	*	*	*	–	–	–	–	–	–	*	*
P. fluviatilis	***	*	**	**	–	–	–	–	–	–	–	*
S. lucioperca	*	*	–	–	–	*	–	–	–	–	–	–
G. cernuus	***	***	***	**	–	**	–	–	–	–	**	**
A. brama	–	*	–	–	–	–	–	–	–	–	–	–
A. bjoerkna	***	*	–	–	–	–	–	–	–	–	–	–
R. rutilus	*	–	–	–	–	–	–	–	–	–	–	–

Values of occurrence and frequency: * = Low, ** = Medium, *** = High, – = Absent.

phic interactions in the ecosystem. We have shown that native predators can quickly assimilate newly occurring exotics into the food chain and so we argue the potential exists to apply fish diet analysis to monitor invasion dynamics and the trophic status of invaders. Furthermore, this and other studies (Van den Brink & Van der Velde 1991, Marguillier et al. 1998, Kelleher et al. 1998) demonstrate the importance of exotic macroinvertebrates for the recent functioning and trophic structure of the ecosystem. However, the rate of non-native species currently entering the lower Rhine system remains high and far exceeds that of indigenous types, indicating that the system has not reached an equilibrium, and is not sufficiently restored to bring about recolonization of native elements. Further research is required in order to determine the implications of this trend to the desired aim of the restoration of populations of native taxa.

ACKNOWLEDGEMENTS

The management of the Gelderland power station gave kind permission to use the facilities. This study was partially funded by an ERASMUS scholarship to B.K. F.W.B. van den Brink and A.D. Buijse are thanked for critically reading the manuscript and D. Nolan for helpful comments. The authors wish to thank B.G.P. Paffen, M.G. Versteeg, R.D. van Anholt for help with collection of samples, and A.E.J. Hanssen for laboratory assistance. The Illustration department of the Faculty of Sciences, University of Nijmegen, prepared the figures.

REFERENCES

Admiraal, W., van der Velde, G., Smit, H. & Cazemier, W.G. 1993. The rivers Rhine and Meuse in the Netherlands: present state and signs of ecological recovery. *Hydrobiologia* 265: 97-128.

Bergers, P.J.M. 1991. Feeding ecology of fishes in the Dutch Rhine branches. Institute for Inland Water Management & Waste Water Treatment, report No.28-1991 of the project *Ecological Rehabilitation of the river Rhine* (in Dutch).

Bergman, E. 1988. Foraging abilities and niche breadths of two percids, *Perca fluviatilis* and *Gymnocephalus cernua,* under different environmental conditions. *J. Anim. Ecol.* 57: 443-453.

Bij de Vaate, A. 1993. Exotic macroinvertebrates in the Dutch part of the River Rhine: causes and effects. In G.M. Van Dijk & E.C.L. Marteijn (eds), *Ecological rehabilitation of the river Rhine, The Netherlands research summary report (1988-1992).* Institute for Inland Water Management & Waste Water Treatment, report No.50-1993 of the project *Ecological Rehabilitation of the rivers Rhine and Meuse.*

Bij de Vaate, A. & Klink, A.G. 1995. *Dikerogammarus villosus* Sowinsky (Crustacea: Gammaridae) a new immigrant in the Dutch part of the Lower Rhine. *Lauterbornia* 20: 51-54.

Bij de Vaate, A., Klink, A.G. & Oosterbroek, F. 1992. The mayfly *Ephoron virgo* Olivier, back in the Dutch parts of the rivers Rhine and Meuse. *Hydrobiol. Bull.* 25: 237-240.

Biró, P. 1974. Observations on the food of eel (*Anguilla anguilla* L.) in Lake Balaton. *Annal. Biol. Tihany* 41: 133-152.

Bronmark, C. & Klosiewski, S.P. 1992. Indirect effects of predators in a freshwater benthic food-chain. *Ecology* 73: 1662-1674.

Capel, P.D., Giger, W., Reichart, P. & Wanner, O. 1988. Accidental input of pesticides into the river Rhine. *Environ. Sci. Tech.* 22: 992-997.

Conlan, K.E. 1994. Amphipod crustaceans and environmental disturbance: a review. *J. Nat. Hist.* 28: 519-554.

Costello, M.J. 1990. Predator feeding strategy and prey importance: a new graphical analysis. *J. Fish Biol.* 36: 261-263.

Den Hartog, C., Van den Brink, F.W.B. & Van der Velde, G. 1989. Brackish water invaders in the River Rhine. A bioindication for increased salinity level over the years. *Naturwissenschaften* 76: 80-81.

Den Hartog, C., Van den Brink, F.W.B. & Van der Velde, G. 1992. Why was the invasion of the River Rhine by *Corophium curvispinum* and *Corbicula* species so successful? *J. Nat. Hist.* 26: 1121-1129.

Drake, J.A., Mooney, H.A., Di Castri, F., Groves, R.H., Kruger, F.J., Rejmánek, M. & Williamson, M. (eds) 1989. *Biological invasions. A Global Perspective.* Chichester: Wiley.

Eie, J.A. & Borgstrøm, R. 1981. Distribution and food of roach [*Rutilus rutilus* (L.)] and perch (*Perca fluviatilis* L.) in the eutrophic Lake Årungen, Norway. *Verh. Internat. Verein. Limnol.* 21: 1257-1263.

Fahnenstiel, G.L., Lang, T.L., Bridgeman, G.A., McCormick, M.J. & Nalepa, T.F. 1995. Phytoplankton productivity in Saginaw Bay, Lake Huron: effects of zebra mussel (*Dreissena polymorpha*) colonization. *J. Great Lakes Res.* 21: 465-475.

Hadderingh, R.H., Van Aerssen, G.H.F.M., Groeneveld, L., Jenner, H.A. & Van der Stoep, J.W. 1983. Fish impingement at power stations located along the rivers Rhine and Meuse in The Netherlands. *Hydrobiol. Bull.* 17: 129-141.

Harris, R.R & Bayliss, D. 1990. Osmoregulation in *Corophium curvispinum* (Crustacea: Amphipoda), a recent coloniser of fresh water. III Evidence for adaptive changes in sodium regulation. *J. Comp. Physiol. B* 160: 85-92.

Hyslop, E.J. 1980. Stomach contents analysis – a review of methods and their application. *J. Fish Biol.* 17: 411-429.

Johnsen, P. 1965. Studies on the distribution and food of the ruffe (*Acerina cernua* L.) in Denmark, with notes on other aspects. *Meddelelser fra Danmarks Fiskeri og Havundersøgelser* 4(6): 137-156.

Kelleher, B., Bergers, P.J.M., Van den Brink, F.W.B., Giller, P.S., Van der Velde, G. & Bij de Vaate, A. (1998). Effects of exotic amphipod invasions on fish diet in the Lower Rhine. *Arch. Hydrobiol.* 143: 363-382.

Marguillier, S., Dehairs, F., Van der Velde, G., Kelleher, B. & Rajagopal, S. 1998. Trophic relationships based on *Corophium curvispinum* in the Rhine using stable isotopes: first results. In P.H. Nienhuis, R.S.E.W. Leuven & A.M.J. Ragas (eds), *New Concepts for Sustainable Management of River Basins*: 171-177. Leiden: Backhuys Publishers.

Mills, E.L. & Forney, J.L. 1988. Trophic dynamics and development on freshwater pelagic food webs. In S.R. Carpenter (ed.), *Complex interactions in lake communities*, 11-31. New York: Springer-Verlag.

Muskó, I.B. 1992. Life history of *Corophium curvispinum* G.O. Sars (Crustacea, Amphipoda) living on macrophytes in Lake Balaton. *Hydrobiologia* 243/244: 197-202.

Muskó, I.B. & Russo, A.R. 1999. Importance of *Corophium curvispinum* in Lake Balaton (Hungary): a colonization study. In F.R. Schram & J.C. von Vaupel Klein (eds), *Crustaceans and the biodiversity crisis*: 445-456. Leiden: Brill.

Pinkster, S., Scheepmaker, M., Platvoet, D. & Broodbakker, N. 1992. Drastic changes in the amphipod fauna (Crustacea) of Dutch inland waters during the last 25 years. *Bijdr. Dierk.* 61: 193-204.

Petts, G.E., Moller, H. & Rouse, A.L. (eds) 1989. *Historical Changes of Large Alluvial Rivers*. Chichester: Wiley.

Popova, O.A. & Sytina, L.A. 1977. Food and feeding relations of Eurasian perch (*Perca fluviatilis* L.) and pikeperch (*Stizostedion lucioperca* L.) in various waters of the USSR. *J. Fish. Res. Board Can.* 34: 1559-1570.

Van den Brink, F.W.B. & Van der Velde, G. 1991. Macrozoobenthos of floodplain waters of the rivers Rhine and Meuse in The Netherlands: a structural and functional analysis in relation to hydrology. *Regul. rivers: Res. Mgmt.* 6: 265-277.

Van den Brink, F.W.B., Van der Velde, G. & Cazemier, W.G. 1990. The faunistic composition of the freshwater section of the river Rhine in The Netherlands: present state and changes since 1900. In G. Friedrich & R. Kinzelbach (eds), *Die Biologie des Rheins. Limnologie aktuell* 1: 191-216.

Van den Brink, F.W.B., Van der Velde, G. & Bij de Vaate, A. 1993. Ecological aspects, explosive range extension and impact of a mass invader, *Corophium curvispinum* Sars, 1895 (Crustacea: Amphipoda), in the lower Rhine (The Netherlands). *Oecologia* 93: 224-232.

Van den Broek, W.L.F. 1978. Dietary habits of fish populations in the Lower Medway estuary. *J. Fish Biol.* 13: 645-654.

Van der Velde, G., Van den Brink, F.W.B., Van der Gaag, M. & Bergers, P.J.M. 1990. Changes in the numbers of mobile macroinvertebrates and fish in the river Waal in 1987, studied by sampling the cooling-water intakes of a power plant: first results of a Rhine biomonitoring project. In G. Friedrich & R. Kinzelbach (eds), *Die Biologie des Rheins. Limnologie aktuell* 1: 325-342.

Van der Velde, G., Paffen, B.G.P., Van den Brink, F.W.B., Bij de Vaate, A. & Jenner, H.A. 1994. Decline of Zebra mussel populations in the Rhine. Competition between two mass invaders (*Dreissena polymorpha* and *Corophium curvispinum*). *Naturwissenschaften* 81: 32-34.

Van der Velde, G., Rajagopal, S., Van den Brink, F.W.B., Kelleher, B., Paffen, B.G.P., Kempers, A.J. & Bij de Vaate, A. 1998. Ecological impact of an exotic amphipod invasion in the River Rhine. In P.H. Nienhuis, R.S.E.W. Leuven & A.M.J. Ragas (eds), *New Concepts for Sustainable Management of River Basins*: 159-169. Leiden: Backhuys Publishers.

Lessepsian immigration:
Human impact on Levantine biogeography

BELLA S. GALIL
Israel Oceanographic and Limnological Research, National Institute of Oceanography, Haifa, Israel

ABSTRACT

The threats of marine invasions to regional biodiversity, resulting in fundamental changes to ecosystem structure and function, impel us to look at the ecological consequences of the Suez Canal. Despite physical and hydrological impediments, hundreds of Red Sea species settled in the Mediterranean, forming thriving populations along the Levantine coasts. Though the expected outcome of invasion is reduction in diversity, we witness increased diversity – it is the unique history of the easternmost Mediterranean, that left it warm, salty and impoverished, that is at the base of a singular synergy between anthropogenic and environmental factors.

1 INTRODUCTION

Invasions by allochthonous species occur in aquatic ecosystems throughout the world, leading to significant, and sometimes severe, biological repercussions and economic effects. The cumulative outcome of invasions dominates anthropogenic global change, harming ecosystem structure and function, and reducing biodiversity (Cohen & Carlton 1995). The realization that human impact may irreversibly alter the marine environment, and threats of marine invasions to regional biodiversity, impel us to look at the ecological consequences of the largest marine engineering enterprise of the last century. The opening of the Suez Canal joined two seas, initiating a remarkable faunal movement. Despite physical and hydrological impediments, hundreds of Red Sea species settled in the Mediterranean, forming thriving populations along the Levantine coasts – this extraordinary movement was named after the canal builder 'Lessepsian migration' (Por 1978).

2 HISTORICAL BACKGROUND

The construction of a sea-level waterway between the Levantine Mediterranean and the Gulf of Suez, linked the Atlantic-Mediterranean biogeographic province with the Indo-Pacific. The deep-water communication between the Palaeo-Mediterranean and the Indian Ocean was breached in the early Miocene (Robba 1987), though intermittent marine contacts continued well into the Messinian (Sonnenfeld 1985), and species of Indo-Pacific origin still inhabited the Pliocene Mediterranean (Sorbini 1988). Early in the Pliocene, the Isthmus of Suez was elevated and the two seas separated altogether. Thus, the present day canal is an iteration of an earlier connection.

The Suez Canal, 163 km long, traverses a series of shallow lakes, the waters of which ranged from brackish to hypersaline. In the first decades following the opening of the canal, presumably only the hardiest organisms could migrate, due to the barrier presented by salinity as high as 16.1% in the Bitter Lakes (Vadiya & Shenuda 1985). As canal water flowed through the Bitter Lakes, the salinity fell, and measurements taken a century later recorded salinity not much higher than in the northern Gulf of Suez (Morcos 1960). Another obstacle to invasion has been removed with the damming of the Nile at Aswan. The Nile floods coincided with the summer current inversion and the inflow of Mediterranean water into the canal, resulting in markedly decreased salinity, as low as 2.6%, in the northern portion of the canal and in the adjacent Mediterranean (Vadiya & Shenuda 1985). Since the completion of the Aswan high dam, the Nile flow has been reduced drastically. Deepened and widened several times, the canal now is 2.5 times its size in 1975, 4 times its size in 1956, and 14 times its size at 1869 (http:/www.sis.gov.eg). Each increase in volume lessened temperature fluctuations within the canal, whereas increasing depth has probably facilitated passage of inhabitants of the infralittoral.

3 LESSEPSIAN MIGRANTS

Lessepsian immigrants founded thriving populations along the Levantine coasts. A considerable number reaches the southern Turkish coast, but declines west of Rhodes. Along the north African coast Red Sea species have spread as far west as Malta, Sicily and Tunis (Chakroun 1966, Ariani & Sarra 1969, Ktari-Chakroun & Bouhal 1971, Stevcic 1979, Ben Tuvia 1985).

Some migrants had become so abundant as to be exploited commercially, their prominence in the local communities attested to by fishery statistics. Immigrant fish constitute nearly half of the total trawl catches off the Israeli coast (Golani & Ben Tuvia 1995), and nearly half of total fish biomass on hard bottom habitats (Spanier et al. 1989). Gücü & Bingel (1994) estimate that, off the southeastern coast of Turkey, immigrants compose 62% of demersal fish biomass in the Gulf of Iskenderun, 34% in Mersin Bay. An early immigrant into the Mediterranean, *Portunus pelagicus* was among the first to appear in local markets: Fox (1927) reported '*Neptunus pelagicus* is fished for sale in the markets at Alexandria and at Haifa' – and it is still in great demand today. Two immigrant penaeid prawns, *Penaeus japonicus* and *Metapenaeus monoceros*, make up most of the shrimp catches along both the Egyptian and Israeli coasts (Galil 1986).

3.1 *Types of migrants*

Many immigrants have become ubiquitous, but by no means, all. In fact, three modes of population gain can be discerned:

1. Persevering,
2. Exploding, and
3. Ubiquitous.

Ixa monodi was first recorded in the Mediterranean by Holthuis & Gottlieb (1958); *Alpheus migrans* was described by Lewinsohn & Holthuis (1978) – both are considered rather rare residents. Many, if not most, of the immigrants have established small, but stable populations off the Levantine coast.

Rather more attention is given those instances of explosive population gain followed by sharp decline. In the late 1940s the immigrant goldband goatfish, *Upeneus moluccensis*, made up 10-15% of the total mullid catches. Following the exceptionally warm winter of 1954-1955, its percentage in the catch increased to 83% (Oren 1957). Its share has since been reduced to 30% of the catch (Ben Tuvia 1973). Following that same winter, the brushtooth lizardfish, *Saurida undosquamis*, became a commercially important fish and its proportion in trawl fisheries catches rose to 20% in the late 1950s. The population then diminished and catches stabilized at about 5% of the total trawl catch (Ben Yami & Glazer 1974). Similarly, the gastropod *Rhinoklavis kochi*, first reported in Haifa Bay in the mid-1960s, spread rapidly to become, by the late 1970s, one of the dominant species on sandy-mud bottoms between 20 and 60 m (Galil & Lewinsohn 1981, Tom & Galil 1990). Samples taken a decade later consisted mostly of empty shells (Galil 1993).

Then, there are immigrants that are common and abundant. *Strombus decorus persicus*, first sighted in Mersin Bay, Turkey, in 1978 (Nicolay & Romagna Manoja 1983) and then in quick succession off southern Turkey (Raybaudi-Massilia 1983), Rhodes (Verhecken 1984, Barash & Danin 1989), Cyprus (Bazzocchi 1985, Robinson 1989) and Israel. The shallow sandy littoral is littered with them: 'one can speak of an invasion... hundreds of dead shells on the beaches and shoals of live *Strombus*, of all sizes, colors and patterns, feeding on the sea floor up to 20 m depth' (Curini-Galletti 1988). Another immigrant that proliferated in astonishingly short space of time is the nomadic jellyfish, *Rhopilema nomadica* (Galil et al. 1990). It was first collected in the Mediterranean in 1977, by the mid-1980s huge swarms would appear each summer along the southeastern Levant coast, and by 1995 also off the southeastern coast of Turkey (Kideys & Gücü 1995) and Cyprus. The massive swarms, the sizable biomass of these voracious planktotrophs must play havoc with the meager sources of this oligotrophic sea, and when those shoals draw nearer shore, they impact fisheries, coastal installations and tourism.

3.2 *Consequences of invasion*

Invasion is often followed by competition for resources or direct interference between autochthonous and allochthonous species, the latter outcompeted wholly, or in part, of their habitat space. Por (1978) maintained that 'Other than the case of *Asterina gibbosa* there is no known case in which a Lessepsian migrant species has completely replaced a local one'. Indeed, the decimation of the indigenous sea star, *A. gibbosa* populations from the Israeli coast paralleled the rapid advent of its Red Sea congener *A. wega* (Achituv 1973). However, it is far from singular: there are other documented instances of an extreme change in abundance that can be attributed to the new competition. A native penaeid prawn, *Penaeus kerathurus*, was 'very commonly caught by trawlers on Israel coastal shelf especially on sandy or sandy mud bottoms' according to Holthuis & Gottlieb (1958), and supported a commercial fishery throughout the 1950s. It has since nearly disappeared and its habitat overrun by the Lessepsian penaeid prawns. Geldiay & Kocatas (1972) reported that off the southern coast of Turkey too, *P. japonicus* is replacing *P. kerathurus*, and the rapid advent of *Metapenaeus monoceros* into the Gulf of Gabes, Tunis, has raised concerns over the fate of *P. kerathurus* fisheries there (Chaouachi et al. 1998). The immigrant snapping shrimps *Alpheus inopinatus* and *A. edwardsii* are more common now in the rocky littoral than the native *A. dentipes* (cf. Lewinsohn & Galil 1982). The decrease in the numbers of the previously prevalent indigenous jellyfish, *Rhizostoma pulmo*, have coincided with the massive presence of *Rhopilema nomadica*, and may also be a case of competitive displacement.

Competition might also take the form of bathymetric adjustment. Such competitive displacement probably occurs between the local red snapping shrimp, *Alpheus glaber*, and the intruding congener *A. rapacida*; the former appears mainly between 35 and 145 m and the latter between 15 and 50 m (Galil 1989). The local red mullet, *Mullus barbatus*, and the native hake, *Merluccius merluccius*, were both displaced into deeper, cooler waters by their respective Red Sea competitors: *Upeneus moluccensis* and *Saurida undosquamis* (Oren 1957). But depth-shifting among Red Sea immigrants could be ascribed also to the physical characteristics of the environment. Red Sea migrants that are rock and rubble dwellers are confined in the southeastern Levant to the littoral for want of hard substrate in deeper waters. Among soft bottom immigrants, some occupy in the Levant depths similar to those in the Red Sea, while others seem to shift their depth range to 20-50 m. As temperature is considered the most important single factor for Red Sea immigrants, it is assumed that the winter cooling of the shallow waters is the cause for the bathymetric shifting (Galil 1989).

However, the immigrants' ascendancy resulted not only in displacement of some indigenous species but, increasingly, in displacement among the immigrant species themselves. *Trachypenaeus curvirostris* was first recorded in the Mediterranean in the late twenties (Steinitz 1929), already so abundant, it was sold on the Haifa fish market and was the most common penaeid on sandy-mud bottoms (Galil 1986). In 1987 another immigrant, *Metapenaeopsis aegyptia* joined it on the sandy-mud bottoms (Galil & Golani 1990). In samples collected along the central Israeli coast at depth of 35 m, *M. aegyptia* outnumbered *T. curvirostris* 3 to 1 by 1993, and 25 to 1 by 1996. That same year yet another immigrant penaeid was recorded on the sandy-mud bottoms, *Metapenaeopsis mogiensis consobrina* (Galil 1997).

Charybdis longicollis, was first recorded in the Mediterranean in the mid 1950s (Holthuis 1961), and has since dominated the macrobenthic fauna on silty sand bottoms off the Israeli coast, forming up to 70% of the biomass in places (Galil 1986). Of the thousands of specimens collected over three decades, none was parasitized till, in 1992, few parasitized crabs were collected, the parasite identified as a sacculinid rhizocephalan, *Heterosaccus dollfusi* – a Red Sea immigrant itself (Galil & Lützen 1995). Within three years it spread as far as the eastern Anatolian coast and infection rate at Haifa Bay rose to 77.0%. The rapid increase and the high prevalence of *H. dollfusi* infestation is ascribed to the dense population of the host and the year-round reproduction of the parasite, facilitating re-infection. In the year just past, we have found, for the first time since the invasion of *Charybdis*, large numbers of the indigenous *Liocarcinus vernalis* – has the toll exacted by the parasite paved the way for the return of the native species?

4 CONCLUSIONS

Lessepsian immigration is a dynamic, ongoing, surprising process, repeatedly upsetting pet assumptions. Contrary to views held a decade ago that 'none of the many zooxanthellae bearing organisms of the Red Sea could get a foothold in the Mediterranean' (Por 1989) – a zooxanthellate jellyfish, *Cassiopea andromeda*, was noted in the Mediterranean as early as the beginning of the century (Maas 1903), and is known from Israel (Spanier 1989), Lebanon (Goy et al. 1988), and the southern Aegean (Schafer 1955). Similarly, it was posited that 'typical coral reef families... cannot colonize the Levant prior to the settlement of the reef building coelenterates' (Ben Tuvia 1985), yet a pomacentrid fish, *Abudefduf vaigiensis*, commonly associated with coral reefs from Australia to the Red Sea, was recently found to inhabit rocky knolls off the northern Mediterranean coast of Israel (Goren & Galil 1998).

The large contingent of successful Red Sea colonists evinces that habitats suitable for tropical species are available in this region. The lower species diversity in the Levant (Fredj 1974) is related to the comparatively late recolonization of the region after the Messinian crisis. Moreover, the summer water temperature is tropical but the present day Levantine fauna, due to the Pleistocene extinctions, lacks the many subtropical and tropical Atlantic species, whose niche is only partially occupied (Türkay 1989), with many of the species present presumably at the limit of their ecological tolerance – leaving it vulnerable to invasion.

The Red Sea, part of the tropical ocean, enjoys higher species diversity than the Levantine Mediterranean, and species from a richer ecosystem are considered competitively superior. In addition, Red Sea species are better adapted to the higher temperature and salinity. The sizable assemblage of Red Sea species that has taken residence along the Levantine infralittoral, modifying the composition and structure of the biota, enhancing its tropical affinities, has disrupted the zoogeographic unity of the Mediterranean fauna and turned the Levant into a 'semi-tropical' province. Though the expected outcome of invasion is reduction in diversity, we witness an invasion that increases diversity. The unique history of the easternmost Mediterranean, that left it warm, salty and impoverished, is at the base of a singular synergy between anthropogenic and environmental factors, past and present. As for the future, the planned expansion of the Suez Canal, the increasing salinity of the eastern Mediterranean (Roether & Klein 1998), and the predicted global warming favor new cohorts of immigrants.

REFERENCES

Achituv, Y. 1973. On the distribution and variability of the Indo-Pacific sea star *Asterina wega* (Echinodermata: Asteroidea) in the Mediterranean Sea. *Mar. Biol.* 18: 333-336.
Ariani, A.P. & Sarra, V. 1969. Sulla presenza del *Portunus pelagicus* (L.) in acque Italiane, con osservazioni sulla morfologia della specie (Crustacea Decapods). *Arch. Bot. Biogeogr. Ital.* 45: 186-206.

Barash, A. & Danin, Z. 1989. Marine Mollusca at Rhodes. *Isr. J. Zool.* 35: 1-74.

Bazzocchi, P. 1985. Prima segnalazione di *Strombus* (*Conomurex*) *decorus raybaudii* Nicolay and Romagna Manoja, 1983 per l'isola di Cipro. *Boll. Malacologico* 21: 64.

Ben Tuvia, A. 1973. Man-made changes in the Eastern Mediterranean sea and their effect on the fishery resources. *Mar. Biol.* 19: 197-203.

Ben Tuvia, A. 1985. The impact of the Lessepsian (Suez Canal) Fish Migration on the Eastern Mediterranean Ecosystem. In M. Moraitu-Apostolopoulou & V. Kiortsis (eds), *Mediterranean Marine Ecosystems*, pp 197-203. New York: Plenum Press.

Ben Yami, M. & Glaser, T. 1974. The invasion of *Saurida undosquamis* (Richardson) into the Levant Basin – An example of biological effect of interoceanic canals. *Fishery Bull.* 72: 359-373.

Chakroun, F. 1966. Captures d'animaux rares en Tunisie. *Bull. Inst. Nat. Sci. Tech. Océanogr. Pêche Salammbo* 1: 75-79.

Chaouachi, B., Dos Santos, A. & Ben Hassine, O.K. 1998. Occurrence of the shrimp *Metapenaeus monoceros* (Fabricius 1798) in the Gulf of Gabes: A first record for Tunisian coasts. *Proceedings and Abstracts of the 4th International Crustacean Congress, Amsterdam, 20-24 July 1998*, p. 80.

Cohen, A.N. & Carlton, J.T. 1995. Nonindigenous aquatic species in a United States estuary: A case study of the biological invasions of the San Francisco Bay and delta. US Fisheries and Wildlife Service, 246 pp.

Curini Galletti, M. 1988. Notes & Tidings. *La Canchiglia* 20(232-233): 14.

Fox, H.M. 1927. Appendix to the Report on the Crustacea Decapoda (Brachyura). Zoological Results of the Cambridge Expedition to the Suez Canal, 1924. *Trans. Zool. Soc. London* 22: 217-219.

Fredj, G. 1974. Stockage et exploitation des données en écologie marine. C. Considérations biogéographiques sur le peuplement benthique de la Méditerranée. *Mém. Inst. Océanogr. Monaco* 7: 1-88.

Galil, B.S. 1986. Red Sea Decapods along the Mediterranean Coast of Israel, Ecology and distribution. In Z. Dubinsky & Y. Steinberger (eds), *Environmental Quality and Ecosystem Stability*, Vol. III/A, pp. 179-183. Bar Ilan University Press.

Galil, B.S. 1989. Bathymetric distribution and habitat preferences of Lessepsian migrant Decapoda along the Mediterranean coast of Israel, or: do decapods have cold feet? In E. Spanier, Y. Steinberger & M. Luria (eds), *Environmental Quality and Ecosystem Stability*, Vol. IV/V pp. 147-153. Jerusalem: ISEEQS.

Galil, B.S. 1993. Lessepsian migration: New findings on the foremost anthropogenic change in the Levant Basin fauna. In N.F.R. Della Croce (ed.), *Symposium Mediterranean Seas 2000*, pp. 307-318.

Galil, B.S. 1997. Two Lessepsian migrant decapods new to the coast of Israel. *Crustaceana* 70: 111-114.

Galil, B.S. & Golani, D. 1990. Two new migrant decapods from the eastern Mediterranean. *Crustaceana* 58: 229-236.

Galil, B.S. & Lewinsohn, Ch. 1981. Macrobenthic Communities of the Eastern Mediterranean Continental Shelf. *Mar. Ecol.* 2: 343-352.

Galil, B.S. & Lützen, J. 1995. Biological observations on *Heterosaccus dollfusi* Boschma (Cirripedia: Rhizocephala), a parasite of *Charybdis longicollis* Leene (Decapoda: Brachyura), a Lessepsian migrant to the Mediterranean. *J. Crust. Biol.* 15: 659-670.

Galil, B.S., Spanier, E. & Ferguson, W.W. 1990. The Scyphomedusae of the Mediterranean coast of Israel, including two Lessepsian migrants new to the Mediterranean. *Zoöl. Meded. (Leiden)* 64: 95-105.

Geldiay, A. & Kocatas, A. 1972. A report on the occurrence of Penaeidae (Decapoda Crustacea) along the coast of Turkey from Eastern Mediterranean to the vicinity of Izmir, as a result of migration and its factors. *17th Congr. Intern. Zool.* Monte Carlo, 7 pp.

Golani, D. & Ben-Tuvia, A. 1995. Lessepsian migration and the Mediterranean fisheries of Israel. In N.B. Armantrout (ed.), *Conditions of the world's aquatic habitats. Proc. World Fish. Cong.*, pp. 279-289. New Delhi.

Goren, M. & Galil, B.S. 1998. First record of the Indo-Pacific reef fish *Abudefduf vaigiensis* (Quoy and Gaimard 1825) in the Levant. *Isr. J. Zool.* 44: 57-59.

Goy, J., Lakkis, S. & Zeidane, R. 1988. Les Meduses de la Méditerranée orientale. *Rapp. Comm. Int. Mer. Médit.* 31: 299.

Gücü, A.C. & Bingel, F. 1994. Trawlable species assemblages on the continental shelf on the Northeastern Levant Sea (Mediterranean) with an emphasis on Lesseptian migration. *Acta Adriat.* 35: 83-100.

Holthuis, L.B. 1961. Report on a collection of Crustacea Decapoda and Stomatopoda from Turkey and the Balkans. *Zoöl. Verh. (Leiden)* 47: 1-67.

Holthuis, L.B. & Gottlieb, E. 1958. An annotated list the Decapod Crustacea of the Mediterranean coast of Israel, with an appendix listing the Decapoda of the Eastern Mediterranean. *Bull. Res. Counc. Isr.* 7B: 1-126.

Kideys, A.E. & Gücü, A.C. 1995. *Rhopilema nomadica*: a Lessepsian scyphomedusan new to the Mediterranean coast of Turkey. *Isr. J. Zool.* 41: 615-617.

Ktari-Chakroun, F. & Bouhlal, M. 1971. Capture de *Siganus luridus* (Ruppell) dans le golfe de Tunis. *Bull. Inst. Nat. Sci. Tech. Océanogr. Pêche Salammbo* 2: 49-52.

Lewinsohn, Ch. & Galil, B. 1982. Notes on species of *Alpheus* (Crustacea Decapoda) from the Mediterranean coast of Israel. *Quad. Lab. Tecno. Pesca. Ancona* 3(2-5): 207-210.

Lewinsohn, Ch. & Holthuis, L.B. 1978. On a new species of *Alpheus* (Crustacea, Decapoda, Natantia) from the eastern Mediterranean. *Zoöl. Meded. (Leiden)* 53: 75-82.

Maas, O. 1903. Die Scyphomedusen der Siboga Expedition. *Siboga Exped. Mon.* 11: 1-91.

Morcos, S.A. 1960. Die Verteilung des Salzgehaltes im Suez-Kanal. *Kiel. Meeresforsch.* 16: 133-154.

Nicolay, K. & Romagna Manoja, L. 1983. *Strombus (Conomurex) decorus raybaudii* n. sp. *La Conchiglia* 15: 176-177.

Oren, O.H. 1957. Changes in the temperature of the eastern Mediterranean Sea in relation to the catch of the Israel trawl fishery during the years 1954/55 and 1955/56. *Bull. Inst. Océanogr. Monaco* 1102: 1-12.

Por, F.D. 1978. Lessepsian Migration – The Influx of Red Sea Biota into the Mediterranean by Way of the Suez Canal. *Ecological Studies, 23*. Heidelberg-New York: Springer-Verlag.

Por, F.D. 1989. *The Legacy of Tethys – an Aquatic Biogeography of the Levant*. Dordrecht: Kluwer.

Raybaudi-Massilia, L. 1983. A *Strombus* from the Mediterranean. *La Conchiglia* 15: 174-175.

Robba, E. 1987. The final occlusion of Tethys: Its bearing on Mediterranean benthic molluscs. In K.G. McKenzie (ed.), *Shallow Tethys 2'*, pp. 405-426. Rotterdam: Balkema.

Robinson, J. 1989. Recent finds, *Strombus decorus persicus*. *Hawaiian Shell News* 37: 10.

Roether, W. & Klein, B. 1998. The great eastern Mediterranean deep water Transient. *Rapp. Comm. Int. Mer. Médit.* 35: 12-16.

Schafer, W. 1955. Eine Qualle aus dem Indischen Ozean in der Ägäis. *Natur u. Volk.* 85: 241-245.

Sonnenfeld, P. 1985. Models of Upper Miocene Evaporite Genesis in the Mediterranean Region. In D.J. Stanley & F.C. Wezel (eds), *Geological Evolution of the Mediterranean Basin*, pp. 323-346. Heidelberg-New York: Springer Verlag.

Sorbini, L. 1988. Biogeography and climatology of Pliocene and Messinian fossil fish of Eastern-Central Italy. *Bull. Mis. Civ. St. nat. Verona* 14: 1-85.

Spanier, E. 1989. Swarming of Jellyfishes along the Mediterranean coast of Israel. *Israel J. Zool.* 36: 55-56.

Spanier, E., Pisanty, S., Tom, M. & Almog-Shtayer, G. 1989. The fish assemblages of the coralligenous shallow shelf of the Mediterranean coast of Israel. *J. Fish. Biol.* 35: 641-649.

Steinitz, W. 1929. Die Wanderung indopazifischer Arten ins Mittelmeer seit Beginn der Quartärperiode. *Inst. Rev. ges. Hydrobiol. Hydrogr.* 22: 1-90.

Stevcic, Z. 1979. Contribution á la connaissance des Crustacés Décapodes de Malta. *Rapp. Comm. Int. Mer. Médit.* 25/26: 127-128.

Tom, M. & Galil, B.S. 1990. The Macrobenthic Associations of Haifa Bay, Mediterranean Coast of Israel. *Mar. Ecol.* 12: 75-86.

Türkay, M. 1989. Subtropische und tropische Elemente in der Fauna des ostlichen Mittelmeer. *Natur Mus.* 119: 183.

Vadiya, V. & Shenuda, S. 1985. Role of the Suez Canal and flow from the Nile river in changing the salinity and fauna of the Mediterranean Sea. *Ichtyol.* 25: 134-135.

Verhecken, A. 1984. *Strombus decorus raybaudii* in de Middellandse Zee. *Gloria Maris* 23(4): 79-88.

Immigration history and present distribution of alien crustaceans in Polish waters

KRZYSZTOF JAŹDŻEWSKI & ALICJA KONOPACKA
Department of Invertebrate Zoology & Hydrobiology, University of Łódź, Poland

ABSTRACT

The authors present the story of the introduction of several malacostracan crustaceans into Polish Baltic and inland waters. These include: two American crayfish, *Orconectes limosus* (introduced in 1890) and *Pacifastacus leniusculus* (introduced around 1980); and two brackish water crabs, *Eriocheir sinensis* (introduced around 1920) and *Rhithropanopeus harrisii* (introduced around 1945), probably ships' ballast water newcomers. Crustaceans entering Polish waters due to the interlinking of formerly separated river basins are the Ponto-Caspian amphipods, *Corophium curvispinum* and *Chaetogammarus ischnus*, which appeared in the Vistula and Oder basins at the beginning of this century. Quite recently new migrants to Polish waters have been recorded: *Gammarus tigrinus* (in 1995), coming from western Europe, and two further Ponto-Caspian gammarids, *Pontogammarus robustoides* (in 1996) and *Dikerogammarus haemobaphes* (in 1997). Possible arrival routes are discussed.

1 INTRODUCTION

Natural changes in the geographical distributions of plants and animals are usually achieved rather slowly. However, the rapid population growth of *Homo sapiens* and the development of civilization has created a new situation whereby range extensions of organisms have begun to fluctuate over shorter time spans. Some species have been and still are quickly exterminated, while others are transplanted, gaining new territories, sometimes attaining incredible success with the elimination of native competitors. Intercontinental human migrations and the trade and transport of goods, so quickly developed after the 15th century, have transported many successful biological invaders from remote countries. These were not always purposely carried by man but perhaps more often accompanied humans inadvertently. Di Castri (1989) presents an interesting review of human driven biological invasions.

The Baltic Sea and the network of northern European inland waterways were formed after the last Würm glaciation. The present temperate brackish and freshwater fauna of this region is therefore very young, and the role of human activity has been a major agent in creating the flora and fauna of this area. Jaźdźewski (1980) summa-

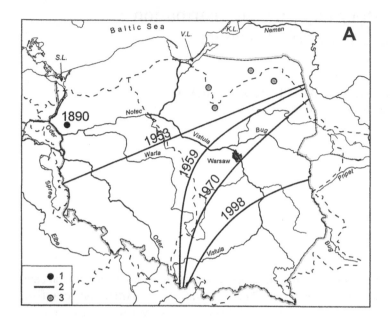

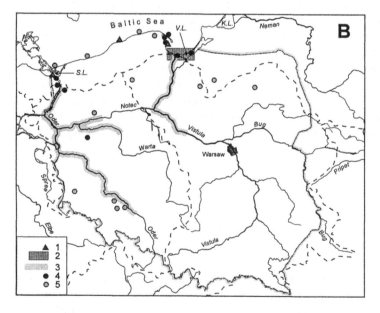

Figure 1. Distribution of alien Decapoda in Poland. A) 1 = *Orconectes limosus*, first introduction in 1890, 2 = idem, southern limit of occurrence, 3 = *Pacifastacus leniusculus*, introduction attempts. B) 1 = *Rhitropanopeus harrisii*, single records, 2 = idem, common occurrence, 3 = *Eriocheir sinensis*, occurrence in river flows before 1939, 4 = idem, records after 1968, 5 = idem, single inland records before 1939. S.L. = Szczecin Lagoon, V.L. = Vistula Lagoon, K.L. = Kuronskij Lagoon. Broken line indicates borders of drainage systems.

rized the situation with regard to acclimatization of amphipod crustaceans to European inland waters and indicated that the dispersal of these animals was enhanced by the construction of canals joining different river systems.

The authors describe here the history of several decapod and amphipod crustaceans introduced into Polish offshore and inland waters. Even thought the area is well studied, some of these species were discovered only very recently, and sometimes their populations are rather dense, indicating that these newcomers are competing successfully with native crustaceans. The area of concern consists mostly of the drainage areas of the Vistula and Oder (Fig. 1). Therefore, when we speak of Polish waters we mean the physiographic entities – the Vistula and Oder basins.

2 DECAPODA

The first alien decapod crustacean ever noted on Polish territory was *Orconectes limosus*, introduced for the first time in Europe in 1890 by Max von Borne to a small pond in the drainage system of the Myśla, tributary of the Oder. The aim was to introduce *O. limosus* as a replacement for the populations of native crayfishes (*Astacus astacus* and *A. leptodactylus*) that were depleted by crayfish plague caused by a fungus, *Aphanomyces astaci*, also an alien invading pathogen probably of American origin. Introduction of *O. limosus* was very successful.

During the past hundred years the species has expanded its distribution in Poland and is still gaining grounds, replacing native crayfish species less adapted to man made changes of the environment and very sensitive to the *Aphanomyces* (Leńkowa 1962). *O. limosus* occupied the western and northern parts of Poland in 1950s. Now it occurs in most of Polish territory (own data and W. Struzynski, pers. comm.). It is definitely more common in the areas with an abundant network of lakes and rivers, i.e., the northern and western lowland parts of Poland (Fig. 1A).

The unintentional introduction of the Chinese mitten crab, *Eriocheir sinensis*, was due perhaps to the emptying of ballast tanks of ships from the Far East into European harbors. First recorded in the Weser River basin in 1912, it spread quickly in the brackish offshore waters of the North Sea and Baltic Sea as well as in the freshwaters of their drainage systems, entering the rivers upstream some hundreds of kilometers from the sea. The species cannot reproduce in freshwater, but adults migrate upstream in masses. In the late 1920s and early 1930s, *E. sinensis* became a true pest; its burrowing habits caused serious damage to hydrotechnical facilities (Peters 1933). However, before World War II this population explosion was over. Although the species is still noted from time to time along West European Atlantic shores from southern France to Great Britain and Germany, and also along the Polish Baltic coast, usually there are only single individuals (Grabda 1973, Gledhill et al. 1992, Vigneux et al. 1993). In Poland, the last deep inland record was in 1972 (Grabda 1973). All other records concern offshore Baltic localities. In the last decade, usually several specimens per year have been brought by fishermen to biological stations of the University of Gdańsk. However, last year in the Szczecin Lagoon (S.L.) a rather abundant occurrence was noted (K. Skóra, pers. comm., E. Skorkowski, pers. comm.) (Fig. 1B). Zettler (1998) recently collected *E. sinensis* in many new inland localities in northeastern Germany (Mecklenburg area) in the region neighboring the Szczecin Lagoon.

This observation accords with the information by Adema (1991) on the unpredictable population fluctuations of *E. sinensis*.

Another immigrant brachyuran in Polish waters is *Rhithropanopeus harrisii*, a native of North American Atlantic brackish waters. Trans-Atlantic transport was obviously the cause of its implantation in European waters. In Poland, it was first noted by Demel (1953) in the brackish deltaic region of the Vistula. It attained a very high population density in the 1950s and 1960s, becoming a major component of the zoobenthos. As with *E. sinensis*, the abundance of *R. harrisii* suddenly decreased in the 70s and 80s. Now the species is moderately common but not abundant in the brackish waters of the delta of the Vistula and in the Vistula Lagoon (V.L.) (Fig. 1B).

The last alien decapod in Polish waters is a signal crayfish *Pacifastacus leniusculus*, originating from North America. In early 1960s and 1970s, it was successfully introduced to Scandinavia and France, where it became a common species used profitably in Swedish and French astaciculture (Laurent 1981, Vigneux et al. 1993). In Poland, attempts to introduce it started in the late 1970s and early 1980s, initially with poor results. Four localities are listed as introduction places with a possible present occurrence (J. Kossakowski, pers. comm.). According to the most recent information from Dr Krzywosz of the Fishery Institute in Olsztyn new attempts at *P. leniusculus* cultivation are now under way in the Lake Masurian region (Fig. 1A).

3 AMPHIPODA

The oldest amphipod immigrant noted in Polish waters is *Corophium curvispinum*. This Ponto-Caspian species originally lived in large rivers emptying into the Black and Caspian Seas (Volga, Dnieper, Dniester, Danube and many others). The earliest record of *C. curvispinum* outside the Ponto-Caspian drainage area was that of Wundsch (1912), already then as far as the Spree-Havel system near Berlin in Germany. The species was first described as new to science, but it soon became obvious that *C. curvispinum* had arrived in the Baltic Sea and North Sea drainage systems through the 18th century canals joining the Dnieper, Vistula, Oder, and Elbe basins. Thus, the species evidently first crossed Polish territory and perhaps was present in the middle Vistula and Noteć rivers system well before its discovery in Poland in the 1920s (Kulmatycki 1930, Wolski 1930).

C. curvispinum is common now and abundant in many localities distributed along the Vistula and Oder flows, as well as in waters connecting these two systems (Fig. 2B). The species is widely distributed and sometimes in mass occurrences (Van den Brink et al. 1989, 1993b, Schöll 1990, Pinkster et al. 1992). Its far western extension in Great Britain was recorded already in the early 1930s by Crawford (1935).

Chaetogammarus ischnus is also a rather old invader of Ponto-Caspian origin. The earliest record of this species outside the drainage systems of the Black and Caspian Seas was that of Jarocki and Demianowicz (1931) from the Vistula river, where the species was first caught in 1928. These authors noted *C. ischnus* along the whole middle and lower Vistula flow below Warsaw. The presence of this species at the mouth of the Vistula was then confirmed by Jaźdżewski (1975). This year (1998) sampling in the middle and lower Vistula and in its tributary, the Bug, showed that *C. ischnus* still occurs in these rivers (Fig. 2B), but is now clearly outnumbered by the fresh Ponto-Caspian incomer *Dikerogammarus haemobaphes*.

Nevertheless, *C. ischnus* is evidently an expansive species. In the late 1970s, it was recorded in the canals joining the Elbe, Weser, and Ems (Herhaus 1978, Herbst 1982). Somewhat later it was found also in some Mecklenburgian and Pommeranian lakes (Waterstraat & Köhn 1989, Jaźdźewski & Konopacka 1990, Köhn & Waterstraat 1990, Zettler 1998), for the first time in stagnant waters outside its natural Ponto-Caspian distribution area. In the home territory, it occurs in several onshore Black Sea lakes and, artificially introduced, in some Ukrainian dam-reservoirs (Jaźdźewski 1980). Finally, the species also arrived into the Lower Rhine (Schöll 1990, Van den Brink et al. 1993a).

Recently, *C. ischnus* crossed the Atlantic. Witt et al. (1997) discovered the species in 1995 in the Great Lakes basin, where it was somehow inadvertently introduced by humans.

A fresh invader of Polish waters is *Gammarus tigrinus*. Gruszka (1995) discovered it in 1994 in the oligohaline Szczecin Lagoon. This year a search for gammarids in the Vistula Lagoon also uncovered abundant *G. tigrinus*, a species which co-occurs in the shallow littoral of this brackish water body with *G. duebenii* and *G. zaddachi* (Fig. 2A).

The spread of *G. tigrinus* in continental European waters after its introduction in late 1950 was closely watched by German and Dutch zoologists (Schmitz 1960, Nijssen & Stock 1966, Ruoff 1968, Pinkster et al. 1977, 1980, 1992, Bulnheim 1985, Platvoet & Pinkster 1985, Platvoet et al. 1989). In Mecklenburg adjacent to Poland, *G. tigrinus* was quite recently discovered by Rudolph (1994a, b); its spread along the Baltic Sea shores and many inland localities of northeast Germany was documented by Zettler (1995, 1998). Due to its high ecological potency and reproductive capacity, the past predictions concerning the eventual range extension of *G. tigrinus* (Bulnheim 1976, Pinkster et al. 1977) have come true. In some twenty years, the distribution of *G. tigrinus* in central Europe along the Baltic coasts moved eastward by some 1000 km (from the Schlei estuary – discovery by Bulnheim in 1975 to the Vistula Lagoon – by present authors in 1998) (Fig. 2A).

The relative abundance in samples of gammarids collected in spring 1998 in the Vistula Lagoon and containing hundreds of specimens was as follows: *G. duebenii* 55%, *G. zaddachi* 3%, *G. tigrinus* 42%. So it seems that *G. tigrinus* successfully competes with indigenous species.

Pontogammarus robustoides originally occurred in the lower courses of large Ponto-Caspian rivers (Volga, Don, Dnieper, Dniester and Danube) as well as in some freshwater and brackish Black Sea onshore lakes (Jaźdźewski 1980). In the 1960s, it was introduced with success to many Ukrainian, Caucasian, and Lithuanian artificial lakes; in the latter case, naturalized in the Neman drainage system, including the Kuronskij Lagoon (K.L.), the species has reached a new sea basin – namely the Baltic Sea (Gasjunas 1968, Jaźdźewski 1980). Recent records of *P. robustoides* come from northeast Germany, namely Mecklenburg (Rudolph 1997, Zettler 1998) and from northwestern Poland (Szczecin Lagoon) (P. Gruszka, pers. comm.). It is possible that the species reached the Oder delta and Mecklenburg via the Baltic from the Neman system and Kuronskij Lagoon. However, recent discoveries of the species in the lower Vistula and in the Vistula Lagoon (Konopacka 1998, and present authors' 1998 samplings) indicate that *P. robustoides* could have followed the dispersal route of

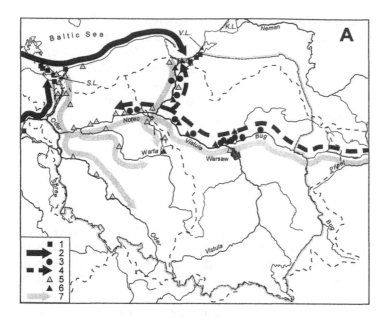

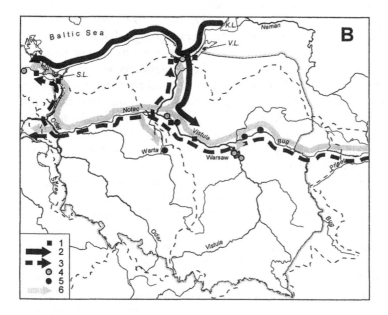

Figure 2. Distribution of alien Amphipoda in Poland. A) 1 = *Gammarus tigrinus*, literature and authors' records, 2 = idem, probable immigration route, 3 = *Dikerogammarus haemobaphes*, recent discovery, 4 = idem, probable immigration route from Dnieper basin, 5 = *Corophium curvispinum*, literature records, 6 = idem, recent authors' records, 7 = idem, probable immigration route from Dnieper basin. B) 1 = *Pontogammarus robustoides*, literature and authors' records, 2 = idem, probable immigration route from Neman basin, 3 = idem, probable immigration route from Dnieper basin, 4 = *Chaetogammarus ischnus*, literature records, 5 = idem, authors' recent records, 6 = idem, probable immigration route from Dnieper basin. S.L. = Szczecin Lagoon, V.L. = Vistula Lagoon, K.L. = Kuronskij Lagoon. Broken line indicates borders of drainage systems.

C. curvispinum and *C. ischnus* via the Pripet-Bug canal from Dnieper into the Vistula (Fig. 2B).

In the 1960s, Morduchaj-Boltovskoj (1964) predicted that *Dikerogammarus haemobaphes* would soon penetrate another basin beyond the Ponto-Caspian system. However, only recently has this species, moving up the Danube, crossed this river system reaching the North Sea basin by entering the Rhine drainage system (Schleuter et al. 1994, Schöll 1995, Leuchs & Schleuter 1996).

D. haemobaphes was discovered in 1977 in Poland (Konopacka 1998). In 1998, sampling along the whole lower Vistula revealed that the species is a dominant gammarid in this river, co-occurring with the less numerous *P. robustoides* and the scarce *C. ischnus* at the mouths of small Vistula tributaries – sometimes also with *Gammarus roeselii* and/or *G. fossarum*. Furthermore, search of the Bug River and the river connecting the Vistula system with Oder basin (Noteć River) enabled us to demonstrate that *D. haemobaphes* is present and sometimes abundant in both these rivers as well (Fig. 2A). In the Noteć River *D. haemobaphes* co-occurs with the very abundant *G. roeselii* as well as with the less numerous *G. varsoviensis*. In the Vistula and Bug rivers, *C. curvispinum* was usually also present in gammarid samples.

The common occurrence and sometimes dominance of *D. haemobaphes* in the Vistula river is quite clear. There is no question that the species is at present in a kind of active expansion phase; it has been recently noted also as far north as near Moscow in the Moskva river (L'vova et al. 1996).

4 CONCLUSIONS

This short review of alien malacostracan crustaceans in Poland demonstrates the enrichment of the local fauna of the Oder and Vistula basins by 9 species that have invaded or been introduced during the last century. Four species are of North American origin, four of Ponto-Caspian, and one comes from China. These invaders make up 10% of the total number of 86 Malacostraca species recorded so far in Polish freshwaters and in the southern Baltic Sea along Polish state borders (Jaźdżewski & Konopacka 1993, 1995, Gruszka 1995, Konopacka 1998). However, if we exclude the typically Baltic malacostracans occurring outside its lagoons and river estuaries this percentage of alien species would increase twofold.

The history of these intentional and unintentional introductions of crustaceans into a medium-sized European country shows how quickly man can influence the composition of local faunas. Scientists have rather limited means to prevent these mostly undesirable changes, but at least we can and we should thoroughly observe these rarely fortunate but otherwise interesting effects of human activity.

REFERENCES

Adema, J.P.H.M. 1991. *De krabben van Nederland en België*. Leiden: Nationaal Natuurhistorisch Museum.
Bulnheim, H.-P. 1976. *Gammarus tigrinus*, ein neues Faunenelement der Ostseeförde Schlei. *Schr. Naturw. Ver. Schlesw.-Holst.* 46: 79-84.

Bulnheim, H.-P. 1985. Genetic differentiation between natural populations of *Gammarus tigrinus* (Crustacea, Amphipoda) with reference to its range extension in European continental waters. *Arch. Hydrobiol.* 102(3): 273-290.

Crawford, G.I. 1935. *Corophium curvispinum* G.O. Sars var. *devium* Wundsch, in England. *Nature* 136: 685.

Demel, K. 1953. Nowy gatunek w faunie Baltyku. *Kosmos*, Ser. Biol. 2, 1(2): 105-106.

Di Castri, F. 1989. History of Biological Invasions with Special Emphasis on the Old World. In J.A. Drake, H.A. Moony, F. Di Castri, R.H. Groves, F.J. Kruger, M. Rejmánek & M. Williamson (eds), *Biological Invasions, A global Perspective, Scope* 37: 1-30. Chichester, New York, Brisbane, Toronto, Singapore: John Wiley & Sons.

Gasjunas, I.I. 1968. Akklimatizacija vysshikh rakoobraznykh kaspijskogo kompleksa v ozerach Litvy. *Limnologija* 3(1): 42-48.

Gledhill, T., Sutcliffe, D.W. & Williams, W.D. 1993. British freshwater Crustacea Malacostraca: A key with ecological notes. *Freshwater Biological Association*, Sc. Pub. 52: 1-173.

Grabda, E. 1973. Krab welnistoszczypcy, *Eriocheir sinensis* Milne-Edwards, 1853 w Polsce. *Prz. Zool.* 17: 46-49.

Gruszka, P. 1995. *Gammarus tigrinus* Sexton, 1939 (Crustacea: Amphipoda) – nowy dla fauny Polski gatunek w estuarium Odry. In I Konf. *Przyrodnicze aspekty badania wód estuarium Odry i wód jeziornych województwa szczecińskiego*, Mat. Konf. No. 7, Uniw. Szczecin: 44.

Herbst, V. 1982. Amphipoden in salzbelasteten niedersächsischen Oberflächengewässern. *Gewässer und Abwässer* 68/69: 35-40.

Herhaus, K.F. 1978. Die ersten Nachweise von *Gammarus tigrinus* Sexton, 1939, und *Chaetogammarus ischnus* (Stebbing, 1906) (Crustacea, Amphipoda, Gammaridae) in Einzugsgebiet der Ems und ihre verbreitungsgeschichtliche Einordung. *Natur und Heimat* 38(3): 99-102.

Jarocki, J. & A. Demianowicz 1931. Über das Vorkommen des ponto-kaspischen Amphipoden *Chaetogammarus tenellus* (G.O.Sars) in der Wisła (Weichsel). *Bull. Int. Acad. Pol., Cl. Math. Nat.* B(II), 1931: 513-530.

Jaźdżewski, K. 1975. Morfologia, taksonomia i wystepowanie w Polsce kiełzy z rodzajów *Gammarus* Fabr. i *Chaetogammarus* Mart. (Crustacea, Amphipoda). *Acta Univ. Łódź*, Łódź, 185 pp.

Jaźdżewski, K. 1980. Range extensions of some gammaridean species in European Inland waters caused by human activity. *Crustaceana Suppl.* 6: 84-107.

Jaźdżewski, K. & Konopacka, A.1991. Nowe, interesujace stanowisko ponto-kaspijskiego kielza *Echinogammarus ischnus* (Stebbing, 1898) (Crustacea, Amphipoda) w Polsce. *Prz. Zool.* 34: 101-111.

Jaźdżewski, K. & Konopacka, A. 1993. Survey and distribution of Crustacea Malacostraca in Poland. *Crustaceana* 65: 176-191.

Jaźdżewski, K. & Konopacka, A. 1995. Pancerzowce – Malacostraca (prócz Oniscoidea). *Katalog Fauny Polski*, Vol. XIII, 1, 165 pp.

Konopacka, A. 1998. Nowy dla Polski gatunek kielza, *Dikerogammarus haemobaphes* (Eichwald, 1841) (Crustacea, Amphipoda) oraz dwa inne rzadkie gatunki skorupiaków obunogich w Wiśle. *Prz. Zool.* 42: 211-218.

Köhn, J. & Waterstraat, A.1990. The amphipod fauna of Lake Kummerow (Mecklenburg, GDR) with reference to *Echinogammarus ischnus* Stebbing, 1899. *Crustaceana* 58: 74-82.

Kulmatycki, W.J. 1930. Ueber das Vorkommen von *Corophium curvispinum* G.O.Sars f. *devium* Wundsch sowie *Carinogammarus roeselii* (Gervais) im Gebiet des Noteć-Flusses. *Fragm. Faun. Mus. Zool. Pol.* 1: 123-134.

Laurent, P.J. 1991. Introductions d'écrevisses. In J.-L. Michelot (ed.), *Les réintroductions animales en Rhône-Alpes*: 199-202. Villeurbanne: Federation Rhône-Alpes de Protection de la Nature, Région Rhône-Alpes.

Leńkowa, A. 1962. Badania nad przyczynami zaniku, sposobami ochrony i restytucja raka szlachetnego *Astacus astacus* (L.) w zwiazku z rozprzestrzenianiem sie raka amerykánskiego *Cambarus affinis* Say. *Ochr. Przyr.* 28: 1-38.

Leuchs, H. & Schleuter, A.1996. *Dikerogammarus haemobaphes* (Eichwald, 1841), eine aus der Donau stammende Kleinkrebsart (Gammaridae) im Neckar. *Lauterbornia* 25: 139-141.

L'vova, A.A., Palij, A.V. & Sokolova, N.U. 1996. Ponto-kaspijskie vselency v reke Moskve v cherte g. Moskvy. *Zool. Zhurnal* 75: 1273-1274.

Morduchaj-Boltovskoj, F.D. 1964. Caspian Fauna beyond the Caspian Sea. *Int. Revue ges. Hydrobiol.* 49: 139-176.

Nijssen, H. & Stock, J.H. 1966. The amphipod, *Gammarus tigrinus* Sexton, 1939, introduced in the Netherlands (Crustacea). *Beaufortia* 13: 197-206.

Peters, N. 1933. Einschleppung und Ausbreitung in Europa. In N. Peters & A. Panning (eds), Die Chinesische Wollhandkrabbe (*Eriocheir sinensis* H.Milne-Edwards) in Deutschland. *Zool. Anz.* 104: 59-156.

Pinkster, S., Dieleman, J. & Platvoet, D. 1980. The present position of *Gammarus tigrinus* Sexton, 1939, in the Netherlands, with the description of a newly discovered amphipod species, *Crangonyx pseudogracilis* Bousfield, 1958 (Crustacea, Amphipoda). *Bull. zoöl. Mus. Univ. Amsterdam* 7: 33-45.

Pinkster, S., Scheepmaker, M., Platvoet, D. & Broodbaker, N. 1992. Drastic changes in the amphipod fauna (Crustacea) of Dutch inland waters during the last 25 years. *Bijdr. Dierk.* 61: 193-204.

Pinkster, S., Smit, H. & Brandse-de Jong, N. 1977. The introduction of the alien amphipod *Gammarus tigrinus* Sexton, 1939, in the Netherlands and its competition with indigenous species. *Crustaceana Suppl.* 4: 91-107.

Platvoet, D. & Pinkster, S. 1985. The present position of the alien amphipods *Gammarus tigrinus* and *Crangonyx pseudogracilis* in the Netherlands (Crustacea, Amphipoda). *Bull. zoöl. Mus. Univ. Amsterdam* 10: 125-128.

Platvoet, D, Scheepmaker, M. & Pinkster, S. 1989. The position of two introduced amphipod crustaceans, *Gammarus tigrinus* and *Crangonyx pseudogracilis* in the Netherlands during the period 1987-1988. *Bull. zoöl. Mus. Univ. Amsterdam* 11: 197-202.

Rudolph, K. 1994a. Erstnachweis des Amphipoden *Gammarus tigrinus* Sexton, 1939 (Crustacea: Gammaridea) im Peenestrom und Achterwasser (südliche Ostseeküste). *Naturschutzarbeit in Mecklenburg-Vorpommern* 37: 23-29.

Rudolph, K. 1994b. Funde des Amphipoden *Gammarus tigrinus* Sexton, 1939 in zwei Havelseen der Region Berlin/Brandenburg (Crustacea: Amphipoda: Gammaridae). *Faun. Abh. Staatl. Mus. Tierkd.* 19(7): 129-133.

Rudolph, K. 1997. Zum Vorkommen des Amphipoden *Pontogammarus robustoides* Sars, 1894, im Peenemündungsgebiet. *Natur und Museum* 127: 306-312.

Ruoff, K. 1968. Experimentelle Untersuchungen über den in die Weser eingebürgerten amerikanischen Bachflohkrebs *Gammarus tigrinus* Sexton. *Arch. Fisch. Wiss.* 19: 134-158.

Schleuter, M., Schleuter, A., Potel, S. & Banning, M. 1994. *Dikerogammarus haemobaphes* (Eichwald, 1841) (Gammaridae) aus der Donau erreicht über den Main-Donau-Kanal den Main. *Lauterbornia* 19: 155-159.

Schmitz, W. 1960. Die Einbürgerung von *Gammarus tigrinus* Sexton auf dem europäischen Kontinent. *Arch. Hydrobiol.* 57: 223-225.

Schöll, F. 1990. Zur Bestandssituation von *Corophium curvispinum* Sars im Rheingebiet. *Lauterbornia* 5: 67-70.

Schöll, F., Becker, C. & Tittizer, T. 1995. Das Makrozoobenthos des schiffbaren Rheins von Basel bis Emmerich 1986-1995. *Lauterbornia* 21: 115-137.

Van den Brink, F.W.B., Van der Velde, G. & Bij de Vaate, A. 1989. A note on immigration of *Corophium curvispinum* Sars, 1895 (Crustacea: Amphipoda) into the Netherlands via the River Rhine. *Bull. zoöl. Mus. Amsterdam* 11: 211-213.

Van den Brink, F.W.B., Paffen, B.G.P., Oosterbroek, F.M.J. & van der Velde, G. 1993a. Immigration of *Echinogammarus* (Stebbing, 1899) (Crustacea: Amphipoda) into the Netherlands via the lower Rhine. *Bull. zoöl. Mus. Amsterdam* 13: 167-170.

Van den Brink, F.W.B., Van der Velde, G. & Bij de Vaate, A. 1993b. Ecological aspects, explosive range extension and impact of a mass invader, *Corophium curvispinum* Sars, 1895 (Crustacea: Amphipoda), in the Lower Rhine. *Oecologia* 93: 224-232.

Vigneux, E., Keith, P. & Noël, P. 1993. Atlas préliminaire des crustacés décapodes d'eau douce de France. Paris: Muséum Nationale d'Histoire Naturelle.

Waterstraat, A.K. & Köhn, J. 1989. Ein Beitrag zur Fauna des Kummerower Sees, Erstnachweis des Amphipoden *Echinogammarus ischnus* Stebbing, 1899, in der DDR. *Arch. Freunde Naturg. Mecklb.* 29: 93-106.

Witt, J.D.S., Hebert, P.D.N. & Morton, W.B. 1997. *Echinogammarus ischnus*: another crustacean invader in the Laurentian Great Lakes basin. *Can. J. Fish. Aquat. Sci.* 54: 264-268.

Wolski, T. 1930. *Corophium curvispinum* G.O. Sars in der Prypeć und in der Warschauer Wasser-leitungsanlagen. *Fragm. Faun. Mus. Zool. Pol.* 1: 152-159.

Wundsch, H.H. 1912. Eine neue Species des Genus *Corophium* Latr. aus dem Müggelsee bei Berlin. *Zool. Anz.* 39: 729-738.

Zettler, M.L. 1995. Erstnachweis von *Gammarus tigrinus* Sexton, 1939 (Crustacea: Amphipoda) in der Darß-Zingster Boddenkette und seine derzeitige Verbreitung an der deutschen Ostseeküste. *Arch. Freunde Naturg. Mecklenburg* 34: 123-140.

Zettler, M. L. 1998. Zur Verbreitung der Malacostraca (Crustacea) in den Binnen- und Küsten-gewässer von Mecklenburg-Vorpommern. *Lauterbornia* 32: 49-65.

Biogeography of trapeziid crabs (Brachyura, Trapeziidae) symbiotic with reef corals and other cnidarians

PETER CASTRO

Biological Sciences Department, California State Polytechnic University, Pomona, USA

ABSTRACT

Like their hosts, most species of *Hexagonalia, Quadrella, Tetralia, Tetraloides,* and *Trapezia* are restricted to the Indo-west Pacific region. Most (60%) are widely distributed across the region. Three species cross the East Pacific barrier and are found in the eastern Pacific. In contrast, ten species are peripheral endemics in the southeastern Pacific, Hawaiian archipelago, eastern Pacific, and southwestern Indian Ocean. It is suggested that the geographic distribution of most of these species is best explained by long-distance larval dispersal. Some of the peripheral endemics are perhaps the result of vicariance.

1 INTRODUCTION

Trapeziid crabs are obligate symbionts of cnidarians, mostly reef corals, in tropical environments. The host provides food (mucus and tissue) as well as shelter (Castro 1976). Although nine genera are sometimes included in the family (Serène 1984), the systematic position of *Calocarcinus, Domecia, Jonesius,* and *Palmyria* remains questionable; therefore these genera are not included here. A reexamination of the systematic status of the species included in the five remaining genera (*Hexagonalia, Quadrella, Tetralia, Tetraloides,* and *Trapezia*) (Castro 1996, 1997a, b, 1998a, b, c and d) has permitted an analysis of their geographic distribution.

All 22 species of *Trapezia* are sympatric symbionts of pocilloporid corals (*Pocillopora, Seriatopora, Stylophora*), zooxanthellate scleractinians widely distributed in the Indo-west and eastern Pacific regions. The eight species of *Tetralia* and *Tetraloides* are sympatric symbionts that are restricted to Indo-west Pacific species of the zooxanthellate scleractinian coral *Acropora.* Some of the species belonging to these three genera may be referred to as sibling species since most are morphologically very close to other congeners and are best distinguished from each other by their color patterns. Species of *Quadrella* are symbionts of several groups of anthozoans, while the rarely collected species of *Hexagonalia* are mostly known from stylasterid corals. Little is know about the biology of the last two genera.

2 LIST OF SPECIES AND THEIR GEOGRAPHIC DISTRIBUTION

Genus *Hexagonalia* Galil, 1986
Hexagonalia brucei (Serène, 1973): western Indian Ocean. Hosts: stylasterid corals and gorgonians.
Hexagonalia laboutei Galil, 1997: southeastern Pacific. Host: stylasterid corals.

Genus *Quadrella* Dana, 1851
Quadrella boopsis Alcock, 1898: southwestern Indian Ocean (Mozambique and Madagascar), Pacific Ocean (Japan to southeastern Pacific). Hosts: dendrophylliid and other ahermatypic corals.
Quadrella coronata Dana, 1852: Red Sea, Persian Gulf, Indian Ocean, western Pacific Ocean (Japan to Coral Sea). Hosts: alcyonaceans, antipatharians, gorgonians.
Quadrella maculosa Alcock, 1898: Red Sea, Indian Ocean, Pacific Ocean (Japan to southeastern Pacific). Hosts: antipatharians.
Quadrella nitida Smith, 1869: eastern Pacific (southern Gulf of California to Galápagos Islands). Hosts: antipatharians.
Quadrella reticulata Alcock, 1898: Red Sea, Indian Ocean, western Pacific Ocean (Japan to Indonesia). Hosts: antipatharians.
Quadrella serenei Galil, 1986: Indian Ocean, Pacific Ocean (Japan to southeastern Pacific). Hosts: antipatharians.

Genus *Tetralia* Dana, 1851
Tetralia cavimana Heller, 1861: Red Sea, Persian Gulf and northwestern Indian Ocean.
Tetralia cinctipes Paulson, 1875: across Indo-west Pacific except Hawaiian Islands.
Tetralia fulva Serène, 1984: across Indo-west Pacific except Hawaiian Islands.
Tetralia nigrolineata Serène & Dat, 1957: western Indian Ocean to western and central Pacific Ocean (Japan to Marshall Islands and Tonga).
Tetralia rubridactyla Garth, 1971: across Indo-west Pacific except Red Sea and Hawaiian Islands.
Tetralia vanninii Galil & Clark, 1988: across Indo-west Pacific except Red Sea and Hawaiian Islands.

Genus *Tetraloides* Galil, 1986
Tetraloides heterodactyla (Heller, 1861): across Indo-west Pacific except Hawaiian Islands.
Tetraloides nigrifrons (Dana, 1852): across Indo-west Pacific except Red Sea, and Hawaiian Islands.

Genus *Trapezia* Latreille, 1828
Trapezia areolata Dana, 1852: southeastern Pacific.
Trapezia bella Dana, 1852: southeastern Pacific.
Trapezia cheni Galil, 1983: Taiwan.
Trapezia corallina Gerstaecker, 1857: eastern Pacific (southern Gulf of California to Galápagos Islands).
Trapezia cymodoce (Herbst, 1801): across Indo-west Pacific except central Pacific east of Caroline Islands.
Trapezia digitalis Latreille, 1828: across Indo-west Pacific and eastern Pacific.
Trapezia ferruginea Latreille, 1828: across Indo-west Pacific and eastern Pacific.
Trapezia flavopunctata Eydoux & Souleyet, 1842: across Indo-west Pacific except northwestern Indian Ocean.
Trapezia formosa Smith, 1869: across Indo-West Pacific (except Hawaiian Islands) and eastern Pacific.
Trapezia garthi Galil, 1983: western Pacific Ocean (Japan to Moluccas and Niue).
Trapezia globosa Castro, 1997: southeastern Pacific.
Trapezia guttata Rüppell, 1830: across Indo-west Pacific except Hawaiian Islands.
Trapezia intermedia Miers, 1886: Hawaiian Islands, Wake Island, Johnston Atoll, Marshall Islands.

Trapezia lutea Castro, 1997: across Indo-west Pacific except Hawaiian Islands.
Trapezia punctimanus Odinetz, 1984: southeastern Pacific.
Trapezia punctipes Castro, 1997: Andaman Sea to western Pacific Ocean (Mariana Islands and Fiji).
Trapezia richtersi Galil & Lewinsohn, 1983: Indian Ocean (Somalia and southwestern Indian Ocean to Andaman Sea).
Trapezia rufopunctata (Herbst, 1799): across Indo-west Pacific.
Trapezia septata Dana, 1852: Indian Ocean (east of Sri Lanka) to Pacific Ocean (Japan to central Pacific west of Marshall Islands and Samoa).
Trapezia serenei Odinetz, 1984: Pacific Ocean (Japan and Moluccas to Line Islands and southeastern Pacific).
Trapezia speciosa Dana, 1852: Indian Ocean (Seychelles to Chagos Archipelago) and South China Sea to southeastern Pacific.
Trapezia tigrina Eydoux & Souleyet, 1842: across Indo-west Pacific.

3 BIOGEOGRAPHY

Most trapeziids, 23 species out of 38 thus far described (60%), are wide ranging, or eurytypic, across the Indo-west Pacific region (Table 1). Three of the 11 widespread species of *Trapezia* (*T. digitalis*, *T. ferruginea*, and *T. formosa*) are also found in the eastern Pacific region and thus show an amphi-Pacific distribution (Fig. 1). Kay (1984) indicated that most of the species (about 37%) of several groups of Indo-west Pacific invertebrates and fishes showed a wide distribution throughout the region in contrast to those found only in the western Pacific, Indian Ocean, or Pacific Ocean. About 60% of the xanthid crabs of her inventory showed a similar distribution, while

Table 1. Geographic distribution of trapeziid crabs. Figures for the four divisions of the Indo-west Pacific region and for the eastern Pacific region refer to number of endemic species.

Genus	No. species	Widespread Indo-west Pacific[1]	Indian Ocean[2]	West-central Pacific[3]	South-eastern Pacific[4]	Hawaiian Islands and north-central Pacific[5]	Eastern Pacific[6]
Trapezia	22	11 (50%)	1 (4%)	4 (18%)	4 (18%)	1 (4%)	1 (4%)
Quadrella	6	5 (83%)	0	0	0	0	1 (17%)
Tetralia	6	5 (83%)	1 (17%)	0	0	0	0
Hexagonalia	2	0	1 (50%)	0	1	0 (50%)	0
Tetraloides	2	2 (100%)	0	0	0	0	0
Total	38	23 (60%)	3 (8%)	4 (10%)	5 (13%)	1 (3%)	2 (5%)

[1]Red Sea to KwaZulu-Natal coast of South Africa, east to Hawaiian Islands and Easter Island. [2]Gulf of Aden to KwaZulu-Natal coast of South Africa east to Andaman Sea and Western Australia. [3]Southern Japan to South Coral Sea east to Line and Cook Islands. [4]French Polynesia, Pitcairn Island and Easter Island. [5]Hawaiian Islands, Wake Island, Marshall Islands and Johnston Atoll. [6]Southern Gulf of California to Galápagos Islands.

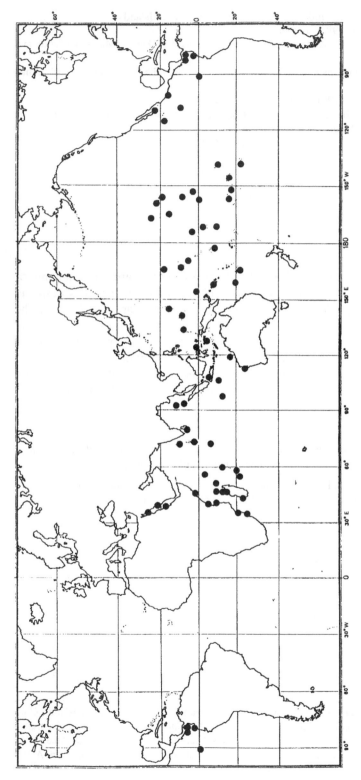

Figure 1. Geographic distribution of *Trapezia ferruginea*, a wide-ranging trapeziid crab.

the figure was only 20% in cryptochiriid crabs. About 69% of the brachyurans of French Polynesia (Forest & Guinot 1962) and about 52% of the western Indian Ocean species (Guinot 1967) showed a wide Indo-west Pacific distribution.

3.1 *Pacific tectonic plate*

The Pacific tectonic plate has been acknowledged as having a relatively large number of endemic species (Springer 1982). Nevertheless, only three species of *Trapezia* (*T. bella*, *T. globosa*, and *T. intermedia*) and one of *Hexagonalia* (*H. laboutei*) are strict endemics of the Pacific Plate. Two additional species (*T. areolata* and *T. punctimanus*) also occur in Easter Island, which is on an adjacent plate. All of these six species are endemic to the southeastern Pacific or to the Hawaiian archipelago and the north-central Pacific, and none is widespread through the plate. A smaller proportion of Pacific Plate endemics (40%) against widespread Indo-west Pacific species (60%) was recorded by Kay (1984) for species belonging to four groups of invertebrates and fishes. The difference is more marked among trapeziids. Of the 32 species found in the Pacific Plate, 6 (19%) are endemic and 26 (81%) are also found elsewhere, of which 21 (66% of the total) are widespread Indo-west Pacific species.

The view that the Indo-Malayan region has served as the center of diversification from which other Indo-west Pacific species became adapted to less favorable conditions in peripheral regions (Ekman 1953, Briggs 1974, 1992) is not supported by the distribution of trapeziids. Only four species, all of *Trapezia*, can be grouped as endemics of an ill-defined but convenient west-central Pacific region (Table 1). The patterns of distribution of these four species, however, vary widely. *T. cheni* is so far known only from Taiwan, while *T. serenei* occurs from Japan and Indonesia to the Line Islands and southeastern Pacific. Species diversity of trapeziids in Indonesia, 22 species of the five genera treated here (Castro 1998b), is similar to the 23 species found in French Polynesia (Castro 1997b). Ladd (1960) and Kay (1984) proposed that the high species diversity of the western Pacific is the result of the accumulation of species from separate sources rather than a center of dispersal. Patterns of distribution, however, should vary among various groups and the hypotheses presented to explain the role of the Indo-Malayan region are not mutually exclusive (Paulay 1997).

3.2 *Hawaiian archipelago*

Only six species of *Trapezia* have been recorded from the Hawaiian archipelago. One species (*T. rufopunctata*) is only known by one confirmed record of one heterosexual pair (Castro 1998c). Five of these six species are widespread in the Indo-west Pacific; one (*T. intermedia*) is endemic to the archipelago and the north-central Pacific (Fig. 2). The rare presence of *Acropora*, relegated to a few islands in the northwestern chain of the archipelago (Grigg et al. 1981, Grigg 1981), explains the apparent absence of *Tetralia* and *Tetraloides*.

Two widespread Indo-west Pacific species of *Trapezia* (*T. guttata* and *T. speciosa*) have been recorded from Johnston Atoll, only 687 km southwest of the Hawaiian Islands. Two other widespread species (*T. lutea* and *T. serenei*) are known from Palmyra Atoll in the Line Islands 1600 km southwest of the archipelago, while other two

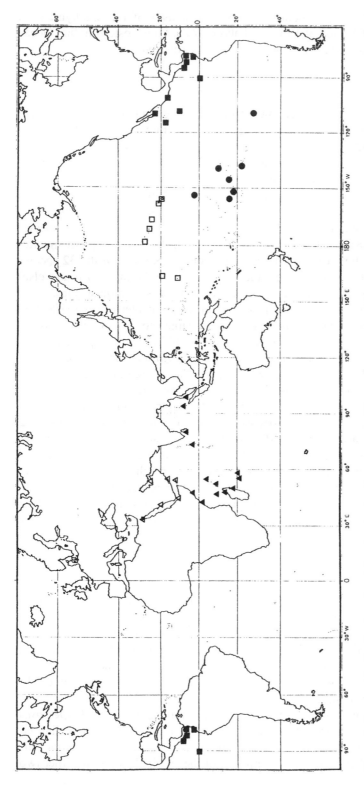

Figure 2. Geographic distribution of peripheral endemics: *Trapezia corallina* (eastern Pacific) (■), *Trapezia intermedia* (Hawaiian Islands and north-central Pacific) (□), *Trapezia areolata*, *T. bella*, *T. globosa*, *T. punctimanus*, and *Hexagonalia laboutei* (southeastern Pacific) (●), *Tetralia cavimana* (△) and *Trapezia richtersi* (▲) (Indian Ocean).

(*T. bella*, a southeastern Pacific endemic, and *T. formosa*, a widespread species) are known from Kiritimati (Christmas) Island in the Line Islands 1850 km south of the archipelago. Conversely, *T. intermedia*, the most common species in the archipelago, is so far only known from Wake Island, the Marshall Islands, and Johnston Atoll, but not from the Line Islands. The remaining five Hawaiian species are found in both Johnston Atoll and the Line Islands.

The shallow-water marine fauna of Hawaii represents an impoverished Indo-west Pacific biota with an endemic element, around 32% of the species among the invertebrate groups tabulated by Kay & Palumbi (1987). Impoverishment, at least in the case of reef corals, seems to be mainly related to distance (Jokiel 1987). The absence of several widespread species of trapeziids, some of which extend not far from the archipelago, is not surprising but puzzling. It is possible that some of these species are present but are very rare and seldom collected (as in *T. rufopunctata*), that larvae cannot reach the islands, or that larvae arrive but are unsuccessful in colonizing corals. The absence of particular hosts is not a factor since Hawaii and Johnston Atoll share the same species of *Pocillopora* (Maragos & Jokiel 1986). Another possibility is that a higher number of species of *Trapezia* were once present in the islands but eventually became extinct as a result of sea-surface cooling and sea-level fluctuations during the Pleistocene, as in the case of *Acropora* (Grigg 1981) and several species of molluscs (Kay & Palumbi 1987). At least some of the species now present in the islands may have subsequently arrived by long-distance larval dispersal from the central Pacific, as in the case of *T. intermedia*, which is distributed from the Marshall Islands to Hawaii. Some of the species may have continued to exist in Hawaii as relics and *T. intermedia* could have colonized the Marshall Islands from Hawaii. Huber (1985) used electrophoresis to analyze several species of *Trapezia*, five from Hawaiian populations and three from Enewetak, Marshall Islands. It was shown that the genetic differences between the Hawaii and Enewetak populations of *T. ferruginea* were similar to those between the remaining seven species. A Pleistocene radiation as a result of extinction and subsequent recolonization of coral habitats was suggested to explain the low genetic differences between the eight species that were analyzed.

3.3 *Eastern Pacific*

The eastern Pacific region has two endemic trapeziids, *Quadrella nitida* and *Trapezia corallina*. The three other species known from the region (*Trapezia digitalis*, *T. ferruginea*, and *T. formosa*) are widespread Indo-west Pacific species that stretch as far as the western Indian Ocean. While only five species (two endemic) belonging to two genera are found in the eastern Pacific, five genera and 23 species (five endemic) are known from the southeastern Pacific. As in Hawaii, one reason for this low diversity is the very rare presence, perhaps as temporary populations, of *Acropora*, the host of *Tetralia* and *Tetraloides*.

The reef associated biota of the eastern Pacific is characterized by low diversity, low endemicity and strong affinities to the Indo-west Pacific. Its possible origin has been much discussed (reviews by Veron 1995 and Paulay 1997), particularly the significance of the East Pacific barrier between the Indo-west Pacific and eastern Pacific (Grigg & Hey 1992). Indo-west Pacific colonizers with better means of dispersal are more diverse in the eastern Pacific than groups with poorly dispersed larvae (Paulay

1997). The crossing of the barrier by long lived, or teleplanic, larvae was suggested for trapeziids by Garth (1974) and demonstrated in some decapod larvae by Johnson (1974) and Scheltema (1988). While the length of larval development in trapeziids remains unknown, coral rafting (Jokiel 1984) may carry juvenile stages of *Trapezia* that have metamorphosed from megalopae in the plankton. Jokiel (1990) found that *Pocillopora* was the most common coral in drifting pumice. Fast, warm currents during El Niño-Southern Oscillation events may also help larvae cross the eastern Pacific barrier (Richmond 1990, Lessios et al. 1998).

Huber (1985) showed that there is little genetic differentiation between eastern Pacific and Hawaii populations of *T. ferruginea*. This, the possibility of gene flow or recent immigration, has also been demonstrated in fishes (Rosenblatt & Waples 1986). Genetic differentiation between eastern Pacific and Marshall Islands populations of *T. formosa*, however, was found to be similar to that between other species of *Trapezia* (Huber 1985), a result not supported by morphological characters (Castro 1998a).

The only eastern Pacific endemic, *T. corallina*, may be a relict from the reef associated biota prior to the Pliocene closure of the Isthmus of Panama (Abele 1982). *Trapezia* and *Tetralia* inhabited Tethys Sea reefs during the Eocene (Müller 1984, Müller & Collins 1991). Pocilloporid hosts, however, may have disappeared from the eastern Pacific during the Pleistocene (Dana 1975).

3.4 *Southeastern Pacific*

High species diversity (23 species, or 60% of the total) coupled with relatively high endemicity (five species, or 22%; Fig. 2) characterize the trapeziids of the island groups of French Polynesia at the southeastern limit of the Indo-west Pacific. Only 8% of the brachyurans listed by Forest & Guinot (1962) were considered endemic to the southeastern Pacific. Although not as isolated as Hawaii, the southeastern Pacific is relatively isolated from the rest of the Indo-west Pacific by distance and prevailing ocean currents. In contrast to Hawaii and the eastern Pacific, the high species diversity of southeastern Pacific trapeziids may be in part explained by the high diversity of acroporid and pocilloporid corals in the area. High host diversity plus a reduced but not restrictive isolation may explain the high diversity and high endemicity of trapeziids. Only one of the five southeastern Pacific endemics reaches the Line Islands and none is found in Hawaii, in contrast to the amphitropical distribution observed in some barnacles (Newman 1986).

3.5 *Other peripheral endemics*

Some peripheral endemism is also present on the western limits of the Indo-west Pacific (Fig. 2): the Red Sea and the northwestern Indian Ocean (*Tetralia cavimana*), southwestern Indian Ocean (*Hexagonalia brucei*), and the western and northern Indian Ocean (*Trapezia richtersi*). Another apparent peripheral endemic, *Trapezia cheni*, is known only from Taiwan. It is highly probable that these and other assumed endemic species may ultimately be discovered elsewhere.

4 CONCLUSIONS

Practically all trapeziids are inhabitants of the Indo-west Pacific region, which is recognized as a single biogeographic entity as a result of a high commonality of species (Briggs 1974, Paulay 1997). The geographic distribution of the group, which is characterized by a large number of sympatric widespread species that co-inhabit with endemics in peripheral regions, presents some vexing questions. Of particular significance is how speciation occurs in seemingly high-dispersal crustaceans with high fecundity, rapid gene flow and large populations that inhabit a vast region (see Palumbi 1992).

Long-distance larval dispersal seems to be the most logical mechanism that explains the biogeography of most trapeziids. Unfortunately, larval competence among trapeziids is unknown. Results among several species raised under laboratory conditions in Japan indicate a long larval development (N. Shikatani, personal communication). Nevertheless, vicariance cannot be ruled out as an explanation of the existence of some of the peripheral endemics. These may be instances of relictual endemism as in some corals in high-latitude localities (Veron 1995).

Several mechanisms of speciation can be proposed as having been in effect. One possibility is the development of reproductive isolation due to host preferences. This may have been the case in some of the species of *Quadrella. Q. boopsis*, for example, seems to be restricted to dendrophylliid and other ahermatypic corals. Species of *Trapezia*, however, do not show narrow host specificity. Several species are typically found sympatrically in a single colony, although species characterized by small individuals, such as *T. formosa*, are more common in smaller coral colonies. The opposite applies to large-size species (*T. flavopunctata* and *T. rufopunctata*). This nevertheless is not the result of host specificity since *Pocillopora* species may show wide variation in the size and morphology of their colonies depending on habitat (Veron 1995). Some species of *Tetralia*, on the other hand, show host-specific patterns of local distribution (T. Sin, in press).

Differences in food preferences or microhabitat within a host may have lead to habitat segregation. There is no evidence of such differences other than the observation that in large colonies small-size species live deeper in the host. The possibility of differences in the length of larval development is an intriguing question. Regions of high diversity (such as the southeastern Pacific in *Trapezia*) were predicted to be populated by a larger number of endemic, narrowly distributed species with larval development that is shorter than widespread species (Vermeij 1978). That at least differences in reproductive effort exist is hinted by the very early sexual maturity in small-size species such as *T. globosa*, a southeastern Pacific endemic (Castro 1997b). Information on reproduction and larval development in trapeziids is unfortunately lacking.

Mate selection and recognition in the sympatric trapeziids may have played a significant role in speciation. There is evidence that species-specific behaviors and chemical detection are involved in maintaining reproductive isolation among some sibling marine species (Palumbi 1994). The demonstration of assortative mating among color variants of *T. digitalis* in Hawaii (Huber 1987) suggests that mate preference could have lead to reproductive isolation among sympatric species of *Trapezia*.

Molecular and genetic studies will help elucidate the mechanisms of speciation that may have been in effect in trapeziids. Such studies will also clarify the nature of genotypic variation, particularly in the widespread species. Slight variations in color patterns among some of these species may demonstrate the existence of additional species, a possibility already suggested by electrophoresis (Huber 1985).

ACKNOWLEDGEMENTS

I am most grateful for the comments made to the manuscript by Dr A. Kay (University of Hawaii) and Dr G. Paulay (University of Guam).

REFERENCES

Abele, L.G. 1982. Biogeography. In L.G. Abele (ed.), *The Biology of Crustacea*, vol. 1, *Systematics, The Fossil Record, and Biogeography*: 241-304. New York: Academic Press.

Briggs, J.C. 1974. *Marine Biogeography*. New York: McGraw-Hill.

Briggs, J.C. 1992. The marine East Indies: centre of origin? *Glob. Ecol. Biogeogr. Letters* 2: 149-156.

Castro, P. 1976. Brachyuran crabs symbiotic with scleractinian corals: a review of their biology. *Micronesica* 12: 99-110.

Castro, P. 1996. The eastern Pacific species of *Trapezia* (Crustacea, Brachyura: Trapeziidae), sibling species symbiotic with reef corals. *Bull. Mar. Sci.* 58: 531-554.

Castro, P. 1997a. Trapeziid crabs (Brachyura: Xanthoidea: Trapeziidae) of New Caledonia, eastern Australia, and the Coral Sea. In B. Richer de Forges (ed.), Les fonds meubles des lagons de Nouvelle-Calédonie (Sédimentologie, Benthos). *Études & Thèses* 3: 59-107.

Castro, P. 1997b. Trapeziid crabs (Brachyura: Xanthoidea: Trapeziidae) of French Polynesia. In B. Richer de Forges (ed.), Les fonds meubles des lagons de Nouvelle-Calédonie (Sédimentologie, Benthos). *Études & Thèses* 3: 109-139.

Castro, P. 1998a. Systematic status and geographic distribution of *Trapezia formosa* Smith, 1869 (Crustacea, Brachyura, Trapeziidae), a symbiont of reef corals. *Zoosystema* 20: 177-181.

Castro, P. 1998b. The Trapeziidae (Crustacea: Brachyura: Xanthoidea) of Indonesia. Results of the Rumphius Biohistorical Expedition to Ambon (1990), part 7. *Zool. Med.* 73: 27-61.

Castro, P. 1998c. The Hawaiian species of *Trapezia* (Crustacea, Brachyura, Trapeziidae), symbionts of *Pocillopora* (Scleractinia). *Bishop Mus. Occ. Pap.* 55:73-760.

Castro, P. 1998d. Trapeziid crabs (Crustacea, Brachyura, Trapeziidae) of the Indian Ocean and the Red Sea. *Zoosystema* 21: 93-120.

Dana, T.F. 1975. Development of contemporary Eastern Pacific coral reefs. *Mar. Biol.* 33: 355-374.

Ekman, S. 1953. *Zoogeography of the sea*. London: Sidgwick & Jackson.

Forest, J. & Guinot, D. 1962. Remarques biogéographiques sur les crabes des archipels de la Société et des Tuamotu. *Cah. Pacif.* 4: 41-75.

Garth, J.S. 1974. On the occurrence in the eastern tropical Pacific of Indo-west Pacific decapod crustaceans commensal with reef-building corals. *Proc. Sec. int. Coral Reef. symp., Brisbane* 1: 397-404.

Grigg, R.W. 1981. *Acropora* in Hawaii. Part 2. Zoogeography. *Pac. Sci.* 35: 15-24.

Grigg, R.W., J.W.Wells & C. Wallace 1981. *Acropora* in Hawaii. Part 1. History of the scientific record, systematics, and ecology. *Pac. Sci.* 35: 1-13.

Grigg, R.W. & R. Hey 1992. Palaeoceanography of the tropical eastern Pacific Ocean. *Science* 255: 172-178.

Guinot, D. 1967. La faune carcinologique (Crustacea Brachyura) de l'Océan Indien Occidental et de la Mer Rouge. Catalogue, remarques biogéographiques et bibliographie. Réunion de spécialistes C.S.A. sur les Crustacés, Zanzibar 1964. *Mém. Inst. Fond. Afr. Noire* no. 77(2): 237-352.

Huber, M.E. 1985. Population genetics of eight species of *Trapezia* (Brachyura: Xanthidae), symbionts of corals. *Mar. Biol.* 85: 23-36.

Huber, M.E. 1987. Phenotypic assortative mating and genetic population structure in the crab *Trapezia digitalis. Mar. Biol.* 93: 509-515.

Johnson, M.W. 1974. On the dispersal of lobster larvae into the East Pacific barrier (Decapoda, Palinuridea). *Fish. Bull.* 72: 639-647.

Jokiel, P.L. 1984. Long distance dispersal of reef corals by rafting. *Coral Reefs* 3: 113-116.

Jokiel, P.L. 1987. Ecology, biogeography and evolution of corals in Hawaii. *Trends Ecol. Evol.* 2: 179-182.

Jokiel, P.L. 1990. Transport of reef corals into the Great Barrier Reef. *Nature* 347: 665-667.

Kay, E.A. 1984. Patterns of speciation in the Indo-west Pacific. In: F.J. Radovsky, P.H. Raven & S.H. Sohmer (eds), Biogeography of the Tropical Pacific. *Bishop Mus. Spec. Publ.* 72: 15-31.

Kay, E.A. & S.R. Palumbi 1987. Endemism and evolution in Hawaiian invertebrates. *Trends Ecol. Evol.* 2: 183-186.

Ladd, H.S. 1960. Reef building. *Science* 134: 703-715.

Lessios, H.A, B.D. Kessing & D.R. Robertson 1998. Massive gene flow across the world's most potent marine biogeographic barrier. *Proc. Royal. Soc. London, Biol. Sci.* ser. B, 265: 583-588.

Maragos, J.E. & P.L. Jokiel 1986. Reef corals of Johnston Atoll: one of the world's most isolated reefs. *Coral Reefs* 4: 141-150.

Müller, P. 1984. Decapod Crustacea of the Badenian. *Geolog. Hungar.,* ser. paleont. 42: 1-317.

Müller, P. & J.S.H. Collins 1991. Late Eocene coral-associated decapods (Crustacea) from Hungary. *Contr. Tert. Quatern. Geol.* 28: 47-92.

Newman, W.A. 1986. Origin of the Hawaiian marine fauna: Dispersal and vicariance as indicated by barnacles and other organisms. *Crust. Iss.* 4: 21-49.

Palumbi, S.R. 1992. Marine speciation on a small planet. *Trends Ecol. Evol.* 7: 114-118.

Palumbi, S.R. 1994. Genetic divergence, reproductive isolation, and marine speciation. *Ann. Rev. Ecol. Syst.* 25: 547-572.

Paulay, G. 1997. Diversity and distribution of reef organisms: 298-353. In C. Birkeland (ed.), *Life and death of coral reefs.* New York: Chapman & Hall.

Richmond, R.H. 1990. The effects of El Niño/Southern Oscillation on the dispersal of corals and other marine organisms: 127-140. In P.W. Glynn (ed.), *Global ecological consequences of the 1982-83 El Niño-Southern Oscillation.* Amsterdam: Elsevier.

Rosenblatt, R.H. & R.S. Waples 1986. A genetic comparison of allopatric population of shore fish species from the eastern and central Pacific Ocean: dispersal or vicariance? *Copeia* 1986: 275-284.

Serène, R. 1984. Crustacés Décapodes Brachyoures de l'Océan Indien occidental et de la Mer Rouge, Xanthoidea: Xanthidae et Trapeziidae. *Faune Trop.* 24: 1-349.

Scheltema, R.S. 1988. Initial evidence for the transport of teleplanic larvae of benthic invertebrates across the east Pacific barrier. *Biol. Bull.* 174: 145-152.

Sin, T. in press. Distribution and host specificity in *Tetralia* crabs (Crustacea: Brachyura) symbiotic with corals in the Great Barrier Reef, Australia. *Bull. Mar. Sci.*

Springer, V.G. 1982. Pacific plate biogeography, with special reference to shorefishes. *Smithson. Contrib. Zool.* no. 367: i-iv, 1-182.

Vermeij, G.J. 1978. *Biogeography and Adaptation. Patterns of Marine Life.* Cambridge: Harvard University Press.

Veron, J.E.N. 1995. *Corals in Space and Time.* Ithaca, N.Y.: Comstock/Cornell.

Decapod crustaceans associated with unbleached and bleached colonies of the coral, *Seriatopora hystrix* from near the thermal discharge of a nuclear power plant in Taiwan

JENG MING-SHIOU
Institute of Zoology, Academia Sinica, Taipei, Taiwan

ABSTRACT

Seriatopora hystrix is a stony coral commonly found in shallow waters of Nanwan Bay in southern Taiwan. A total of 32 unbleached and 10 bleached colonies of *S. hystrix* were collected near the warm effluent of a nuclear power plant in the western part of the Bay. Altogether 2148 specimens of 85 species of coral inhabiting animals were found in these colonies. Among them, 53 species (1860 specimens, 86.6% of total) were decapod crustaceans. Of these, 32 species and 1629 individuals were found from unbleached colonies; 27 species and 231 individuals were found on bleached colonies. The ten most abundant species were: *Trapezia lutea, T. guttata, Calcinus seurati, Alpheus lottini, T. areolata, A. gracillipes, A. pachychirus, Lissoporcellana spinuligera, Synalpheus charon*, and *S. tumidomanus*. Of these species, two alpheid shrimps (*A. lottini* and *S. charon*) and three xanthid crabs (*T. lutea, T. areolata*, and *T. guttata*) are possibly obligatory symbionts of *S. hystrix*, as they were found only on unbleached colonies.

1 INTRODUCTION

Decapod crustaceans are among the most abundant and diverse animals in tropical coral reef communities. Serène (1972) estimated that more than 500 of approximately 2000 species of Indo-West Pacific brachyurans are associated with living corals or are regular inhabitants of the dead parts of reefs. The general behavior of tropical decapod crustaceans has been little studied and, in general, only casual observations of isolated species are available. It is clear that the behavior of decapods is important in maintaining the association with coral hosts. However, the general behavior of commensal symbiotic species closely resembles that of free-living decapods, and it is probable that photophobic and thigmotactic responses of the latter have been an important factors in the evolution of symbiotic habits (Bruce 1976). Thermal effects to symbionts on branching corals, however, are unknown.

Head forming branching corals provide discrete habitats that contain a distinct fauna of small shrimps, crabs, and fishes. Patton (1994) showed that the shrimps and crabs feed on coral mucus and, to lesser and varying degrees, zooplankton, suspended matter, and coral tissue. Generally, host corals provide crustaceans with shelter form predators and with mucus as a food source; resident crustaceans protect their coral host against corallivores. Glynn (1980) demonstrated that the symbiontic shrimp *Alpheus lottini* and crab *Trapezia* spp. protect their host coral (*Pocillopora elegans*) by detecting an approching predative sea star (*Acanthaster planci*) by chemical cues. These crustaceans are obligate symbionts of pocilloporid corals and feed primarily on the host coral's mucus (Knudsen 1967, Patton 1974).

Coral bleaching (loss of pigmentation) is a widespread phenomenon in coral reef ecosystems. Despite this, the effects of bleaching on decapods associated with host corals are poorly understood. More than 250 species of coral grow very well in Nanwan Bay, southern Taiwan. The Third Nuclear Power Plant, located in the western side of the bay became operative in May 1984. Thermal effluent is discharged into the bay (Fig. 1). In early July 1987 the surface water temperature often reached 32°C or even higher, and almost all the corals close by at depths < 3 m were bleached (Fan 1988). Only one species below 6 m, *Seriatopora hystrix,* was bleached. Decapods die in unfavorable habitats and survive in more favorable ones, a better description of what might be happening. This behavioral trait appears to be of major importance for the survival of decapods associated with corals, yet so far nothing is known about it. This study determined the incidence of decapod species associated with unbleached and bleached *Seriatopora hystrix* in the vicinity of the nuclear power plant.

2 MATERIAL AND METHODS

2.1 *Study area*

The study area is located at Nanwan Bay on the southern tip of Taiwan (21°55′N, 120°45′E) (Fig. 1). The tropical climate of southern Taiwan is characterized by a marked seasonal variation in sea temperatures ranging from 20.3 to 29.2°C. Sea temperatures of southern Taiwan are relatively uniform in terms of different localities

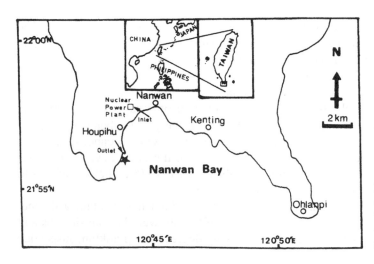

Figure 1. Map showing the study area in southern Taiwan. The location with bleached corals is marked ★.

and depths (Dai 1991). The marine fauna of the study area is strongly influenced by the warm Kuroshio ('the black current'), and it has a strong affinity with those of the Philippines and the Ryuku Islands. Southern Taiwan is surrounded by well-developed fringing reefs, with 235 species of scleractinians and 47 species of alcyonaceans; the coral fauna is very rich and comparable to the richest areas in the western Pacific in terms of species diversity (Dai 1988, 1991). *Seriatopora hystrix* is a stony coral commonly found in shallow water.

2.2 *Collection of specimens*

Seriatopora hystrix coral heads and their associated macrofauna from Nanwan Bay were either enclosed in plastic bags and removed, or broken off and put in a bucket and taken to the surface where all macroorganisms were removed. A total of thirty-two unbleached colonies were collected from Nanwan Bay, and ten of these were bleached coral heads collected in the warm effluent of a nuclear power plant in the western part of the bay at 3 m depth. The macrofauna of these coral heads consisted almost entirely of decapod crustaceans.

Dry weight was used as a measure for the size of the coral heads. The dead base and substrates were not processed, nor were animals collected from these portions. Decapod crustaceans were sorted into taxonomic groups and identified to species level whenever possible. Specimens were deposited in the Institute of Zoology, Academia Sinica, Taipei, Taiwan.

3 RESULTS

A total of 2148 macrofaunal individuals was collected and sorted into 85 species categories. Some 86.6% of all individuals were decapods and they consisted of shrimps, hermit crabs, and crabs. Among 53 decapod species (1860 individuals), 32 species (1629 individuals) were associated with unbleached colonies, whereas 27 species (231 individuals) were associated with bleached colonies. The remaining 13.4% of the fauna consisted primarily of molluscs, echinoderms, polychaetes, barnacles, and fishes. There were ten major decapod species whose individual numbers were more than 1% of the 1860 total collected decapod animals (Table 1). The numbers of alpheid shrimps (belonging to 16 species) that occurred with each unbleached colony are listed in Table 2. There appears to be different factors in the way that species perceive each coral head as habitat.

Comparing the ten major species in Table 3, the number of individuals of each species exceeded 1% of the 1860 total collected decapod individuals, and these species cumulatively accounted for 55.2% of total numbers. Comparing the species composition from unbleached and bleached coral heads revealed some differences in the biology of the two faunas. In the unbleached coral head, the xanthid crab *Trapezia lutea* (Castro 1997), was the most abundant species in each colony. In unbleached colonies *T. guttata*, the snapping shrimp *Alpheus lottini,* and *A. gracillipes* were also commonly found. However, the above species were represented by one or a few individuals so that they may occur at bleached coral head but were not yet collected. Es-

Table 1. Comparison of species composition, numbers of individuals, relative abundance, and frequency of occurrence of decapod crustaceans associated with *Seriatopora hystrix* between unbleached and bleached coral colonies.

	Associate species	Number of individuals		Relative abundance		Frequency of occurence (%)	
		Unbleached	Bleached	Unbleached	Bleached	Unbleached	Bleached
1	*Trapezia lutea*	252	0	0.1547	0	100.0	0
2	*Trapezia guttata*	187	0	0.1148	0	96.9	0
3	*Calcinus seurati*	134	35	0.0823	0.1515	75.0	40.0
4	*Alpheus lottini*	123	0	0.0755	0	93.8	0
5	*Trapezia areolata*	92	0	0.0565	0	87.5	0
6	*Alpheus gracillipes*	49	14	0.0301	0.0606	62.5	60.0
7	*Alpheus pachychirus*	44	8	0.0270	0.0346	37.5	50.0
8	*Lithoporcellana spinuligera*	42	5	0.0258	0.0216	46.9	40.0
9	*Synalpheus charon*	21	0	0.0129	0	31.3	0
10	*Synalpheus tumidomanus*	18	7	0.0110	0.0303	40.6	30.0

For the unbleached colonies $n = 32$, and total number of individuals is 1629; for bleached colonies $n = 10$, and total number of individuals is 231.

pecially *T. lutea, T. guttata, T. areolata, A. lottini,* and *Synalpheus charon* were not collected form bleached coral head.

4 DISCUSSION

During the summer of 1988 there was widespread damage in the area, with 100% bleaching (118 coral species) to a depth of 3 m and 50% bleaching at 3~6 m. One species, *S. hystrix,* bleached at 6~10 m. This revealed that it was more sensitive to temperature stress (Jeng, unpublished data). *S. hystrix* did not bleach when exposed to 31°C in separate experiments, but bleached if it was exposed to 32°C for 24 h. Results from the survey conducted from 1987 to 1996 at the western side of the mouth of the Third Nuclear Power Plant discharge canal showed that most corals at < 3 m depth bleached during the summer seasons.

The bushing coral *Seriatopora hystrix* bleaches rapidly when exposed to water temperatures > 30°C. Hoegh-Guldberg & Smith (1989) showed that bleached corals had reduced densities of zooxanthellae despite the normal pigment contents of zooxanthellae. Present evidence suggests that many corals are unable to adapt physiologically or genetically to such marked and rapid increased temperatures (Glynn 1993).

In the unbleached coral colonies of *S. hystrix, Trapezia lutea, T. areolata, Apheus lottini,* and *Synalpheus charon* all appear to be obligate associates, and they were not found on bleached colonies. The majority of the species is considered to be obligate commensals of living corals, although there were a few facultative commensal species that also occurred on bleached coral. Only five of these species were common. This were the alpheid shrimps *A. gracilipes, A. pachychirus, Synalpheus tumidomanus,* the hermit crab *Calcinus seurati* and the porcelain crab *Litho-*

Table 2. List of Alpheidae species and number of individuals collected from 32 unbleached colonies of *Seriatopora hystrix* of different sizes.

Coral colony no.	Colony size (kg)	*Athanas areteformis*	*Athanas grantii*	*Alpheopsis equalis*	*Alpheopsis sp.*	*Alpheus diadema*	*Alpheus frontalis*	*Alpheus gracilipes*	*Alpheus latipes*	*Alpheus lottini*	*Alpheus pachychirus*	*Metalpheus paragracilis*	*Synalpheus charon*	*Synalpheus bituberculatus*	*Synalpheus hastilicrassus*	*Synalpheus pococki*	*Synalpheus tumidomanus*	Total no. of alpheid individuals	Total no. of decapod individuals
1	0.5	1								2								3	19
2	0.5					1		2		2	3		2				1	11	55
3	0.6							2		3			2					7	27
4	0.6					2		2			1					2		7	33
5	0.6					3		1		6							1	11	67
6	0.7							3									1	4	42
7	0.7		1					1		5			2			1		10	43
8	0.8									3			2				2	7	32
9	0.8									4			2				1	7	65
10	0.8									4						1		5	53
11	0.8									2								2	49
12	0.9				1	2		1		2	3				2		2	13	48
13	0.9									3					1			4	42
14	1.0					1				2								3	21
15	1.1							1		3							2	6	24
16	1.1							4		6	6	2	5			1	2	26	78
17	1.2					2				1								3	49
18	1.2							2		4								6	37
19	1.2						2	1		1								4	40
20	1.3							1		1	2		2		2		1	9	39
21	1.4					2	3			1								6	41
22	1.4							5		7								12	79
23	1.5			2				1		9	3		1				2	18	69
24	1.5							10		10	11				2		1	34	80
25	1.6									5	3							8	46
26	1.8							1	1	3								5	69
27	1.8							4		14	5		2		4	3		32	85
28	1.9					2		2		9	4							17	43
29	2.0							2		3				1				6	44
30	2.4									6	2		1				1	10	89
31	2.6									2	1							3	47
32	2.6					2		3							2		1	8	74
Total no. individuals		1	1	2	1	17	5	49	1	123	44	2	21	1	13	8	18	307	1629

Table 3. Decapod crustacean species and number found on unbleached and bleached colonies of *Seriatopora hystrix* (*n* = 42). Ten species are listed whose individual number was >1% of 1860 total collected decapod organisms.

Associate species	No. of inhabited Coral colonies	No. of individuals	Percent total no. of decapod individuals	Cumulative %
Trapezia lutea	32	252	13.5	13.5
Trapezia guttata	31	187	10.0	23.5
Calcinus seurati	28	169	9.1	32.6
Alpheus lottini	30	123	6.6	39.2
Trapezia areolata	28	92	4.9	44.1
Alpheus gracillipes	26	63	3.4	47.5
Alpheus pachychirus	17	52	2.8	50.3
Lithoporcellana spinuligera	19	47	2.5	52.8
Synalpheus tumidomanus	16	25	1.3	54.1
Synalpheus charon	10	21	1.1	55.2

porcellana spinuligera. However, healthy (= unbleached) colonies of *S. hystrix* do not only provide their decapod inhabitants shelter and room, but also, and much more importantly, produce mucus as a food source for these symbiotic crustaceans.

Currently little is know about the impact of bleaching on the remainder of the reef community and the long-term effects on competition, predation, symbiosis, bioerosion, and substrate condition. This are all factors that can influence the long-term recovery and stability of coral reefs. Moreover, bleaching disturbances are not confined solely to symbiotic algae, but also affect some corals and sponges without endosymbionts (Williams & Bunkley-Williams 1990) and numerous other species that depend on live coral for shelter, food, and other requisites (Glynn 1990).

A general survey of crustaceans associated with branching corals has been conducted in the waters of Taiwan (Soong & Chang 1983, Jeng & Chang 1985, Chang et al. 1987, Jeng 1994). A rich crustacean fauna inhabits colonies of several branching coral species. In this paper, I examined a relatively recently discovered mutualism involving crustaceans and a coral host. Of these species, two alpheid shrimp (*Alpheus lottini* and *Synalpheus charon*) and three xanthid crabs (*Trapezia lutea, T. areolata,* and *T. guttata*) are likely to be obligatory symbionts of *Seriatopora hystrix,* because they only inhabit unbleached colonies. These species can defend their coral hosts against corallivores, and the mutualistic associations are assumed to be coevolutionary (Glynn 1983).

Most common obligate crustacean symbionts of corals belong to four families: Xanthidae, Palaemonidae (Pontoniinae), Alpheidae, and Hippolytidae (Patton 1966). Branching pocilloporid corals of the genera *Pocillopora, Stylophora,* and *Seriatopora* are hosts for *Trapezia* and *Alpheus.* However, host specificity among the many pocilloporid species has not been analyzed (Castro 1976). In most average sized coral colonies, only one symbiont species is present. However, more than one species may be found on large colonies. Most symbionts are found on the central branches with their head ends facing outwards. Smaller crabs and shrimps (adults as well as juveniles) are more common at the base of the colony.

The establishment of territories among symbionts is a consequence of pairing. Preston (1973) indicated that interferential competition resulting from random encounters during host selection, may explain the distribution patterns observed in the species. It was found that almost all movements outside of the coral took place during the night. The migration of *Trapezia ferruginea* among coral colonies illustrates the host selection by adults, which appears to be elicited by the lack of a suitable partner and the inadequate size of their established territory in the colony (Castro 1978). It is still not known, however, how often these migrations occur under natural conditions. The most sensitive coral and crustacean species are affected by thermal discharges. However, crustaceans can often escape before coral bleaching occurs.

The number of symbiotic decapod species (and individuals) on *Seriatopora hystrix*, increases with the size of the coral colony. This corroborates with results of Abele and Patton (1976) on crustacean communities that inhabit the branching coral *Pocillopora damicornis*. Higher species diversity was observed in bleached colonies of *S. hystrix* than was in unbleached ones. It appears that more obligate symbiotic decapods escape from unbleached colonies than recruit onto bleached colonies.

Seriatopora hystrix is highly sensitive to elevated water temperatures. In addition to this, the broad distribution and fast growth rate makes *S. hystrix* an ideal indicator species for monitoring marine environments (Jeng, unpubl. data). Moreover, from the inhabiting decapods of *S. hystrix,* it is obvious that at least two alpheid shrimp and three xanthid crabs have close symbiotic relationships with their hosts.

ACKNOWLEDGMENTS

This research was supported by the Kenting National Park and the Institute of Zoology, Academia Sinica, Taiwan.

REFERENCES

Abele, L.G. & Patton, W.K. 1976. The size of coral heads and the community biology of associated decapod crustaceans. *J. Biogeog.* 3: 35-47.

Bruce, A.J. 1976. Shrimps and prawns of coral reefs, with special reference to commensalism. In O.A Jones. & R. Endean (eds), *Biology and geology of coral reef* Vol. III: 37-94. New York: Academic Press.

Castro, P. 1976. Brachyuran crabs symbiotic with scleractinian corals: a review of their biology. *Micronesica* 12: 99-110.

Castro, P. 1978. Movements between coral colonies in *Trapezia ferruginea* (Crustacea: Brachyura), an obligate symbiont of scleratinian corals. *Mar. Biol.* 46: 237-245.

Castro, P. 1997. Trapeziid crabs (Brachyura: Xanthoidea: Trapeziidae) of New Caledonia, eastern Australia, and the Coral Sea. In B. Richer De Forges (ed.), Les fonds meubles des lagons de Nouvelle-Calédonie (Sédimentologie, Benthos). *Études & Thèses ORSTOM* 3: 59-107.

Chang, K.H., Chen, Y.S. & Chen, C.P. 1987. Xanthid crabs in the corals, *Pocillopora damicornis* and *P. verrucosa* of southern Taiwan. *Bull. Mar. Sci.* 41: 214-220.

Dai, C.F. 1988. Coral communities of southern Taiwan. *Proc. 6th int. Coral Reef Symp.* 2: 647-652.

Dai, C.F. 1991. Reef environment and coral fauna of southern Taiwan. *Atoll Res. Bull.* 354: 1-24.

Fan, K.L. 1988. The thermal effluent incident of the Third Nuclear Power Plant in southern Taiwan. Acta Oceanogr. *Taiwanica* 20: 117-125.

Glynn, P.W. 1980. Defense by symbiotic Crustacea of host corals elicited by chemical cues from predator. *Oecologia* 47: 289-290.

Glynn, P.W. 1983. Crustacean symbionts and the defense of corals: Coevolution on the reef? In M.H. Nitecki (ed.), *Coevolution*: 111-178. Chicago: University of Chicago Press.

Glynn, P.W. 1990. Coral mortality and disturbances to coral reefs in the tropical eastern Pacific. In P.W. Glynn (ed.) *Global ecological consequences of the 1982-83 El Niño-Southern Oscillation*: 55-126. Amsterdam: Elsevier.

Glynn, P.W. 1993. Coral reef bleaching: ecological perspectives. *Coral Reefs* 12: 1-17.

Hoegh-Guldberg, O. & Smith, G.J. 989. The effect of sudden changes in temperature, light and salinity on tha population density and export of zooxanthellae from tha reef corals *Stylophora pistillata* Esper and *Seriatopora hystrix* Dana. *J. Exp. Mar. Biol. Ecol.* 129: 279-303.

Jeng, M.S. 1994. Newly recorded symbiotic crabs (Crustacea: Decapoda: Brachyura) from southern Taiwan coral reefs. *Zool. Stud.* 33: 314-318.

Jeng, M.S. & K.H. Chang 1985. Snapping shrimps (Crustacea: Decapoda: Alpheidae) of Taiwan. *Bull. Inst. Zool. Acad. Sinica* 24: 241-256.

Knudsen, J.W. 1967. *Trapezia* and *Tetralia* (Decapoda, Brachyura, Xanthidae) as obligate ectoparasites of pocilloporid and acroporid corals. *Pacific Sci.* 21: 51-57.

Patton, W.K. 1966. Decapod Crustacea commensal with Queensland branching corals. *Crustaceana* 10: 271-295.

Patton, W.K. 1974. Community structure among the animals inhabiting the coral *Pocillopora damicornis* at Heron Island, Australia. In W.B. Vernberg (ed.), *Symbiosis in the sea*: 219-243. Columbia: University of South Carolina Press.

Patton, W.K. 1994. Distribution and ecology of animals associated with branching corals (*Acropora* spp.) from the Great Barrier Reef, Australia. *Bull. Mar. Sci.* 55: 193-211.

Preston, E.M. 1973. A computer simulation of competition among five sympatric congeneric species of xanthid crabs. *Ecology* 54: 469-483.

Serène, R. 1972. On the brachyuran fauna of the Indo-Pacific coral reefs. *Proc. Symp. Corals and Coral Reefs, Mar. Biol. Ass. India*: 419-424.

Soong, K.Y. & Chang, K.H. 1983. The coral-inhabiting barnacles (Crustacea: Thoracia: Pyrgomatidae) from southernmost coast of Taiwan. *Bull. Inst. Zool. Acad. Sinica* 22: 243-252.

Williams, E.H. & Bunkley-Williams, L. 1990. The world-wide coral reef bleaching cycle and related sources of coral mortality. *Atoll Res. Bull.* 335: 1-71.

The introduction of alien crayfish species into Britain for commercial exploitation – An own goal?

DAVID M. HOLDICH
School of Biological Sciences, The University of Nottingham, Nottingham, UK

ABSTRACT

Since the early 1970s various alien species of freshwater crayfish have been introduced into Britain for aquacultural purposes and also for the restaurant and aquarist trades. Escapes and deliberate implants have led to many populations developing in natural waters. In the mid-1990s the harvest from these populations rivalled that produced from aquaculture. Some of the established populations of alien crayfish are having an adverse physical and biological impact on the freshwater environment through their burrowing and trophic activities. Some also harbour a fungal disease, crayfish plague, to which the single native crayfish species in Britain, *Austropotamobius pallipes*, is totally susceptible. The disease has resulted in the loss of many populations since the early 1980s. As a result of these problems legislation has been introduced, firstly designating some of the alien crayfish as pests in 1992, secondly banning the keeping of all alien, temperate crayfish in 1996 except for the main species being farmed, *Pacifastacus leniusculus*, and then only outside specified no-go areas. Those involved in supplying crayfish for food such as fish markets and restaurants are exempt from the ban, although they must adhere to strict guidelines. The introduction of alien crayfish for commercial purposes has therefore scored an own goal resulting in the industry being subjected to stringent legislation and facing stiff competition from natural stocks of the same species they seek to cultivate. The British crayfish example highlights what is a world-wide problem concerning the movement of animals outside their native ranges and the resulting legislation required to try and alleviate the subsequent problems when they become established.

1 INTRODUCTION

The British Isles has a single species of native crayfish, the white-clawed crayfish, *Austropotamobius pallipes* (Lereboullet) (Plate 1). It has a widespread distribution in England, Northern Ireland, the Irish Republic and Wales, where it inhabits streams, rivers, lakes and reservoirs, most often in areas with mineral-rich waters (Holdich & Rogers 1997, Reynolds 1997, Holdich et al. 1999b). It is naturally absent from Scotland although one introduced population exists in the north-west (Fig. 1). Although catching and eating crayfish was a popular past-time prior to this century, since then there has not been much of a tradition for these activities except at a local level and no native crayfish fishery as such exists, unlike in mainland Europe (Holdich 1993, Ackefors 1998). In the 1970s various foreign species of crayfish started to be imported into England and Wales for the restaurant and aquarist trades and for aquacultural purposes (Holdich & Reeve 1991). No such introductions were made into Ireland as legislation existed to stop them (Holdich et al. 1999b). Although, *A. pallipes* has had to face many threats to its existence in the last few hundred years, mainly due to human activities resulting in pollution and habitat destruction, it is the recent influx of alien crayfish and an associated fungal disease, crayfish plague, which has had most impact, particularly in England. The problem and the resulting legislation introduced to try and alleviate it are outlined below. This is then put in context with the worldwide problem of crayfish introductions.

2 CRAYFISH PLAGUE

Since the 1980s the range of *Austropotamobius pallipes* in England and Wales has diminished, particularly in southern parts, although the species is still relatively widespread (Fig. 1) (Holdich & Rogers 1997, Holdich et al. 1999b). The diminution is largely the result of the disease, crayfish plague or aphanomycosis, caused by the fungus, *Aphanomyces astaci* (Schikora). The plague fungus is endemic to North America where it is associated with astacid and procambarid crayfish which are largely immune to its effects. Non-North American crayfish are, however, very susceptible to it (Unestam 1975). The disease entered Europe in the mid-1800s and rapidly spread through many countries decimating native crayfish stocks (Alderman 1996), especially those of the noble crayfish, *Astacus astacus* (L.), which was being harvested in large quantities at the turn of the century, particularly in Sweden and Finland (Brinck 1983, Westman 1991, Ackefors 1997). In the 1980s it affected populations of the narrow-clawed crayfish, *Astacus leptodactylus* Eschscholtz, in Turkey and severely upset the trade to Western Europe (Köksal 1988, Baran & Soylu 1989, Ackefors 1998). The disease is still prevalent in mainland Europe (Westman 1995, Diéguez-Uribeondo et al. 1997, Machino & Diéguez-Uribeondo 1998), and is likely to continue to be so as long as the North American crayfish vectors are present.

In England and Wales, crayfish plague affected many populations of *A. pallipes* between 1981-1993 (Alderman 1993) although there have been very few reported cases since 1993 (Holdich & Rogers 1997). Most populations did not recover unless restocking programmes were initiated (Holdich et al. 1999b). Some lakes in central Ireland were also badly affected in the mid-1980s but due to restocking they are now recovering (Reynolds 1988, Reynolds 1997). The outbreak of the disease in Ireland is thought to have been due to fungal spores being brought in on contaminated fishing equipment from mainland Europe (Reynolds 1988). However, in England there is evidence (Lilley et al. 1997) to link outbreaks with the importation (in the 1970s and 1980s) of the North American signal crayfish, *Pacifastacus leniusculus* (Dana), which is a proven vector of crayfish plague in Europe (Smith & Söderhäll 1986, Alderman et al. 1990, Diéguez-Uribeondo et al. 1997).

3 INTRODUCTIONS OF ALIEN CRAYFISH INTO EUROPE

In order to compensate for the loss of crayfish stocks due to crayfish plague various North American crayfish species have been introduced into European waters. Firstly, the striped crayfish, *Orconectes limosus* (Rafinesque) (Plate 2) was introduced into Germany in the 1890s, then *Pacifastacus leniusculus* (Plate 3) into Sweden in the 1960s, and finally the red swamp crayfish, *Procambarus clarkii* (Girard) (Plate 4) into Spain in the 1970s (Lowery & Holdich 1988). These all became widespread in Europe, usually due to human intervention, and have to some extent made up for the loss of production (Brinck 1983, Ackefors & Lindqvist 1994, Westman 1995, Ackefors 1997, 1998), although *O. limosus* is rarely utilised. *P. leniusculus* was considered to be an ecological and gastronomic homologue to *A. astacus* (Plate 5). *P. clarkii* has been particularly successful and large harvests have been made from Spanish populations and exported, mainly to Scandinavia (Gaudé 1986, Ackefors & Lindqvist 1994, Ackefors 1998). All three species have, however, been shown to be vectors of crayfish plague (*O. limosus* – Vey et al. 1983, *P. clarkii* – Diéguez-Uribeondo & Söderhäll 1993, *P. leniusculus* – see above) and so the situation has only been made worse for the native crayfish. In addition, the three North American crayfish chosen for introduction seem to have done particularly well in Europe. This is largely due to their immunity to crayfish plague, fast growth, high fecundity, invasive capabilities and physiological tolerance which makes them superior competitors to the native European crayfish (Huner & Lindqvist 1995, Lindqvist & Huner 1999). Not all North American crayfish populations in Europe harbour the crayfish plague fungus, however, and some mixed populations of North American and European crayfish have developed (Holdich 1999). More often than not though the North American species come to dominate and displace the native species (Holdich & Domaniewski 1995, Söderbäck 1995).

In addition to the impact North American crayfish are having on native European crayfish, they are also having an impact on the freshwater environment (Holdich 1999). Often their populations become very large and they consume large amounts of animal and plant material to the detriment of other organisms (Guan & Wiles 1996, 1997, Nyström 1999). Although there is nothing in the literature to suggest that *P. leniusculus* is a burrowing species, in Britain, at least, it is an extensive burrower and can cause considerable damage to lake and river banks (Plate 6) (Guan 1994, Holdich 1999). *P. clarkii* is also a burrower and causes damage in rice fields both to the irrigation structures, and the crop by increasing turbidity and eating shoots (Gaudé 1986, Fonseca et al. 1997).

Various Australian crayfish belonging to the genus *Cherax* have been introduced into Europe for aquacultural trials and aquarist purposes (Holdich 1999). A wild population of the yabby, *C. destructor* Clark, has become established in Spain and this species is being cultivated in Italy (see papers in Gherardi & Holdich 1999). Australian crayfish are susceptible to crayfish plague but the yabby has many of the characteristics of the North American crayfish introduced into Europe, including the fact that it burrows. If it becomes widespread then it could pose a serious threat to native crayfish as well as the freshwater environment. The tropical redclaw, *Cherax quadricarinatus* (Von Martens) (Plate 7), is sometimes kept by aquarists in Europe but it is thought unlikely it will become established if it escapes – at least in northerly temperate regions.

Astacus leptodactylus (Plate 8) is considered by some to be an alien crayfish in much of Europe as its native range is mainly the Ponto-Caspian basin (Holdich 1999). It has gradually spread westwards, often aided by humans to supplement stocks of European crayfish eliminated by crayfish plague, and by the building of canals. For example, it became commercially important in the former USSR after *A. astacus* stocks were affected by crayfish plague (Cukerzis 1988). As it moved westwards it has displaced other crayfish species, e.g. Huxley (1880) stated that '... the invading *A. leptodactylus* is everywhere overcoming and driving out *A. nobilis* in the struggle for existence, apparently in virtue of its more rapid multiplication.' Huxley was referring to the displacement of *A. astacus* in the White Sea region of Russia and the fact that *A. leptodactylus* had probably reached this region via canals connecting its rivers to the River Volga.

4   THE BRITISH SITUATION

4.1   *Introduction, establishment and farming of alien crayfish*

Prior to the 1970s Britain only had its single native species, *Austropotamobius pallipes*, and now in the 1990s there are five alien species in the wild and one kept by aquarists.

In the 1970s, *Astacus leptodactylus* from continental Europe was on sale in fish markets (Wickens 1982) and soon escaped into wild waters or was deliberately introduced. In the 1990s many populations inhabit canals, lakes, reservoirs and water-filled gravel pits in and around London, some populations are very large and can sustain harvesting (Holdich & Reeve 1991, Rogers & Holdich 1995, Holdich & Rogers 1997, Holdich et al. 1999b). In one instance, a riverine population has developed in eastern England and has extended downstream into the tidal parts. *A. leptodactylus* is highly tolerant of saline conditions and may become established in estuarine environments in Britain as it has done in Eastern and Northern Europe (Holdich et al. 1997). Only one farm (in central England) currently exists for *A. leptodactylus* but this has not been a success. Although *A. leptodactylus* does not carry crayfish plague it is a threat to *A. pallipes* and the freshwater environment because of its fast population growth, high fecundity and invasive capabilities – like many North American crayfish (Holdich 1999).

In the 1980s, *Astacus astacus* was introduced into south west England from Bavaria (Holdich & Reeve 1991, Holdich & Rogers 1997). It is farmed at one site from which it has escaped into the wild in a catchment where *A. pallipes* exists. Although *A. astacus* does not carry crayfish plague it is a threat to *A. pallipes* because of its larger size and higher fecundity. It is likely that in a competitive situation the native crayfish would be displaced by the alien crayfish.

In the 1980s *Procambarus clarkii* started to appear for sale in aquarist centres as well as being on the menus in some restaurants. As with other crayfish species, *P. clarkii* is a master of escape and there were soon various reports of individuals being found in the wild, some carrying eggs, although only one population as such was reported (in a London lake) and it has yet to be proved that it has become established, i.e. breeding (Holdich & Rogers 1997, Holdich et al. 1999b).

There have been two reports in the 1990s of *Orconectes limosus* occurring in England, although it is not known if they have become established (Holdich et al. 1999b). If this proves to be the case this could be a very serious development and a severe threat to *A. pallipes* because of *O. limosus*' characteristics (see above).

Various *Cherax* species have been imported into Britain in recent years but no serious attempts have been made to farm them (Holdich 1999). They are more often utilised as an aquarium pet, particularly *C. quadricarinatus* (see Section 4.2).

Pacifastacus leniusculus was introduced into Britain in the mid-1970s supposedly to replace *A. pallipes* (Plate 9) with a larger, more vigorous species which was immune to crayfish plague and which could provide a valuable harvest. Importers were encouraged to do this under a government-sponsored agricultural diversification scheme. Suggestions were made that *A. pallipes* was on the decline having been decimated by crayfish plague. There was no evidence for this and crayfish plague probably did not reach England until the 1980s, after the introduction of *P. leniusculus* – a known vector of the disease (see above). Neither was *A. pallipes* on the decline as subsequent studies showed it to be very abundant, particularly in central and northern England and the Welsh borderlands (Holdich & Rogers 1997, Holdich et al. 1999b).

Despite the concerns expressed by scientists, *P. leniusculus* was distributed widely after successful growth trials to potential crayfish farmers in England, Scotland and Wales in the 1970s and 1980s (Holdich & Reeve 1991). In the region of 300 implants were made although not all became established. In the 1990s, however, there are still extant many populations, some are in natural waters and others are on so-called crayfish farms (Fig. 2). The current distribution of *P. leniusculus* overlaps that of *A. pallipes* in much of central and southern Britain (cf. Figs 11 and 12). The new industry met with mixed success as some populations developed but others did not mainly due to a lack of proper management. Sufficient production was achieved for a British Crayfish Marketing Association to be formed with the aim of controlling prices and quality. However, this soon collapsed in the face of independent producers and sellers (Rogers & Holdich 1995). In 1990 there were 68 sites holding *P. leniusculus* registered with the Ministry of Agriculture, Fisheries & Food. This number rose to 82 in 1992 and 99 in 1994. However, production was never very high and was estimated at 0.5 tons in 1984 rising to a peak in 1992 of 6.5 tons (Auchterlonie 1993, Rogers & Holdich 1995). Holdich & Rogers (1992) and Rogers & Holdich (1995) have described the various types of facilities used for culturing crayfish, most being existing fish ponds or extensive lakes, very few purpose-built facilities having been constructed. Realistic production figures for semi-intensive facilities have been estimated to be 400 kg ha^{-1} and 200 kg ha^{-1} for more extensive systems, which compare well with those obtained for *P. leniusculus* in other countries but are well below those for some *Cherax* and *Procambarus* species (Rogers & Holdich 1995). After 1996, following the introduction of new legislation (see below), and in keeping with the registration of (fin) fish farms, many sites were deregistered by MAFF as they were not considered to be farms because they were not producing juveniles or rearing crayfish by feeding, they were simply harvesting crayfish from the wild. What was the 'wild' had never satisfactorily defined until recently and, as far as crayfish are concerned,

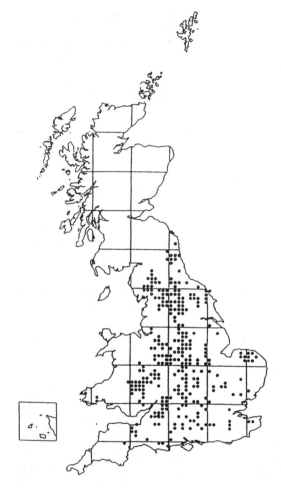

Figure 1. Distribution of *Austropotamobius pallipes* (Lereboullet) in Britain for the period 1990-1996 at the 10-km square level.

MAFF have defined it as – any body of water from which crayfish could move to a natural water course. The majority of crayfish farms subsequently were considered to be the 'wild' and the number of registered crayfish farms has now reduced to 10 in 1998 (A. Scott, pers. comm.).

Very few people are, therefore, making money out of farming crayfish and the problem is compounded by the fact that populations of some alien crayfish from natural waters are now so large that they can be harvested, making farming unnecessary. As most of these natural populations are derived from escapes from holding facilities then the industry has scored an own goal in this respect. Virtually nothing was harvested from the natural waters before 1990, but by 1994 it has been estimated that in the region of 8 tons were being taken (Rogers & Holdich 1995). This was not entirely based on *P. leniusculus* but also *A. leptodactylus*. Unfortunately, the lack of management on previous farm sites means that more escapes into the wild will occur (despite the legislation) thus making the situation for the native crayfish worse as, even if *P. leniusculus* is free of crayfish plague, it is known that it can out compete and predate on *A. pallipes* (Plate 10) (Holdich et al. 1995).

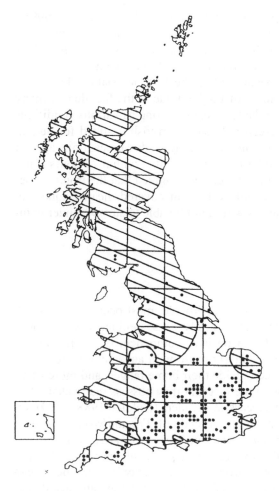

Figure 2. Distribution of *Pacifastacus leniusculus* (Dana) in Britain for the period
1990-1996 at the 10-km square level superimposed on no-go areas (shaded) where the
keeping of all temperate alien crayfish without a licence is banned (see text for exemptions).

4.2 *Legislation*

It became clear in the 1980s and 1990s that the future survival of *A. pallipes* was being threatened by established populations of alien crayfish, particularly *P. leniusculus*
which grows to a much larger size (Plate 9), is more fecund, and may predate on *A.
pallipes*. In 1986 *A. pallipes* was made a protected species and it became illegal to
take or sell it under the terms of the Wildlife & Countryside Act 1981. It was also
clear that alien crayfish were causing both physical and biological damage to the
freshwater environment. These facts prompted the British Government to designate
P. leniusculus, *A. leptodactylus* and *A. astacus* as pests in 1992 under the Wildlife &
Countryside Act, thus making illegal to allow them to escape into natural waters.
However, the threat continued and in 1996 a law was implemented making it illegal
to keep all alien crayfish over the whole of England, Scotland and Wales, except by
licence, under a Prohibition of Keeping of Live Fish (Crayfish) Order of the Import
of Live Fish Act. In effect, much of Britain became a no-go area for alien crayfish

(Fig. 2) (Holdich & Rogers 1997, Rogers & Holdich 1997, Holdich et al. 1999b). Current legislation will make any new crayfish farming ventures virtually impossible unless they are in escape-proof facilities, but considering how easy it now is to harvest alien crayfish from natural waters, such ventures would not seem economically worthwhile! There were three notable exemptions to the 1996 legislation. First, it is permitted to keep *P. leniusculus* in the south of England due to the fact that so many natural populations exist there (Fig. 2, unshaded areas), as long as various conditions are adhered to, particularly relating to escapes. Second, crayfish destined for human consumption are allowed to be kept in restaurants as long as due precautions against escapes are taken and they are killed or returned to the supplier within 72 hours if not used. Third, in order to satisfy the aquarium trade, it was agreed that the tropical species, *Cherax quadricarinatus* (Plate 7) could be kept in covered, heated aquarium tanks, as it was considered that any escapees would not be able to survive under natural conditions in Britain.

5 DISCUSSION

There are some 500 species of crayfish known worldwide, most occurring in North America and Australia (Hobbs Jr. 1988). They rarely cause problems within their home ranges and are usually considered as a positive component of the freshwater environment. Some species, however, particularly those with aquaculture and stocking potential, have been widely distributed outside their home ranges, and more often than not they have caused problems due to the fact that they possess many characteristics of r-selected species making them successful invaders (Holdich 1988, 1999). Of particular concern are the spread (or potential spread) of *Procambarus clarkii* in Europe, Africa and Asia; *Pacifastacus leniusculus* in Europe; *Cherax destructor* in Africa; *Cherax quadricarinatus* in South America; and *Orconectes* species in Europe.

Not all problems result from deliberate introductions and crayfish may be spread outside their home ranges by the building of canals, floods, predators, and accidental release from holding facilities, or even from bait used by anglers (Holdich 1999); however, the resulting problems are usually the same. *Orconectes rusticus* (Girard), for example, has been widely distributed within North America by the latter method and has become a pest in many waterbodies (Lodge et al. 1985). It appears to be a superior competitor to all other species of crayfish it encounters, and its ability to quickly build up dense populations means that it has significant detrimental impacts on the native biota when it becomes established (Lodge & Hill 1994). It has been implicated in causing environmental damage through competitive exclusion of other crayfish species, predation on fish eggs, reduction in fish stocks, and consumption of large amounts of macrophytes and invertebrates (see Magnuson et al. 1975, Hobbs III et al. 1989, Momot 1997). Legislation has been passed in two states banning its possession (Taylor et al. 1996).

Moyle et al. (1986) have stated that introductions are often made to solve some local or regional problem but, if the broad-scale consequences of each introduction are not considered, then it may ultimately cause more problems than it solves, i.e. plans to improve on nature are not always beneficial. They described this as the Frankenstein Effect and illustrated how it could be applied to fish introductions in North

America. There are many examples of this phenomenon in biology, e.g. the wide-ranging, negative, cascading effects within food webs caused by the introduction of the opossum shrimp, *Mysis relicta*, into some North American lakes and rivers (Spencer et al. 1991). Whilst there have been undoubted positive commercial, socio-economic, recreational and biological effects to crayfish introductions in Europe (Ackefors 1999) there have also been many negative aspects to such introductions and conservationists, at least, would consider their introduction as yet another example of the Frankenstein Effect (Holdich 1999, Holdich et al. 1999b). A good example of this is what has happened in Britain since the mid-1970s.

Attempts to develop a crayfish farming industry in Britain have not been successful but the legacy it has left behind is likely to have far-reaching environmental consequences. The industry has scored an own goal due to the fact that alien crayfish used for aquacultural, culinary and aquarist purposes, particularly *Pacifastacus leniusculus*, have escaped into natural waters (or been deliberately implanted) where they are causing damage to stocks of native crayfish through the transfer of disease and competition, upsetting the ecological balance of many waterbodies due to their trophic activities, and causing physical damage by burrowing. This has prompted the British Government to introduce stringent legislation to reduce their impact which will make future crayfish farming developments almost impossible. In addition, the establishment of *P. leniusculus* and *Astacus leptodactylus* in natural waters has led to populations developing which are now large enough to sustain harvesting to a level which is greater than that produced from farming (Rogers & Holdich 1995), thus making farming them uneconomic.

One of the original aims of crayfish farmers in Britain was to produce crayfish for export to the lucrative Scandinavian market where they fetch a high price (Holdich 1993). This did not happen to any extent until recently and ironically most of the exports are now being made with crayfish harvested from natural waters in Britain not from farming! Such crayfish are being exported to Sweden which has lifted all restrictions on the import of live crayfish since the lowering of trade barriers in the European Union (L. Edsman, pers. comm.). In theory this could allow crayfish carrying crayfish plague to be disseminated in Sweden and seriously threaten the remaining stocks of its native crayfish, *Astacus astacus*. Members of the EU apparently cannot prevent live crayfish imports from other members of the EU because of GATT and free trade agreements – as Germany found out recently (European Court of Justice 1994). Britain is also affected by such regulations but it has got round this by introducing legislation to control the destination of alien crayfish once they enter the country. A licence is also needed in Britain to introduce live crayfish from outside Europe and this is currently only granted for one species – *Cherax quadricarinatus* from Australia.

The problem of what to do about established populations of alien crayfish in natural waters remains in Britain. The Government is now keen to try and eradicate those populations causing problems. Very few attempts have been made in any country to try and eradicate nuisance crayfish populations, other than in relatively small, enclosed waterbodies, and certainly not in rivers. Holdich et al. (1999a) have reviewed those methods which have been attempted and have come to the conclusion that, whilst it might be possible to control nuisance crayfish populations by using intensive trapping of adults coupled with predatory fish species to keep juvenile numbers low, it would probably be impossible to eradicate them completely by such methods. This particularly applies to burrowing species such as *Cherax destructor*, *Pacifastacus leniusculus* and *Procambarus clarkii*, which may be able to escape the attentions of predators, and can certainly survive long periods in their burrows if the water level is lowered to try and enhance trapping. The use of organophosphate and organochlorine pesticides has been found to be effective in some cases but their use is not to be recommended due to the environmental impact they have. Piscicides such as rotenone have been found to have little effect on crayfish except when used at very high concentrations, e.g. > 50 mg l^{-1}, compared to those needed to kill some fish species, e.g. 0.5 mg l^{-1} (Holdich, unpublished results). Fourth generation pyrethroids are very effective against crayfish and, if authorities allow their use, may be the only answer for eradicating nuisance populations.

Hobbs III et al. (1989) stated that introducing crayfish may impose considerable environmental stress on the freshwater ecosystem such that irreparable shifts in species diversity may occur. They recommended that 'every effort should be taken to prevent unsolicited transplantations of any crayfish outside of its native range' and that 'resource managers should realise that successful crayfish transfers for aquacultural purposes will almost certainly lead to the establishment of breeding populations in the natural environment.' Crayfish species will continue to cause problems when introduced outside their home ranges unless this advice is taken – many countries are counting the cost of not listening to people who know the potential dangers.

ACKNOWLEDGEMENTS

Thanks are due to the organisers of the 4th International Crustacean Congress (Amsterdam 1998) for inviting me to give the plenary lecture on which this paper is based, to David Fox for taking the photographs for Plates 1, 3, 4, 7, 8 and to Dr R. Berg for supplying the colour original from which Plate 2 has been reproduced.

REFERENCES

Ackefors, H. 1997. The development of crayfish culture in Sweden during the last decade. *Freshwater Crayfish* 11: 627-654.
Ackefors, H. 1998. The culture and capture crayfish fisheries in Europe. *World Aquaculture* 29(2): 18-24, 64-67.
Ackefors, H. 1999. The positive effects of established crayfish introductions in Europe. In F. Gherardi & D.M. Holdich (eds), *Crayfish in Europe as alien species: how to make the best of a bad situation*?: 49-61. Crustacean Issues 11. Rotterdam: Balkema.
Ackefors, H. & O.V. Lindqvist, 1994. Cultivation of freshwater crayfish in Europe. In J. V. Huner (ed.), *Freshwater crayfish aquaculture in North America, Europe and Australia*: 157-216. New York: The Haworth Press, Inc.
Alderman, D.J. 1993. Crayfish plague in Britain, the first twelve years. *Freshwater Crayfish* 9: 266-272.
Alderman, D.J. 1996. Geographical spread of bacterial and fungal diseases of crustaceans. *Revue Scientifique et Technique, Office International des Epizooties* 15(2): 603-632.
Alderman, D.J., D.M. Holdich & I.D. Reeve 1990. Signal crayfish as vectors in crayfish plague in Britain. *Aquaculture* 86: 3-6.
Auchterlonie, N. 1993. Survey of crayfish farms 1992. *Trout News* 16: 10.

Baran, I. & E. Soylu 1989. Crayfish plague in Turkey. *Journal of Fish Diseases* 12: 193-197.

Brinck, P. 1983. An ecologist's approach to dealing with the loss of *Astacus astacus*. *Freshwater Crayfish* 5: 21-34.

Cukerzis, J. 1988. *Astacus astacus* in Europe. In D.M. Holdich & R.S. Lowery (eds), *Freshwater crayfish: biology, management and exploitation*: 309-340. London: Croom Helm (Chapman & Hall).

Diéguez-Uribeondo, J. & K. Söderhäll 1993. *Procambarus clarkii* Girard as a vector for the crayfish plague fungus, *Aphanomyces astaci* Schikora. *Aquaculture and Fisheries Management* 24: 761-765.

Diéguez-Uribeondo, J., C. Temiño & J.L. Múzquiz 1997. The crayfish plague fungus (*Aphanomyces astaci*) in Spain. *Bulletin Français de la Pêche et de la Pisciculture* 347: 753-763.

European Court of Justice 1994. Case C-131/93, Free movement of goods. Commission of the European Communities *vs* Federal Republic of Germany.

Fonseca, J.C., J.C. Marques & V.M.C. Madeira 1997. Oxygen uptake inhibition in *Procambarus clarkii*, red swamp crayfish by biogradable surfacants: An ecotechnological approach for population control in rice fields. *Freshwater Crayfish* 11: 235-242.

Gaudé, A.P. 1986. Ecology and production of Louisiana red swamp crawfish *Procambarus clarkii* in southern Spain. *Freshwater Crayfish* 6: 111-130.

Gherardi, F. & D.M. Holdich (eds) 1999. *Crayfish in Europe as alien species: how to make the best of a bad situation*? Crustacean Issues 11. Rotterdam: Balkema.

Guan, R. 1994. Burrowing behaviour of signal crayfish, *Pacifastacus leniusculus* (Dana), in the River Great Ouse, England. *Freshwater Forum* 4: 155-168.

Guan, R. & P.R. Wiles 1996. Growth, density and biomass of crayfish, *Pacifastacus leniusculus*, in a British lowland river. *Aquatic Living Resources* 9: 265-272.

Guan R. & P.R. Wiles 1997. Ecological impact of introduced crayfish on benthic fishes in a British lowland river. *Conservation Biology* 11(3): 641-647.

Hobbs, H.H. Jr.1988. Crayfish distribution, adaptive radiation and evolution. In D.M. Holdich & R.S. Lowery (eds), *Freshwater crayfish: biology, management and exploitation*: 52-82. London: Croom Helm (Chapman & Hall).

Hobbs III, H.H., J.P. Jass & J.V. Huner 1989. A review of global crayfish introductions with particular emphasis on two North American species (Decapoda, Cambaridae). *Crustaceana* 56(3): 300-316.

Holdich, D.M. 1988. The dangers of introducing alien animals, with particular reference to crayfish. *Freshwater Crayfish* 7: xv-xxx.

Holdich, D.M. 1993. A global review of astaciculture – freshwater crayfish farming. *Aquatic Living Resoures* 6(3): 307-317.

Holdich, D.M. 1999. The negative effects of established crayfish introductions. In F. Gherardi & D. M. Holdich (eds), *The introduction of alien crayfish species in Europe – how to make the best of a bad situation*?: 31-47. Crustacean Issues 11. Rotterdam: Balkema.

Holdich, D.M. & J.C.J. Domaniewski 1995. Studies on a mixed population of the crayfish, *Austropotamobius pallipes* and *Pacifastacus leniusculus*. *Freshwater Crayfish* 10: 37-45.

Holdich, D.M., R. Gydemo & W.D. Rogers 1999a. A review of possible methods for controlling alien crayfish populations. In F. Gherardi & D. M. Holdich (eds), *Crayfish in Europe as alien species: how to make the best of a bad situation*?: 245-270. Crustacean Issues 11. Rotterdam: Balkema.

Holdich, D.M., M.M. Harlioğlu & I. Firkins 1997. Salinity adaptations of crayfish in British waters with particular reference to *Austropotamobius pallipes*, *Astacus leptodactylus* and *Pacifastacus leniusculus*. *Estuarine, Coastal and Shelf Science* 44: 147-154.

Holdich, D.M. & I.D. Reeve 1991. The distribution of freshwater crayfish in the British Isles with particular reference to crayfish plague, alien introductions and water quality. *Aquatic Conservation* 1(2): 139-158.

Holdich, D.M. & W.D. Rogers 1992. Crayfish populations in the British Isles, farming, legislation, conservation and management. *Finnish Fisheries Research* 14: 23-32.

Holdich, D.M. & W.D. Rogers 1997. The white-clawed crayfish, *Austropotamobius pallipes*, in Great Britain and Ireland with particular reference to its conservation in Great Britain. *Bulletin Français de la Pêche et de la Pisciculture* 347(4): 597-616.

Holdich, D.M., W.D. Rogers, J.P. Reader & M.M. Harlioğlu 1995. Interactions between three species of freshwater crayfish *(Austropotamobius pallipes, Astacus leptodactylus* and *Pacifastacus leniusculus)*. *Freshwater Crayfish* 10: 46-56.

Holdich, D.M., W.D. Rogers & J. D. Reynolds 1999b. Native and alien crayfish in the British Isles. In F. Gherardi & D. M. Holdich (eds), *Crayfish in Europe as alien species: how to make the best of a bad situation?*: 221-235. Crustacean Issues 11. Rotterdam: Balkema.

Huner, J.V. & O.V. Lindqvist 1995. Physiological adaptations of freshwater crayfishes that permit successful aquacultural enterprises. *American Zoologist* 35: 12-19.

Huxley, T. H. 1880. *The crayfish. An introduction to the study of zoology*. London: Keegan Paul.

Köksal, G. 1988. *Astacus leptodactylus* in Europe. In D.M. Holdich & R.S. Lowery (eds), *Freshwater crayfish: biology, management and exploitation*: 365-400. London: Croom Helm (Chapman & Hall).

Lilley, J.H., L. Cerenius & K. Söderhäll 1997. RAPD evidence for the origin of crayfish plague outbreaks in Britain. *Aquaculture* 157: 181-185.

Lindqvist, O.V. & J.V. Huner 1999. Life history characteristics of crayfish – what makes some of them good colonizers? In F. Gherardi & D. M. Holdich (eds), *Crayfish in Europe as alien species: how to make the best of a bad situation?*: 23-30. Crustacean Issues 11. Rotterdam: Balkema.

Lodge, D.M. & A.M. Hill 1994. Factors governing species composition, population size, and productivity of cool-water crayfishes. *Nordic Journal of Freshwater Research* 69: 111-136.

Lodge, D.M., A. Beckel, & J.J. Magnuson 1985. Lake-bottom tyrant. *Natural History* 94(8): 32- 37.

Lowery, R.S. & D.M. Holdich 1988. *Pacifastacus leniusculus* in North America and Europe, with details of the distribution of introduced and native crayfish species in Europe. In D.M. Holdich & R.S. Lowery (eds), *Freshwater crayfish: biology, management and exploitation*: 283-308. London: Croom Helm (Chapman & Hall).

Machino, Y. & J. Diéguez-Uribeondo 1998. Un cas de peste des écrevisses en France dans le bassin de la Seine. *L'Astaciculteur de France* 54: 2-11.

Magnuson, J.J., G.M. Capelli, J.G. Lorman & R.A. Stein 1975. Consideration of crayfish for macrophyte control. In P.L. Bezonik & J.L. Fox (eds), *The Proceedings of a Symposium on Water Quality Management Through Biological Control*, 66-74. Gainesville: University of Florida.

Momot, W.T. 1997. History of the range extension of *Orconectes rusticus* into northwestern Ontario and Lake Superior. *Freshwater Crayfish* 11: 61-72.

Moyle, P.B., H.W. Li & B.A. Barton 1986. The Frankenstein Effect: impact of introduced fishes on native fishes in North America. In R.H. Stroud (ed.) *Fish culture in fisheries management*: 415-426. Bethesda, MD: American Fisheries Society.

Nyström, P. 1999. Trophic aspects of crayfish introductions in Europe. In F. Gherardi & D. M. Holdich (eds), *Crayfish in Europe as alien species: how to make the best of a bad situation?*: 63-86. Crustacean Issues 11. Rotterdam: Balkema.

Reynolds, J.D. 1988. Crayfish extinctions and crayfish plague in Ireland. *Biological Conservation* 45: 279-285.

Reynolds, J.D. 1997. The present status of freshwater crayfish in Ireland. *Bulletin Français de la Pêche et de la Pisciculture* 347(4): 693-700.

Rogers, W.D. & D.M. Holdich 1995. Crayfish production in Britain. *Freshwater Crayfish* 10: 583-596.

Rogers, W.D. & D.M. Holdich 1997. New legislation to conserve the native crayfish in Britain – will it work? *Freshwater Crayfish* 11: 619-626

Smith, V.J. & K. Söderhäll 1986. Crayfish pathology: an overview. *Freshwater Crayfish* 6: 199- 211.

Spencer, C.N., B.R. McClelland, & J.A. Stanford 1991. Shrimp stocking, salmon collapse, and eagle displacement. *BioScience* 41(1): 14-21.

Söderbäck, B. 1995. Replacement of the native crayfish *Astacus astacus* by the introduced species *Pacifastacus leniusculus* in a Swedish lake: possible causes and mechanisms. *Freshwater Biology* 33: 291-304.

Taylor, C.A., M.L. Warren Jr, J.F. Fitzpatrick Jr, H.H. Hobbs III, R.F. Jezerinac, W.L. Pflieger & H.W. Robinson 1996. Conservation status of crayfishes of the United States and Canada. *Fisheries* 21(4): 25-38

Unestam, T. 1975. Defense reactions in and susceptibility of Australian and New Guinean freshwater crayfish to European crayfish plague fungus. *Australian Journal of Experimental Medicine and Biological Sciences* 53: 349-359.

Vey, A., K. Söderhäll, & R. Ajaxon 1983. Susceptibility of *Orconectes limosus* Raff. to the crayfish plague, *Aphanomyces astaci* Schikora. *Freshwater Crayfish* 5: 284-291.

Westman, K. 1991. The crayfish fishery in Finland – its past, present and future. *Finnish Fisheries Research* 12: 187-216.

Westman, K. 1995. Introduction of alien crayfish in the development of crayfish fisheries; experience with signal crayfish (*Pacifastacus leniusculus* (Dana)) in Finland and the impact on the native noble crayfish (*Astacus astacus* (L.)). *Freshwater Crayfish* 10: 1-17.

Wickens, J.F. 1982. Opportunities for farming crustaceans in western temperate regions. In J.F. Muir & R.J. Roberts (eds). *Recent advances in aquaculture*: 87-177. London: Croom Helm.

Life history patterns of the red swamp crayfish (*Procambarus clarkii*) in an irrigation ditch in Tuscany, Italy

FRANCESCA GHERARDI, ANDREA RADDI,
SILVIA BARBARESI & GABRIELE SALVI
Dipartimento di Biologia Animale e Genetica 'Leo Pardi', Firenze, Italy

ABSTRACT

The red swamp crayfish, *Procambarus clarkii*, has been successfully introduced into many countries, Italy included, from its native North America. A number of biological traits of a population of this species was studied through a year cycle in an irrigation ditch in the neighbourhood of Florence. Here, we describe several traits of this species' life history (population structure, molting and reproduction) and discuss the relationships among them.

1 INTRODUCTION

The red swamp crayfish, *Procambarus clarkii* (Girard), is the most cosmopolitan species of crayfish. From its native distribution area, ranging from northeastern Mexico to south-central USA (Huner 1988), this crayfish has been introduced into all continents except Australia and Antarctica (Huner 1977, Huner & Avault 1979).

Some of the life history traits of *P. clarkii*, i.e. adaptability to burrow environments, resistance to air exposure, polytrophism, rapid growth, high fecundity and disease resistance (Huner & Lindqvist 1995), make its cultivation commercially feasible. Thus, following the profitable examples of the southern USA (annual production exceeding 60,000 tons, Huner & Lindqvist 1995) and the People's Republic of China (annual production of 40,000 tons, Huner & Lindqvist 1995), this crayfish has been imported into Italy, mostly from Spain. However, *Procambarus*' biology makes also its invasion of the wild extremely successful, especially in Italy where most farmers have failed to take adequate precautions to prevent the crayfish from escaping from farm enclosures.

At present, this species is well acclimatized in northern and central Italian inland waters (Gherardi et al. 1999). In the north, it is in great expansion in some sectors of the Po River (Delmastro 1992) and the Reno River (Mazzoni et al. 1996) drainage basins. In central Italy, it is widespread in: Tuscany (especially in the 112-km^2 Massaciuccoli Lake; Baldaccini 1995, Ercolini et al., in press), Umbria, the Marches (Gabucci et al. 1990), Latium (Welcomme 1988, Delmastro & Laurent 1997) and Abruzzo (G. De Luise, pers. comm.).

The paper will analyze some patterns of the biology of this crayfish in an irrigation ditch in Tuscany. It is part of an extensive research on the distribution and ecology of *P. clarkii*, having the final purposes of a) assessing the environment impact of this species in the Italian freshwater habitats and b) finding the means for better control and management of this invasive species.

2 MATERIALS AND METHODS

Fieldwork was conducted from June 1996 to May 1997 in an irrigation ditch in the neighborhood of Florence. Water depth ranged from 0 to 50 cm; the average pH was 7.7. The population inhabiting a 200-m strip was inspected every second week over a year cycle, using five cylindrical, 80-cm long baited traps (aperture: 28 cm in diameter, mesh size: 2 mm). These were located 40 m apart for 24 hours; their position remained constant through the year.

Collected animals were sexed and their cephalothorax length including the rostrum (CL) measured as an estimate of size. Their molting phase, degree of maturation, ovigerous state if females, and form I or II if males were recorded. Before being released, crayfish were individually marked using plastic tags glued to the first half of their cephalothorax.

Statistical analysis (using Willks' or Binomial tests) was done only when sample size was over 10. The level of significance under which the null hypothesis was rejected is $\alpha = 0.05$.

3 RESULTS

3.1 *Population size*

The size of the population was estimated applying Petersen's formula

$$P = m \ (r + u)/r$$

where m = specimens captured and marked one time, r = specimens captured two times and u = unmarked specimens. This provides the most likely value of the (active) population size, measured as the weight of the catch (the weight of an average-sized crayfish is 40 g) over the studied area (Fig. 1).

3.2 *Population structure compared among seasons*

In both females ($G = 62.829$, $df = 12$, $P < 0.001$) and males ($G = 59.965$, $df = 12$, $P < 0.001$) there were significant differences in the population structure of active crayfish among the seasons (Figs 2 and 3). Although minimum and maximum values of CL did not differ between sexes (females vs males, minimum CL: 16 mm vs 18.3 mm, maximum CL: 70.6 mm vs 70.5 mm), females reached a larger size than males in spring ($G = 14.502$, $df = 4$, $P < 0.001$) and summer ($G = 16.160$, $df = 5$, $P < 0.01$) but not in autumn ($G = 6.846$, $df = 5$, ns) and winter ($G = 3.948$, $df = 4$, ns).

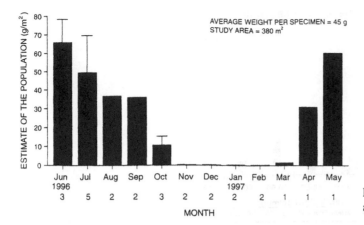

Figure 1. Estimate of the active population size.

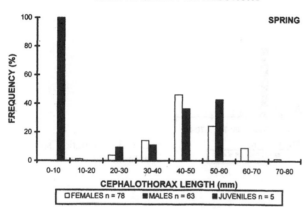

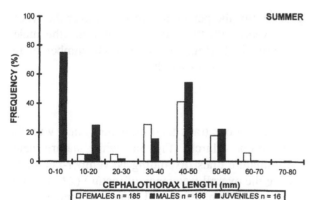

Figure 2. Population structures in spring and summer.

SIZE FREQUENCY DISTRIBUTIONS

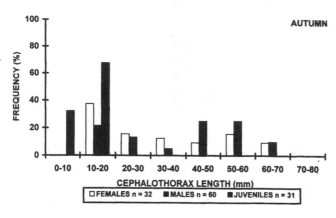

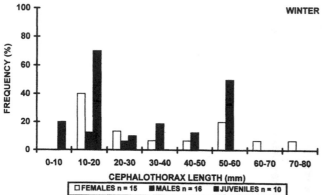

Figure 3. Population structures in autumn and winter.

3.3 *Sex ratio*

If we exclude December-February, sex ratio (the percentage of males over the whole population) was balanced through the year, with the exceptions of June (the males prevailed) and October (the females prevailed) (Fig. 4). The overall number of females (309) was nearly equal to that of males (308, $G = 0.002$, d$f = 1$, ns).

3.4 *Mature vs immature specimens*

Mature females occurred more frequently in our traps than immature ones, with the exception of November (excluding December-March) (Fig. 5a), while mature males were always higher in number than immature ones (excluding November-February) (Fig. 5b). This was also true, comparing the overall frequency of mature (221 females and 250 males) and immature specimens (82 females and 51 males, in females: $G = 66.106$, d$f = 1$, $P < 0.001$, in males $G = 258.014$, d$f = 1$, $P < 0.001$).

Obviously, the size of mature specimens in both females ($G = 180.555$, d$f = 6$, $P < 0.001$) (Fig. 6a) and males ($G = 171.371$, d$f = 6$, $P < 0.001$) (Fig. 6b) was larger than immature crayfish, but the two distributions superimposed in correspondence of size classes 30-40 and 40-50 mm CL.

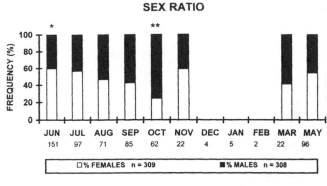

Figure 4. Sex ratio through the year.

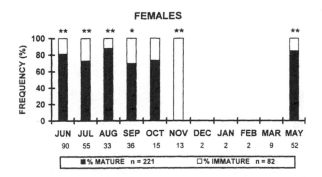

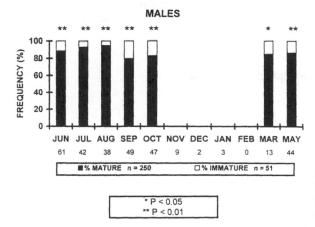

Figure 5. Occurrence of mature and immature specimens through the year in both females (a, top) and males (b, bottom).

3.5 *Ovigerous state*

Ovigerous females and females with hatchlings appeared in our catches in August-September and in September-October, respectively, but in a very low number (5 on 221, $G = 143.133$, d$f = 1$, $P < 0.001$) (Fig. 7). These reproducing individuals belonged to the size distribution of mature females ($G = 0.591$, d$f = 5$, *ns*) (Fig. 8).

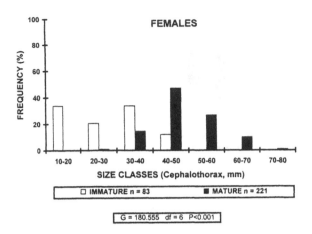

Figure 6. Size distributions of mature and immature females (a, top) and of mature and immature males (b, bottom).

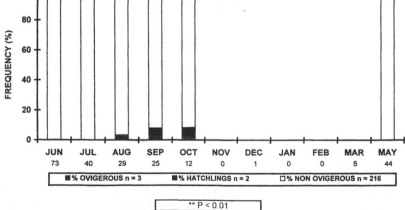

Figure 7. Occurrence of females brooding eggs and hatchlings through the year.

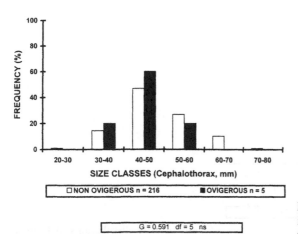

Figure 8. Size distributions of reproducing and not reproducing (but mature) females.

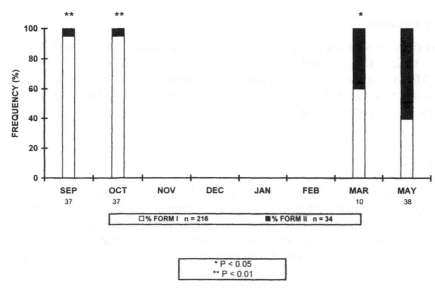

Figure 9. Occurrence of forms I and II males through the year.

3.6 *Form I and II males*

Sexually active, or form I males have a red color laterally and are very dark, often black, dorsally, while sexually inactive but mature males, or form II males, are greenish-brown. The former have more evident secondary sexual characters, such as prominent copulatory hooks at the bases of the third and fourth pairs of walking legs, cornified gonopodia, and rectangular and greatly inflated chelae (Huner 1988).

Form I males were more abundant in September and October than form II (excluding November-February) and were on overall more frequent (216 vs 34, $G =$ 147.461, d$f = 1$, $P < 0.001$) (Fig. 9). The two forms belong to the same size-class distribution ($G = 48.58$, d$f = 4$, *ns*) (Fig. 10).

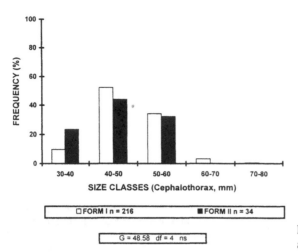

Figure 10. Size distributions of forms I and II males.

3.7 *Molting specimens*

Soft-shelled crayfish were always less abundant in our catches than hard-shelled ones (females: 10 vs 297, $G = 336.888$, $df = 1$, $P < 0.001$; males: 7 vs 300, $G = 358.237$, $df = 1$, $P < 0.001$) (Fig. 11). The time of molting was September-November and May-June in females, being in males more diffused through the year (molting males were found in June, August, October, and December).

4 DISCUSSION

Figure 12 illustrates the biological cycle of *P. clarkii* in the studied habitat at this latitude, as hypothesized on the basis of our study. It remains dormant inside burrows in winter. Mating takes place all over the active period, as shown by the constant occurrence of form I males in the active phase. Females spawn in April (G. Salvi, pers. comm.) and August (this study), but the number of females brooding first eggs and then hatchlings was very small in our catches. Males molt through the whole active period, while peaks of soft-shelled females appear after the hatchlings become independent. Maturity is reached soon, as suggested by the higher frequency of mature specimens in both sexes. From the management point of view, an important aspect to emerge from this study is that this introduced crayfish appears well adapted to a fluctuating aquatic system, e.g. an irrigation ditch, displaying a life cycle similar to the one shown in their natural environment, the swamplands in Louisiana (Gutiérrez-Yurrita et al. 1999).

In our catches only a small number of both ovigerous females and molting specimens were found, possibly because spawning, brooding eggs (and then hatchlings) and molting occur inside burrows. The red swamp crayfish, considered a tertiary burrower by Huner & Barr (1984), constructs several different types of burrows from short, submerged tunnels to one vertical, undulating tube which terminates in an enlarged chamber normally below the water table to U-shaped structures, enlarged at the bottom with chambers and covered by chimneys or mud plugs. As shown by this

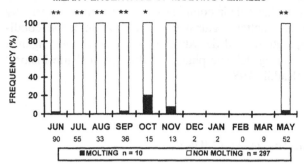

MEAN PERCENTAGE OF MOLTING FEMALES

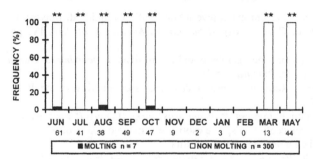

MEAN PERCENTAGE OF MOLTING MALES

Figure 11. Occurrence of soft- and hard-shelled females (a, top) and males (b, bottom).

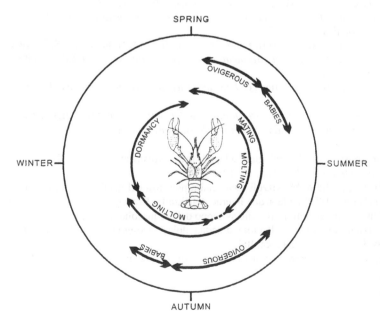

Figure 12. Biological cycle of *Procambarus clarkii* in the studied irrigation ditch.

study (as well as in: Hobbs, 1981 and Huner & Barr 1984), burrows play a central role in *P. clarkii*'s biology. This explains their complexity and the energy spent by these crayfish to dig them: they are shelters against extreme environmental conditions, such as low and high temperatures and dehydration, but also refuges against predators (conspecifics included) during delicate phases of this species' life history, such as reproducing and molting (Raddi 1998).

ACKNOWLEDGMENTS

This study was funded by Regione Toscana and MURST.

REFERENCES

Baldaccini, G.N. 1995. Considerazioni su alcuni macroinvertebrati dell'area umida di Massaciuccoli (Toscana). In P.E. Tomei & E. Guazzi (eds), *Il bacino del Massaciuccoli* Vol. IV: pp. 91-103. Pisa: Pacini.

Delmastro, G.B. 1992. Sull'acclimatazione del gambero della Lousiana *Procambarus clarkii* (Girard, 1852) nelle acque dolci italiane. *Pianura* 4: 5-10.

Delmastro, G.B. & P.J. Laurent 1997. Récentes avancées des écrevisses américaines en Italie. *L'Astaciculteur de France* 50: 2-6.

Ercolini, P., G.N. Baldaccini & M. Mattioli, in press. *Procambarus clarkii* (Girard) (Crustacea, Decapoda) nella zona umida di Massaciuccoli: una specie esotica infestante o una risorsa da sfruttare? Atti del Convegno 'I biologi e l'ambiente oltre il duemila'.

Gabucci, L., R. Para & M. Poselli 1990. *Pesci e Crostacei d'acqua dolce della Provincia di Pesaro-Urbino*. Villa Verrucchio: TPL La Pieve.

Gherardi, F., G.N. Baldaccini, S. Barbaresi, P. Ercolini, G. De Luise, D. Mazzoni, & M. Mori 1999. The situation in Italy. *Crustacean Issues* 11: 107-128.

Gutiérrez-Yurrita, P.J., J.M. Martinez, M. Ilhéu, M.A'. Bravo-Utrera, J.M. Bernardo & C. Montes 1999. The status of crayfish populations in Spain and Portugal. *Crustacean Issues* 11: 161-192.

Hobbs, H.H., Jr. 1981. The crayfishes of Georgia. *Smithsonian Contrib. Zool.* 318: 1-549.

Huner, J.V. 1977. Introductions of the Louisiana red swamp crayfish, *Procambarus clarkii* (Girard): an update. *Freshwater Crayfish* 3: 193-202.

Huner, J.V. 1988. *Procambarus* in North America and elsewhere. In D.M. Holdich & R.S. Lowery (eds), *Freshwater crayfish. Biology, management and exploitation*: pp. 239-261. London: Chapman & Hall (Croom Helm).

Huner, J.V. & J.W. Avault Jr 1979. Introductions of *Procambarus* spp. *Freshwater Crayfish* 4: 191-194.

Huner, J.V. & J.E. Barr 1984. *Red swamp crawfish: biology and exploitation*. Baton Rouge, Louisiana: Louisiana Sea Grant College Program, Center for Wetland Resources, Louisiana State University.

Huner, J.V. & O.V. Lindqvist 1995. Physiological adaptations of freshwater crayfish that permit successful aquacultural enterprises. *Am. Zool.* 35: 12-19.

Mazzoni, D., G. Minelli, F. Quaglio & M. Rizzoli 1996. Sulla presenza del gambero della Lousiana *Procambarus clarkii* (Girard, 1852) nelle acque interne dell'Emilia-Romagna. Atti del Convegno Nazionale 'Il contributo dei progetti di ricerca allo sviluppo dell'acquacoltura nazionale', 75-82.

Raddi, A. 1998. Introduzione di gamberi alloctoni d'acqua dolce: eco-etologia di *Procambarus clarkii*. Master thesis, University of Florence, Italy.

Welcomme, R.L. 1988. International introductions of inland aquatic species. FAO Fisheries Technical Paper 294.

Dispersal in the desert rock pool anostracan *Branchipodopsis wolfi* (Branchiopoda: Anostraca)

LUC BRENDONCK
Royal Belgian Institute of Natural Sciences, Freshwater Biology, Brussels, Belgium
Present adress: Laboratory of Aquatic Ecology, Leuven, Belgium

BRUCE J. RIDDOCH
University of Botswana, Department of Biological Sciences, Gaborone, Botswana

ABSTRACT

The freshwater anostracan *Branchipodopsis wolfi* inhabits clusters of desert rock pools in southeastern Botswana. These rock pool sites are rare and occur in a matrix of unsuitable habitat for establishment. The temporal and spatial variability of such (partially isolated) habitats is expected to favour selection for dispersal. This paper reviews recent work on dispersal strategies in *B. wolfi*. By using egg traps at different sites and heights, direct proof was gathered that at least part of the egg bank of *B. wolfi* is dispersed by wind. Wind-borne dispersal of resting eggs could, however, only be confirmed at one site with shallow basins and with little vegetation cover and this only at saltation height. Long-range dispersal by wind is considered a rare event and mal-adaptive for rock pool inhabitants. To reduce loss of eggs by landing in unsuitable places, short-range dispersal by means of drought-resistant resting eggs is expected to be the selectively superior strategy, very much as atelechory in desert plants and reduced dispersal ability in animals and plants of oceanic islands. After inundation, part of the egg bank of *B. wolfi* floats and is considered to be an efficient means of short-range dispersal. Analysis of the genetic structure of three rock pool metapopulations supports our hypothesis on the dominance of short-range in comparison with long-range dispersal in *B. wolfi*. The production of sticky eggs that collect dirt within seconds after deposition is considered a case of anti-telechory to inhibit long-range dispersal. The generation of multi-year egg banks and mixed dispersal strategies allows *B. wolfi* to respond to unpredictable habitat availability/-suitability in both time and space.

1 INTRODUCTION

Anostracans are a conspicuous component of the biological communities of temporary aquatic habitats. They produce resistant resting eggs that synchronise their life cycle with favourable habitat phases and that form egg banks to buffer against catastrophic population crashes due to early drying of pools. The temporal and spatial variability of such (partially isolated) habitats is expected to favor selection for dispersal (McPeek & Holt 1992, Perry & Gonzalez-Andujar 1993), which in anostracans is supposed to be mainly effected by resting eggs. The effectiveness of passive dispersal is indirectly demonstrated by their ability to colonise even remote oceanic islands (Brendonck et al. 1990, Peck 1994), and by the broad distributions of some species (Banarescu 1990, Hamer et al. 1994a, b). Wind (Riddoch et al. 1994) and animal vectors (Proctor 1964, Proctor et al. 1967, Belk & Lindberg 1979, Brendonck et al. 1990, Saunders et al. 1993) have been suggested as important dispersal agents, but few, if any, direct measurements have been made in anostracans or any other zooplankton groups.

Favorable sites with temporary pools often occur in clusters in a large matrix of unsuitable habitat. Under these conditions, long-range dispersal may cause a high mortality cost due to frequent transfer of eggs to uninhabitable places. Several hypotheses have been formulated in relation with the dispersal capacities of anostracan resting eggs. In contrast to the hypothesis that the specific sculpturing of resting eggs enhances the dispersal by wind, very much like a golf ball (Richard Hill, pers. comm.), others (e.g. Bill Williams pers. comm.), suggested that the egg shell morphology may increase the surface to allow more particles to attach and as such to increase the specific weight of the eggs. Brendonck et al. (1992) observed that the tetrahedral eggs characteristic of the streptocephalid subgenus *Parastreptocephalus* were difficult to remove from the bottom of petridishes and considered this also as an inhibition to wind dispersal.

The southern African fairy shrimp *Branchipodopsis wolfi* Daday, 1910 is particularly abundant in desert rock pools that usually occur as clusters on isolated granite outcrops. These rocky domes are rare and are separated by stretches of between 2 and 50 km of unfavorable habitat (Fig. 1: top). In southeastern Botswana, each such rock pool site consists of between 10 and 20 pools that are usually less than 10 m apart (Fig. 1: bottom). Very much as in desert plants (Ellner & Shmida 1981) or in island plant populations (Cody & Overton 1996), we expect that in the above geographic setting with fragmented populations, short range dispersal (atelechory or anti-telechory sensu Ellner & Shmida 1981) of eggs will be the selectively superior strategy reducing the mortality cost of landing in unsuitable places.

With the present contribution we want to review recent field work and laboratory studies on the dispersal potentials of resting eggs of *B. wolfi* in rock pools in southeastern Botswana. In these studies it was assessed to what extent eggs are apt for dispersal by wind (considered here as a long-range vector) or are more suitable for short-range dispersal. Evidence for our hypothesis was gathered directly from field observations, or indirectly from the morphology or behavior of eggs, and the genetic structure of the populations.

2 STUDY MATERIAL

2.1 *Study species*

Branchipodopsis species are distributed widely in southern and eastern Africa (Barnard 1929, Hamer & Appleton 1996) and are the only anostracans in the region to persist in short-lived rock pools. *Branchipodopsis wolfi* is the most widely distributed and morphologically variable one of the genus (Hamer & Appleton, 1996). The spe-

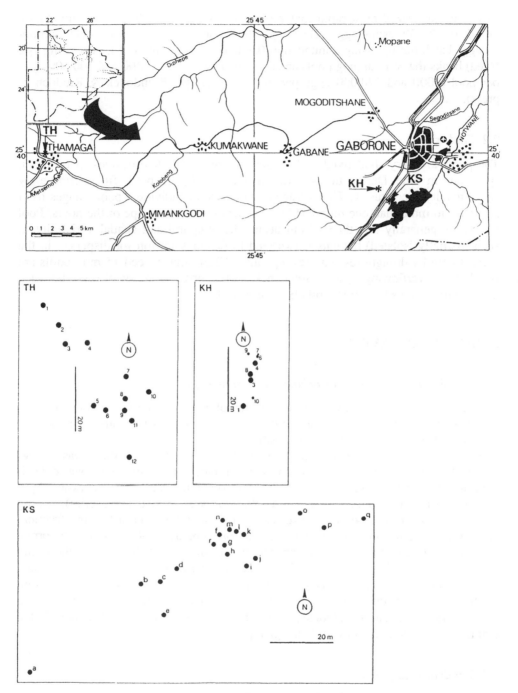

Figure 1. Rock pool sites of *Branchipodopsis wolfi* Daday, 1910 in southeastern Botswana (see in-set) to illustrate relative distances between sites (top) and basins at the three different sites (bottom). KS: Kgale Siding, KH: Kgale Hill, Th: Thamaga.

cies is characterized by a rapid maturation (3-4 days) and a high fecundity (daily broods of 30-150 resting eggs). Hatching is light and temperature dependent, and not all eggs hatch during a single inundation (Brendonck et al., in press), generating egg (seed) banks that vary among pools and seasons between about 200 and 500,000 eggs per pool (1000 and 220,000 eggs per m^2, respectively) (Brendonck & Riddoch in press).

2.2 *Study sites*

The pools are very short-lived with largely unpredictable hydrocycles that vary in space and in time. During the wet season (October-April), pools fill intermittently and often dry prematurely. The average hydroperiod of individual pools ranges from five days to more than one month (Fig. 2) depending on the shape of the pools. Pool basins are generally less than 15 cm deep. The temporal and spatial variability of such (partially isolated) habitats is expected to favour selection for dispersal, in the present case by drought resistant resting eggs. When rains exceed 15 mm, pools are filled till overflowing (Luc Brendonck pers. obs.) and overflows between neighbouring pools have frequently been observed.

3 RESULTS, OVERVIEW

3.1 *Morphology and buoyancy behavior of resting eggs*

All eggs have a similar external sculpturing of the egg shell consisting of ribs surrounding polygonal depressions, but some are smooth while others are sticky, collecting dirt within seconds after deposition (Fig. 3).

In Brendonck et al. (1998) it was shown that after inundation there was always part of the egg bank that floated. This behavior of part of the resting eggs was considered an adaptation to escape from the mud in order to receive the necessary light stimulus to hatch. In another study (Luc Brendonck submitted manuscript), resting eggs produced in the laboratory were separated in a sinking and a floating fraction. Eggs were dried for three and a half weeks before separation of fractions. To correct for empty eggs, part of each fraction was decapsulated with bleach. Of the entire batch of eggs 28.6% were floating. Both fractions were dry-stored for an additional two weeks to determine any changes in buoyancy behaviour. Of the original floating fraction, 55.1% became sinking, while of the original sinking fraction, 43.8% became floating after the additional storage period. There were no structural differences in the diameter or chorion thickness of both egg types.

3.2 *Windborne dispersal*

In a former study during the dry season (Brendonck & Riddoch 1999), egg traps (sticky quadrats) were mounted at approximately 1-metre intervals around a total of 33 dry pool margins at three rock pool sites to estimate the rate of outgoing and to a lesser extent of incoming eggs due to wind action. The distance between traps and each egg bank varied with the depth of the pool basin but was usually more than 10

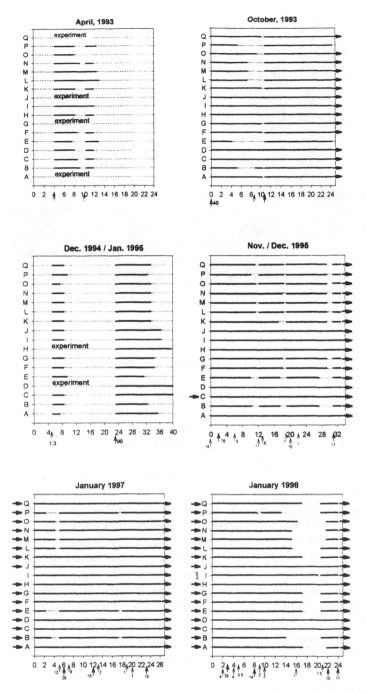

Figure 2. Hydrocycles of rock pools at one metapopulation (Kgale Siding, KS) at different years, illustrating temporal and spatial variability in the pools' suitability for reproduction. Thick horizontal lines show the duration of each hydration period of the pools (KSA-KSQ). Vertical arrows at *x*-axis show rainfall events with their volume in ml/m² (when available). Horizontal arrows at *y*-axis indicate that pools were immersed at the start/end of the observation period. 'experiment' = hydrocycle not presented due to manipulation.

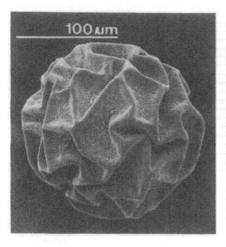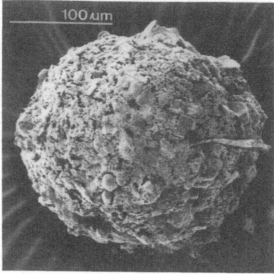

Figure 3. External morphology of the resting egg of the fairy shrimp *Branchipodopsis wolfi* Daday, 1910 with smooth (left) and sticky (right) surface.

cm. Some additional quadrats were mounted on the rocky surface between pools to trap resting eggs that are dispersed by wind in successive short jumps, or to catch so-called 'egg rains', eggs that are carried by wind over longer distances and at greater altitude. A total of 8 viable eggs, of which two of the sticky type, were trapped around pool margins, mainly at one site with shallow basins and limited vegetation cover. With the exception of one pool, all eggs were captured around unvegetated basins. No egg trap contained more than one egg. No eggs were collected at the rock pool site with deepest basins and most vegetation, nor on any of the quadrats mounted between the pools.

In a recent study (unpublished) egg traps were mounted at different levels from the rocky substrate at one rock pool site: 109 Tanglefoot™ yellow sticky traps at 1-m intervals around the margin of 13 dry pools (max 15 cm deep), three wooden palets each with 80 Tanglefoot™ yellow sticky traps for detecting any saltation of eggs more than 13 cm above the ground and/or of any settlement of airborne eggs, 131 egg traps (inverted test tubes smeared with brush-on tangle trap™ insect trap coating) around the entire rock pool site to trap any eggs transported by wind about 1m above the ground. All egg traps were left in the field for about three weeks. With the exception of two viable eggs around individual pools no other resting eggs were collected on any of the other devices (Luc Brendonck & Bruce Riddoch unpubl. data).

3.3 *Genetic structure*

In a study of the genetic structure of three metapopulations of *B. wolfi* (Brendonck et al. in press), a total of seventeen pools from three rockpool sites (KS, KH and Th), varying in morphometry and representing a range of geographic distances (Fig. 1) were selected. Samples from whole organisms were electrophoresed on Titan III cellulose acetate gels (Helena Laboratories) using the methodology of Hebert & Beaton (1989). After screening for polymorphy, 4 enzyme loci were retained: PGM (Phosphoglucose mutase), PGI (phosphoglucose isomerase), APK (arginine phosphokinase), and GOT (Glutamate-oxaloacetate transferase). Nearly all populations showed a net deficiency of heterozygotes for pooled genotypes and in about 50% there was a significant deviation from H-W equilibrium. Clustering of all populations performed with UPGMA based on Rogers' (1972) genetic similarities (S) showed clearly differentiated metapopulations, but levels of differentiation did not correspond with geographic distance. The KH and KS rock pool sites separated by only 2 km were genetically more differentiated than KH and Th sites that are about 50 km apart. At the individual sites some effect of distance was revealed only at the KS site with deeper pool basins and where two pools were more isolated (about 20 m) than the rest.

4 DISCUSSION

In southeastern Botswana, the rock pool habitat of the freshwater anostracan *Branchipodopsis wolfi* shows temporal and spatial heterogeneity. The different patches of the environment are of different quality at any given time. In such an environment dispersal of part of the egg (seed) bank is generally favored (Venable & Brown 1988, McPeek & Holt 1992, Perry & Gonzalez-Andujar 1993). As favorable sites (rock domes) are rare and occur scattered in a large matrix of habitat unsuitable for establishment, short-range dispersal will reduce the mortality cost of transporting eggs to unsuitable places and will therefore be the selectively superior strategy. This corresponds with atelechory (seeds lacking dispersal-enhancing characters) which is the dominant strategy in plants in the open deserts of the Old World (Ellner & Shmida 1981). Strong reductions in dispersal potential are also characteristic of many diverse taxa, both plants and animals, on oceanic islands and reflect strong selection for reduced dispersal potential in small and isolated natural populations (Cody & Overton 1996). After rehydration, part of the egg banks of *B. wolfi* populations floats (Brendonck et al. 1998). As floating eggs are more likely to be dispersed by means of overflows than the sinking eggs, and as overflows are very common between the shallow rock pools, and more important than other possible dispersal agents, Brendonck (submitted manuscript) stated that floating eggs are more dispersed than the sinking eggs and form an effective means of short-range dispersal. The production of eggs with sticky surface that collect particles within seconds after their deposition (Fig. 3) may be an example of antitelechory (sensu Ellner & Shmida 1981) to hinder dispersal.

Besides dispersal by overflows, additional short-range vectors may be cattle and amphibians (Luc Brendonck, pers. obs.), but the survival of *B. wolfi* resting eggs after defecation by these organisms has not been tested. Also with these cases of zoochory, floating eggs may be more easily attached to animal coat or ingested by drinking. Only when sinking eggs would be stickier than floating ones would they be more easily dispersed in plumes and on pats of waterfowl. Not only is this vector of lower importance, there is also no indication for differential stickiness between sinking and floating egg types in *B. wolfi*.

Windborne dispersal of resting eggs could only be shown to some extent from one site with shallow basins (Th) and this mainly around basins with little vegetation (Brendonck & Riddoch 1998). In an additional experiment, no eggs were captured at a height of more than 15 cm above the ground. These observations should, therefore, mainly be considered as saltation of eggs by wind action rather than long-range dispersal on wind currents.

The genetic analysis of the population structure (Brendonck et al. in press) gives indirect support for the inefficiency of long-range dispersal. The three metapopulations (KS, KH and Th) are genetically strongly differentiated without clear geographic pattern. The two sites being separated by less than 2 km (KS and KH) were even more different than the KH and Th sites being about 50 km apart. This lack of pattern in the isolation by distance may reflect the stochastic nature of past colonisation events by rare long-range dispersal. In individual metapopulations, a significant positive relationship between geographic and genetic distance was only clear at one site (KS) with deeper pools. Pools separated by more than 20 meters were clearly differentiated (Brendonck et al. in press). A similar pattern was revealed by Riddoch et al. (1994), and gives additional support to the above hypothesis on the importance of short-range dispersal in the fragmented rock pool sites in consideration.

All populations except one were heterozygote-deficient and in about 50% of populations a significant deviation from H-W equilibrium was observed. Heterozygote deficiencies are usually explained in terms of selection, non-random mating, inbreeding, or population subdivision (Wahlund effect). The most likely cause of the observed heterozygote deficiency is a Wahlund effect resulting from the simultaneous hatching of eggs from different generations, and/or from mixing of eggs from several pools by overflows after heavy rains promoting the mixing of populations.

The predominance of short- rather than long-range dispersal in *B. wolfi*, may be an important factor in explaining the high morphological variability between regions observed in this species (Hamer & Appleton 1996). The strong regional patterning of gene pools in *B. wolfi* is reminiscent of the pronounced regional divergence in some members of the *Daphnia carinata* complex, and may lead to allopatric speciation (Hebert & Wilson 1994). Local groups of populations communicating by exchange of resting eggs could be considered as important units of evolution. Their potential for divergence is enhanced because of their isolation and subdivided population structure (Templeton 1982). The generation of multi-year egg banks and the production of egg types with different dispersibility and hatchability make *B. wolfi* an extreme bethedger (spreading of risk) with means to respond to unpredictable habitat availability/suitability in both time and space.

ACKNOWLEDGEMENTS

L.B. is a postdoctoral researcher with the Fund for Scientific Research (F.W.O.-Flanders, Belgium). This study was partly financed by the research grant G.0260.97 provided by the F.W.O-Flanders. I am most grateful to Luc De Meester for his help with the electrophoresis of samples and Richard Hill is thanked for the constructive discussions on egg dispersal in anostracans. Mrs V. Arkosi kindly corrected the English.

REFERENCES

Banarescu, P. 1990. Vol. 1. General distribution and dispersal of freshwater animals. *Zoogeography of Fresh Waters*. Wiesbaden: Aula-Verlag.

Barnard, K.H. 1929. A revision of the South African Branchiopoda (Phyllopoda). *Ann. S. Afr. Mus.* 29: 181-272

Belk, D. & D.R. Lindberg 1979. First freshwater animal reported for Isla de Guadelupe represents a southern range extension for *Branchinecta lindahli* (Crustacea: Anostraca). *Southwest. Nat.* 24: 371-396.

Brendonck, L., submitted. Floating resting eggs as an efficient means of short range dispersal in the rock pool anostracan *Branchipodopsis wolfi* (Crustacea: Branchiopoda).

Brendonck, L. & B. Riddoch 1999. Proof for egg dispersal by wind in anostracans (Crustacea: Branchiopoda). *Biol. J. Linn. Soc.* 67: 87-95.

Brendonck, L. & B. Riddoch in press. Egg bank dynamics in anostracan desert rock pool populations. *Arch. Hydrobiol.*

Brendonck, L., L. De Meester & B.J. Riddoch in press. Regional structuring of genetic variation in short-lived rock pool populations of *Branchipodopsis wolfi* (Crustacea: Anostraca). *Oecologia.*

Brendonck, L., M. Hamer & A. Thiery 1992. The occurrence of tetrahedral eggs in the Streptocephalidae (Crustacea: Branchiopoda: Anostraca) with descriptions of a new subgenus: *Parastreptocephalus* and a new species: *Streptocephalus (Parastreptocephalus) zuluensis* Brendonck and Hamer. *J. Crust. Biol.* 12: 282-297.

Brendonck, L., A. Thiery & A. Coomans 1990. Taxonomy and biogeography of the Galapagos branchiopod fauna (Anostraca, Notostraca, Spinicaudata). *J. Crust. Biol.* 10: 676-694.

Brendonck, L., B.J. Riddoch, V. van de Weghe & T. Van Dooren 1998. The maintenance of egg banks in very short-lived pools – a case study with anostracans (Branchiopoda). *Arch. Hydrobiol.* (Spec. Issue) 52: 141-161.

Cody, M.L. & J. McC. Overton 1996. Short-term evolution of reduced dispersal in island plant populations. *J. Ecology* 84: 53-61.

Ellner, S. & A. Shmida 1981. Why are adaptations for long-range seed dispersal rare in desert plants? *Oecologia* 51: 133-144.

Hamer, M. & C.C. Appleton 1996. The genus *Branchipodopsis* (Crustacea, Branchiopoda, Anostraca) in southern Africa. *Ann. S. Afr. Mus.* 104: 311-377.

Hamer, M., L. Brendonck, A. Coomans & C.C. Appleton 1994a. A review of African Streptocephalidae (Crustacea: Branchiopoda: Anostraca). Part 1: South of Zambezi and Kunene rivers. *Arch. Hydrobiol.* 3: 235-277.

Hamer, M., L. Brendonck, A. Coomans & C.C. Appleton 1994b. A review of African Streptocephalidae (Crustacea: Branchiopoda: Anostraca). Part 2: North of Zambezi and Kunene rivers, and Madagascar. *Arch. Hydrobiol.* 3: 279-311.

Hebert, P.D.N. & M.J. Beaton 1989. *Methodologies for allozyme analysis using cellulose acetate electrophoresis*. Beaumont, Texas: Helena Laboratories.

Hebert, P.D.N. & C.C.Wilson 1994. Provincialism in plankton: endemism and allopatric speciation in Australian *Daphnia. Evolution* 48: 1333-1349.

McPeek, M.A. & R.D. Holt 1992. The evolution of dispersal in spatially and temporally varying environments. *Amer. Nat.* 140: 1010-1027.

Peck, S.B. 1994. Diversity and zoogeography of the non-oceanic Crustacea of the Galapagos Islands, Ecuador (excluding terrestrial Isopoda). *Can. J. Zool.* 72: 54-69.

Perry, J.N. & J.L. Gonzalez-Andujar 1993. Dispersal in a metapopulation neighbourhood model of an annual plant with a seedbank. *J. Ecol.* 81: 453-463.

Proctor, V.W. 1964. Viability of crustacean eggs recovered from ducks. *Ecology* 45: 656-658.

Proctor, V.W., E.R. Malone & V.L. DeVlaming 1967. Dispersal of aquatic organisms: viability of disseminules recovered from the intestinal tract of captive killdeer. *Ecology* 48: 672-676.

Riddoch, B.J., S.W. Mpoloka, & M. Cantrell 1994. Genetic variation and localised gene flow in the fairy shrimp, *Branchipodopsis wolfi* in temporary rainwater pools in south-eastern Botswana. In A. Beaumont (ed.), *Genetics and Evolution of Aquatic Organisms:* 96-102. London: Chapman & Hall.

Rogers, J.S. 1972. Measures in genetic similarity and genetic distance. Studies in genetics VII. University of Texas Publication 7213: 145-153.

Saunders III, J.F., D. Belk, & R. Dufford 1993. Persistence of *Branchinecta paludosa* (Anostraca) in southern Wyoming, with notes on zoogeography. *J. Crust. Biol.* 13: 184-189.

Templeton, A.R. 1982. Genetic architectures of speciation, pp. 105-121. In C. Barigozzi (ed.), *Mechanisms of speciation.* New York: Alan R. Liss, Inc.

Venable, D.L. & J.S. Brown 1988. The selective interactions of dispersion, dormancy, and seed size as adaptations for reducing risk in variable environments. *Amer. Nat.* 131: 360-384.

Bathymetric distribution of decapods associated with a *Posidonia oceanica* meadow in Malta (Central Mediterranean)

JOSEPH A. BORG & PATRICK J. SCHEMBRI
Department of Biology, University of Malta, Msida, Malta

ABSTRACT

To study the bathymetric distribution of decapods associated with a *Posidonia oceanica* meadow, samples were collected from four stations at depths of 6, 11, 16, and 21 m over a one year period, between August 1993 and September 1994, using a suction sampler and a hand held net. A total of 4607 individuals belonging to 41 species were collected. The hermit crab *Cestopagurus timidus* had the highest abundance in the suction samples, whilst the shrimp *Hippolyte inermis* had the highest abundance in the net samples. The suction sampler collected more than twice the number of species collected by the hand net; no brachyurans were collected by the hand held net. The application of Multidimensional Scaling to the decapod species-abundance data obtained from the suction samples gave three groups corresponding to three depth zones: 6 m, 11-16 m and 21 m. Shannon-Wiener's index of diversity was highest for the decapod assemblage inhabiting the middle part of the meadow, at a depth of 16 m. This part of the meadow also had the thickest 'matte'. Species richness and abundance of decapods varied significantly with bathymetry, and appeared to be correlated with changes in the morphology of the *Posidonia* meadow coincident with the depth gradient, particularly shoot density and structural complexity of the 'matte'. While some temporal changes in the structure of the decapod assemblages were evident, these were not significant.

1 INTRODUCTION

Decapods constitute one of the most conspicuous faunal groups inhabiting seagrass meadows and, together with other crustaceans, are important intermediaries in the transfer of energy in these ecosystems (Bell & Harmelin Vivien 1983, Chessa et al. 1983). As a result, they have received considerable attention in studies on seagrass associated fauna (e.g. Heck & Thoman 1984, Heck et al. 1989), in those on the relationship between meadow complexity and abundance of the associated fauna (e.g. Bell & Westoby 1986a, b, c), and those on predator-prey relationships in seagrass habitats (e.g. Heck & Thoman 1981, Nelson 1981, Leber 1985).

In the Mediterranean, extensive meadows are formed by the endemic *Posidonia oceanica*, which occurs from ca 1 m down to ca 40 m (Mazzella et al. 1992). *P. oceanica* meadows are very important coastal biotopes supporting a high diversity of associated macrofauna, including decapods (Mazzella et al. 1989, Gambi et al. 1992, Mazzella et al. 1992, Scipione et al. 1996). However, there are relatively few published studies dealing specifically with the total decapod assemblage of these meadows (García Raso, 1990, Zupo 1990, García Raso et al. 1996). Most of the available information has been derived from general studies on the macroinvertebrates associated with this seagrass (e.g. Ledoyer 1962, Harmelin 1964, Scipione et al. 1983, Templado 1984, Mazzella et al. 1989, Cozzolino et al. 1992, Gambi et al. 1992, Pessani et al. 1995, Scipione et al. 1996); or from studies on particular sub-habitats of the *Posidonia* system (the leaf stratum, e.g. Falciai 1985-86; the algal epiphytes of the root-rhizome stratum e.g. García Raso 1988, López de la Rosa & García Raso 1992); or from others on particular decapod taxa (e.g. Vadon 1981, Manjón Cabeza & García Raso 1994, 1995). Moreover, much of this work has been carried out in the western Mediterranean but there is little information from the south-central and eastern regions of this sea.

The structure of *Posidonia* meadows changes with depth (Mazzella et al. 1992) and such changes have been shown to affect the associated macrofaunal assemblages, establishing a clear bathymetric zonation (Falciai 1985-86, Mazzella et al. 1989, Cozzolino et al., 1992, Gambi et al. 1992). However, these studies employed a sampling technique that is known to preferentially collect the vagile macrofauna of the leaf stratum (Borg 1991). Studies in which the vagile fauna of both the root-rhizome and the leaf strata was adequately sampled are lacking.

Dense meadows of *P. oceanica* occur over an extensive depth range in the Maltese Islands (Borg & Schembri 1995). We studied the distribution and abundance of decapods associated with one such meadow along a bathymetric gradient and during different times of the years, using two different techniques which together sample both the leaf and the root-rhizome strata. Our main objectives were to extend previous observations on the decapod assemblages of *Posidonia* meadows to the south central Mediterranean area and to the total decapod community of the meadow, and to correlate any patterns observed to meadow morphology.

2 MATERIALS AND METHODS

2.1 *Study area*

The *P. oceanica* meadow studied was located off the White Tower headland (latitude 35°59.89′N, longitude 14°21.83′E) at the northern tip of the island of Malta (Fig. 1) in an area that has some of the most extensive and healthy seagrass meadows in the Maltese Islands (Borg & Schembri 1995). Here, the bottom slopes gently and supports a continuous *P. oceanica* meadow between depths of ca 5 and 25 m. In the 10-18 m depth range, the *P. oceanica* shoots grow on a thick matte whose upper 2-5 cm consist of a dense root-rhizome lattice in which the interstices are relatively free of sediment. The exposed roots and rhizomes support a rich, sessile epibiota dominated by calcareous algae and bryozoans. The absence of interstitial sediment from the surface layers of the matte is probably the result of winnowing by the strong cur-

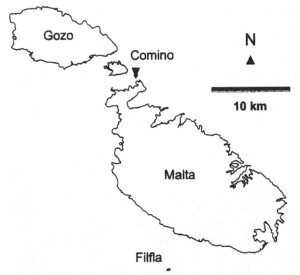

Figure 1. Map of the Maltese Archipelago showing the location of the study
area (arrow). The islands are centered
on latitude 35°55.5′N and longitude
14°23.5′E.

rents that prevail in the area. The network of small spaces provided by the surface
layers of the matte, and by the epibiota it supports, provides a complex microhabitat,
especially for cryptic invertebrates such as decapods. At the shallower end of the
meadow (6 m), the *P. oceanica* shoots mainly grow on medium to coarse sand, while
in the deeper parts (21 m), the shoots grow on a thin matte or on coarse sand.

2.2 *Sampling and data analyses*

Four stations were established at depths of 6, 11, 16, and 21 m, along a shore-normal
transect. For each of the four stations, *Posidonia* shoot density was estimated in
August 1993 by counting the shoots within a 0.125 m^2 quadrat. Five replicate counts
were taken at each station. Shoot density estimation was made once only, as negligible change was expected over a one-year period. Twenty five *P. oceanica* shoots
were collected at random from each of the four stations in August and December
1993 and April 1994, to estimate mean leaf width, mean leaf length and the mean
number of leaves per shoot. Shoot density, Leaf Standing Crop (LSC, dry weight
gm^{-2}) and Leaf Area Index (LAI, m^2 m^{-2}) were used to quantify changes in meadow
structure with depth.

Photographs of the matte surface were taken at each station using a Nikonos V
camera equipped with a 35 mm lens with close-up attachment. A 'matte cavity index'
was estimated from the photographs. This index represents a measure of the proportion of cavities (which appear in the photographs as very dark areas) in the matte surface layer and takes values between zero (no cavities) and one (100% cavities).

Sediment samples were collected from each station for determination of organic
matter content and for granulometry using the method of Buchanan (1984).

The vagile fauna was sampled using a combination of the two most widely used
samplers in studies of the macrofauna associated with *P. oceanica* meadows: the
hand-held net and the suction sampler.

The suction sampler used was based on the design of Vadon (1981) and was fitted with a 1 mm mesh collection bag. A 1 m² sample was collected from each of the four stations in August and December 1993 and April 1994. The hand net samples were collected according to Russo et al. (1985). An area of ca 20 m² was sampled at each of the four stations in September and December 1993, and in March, June and September 1994.

All suction and hand net samples were sorted by hand to separate the decapods, which were transferred to 70% ethanol for preservation and subsequent identification and counting.

Multivariate analyses of the species abundance data from the August and December 1993, and April 1994 suction samples were made using the PRIMER (Plymouth Routines in Multivariate Ecological Research) suite of programs (Clarke & Warwick 1994). The ordination method applied was Non-Metric Multidimensional Scaling (MDS) using the Bray-Curtis coefficient as a similarity measure (Field et al. 1982). Analysis of similarity (ANOSIM) was used to test the significance of differences in the composition of the decapod assemblages from the different depths and seasons. The contribution of the various species to the similarity/dissimilarity within and between stations was investigated using SIMPER (Clarke 1993). The relationship between the composition of the decapod assemblages and environmental parameters was investigated using the BIOENV procedure (Clarke & Ainsworth 1993). The parameters tested were: shoot density, LAI, LSC, 'matte cavity index', % organic matter, and median sediment grain size.

Changes in total abundance of decapods (N), Shannon-Wiener diversity (H') and Pielou's evenness (J) with depth were investigated using univariate methods. No statistical analyses were carried out on the data from the hand net samples as this sampler mainly collected decapods from the leaf stratum (see the results section below).

3 RESULTS

In general, *P. oceanica* shoot density and Leaf Area Index decreased with increasing depth. Leaf Standing Crop decreased with increasing depth in December but higher values were recorded at the 11 and 16 m stations than at the 6 m station during August and April. The mean 'matte cavity index' varied between a minimum of 0.03 at 6 m and a maximum of 0.35 at 16 m. The mean % organic matter content of the sediment (Morgan's ratio) varied between a minimum of 0.77 at 21 m and a maximum of 0.87 at 16 m. Sediment grain size varied between a minimum of 1.0ϕ at 11 m and 1.51ϕ at 16 m (Table 1).

A total of 4607 individual decapods belonging to 41 species were collected by the two sampling methods. Of these, 1768 individuals (38.4%) were collected by the suction sampler and 2839 individuals (61.6%) by the hand net. Of the 41 species recorded, 40 (97.6%) were collected by the suction sampler and 16 (39%) were collected by the hand net (Table 2). Therefore, although a larger number of individuals were collected by the hand net as compared to the suction sampler, less than half the total number of species were collected by the former method. Only a single species (*Anapagurus* cf. *chiroacanthus*) was recorded from the hand net samples only, while not a single brachyuran was collected using the hand net. The hermit crab *Cestopagu-*

Table 1. Mean values (±SD) of 'matte cavity index', % organic matter in the sediment, sediment grain size (graphic mean), shoot density, leaf area index (LAI), and leaf standing crop (LSC) for the four stations. Values are for August 1993, except in the case of LAI and LSC where values for August 1993 (top), December 1993 (middle) and April 1994 (lower) are given.

Station	Matte cavity index ($n = 3$)	% Organic matter ($n = 2$)	Sediment grain size (Mdφ) ($n = 2$)	Shoot density ($n = 5$)	LAI (m^2 m^{-2})	LSC (gdw m^{-2}) ($n = 1$)
6 m	0.03 ± 0.02	0.81 ± 0.01	1.19 ± 0.05	787 ± 19	20.2	710
					14.3	630
					13.1	660
11 m	0.32 ± 0.07	0.82 ± 0.11	1.0*	607 ± 31	15.1	840
					14.5	550
					14.8	780
16 m	0.35 ± 0.1	0.87 ± 0.04	1.51 ± 0.09	486 ± 28	12.1	864
					9.6	350
					11.7	750
21 m	0.13 ± 0.05	0.77 ± 0.01	1.06 ± 0.15	373 ± 27	9.2	622
					8.8	440
					11.0	650

*Single measurement.

rus timidus had the highest abundance in the suction samples, whilst the shrimp *Hippolyte inermis* had the highest abundance in the net samples (Table 2).

Total decapod abundance in the suction samples decreased with increasing depth, with the greatest temporal variation being recorded from the 6 m station. Temporal variation in abundance at the other stations was very small (Fig. 2). Diversity and evenness increased with increasing depth, with maximum values at the 16 m station (Figs 3 and 4). The two-dimensional MDS ordination plot (Fig. 5) of the suction samples showed a distinct grouping of shallow (6 m) samples to the right. The remaining samples did not group as distinctly, although samples collected from intermediate depths (11 and 16 m) tended to group in the middle of the plot, while samples collected from the deep (21 m) station grouped towards the left (Fig. 5).

The ANOSIM analysis showed a significant difference between samples taken from different depths in each sampling session ($r = 0.562$, $p < 0.01$) but there was no significant difference between samples taken from the same depth at different times of the year.

The SIMPER analysis showed that within groups of samples collected from the same station, samples collected from the 6 m station had the highest similarity (70.3%), while samples collected from the 21 m station had the lowest (55.1%). The analysis showed also that between groups of samples collected from different stations, the highest dissimilarity was between the 6 m and the 21 m stations (55.6%), while the lowest dissimilarity was between the 11 m and the 6 m stations (37.2%). While the hermit crab *C. timidus* had an overall significant contribution to 'within' station similarity (12.33% to 20.05%), this species had a much lower contribution to 'between' station dissimilarity (7.89% to 11.70%). Several species, including *Calcinus tubularis*, *Galathea bolivari*, *Pilumnus hirtellus* and *Eurynome aspera*, contributed significantly to 'between' station dissimilarity.

Table 2. List of decapod species collected by the suction sampler and the hand net. Stations = the number of individuals recorded from each station in the following order 6 m + 11m + 16 m + 21 m; N = total number of individuals.

Species	Suction sampler		Hand net	
	Stations	N	Stations	N
1. *Athanas nitescens* var. *laevirhincus* (Risso, 1816)	43+68+40+43	194	0+1+2+1	4
2. *Alpheus dentipes* (Guérin, 1832)	11+14+15+11	51	1+1+1+1	4
3. *Alpheus macrocheles* (Hailstone, 1835)	0+3+3+5	11		
4. *Processa edulis* (Risso, 1816)	0+4+0+0	4	1+0+0+2	3
5. *Hippolyte inermis* (Leach, 1815)	9+2+13+9	33	377+ 307+523+553	1760
6. *Hippolyte garciarasoi* (d'Undekem d'Acoz, 1996)	1+2+0+0	3	2+2+11+21	36
7. *Thoralus cranchii* (Leach, 1817)	2+11+9+8	30	5+3+18+6	32
8. *Palaemon xiphias* (Risso, 1816)	1+0+0+0	1	11+1+10+7	29
9. *Palaemon longirostris* (A. Milne Edwards, 1837)	0+0+0+1	1	0+0+1+5	6
10. *Philocheras fasciatus* (Risso, 1816)	1+0+0+1	2		
11. *Upogebia* cf. *nitida mediterranea* (Noel, 1992)	0+0+1+0	1		
12. *Callianassa* sp.	0+1+0+0	1		
13. *Calcinus tubularis* (Linnaeus, 1767)	113+10+7+2	132	12+8+3+13	36
14. *Cestopagurus timidus* (Roux, 1830)	707+135+25+38	905	137+290+178+280	885
15. *Anapagurus* cf. *breviaculeatus* (Fenizia, 1937)	0+0+1+0	1	1+0+0+0	1
16. *Anapagurus* cf. *chiroacanthus* (Lilljeborg, 1856)			0+0+1+0	1
17. *Pagurus chevreuxi* (Bouvier, 1896)	79+28+23+22	152	0+1+1+8	10
18. *Pagurus anachoretus* (Risso, 1827)	6+5+4+2	17	0+11+9+7	27
19. *Pagurus* cf. *cuanensis* (Bell, 1846)	0+0+1+0	1		
20. *Galathea bolivari* (Zariquiey Alvarez, 1950)	2+7+13+7	29	1+0+0+1	2
21. *Galathea squamifera* (Leach, 1814)	0+1+0+0	1		
22. *Pisidia longimana* (Risso, 1816)	27+17+7+1	52	1+1+0+1	3
23. *Dromia personata* (Linnaeus, 1759)	0+1+0+1	2		
24. *Ethusa mascarone* (Herbst, 1785)	1+0+0+1	2		
25. *Ebalia edwardsi* (O.G. Costa, 1838)	3+11+11+6	31		
26. *Liocarcinus corrugatus* (Pennant, 1777)	0+1+0+1	2		
27. *Pilumnus aestuarii* (Nardo, 1869)	1+0+1+0	2		
28. *Pilumnus* sp.	0+0+1+0	1		
29. *Pilumnus hirtellus* (Linnaeus, 1761)	27+10+1+0	38		
30. *Xantho incisus granulicarpus* (Forest, 1953)	4+1+0+0	5		
31. *Xantho pilipes* (A. Milne Edwards, 1867)	0+2+0+0	2		
32. *Paractea monodi* (Guinot, 1969)	0+0+0+1	1		
33. *Pisa muscosa* (Linnaeus, 1758)	0+1+1+4	6		
34. *Herbstia condyliata* (Fabricius, 1787)	8+10+7+5	30		
35. *Eurynome aspera* (Pennant, 1777)	0+1+5+3	9		
36. *Eurynome spinosa* Hailstone, 1835	0+0+1+0	1		
37. *Acanthonyx lunulatus* (Risso, 1816)	1+0+0+0	1		
38. *Inachus* cf. *phalangium* (Fabricius, 1775)	0+1+0+0	1		
39. *Achaeus cranchii* (Leach, 1817)	1+2+0+5	8		
40. *Achaeus gracilis* (O.G. Costa, 1839)	0+0+1+1	2		
41. *Macropodia czerniavskii* (Brandt, 1880)	1+0+1+0	2		
Total		1768		2839

Total abundance (m⁻²)

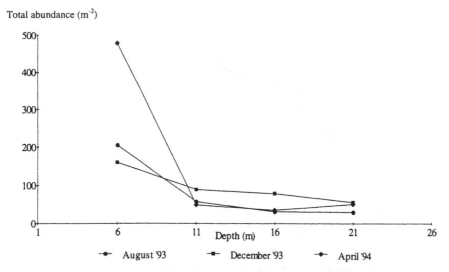

Figure 2. The total abundance of decapods in suction samples from the four stations.

Diversity (H')

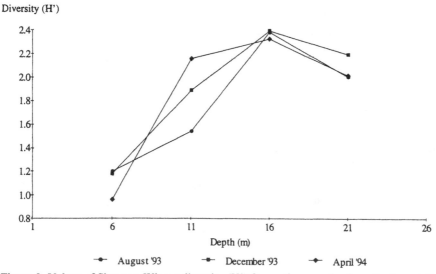

Figure 3. Values of Shannon-Wiener diversity (H') for suction samples from the four stations.

The results of the BIOENV analysis showed that the highest correlation between the species data and environmental parameters (0.501) was for a combination of two variables: shoot density and % organic matter in the sediment.

4 DISCUSSION

Although species richness was high, only twelve species had a relative abundance greater than 1%. Of these, the hermit crab *C. timidus* had the highest value (51.2%).

Evenness (J)

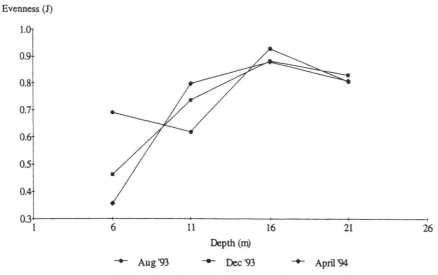

Figure 4. Values of Pielou's evenness (J) for suction samples from the four stations.

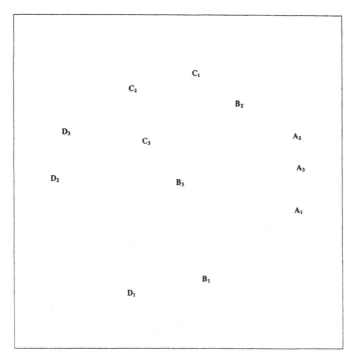

Figure 5. MDS plot for the decapod taxocene from the suction samples based on abundance after double square root transformation and using the Bray-Curtis similarity measure. Stress = 0.15. Letters represent depth stations as follows: A) 6 m, B) 11 m, C) 16 m and D) 21m. The subscript indicates the collection period, 1) August 1993, 2) December 1993, 3) April 1994.

García Raso (1990) and Zupo (1990) obtained similar results from the Western and Central Mediterranean, respectively.

In all sampling sessions, the highest total abundance value recorded in the suction samples came from 6 m. Temporal variation in total abundance was significant at the 6 m station but not at the other depths sampled. The high total abundance recorded at this depth may be related to the high *P. oceanica* shoot density at this station and the proximity of a different habitat – rock with photophilic algae – that extends from mean sea level to the edge of the *Posidonia* meadow at 6 m. High seagrass shoot densities have been shown to be behaviourally selected for by decapods as a refuge from predation (Bell & Westoby 1986b). Dense meadows may provide additional living space and may supply more abundant food than sparser meadows by providing a larger surface area that can be occupied by epiphytes (Heck & Wetstone 1977, Leber 1985). The observed high abundance at the 6 m station was almost entirely due to three small hermit crabs, *Cestopagurus timidus*, *Calcinus ornatus* and *Pagurus chevreuxi*. It may be significant that these species are able to climb up the *Posidonia* leaves and forage in the foliar stratum. The proximity of a different habitat may be enhancing recruitment of some decapods; for example, the hermit crab *Calcinus ornatus* is known to occur also on rocky substrata supporting photophilic algae.

As *Posidonia* shoot density decreases with depth, so does the abundance of decapods. However, diversity was lowest at the 6 m station and highest at intermediate depth (16 m), although even at the deepest station (21 m), diversity was still higher than at the shallowest station. This also appears to be related to habitat complexity. Although shoot density is highest at the shallow station, here *Posidonia* was growing on sand. At the 11 m and 16 m stations, the thick matte present provides additional microhabitats for decapods. Although not as thick as at the intermediate depths, the matte is also present at the deepest station and diversity is still higher than for *Posidonia* growing on sand. It would appear therefore, that the decrease in habitat complexity of the foliar stratum resulting from a decrease in shoot density, is more than made up for by an increase in complexity in the root-rhizome stratum. While the importance of different sub-habitats within seagrass meadows in influencing the species richness and abundance of the associated macrofauna has been pointed out (e.g. Orth et al. 1980), this is an interesting result since studies on the relationship between bathymetry, meadow structural complexity and diversity of associated macrofauna are generally lacking.

Our results suggest that the root-rhizome stratum supports a higher diversity of decapods than the leaf stratum. This is in agreement with García Raso's (1990) findings for meadows from southeastern Spain and with the published observations of several authors on the importance of the root-rhizome matte in influencing diversity (e.g. Harmelin 1964, Templado 1984). However, some decapods cannot be easily classified as inhabitants of the leaf or root-rhizome strata since they occur in both microhabitats. For example, in the Maltese meadow studied, the hermit crab *Cestopagurus timidus* and the shrimp *Thoralus cranchii* occur equally abundantly in both the leaf stratum and the root-rhizome layer.

The decapods from the *P. oceanica* meadow studied could be divided into three sub-assemblages: a shallow water group (6 m), an intermediate group (11-16 m), and a deeper water group (21 m). The occurrence of these sub-assemblages appears to be correlated with changes in meadow structure as depth increases (Mazzella et al. 1989,

Gambi et al. 1992), and with the proximity of different habitats at the shallow and deep ends of the meadow. In turn, meadow structure is dependant on a number of physical factors, including availability of light and the hydrodynamic regime (Mazzella et al. 1989, Gambi et al. 1992). The species composition of the shallow (6 m) and deep (21 m) sub-assemblages is partly influenced by the occurrence of species which are not permanent residents of the *Posidonia* meadow but which migrate to it from nearby habitats. For example, *P. hirtellus,* which is a species typical of the assemblage of photophilic algae growing on rock (Vadon 1981), occurred predominantly at the shallower (6 m) end of the meadow which lies adjacent to exposures of bedrock with photophilic algae. This species may be using the shallow water *Posidonia* meadow as a nursery (Vadon 1981). The sub-assemblage from intermediate depths (11-16 m) had the highest diversity. This may be partly a result of the high structural complexity of the root-rhizome layer here, as already discussed, as well as of the greater availability of food – the intermediate stations had the largest 'matte cavity index' as well as the highest values of sediment organic content. Cryptic detrivorous species, such as *Athanas nitescens, Alpheus dentipes, Thoralus cranchii* and *Galathea bolivari*, were mainly restricted to the middle region of the meadow. The deepest station is different from the others more because of the absence of species that were present at intermediate depths than because of the presence of species restricted to this deeper station. This decrease may be correlated with both the sparse cover of *Posidonia* shoots and to the thinning matte, as well as to a reduction in sediment organic content.

It is very clear from our data that the hand held net was less efficient than the suction sampler in sampling the total decapod assemblage of the *P. oceanica* meadow studied; the hand net sampled mainly the leaf stratum. However, a comparative assessment of the suitability and efficiency of these two samplers is outside the scope of the present report.

ACKNOWLEDGEMENTS

We are grateful to the University of Malta for financial assistance. We thank Mr Hassan M. Howege for help with sample processing and for supplying physio-chemical data. We thank Dr Ashley A. Rowden (University of Plymouth, UK) for help with the data analysis and for many useful discussions. We also thank all the divers, laboratory officers, technicians and students who have, in some way or other assisted us in this study. This paper benefited greatly from the comments of two anonymous referees

REFERENCES

Bell, J.D. & Harmelin Vivien, M.L. 1983. Fish fauna of French Mediterranean *Posidonia oceanica* seagrass meadows. 2. Feeding habits. *Tethys* 11: 1-14.

Bell, J.D. & Westoby, M. 1986a. Importance of local changes in leaf height and density to fish and decapods associated with seagrasses. *J. Exp. Mar. Biol. Ecol.* 104: 249-274.

Bell, J.D. & Westoby, M. 1986b. Abundance of macrofauna in dense seagrass is due to habitat preference, not predation. *Oecologia* 68: 205-209.

Bell, J.D. & Westoby, M. 1986c. Variation in seagrass height and density over a wide spatial scale: effects on common fish and decapods. *J. Exp. Mar. Biol. Ecol.* 104: 275-295.

Borg, J.A. 1991. Species richness and abundance of decapod crustaceans associated with a Maltese *Posidonia oceanica* (L.) Delile meadow. Unpublished MSc dissertation, vi + 144 pp. Malta: University of Malta.

Borg, J.A. & Schembri, P.J. 1995. The state of *Posidonia oceanica* (L.) Delile meadows in the Maltese Islands (Central Mediterranean). Rapp. Comm. int. Mer Médit. 34: 123.

Buchanan, J.B. 1984. Sediment Analysis. In N.A. Holme & A.D. McIntyre (eds), *Methods for the study of marine benthos*: 41-65. Oxford, London, Edinburgh, Boston, Palo Alto, Melbourne: Blackwell Scientific.

Chessa, L.A., Fresi, E. & Soggiu, L. 1983. Preliminary data on consumer food web in a *Posidonia oceanica* (L.) Delile bed. *Rapp. Comm. int. Mer Médit.* 28(3): 159-160.

Clarke, K.R. 1993. Non-parametric multivariate analyses of changes in community structure. *Aus. J. Ecol.* 18: 117-143.

Clarke, K.R. & Ainsworth, M. 1993. A method of linking multivariate community structure to environmental variables. *Mar. Ecol. Prog. Ser.* 92: 205-219.

Clarke, K.R. & Warwick, R.M. 1994. *Change in marine communities: an approach to statistical analysis and interpretation*; 144 pp. UK: National Environmental Research Council.

Cozzolino, G.C., Scipione, M.B., Lorenti, M. & Zupo, V. 1992. Migrazioni nictemerali e struttura del popolamento a crostacei peracaridi e decapodi in una prateria a *Posidonia oceanica* dell'Isola d'Ischia (Golfo di Napoli). *Oebalia* XVII: 343-346.

Falciai, L. 1985-86. Fauna vagile di una prateria a *Posidonia oceanica*: I crostaci decapodi. *Quaderni Ist. Idrobiol. Acquacolt. Brunelli* 5-6: 74-84.

Field, J.G., Clarke, K.R. & Warwick, R.M. 1982. A practical strategy for analysing multispecies distribution patterns. *Mar. Ecol. Prog. Ser.* 8: 37-82.

Gambi, M.C., Lorenti, M., Russo, G.F., Scipione, M.B. & Zupo, V. 1992. Depth and seasonal distribution of some groups of the vagile fauna of the *Posidonia oceanica* leaf stratum: structural and trophic analyses. *P.S.Z.N.I. Mar. Ecol.* 1: 17-39.

García Raso, J.E. 1988. Consideraciones generales sobre la taxocenosis de crustáceos decápodos de fondos de concrecionamiento calcáreo superficial del alga *Mesophyllum lichenoides* (Ellis & Sol.) Lemoine (Corallinaceae) del mar de Alborán. *Inv. Pesq.* 52: 245-264.

García Raso, J.E. 1990. Study of a crustacea decapoda taxocoenosis of *Posidonia oceanica* beds from the southeast of Spain. *P.S.Z.N.I. Mar. Ecol.* 11: 309-326.

García Raso, J.E., López de la Rosa, I. & Rosales, J.M. 1996. Decapod crustacean communities from calcareous seaweed and *Posidonia oceanica* (rhizome stratum) in shallow waters. *Ophelia* 45: 143-158.

Harmelin, J.G. 1964. Etude de l'endofaune des 'mattes' d'herbiers de *Posidonia oceanica*. *Rec. Trav. Stn. Mar. Endoume* 35(51): 43-106.

Heck, K.L., Able, K.W., Fahay, M.P. & Roman, C.T. 1989. Fishes and decapod crustaceans of Cape Cod eelgrass meadows: species composition, seasonal abundance patterns and composition with unvegetated substrates. *Estuaries* 12(2): 59-65.

Heck, K.L. & Thoman, T.A. 1981. Experiments on predator-prey interactions in vegetated aquatic habitats. *J. Exp. Mar. Biol. Ecol.* 53: 125-134.

Heck, K.L. & Thoman, T.A., 1984. The nursery role of seagrass meadows in the Upper and Lower reaches of the Chesapeake Bay. *Estuaries* 7(1): 70-92.

Heck, K.L. & Wetstone, G.S., 1977. Habitat complexity and invertebrate species richness and abundance in tropical seagrass meadows. *J. Biogeogr.* 4: 135-142.

Leber, K.M. 1985. The influence of predatory decapods, refuge, and microhabitat selection on seagrass communities. *Ecology* 66: 1951-1964.

Ledoyer, M. 1962. Étude de la faune vagile des herbiers superficiels de zosteracees et de quelques biotopes d'algues littorales. *Rec. Trav. Stn. Mar. Endoume* 25(39): 173-216.

López De La Rosa, I. & García Raso, J.E. 1992. Crustáceos decápodos de fondos de concrecionamientos calcáreos asociados a *Posidonia oceanica* de sur de España (Almeria). *Cah. Biol. Mar.* 33: 55-74.

Manjón-Cabeza, M.E. & García Raso, J.E. 1994. Estructura de una poblacion del cangrejo ermitano *Cestopagurus timidus* (Crustacea, Decapoda, Anomura) de fondos de *Posidonia oceanica* del SE de España. *Cah. Biol. Mar.* 35: 225-236.

Manjón-Cabeza, M.E. & García Raso, J.E. 1995. Study of a population of *Calcinus tubularis* (Crustacea, Diogenidae) from a shallow *Posidonia oceanica* meadow. *Cah. Biol. Mar.* 36: 277-284.

Mazzella, L., Scipione, M.B. & Buia, M.C. 1989. Spatio-temporal distribution of algal and animal communities in a *Posidonia oceanica* meadow. *P.S.Z.N.I. Mar. Ecol.* 10(2): 107-129.

Mazzella, L., Buia, M.C., Gambi, M.C., Lorenti, M., Russo, G.F., Scipione, M.B. & Zupo, V. 1992. Plant-animal trophic relationships in the *Posidonia oceanica* ecosystem of the Mediterranean Sea: a review. In D.M. John, S.J. Hawkins & J.H. Price (eds) *Plant-Animal interactions in the marine benthos*: 165-187. Oxford: Clarendon Press.

Nelson, W.G. 1981. Experimental studies of decapod and fish predation on seagrass macrobenthos. *Mar. Ecol. Prog. Ser.* 5: 141-149.

Orth, R.J., Heck, K.L. & Van Montfrans, J. 1984. Faunal communities in seagrass beds: a review of the influence of plant structure and prey characteristics on predator-prey relationships. *Estuaries* 7(4A): 339-340.

Pessani, D., Poncini, F., Sperone, P. & Vetere, M. 1995. Fauna vagile della prateria di *Posidonia oceanica* di Diano Marina (Liguria Occidentale): molluschi e decapodi. *Biol. Mar. Medit.* 2: 401-403.

Russo, G.F., Fresi, E. & Vinci, D. 1985. The hand-towed net method for direct sampling in *Posidonia oceanica* beds. *Rapp. Comm. Int. Mer Medit.* 29(6): 175-177.

Schembri, P.J. & Lanfranco, E. 1984. Marine Brachyura (Crustacea: Decapoda: Brachyura) from the Maltese Islands and surrounding waters (Central Mediterranean). *Centro* [Malta] 1(1): 21-39.

Scipione, M.B., Fresi, E. & Wittman, K.J. 1983. The vagile fauna of *Posidonia oceanica* (L.) Delile foliar stratum: a community approach. *Rapp. Comm. int. Mer Medit.* 28(3): 141-142.

Scipione, M.B., Gambi, M.C., Lorenti, M., Russo, G.F. & Zupo, V. 1996. Vagile fauna of the leaf stratum of *Posidonia oceanica* and *Cymodocea nodosa* in the Mediterranean Sea. In J. Kuo, R.C. Phillips, D.I. Walker & H. Kirkman (eds) *Seagrass Biology: Proceedings of an International Workshop, Rottnest Island, Western Australia, 25-29 January 1996*: 249-260.

Templado, J. 1984. Las praderas de *Posidonia oceanica* en el sureste Espanol y su biocenosis. In C.F. Boudouresque, A. Jeudy de Grissac & J. Olivier (eds) *International workshop on Posidonia oceanica beds 1*: 159-172. France: GIS Posidonie.

Vadon, C. 1981. Les brachyoures des herbiers de Posidonies dans la région de Villefranche-sur-mer: biologie, écologie et variations quantitatives des populations. Unpublished PhD thesis, 227 + XII pp. Paris: Universite Pierre et Marie Curie.

Zupo, V. 1990. I decapodi delle praterie di *Posidonia oceanica*: confronto tra metodiche di campionamento in rapporto alla zonazione del taxon. *Oebalia* 16(2): 817-822.

Distribution patterns of two species of swimming crabs (Portunidae), *Liocarcinus depurator* (L.) and *Macropipus tuberculatus* (Roux), in the southwestern Adriatic Sea (Mediterranean Sea)

NICOLA UNGARO, GIOVANNI MARANO & ALESSANDRO VLORA
Laboratorio Provinciale di Biologia Marina, Bari, Italy

GIUSEPPE PASSARELLA
C. N. R., Istituto Ricerca Sulle Acque, Bari, Italy

ABSTRACT

Quantitative data and length distribution data on the blue-leg swimming crab, *Liocarcinus depurator*, and the knobby swimming crab, *Macropipus tuberculatus*, were collected during eight seasonal trawl surveys carried out from 1991 to 1995 in the Southwestern Adriatic Sea. The spatial distributions were analysed by means of geostatistics, and the population demography was described. Referring to the kriging per species, *L. depurator* showed the highest abundance on the shelf bottoms (mostly in the northern zone of investigated area) and it appears to be more homogeneously distributed during autumn season. *M. tuberculatus* showed the highest abundance on the lower shelf border and on the slope bottoms. Moreover, the presence of a 'spatial competition' rate between the species in the areas of bathymetric overlapping could be underlined as it was shown by seasonal kriging. The carapace length ranged from 13 to 40 mm in the blue-leg swimming crab, and from 12 to 30 mm in the knobby swimming crab.

1 INTRODUCTION

Portunid crabs – in particular the blue-leg swimming crab, *Liocarcinus depurator* (Linnaeus, 1758), and the knobby swimming crab, *Macropipus tuberculatus* (Roux, 1830) – represent more than 15% of total crustacean catches on Southwestern Adriatic trawlable bottoms (AA.VV. 1993, AA.VV. 1996) but these crustaceans are usually discarded by fishermen.

Scientific information about *Liocarcinus depurator* and *Macropipus tuberculatus* is very poor in the just mentioned area of the Adriatic Sea; in the present paper the authors report preliminary data on spatial distribution of the species, analysed by means of GIS method, a statistical technique recently used also for crustacean populations in the Mediterranean basin (Maynou et al. 1995, Lembo et al., in press). Some information about the demographic structure of the populations is also reported.

2 MATERIALS AND METHODS

Biological data referring to swimming crabs (biomass as g h^{-1} trawling and abundance as n h^{-1} trawling) were collected during 8 trawl surveys (1991-1995) carried out in the Southwestern Adriatic Sea (Lat. 40°-42°N; Long. 16°-19°E) (Fig. 1). The data came from a total of 252 hauls (126 'Spring' hauls during the '91, '92, '94, and '95 surveys; 126 'Autumn' hauls during the '91, '92, '94, '95 surveys). Moreover, during the autumn 1995 survey length measurements by sex (carapace length in mm) were also collected in order to obtain length frequency distributions.

The sampling gear was an 'Italian' bottom trawl net (cod-end stretched mesh = 36 mm); the net wing spread was 14-20 m (it increases with depth) and the swept area for each haul (one trawling hour) was approximately 0.1 km^2. The sampling design was random stratified. Five bathymetric strata were chosen on the ground of the main demersal species assemblages: 10-50 m, 51-100 m, 101-200 m, 201-450 m, 451-800 m). The samples are summarized in Table 1.

The species distribution was analysed by means of geostatistical tools (Royle et al. 1981, Burrough 1986). In order to improve the representativeness of species distribution, the structural analysis was performed on the cumulative sample distributions for 'Spring' surveys and 'Autumn' ones (Ungaro et al. 1998). After a preliminary statistical analysis of data, several outlier values (probably due to measurement errors) were eliminated from the data set (Isaaks & Srivastava 1989). The variogram models were cross validated (Journel & Huijbregts 1989) and kriging estimates and variances were mapped (Oliver 1990).

Length distribution ranges and the mean size were computed by sex for both species while sex ratios were statistically analysed by means of chi-squared test.

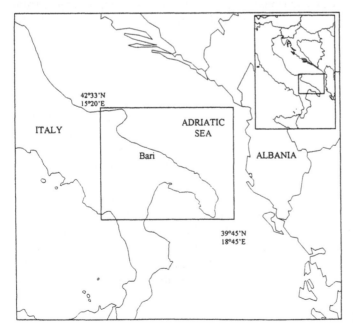

Figure 1. Investigated area.

Table 1. Number of hauls per survey and depth stratum.

Depth stratum Survey	Spring '91	Autumn '91	Spring '92	Autumn '92	Spring '94	Autumn '94	Spring '95	Autumn '95
10-50 m	6	6	6	6	6	6	6	6
51-100 m	6	6	6	6	9	9	9	9
101-200 m	5	5	5	5	13	13	13	13
201-450 m	5	5	5	5	5	5	5	5
451-800 m	3	3	3	3	5	5	5	5

Table 2. Variograms parameters (exponential model).

Species	Season	Nugget	Sill	Range
L. depurator	Autumn	0	10,000	0.480
L. depurator	Spring	0	27,000	0.209
M. tuberculatus	Autumn	0	26,000	0.165
M. tuberculatus	Spring	0	20,000	0.360

3 RESULTS

The swimming crabs *L. depurator* and *M. tuberculatus* were collected from 13 m to 680 m depth (blue-leg swimming crab from 13 m to 560 m, knobby swimming crab from 19 m to 680 m). Raw abundance values from net catches showed a high variability rate for both species – probably due to the aggregated distributions of the crabs. All data were standardized to on hour of trawling and analysed by means of geostatistical method.

The omnidirectional variogram models resulting from the structural analysis (per species and season) showed in any case a low spatial correlation. The abundance and biomass distributions were quite similar (for the reason stated above we choose the first one to be represented). The theoretical exponential model appeared to represent the spatial distribution of data the best (Table 2) (Fig. 2).

Referring to the kriging per species, *L. depurator* showed the highest abundances on the shelf bottoms (mostly in the northern zone of investigated area) and it appears to be more homogeneously distributed during autumn season (Fig. 3). *M. tuberculatus* showed the highest abundance on the lower shelf border and on the slope bottoms (Fig. 4).

The estimated kriging variance appeared to be rather large in both species and seasons. However, the estimation can be considered acceptable in a wide part of the investigated area, as it is shown by the position of the interpolation significance limits (Figs 3 and 4).

The carapace length of the blue-leg swimming crab ranged from 13 to 40 mm (mostly from 17 to 23 mm in the whole area during the Autumn 1995 survey). The carapace length of the knobby swimmming crab ranged from 12 to 30 mm (mostly from 15 to 21 mm in the whole area, Autumn 1995 survey) (Fig. 5). The length of the blue-leg swimming crab ranged from 13 to 40 mm in males and from 14 to 36 mm in females. The length of the knobby swimming crab ranged from 14 to 30 mm in males

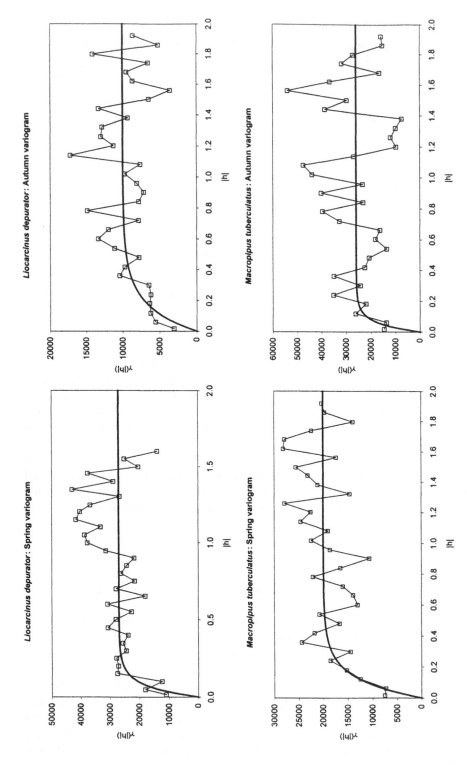

Figure 2. Experimental and theoretical variograms per species and seasons (h = distance referred to geographic co-ordinates).

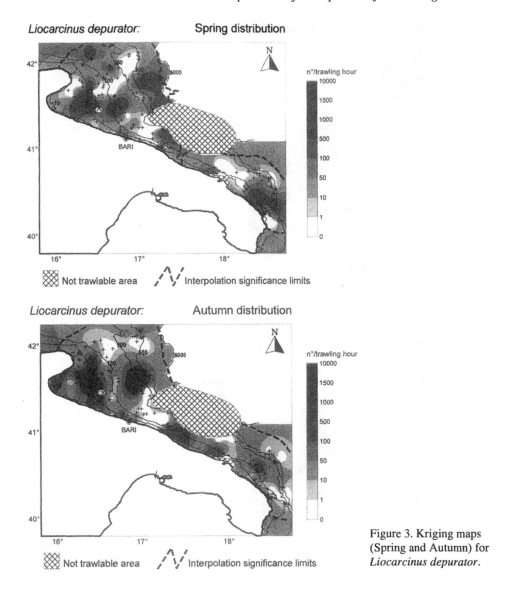

Figure 3. Kriging maps (Spring and Autumn) for *Liocarcinus depurator*.

and from 12 to 25 mm in females (Table 3). The overall sex ratio for *L. depurator* in our samples was not significantly different (chi-squared test) from the expected 1:1, whereas the sex ratio for *M. tuberculatus* was significantly biased towards the males (2.01:1; chi-squared test, $p < 0.01$).

4 DISCUSSION AND CONCLUSION

The depth distribution of *L. depurator* in the Southwestern Adriatic was quite similar as in other Mediterranean Sea districts (Mori & Zunino 1987, Abellò et al. 1988). In the investigated area, *M. tuberculatus* was found at less deep bottoms compared to

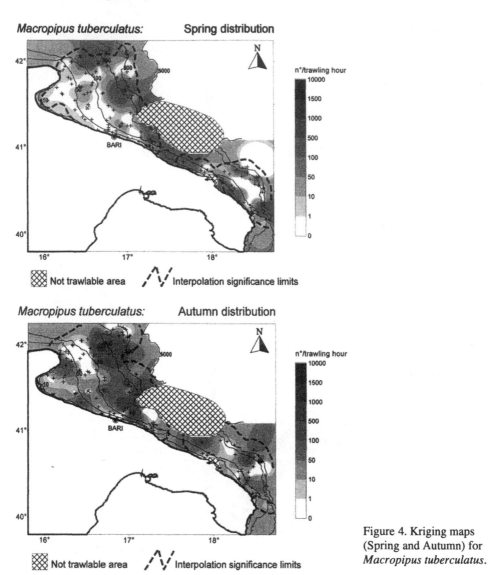

Figure 4. Kriging maps (Spring and Autumn) for *Macropipus tuberculatus*.

other Mediterranean Sea districts (Mori 1987, Abellò et al. 1988). As was shown by geostatistical analysis, generally the distribution patterns of the two species in the Southwestern Adriatic Sea were in agreement with results from other Mediterranean areas (Mori 1987, Mori & Zunino 1987, Abellò et al. 1988, Abellò 1989a, 1989b). In fact, the blue-leg swimming crab showed the highest abundance on the shelf bottoms, whereas the knobby swimming crab showed the highest abundance on the lower shelf border and on the slope bottoms.

The two swimming crab species showed different estimated 'aggregation intensity' in relationship with the spring and autumn seasonal kriging maps. These reported differences between the spatial distributions of the two species were probably due to environmental factors, mostly depth, but also bottom steepness or sediment

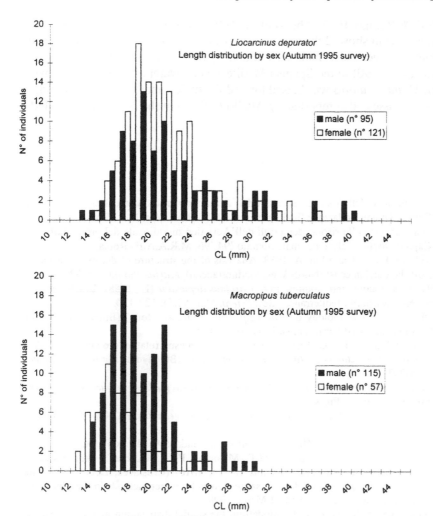

Figure 5. Length distributions for *Liocarcinus depurator* and *Macropipus tuberculatus*.

Table 3. Length distribution (Carapace length, mm) ranges and mean sizes per species and sex.

Species	Sex	Minimum	Maximum	Mean
L. depurator	Male	13	40	22.8* (s.d. 4.5)
L. depurator	Female	14	36	20.9* (s.d. 2.8)
M. tuberculatus	Male	14	30	18.8* (s.d. 3.3)
M. tuberculatus	Female	12	25	16.5* (s.d. 3.7)

*After log-trasformation of data to correct distribution asimmetry.

composition (Abellò et al. 1988). However, the presence of a seasonal 'spatial competition' rate between the species in the bathymetrically overlapping areas, seemed to be underlined by the kriging representations.

The length distribution of the blue-leg swimming crab was quite similar to literature data (Mori & Zunino 1987, Abellò et al. 1990). The length distribution of the knobby swimming crab showed the absence of specimen larger than 30 mm carapace length. In contrast, larger individuals were found by Mori (1987) in the Ligurian Sea and by Abellò et al. (1990) in the Spanish Mediterranean waters. Moreover, the overall sex ratio in *M. tuberculatus* was biased toward the males, that were more abundant in the larger size classes, also reported by Abellò (1989b) in the Northwestern Mediterranean.

REFERENCES

AA.VV. 1993. Valutazione delle risorse demersali dell'Adriatico meridionale dal promontorio del Gargano al Capo d'Otranto: relazione finale triennio '90-'93. Min. Mar. Mer., Roma.

AA.VV. 1996. Valutazione delle risorse demersali dell'Adriatico meridionale dal promontorio del Gargano al Capo d'Otranto: relazione finale triennio '94-'96. M.R.A.A.F., Roma.

Abellò, P., Valladares, F.J. & Castellon, A. 1988. Analysis of the structure of decapod crustacean assemblages off the Catalan coast (North-West Mediterranean). *Marine Biology* 98: 39-49.

Abellò, P. 1989a. Reproduction and moulting in *Liocarcinus depurator* (L., 1758) (Brachiura: Portunidae) in the Northwestern Mediterranean Sea. *Scient. Mar.* 53(1): 127-134.

Abellò, P. 1989b. Reproductive biology of *Macropipus tuberculatus* (Roux) (Brachiura: Portunidae) in the Northwestern Mediterranean. *Ophelia* 30(1): 47-53.

Abellò, P., Pertierra, J.P. & Reid, D.G. 1990. Sexual size dimorphism, relative growth and handedness in *Liocarcinus depurator* and *Macropipus tuberculatus* (Brachyura: Portunidae). *Scient. Mar.* 54(2): 195-202.

Burrough, P.A. 1986. *Principles of geographical information system for land resources assessment.* New York: Oxford University Press.

Isaaks, E.H. & Srivastava, R.M. 1989. *An introduction to applied geostatistics.* New York: Oxford University Press.

Journel, A.G. & Huijbregts, C. J. 1989. *Mining Geostatistics.* London: Academic Press.

Lembo, G., Tursi, A., D'Onghia, G., Spedicato, M.T., Maiorano, P. & Silecchia, T. (in press). Spatio-temporal distribution of *Aristeus antennatus* (Risso, 1816) (Crustacea, Decapoda) in the northwestern Ionian Sea. Assessment of demersal resources by direct methods in the Mediterranean and the adjacent Seas. Pisa (Italy) 18-21 March 1998.

Maynou, F., Conan, G.Y. & Sardà, F. 1995. Modelling the spatial distribution and assessment of *Nephrops norvegicus* (L.) by geostatistics. *Rapp. Comm. Int. Mer Médit.* 34: 35.

Mori, M. 1987. Observations on reproductive biology, and diet of *Macropipus tuberculatus* (Roux) in the Ligurian Sea. *Inv. Pesq.* 51(Supp. 1): 147-152.

Mori, M. & Zunino, P. 1987. Aspects of the biology of *Liocarcinus depurator* (L.) in the Ligurian Sea. *Inv. Pesq.* 51(Supp. 1): 135-145.

Oliver, M.A. 1990. Kriging: a method of interpolation for geographical information systems. *Int. Journ. of Geog. Inf. Sys.* 4(4): 313-332.

Royle, A. G., Clausen, F. L. & Frederksen, P. 1981. Practical universal kriging and automatic contouring. *Geoprocessing* 1: 377-394.

Ungaro, N., Marano, G. & Martino, M. 1998. Mapping of fourspotted megrim, *Lepidorhombus boscii* (Risso, 1810), resource on South-Western Adriatic Sea trawlable bottoms. *Rapp. Com. Int. Mer Médit.* 35: 496-497.

Diamysis bahirensis: A mysid species new to the Portuguese fauna and first record from the west European coast

MARINA RIBEIRO DA CUNHA & MARIA HELENA MOREIRA
Departamento de Biologia, Universidade de Aveiro, Aveiro, Portugal

JEAN CLAUDE SORBE
Laboratoire d'Océanographie Biologique, UMR 5805 (CNRS-UB1), Arcachon, France

ABSTRACT

A population belonging to the *Diamysis bahirensis* species group is reported from the Atlantic lagoon Ria de Aveiro (NW Portugal). The specimens were collected with a sled-type sampler among macrophytes (1.0-2.5 m depth), during a study of the suprabenthic fauna (June 1995-June 1996). A maximum density of 51 ind. m^{-2} was recorded in June 1995 (salinity = 26‰). Ovigerous females (9-14% of the population) occurred at the end of spring. This is the westernmost record of *Diamysis*. This genus is considered to be of Tethyan origin and *D. bahirensis* is consistently mentioned as a Mediterranean endemic species. Information on the morphology, reproductive biology and ecology of the population living in Ria de Aveiro is given, and its presence in this Atlantic coastal system is discussed as probably due to ballast water transport from a western Mediterranean locality.

1 INTRODUCTION

The genus *Diamysis* was established by Czerniavsky (1882) along with *Euxinomysis* and *Potamomysis*. Zimmer (1915a), Derjavin (1924) and Martynov (1924) contributed to put the three genera under *Diamysis*. The origin of the genus *Diamysis*, subject of discussion since 1940, has been associated with the ancient seas of Tethys and Parathetys and its evolution and speciation have been related to the origin and evolution of the ponto-caspian and Mediterranean basins (Bacescu 1940, 1954, 1980, Ariani 1979, 1980).

In his taxonomic list of Mysidacea, Mauchline (1980) mentions four species within the genus *Diamysis*: *D. bahirensis* (G.O. Sars, 1877) from the Mediterranean and the Black Sea (designated by some authors as the subspecies *D. b. mecznikowi*); *D. frontieri* Nouvel, 1965 from Madagascar; *D. pengoi* (Czerniavsky, 1882), a freshwater mysid from the rivers of the Black Sea region; *D. pusilla* G.O. Sars, 1907, restricted to the Caspian Sea. The complete synonymy and references on these species can be found in the catalogues of Gordan (1957), Mauchline & Murano (1977) and Müller (1993). Other closely related species are *Surinamysis americana* (= *Antromysis americana*) from Surinam and *Mysidium columbiae* with a wide distribution in

Central America, that have been described respectively by Tattersall (1951) as *Diamysis americana* and by Zimmer (1915b) as *D. columbiae. Potamomysis assimilis* described by Tattersall (1908) from India was under *Diamysis* when the synonymy of the two genera was established but was later removed to the genus *Gangemysis*.

Diamysis frontieri has only been reported again by Bacescu (1973) as also occurring in the Red Sea, but *D. pengoi* and *D. pusilla* are relatively well known in their areas of distribution. Since it was first described by Sars as *Mysis bahirensis* from the 'lake' El Bahira (Tunis), *D. bahirensis* has been reported frequently from several areas all around the Mediterranean.

Several authors have recognised a large degree of morphological variability in the specimens referred to *Diamysis bahirensis* sensu lato (e.g. Spandl 1926, Bacescu 1940, Holmquist 1955, Ariani 1979) and various infraspecific entities have been proposed on the basis of morphological, ecological and biogeographical features. This systematic issue is not yet satisfactory but a thorough taxonomic revision is being prepared (Wittmann, pers. comm.) by Prof. A. Ariani (Istituto di Zoologia dell' Università di Napoli) and Prof. K. Wittmann (Institut für Allgemeine Biologie, Wien).

During a benthos sampling program carried out by Marques et al. (1993) a few damaged mysid specimens were collected with a Van Veen grab, in the Canal de Ovar, Ria de Aveiro. A full identification was not possible at that time but it was clear that these specimens did not belong to any of the known species of the Ria. Two years later the same area was sampled with a suprabenthic sled with the purpose of studying the peracaridan fauna of the Ria de Aveiro (Cunha et al. in press). On this occasion, numerous specimens were collected and assigned to *Diamysis bahirensis* sensu lato. Prof. Wittmann kindly examined this new material and confirmed our identification. He also pointed out that the specimens from Ria de Aveiro correspond perfectly to those from Sardinia and the Mediterranean coast of France which is somewhat surprising, considering the high diversity among Mediterranean populations (Wittmann, pers. comm.).

The purpose of this paper is to give some information on the morphology, reproductive biology and ecology of the new *Diamysis* population in comparison to the known Mediterranean populations and to hypothesize on its presence in this Atlantic coastal system.

2 MATERIAL AND METHODS

2.1 *Study area*

The Ria de Aveiro is a shallow coastal lagoon on the west coast of Portugal (Fig. 1) which can be considered as a bar-built estuary, according to the classification of Pritchard (1967). It is a very recent geological feature, resulting from sediment transport along the coast and sedimentation of materials carried by rivers. These mechanisms, initiated in the XI-XII century, have not yet attained an equilibrium and the present trend is to silt up. From the XII to the XVI centuries the formation of a sand bar separated a former bay from the sea and created a wide and sheltered harbour frequently used by ocean-going vessels. During the XVII century the evolution of the sand-bar made the communication with the sea difficult, and by the end of the XVII century

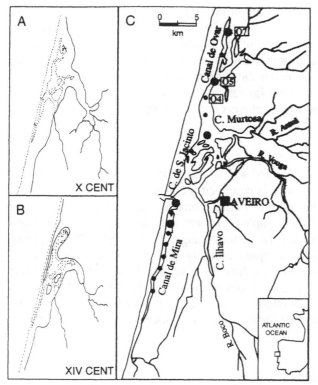

Figure 1. Evolution of Ria de Aveiro. (A) coastline before the Xth century; (B) coastline in the XIVth century; (C) actual map of Ria de Aveiro with location of the sampling stations. Large symbols indicate the stations sampled monthly, small symbols indicate stations sampled every third month. The three stations where *Diamysis* specimens occurred are labelled.

the Ria was transformed in a closed swampy lagoon (Abecasis 1961, Oliveira 1988).

The present mouth of the lagoon was artificially opened in 1808 for maritime, commercial and fishing purposes. The main channel of navigation starts at the mouth and extends for 9 km ending near the city of Aveiro. The Port is located alongside this channel and includes long-distance and coastal fishing, commercial and chemical terminals. The Port activity has been increasing since the early 1950s. In the early 1980s the total traffic of goods was stabilised in 500×10^3 t. With the General Plan of Development of the Port and Ria de Aveiro in 1988, the traffic increased more than five times, reaching approximately 2.7×10^6 t and 1300 registered ship arrivals in 1997.

The topography of Ria de Aveiro is rather complex with three main channels that radiate from the mouth with several branches, islands and mudflats. With a length of 45 km and a maximum width of 10 km, the lagoon covers an area of 43 to 47 km^2 at low and high tide, respectively. The hydrological circulation is dominated by sea water. The volume exchanged during a tidal cycle varies from 2.5×10^7 to 7.0×10^7 m^3 for tides with 1.0 to 2.5 m range, respectively, whilst the mean total river discharge is only 1.8×10^6 m^3. About 40% of the tidal prism of a flowing tide is diverted to the Canal de S. Jacinto, a channel that runs parallel to the coast and extends for 27 km from the mouth of the Ria (Vicente 1985). The Canal de Ovar is the northern part of this channel. It forms a 'dead end' with an insignificant supply of freshwater ex-cept during the rain season. Due to these conditions, the Canal de Ovar behaves, in like manner, as the small coastal systems of the Mediterranean: the salinity is rather high throughout the year but it can drop almost to 0 during periods of heavy rainfall; the temperature of the water may reach 28°C or more in the summer months; the shallow and muddy bed of the channel favours the growth of macrophytes (*Zostera noltii*, *Ruppia cirrhosa*, *Potamogeton pectinatus* and macroalgae).

2.2 *Sampling*

The material reported herein was collected with a small suprabenthic sled towed over the bottom with a motor boat. The sled is equipped with one net (0.5 mm mesh size) and allows quantitative sampling of the motile fauna in the 0-50 cm water layer above the bottom. For each haul the sampled area was calculated from the net width and the estimated towed distance on the bottom (range: 12.5-25 m^2). The sampling programme included five stations sampled monthly and eleven stations sampled every three months during the period from June 1995 to June 1996, in Canal de Ovar and Canal de Mira. Sampling was carried out during the day, at flood. The results presented herein concern only the stations where *Diamysis* specimens were collected (stations O4, O5 and O7, in Canal de Ovar, respectively located at 17, 20 and 26 km from the mouth of the Ria).

2.3 *Material examined and methods*

The material from Ria de Aveiro was preserved with 10% neutral formalin and later kept in 70% ethanol. The density was expressed as number of individuals per m^2.

The individuals were classified into the following demographic categories: juve-niles; immature males; mature males, with setose lobus masculinus and well devel-oped endopod of fourth pleopod; immature females; incubating females, with fully developed and setose marsupium carrying eggs or larvae; empty females, with fully developed but empty marsupium.

The following measurements were made in some specimens with a calibrated ocular micrometer under a dissecting microscope: total body length, measured from the tip of the rostrum to the end of the telson, not including the spines; cephalotho-racic length, from the tip of the rostrum to the posterior margin of the carapace, in lateral view. The number of lamellar processes on apical cleft and of lateral spines of the telson, including the apical ones was counted. The number of intra-marsupial in-dividuals was counted in the incubating females with undamaged marsupium. The developmental stage was determined considering the following three categories (Mauchline 1980, Wittmann 1981): stage 1 (embryonic stage), rounded embryos; stage 2 (nauplioid stage), elongated larvae with developing appendages; stage 3 (postnauplioid stage), larvae with pedunculated and pigmented eyes.

Some material from the collection of the Natural History Museum of London (NHML) was also examined:

#1. *Diamysis bahirensis*; Norman collection, 1911.11.8; 3535-3539; Adriatic (Coll.:Prof. Claus),

#2. *Diamysis bahirensis*; 1964:1,21,846-860; Egypt (Coll.: Tattersall),

#3. *Diamysis bahirensis*; 1964; River Kishon (Coll.: Tattersall, 1924),

#4. *Diamysis* (unpublished species); 1997:208-217; Adriatic, coast of Apulia, Italy (Coll.: Ariani & Wittmann, 1995).

Some specimens or interesting parts thereof, were dehydrated in an ascending acetone gradient and critical point dried. The material was glued onto a stub with Araldite, sputter-coated with gold-palladium and observed with a Hitachi S-2500 scanning electron microscope.

3 RESULTS

3.1 *Morphology*

The maximum body length observed in *Diamysis* from Ria de Aveiro is 7.7 mm and 6.3 mm for the females and males, respectively, but in most mature specimens the total length is about 5.0-5.5 mm for the females and around 4.5 mm for the males. The body is not conspicuously pigmented.

In the female, the second article of the antennular peduncle has one large plumose seta at its inner distal corner. The third article has at its distal inner corner one large plumose seta (about the same length as the one in the second article) and two smaller ones; one of these latter is more sparsely plumose (Plate 1A). In the male the *lobus masculinus* is posteriorly directed (Plate 1B).

The fourth pleopod of the male is typical of the genus. The first article of the exopod has a bare seta at the distal end.

Mature specimens have 9-11 lateral spines (including the apical one) on each margin of the telson (Fig. 2). Occasionally, in the larger females, 12 lateral spines can occur. Most specimens have 15 apical lamellar processes with little variation: 14 or 16 lamellar processes were counted in some specimens and exceptionally 19-22 in large females and 18 in large males.

From the collection of the NHML the specimens reported as #1 (Adriatic) were all immature females. The mature specimens #2 and #3, from Port Said and Kishon

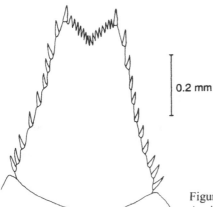

0.2 mm

Figure 2. *Diamysis bahirensis* (sensu lato) from Ria de Aveiro. Telson.

River (Tattersall 1927) have about the same body length as the specimens of Ria de Aveiro. The number of lateral spines ranges from 8 to 11 and 6 to 9, respectively. The variation observed in the number of apical processes (12-24, 18 in average) is however larger than in the specimens of Ria de Aveiro. The recently collected specimens from the Adriatic (#4) were larger and stouter than the Aveiro specimens and the body was conspicuously pigmented. The proportions of the telson were also different. The apex was wider with a slighter incision and about 30 lamellar processes. The number of lateral spines ranged from 8 to 10.

3.2 *Reproductive biology*

During the sampling period, incubating females (9 to 14% of the population) occurred from June to August 1995 and again from May to June 1996. Empty females and small sized juveniles were observed at the same periods. The sex ratio was generally close to 1 except in some samples with a few individuals (global mean: 1.08). The brood size (Fig. 3) ranged from 4-5 in small females (4.1 mm long) to 19-20 in large females (about 7 mm long). From a total of 173 females observed 23.1% had intra-marsupial individuals at stage 1, 54.4% at stage 2 and the remaining 22.5% at stage 3. Embryos at stage 1 have a diameter of 0.35-0.40 mm. The total length of nauplioid (stage 2) and postnauplioid larvae (stage 3) ranges from 0.65 to 0.9 mm and from 0.9 to 1.1 mm, respectively. Larger females have, in general, larger larvae.

3.3 *Ecology*

The specimens of *Diamysis* from Ria de Aveiro were found only in three stations in the upper reaches of Canal de Ovar (O4, O5, O7). From the 1537 specimens examined, about 2/3 were collected at O7 and only two individuals were collected at O4.

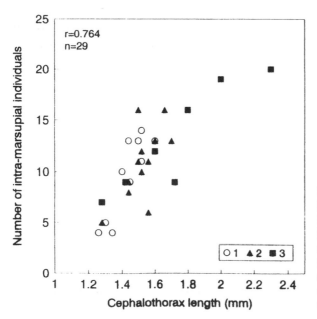

Figure 3. Fecundity of *Diamysis* in Ria de Aveiro. Relation between the brood size and the cephalothorax length of incubating females. The symbols indicate different developmental stages. 1 = stage I, rounded embryos; 2 = stage II, nauplioid larvae; 3 = stage III, post-nauplioid larvae.

Table 1. Density (ind. m^{-2}), abundance (total number of specimens collected) and demographic composition (number of individuals) of *Diamysis bahirensis* in the Canal de Ovar, Ria de Aveiro (only positive hauls are mentioned for the last two stations).

Date	Density	Abun.	J	Fi	Fine	Fe	Mi	Mm	M/F
Station O7									
21 Jun 95	51.47	772	279	86	110	62	26	209	0.91
24 Jul 95	0.00	0							
30 Aug 95	1.44	23	16	0	3	4	0	0	–
19 Sep 95	0.00	0							
18 Oct 95	0.00	0							
17 Nov 95	0.07	1	1	0	0	0	0	0	–
19 Dec 95	0.20	3	3	0	0	0	0	0	–
17 Jan 96	0.53	8	4	2	0	0	2	0	1.00
15 Feb 96	0.33	5	2	2	0	0	0	1	0.50
14 Mar 96	0.00	0							
11 Apr 96	0.00	0							
14 May 96	4.60	69	20	15	9	1	3	21	0.96
13 Jun 96	9.27	139	35	20	19	13	6	46	1.00
Station O5									
21 Jun 95	28.97	507	180	52	47	29	19	180	1.55
24 Jul 95	0.48	6	1	1	1	2	0	1	0.25
19 Sep 95	0.13	2	2	0	0	0	0	0	–
Station O4									
21 Jun 95	0.08	2	2	0	0	0	0	0	–

J = juveniles; Fi = immature females; Finc = incubating females; Fe = empty females; Mi = immature males; Mm = mature males; M/F = male-female ratio.

The maximum density of 51.5 ind. m^{-2} was observed in June 1995 (Table 1) but then it decreased to less than 1.5 ind. m^{-2}. During the winter density was very low (no animals were found in October 1995 and March and April 1996) but in spring *Diamysis* reappeared with a significant density, however lower than in the preceding year and only at the northern station.

The sampling stations where *Diamysis* was collected are shallow (1.0-2.5 m depth) and the bottom is covered by macrophytes. Drifting macroalgae and filamentous algae may also occur in large quantity. The sediment is mainly sandy mud with a high content in organic matter (on average 7% of the sediment dry weight). Water temperature and salinity are rather high almost all year round. Higher densities of *Diamysis* were recorded at the end of spring with water temperature around 22-25°C and salinity around 26. In the spring of 1996 long periods of rainfall occurred and salinity values were lower than those observed in 1995. Parallel to the lower salinity observed in June 1996, a lower number of *Diamysis* specimens was collected.

Table 2 shows the main environmental characteristics of station O7, where *Diamysis* was collected more frequently.

The suprabenthic community in the area of occurrence of *Diamysis* is dominated by amphipods (mainly *Corophium acherusicum*, *C. insidiosum*, *Ampithoe valida*, *Gammarus insensibilis*, *Melita palmata*) and isopods (mainly *Lekanesphaera hookeri*, *L. levii*, *Idotea chelipes*). The mysids *Praunus flexuosus*, *Paramysis bacescoi* and

Table 2. Main characteristics of the station O7 during the sampling period in Canal de Ovar, Ria de Aveiro.

Date	Time (h:m)	Depth (m)	Water Temp (°C)	Sal. ‰	Surficial sediments Gv %	Cs (%)	Ms (%)	Fs (%)	SC (%)	Mz (mm)	Md (mm)	OM (%)
21 Jun 95	16:30	1.5	22.0	26	8.8	2.2	8.7	8.0	72.3	0.041	0.067	9.2
24 Jul 95	17:00	1.5	27.0	25	0.5	2.9	41.4	6.1	49.1	0.072	0.086	6.7
30 Aug 95	15:40	1.0	25.0	32	0.5	2.3	17.9	2.2	77.1	0.038	0.050	9.6
19 Sep 95	17:25	1.0	20.5	32	0.1	0.9	9.5	7.4	82.1	0.036	0.041	10.0
18 Okt 95	15:45	1.5	21.0	23	0.7	3.0	23.6	3.1	69.6	0.043	0.056	9.2
17 Nov 95	13:30	1.5	16.5	23	2.3	2.4	23.4	3.7	68.2	0.043	0.060	9.4
19 Dec 95	18:10	1.5	11.5	18	3.4	11.3	76.3	1.1	7.9	0.268	0.293	1.3
17 Jan 96	15:45	2.0	13.0	5	1.5	13.5	82.7	0.6	1.7	0.277	0.308	0.3
15 Feb 96	15:30	1.0	11.0	1	0.7	2.6	26.8	5.2	64.7	0.046	0.066	7.3
14 Mar 96	14:40	1.5	13.5	12	1.9	5.6	48.9	5.7	37.9	0.149	0.115	5.6
11 Apr 96	13:40	1.5	19.5	7	0.7	1.8	9.9	2.0	85.6	0.035	0.043	10.8
14 May 96	14:45	1.0	22.0	14	2.8	18.7	67.9	1.3	9.3	0.277	0.310	1.7
13 Jun 96	16:30	1.0	24.5	13	3.3	3.2	38.5	5.8	49.2	0.067	0.087	9.2

Temp. = temperature; Sal. = salinity; Gv = gravel; Cs = coarse sand; Ms = medium sand; Fs = fine sand; SC = silt and clay; Mz = granulometric mean; Md = granulometric median; OM = organic matter content.

Mesopodopsis slabberi occurred in high densities. Some specimens of *Siriella clausii* and *Neomysis integer* were also collected. *Carcinus maenas, Palaemon* sp., *Cardium glaucum, Hydrobia ulvae*, ascidians, *Sygnathus* sp. and *Anguilla anguilla* were some of the most frequent species among other constituents of the fauna.

4 DISCUSSION

The zoogeographical distribution of the majority of mysid species is either unknown or has not been adequately analysed. The coastal and littoral nature of the habitats of many species and the difficulty to sample the mysid fauna resulted in unrepresentative sampling in many regions (Mauchline 1980). Figure 4 shows the known distribution of *Diamysis bahirensis* (sensu lato) in the Mediterranean and Black Sea based on the available published data.

According to Mauchline (1980) the mysid fauna of the Mediterranean consists of more than 20 endemic species and about 30 other species that also occur in the northeastern Atlantic but, in most cases, do not penetrate the eastern Mediterranean. *Diamysis bahirensis* is the only Mediterranean endemic species that also occurs in the Black Sea (*D. b. mecznikowi*). The mapping of records (Fig. 4) shows that *Diamysis* has not been found in the westernmost part of the African coast of the Mediterranean between Tunis and Ceuta. Only recently it has been collected in the coast of Spain in the Ebro Delta (San Vicente, pers. comm.) and in the tidal channels of the Bay of Cadiz (Drake et al. 1997). The Strait of Gibraltar does not represent an important zoogeographical boundary and the Mediterranean is not a distinct faunal unit but en-

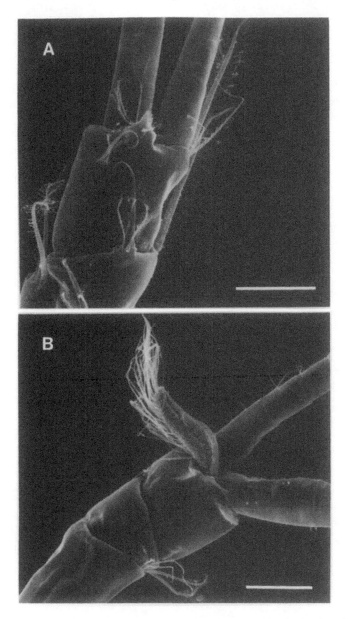

Plate 1. *Diamysis bahirensis* (sensu lato) from Ria de Aveiro. SE-micrographs of the antennula (details). (A) female, dorsal view; (B) male, ventral view. Scale bar: 0.1 mm.

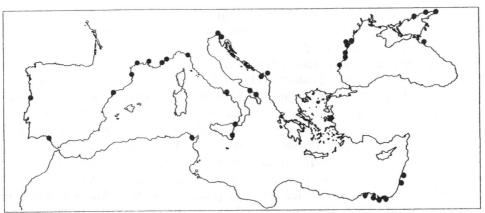

Figure 4: Known distribution of *Diamysis bahirensis* (sensu lato). Data were gathered from Drake et al. (1997) and San Vicente (pers. comm.) for the coast of Spain; Bacescu (1941), Gourret (1892), Walker (1901) for the coast of France; Ariani et al. (1983), Genovese (1956), Sars (1877) for the west coast of Italy and Sicily; Ariani (1966, 1967, 1981), Ariani et al. (1981, 1982, 1983), Claus (1884), Colosi, (1922), De Matthaeis et al. (1982), Graeffe (1902), Hoenigman (1963), Holmquist (1955, 1972), Spandl (1926) for the Adriatic coast and Lake Scutari; Katagan (1985), Katagan & Ledoyer (1979) for the Aegean coast; Almeida Prado-Por (1981), Almeida Prado-Por et al. (1981), Tattersall (1927), Zimmer (1936) for the Levantine coast and Egypt; Ariani (1981), Ariani et al. (1981, 1983), Sars (1877) for the coast of Tunisia; Antonescu (1934), Bacescu (1934, 1940, 1954), Sowinsky (1898), Valkanov (1936) for the Black and Azov seas.

ters into a wider one that includes the Lusitanian and the Mauretanian regions (Ekman 1953). It is therefore difficult to establish if the known distribution of *Diamysis* results from biogeographical causes or simply from non-comprehensive sampling of mysids in these regions.

Several authors agree that *Diamysis,* as well as a group of closely related genera like *Surinamysis* and *Mysidium* from the Caribbean Sea or *Gangemysis* and *Indomysis* from the Indian Ocean, evolved from a common ancestor inhabiting the ancient Sea of Tethys (Ariani 1980, Bacescu 1940, 1954, 1980, Por 1989). During the second half of the Miocene, the Mediterranean lost contact with the Indian Ocean and gradually also with the brackish Parathetys in the north. At the start of the Messinian period (seven million years ago) a massive and rapid glaciation or a tectonic event lowered the sea level and isolated the Mediterranean also from the Atlantic. During a period of 600 000 years the Mediterranean had not a permanent connection with the Atlantic and was reduced to a series of lagoons. The re-establishment of the permanent connection with the Atlantic through the Strait of Gibraltar marked the start of the Pliocene (Por 1989). According to the biogeographic hypothesis of Ariani (1980) the differentiation of *Diamysis* was probably initiated during the Messinian salinity crisis and different morphological and ecological forms of *D. bahirensis* are found today in the corresponding limits of the Messinian lagoons.

In fact, a variety of subspecies or infra-specific categories of *Diamysis bahirensis* have been proposed on the grounds of morphological and ecological differences. Bacescu (1940, 1954) suggested that the specimens from the freshwater lake Scutari should be considered as a separate entity (*D. b.* mod. *lacustris*) and that the species

D. mecznikowi known from the Black Sea should be regarded only as a geographical race of *D. bahirensis*. Ariani (1980), on the other hand, considers that *D. b. mecznikowi* is a separate species and that the specimens from some brackish water environments in the Adriatic (unnamed subspecies) are different from the type subspecies (*D. b. bahirensis*). Almeida Prado-Por (1981) describes two other subspecies from the coasts of Israel (*D. b. hebraica*) and Sinai (*D. b. sirbonica*). Wittmann (pers. comm.) found different morphological characters in the specimens from the south of France and in the specimens from the type locality (Tunis). *Diamysis bahirensis* (sensu lato) has been reported from environments with salinity ranging from freshwater (lake Scutari) to 70 (Bardawil lagoon, Sinai). However, *D. bahirensis* sensu lato is frequently found in more or less enclosed littoral areas and harbours. In the Adriatic it was sometimes collected off shore (Hoenigman 1963, Ariani 1980). The morphological differences are mainly based in the proportions and armature of the telson, in the number of propodal articles in the pereopods, in the armature of the maxillary palpus and in the characteristics of the *lobus masculinus*. However, the systematic value of some of these features has been questioned by Holmquist (1955) and Ariani (1980). Almeida Prado-Por (1981) presents a table, comparing the best known *Diamysis* forms, that confirms the diversity in the morphology and habitat of this mysid.

The finding of *Diamysis* in Ria de Aveiro raises different questions. If the mysid was an inhabitant of this area before the Messinian crisis, shouldn't it have evolved into a different morphological type like, apparently, it happened in different areas of the Mediterranean? Why are the specimens collected in Ria de Aveiro so similar to the ones in the south of France? Could the individuals have dispersed from the south of France to the Atlantic waters? Which modes of dispersion might have been used considering that these populations inhabit more or less enclosed brackish environments? Was the coast of Aveiro a suitable environment for the colonisation of this mysid before the Xth century when the formation of Ria de Aveiro was initiated?

Our knowledge of the fauna of Ria de Aveiro is recent. The most comprehensive work, from the beginning of the century (Nobre et al. 1915), mentions only one mysid species, *Neomysis integer* (recorded as *Mysis vulgaris*). Recently some benthic and zooplankton surveys were carried out (Pinho et al. 1992, Marques et al. 1993, Moreira et al. 1993, Cunha & Moreira 1995, Morgado 1998) but representative sampling of the mysid fauna was only accomplished recently (Cunha et al. in press). Due to the little biological research undertaken in the past in Ria de Aveiro it is not possible to determine whether or not *Diamysis* is a recent inhabitant of this coastal ecosystem. However, when we consider the high diversity observed in Mediterranean populations, the low morphological variability of the specimens collected in Ria de Aveiro may be indicative of a recent colonisation.

Since the morphology of the Portuguese specimens corresponds perfectly to those from Sardinia and the Mediterranean coast of France (Wittmann, pers. comm.), the likelihood of the hypothesis of unintentional introduction has been investigated. Cases of introduced mysids are not completely unknown: *Hemimysis anomala*, an endemic freshwater species from the ponto-caspian region, has been unintentionally introduced in the Baltic region and now dominates the diet of fishes in the Couronian lagoon in Lithuania (Elmgren & Hill 1997). In their interesting work on the role of ballast water for the introduction of non-indigenous species in Australia, Williams et al. (1988) report three mysid species (*Acanthomysis shrencki, Archaeomysis grebnitzkii* and *Neomysis japonica*) collected from the ballast water and from sediment remaining at the bottom of drained ballast tanks.

Ballast water is the principal mode of unintentional introduction of non-native species all over the world (Pullen 1997). The off-load of ballast water is forbidden inside the Ria de Aveiro but the Port Authorities have recognised their incapability to control this procedure. With the increasing portuary activity since 1950, the probability of introduction of non-native species in Ria de Aveiro must be considered and ballast water transport seems to be a likely mode of dispersion by which *Diamysis* might have been carried from the Mediterranean to Ria de Aveiro. In fact, most ships arriving from the Mediterranean to the Port of Aveiro come from Italy (Naples and Piombino) and from France (Marseille, Fos, Sète and La Nouvelle). These French ports are located in the close neighbourhood of the coastal systems (e.g. Étang de Lavalduc, Étang des Eaux-Blanches) where *Diamysis* is known to occur (Gourret 1892, Bacescu 1941) enhancing the likelihood of the introduction hypothesis. Recently *Diamysis* specimens were collected at Port de Fos (Wittmann, pers. comm.).

Unintentional introduction linked to intentional introduction of species for fish farming or aquaculture is quite improbable since, to our knowledge, no Mediterranean species have been brought to Ria de Aveiro for such purposes. The carriage of *Diamysis* amongst fouling on ship' hulls is also unlikely.

The ecological features of the upper reaches of Canal de Ovar appear to be especially suitable for the establishment of a *Diamysis* population. The relatively high temperatures combined with relatively high salinity values, the presence of macrophytes and calm hydrodynamic conditions create an environment somehow similar to the French Mediterranean lagoons. The observed density of *Diamysis* (at least at the end of spring in two consecutive years) as well as the presence of incubating females and juveniles is indicative of a wealthy population. The possibility that *Diamysis* has a wider distribution along the Iberian and African Atlantic coasts in similar environments can not be disregarded: they may have been overlooked in the past, because of unrepresentative sampling of the mysid fauna or difficulties in taxonomic determination. The hypothesis of unintentional introduction by cargo vessel ballast water is a probable explanation for the occurrence of *Diamysis* in Ria de Aveiro. However, thorough morphological and molecular research on *Diamysis bahirensis* (sensu lato) is needed to clarify the taxonomy and biogeography of this polilypical species.

ACKNOWLEDGEMENTS

We wish to thank the Natural History Museum of London, specially Chris Jones (SEM Unit) and Paul Clark (Curator for Crustacea). We are indebted to Prof. Wittmann for his interest and helpful comments on our material. Thanks are due to Mr Rui Marques for his assistance with sampling and to Dr Oliveira from the Junta Autónoma do Porto de Aveiro, who kindly helped to gather all the information about the Port of Aveiro. This research was partially supported by Centro das Zonas Costeiras e do Mar (CZCM), Universidade de Aveiro.

REFERENCES

Abecasis, C.K. 1961. *As formações lagunares e seus problemas de engenharia litoral.* Lisboa.

Almeida Prado-Por, M.S. 1981. Two new subspecies of the *Diamysis bahirensis* Sars species group (Crustacea: Mysidacea) from extreme salinity environments on the Israel and Sinai coasts. *Isr. J. Zool.* 30: 161-175.

Almeida Prado-Por, M.S., Ortal, R. & Por, F.D. 1981. *Diamysis* from the brackish river Nahal Taninim in Israel and its associated fauna. *Rapp. Comm. int. Mer Médit.* 27(4): 181-182.

Antonescu, C.S. 1934. Ueber Mysideen aus dem Süss- und Brackwasser Rumäniens. *Notationes Biologicae* 2(1): 32-36.

Ariani, A.P. 1966. Su una forma di *Diamysis bahirensis* (G.O. Sars) rinvenuta in territorio pugliese. *Boll. Zool.* 33: 227-228.

Ariani, A.P. 1967. Osservazioni su Misidacei della costa adriatica pugliese. *Annar. Ist. Mus. Zool. Univ. Napoli* 18(5): 1-38.

Ariani, A.P. 1979. Contribution à l'étude écotaxonomique et biogéographique des *Diamysis* d'eau saumâtre de la Méditerranée. *Rapp. Comm. int. Mer Médit.* 25-26(3):159-160.

Ariani, A.P. 1980. Systématique du genre *Diamysis* et paléogéographie de la Méditerranée. *Journées Étud. Systém. et Biogéogr. Médit.*: 121-130.

Ariani, A.P. 1981. Experiences d'hybridation entre populations méditerranéennes du genre *Diamysis*. *Rapp. Comm. int. Mer Médit.* 27(4): 177-180.

Ariani, A.P., Marmo, F., Balsamo, G. & Franco, E. 1981. Vaterite in the statoliths of a mysid crustacean (*Diamysis bahirensis*). *Annuar. Ist. Mus. Zool. Univ. Napoli* 24: 69-78.

Ariani, A.P., Marmo, F., Balsamo, G., Cesaro, G. & Maresca, N. 1982. Prime osservazioni sullo sviluppo degli statoliti di crostacei misidacei. *Annuar. Ist. Mus. Zool. Univ. Napoli* 25: 326-341.

Ariani, A.P., Marmo, F., Balsamo, G., Franco, E. & Wittmann, K. 1983. The mineral composition of statoliths in relation to taxonomy and ecology in mysids. *Rapp. Comm. int. Mer Médit.* 28(6): 333-336.

Bacescu, M. 1934. Contributions à l'étude des mysidés de la mer Noire ainsi que des limans et des lacs en relation avec la mer ou avec la Danube. *Ann. Sci. Univ. Jassy* 19: 331-338.

Bacescu, M. 1940. Les Mysidacés des eaux roumaines (étude taxonomique, morphologique, biogéographique et biologique). *Ann. sci. Univ. Jassy* 26: 453-804.

Bacescu, M. 1941. Les Mysidacés des eaux méditerranéennes de la France (spécialement de Banyuls) et des eaux de Monaco. *Bull. Inst. océanogr. Monaco* 795: 1-46.

Bacescu, M. 1954. Crustacea: Mysidacea. *Fauna Republicii Populare Romîne* 4(3):1-126.

Bacescu, M. 1973. Contribution à la connaissance des Mysidés benthiques de la mer Rouge. *Rapp. Comm. int. Mer Médit.* 21: 643-646.

Bacescu, M. 1980. Problèmes de systématique évolutive concernant quelques crustacés de la Mer Noire. *Journées Étud. Systém. et Biogéogr. Medit.*: 85-88.

Claus, C. 1884. Zur Kenntnis der Kreislauforgane der Schizopoden und Decapoden. *Arb. Zool. Inst. Univ. Wien, Zool. Sta. Triest* 6: 271-318

Colosi, G. 1922. Eufausiacei e Misidacei dello Stretto di Messina. *Mem. R. Comitato Talassografico Italiano* 98: 1-22.

Cunha, M.R. & Moreira, M.H. 1995. Macrobenthos of *Potamogeton* and *Myriophyllum* beds in the upper reaches of Canal de Mira (Ria de Aveiro, NW Portugal): community structure and environmental factors. *Neth. J. Aqua. Ecol.* 29(3-4): 377-390.

Cunha, M.R., Sorbe, J.C. & Moreira, M.H. (in press). Spatial and seasonal changes of brackish suprabenthos: peracaridean assemblages and their relation to environmental variables in two tidal channels of Ria de Aveiro (NW Portugal). *Mar. Ecol. Prog. Ser.*

Czerniavsky, V. 1882. Monographia Mysidarium imprimis imperii Rossici. Fasc.I. *Trav. Soc. Imper. Natur. St. Petersburg* 12: 1-170.

De Mattheis, E., Colognola, R., Sbordoni, V., Cobolli-Sbordoni, M. & Pesce, G.L. 1982. Genetic differentiation and variability in cave dwelling and brackish water populations of Mysidacea (Crustacea). *Zeitschrift für Zoologische Systematik und Evolutionsforschung* 20(3): 198-208.

Derjavin, A.N. 1924. Freshwater Peracarida from the coast of the Black Sea of the Caucasus. *Russische Hydrobiol. Zeitschr.* 3: 113-129.

Drake, P., Arias, A.M. & Conradi, M. 1997. Aportación al conocimiento de la macrofauna supra y epibentónica de los caños mareales de la bahía de Cádiz (España). *Publ. Esp. Inst. Esp. Oceanog.* 23: 133-141.

Ekman, S. 1953. *Zoogeography of the sea.* London: Sidgwick and Jackson Limited.

Elmgren, R. & Hill, C. 1997. Ecosystem function at low biodiversity – the Baltic example. In R.F.G. Ormond, J.D. Gage & M.V. Angel (eds), *Marine biodiversity*, pp 319-336. Cambridge: Cambridge University Press.

Genovese, S. 1956. Su due Misidacei dei laghi di Ganzirri e di Faro. Boll. Zool. 23: 177-197.

Gordan, J. 1957. A bibliography of the order Mysidacea. *Bull. Amer. Mus. nat. Hist.* 112: 279-394.

Gourret, P. 1892. Notes zoologiques sur l'étang des Eaux-Blanches (Cette). *Ann. Mus. Hist. Nat. Marseille 4 mém.* 2:1-26.

Graeffe, E. 1902. Übersicht der Fauna des Golfes von Triest nebst Notizen über Vorkommen, Lebensweise, Erscheinungs- und Laichzeit der einzelnen Arten. Crustacea. *Arb. Zool. Inst. Univ. Wien* 13: 33-80.

Hoenigman, J. 1963. Mysidacea de l'expédition 'Hvar' (1948-49) dans l'Adriatique. *Rapp. Comm. int. Mer Médit.* 17: 603-616.

Holmquist, C. 1955. Die '*Mysis relicta*' aus dem Mittelmeergebiet (=Diamysis bahirensis (G.O. Sars,1877) und *Paramysis helleri* (G.O. Sars, 1877)). Zool. Anz. 154: 277-288.

Holmquist, C. 1972. Mysidacea. *Binnengewässer* 26: 247-256.

Katagan, T. 1985. Mysidacés et cumacés des côtes Egéennes de Turquie. *Rapp. Comm. int. Mer Médit.* 29(5): 287-288.

Katagan, T. & Ledoyer, M. 1979. Crustacea Mysidacea des côtes de Turquie et deux nouvelles espèces pour la Méditerranée orientale. *Téthys* 9: 129-131.

Marques, L.F., Cruz, M.M., Moreira, M.H., Rua, J.M., Rebelo, J.E., Luís, A.S., Cunha, M.R., Barroso, C.M., Pinho, R.F. & Marques, A.M. 1993. *Estudo de impacte ambiental do projecto de dessassoreamento da Ria de Aveiro. Relatório de síntese.* Junta Autónoma do Porto de Aveiro.

Martynov, A. 1924. On some interesting Malacostraca from freshwaters of European Russia. *Russische Hydrobiol. Zeitschr.* 3: 210-216.

Mauchline, J. 1980. The biology of mysids and euphausiids. *Advances in Marine Biology*, 18: 1-681.

Mauchline, J. & Murano, M. 1977. World list of the Mysidacea, Crustacea. *J. Tokyo Univ. Fish.* 64: 39-88.

Moreira, M.H., Queiroga, H., Machado, M.M. & Cunha, M.R. 1993. Environmental gradients in a southern Europe estuarine system: Ria de Aveiro, Portugal. Implications for soft bottom macrofauna colonization. *Neth. J. Aqua. Ecol.* 27 (2-4): 465-482.

Morgado, F. 1998. Ecologia do zooplâncton da Ria de Aveiro. PhD thesis. Universidade de Aveiro.

Müller, H.-G. 1993. *World Catalogue and Bibliography of recent Mysidacea.* Wetzlar: Laboratory for Tropical Ecosystems, Research & Information Service.

Nobre, A., Afreixo, J. & Macedo, J. 1 915. *A Ria de Aveiro. Relatório oficial do Regulamento da Ria.* Lisboa: Imprensa Nacional.

Oliveira, O. 1988. Origens da Ria de Aveiro. Aveiro: Câmara Municipal de Aveiro.

Pinho, P., Moreira, M.H., Costa, C., Beato, C., Margalha, J., Rebelo, J.E., Luís, A.S., Cunha, M.R., Barroso, C.M. & Pinho, R.F. 1992. Estudo de impacte ambiental do estudo prévio da marina da Barra. Análise dos recursos biológicos, qualidade do ambiente e infraestruturas. Departamento de Biologia, Universidade de Aveiro, Paulo Pinho, Estudos e Projectos de Planeamento e Ambiente Lda.

Por, F.D. 1989. *The legacy of Tethys. An aquatic biogeography of the Levant.* Dordrecht: Kluwer Academic Publishers.

Pritchard, D.W. 1967. What is an estuary: physical viewpoint. In G.E. Lauff (ed.), *Estuaries*: 3-5. Washington: Association for the Advancement of Science.

Pullen, J.S.H. 1997. Protecting marine biodiversity and integrated coastal zone management. In R.F.G.Ormond, J.D. Gage & M.V. Angel (eds), *Marine biodiversity*, pp 394-414. Cambridge: Cambridge University Press.

Sars, G.O. 1877. Nye Bidrag til Kundskaben om Middelhavets Invertebrat fauna. I. Middelhavets Mysider. *Arch Math. Naturv.* 2: 10-119.

Sowinsky, W. 1898. Scientific results of the 'Atmanai' expedition. Crustacea, Malacostraca of the Azov sea. *Bull. Acad. Imp. Sci. St. Petersbourg*, ser. 5, 8: 359-398.

Spandl, H. 1926. Beiträge zur Kenntnis der im Süsswasser Europas vorkommenden Mysidaceen. *Int. Rev. ges. Hydrobiol.* 15: 358-375.

Tattersall, W.M. 1908. The fauna of brackish ponds at Port Canning, Lower Bengal. XI. Two new Mysidae from brackish water in the Ganges delta. *Rec. Indian Mus.* 2(3): 233-239.

Tattersall, W.M. 1927. The Cambridge Expedition to the Suez Canal. Report on the Crustacea Mysidacea. *Trans. Zool. Soc. London* 22: 185-199.

Tattersall, W.M. 1951. A review of the Mysidacea of the United States National Museum. *Bull. US Nat. Mus.* 201: 1-292.

Valkanov, A. 1936. Über einige Mysiden aus Bulgarien. *Zool. Anz.* 115: 25-27.

Vicente, C.M. 1985. Caracterização hidraulica e aluvionar da Ria d Aveiro. Utilização de modelos hidráulicos no estudo de problemas da Ria. In *Jornadas da Ria de Aveiro*. Vol. III: 41-58. Aveiro: Câmara Municipal de Aveiro.

Walker, A.O. 1901. Contributions to the malacostracan fauna of the Mediterranean. *J. Linn. Soc. Zool. London* 28: 294.

Williams, R.J., Griffiths, F.B., Van der Wal, E.J. & Kelly, J. 1988. Cargo-vessel ballast water as a vector for the transport of non-indigenous marine species. *Est. Coast. Shelf Sci.* 26: 409-420

Wittmann, K.J. 1981. Comparative biology and morphology of marsupial development in Leptomysis and other Mediterranean Mysidacea (Crustacea). *J. exp. mar. Biol. Ecol.* 52: 243-270.

Zimmer, C. 1915a. Die Systematik des Tribus Mysini H.J. Hansen. *Zool. Anz.* 46: 202-216.

Zimmer, C. 1915b. Schizopoden des Hamburger Naturhistorischen (Zoologischen) Museums. *Mitt. Nat. Mus. Hamburg* 32: 159-182.

Zimmer, C. 1936. Appendix: Schizopoda. In H. Balss, Decapoda. The Fishery grounds near Alexandria. *Notes Fish. Res. Directorate, Egypt* 15: 1-46.

An occasional record of *Funchalia woodwardi* (Penaeidae) in the Sardinian Channel, eastern Mediterranean

MATTEO MURENU, MARCO MURA & ANGELO CAU
Department of Animal Biology and Ecology, University of Cagliari, Cagliari, Italy

ABSTRACT

The capture of some specimens of *Funchalia woodwardi* off the southern coast of Sardinia (Western Mediterranean) during commercial trawling operations is reported on. The main biometric parameters were verified, as was the presence of spermatophores in the telicum, the size of the female ovary and the presence of spermatophores in the terminal ampullae of the male gonoducts. Whenever possible stomach content was analysed. The finding of these examples, captured by vessels not suitably equipped to catch pelagic species such as *F. woodwardi*, represents more than 50% of the total number of examples captured in the whole of the Mediterranean Basin. Therefore, of all of the areas in the Mediterranean that are suitable for trawler fishing, the middle slope of this area is of extreme scientific interest in the study of this species, with possible economic advantages in the exploitation of alternative alieutic resources.

1 INTRODUCTION

Funchalia woodwardi is a pelagic species with a wide geographic distribution including the Atlantic Ocean and the Mediterranean Sea (Zariquiey Alvarez, 1968). No information is available on its behavior and biology in the Mediterranean. Moreover, even sure findings in this area are few and refer to a limited number of specimens (Relini Orsi & Costa 1975; Grippa 1987).

Between 1908 and 1910, during the Thor expedition eight specimens were found in seven different places in the easternmost part of the western Mediterranean, between the depths of 75 and 3700 m (Stephensen, 1923). In 1938, an individual of the species that had been captured off the coast of Nice was identified in the material found at the Zoological Station of Villefranche-Sur-Mer. Relini Orsi & Costa (1975) registered a male, captured on the surface in the Gulf of Genoa. Grippa (1976) recorded the presence of three specimens of *F. woodwardi* in the decapod collections of the Civic Museum of Natural History of Milan. These specimens were found between 1921 and 1924 off the coast of Messina. Grippa (1987) reported on five other specimens captured during four fishing expeditions within the framework of a research

Table 1. Mediterranean specimens (data from Relini and Costa (1975) and Grippa (1987)).

Museum	Month	Year	Place	Specimens
MCSNM[1]	September	1921-24	Messina	3 f
MMF[2]	May	1955	Madeira	7 f, 2 m
MMF[2]	September	1955	Madeira	1 m
MMF[2]	September	1955	Madeira	5 f, 1 m
MMF[2]	April	1957	Madeira	1 m
MUG[3]	July	1975	Liguria	1 m
USNM[4]	September	1970	39°58′N – 12°24′E	2 juv
USNM[4]	September	1970	40°06′N – 12°32′E	1 m
USNM[4]	September	1970	40°17′N – 12°33′E	2 juv
USNM[4]	September	1970	41°18′N – 06°14′E	1 m
Thor spedition	January	1909	40°14′N – 12°23′E	1 juv
Thor spedition	January	1909	40°45′N – 09°30′E	1 juv
Thor spedition	January	1909	40°53′N – 13°43′E	1 juv
Thor spedition	July	1909	38°57′N – 09°47′E	2m
Thor spedition	July	1909	39°35′N – 11°21′E	1 f
Thor spedition	July	1910	34°31′N – 18°40′E	1 f
Thor spedition	August	1910	38°33′N – 15°29′E	1 juv
Villefranche coll.	May	1926	off of Nizza	2 f
DBAE[5]	September	1997	South Eastern Sardinia	15 f, 5 m

[1]Civil Museum of Natural History of Milan, [2]Museum on Funchal in Madeira, [3]University of Genova Museum, [4] National Museum of Natural History – Smithsonian Institution, Washington, [5]Department of Animal Biology and Ecology – University of Cagliari.

campaign by the Smithsonian Institute in the Mediterranean.

In this work we report on twenty adult specimens found on the mesobathyal bottoms of southern Sardinia by two commercial trawlers engaged in deep-shrimp fishing. Considering the exceptional nature of this event in the Mediterranean, it was deemed opportune to report the finding. We also give some biometric data for the determination of ratios (e.g. between the length of the final abdominal segment and the length of the telson and between the length of the uropods and the length of the telson) that distinguish this species. These also are needed to compare adult individuals of the same sex (e.g. between the length of the dactyl and that of the propodite of the fourth pereiopodites) (Grippa 1987). We also give the conditions of maturity and observations on stomach content of specimens examined.

2 MATERIALS AND METHODS

The material was found in September 1997 during trawling operations by two different trawlers that habitually practice the bentho-bathyal fishing known as 'deep-shrimp fishing'. Both trawlers were using an Italian-type dragnet, with a 450 mesh pocket from 15 to 20 mm, and trawl doors of the multipurpose type made of iron.

Twenty specimens (15 females and 5 males) were found off Capo Carbonara and Capo Boi (southeastern Sardinia) in the Gulf of Cagliari (Central Western Mediterranean Sea) not far from one another (a maximum of 10 knots). The bottoms, with depths between 450 and 700 m, are characterized by *Isidella elongata* facies (Fig. 1).

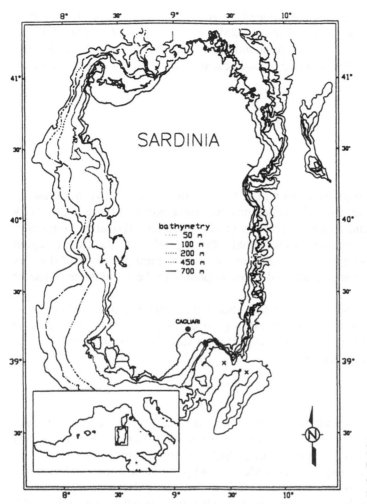

Figure 1. Fishing area
(X = occurence of *Funchalia woodwardi*, Johnson 1867).

The material gathered was preserved in neutral formalin and taken to the laboratory. There, after ascertaining the sex of each specimen, the following biometric parameters were measured to a precision of ± 0.1 mm:

- CL: Carapace length, from the orbital corner to the posterior edge;
- L.A.S.: Last abdominal somite length;
- Ltel: Telson length;
- Lpro: Propod length of the fourth pereiopods;
- Ldat: Dactyl length of the fourth pereiopods;
- Luro: Uropods length.

The rostral formula was determined for each individual as well. In fresh female specimens, the presence of spermatophores in the telicum was verified and the size of the ovary was noted. In males, the presence of spermatophores in the terminal ampullae of the gonoducts was verified.

Since the stomach content sample was limited, qualitative aspects were given priority in the analysis. Foregut contents were preserved in alcohol (70% vol.) and later examined. Fullness of foregut was estimated as: empty or nearly empty (f1 ≤ 25%), partially full (f2 > 25% and ≤ 50%), full (f3 ≤ 75%) and very full (f4 > 75%). Foregut contents were analysed and prey items identified to the lowest possible taxonomic level. Prey were classified as digested or undigested on the basis of Cartes' proposal (1993a).

3 RESULTS

3.1 *General aspects*

Specimen sizes ranged from 31 to 45 mm of CL in females, and between 23 and 29 mm of CL in males (Fig. 2). All females had spermatophores in the telicum. Only two females (CL 32.2 and 33.8 mm) presented thin and collapsed ovaries. In the others the ovaries were voluminous, with frontal lobes extending to the frontal region and with very wide caudal lobes extending to the third abdominal somite. All males presented a well-developed petasma and spermatophores in the terminal ampullae of the gonoducts.

The relation between length of the last abdominal somite and the telson (r^2 = 0.995), between length of uropods and telson (r^2 = 0.994), and between the propod and dactyl of the fourth pereiopods (r^2 = 0.914) were statistically significant (Fig. 3a, b, c). Rostral teeth varied from 10 to 12.

3.2 *Stomach content*

The stomachs of only three individuals were found to be empty. In the others, the degree to which they were full was high (f3 = 8, f4 = 4). All contents were already partly digested. There was a prevalence of tentacles of small cephalopods (in 6 individuals), remains of bony fish, [scales, vertebras and parts of tails (in 4 individuals), antennae, legs and telsons of crustaceans (in 2 individuals)]. Because of partial digestion of the cephalopods, it was possible to identify prey as belonging to the genus *Sepietta* in only two cases. The bony fishes belonged to the genus *Cyclothone* sp.

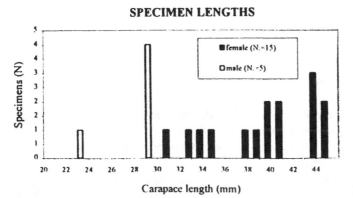

Figure 2. Specimens lengths.

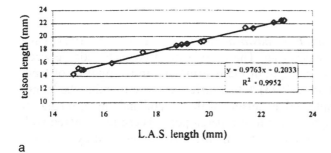

a

b

c

Figure 3. Linear regression between length of: last abdominal somite and telson (a) propod and dactyl of the fourth pereiopods (b) uropods and telson (c).

(1 case) and to the family Myctophidae (2 cases). The crustaceans were Euphausiacea (1 case).

In every full stomach, only one kind of prey was found, with the exception of the case of one of the largest females of *F. woodwardi* (CL = 43.6 mm) in the stomach of which the tails of 2 bony fish were recognized. Moreover, no detritus, foraminiferans, bivalve residues or non-assimilable fragments were found, as has been reported for other mesopelagic species of Mediterranean Decapods (Lagardere 1975; Cartes 1993b, 1993c).

4 DISCUSSION

To date, two species of *Funchalia*, *F. woodwardi* and *F. villosa* (Grippa, 1987) have been found in the Western Mediterranean. The most distinct characteristic between the two species is the difference in length of the rostrum and in the number of teeth on its upper edge (Crosnier & Forest, 1973). Other differences are found in the length ratios of the last abdominal somite and the telson, of the uropod and the telson and of the dactylus and that of the propod of the fourth pereiopods. These ratios agree with those reported by Grippa (1987). The other taxonomic features reported by the same author confirm that our specimens belong to *Funchalia woodwardi*.

Although the number of stomachs examined cannot be deemed sufficient to produce reliable information on the feeding habits of the species, estimates of the degree of digestion of the stomach content and the high values for stomach fullness suggest a nocturnal feeding period. The sole component of the diet of the specimens examined is represented by small pelagic organisms such as cephalopods, bony fish (mesopelagic) and Euphausiacea. *F. woodwardi* most likely behaves as an active predator of small pelagic prey, thus confirming what is found in the literature on the behavior of this species.

The fragmentary information on the distribution of the species justifies the importance of this report. The number of specimens reported on here comprise more than 50% of the total number of specimens fished within the whole of the Mediterranean basin. Moreover, since *F. woodwardi* is well known to be a pelagic species (Holthuis, 1987), and the capture took place with a net unsuitable for such specimens, it is to be hoped that *ad hoc* thalassographic campaigns will now be undertaken to improve our knowledge concerning this species.

REFERENCES

Cartes, J.E. 1993a. Day-night feeding by decapod crustaceans in a deep-water bottom community in the western Mediterranean. *J. Mar. biol. Ass. UK* 73: 795-811.

Cartes, J.E. 1993b. Feeding habits of oplophotid shrimps in the deep Western Mediterranean. *J. Mar. biol. Ass. UK* 73: 193-206.

Cartes, J.E. 1993c. Feeding habits of pasiphaeid shrimps close to the bottom on the Western Mediterranean slope. *Mar. Biol.* 117: 459-468.

Crosnier, A & Forest, 1973. Les Crevettes profondes de l'Atlantique Oriental Tropical. *Fauna Tropicale* 19, 296-297.

Grippa, G. 1987. A revision of gen. *Pelagopenaeus* Burkenroad, 1934 and Funchalia Johnson, 1987 (Crustacea, Decapoda, Peneidea). *Inv Pesq.* 51(1): 73-85.

Holthuis, L.B. 1987. Crevettes. In W. Fischer, M.L. Bauchot & M. Schneider (eds), *Fiches Fao d'identification des espèces nouvelles pour les besoins de la pêche; Méditerranée et Mer Noire*: 191-367.

Lagardere, J.P. 1975. Recherches sur l'alimentation des crevettes bathypelagiques du talus continental du Golfe de Gascogne. *Rev. Trav. Inst. Pêches marit.* 39(2): 213-229.

Relini-Orsi, L. & Costa, M.R. 1975. Segnalazione di *Funchalia woodwardi* Johnson in Mar Ligure (Crustacea, Decapoda, Penaeidae). *Boll. Mus. Ist. Univ. Genova* 43: 33-39.

Stephensen, K. 1923. Decapoda Macrura excl. Sergestidae. *Rep. Danish Oceanogr. Exped. (1908-10) Medit. 2, Biology.*

Zariquiey Alvarez, R. 1968. Crustaceos Decapodos Ibéricos. *Inv. Pesq.* 32: 55-56.

2 Ecology and behavior

The role of infochemicals in lake plankton crustacean behavior and predator-prey relations

J. RINGELBERG

Department of Aquatic Ecology, University of Amsterdam, NIOO – Centre for Limnology, Nieuwersluis, Netherlands

ABSTRACT

Defenses in *Daphnia* (Cladocera) against invertebrate and vertebrate predation are predominantly conditionally induced. Chemical substances or kairomones, associated with the predators, suffice to cause changes in morphology, life history or behavior. The extent of these adaptations is variable and genotype specific. Trade-offs and a variable environment are important to understand why induced behavior is common. An example of diel vertical migration (DVM) and the underlying photobehavior is discussed. In the given example, DVM occurs during a short period of heavy predation by juvenile fish. Costs are high and explain why it is conditionally induced. The *Daphnia* population complex consists of many clones of which dominance changes rapidly. Infochemicals, like kairomones, are important factors that determine food web relations and community composition.

1 INTRODUCTION

1.1 *Predator-induced defenses*

The herbivorous zooplankton are important links in pelagic food webs of oceans and lakes. Calanoid copepods, in lakes together with cladocerans, are staple food for both invertebrate and vertebrate predators. Heavily preyed upon, they consequently have evolved various kinds of defense mechanisms. The freshwater genus *Daphnia* was studied extensively, and morphological, life history and behavioral adaptations have been described. The traits in concern might be genetically fixed, as the transparency of *Daphnia* or the jumping behavior of calanoids seems to be, but most of them are phenotypically induced in the presence of predators. In fact, predator-correlated chemicals or kairomones suffice (see for terminology Dicke & Sabelis 1988) to bring about these defenses. Of course, the property of phenotypic induction as such is genetically determined, the mechanism of sensory perception of the chemicals and of the processing of the obtained information by the central nervous system, included.

A well-studied example of a morphological adaptation is the induction of small protuberances or neck teeth in several juvenile instars of *Daphnia pulex* by kairomones from the phantom midge *Chaoborus*. These small spines develop on the dorsal anterior margin of the head carapace in pre-reproductive instars smaller than 1200 µm (Vuorinen et al. 1989, Havel & Dodson 1987). The expression (number, length) depends on concentration of the kairomone (Tollrian 1995). According to Havel & Dodson (1987), individuals with spines escape an attack by *Chaoborus* more often than specimens without spines, but Lüning (1992) casts some doubt on the direct effectiveness of the structure and suggested that associate parameters of escape behavior could be responsible. Spined morphs have a lower growth rate and a reduced reproduction (Havel & Dodson 1987, Lüning 1992). The expression of costs may depend, however, on food supply since Hanazato (1991) found that neither body size, brood size, nor intermolt time in *D. ambigua* differed significantly from controls at a high food concentration. The induction of this change in morphology is phenotypically plastic and genotype-dependent (Lüning 1992, Black 1993, Spitze 1992).

Life history adaptations are properties of characters such as growth, maturation, number and size of eggs profitable in a particular environmental situation. In *Daphnia* again, extensive studies were made of effects of predator kairomones. In the presence of fish-associated exudates, pelagic daphnids, like *D. galeata* and *D. hyalina*, mature earlier at a smaller size and the females have larger clutches of small eggs (Machacek 1993, Reede & Ringelberg 1995, Stibor 1992). As a defense against visually predating fish this is considered an advantage. However, the small-sized neonates, born from these small eggs, are less resistant to starvation (Cowgill et al. 1984, Tessier & Consolatti 1989). Food concentration is a very important factor to influence life-history: as it diminishes, females may increase egg size, thus enhancing survival of the neonates (Glazier 1992, Guisande & Gliwicz 1992). Food quality (for review see Gulati & DeMott 1997) likewise determines the configuration of life-history traits (Weers & Gulati 1997, Tessier & Consolatti 1991). The interacting effects of food and fish kairomone is complicated (Reede 1997) and as yet poorly understood. All these influencing factors make the application of life history results to a particular field situation difficult. Moreover, the change in traits is clone dependent and correlated with behavior. Unless planned with a particular situation in mind, collecting data from life history experiments provide a picture of expectations only.

A widespread phenomenon in oceans and lakes is diel vertical migration (DVM) (for reviews see Cushing 1951, Segal 1970). Before sunrise, zooplankton descend to escape predation by visually hunting predators, and after sunset they return to the surface layers, rich in algal food, to graze. The basic behavior in *Daphnia* is a negatively and positively phototactic response to the relative increases and decreases in light intensity of dawn and dusk, respectively (Ringelberg 1993, 1995b). This photobehavior is enhanced by fish kairomone and might lead to extensive migrations. The many-sided aspects of diel vertical migration make this phenomenon an excellent subject to discuss the role of infochemicals in behavior and predator-prey relations and attention will be focused on this subject.

1.2 *Advantage of mechanisms*

Why are induced predator defence mechanisms widespread and obviously of advantage over constitutive traits in *Daphnia* and certainly other zooplankton organisms? Several arguments have been listed: reliable predator cues, short response times, unpredictable strong predation pressure, associated costs (Adler & Harvell 1990, Pya-nowska 1993, Harvell 1990, Gliwicz & Pijanowska 1988, De Meester et al. 1999). The first two are, in fact, conditions rather than arguments in answer to the 'why' question and predictability is a cost aspect. So answering the question boils down to a costs-benefits balance. Predictability has to do with the frequency and regularity of occurrence in relation to the time-scale of an observer. The occurrence of large shoals of 0+ fish in early summer is highly predictable to a human observer: in Lake Maarsseveen (Netherlands) juvenile perch (*Perca fluviatilis*) start eating *Daphnia* at the end of May, as we have observed for ten years now. With regard to *Daphnia*, only few of many yearly generations are actually confronted with these predators. There-fore, to *Daphnia*, the early summer peak in predation may be unpredictable. Costs are usually assessed as a reduction in some fitness component. Vertical migrating ani-mals spent most of the day in food-poor and cold water layers. Consequently, fecun-dity is decreased and development rate is increased, which results in lower birth rates. The payoff is a lower mortality due to predation. To keep inducement in existence, different fitness aspects must be negatively correlated. If no trade-offs were present, the traits responsible for migration behavior would become constitutive because the genotype with these traits would always have a superior fitness. Of course this is de-pendent on whether the environment is variable with regard to the relevant factors. On the time scale of a few weeks of heavy predation, as in Lake Maarsseveen, it is quite possible that costs for some genotypes are larger than the benefits. The result is a frequent change in genotype composition. The system of inducible defences can only be understood if the presence of trade-offs and a frequently changing environ-ment is considered.

Infochemicals are material carriers of information. The information in kairomones is about the presence of predators. It is characteristic of information carriers that the material base represents a low energy content in relation to the energetic outburst of the organisms' response. A chemical sense must be involved, but the perception mechanism of kairomones is unknown in copepods and cladocerans. In addition, the chemical nature of these aquatic invertebrate and vertebrate predator kairomones is unknown. After perception, information is translated and sent to the central nervous system. Successive processing leads to messages to effector organs. We know next to nothing of these pathways between signal perception and responsive behavior, al-though insight might certainly increase our knowledge of evolutionary routes. For in-stance, kairomone concentration seems to have to surpass a threshold value before being effective (Loose et al. 1993, Van Gool & Ringelberg 1998) but where the high-est threshold resides, at the receptor side or more central, cannot be told when be-havior is studied only. As long as attention is predominantly fixed on functional, ul-timate aspects, deeper understanding of the evolutionary process will escape us.

'Cued' defenses are not simple 'on-off' effects. The reaction can be modulated, depending on the kairomone concentration and thus on the density of the predator. The variety in response, or reaction norm of the phenotype, might be different for dif-ferent genotypes. Therefore, selection must act on plasticity and a broad combination of induced defense properties realizes highest fitness for a particular genotype in a certain configuration of environmental circumstances. As will be demonstrated by the example of *Daphnia* in Lake Maarsseveen, the system is highly dynamic.

2 INFOCHEMICAL-INDUCED BEHAVIOR IN *DAPHNIA*

Extensive studies were done in Lake Maarsseveen, a mesotrophic lake in the center of the Netherlands (see for detailed description: Swain et al. 1987). This man-made lake has a surface area of 70 ha. with a maximum depth of about 30 m. Morphology and hydrology are simple: the lake has a nearly rectangular shore line with a small littoral zone and is fed by precipitation and ground water seepage from diluvial hills only. One small outlet regulates water level. This simplicity makes the lake ideal for studies in aquatic ecology. The population of *Daphnia* consists of *D. pulicaria*, *D. hyalina*, *D. galeata*, and the hybrid *D. galeata* × *hyalina*. The last two taxa dominate. As allozyme analysis has revealed, relative dominance changes from year to year, as well as within years. For example: in April 1989, when this kind of analysis was first performed, only the hybrid was found. However, from July onwards *D. galeata* increased in numbers and reached a maximum of 56% in July next year. Then the percentage gradually dropped and the species was only rarely observed in 1991 (Spaak & Hoekstra 1993). Obviously, selecting forces operate continuously but the resulting direction of selection changes.

Diel vertical migration is a yearly returning, conspicuous phenomenon in the lake. In the last week of May, DVM starts and will last until mid-July (Fig. 1). The period coincides with the presence of large shoals of 0+ perch (*Perca fluviatilis*). Perch are visually hunting fish that need a sufficiently high light intensity to predate. Not all *Daphnia* in the population migrate. During daytime of the first 2 weeks of the migration period, part of the population is still present in the upper 5 m (Fig. 2). The animals are heavily preyed upon and rapidly decline in numbers. In the first and second week of the period in 1990, the number of these non-migrating adult *Daphnia* decreased with 83% and 96%, respectively. During a short time in late evening and early morning, also the migrating part of the population is prone to predation, and, in these two first weeks, the total population plummeted with 47% and 27%, respectively. Nevertheless, the population was sustained, and, in some years, might even increase in size near the end of the predation period. In spite of imperfections and disadvantages, diel vertical migration is a successful strategy to cope with visual predation.

However, costs are involved. A sojourn during the larger part of the day in cold and food-poor, deepwater layers effectuates the main drawback. Fecundity drops and egg development time is increased and especially the latter factor is responsible for a lower reproduction and birth rate. To illustrate the nature and extent of costs, some calculations were done for the situation in Lake Maarsseveen on 10 June 1992. The amplitude of migration was large, with an epilimnion nearly empty in daytime, as was the hypolimnion during the night (Fig. 3). If we assume that all individuals are mixed and that all migrate with an immediate shift between the two habitats, a time/depth-weighted population, mean egg development time of 10.17 days can be calculated. This is twice as long as the development time of 4.67 days that holds for a non-migrating population with a night-time vertical distribution as in Figure 3. This vertical distribution is comparable with distributions found outside the migration period. The longer egg development time, or the resulting decreased birth rate, must be less than the high predation mortality that a non-migrating *Daphnia* would experience. However, the assumption of a well-mixed population does not hold. Clonal

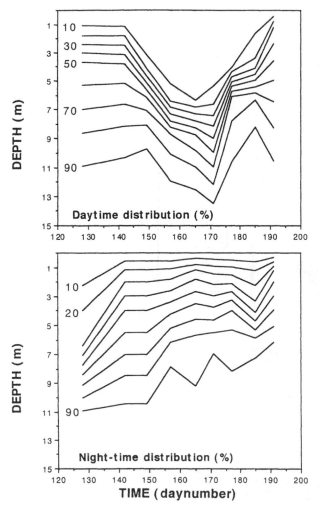

Figure 1. The seasonal period of Diel Vertical Migration in Lake Maars-seveen (Netherlands) in 1990. The isoplethes represent percentages of the total number of adult *Daphnia galeata* and the hybrid *D. galeata* × *hyalina* encountered from the water surface to the depth given by the isopleth.

composition is different at different depth but to what extent the various types stay together during vertical migration is unknown. If it is now assumed, that the individuals of each daytime depth shifted an equal number of sampling layers upward during the night – layers 7 and 8 exempted, that were both thought to migrate to the 6.25 m sampling layer – the night-time percentage distribution is approximated. These shifts in sampling layer are indicated in the right hand side column of Figure 3. The assumption implies, that all individuals (clones) have similar migration amplitudes but at different depths; this is not necessarily true. Egg development time was calculated using the polynomial function in Bottrell et al. (1976) and the temperatures of the shown curve. The values range from 2.98 days, at 1.25 m for eggs of a non-migrating individual, to 14.77 days, for eggs of a daphnid migrating from 8.75 m to 21.25 m. The reproductive rate, expressed as the number of neonates per egg leaving

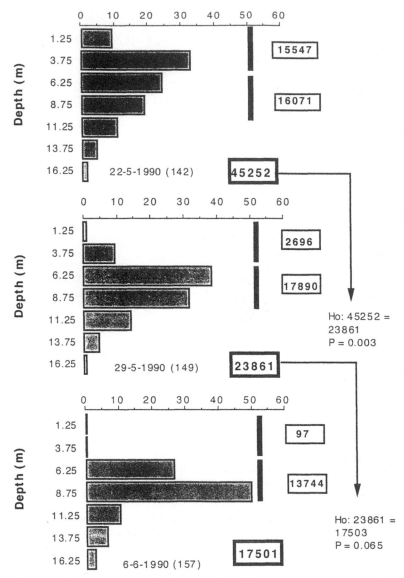

Figure 2. Vertical daytime distribution (%) of *Daphnia galeata* and the *D. galeata* × *hyalina* in Lake Maarsseveen during the first three weeks of a migration period in 1990. Framed numbers indicate total abundance in the water column below 1 m² and numbers in the 0-5 m and 5-10 m water-layers. The probalities of equal total abundances between successive dates is given. (Unpublished data by Flik & Ringelberg).

the brood chamber per day, is also given in Figure 3. The disadvantage for a clone, migrating to 21.25 m, as compared with a non-migrating one at 1.25 m, is obvious. Note that a reproductive rate for a migratory situation, ranging from 11.25 m to 1.25 m, is equal to a stationary situation at 8.75 m. In most years, during the day a high concentration of animals was found in the two sampling depths of 8.75 m and 11.25 m.

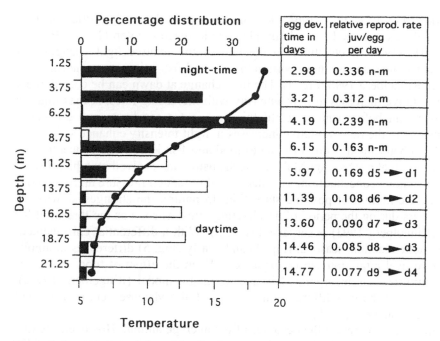

Percentage distribution			egg dev. time in days	relative reprod. rate juv/egg per day
1.25 night-time			2.98	0.336 n-m
3.75			3.21	0.312 n-m
6.25			4.19	0.239 n-m
8.75			6.15	0.163 n-m
11.25			5.97	0.169 d5 → d1
13.75			11.39	0.108 d6 → d2
16.25	daytime		13.60	0.090 d7 → d3
18.75			14.46	0.085 d8 → d3
21.25			14.77	0.077 d9 → d4

Figure 3. Some characteristics of the *Daphnia* population in Lake Maarsseveen, 10 June 1992. The vertical distribution during the night and the day is given in percentages. The curve represents the temperature profile. To calculate the egg development time for a stationary situation, temperature at the corresponding water layer was used. The relative reproductive rate for a stationary situation (n-m) and for a migrating situation is also given (the arrow indicates migration transition from a deeper layer towards a higher water layer). (Unpublished data by Flik & Ringelberg).

During the beginning of a migration period, when some *Daphnia* are still present in the epilimnion, females were collected around noon at depths of 0-5 m and 15-20 m. The first were considered non-migrating and the second migrating genotypes. Individuals from both groups were cultured as clones to study the effect of fish kairomones on life history and photobehavior. The experimental results indicate that the non-migrating genotypes are more apt to change life history characteristics (Reede-Dekker 1998), while migrating clones seem to have a stronger reaction to changes in light intensity (Van Gool & Ringelberg 1998). For instance, length at maturity in the non-migrating group became significantly smaller than in the migrating group in the presence of fish kairomone. In addition, these clones from the surface layers changed life history characteristics at a lower kairomone concentration (Reede & Ringelberg 1995) than the migrating ones.

Fish kairomone and food conditions are intertwined factors, influencing the amplitude of migration. From an adaptive point of view this is to be expected. If animals find insufficient food in the epilimnion, the probability of death by starvation might be higher than the probability of death by predation for a non-migrating individual. Assessment of both factors and attuned behavior is important. Therefore, photobehavior, thought to be at the base of diel vertical migration (Ringelberg 1964), was studied with regard to this aspect. Reactions included both a downward and an upward swimming caused by relative increases and decreases in light intensity, respectively (Clarke 1932, Ringelberg 1964). Kairomone and food have but little effect on the intensity of this so-called secondarily phototactic reaction (Van Gool & Ringelberg 1997). However, accelerations in the rate of the relative light changes can strongly enhance the response in proportion to food and fish kairomone concentration (Van Gool & Ringelberg 1998; Fig. 4). The light change at dawn can be considered a sequence of relative increases in light intensity with increasing rates. A maximum is reached 40-45 min before sunrise. Until this time, rates accelerate and especially these expansions in the rate of relative increases in light intensity enhance swimming velocity. The enhancement is proportional to food and fish kairomone concentration. It was suggested, that a 'Decision-Making-Mechanism' was evolved by which *Daphnia* assess kairomone concentration, as a measure of predation pressure and food concentration, as a criterium of possible starvation. In nature, the strongest downward migration occurs during the period of accelerating rates, as soon as the threshold for phototaxis (Ringelberg 1993) is surpassed (Fig. 5). With a differential enhancement of the phototactic reaction, seasonal (Fig. 1) and yearly (Fig. 5) differences in amplitude of diel vertical migration can be explained. When the effect of food and kairomone concentration was incorporated in a mechanistic model (Ringelberg 1995a), timing and depth of the two different migrations in Lake Maarsseveen, of 1990 and 1992, could be simulated (Fig. 5).

These simulations represent the behavior of the average animal. However, the different behavior of genotypes in the population complex is an essential aspect of DVM and of importance to understand the evolution of phototactic behavior. In Figure 6, a more detailed picture of downward migration for the two years is presented. In 1990 (Fig. 6A), a considerable part of the population remained at the lower edge of the potential predation zone. This is the area above the broken line in the figure (delimiting the maximum operating depth of the young perch), and to the right of the isoplethes of high enough light intensity to make visual predation possible. Other individuals moved into the hypolimnion. As yet, insufficient knowledge of relevant genetics is available to state that we deal with clones with different traits for phototaxis and migration. In 1992, all animals migrated much deeper (Fig. 6B). This might be ascribed to a different clone composition or to a higher kairomone concentration

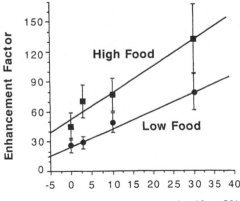

Figure 4. Enhancement of the phototactic reaction caused by accelerations in rate of the relative increase in light intensity as a function of kairomone level and food concentration. (Slightly modified from Figure 3 in Van Gool & Ringelberg 1998).

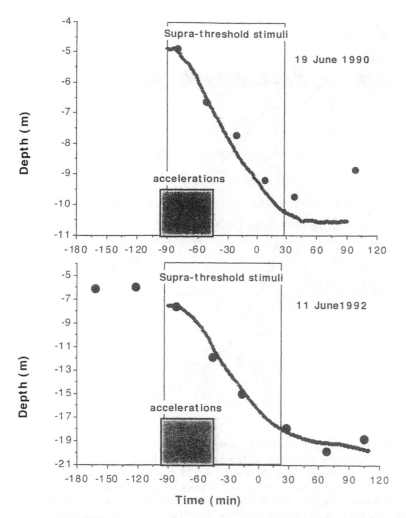

Figure 5. A comparison of two downward migrations of *Daphnia* in Lake Maarsseveen (dots) with model simulations (drawn lines). The phototactic response model was run with a similar food concentration mostly found in the lake around that time of the year. For 1989, the fish kairomone level used in the simulation was ten times lower than the one used in 1992. The natural light signal of dawn was input for the phototactic reaction. Note that the scale of the vertical axes is different. Indicated are also the periods of relative light intensity increases larger than the threshold for phototaxis and the period of accelerations in these relative increases. (Unpublished data by Flik, Ringelberg and Van Gool).

due to a possible larger year class strength of the perch, or both (Ringelberg & Flik 1994).

The clonal differences in life history and behavior, as found in the laboratory, are small and we deal with ranges of overlapping reaction norms. The ever changing combination of selecting forces slides continuously over this field of ranges and causes shifts in clonal composition. The change can be rapid in parthenogenetic animals with a high reproductive potential. In Figure 7 the composition of some allo-

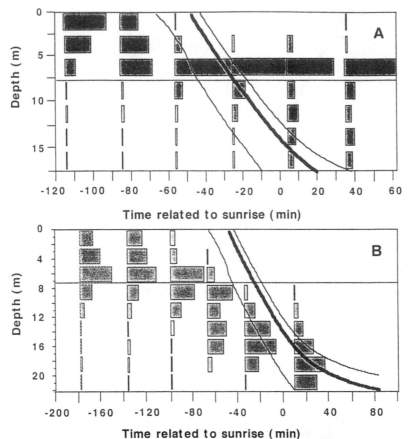

Figure 6. Early morning, downward migrations of *Daphnia* in Lake Maarsseveen on A) 30 May 1990 and B) 11 June 1992. Time is with regard to sunrise, at 5.25 h, and 5.17 h, local time, respectively. Mind different depth and time scales. The curved isolumes indicate light intensities at which 0+ perch eat 10%, 50%, 90% of *Daphnia* in an experimental situation. The horizontal line indicates depth below 0+ perch were never caught in the lake. (Unpublished data by Flik & Ringelberg).

zyme-types before, during and after DVM in Lake Maarsseveen is presented. Samples were taken around noon at 3.75 m with the exception of 11 June, when they were obtained at 21 h. This is nearly two hours before the larger part of the *Daphnia* population arrives near the surface and might be considered, more or less, a daytime situation. Within two weeks, the fairly diverse clonal composition changed to a complete dominance of the MMMF genotype at this depth. On 29 July, three weeks after the migration was over, the clonal composition seems to diversify again.

3 DISCUSSION

For an ecosystem, like Lake Maarsseveen, it is obvious that migration behavior could never become genetically fixed. Ecological costs are high and DVM does not pay in

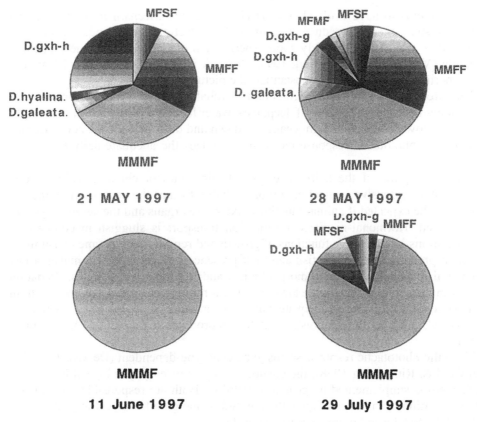

Figure 7. Species and allozyme type composition of *D. galeata*, *D. hyalina*, the *D. galeata* × *hyalina* with back crosses in Lake Maarsseveen around the period of Diel Vertical Migration lasting from the end of May until the beginning of July. (Unpublished data by Ringelberg, Van Gool & Brehm).

the absence of heavy predation. Non-migrating genotypes would take over. Seasonal studies of DVM in other lakes are scarce and a comparison is impossible. In Lake Constance, Stich studied migration for a longer period and found migration present from June until November (Stich 1989, Stich & Lampert 1981).

Photobehavior is at the base of DVM and the phototactic reaction to relative changes in light intensity seems to be of a genetically fixed character. Phototaxis to changes in light intensity has also been described for marine copepods (Johnson 1938, Stearns & Forward 1984). It probably serves another, more permanently present, goal. My guess is that it evolved to protect against harmful UV-B radiation (Hairston 1976, Siebeck et al. 1994). In most lakes, an early morning descent of 1.5-2 m would suffice (Ringelberg & Flik 1984). Superimposed on this constitutive trait evolved the enhancement of phototactic reactions triggered by accelerations in rate of relative changes in light intensity. The mechanism is induced in the presence of fish-mediated kairomones only, and the reaction intensity depends on kairomone and food concentration (Van Gool & Ringelberg 1997). We deal with a genotype-dependent, plastic response that can be considered a component of genetic specialization. It seems to have evolved as a 'Decision-Making-Mechanism' of the central nervous system, meant to reduce costs of DVM. Selection acts strongly on the cost aspect, of course. Although fairly good experimental evidence for this mechanism is available (Van Gool & Ringelberg 1997, 1998), it is still informed speculation and new experiments are needed to substantiate the picture.

The information-conveying kairomone is a reliable cue for fish presence but not for a timely retreat to the refuge of deep, dark water layers or a safe return when light intensity is low enough. The light change of dawn and dusk provide the adequate signals and the phototactic response mechanism realizes the adequate daily timing aspect.

After perception of the kairomone, phototactic behavior changes within hours, which enables extensive vertical migration probably to start the next dawn. A fast response can be expected if animal functions, like sense organs and the nervous system, are involved. Hormonal-mediated information transport is sluggish in comparison and alterations in vegetative functions of growth and reproduction is time consuming. The high predation rate of 0+ fish and their presence in large numbers makes a fast behavioral response superior to morphological and life history alterations. Predation rates of invertebrate predators are lower and the formation of spines and neck teeth in the next generation might be adequate. On the other hand, ecological costs, associated with DVM, such as the diminished relative reproduction rate (Fig. 3), are much higher.

Since the phototactic response seems to be genotype-dependent (De Meester 1993, Van Gool & Ringelberg 1998), the phenotypically induced physiological-behavioral mechanism is supplemented by genetic variability. Both are responsible for a differential migration of the population with amplitudes changing from year to year (Figs 5 and 6), lake to lake and within a season (Fig. 1).

The frequently occurring shifts in clonal dominance in the lake (Fig. 7) indicate that selecting forces are very active. DVM is a conspicuous phenomenon triggered by an outburst of predating juvenile fish. Nevertheless, the presumption that fish predation is an important, selective factor, although obvious, is not necessarily true. Juvenile perch do not selectively feed on genotypes. On the other hand, after sunset, these fish are the first to arrive in the open water zone and start feeding on non-migrating daphnids that are still present in the epilimnion. Successively, feeding is on those daphnids that migrate gradually into the danger zone. This pseudoselective predation might lead to shifts in genotype composition. An illustration of pseudoselective predation can be found in our observation that allozyme types in the stomachs of fishes caught around 23.30 h were identical to the types at the 3.75 m sampling layer in the lake, but completely different from that composition at 1.25 m at that time. Figure 7 probably gives a glimpse of changing directions within the mosaic of selecting forces. It cannot be inferred that the juvenile perch are responsible for the diminished diversity because cogent facts are absent. Juvenile perch initiate different behavior in *Daphnia* and this triggers a cascade of changing environmental circumstances. Before migration started, the individuals were more or less evenly distributed over the first 10 m (Fig. 1). During DVM, they are at low temperatures, which depress metabolism and prolong development time. They also have less food deep down in the water column. At night, they feed for a few hours near the surface and compete more heavily than before. On the other hand, the number of daphnids diminishes due to predation and, if so, competition might be released perhaps. The fact that in some years the population decreases, in other years increases, indicate that apart from fish other factors are important to determine the wax and wane of the population.

Attention was focussed on the effect of fish kairomone on photobehavior and population dynamics in *Daphnia*. However, the calanoid copepod *Eudiaptomus gracilis* starts to migrate simultaneously in the lake and we know nothing about the changing relations between both herbivores. Apart from fish kairomones, other infochemicals are involved. For instance, *Daphnia* produces a chemical substance that influences morphology in algae, thereby probably diminishing edibility (Hessen & Van Donk 1993, Lampert et al. 1995; Van Donk et al. 1999). *Daphnia* also differentiate between chemical substances produced by different algal species and is able to change its behavior accordingly (Van Gool & Ringelberg 1996). A picture unfolds of a pelagic ecosystem where various infochemicals determine behavior and feeding and thus the transfer of matter between populations. Future research will certainly reveal that food web structure and community composition is strongly influenced by information flows.

ACKNOWLEDGEMENTS

A seasonally recurring intensive sampling program, conducted from 1989 until 1998 by Ben Flik, Dick Lindenaar and Koen Royackers (Department of Aquatic Ecology, University of Amsterdam), provided the data on diel vertical migration and population dynamics. I also drew data from thesis work of Tineke Reede and Erik van Gool. The many discussions with Erik van Gool increased our understanding of the mechanism of phototaxis. The allozyme study was subsidized by the Beyerinck-Popping Fund of the Royal Netherlands Academy of Sciences and Arts. Analysis was performed by Michaela Brehm.

REFERENCES

Adler, F.R. & Harvell, C.D. 1990. Inducible defenses, phenotypic variability and biotic environment. *TREE* 5: 407-410.

Black, A.W. 1993. Predation-induced phenotypic plasticity in *Daphnia* pulex: Life history and morphological responses to *Notonecta* and *Chaoborus*. *Limnol. Oceanogr.* 38: 986-996.

Bottrell, H.H., Duncan, A., Gliwicz, Z.M., Grigierek, E., Herzig, A., Hillbricht-Ilkowska, A., Kurasawa, H., Larsson, P. & Weglenska, T. 1976. A review of some problems in zooplankton production studies. *Norw. J. Zool.* 24: 419-456.

Clarke, G.L. 1932. Quantitative aspects of the change in phototropic signs in *Daphnia*. *J. Exp. Biol.* 9: 180-211.

Cowgill, U.M., Williams, D.M. & Esquivel, J.B. 1984. Effects of maternal nutrition on fat content and longevity of neonates of *Daphnia* magna. *J. Crust. Biol.* 4: 173-190.

Cushing, D.H. 1951. The vertical migration of planktonic Crustacea. *Biol. Rev.* 26: 158-192.

De Meester, L. 1993. Genotype, fish-mediated chemicals, and phototactic behavior in *Daphnia magna*. *Ecology* 74: 1467-1474.

De Meester, L., Dawidowicz, P., Van Gool, E. & Loose, C.J. 1999. Ecology and evolution of predator-induced behavior of zooplankton: depth selection and diel vertical migration in *Daphnia*. In R. Tollrian & C.D. Harvell (eds), *The Ecology and Evolution of Inducible Defences*. Princeton: Princeton University Press.

Dicke, M. & Sabelis, M.W. 1988. Infochemical terminology: based on cost-benefit analysis rather than origin of compounds? *Functional Ecology* 2: 131-139.

Glazier, D.S. 1992. Effects of food, genotype, and maternal size and age on offspring investment in *Daphnia magna. Ecology* 73: 910-926.

Gliwicz, Z.M. & Pijanowska, J. 1988. Effect of predation and resource depth distribution on vertical migration of zooplankton. *Bull. Mar. Sci.* 43: 695-709.

Guisande, C. & Gliwicz, Z.M. 1992. Egg size and clutch size in two *Daphnia* species grown at different food levels. *J. Plankton Res.* 14: 997-1007.

Gulati, R.D. & DeMott, W.R. 1997. The role of food quality for zooplankton: remarks on the state-of-the-art, perspectives and priorities. *Freshw. Biol.* 38: 753-768.

Hairston Jr, N.G. 1976. Photoprotection by carotenoid pigments in the copepod *Diaptomus nevadensis. Proc. Nat. Acad. Sci. USA* 73: 971-974.

Hanazato, T. 1991. Influence of food density on the effects of a *Chaoborus*-released chemical on *Daphnia ambigua. Freshw. Biol.* 25: 477-483.

Harvell, C.D. 1990. The ecology and evolution of inducible defenses. *Quart. Rev. Biol.* 65: 322-340.

Havel, J.E. & Dodson, S.I. 1987. Reproductive costs of *Chaoborus*-induced polymorphism in *Daphnia pulex. Hydrobiologia* 150: 273-281.

Hessen, D. & Van Donk, E. 1993. Morphological changes in *Scenedesmus* induced by substances released from *Daphnia. Arch. Hydrobiol.* 127: 129-140.

Johnson, W.H. 1938. The effect of light on the vertical movement of *Acartia clausi* (Giesbrecht). *Biol. Bull.* 75: 106-118.

Lampert, W., Rothhaupt, K.O. & Von Elert, E. 1995. Chemical induction of colony formation in a green alga (*Scenedesmus acutus*) by grazers (*Daphnia*). *Limnol. Oceanogr.* 39: 1543-1550.

Loose, C.J., Von Elert, E. & Dawidowicz, P. 1993. Chemical-induced diel vertical migration in *Daphnia*: a new bioassay for kairomones exuded by fish. *Arch. Hydrobiol.* 126: 329-337.

Lüning, J. 1992. Phenotypic plasticity of *Daphnia pulex* in the presence of invertebrate predators: morphology and life history responses. *Oecologia* (Berlin) 92: 383-390.

Machacek, J. 1993. Comparison of the response of *Daphnia galeata* and *Daphnia obtusa* to fish-produced chemical substance. *Limnol. Oceanogr.* 38: 1544-1550.

Pyanowska, J. 1993. Diel vertical migration in zooplankton: fixed or inducible behavior? *Arch. Hydrobiol. Beih. Ergebn. Limnol.* 39: 89-97.

Reede, T. 1997. Effects of neonate size and food concentration on the life history responses of a clone of the hybrid *D. galeata* × *hyalina* to fish kairomones. *Freshw. Biol.* 37: 389-396.

Reede-Dekker, T. 1998. Life-history adaptations of *Daphnia* during a diel vertical migration period. Ph-D Thesis, University of Amsterdam.

Reede, T. & Ringelberg, J. 1995. The influence of a fish exudate on two clones of the hybrid *D. galeata* × *hyalina. Hydrobiologia* 37: 207-212.

Ringelberg, J. 1964. The positively phototactic reaction of *Daphnia magna* Straus: A contribution to the understanding of diurnal vertical migration. *Neth. J. Sea Res.* 2: 319-406.

Ringelberg, J. 1993. Phototaxis as a behavioral component of diel vertical migration in a pelagic *Daphnia. Arch. Hydrobiol. Beih. Ergebn. Limnol.* 39: 45-55.

Ringelberg, J. 1995a. An account of a preliminary mechanistic model of swimming behavior in *Daphnia*: its use in understanding diel vertical migration. *Hydrobiologia* 307: 161-165.

Ringelberg, J. 1995b. Changes in light intensity and diel vertical migration: a comparison of marine and freshwater environments. *J. Mar. Biol. Assoc. UK* 75: 15-25.

Ringelberg, J. & Flik, B.J.G. 1984. The mortality effect of ultraviolet radiation in a translucent and in a red morph of *Acanthodiaptomus denticornis* (Crustacea, Copepoda) and its possible ecological relevance. *Hydrobiologia* 112: 217-222.

Ringelberg, J. & Flik, B.J.G. 1994. Increased phototaxis in the field leads to enhanced diel vertical migration. *Limnol. Oceanogr.* 39: 1855-1864.

Segal, E. 1970. Light. In O. Kinne (ed.), *Marine ecology*: 159-211. London: Wiley.

Siebeck, O., Vail, T.L., Williamson, C.E., Vetter, R., Hessen, D., Zagarese, H. & Little, E. 1994. Impact of UV-B radiation on zooplankton and fish in pelagic freshwater ecosystems. *Arch. Hydrobiol. Beih. Ergebn. Limnol.* 43: 101-114.

Spaak, P. & Hoekstra, J.R. 1993. Clonal structure of the *Daphnia* population in Lake Maarsseveen: its implications for diel vertical migration. *Arch. Hydrobiol. Beih. Ergebn. Limnol.* 39: 157-165.

Spitze, K. 1992. Predator mediated plasticity of prey life history and morphology: *Chaoborus americanus* predation on *Daphnia pulex*. *Amer. Nat.* 139: 229-247.

Stearns, D.E. & Forward Jr, R.B. 1984. Photosensitivity of the calanoid copepod *Acartia tonsa*. *Mar. Biol.* 82: 85-89.

Stibor, H. 1992. Predator induced life-history shifts in a freshwater cladoceran. *Oecologia* 92: 162-165.

Stich, H.-B. & Lampert, W. 1981. Predator evasion as an explanation of diurnal vertical migration. *Nature* 293: 396-398.

Stich, H.B. 1989. Seasonal changes of diel vertical migrations of crustacean plankton in Lake Constance. *Arch. Hydrobiol.* 83: 355-405.

Swain, W.R., Lingeman, R. & Heinis, F. 1987. A characterization and description of the Maarsseveen Lake system. *Hydrobiol. Bulletin* 21: 5-16.

Tessier, A.J. & Consolatti, N.L. 1989. Variation in offspring size in *Daphnia* and consequences for individual fitness. *Oikos* 56: 269-276.

Tessier, A.J. & Consolatti, N.L. 1991. Resource quality and offspring quality in *Daphnia*. *Ecology* 72: 468-478.

Tollrian, R. 1995. Predator-induced morphological defenses: costs, life history shifts, and maternal effects in *Daphnia pulex*. *Ecology* 76: 1691-1705.

van Donk, E., Lürling, M. & Lampert, W. 1999. Consumer-induced changes in phytoplankton: inducibility, costs, benefits and the impact on grazers. In R. Tollrian & C.D. Harvell (eds), *The Ecology and Evolution of Inducible Defences*. Princeton: Princeton University Press.

van Gool, E. & Ringelberg, J. 1996. Daphnids respond to algae-associated odors. *J. Plankton Res.* 18: 197-202.

van Gool, E. & Ringelberg, J. 1997. The effect of accelerations in light increase on the phototactic downward swimming of *Daphnia* and the relevance to diel vertical migration. *J. Plankton Res.* 19: 2041-2050.

van Gool, E. & Ringelberg, J. 1998. Light-induced migration behavior of *Daphnia* modified by food and predator kairomones. *Animal Behavior* 56: 741-747.

Vuorinen, I., Ketola, M. & Walls, M. 1989. Defensive spine formation in *Daphnia pulex* Leydig and induction by *Chaoborus crystallinus* De Geer. *Limnol. Oceanogr.* 34: 245-248.

Weers, P.M.M. & Gulati, R.D. 1997. Effect of the addition of polyunsaturated fatty acids to the diet on the growth and fecundity of *Daphnia galeata*. *Freshw. Biol.* 38: 721-729.

Population characteristics and feeding parameters of *Aristaeomorpha foliacea* and *Aristeus antennatus* (Decapoda: Aristeidae) from the Ionian Sea (Eastern Mediterranean)

K. KAPIRIS, M. THESSALOU-LEGAKI & M. MORAITOU-APOSTOLOPOULOU
Department of Zoology-Marine Biology, School of Biology, University of Athens, Greece

G. PETRAKIS & C. PAPACONSTANTINOU
National Centre for Marine Research, Institute of Marine Biological Resources, Athens, Greece

ABSTRACT

Aristaeomorpha foliacea and *Aristeus antennatus* were collected from deep waters (473-728 m) of the SE Ionian Sea in March, April and May 1997. Size distributions, sex ratio, percentage of empty stomachs, fullness indices (Hureau's FI, % BW Wet, % BW Dry) and food quality indices (% DW and % AFDW of stomach contents) from 577 specimens of *A. foliacea* and 363 specimens of *A. antennatus* were examined. Both sexes of *A. foliacea* showed an increased size compared to that of *A. antennatus*. In both sympatric species, females were larger than males. Numerical predominance of females over males increased towards the end of spring in *A. antennatus*, while both sexes were equally represented in the three surveys in *A. foliacea*. In most cases, no difference was found between sexes of both species in the percentage of empty stomachs of individuals of the same month and depth zone. In addition, when pooled data were used, the values of the fullness indices did not differ between sexes of each species. In *A. antennatus* both sexes showed an increased percentage of empty stomachs in the deeper zone. The comparison of fullness indices proved that both sexes of *A. foliacea* were better fed than in *A. antennatus*. Nevertheless, *A. foliacea* fullness indices decreased from March to May, while depth did not affect the stomach fullness. On the other hand, in males of *A. antennatus* no consistent pattern was evident for the effect of the two factors, although in some cases a significant effect has been obtained. Neither time or depth significantly affected the fullness indices of the females of the latter species. Regarding food quality indices, both sexes of *A. foliacea* and males of *A. antennatus* showed similar % DW and % AFDW of stomach contents, while the females of *A. antennatus* exhibited significantly higher % DW and lower % AFDW in their stomach contents. Finally, the % organic matter in the diet of *A. foliacea* did not change with depth or time, whereas significant variations according to depth and time were observed in *A. antennatus*.

1 INTRODUCTION

Aristeid shrimps (Decapoda: Penaeoidea) include tropical and subtropical deep-water species with a wide geographical distribution. The rose shrimp *Aristeus antennatus*

177

occurs in the Eastern Atlantic from Portugal to Cape Verde Islands and Azores, the Mediterranean and the Indian Ocean (Pérez Farfante & Kensley 1997). In the Western Mediterranean its depth range extends from 300 to 2200 m (Sardà et al. 1994). It is a dominant benthic species of the upper and middle continental slope (Cartes & Sardà 1992, 1993, Sardà & Cartes 1993a, b). Because of its great ecological and economical interest, a wealth of information exists on the ecology and biology of the species from the Western and Central Mediterranean (Relini Orsi & Relini 1979, Sardà & Demestre 1987, Demestre & Fortuño 1992, Ragonese & Bianchini 1992, Bianchini & Ragonese 1994 and the references within, Sarda & Cartes 1997, Sardà et al. 1997) as well as on population dynamics and fisheries aspects (Relini & Orsi Relini 1987, Tobar & Sardà 1987, Sardà 1993, Demestre & Lleonart 1993, Demestre & Martín 1993, Sardà et al. 1994, Ragonese & Biancini 1996). *A. antennatus* exhibits a large-scale, seasonally, well-defined, mobility pattern that results in an important sex and size segregation: males and juveniles inhabit marine canyons during autumn and early winter whereas adult females are present in the upper and middle slope as well as canyons all year around (Sardà et al. 1997).

The red shrimp *Aristaeomorpha foliacea* exhibits a wider distribution in the Western and Eastern Atlantic, the Mediterranean and the Indo-Pacific (Pérez Farfante & Kensley 1997). It has been reported from 150 to 1850 m depth with a peak in abundance in the range 300-700 m. This species is, or has been, heavily exploited in the Western Mediterranean, and is currently fished in the Central Mediterranean; in the Eastern Mediterranean its stocks are pristine (Bianchini & Ragonese 1994). The relative importance of the two aristeids in deep-water fisheries changes on the west-east axis of the Mediterranean with an increasing importance of *A. foliacea* in the Central Mediterranean (Ragonese 1995). Although scanty information exists on the occurrence of both species in the Eastern Mediterranean, records suggest that in the area *A. foliacea* is more common than *A. antennatus* (Thessalou-Legaki 1994, Mytilineou & Politou 1997). In contrast to *A. antennatus*, little is known about the red shrimp in the Mediterranean. Most information comes from Italian waters (Cau et al. 1982, Spedicato et al. 1994, Ragonese et al. 1994, Ragonese 1995, Ragonese & Biancini 1995, as well as reports in Bianchini & Ragonese 1994).

As deep-water shrimps can be regarded as an important link in the food web of the deep-water Mediterranean ecosystem, the study of their feeding ecology could give evidence on the resource partitioning by the two sympatric species. The study of the feeding ecology of *A. antennatus* started with listing of identified prey organisms in the stomach contents (Maurin & Carries 1968, Lagerdère 1972, Relini Orsi & Wurtz 1977) and evolved to intraspecific comparisons of dietary aspects in relation to factors such as sex, season, depth of occurrence and daily ration (Cartes & Sardà 1989, Cartes 1994, Maynou & Cartes 1997) as well as to interspecific comparisons of deep-water decapod feeding strategies (Cartes, 1998). On the other hand, the feeding ecology of *A. foliacea* was not thoroughly studied until now. Information on the diet of this shrimp is given in Maurin & Carries (1968), Lagardère (1972) Gristina et al. (1992), Pipitone et al. (1994) and Cartes (1995).

In an attempt to extend the knowledge of the two species in the Eastern Mediterranean, the first data on these deep-water shrimps from the SE Ionian Sea are presented in the present paper. Priority was given to the feeding parameters as predation was regarded as a key factor in the oligotrophic waters of the region.

2 MATERIALS AND METHODS

Shrimps were collected in the SE Ionian Sea (Greece), between the Peloponnisos and Zakynthos Islands in March, April and May 1997 (Fig. 1). Experimental 1h daytime hauls were performed by a commercial bottom trawler equipped with a net of 16 mm mesh cod end. As the depth range of the surveys was 280-400 f, three depth zones were established for the presentation of the results as zone 1: < 300 f (473-546 m), zone 2: 301-350 f (548-637 m) and zone 3: 351-400 f (638-728 m). During the three surveys a total of 24 hauls were made. Both species were found together in 67% of the hauls.

A total of 363 individuals of *A. antennatus* (125 males and 238 females) and 577 individuals of *A. foliacea* (273 males and 304 females) were caught. Sex ratio was calculated as (males/females) × 100. The material was kept at +4°C on board and at –30°C after landing. In the laboratory specimens were thawed in batches, separated by species and sexed. After measuring carapace length (CL, in mm) and body wet weight (BWW), the shrimps were dissected and their stomach contents were weighted as wet (SWW), dry (SDW, at 70°C for 24 h) and ash free dry weight

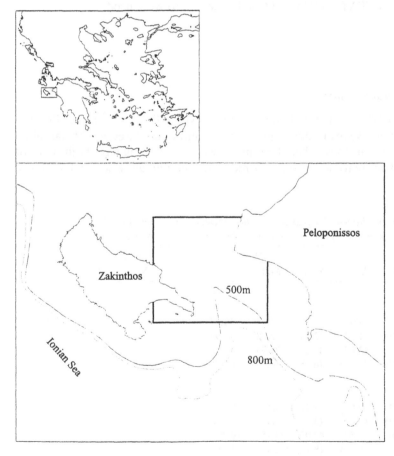

Figure 1. Map of the sampling area.

(SAFDW, as loss on ignition at 450°C for 3 h). All weights were measured with an accuracy of 0.0001 g.

For comparison, three stomach fullness indices were estimated:

a) Hureau's (1969) % of filling FI = (SWW/W) × 100 where SWW is the wet weight (g) of the stomach contents and W is body wet weight (g) of shrimp, without the stomach contents,

b) % BW Wet = (SWW/ BWW) ×100,

c) % BW Dry = (SDW/ BWW) × 100 (Héroux & Magnan 1996).

In order to approach the food quality aspects of the two sympatric aristeid species, two expressions of food quality were taken into account: the % DW = (SDW/SWW) × 100, and the % AFDW = (SAFDW/SDW) × 100.

Length-weight relationships (CL-BWW) for each species and sex were established using linear regression (least squares method) after $\log_{10}$ transformation according to the model $W = a(CL)^b$. The effect of sampling time and depth was tested using two-way ANOVA, while comparisons between samples were made using one way ANOVA. Relationships between two parameters were tested using correlation coefficient (r) and comparisons of correlation coefficients were performed using Z statistic. Comparisons of empty stomachs frequencies and testing of divergence of the sex-ratio from 1:1 were made by chi-square tests with Yates correction. For the statistical analysis the packages STATISTICA and Corel Quattro Pro were used.

3 RESULTS

3.1 *Population characteristics*

The mean CL and BWW during each survey are shown in Table 1 and size distribution of both species and sexes is shown in Figure 2. During the three-month sampling period, the two species and sexes had the same size distribution pattern: both sexes of *A. foliacea* showed an increased size compared to *A. antennatus*, whereas females in

Table 1. *Aristaeomorpha foliacea, Aristeus antennatus*: Mean carapace length (CL), mean body wet weight (BWW) (SD in parenthess) and sex ratio from the three surveys in the SE Ionian Sea.

Species	CL			BWW			Sex ratio		
	March	April	May	March	April	May	March	April	May
Aristaeomorpha foliacea:									
– Male	34.02	33.97	34.12	14.17	12.33	14.67	101.77	64.10	101.92
	(3.73)	(1.97)	(2.78)	(2.78)	(2.92)	(2.42)			
– Female	45.50	43.67	45.89	29.95	24.26	31.77			
	(5.03)	(4.46)	(6.25)	(8.55)	(7.57)	(9.54)			
Aristeus antennatus:									
– Male	27.34	24.61	27.47	8.50	6.77	8.39	37.07	11.57	10.58
	(2.92)	(1.64)	(3.57)	(1.99)	(1.12)	(2.27)			
– Female	40.40	37.53	42.07	24.15	19.40	31.63			
	(8.51)	(6.24)	(8.15)	(12.15)	(7.73)	(10.85)			

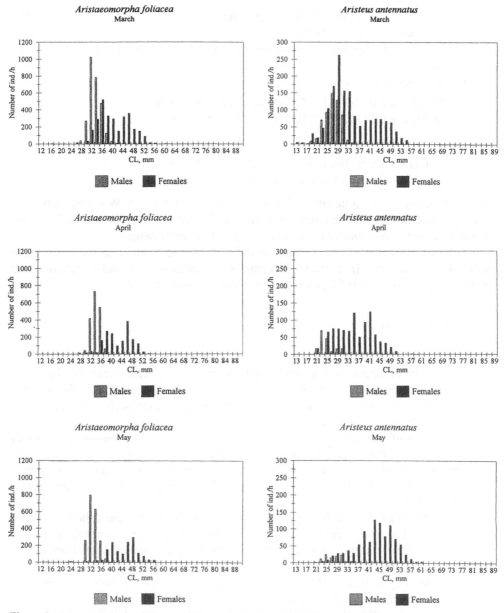

Figure 2. *Aristaeomorpha foliacea, Aristeus antennatus*: CL frequency distribution in the three surveys.

both sympatric species were larger than males. In general, CL of *A. foliacea* ranged from 25.38 to 62.49 mm in females and from 22.48 to 41.60 mm in males. In *A. antennatus* the CL range was 19.52-62.38 mm and 20.14-35.59 mm for females and males respectively.

Body wet weight showed a similar variability as the CL (Table 1). The BWW of *A. foliacea* females ranged from 5.36 to 56.27 g and from 8.66 to 23.81 g for males, whereas in *A. antennatus* the BWW varied from 21.31 to 57.23 g and 4.11 to 13.29 g for females and males respectively.

The CL/BWW regression equation coefficients for the two species from all surveys were:

Aristaeomorpha foliacea:

   females a = 0.00232 b = 2.47035 r^2 = 0.84, N = 306;
   males a = 0.01102 b = 2.02314 r^2 = 0.74, N = 226

Aristeus antennatus:

   females a = 0.00264 b = 2.43867 r^2 = 0.96, N = 226;
   males a = 0.00745 b = 2.12146 r^2 = 0.89, N = 126

In all species-sex cases a negative allometry was evident. CL-BWW correlation coefficients were higher in females of both species (Z = 3.0897 and 2.2925 for *A. foliacea* and *A. antennatus* respectively, $p < 0.05$ in both cases).

The sex ratio (Table 1) in *A. foliacea* did not diverge from equality in the three surveys even in April ($p > 0.05$). In contrast, *A. antennatus* females outnumbered males significantly ($p < 0.05$), increasing towards the end of spring.

3.2 Feeding parameters

3.2.1 Percentage of empty stomachs

The occurrence of empty stomachs (only totally empty stomachs were taken into consideration) in both species and sexes is shown in Figure 3. In general *A. foliacea* exhibited a smaller variation in the percentage of empty stomachs than *A. antennatus* (10-33% and 0-38% respectively).

Comparison of time variations in *A. foliacea* showed that the frequency of empty stomachs was equal between the two sexes during each month ($p > 0.05$). Nevertheless, the increase of the empty stomachs in April, compared to the other two months, was significant ($p < 0.05$) only in females possibly because of the small number of males collected.

In *A. antennatus*, female frequencies were statistically higher than those of males only in April. The time variations (Fig. 3A) of the female empty stomach frequency did not proved to be significant ($p > 0.05$), whereas males showed significant higher percentage of empty stomachs in May ($p < 0.05$).

Regarding depth (Fig. 3B), no differences were detected between sexes of *A. foliacea* from the same depth, nor between the three depth zones in each sex (in all comparisons $p > 0.05$). No differentiation ($p > 0.05$) was also exhibited among the two sexes of *A. antennatus* in each depth zone, while both sexes showed a statistically higher frequency of occurrence of empty stomachs in the deeper zone ($p < 0.05$).

3.2.2 Fullness indices

No significant correlation existed between CL and stomach fullness indices in both sexes of the two species ($p > 0.05$).

One-way ANOVA (Scheffe test, $p < 0.05$) of pooled data from all surveys, showed both sexes of *A. foliacea* were more well fed than those of *A. antennatus* in all three stomach fullness indices examined (mean values in Table 2) except for the compari-

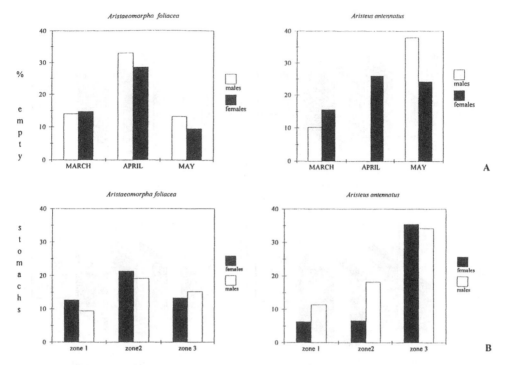

Figure 3. *Aristaeomorpha foliacea, Aristeus antennatus*: Frequency of empty stomachs according to sampling time (A) and depth (B).

son of % BW Dry in males where the increased value in *A. foliacea* was not statistically significant. Comparing the two sexes of the same species no difference was significant in either species ($p > 0.05$).

Regarding the effect of depth and sampling time (Fig. 4), a two-way ANOVA (Table 3) showed that *A. foliacea* fullness indices were significantly affected by sampling period but not by depth. In *A. antennatus* males on the other hand, no consistent pattern is evident for the effect of the two factors, which in some cases proved to be significant, whereas female fullness indices were not significantly affected by sampling time or depth.

3.2.3 *Food quality indices*

One-way ANOVA (Scheffe test, $p < 0.05$) of pooled data from all surveys (Table 2), showed that the females of *A. antennatus* had the highest % DW of stomach contents (51%) that was statistically different from all other species-sex combinations. The males of this species formed a group of not statistically different means with the two sexes of *A. foliacea*. The same comparisons for the % AFDW of stomach contents proved that the above mentioned increase of % DW of stomach contents in *A. antennatus* females corresponded with the lowest % AFDW (48.7%), significantly different from the rest species-sex combinations. The males of the same species showed an increased mean % AFDW (58.0%), not significantly different from both sexes of *A. foliacea*.

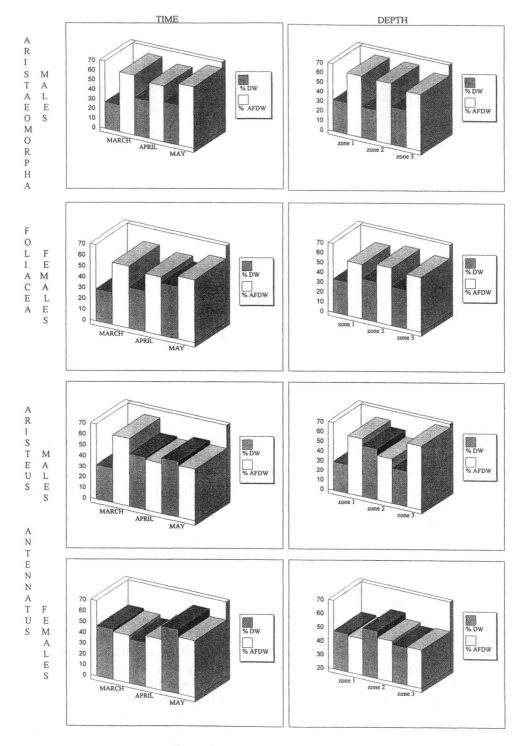

Figure 4. *Aristaeomorpha foliacea, Aristeus antennatus*: Mean values of stomach fullness indices (FI, % BW Wet, % BW Dry) according to sampling time and depth.

Table 2. *Aristaeomorpha foliacea, Aristeus antennatus*: Mean values of stomach fullness (FI, % BW Wet, % BW Dry) and quality indices (% DW, % AFDW).

Species	FI	% BW Wet	% BW Dry	% DW	% AFDW
Aristaeomorpha foliacea:					
– Male	10.30	1.010	0.281	33.54	62.39
– Female	0.978	0.967	0.305	37.75	56.94
Aristeus antennatus:					
– Male	0.716	0.731	0.219	38.76	57.99
– Female	0.514	0.518	0.215	50.98	48.71

Table 3. *Aristaeomorpha foliacea, Aristeus antennatus*: Values of *F*-ratio of the two-way ANOVA tests for the effect of sampling time and depth on the three fullness indices (* = P < 0.05, ** = P < 0.01, *** = P < 0.001, NS = non significant).

Species		Month	Depth
Aristaeomorpha foliacea:			
– Male	FI	16.20^{***}	1.46^{NS}
	% BW Wet	16.43^{***}	1.51^{NS}
	% BW Dry	6.16^{**}	0.41^{NS}
– Female	FI	8.46^{***}	0.21^{NS}
	% BW Wet	7.80^{**}	0.11^{NS}
	% BW Dry	3.29	0.05^{NS}
Aristeus antennatus:			
– Male	FI	0.61^{NS}	5.95^{**}
	% BW Wet	0.14^{NS}	5.95^{**}
	% BW Dry	7.97^{***}	0.58^{NS}
– Female	FI	0.52^{NS}	1.11^{NS}
	% BW Wet	1.22^{NS}	1.42^{NS}
	% BW Dry	2.86^{NS}	1.59^{NS}

Table 4. *Aristaeomorpha foliacea, Aristeus antennatus*: Values of F-ratio of two-way ANOVA tests for the effect sampling time and depth on % DW and % AFDW of stomach contents (* = P < 0.05, ** = P < 0.01, *** = P < 0.001, NS = non significant).

Species	% DW		% AFDW	
	Month	Depth	Month	Depth
Aristaeomorpha foliacea:				
– Male	27.92^{***}	3.63^{**}	0.58^{NS}	1.05^{NS}
– Female	41.33^{***}	3.03^{**}	0.86^{NS}	1.06^{NS}
Aristeus antennatus:				
– Male	10.04^{***}	18.91^{***}	13.09^{***}	10.56^{***}
– Female	6.45^{***}	3.65^{**}	5.15^{***}	0.04^{NS}

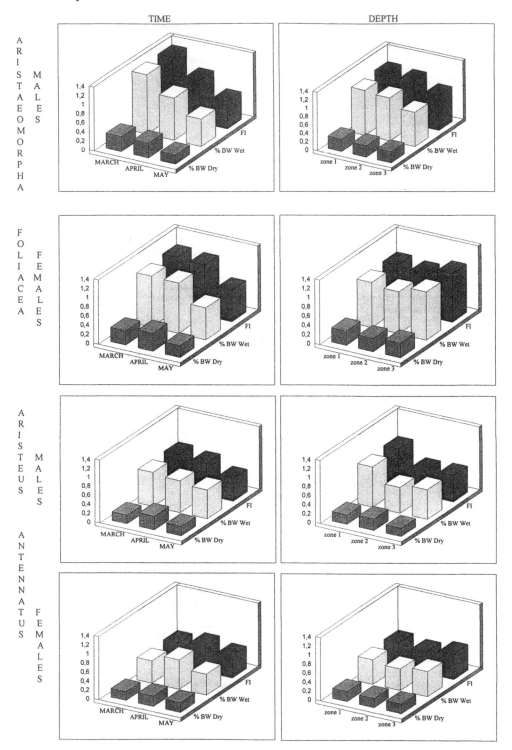

Figure 5. *Aristaeomorpha foliacea, Aristeus antennatus*: Mean values of prey quality indices (% DW, % AFDW) according to sampling time and depth.

Regarding depth and sampling time (Fig. 5), a two-way ANOVA (Table 4) showed significant effects of these two factors on the % DW of stomach contents in both sexes of the two aristeids. On the other hand, the organic content of prey, % AFDW, proved not to be affected by depth of occurrence and sampling time for both sexes of *A. foliacea*, while significant variations according to depth and sampling period were observed in *A. antennatus* in most cases (Table 4).

4 DISCUSSION

The present study provides first data on two sympatric aristeids in the Eastern Mediterranean. Because of the limitations in space and time (small sampling depth range especially for *A. antennatus,* and three surveys during the spring months) the extrapolation of the bio-ecological aspects already observed in the Western Mediterranean is not feasible at the moment. Nevertheless, some tentative comparisons can be made for the population characteristics of the two species from the present area and the Western and Central Mediterranean.

Size ranges of both sexes of the two species in the present study fall within the ranges given in the literature for similar depths (see reports in Biancini & Ragonese 1994). The wider size range usually reported could be attributed to seasonal sampling over more than one year. Nevertheless, when the same sampling period is considered, a well-defined mode of recruits is evident in *A. foliacea* females from the Central Mediterranean (Ragonese et al. 1994, Matarrese et al. 1997) totally missing from our samples. However, Ragonese (1995) reported from the Cicilian Channel that these small-sized individuals were irregularly represented between locations and spring time surveys of different years.

The size distributions in favor of females that were observed in both species during the present study, are considered as prominent features of the two aristeids. The species are markedly sexually dimorphic (Sardà & Demestre 1987, Arrobas & Ribeiro-Cascalho 1987, Demestre & Lleonart 1993, Ragonese 1995, Sardà et al. 1997, Ragonese et al. 1997). In particular males of *A. foliacea* – a species that showed larger size difference between sexes in our surveys – grow linearly and not curvilinearly with age after sexual maturity (Ragonese et al. 1994).

In the Eastern Mediterranean, sampling depth range, females of *A. antennatus* dominated quantitatively, their importance increased towards the end of spring. As already demonstrated from the Western Mediterranean, *A. antennatus* follows a well defined seasonal mobility pattern. It is more abundant on the open slope areas (800 m) in late winter and spring, while later in the end of summer and during autumn, a large part of the population moves to lower depths in the canyons. Important sex and size segregations exist in adult females that are present all year around in different habitats, whereas the abundance of males and juveniles is higher in canyons in autumn and early spring. Females dominate the population all-year around, but form aggregations on the middle slope in spring and summer, this period coinciding with ripening of the gonads (Relini Orsi & Relini 1979, Arrobas & Ribeiro-Cascalho 1987, Sardà & Demestre 1987, Demestre & Fortuño 1992, Sardà et al. 1994, 1997). Therefore, increasing female dominance in the Eastern Mediterranean towards the end of spring conforms with the above pattern.

In *A. foliacea* on the other hand, most references support the equal presence of the two sexes (Spedicato et al. 1994, Ragonese 1995) as was found with the present surveys.

Also consistent with other studies (Arrobas & Ribeiro-Cascalho 1987, Ragonese et al. 1997) is the negative allometry in length-weight relationships as observed in both sexes of the two aristeids.

For both species no difference in empty stomachs percentage was, in most cases, evident between sexes in the same month or depth zone. In addition, when pooled data were used, the values of the fullness indices examined did not differ between sexes of the same species. Cartes & Sarda (1989) also found no difference in stomach fullness in the two sexes of *A. antennatus* from the Western Mediterranean.

The highest percentage of empty stomachs of *A. foliacea* observed in the present study was 33%. In a two-year study in the Catalan Sea, Cartes (1995) found a relatively higher mean percentage of empty stomachs (41.5%) in similar depths to those of the present study. Nevertheless, no direct comparison can be made as the latter value has been derived considering as empty stomachs those between 0-25% fullness. In *A. antennatus* from SE Ionian Sea, the percentage of empty stomachs increases with depth (maximum 35.4%). In the same species in the Catalan Sea a 22.7% of the individuals exhibited stomach fullness lower than 20% at the same depth range. Relatively more empty stomachs and less numerous prey items were found in deeper waters (1000-1900 m) (Cartes 1994).

The comparison of the fullness indices of females of the two sympatric species proved that the stomach fullness of *A. antennatus* is reduced compared to that of *A. foliacea*. Moreover, their food contains less organic matter than that of *A. foliacea*. Qualitative food studies in the two species have shown that *A. foliacea* has the most highly specialised diet among the deep-water penaeoideans in the W Mediterranean. It preys on pelagic or suprabenthic resources (Cartes 1995). *A. antennatus* on the other hand, has a highly diversified diet consisted mainly of endobenthic and epibenthic invertebrates (Cartes & Sarda 1989, Cartes 1994). As time of feeding is concerned, Lagardere (1972) suggests that *A. foliacea* feeds mainly during the day, while Rainer (1992) considers the species both diurnal and nocturnal feeder. A slight dusk-night feeding peak has been found in *A. antennatus* (Maynou & Cartes 1997). Consequently, the above-mentioned differences in stomach content quantity and quality of the present study – which was conducted during day-time – can be attributed to the different feeding strategies of the two sympatric species.

In both sexes of *A. foliacea*, the fullness indices have been significantly reduced from the beginning to the end of spring, while % DW of the stomach contents increased. Nevertheless, animals seemed to keep the quality of their food (organic matter) stable. It is worth to mention that May represents in SE Ionian the beginning of the maturation period (unpubl. data). It could be suggested that the reduction of stomach fullness in May might be related to the onset of the reproductive period. On the other hand, depth seems to affect only the % DW of stomach contents and not the fullness indices or the % AFDW of *A. foliacea*. No clear pattern of the effects of depth and time existed in *A. antennatus* stomach fullness and % AFDW of stomach contents. On the contrary, the % DW of stomach contents showed a similar to *A. foliacea* increase with time.

Although this study presents the first preliminary data on feeding of the two sympatric aristeids in SE Ionian Sea, some indications derived which revealed to a relative predominance of *A. foliacea* as quantitative feeding parameters are concerned. Extension of the observations over time and supplementary studies on qualitative feeding aspects are in progress in order to elucidate the alimentary strategies and provide evidence for comparisons with the W Mediterranean populations.

ACKNOWLEDGMENTS

We thank Mr S. Kavadas, NCMR for his assistance in the map presentation, the captain and the crew of the trawler 'Panagia Faneromeni II' for their collaboration. The surveys were part of the EU FAIR Project CT 95-0655 'Developing deep-water fisheries'. Thanks are also due to the anonymous reviewers for their critical suggestions.

REFERENCES

Arrobas, I. & Ribeiro-Cascalho, A. 1987. On the biology and fishery of *Aristeus antennatus* (Risso, 1816) in the south Portuguese coast. *Inv. Pesq.* 51 (Supl. 1): 233-243.

Bianchini, M.L. & Ragonese, S. (eds) 1994. *Life cycles and fisheries of the deep-water red shrimps* Aristaeomorpha foliacea *and* Aristeus antennatus. *Proceedings of the International workshop held in the Istituto di tecnologia della pesca e del pescato. Mazara del Vallo: N.T.R.-I.T.P.P. Special Publication N. 3*, 87 pp.

Cartes, J.E. 1994. Influence of depth and season on the diet of the deep-water aristeid *Aristeus antennatus* along the continental slope (400 to 2300 m) in the Catalan Sea (Western Mediterranean). *Mar. Biol.* 120: 639-648.

Cartes, J.E. 1995. Diets of, trophic resources exploited by, bathyal penaeoidean shrimps from the Western Mediterranean. *Mar. Freshw. Res.* 46: 889-96.

Cartes, J.E. 1998. Feeding strategies and partition of food resources in deep-water decapod crustaceans (400-2300 m). *J. Mar. Biol. Ass. UK* 78: 509-524.

Cartes, J.E. & Sardà, F. 1989. Feeding ecology of the deep-water aristeid crustacean *Aristeus antennatus. Mar. Ecol. Prog. Ser.* 54: 229-238.

Cartes, J. & Sardà, F. 1992. Abundance and diversity of decapod crustaceans in the deep-Catalan Sea (Western Mediterranean). *J. Nat. Hist.* 26: 1305-1323 .

Cartes, J. & Sardà, F. 1993. Zonation of deep-sea decapod fauna in the Catalan Sea (Western Mediterranean). *Mar. Ecol. Prog. Ser.* 94: 27-34.

Cau, A., Deiana, A.M. & Mura, M. 1982. Nuovi dati sull' accrescimento e sulla maturita sessuale di *Aristaeomorpha foliacea* (Decapoda Penaeidae). *Naturalista Sicil.* IV, 6: 429-434.

Demestre, M. & Fortuño, J.M. 1992. Reproduction of the deep-water shrimp *Aristeus antennatus* (Decapoda: Dendrobranchiata) in the Northwestern Mediterranean. *Sci. Mar.* 57: 183-189.

Demestre, M. & Lleonard, J. 1993. The population dynamics of *Aristeus antennatus* (Decapoda: Dendrobranchiata). *Mar. Ecol. Prog. Ser.* 84: 41-51.

Demestre, M. & Martín, P. 1993. Optimum exploitation of a demersal resource in the Western Mediterranean: the fishery of the deep-water shrimp *Aristeus antennatus* (Risso, 1816). *Sci. Mar.* 57: 175-182.

Gristina, M., Badalamenti, F., Barbera, G., D' Anna, G., & Pipitone, C. 1992. Preliminary data on the feeding habits of *Aristaeomorpha foliacea* (Risso) in the Sicilian Channel. *Oebalia* Suppl. XVII: 143-144.

Héroux, D. & Magnan, P. 1996. In situ determination of food daily ration of fish: review and field evaluation. *Environ. Biol. Fish.* 46: 61-74.

Hureau, J.C. 1969. Biologie comparée de quelques poissons antarctiques (Nototheneidae). *Bull. Inst. Océanogr. Monaco* 68: 1-131.

190   *K. Kapiris et al.*

Lagardère, J.P. 1972. Reserches sur l' alimentation des crevettes de la pente continentale Maro-
caine. *Tethys* 3: 655-675.

Matarrese, A., D'Onghia, G., Tursi, A. & Maiorano P. 1997. Vulnerabilita e resilienza di *Aristaeomorpha foliacea* (Risso, 1827) e *Aristeus antennatus* (Risso, 1816) (crostacei, decapodi) nel Mar
Ionio. *S. It. E. Atti* 18: 535-538.

Maurin, C. & Carries, C. 1968. Note préliminaire sur l' alimentation des crevettes profondes. *Rapp.
Comm. Int. Mer Médit.* 19: 155-156.

Maynou, F. & Cartes, J. 1997. Field estimation of daily ratio in deep-sea shrimp *Aristeus antennatus* (Crustacea: Decapoda) in the Western Mediterranean. *Mar. Ecol. Prog. Ser.* 153: 191-196.

Mytilineou, C. & Politou, C-Y. 1997. New references for the presence of the red shrimp *Aristaeomorpha foliacea* in the Greek waters. *Fifth Hellenic Symposium on Oceanography and Fisheries
held in Kavala, Greece, April 15-18, 1997*, Vol. 2: 87-89.

Perez Farfante, I. & Kensley, B. 1997. Penaeoid and sergestoid shrimps and prawns of the world.
Keys and diagnoses for the families and genera. *Memoires du Museum National d' Histoire
Naturelle*. Tome 175 – Zoologie, Paris, 233 pp.

Pipitone, C. & Andaloro, F. 1994. First observations of *Aristaeomorpha foliacea* and *Aristeus antennatus* along the Eastern Sicily coast. In M.L. Bianchini & S. Ragonese (eds), *Life cycles and
fisheries of the deep-water red shrimps* Aristaeomorpha foliacea *and* Aristeus antennatus: *Proceedings of the International workshop held in the Istituto di Tecnologia della pesca e del pescato (ITTP-CNR), Mazara del Vallo, Italy, 28-30 April.*

Ragonese, S. 1995. Geographical distribution of *Aristaeomorpha foliacea* (Crustacea: Aristeidae) in
the Sicilian Channel (Mediterranean Sea). *ICES Mar. Sci. Symp.* 199: 183-188.

Ragonese, S., Bertolino, F. & Bianchini, M.L. 1997. Biometric relationships of the red shrimp,
Aristeomorpha foliacea Risso 1827, in the Strait of Sicily (Mediterranean Sea). *Sci. Mar.* 61:
367-377.

Ragonese, S. & Bianchini, M.L. 1992. Stima dei parametri di crescita di *Aristeus antennatus* nel
Canale di Sicilia. *Oebalia* 17: 101-107.

Ragonese, S. & Bianchini, M.L. 1995. Size at sexual maturity in red shrimp females, *Aristaeomorpha foliacea*, from the Sicilian Channel (Mediterranean Sea). *Crustaceana* 68: 73-82.

Ragonese, S. & Bianchini, M.L. 1996. Growth, mortality and yield per recruit of the deep-water
shrimp *Aristeus antennatus* of the Strait of Sicily (Mediterranean Sea). *Fish. Res.* 26: 125-137.

Ragonese, S., Bianchini, M.L. & Gallucci, V.F. 1994. Growth and mortality of the red shrimp
Aristaeomorpha foliacea in the Sicilian Channel (Mediterranean Sea). *Crustaceana* 67: 348-361.

Rainer, S.F. 1992. Diets of prawns from the continental slope of north-western Australia. *Bulletin of
Marine Science* 50: 258-274.

Relini Orsi, L. & Relini, G. 1979. Pesca e riproduzione del gambero rosso *Aristeus antennatus* (Decapoda Penaeidae) nel Mar Ligure. *Quad. Civ. Staz. Idrobiol. Milano* 7: 39-62.

Relini Orsi, L. & Wurtz, M. 1977. Aspetti della rete trofica batiale riguardanti *Aristeus antennatus*
(Crustacea, Penaeidae). *Atti del IX Congresso della Societa Italiana di Biologia Marina. Lacco
Ameno d' Ischia 19-22 Maggio 1977*: 389-398.

Relini, G. & Relini Orsi, L. 1987. The decline of red shrimp stocks in the Gulf of Genoa. *Inv. Pesq.*
51 (Suppl 1): 245-260.

Sardà, F. 1993. Bio-ecological aspects of the decapod crustacean fisheries in the Western Mediterranean. *Aquat. Liv. Res.* 6: 299-306.

Sardà, F. & Cartes, J.E. 1993a. Distribution, abundance and selected biological aspects of *Aristeus
antennatus* (Decapoda: Aristeidae) in deep-water habitats in the NW Mediterranean. *BIOS Thessaloniki* 1(1): 59-73.

Sardà, F. & Cartes, J.E. 1993b. Relationship between size and depth in decapod crustacean populations on the deep slope in the Western Mediterranean. *Deep-Sea Res.* 40: 2389-2400.

Sardà, F. & Cartes, J.E. 1997. Morphological features and ecological aspects of early juvenile
specimens of the aristeid shrimp *Aristeus antennatus* (Risso, 1816). *Mar. Freshwater Res.* 48:
73-77.

Sardà, F., Cartes, J.E. & Norbis, W. 1994. Spatio-temporal structure of the deep-water shrimp
Aristeus antennatus population in the Western Mediterranean. *Fish. Bull. US* 92: 599-607.

Sardà, F. & Demestre, M. 1987. Estudio biologico de la gamba *Aristeus antennatus* en el Mar Catalan (NE de Espagna). *Inv. Pesq.* 51(Supl. 1): 213-232.

Sardà, F., Maynou, F. & Tallo, L. 1997. Seasonal and spatial mobility patterns of rose shrimp *Aristeus antennatus* in the Western Mediterranean: results of a long-term study. *Mar. Ecol. Prog. Ser.* 159: 133-141.

Spedicato, M.T., Lembo, G., Carbonara, P. & Silecchia, T. 1994. Biological parameters and dynamics of *Aristaeomorpha foliacea* in Southern Tyrrhenian Sea. In M. L. Bianchini & S. Ragonese (eds), *Life cycles and fisheries of the deep-water red shrimps* Aristaeomorpha foliacea *and* Aristeus antennatus. *Proceedings of the International workshop held in the Istituto di tecnologia della pesca e del pescato (ITTP-CNR), Mazara del Vallo, Italy, 28-30 April:* 35-36.

Tobar, R. & Sardà, F. 1987. Analisis de la evolution de las capturas de gamba rosada *Aristeus antennatus* en los ultimos decenios en Cataluna. *Inf. Tecn. Inst. Invest. Pesq. Barcelona* 142: 1-20.

Thessalou-Legaki, M. 1994. Distribution of *Aristeus antennatus* and *Aristaeomorpha foliacea* in the Eastern Mediterranean Sea. In M. L. Bianchini & S. Ragonese (eds), *Life cycles and fisheries of the deep-water red shrimps* Aristaeomorpha foliacea *and* Aristeus antennatus. *Proceedings of th International workshop held in the Istituto di tecnologia della pesca e del pescato (ITTP-CNR), Mazara del Vallo, Italy, 28-30 April:* 61-62.

Sex and seasonal variation in the stomach content of the red king crab, *Paralithodes camtschaticus* in the southern Barents Sea

J.H. SUNDET
Norwegian Institute of Fisheries and Aquaculture Ltd., Tromsø, Norway

E.E. RAFTER
Polaria, Tromsø, Norway

E.M. NILSSEN
Norwegian College of Fisheries, University of Tromsø, Tromsø, Norway

ABSTRACT

Samples of the red king crab, *Paralithodes camtschaticus*, were taken by trawling and scuba diving on several occasions between May 1994 and September 1996. The stomachs were dissected immediately after capture and fixed for later analyses in the laboratory. The contents were highly fragmented and each item was identified to the lowest possible taxon. The relative frequency of occurrence of each prey category was calculated. The dominant prey groups were bivalves and polychaetes. In addition, many stomachs contained algae, gastropods and several classes of echinoderms. Differences in prey composition were tested using a chi-square test and a two-way ANOVA. Stomach content weight data were submitted to an MDS-analysis, from which the score values were tested by ANOVA. We found significant differences in both food composition and stomach content weight of males and females, particularly for the main prey groups. There were also significant differences in prey composition when comparing spring and autumn values. A majority of large males in our samples, combined with migration patterns and sexual segregation parts of the year may explain the differences due to sex and season. The diversity of prey categories suggests that the king crab in the Barents Sea is an opportunistic predator feeding on what is available.

1 INTRODUCTION

Since the red king crab, *Paralithodes camtschaticus*, also called Kamchatka crab, was introduced by Russian scientists in the Barents Sea during the 1960s (Orlov & Karpevich 1965, Orlov & Ivanov 1978), this species has become abundant in Russian and Norwegian waters (Kuzmin et al.1996). In Norway, the crab population is continuously expanding westwards along the northern coast, invading new coastal areas.

As the king crab is an alien species in these waters, its influence on the ecosystems is unknown. Although the biology of the red king crab has been studied thoroughly (Wallace et al. 1949, Melteff 1990), we had to document its basic biological features in the Barents Sea. In Alaska, the king crab feeds on most invertebrate classes available (Cunningham 1969, Feder & Paul 1980).

Predation may therefore be one of primary impacts of the red king crab as a new species in the Barents Sea. A preliminary study (Rafter et al. 1996) revealed a great variety of benthic groups and species in the diet of the crab in north Norway. Gerasimova (1997) found that the king crab in these areas consumed less than one percent of what is ingested by benthic feeding fish in the same area. However, this study did not take into account seasonal migrations or the fact that growth performances are dependent on both size and sex. Migratory behavior may subject the crabs to different substrata and thereby to different prey organisms (Jewett et al. 1990). In this study we identified the main prey items of the Barents Sea red king crab and estimated the amount of food ingested relative to sex, size and season.

2 MATERIAL AND METHODS

2.1 *Sampling and collection of data*

The material was collected in May and November 1994, and in May and September 1995 and 1996, in the southern part of the Varangerfjord in Finnmark, north Norway. The sampling sites were the same in spring and autumn. In total, 641 stomachs from crabs of 44-200 mm carapace length were analyzed. Crabs were taken at different depths (18-258 m), mainly using a specially designed Agassiz trawl. Some crabs were caught by scuba diving. The sampled crabs varied from newly molted soft-shelled to individuals with very old shells. The stomachs were immediately removed onboard the vessel and stored in buffered 4% formaldehyde. Carapace length and width of each crab were measured using a vernier caliper. After 24 hours in running fresh water, the stomach content was analyzed in the laboratory. The crab breaks up its food before ingesting it; therefore food organisms were identified to phyla and classes only for quantitative analysis. Identification of organisms to species level was made whenever possible. Benthic invertebrates from the crab's habitats in Varanger were collected before the analysis to help in the identification of prey species. Stomach content wet weight was recorded from 460 crabs, whereas the relative frequency of occurrence of each prey group was calculated for all individual stomachs that contained food.

2.2 *Data analysis*

Statistical treatment and graphical presentations were performed using different modules of SYSTAT® and SYGRAHPH® (Wilkinson 1990a, b). Differences in the relative frequency of occurrence between sexes integrated over seasons, and seasons integrated over sexes and the seasonal difference in numbers of empty stomachs, were tested using chi-square tests. Jaccard's dichotomy coefficients (Gower 1985, Krebs 1989) were used to test the influence of parameters as season, sex, carapace length, sampling depth and sampling time. This coefficient relates variables where values represent the presence or absence (1 or 0 respectively) of the different prey categories. A Multi Dimensional Scaling analysis (MDS) was done on the Jaccard coefficients to generate score values in two dimensions (Field et al. 1982), which in turn were tested using a two-way ANOVA (Zar 1984). In the ANOVA the score values were grouped with respect to sex and season, and any effect of sampling depth, carapace length and sampling time were removed by using them as co-variables in the ANOVA. This was done because we knew *a priori* that in our data the overall size of the males would be greater than of the females. Likewise, our sampling showed that the occurrence of females and males depended on both season and depth. The stomach content weights were analyzed using a two-way ANOVA with the same parameters, categories and co-variables as in the analysis above.

3 RESULTS

Of the 641 crab stomachs, 65 were empty (Table 1) and there were no seasonal differences in the number of empty stomachs between sexes (Fig. 1). The proportion of males and females in the samples were the same at all collections (Table 1).

The king crab in Varanger feeds on a whole range of prey categories with polychaetes and bivalves as the most common groups (Figs 1 and 2). Algae, gastropods,

Table 1. Red king crab (*Paralithodes camtschaticus*). Number of empty and filled stomachs from male and female king crabs sampled in spring and autumn annually in 1994, 1995 and 1996, Varangerfjord, north Norway.

	Males		Females		
	Filled	Empty	Filled	Empty	Total
Spring	129	18	79	13	
Autmn	223	20	145	4	
Σ	352	48	224	17	641

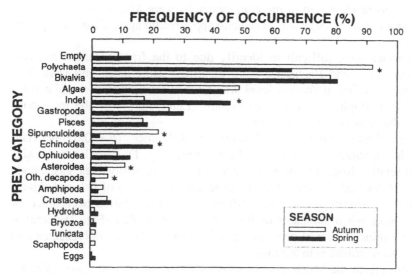

Figure 1. Frequency of occurrence of prey categories in stomachs of red king crab from the Barents Sea during autumn and spring. Males and females are pooled for each season. Asterix indicate significant (*p* < 0.05) seasonal differences for the same food category.

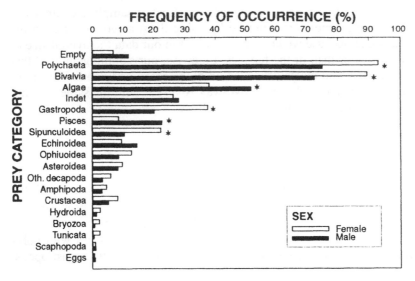

Figure 2. Frequency of occurrence of prey categories in stomachs of male and female red king crab from the Barents Sea. Data from both seasons are pooled for each sex. Asterix indicate significant ($p < 0.05$) sexual differences for the same food category.

echinoderms and fish seem to constitute the second most important group, together with a large amount of not identified stomach content. In addition, 8 other categories of food items were observed of which none occurred in more than 9% of the stomachs.

Bathyarcha glacialis was the most abundant bivalve species, but specimens of the genera *Nucilana*, *Nuculoma* and *Clinocardium* were also quite numerous. Other bivalve species or genera identified in the crab stomachs were *Mytilus edulis*, *Arctica islandica*, *Macoma calcarea*, *Mya* spp., *Yoldiella* spp., *Yoldia* spp., *Cyprina* spp. and *Thyasira* spp.

Polychaeta species were difficult to identify due to the fragmentation, but two genera, *Spiochaeopterus* and *Pectinaria* were identified using setae and jaws. The only echinoderm identified at species level was *Ctenodiscus crispatus,* while fragments of brittle stars (Ophiuridae) and sea urchins (Echinidae) were the most conspicuous among the echinoderm prey group.

Pooled spring and autumn data showed significant differences ($p < 0.05$) between males and females in some of the main food categories (Fig. 1). Females had a significantly larger relative frequency of occurrence of polychaetes, bivalves, gastropods and sipunculids. Males had a significantly larger relative frequency of fish and algae. Likewise, there were significant seasonal differences in the relative occurrence of some prey categories when stomachs of both sexes were pooled (Fig. 2). Echinoderms and undetermined material predominated in spring, whereas polychaetes and sipunculids were most common in autumn.

The score values from the MDS-analysis showed that both sex and season significantly affected the composition of prey categories in the stomachs (Table 2).

Season alone did not have any effect on the variation of the stomach content weight, but combined with sex it was highly significant (Table 3).

Table 2. Red king crab (*Paralithodes camtschaticus*) stomach content. ANOVA with hour of catch, carapace length and depth as covariate. Score values in dimension I and II from the MDS-analysis based on present-absent (Jaccard index) of the different prey categories. Groups are season and sex. Df = degrees of freedom.

	Df	Dimension I		Dimension II	
		F	*p*-value	F	*p*-value
Test groups					
Season	1	0.9	0.331	32.0	< 0.001
Sex	1	14.0	< 0.001	5.5	0.019
Season*Sex	1	0.1	0.817	4.4	0.037
Covariate					
Hour	1	3.4	0.066	0.2	0.657
Carapace lenght	1	18.6	< 0.001	12.5	< 0.001
Depth	1	1.7	0.198	27.0	< 0.001
Error	569				

Table 3. Red king crab (*Paralithodes camtschaticus*). Two-way ANOVA of stomach content weight (Log – transformed values) with hour of catch, log (carapace length) and depth as covariate. Groups are season and sex. Df = degrees of freedom.

Source	Df	F	*p-value*
Test groups			
Season	1	6.8	0.794
Sex	1	80.6	0.005
Season*Sex	1	64.8	0.011
Covariate			
Hour of catch	1	28.9	0.090
Depth	1	47.6	0.030
Log (carapace length	1	93.8	0.002
Error:	459		

4 DISCUSSION

When analyzing the importance of the different prey items in the diet of the crab, certain biases arise. For example, hard fragments in the diet such as bivalve shells or gastropod opercula are less digestible than soft tissues and may thus be retained for a longer time in the digestive system than the soft parts, as shown by Logvinovich (1945) in controlled feeding experiments with previously starved crabs.

Mollusc shells are also heavier than much of the other food material ingested by the crab. This might result in an overestimate of the weight of bivalves and gastropods food for the king crab. However, since all stomach content weights in this study are taken for comparative reasons, we assume the biases due to hard fragments being constant for all samples.

In this study, the relative number of empty stomachs was lower than that found in earlier studies from this area (Rafter et al. 1996, Gerasimova 1997). It was, however, higher than that is found in postlarvae stomach investigations (Feder et al. 1980), but similar to that found in king crabs in Alaska (Feder and Paul 1980, Jewett et al. 1990). Surprisingly, there were no differences in the relative numbers of empty stomachs between sexes and seasons. Other studies have revealed that mating and molting cause large differences in feeding behavior between males and females (Gerasimova 1997). In general, feeding ceases or is at a minimum during the few weeks of the molting-mating period in spring (Kulichova 1955, Cunningham 1969). In our May samples, mature females were almost exclusively in a newly molted, soft-shelled stage.

Many previous studies of the diet of the red king crab concluded that at least one food group and/or species is dominant (Takeuchi 1967, Tarverdieva 1978, Feder & Paul 1980, Jewett et al. 1990). The present investigation corroborated these studies, showing 2-3 dominant prey groups: polychaetes, molluscs and echinoderms. Although the main prey categories were the same as those found in the Bering Sea, the relative occurrences of each category were slightly different (Jewett et al. 1990). Furthermore, there was no large variation in main prey categories due to season as found in the Bering Sea.

The seasonal migration of the king crab is within a narrower spatial range in Norwegian waters than in the Bering Sea (Rodin 1990, Stone & O'Clair 1988). The range of habitats and thereby available prey may therefore be smaller in the Barents Sea. Although there are seasonal differences for each single prey category, the variations in relative occurrence between spring and autumn of each category are negligible.

In our material, there was a majority of males in both spring and autumn (Table 1). The single catches, however, were always dominated by one of the sexes. This was probably due to the known spatial segregation between sexes that occurs in all seasons except for the molting–mating period during spring. It is therefore reasonable to believe that the two sexes have different foraging areas and thereby different abundance of available prey organisms.

Due to the fact that there is a change in prey preference according to size (Tarverdieva 1978), the predominance of large males may be an alternative explanation for the difference in prey composition between the sexes.

It has been suggested that feeding habits of crabs were influenced by morphological changes of feeding-related structures during growth, such as chelae and mouthparts (Elner 1980, Perez & Bellwood 1988). Most of the crabs in our samples were, however, adult specimens, all of which had similar feeding appendage morphology.

There is no consistency in earlier studies in the relationship between the amount of food ingested by the king crab, and its sex (Logvinovich 1945, Cunningham 1969, Tarverdieva 1978, Jewett & Feder 1982). In this study, sex indeed influenced the weight of the stomach content. As well as sampling season combined with sex, contributed significantly to the variations in the stomach content weight.

Growth and production of offspring mainly governed the food intake by the king crab. It is therefore possible that the nutritional needs shortly after the molting–mating process in spring are different than in the autumn. Jewett & Feder (1982) documented greater food intake in the king crabs from Kodiak Island in spring and early summer than in autumn.

Our spring samples were from May, just after the females had molted. At this stage, the crab is probably therefore either in a stage of non-feeding (Kun & Mikulich 1954), or has just started to feed again to compensate losses during ecdysis

(Kulichkova 1955). Male king crabs in the Barents Sea seem to moult earlier in winter (unpublished observations) and may be in a different stage of their annual feeding regime in May.

The diversity of prey items found in the crab stomachs in this study suggests that the king crab in the Barents Sea may be an opportunistic predator as found elsewhere (Gerasimova 1997). In addition, the large relative frequency of fish in the diet may also reflect a scavenging behavior since feeding on live fish is not likely.

This study has identified the dominant prey categories in the diet of the red king crab in the Barents Sea. The results also indicate that the king crab diet in the Barents Sea may be influenced by sex and sampling season. To document the relative importance of these parameters additional and more specially designed investigations are needed.

ACKNOWLEDGEMENTS

This study was partly financed by the Norwegian Research Council. We are thankful to the crew of R/V 'Johan Ruud', who contributed to the practical sampling at sea. Ann Merete Hjelset did much of the tedious stomach content identification.

REFERENCES

Cunningham, D.T. 1969. A study of the food and feeding relationships of the Alaskan king crab *Paralithodes camtschatica*. Master's Thesis. State College, California, San Diego. 84 p.

Elner, R.W. 1980. The influence of temperature, sex, and chela size in the foraging strategy of the shore crab, *Carcinus maenas* (L.). *Marine Behaviour and Physiology* 84: 15-24.

Feder, H.M. & Paul A.J. 1980. Food of the king crab *Paralithodes camtschatica* and the Dungeness crab, *Cancer magister* in Cook Inlet, Alaska. *Proc. Natl. Shellfish Assoc.* 70: 240-246.

Feder, H.M., McCumby, K. & Paul, A.J.1980. The food of post-larval king crab, *Paralithodes camtschatica*, in Kachemak Bay, Alaska (*Decapoda, Lithodidae*). *Crustaceana* 39: 315-318.

Field, J.G., Clarke, K.R. & Warwick, R.M. 1982. A practical strategy for analysing multispecies distribution patterns. *Mar. Ecol. Prog. Ser.* 8: 37-52.

Gerasimova, O.V. 1997. Analysis of king crab (*Paralithodes camtschatica*). Trophic links in the Barents Sea. *ICES. CM 1997/GG: 03.* 21 pp.

Gower, J.C. 1985. Measures of similarity, dissimilarity, and distance. In S. Kotz & N.L. Johnson (eds), *Encyclopedia of statistical sciences*. Vol. 5. New York: John Wiley & Sons Inc.

Jewett, S.C. & Feder, H.M. 1982. Food and feeding habits of the king crab *Paralithodes camtschatica* near Kodiak Island, Alaska. *Mar. Biol.* 66: 243-250.

Jewett, S.C., Gardner, L.A. & Rusanowski, P.M. 1990. Food and feeding habits of red king crab from north-western Norton Sound Alaska. *Proc. Intern. Symp. King Tanner crabs, Univ. Alaska. Alaska Sea Grant Rep.* 90-04: 219-232.

Krebs, C.J. 1989. *Ecological methodology*. New York: Harper Collins Publishers Inc.

Kulichkova, V.A. 1955. Pitanie Kamchatskogo kraba v vesenneletnii period u beregov Kamchatki i Sakhalina [The feeding pattern of the Kamchatka crabs off the coasts off Kamchatka and Sakhalin. *Izvestiya Tikhookeanskogo Nauchno-Issledovatel'skovo Istituta Rybnogo Khoziaistva i Okeanografii* 43: 21-42 (in Russian).

Kun, M.S. & Mikulich, L.V. 1954. Sostav pishchi dalnevostochnykh promy slovikh krabov v letnii period [Diet composition of Far Eastern crabs of commercial quality during the summer]. *Izvestiya Tikhookeanskogo Nauchno-Issledovatel'skovo Istituta Rybnogo Khoziaistva i Okeanografii* 41: 319-332 (in Russian, Japanese, English).

Kuzmin, S., Olsen, S. & Gerasimova, O. 1996. Barents Sea king crab (*Paralithodes camtschaticus*): Transplantation experiments were successful. *High latitude crabs: Biology, management, and economics. Alaska Sea Grant College Program Rep.* 96-02: 649-663.

Logvinovich, D.N. 1945. Akvarial'nya nablyudeniya nad pitaniem Kamchatskogo kraba [Aquarium observations on the feeding of the Kamchatka crab]. *Izvestiya Tikhookeanskogo Nauchno-Issledovatel'skogo Istituta Rybnogo Khozyaistva i Okeanografii* 19: 79-97 (in Russian, Japanese, English).

Melteff, B. (ed.) 1990. *Proc. Intern. King Tanner Crab Symp. November 28-30, 1989, Anchorage, AK. Alaska Sea Grant Coll. Prog. Rep.* 90-04. Fairbanks: University of Alaska.

Orlov, Y.I. & Ivanov, B.G. 1978. On the introduction of the Kamchatka king crab *Paralithodes camtschatica* (Decapoda: Anomura: Lithodidae) into the Barents Sea. *Mar. Biol.* 48: 373-375.

Orlov, Yu., I. & Karpevich, A.F. 1965. On the introduction of the commercial crab *Paralithodes camtschaticus* (Tilesius) into the Barents Sea. In H.A. Cole (ed.), ICES Spec. Meeting 1962 to consider problems in the exploitation and regulation of fisheries for Crustacea. *Rapp. P. -v. Réun. Cons. Int. Explor. Mer* 156: 59-61.

Perez, O.S. & Bellwood, D.R. 1988. Ontogenetic changes in the natural diet of the sandy shore crab, *Matuta lunaris* (Forskaal) (Brachyura: Calappidae). *Aust. J. Mar. Freshwat. Res.* 39: 193-199.

Rafter, E., Nilssen, E.M. & Sundet, J.H. 1996. Stomach content, life history, maturation and morphometric parameters of red king crab, *Paralithodes camtschaticus*, from Varangerfjord area, North Norway. *ICES CM 1996/K* 10.

Rodin, V.E. 1990. Population biology of the king crab *Paralithodes camtschatica* Tilesius in the north Pacific ocean. *Proc. Intern. Symp. King Tanner Crabs. Univ. Alaska Sea Grant Rep.* 90-04: 133-144.

Stone, R.P. & O'Clair, C.E. 1988. Seasonal migration of primiparous and multiparous female red king crab (Paralithodes camtschatica). *Proc. Intern. King Tanner Crab Symp. November 28-30, 1989, Anchorage, AK. Alaska Sea Grant Coll. Prog. Rep.* 90-04: 189-191.

Takeuchi, I. 1967. Food of king crab, *Paralithodes camtschatica*, of the west coast of Kamchatka Peninsula, 1958-1964. *Bull. Hokkaido Reg. Fish. Res. Lab.* 33: 32-44 (in Japanese with English summary).

Tarverdieva, M.I. 1978. Sutochnyi ritm pitaniya Kamchatskogo kraba. *Biologiya Morya* 1978(3): 91-95. [Translated in English as Daily feeding rhythm of the kamchatka crab. *Sov. J. Mar. Biol.* 4(3): 711-714].

Wilkinson, L. 1990a. *SYSTAT: the system for statistics.*, Evanston (Ill.): *Systat Inc.*

Wilkinson, L. 1990b. *SYGRAH: the system for graphics.* Evanston (Ill.): *Systat Inc.*

Zar, J.H. 1984. *Biostatistical analysis.* Prentice-Hall International Inc.

Combined field and laboratory studies on agonistic behavior in shore crabs, *Carcinus maenas*: metabolic consequences of variable oxygen tensions

LYNNE U. SNEDDON, ALAN C. TAYLOR & FELICITY A. HUNTINGFORD
Division of Environmental & Evolutionary Biology, University of Glasgow, UK

ABSTRACT

To bridge the gap between field and laboratory studies in the agonistic behavior of the shore crab, *Carcinus maenas*, the physiological status of crabs after fighting in the field and the laboratory was determined. The behavioral aspects of fights and the physiological consequences were examined using established laboratory techniques and a specially developed field system in which controlled experimentation was possible. The metabolic consequences of aggressive behavior were assessed by analysis of key metabolites (L-lactate, glucose and glycogen) associated with anaerobiosis in pairs of fought male crabs in the laboratory and in the field. Contests were staged in differing levels of water pO_2 since shore crabs regularly experience hypoxia in the field. These experiments demonstrated that the behavioral acts performed by fighting shore crabs were the same in the field as those seen in the laboratory. There was also a similar influence of being a larger relative size and winning in the field and the laboratory. Hypoxic conditions in the field resulted in shorter contest durations and increased concentrations of L-lactate and glucose and a reduction in glycogen in the blood and tissues of fought crabs, when compared with contests staged under normoxic conditions in the field. However, contest duration in the field was much shorter than under laboratory conditions and unlike the laboratory results, there were no differences in metabolite concentrations between fought crabs and crabs which had not fought.

1 INTRODUCTION

Most studies on individual behavior in aquatic crustaceans have been made under tightly controlled laboratory conditions which are necessary when looking into behavioral mechanisms (review in Dingle 1983). However, such experiments cannot demonstrate how relevant this behavior is in the ecology of an individual *in situ*. Without further knowledge of context, the results from controlled experiments cannot be interpreted in functional terms. Non-manipulative studies in semi-natural environments in large aquaria circumvent some of the disadvantages of both highly controlled environments and manipulative field studies. However, even these conditions

201

remain artificial, since it is impossible to imitate realistic emigration and immigration, the threat of predation and fluctuation in environmental factors. Many questions about the incidence and nature of agonistic interactions cannot be addressed without knowledge of behavior in the natural context.

Therefore, the purpose of this study was to obtain observations on the intraspecific agonistic behavior between shore crabs, *Carcinus maenas,* in the field to determine if the aggressive behavior that the crabs display in the laboratory is also shown in the natural environment. Ideally this should be non-manipulative observations, but this was not feasible during the course of the present study. Laboratory studies have already demonstrated that fighting results in an elevation of L-lactate and glucose in the blood and tissues of *C. maenas* and under low oxygen tensions this effect is amplified and is accompanied by a significant reduction in glycogen stores (see Sneddon et al. 1998, 1999). Also the duration of contests between shore crabs were much shorter under very low oxygen tensions. Therefore, the behavioral aspects of fights and the physiological consequences were examined using established laboratory techniques and a specially developed field system in which controlled experimentation was possible. The present work was carried out on the upper shore in a large rock pool. The metabolic consequences of agonistic behavior were assessed by analysis of key metabolites (L-lactate, glucose, glycogen) associated with anaerobiosis in fought crabs. Also the rock pool became hypoxic so observations on the crabs experiencing these conditions in the field were made and the effects of hypoxia on behavior and metabolic physiology of agonistic interactions in the field will be compared with laboratory results.

2 MATERIALS AND METHODS

2.1 *Field study*

Field observations were made on the rocky shore in the vicinity of the University Marine Biological Station, Millport, Isle of Cumbrae during the months of June, July and August in 1996 and 1997. The pool was chosen since *C. maenas* were present (2-4 crabs were present on each sampling occasion), the steep sides would prevent crabs from leaving the pool once released into it and the pool was located high on shore which allowed a longer access time when the tide was out (Maximum length 132 cm, breadth 35 cm and height 47 cm). Plastic mesh (0.5 × 0.5 cm) barriers were placed at either end of the pool to prevent crabs leaving the observation area.

2.1.1 *Experiments under ambient oxygen tensions and under normoxia*
Crabs used in field observations were freshly creeled from the Clyde Sea area and males were marked numerically with nail varnish on the back of the carapace, measured (carapace width and propodus length to 0.1 mm) and placed in separate holding tanks for 24 hours prior to experimentation. The crabs were placed into 10 individual plastic mesh cages (cylindrical shape, 8 cm diameter ×15 cm length) and taken to the rock pool at 10 pm. Observations were made between 2.30 am and 4 am when oxygen tensions in the pool water would be lowest (Hill 1989). The pO_2 of the pool was measured with a calibrated, hand held oxygen meter and was found to lie between 30

and 42 Torr. The cages were split into two groups of five and tied to rocks and placed in the pool at either end. Crabs were chosen at random and 2 individuals released into the pool. Bait was thrown into the pool to encourage the crabs to fight. Baiting individuals to attract them and encourage fights has been used in a variety of animals such as *N. puber* (Smith 1990), and *N. norvegicus* (Bjordal 1986).

2.1.2 *Observations in the field*
The acts of the fighting crabs were observed with the aid of a diving lamp with a red filter, and were recorded on audio tape. A total of 22 fights were taped and the information transferred onto the event recorder by playing back the audio tapes. This provided information on behavioral content and duration of fights (see Sneddon et al. 1997a). The influence of relative size was determined since the crabs had been marked and measured previously. Observations were also made in this rock pool under normoxia by aerating the pool using a cylinder of compressed air and a regulator to which airline tubing and three airstones were attached. The airstones were positioned at either end of the pool and one in the middle and air was gently bubbled through the water one hour prior to behavioral observations. The measured pO_2 was between 145 and 156 Torr.

2.1.3 *Sampling in the field*
Five pairs of crabs which had fought at ambient oxygen conditions and also under normoxia, were chosen at random and immediately removed from the pool after fighting and a blood sample (0.1 ml) taken by piercing the arthrodial membrane at the base of the third pereiopod using a needle (19 g) and syringe (1 ml). The sample was immediately added to 0.1 ml of 0.6 M Perchloric acid (PCA) in an 1.5 ml Eppendorf tube, shaken and then placed into a box containing ice to keep the sample chilled whilst on the shore. The PCA deproteinises the blood and halts all metabolic processes. Another three pairs of crabs, chosen at random, were immediately removed from the pool after fighting and each crab was placed into a Dewar of liquid N_2 to instantly freeze the tissues and halt all metabolic processes. Five crabs, which were treated as described above but were released singly into the pool, were removed after three minutes and a blood sample taken to obtain values of metabolites from crabs that had not engaged in fighting. Three crabs under both oxygen conditions were also treated in this manner but were placed into liquid N_2 after the three minutes. The crabs were relatively inactive after their release into the pool. These samples will be referred to as samples from crabs at rest.

2.2 *Laboratory experiments*

Male shore crabs were obtained (carapace width range 55-80 mm) by creeling, transferred to the laboratory in Glasgow and kept in the conditions described in Sneddon et al. (1999). The crabs were deprived of food for 7 days prior to any experiments since this is known to reduce intra-individual variation in metabolite concentrations (Hill 1989).

The crabs were transferred separately to a partitioned arena (55 × 25 × 30 cm) which was screened from visual disturbance and left them for four hours. During the first hour under normoxia, air was bubbled through air stones that were positioned at opposite ends of the tank. Oxygen electrodes were positioned on opposite sides from the air stones to closely monitor oxygen tensions and these were always within 2% agreement. In the severely hypoxic treatment, the P_{O2} of the water was decreased gradually over the hour to 35 Torr (mean oxygen tension in the field) by altering the percentage composition of a gas mixture (N_2, CO_2, O_2) through a precision gas mixing system. The partition was raised from outside the screening, food extract (whitebait homogenized in seawater which was a chemical stimulus and provided a perceived resource) was injected into the middle of the arena via tubing and the interaction viewed through a hole in the screen. The winner of a fight was the crab that elicited 2 successive retreats from its opponent or successfully climbed on top of the other contestant, the loser (Sneddon et al. 1997a). After the resolution of a fight, when a clear winner and loser was apparent, each crab was removed immediately and a haemolymph sample taken. A further 5 fights were staged with crabs being removed after the resolution of a contest and each placed immediately into a Dewar of liquid N_2.

To obtain blood and tissue samples from crabs that had not fought, crabs were treated as described above and each crab was transferred singly to the arena. The partition was raised and then lowered after the mean duration of contests staged in normoxia (559 sec) and in hypoxia (193 sec); see Sneddon et al. (1998). Then crabs were removed and had a blood sample taken (n = 5) or placed into liquid N_2 (n = 5).

2.3 *Enzymatic analyses*

The haemolymph (0.25 ml) and tissue (50 μg) samples were treated and assayed for L-lactate, glucose and glycogen as described in Sneddon et al. (1999).

2.4 *Statistical analyses*

The behavioral and metabolic data were not normally distributed so non-parametric statistics were used throughout.

3 RESULTS

3.1 *Behavioral content*

The behavioral acts of fighting *C. maenas* in the field is similar to that observed during fights in the laboratory (Table 1; see Sneddon et al. (1997a) for details of laboratory behavior). Fights consisted of a pushing contest where opponents stand high on their 4th and 5th pereiopods, wrap the other walking legs around their opponent and push against one another. This frequently resulted in crabs losing balance and being overturned and the overturned crabs were the eventual loser. Winners perform fewer 'move to' acts and losers perform fewer 'move away' acts, since in the pool contests, fights consisted of one interaction with no re-engagement of an opponent.

3.2 *Duration*

The durations of contests under laboratory conditions were much longer than in the

Table 1. Comparison of the behavioral content of contests observed in the rock pool (n = 34) and the behavior of crabs in laboratory (n = 30) using Kruskal Wallis tests. W: winner; L: loser; To: move to; Away: move away; Display: chelipede display; Contact: strike, grasp, and push.

Act	W/L	Median		H	p-value
		Pool	Lab		
To	W	0.65	1.0	10.00	0.002
To	L	0.15	0.5	5.84	0.016
Away	W	0.00	0.1	0.38	0.536
Away	L	0.60	1.0	8.58	0.003
Display	W	1.20	0.0	0.94	0.332
Display	L	0.40	0.4	2.57	0.109
Contact	W	1.80	2.0	0.09	0.770
Contact	L	1.00	1.0	0.35	0.555

field under hypoxia (H = 9.06, df = 1, p = 0.003) and under normoxia (H = 15.14, df = 1, p < 0.001, Fig. 1). However, fights under normoxic conditions in the field are of similar duration to laboratory contests staged under hypoxia (H = 1.57, df = 1, p = 0.210).

3.3 *Relative size of winners and losers*

A previous study which used food as the contested resource found that crabs had an equal chance of winning when there was a 30% difference in body size and an 18% difference in claw size (Sneddon et al. 1997b) and therefore 86% of fights observed in the field were between equally matched crabs (using mates as a resource, this size effect was amplified and only 33% of fights over mates were between equally matched crabs, L.U. Sneddon, unpub. data). Under normoxic conditions in the field, equal numbers of crabs initiated fights but larger individuals tended to win, for both body (χ^2_{cy} = 1.5, p = 0.21, n = 12) and claw size (χ^2_{cy} = 3.23, p < 0.05, n = 12; Table 2). However this is significant for claw size only. This is similar to the results of laboratory studies in which crabs with larger claws (χ^2 = 4.08; p < 0.01; n = 46; see Sneddon et al. 1997b) tended to win more fights than crabs with larger bodies (χ^2 = 1.6, p > 0.05, n = 46). When conditions were hypoxic in the pool, equal numbers of crabs initiated fights but the majority were won by the larger individual for both body (χ^2_{cy} = 3.45, p < 0.05, n = 22) and claw size (χ^2_{cy} = 5.73, p = 0.023, n = 22). This effect is stronger for weapon size, where 86% of fights were won by crabs with larger claws compared with 77% of fights won by crabs with a larger body size.

3.4 *Metabolic physiology of crabs at rest in the field and laboratory*

Concentrations of L-lactate in the blood of crabs at rest under normoxia were higher in the field than in laboratory conditions (H = 6.82, p = 0.009). Under hypoxic conditions, concentrations of L-lactate in the blood of crabs at rest were similar to those obtained in the laboratory (H = 1.87, p = 0.172). Under both normoxic (H = 1.09, p = 0.297) and hypoxic conditions (H = 1.80, p = 0.180) in the field, concentrations of L-lactate in the tissues of resting crabs, were similar to those found under laboratory

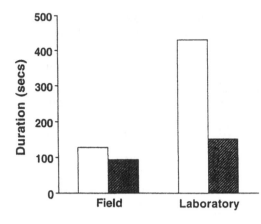

Figure 1. The median duration of fights under normoxia (empty bars; field n = 12; lab n = 15) and hypoxia (shaded bars; field n = 22; lab n = 15) in both field and laboratory observations.

Table 2. Numbers of smaller and larger crabs initiating fights and winning fights in the rock pool under normoxic (n = 12) or hypoxic conditions (n = 22). Results are shown for both carapace width and chela length.

	Carapace width		Chela lenght	
	Smaller	Larger	Smaller	Larger
Normoxia:				
– Initiate	7	5	6	6
– Win	3	9	1	11
Hypoxia:				
– Initiate	12	10	11	11
– Win	5	17	3	19

conditions. Under both normoxic (H = 6.82, p = 0.009) and hypoxic conditions (H = 5.77, p = 0.016) in the field, glucose concentrations in the blood of crabs at rest were higher than concentrations obtained from crabs at rest in the laboratory. Tissue glucose concentrations in the field were similar to those found in the laboratory for crabs at rest under both normoxia (H = 1.84, p = 0.175) and hypoxia (H = 0.56, p = 0.456). Under both oxygen tensions, crabs at rest in the field had similar glycogen concentrations to those in the laboratory (Normoxia H= 0.56; p = 0.456; Hypoxia H = 0.56, p = 0.456).

3.5 *Metabolic physiology of fought crabs*

Concentration of L-lactate in the blood of crabs after fighting under normoxia were higher in the field than in laboratory conditions (H = 14.3, p < 0.001). However, under hypoxia, blood L-lactate concentrations from fought crabs were similar to those obtained in the laboratory (H = 0.76, p = 0.384). Under both normoxic (H = 1.43, p = 0.232) and hypoxic conditions (H = 0.11, p = 0.735) in the field, concentrations of L-lactate in the tissues of fought crabs, were similar to those found under laboratory conditions. Under both oxygen tensions in the field, glucose concentrations in the blood of crabs after fighting were higher than concentrations obtained from crabs in the laboratory (Normoxia H = 8.48, p = 0.004; Hypoxia H = 6.23, p = 0.013). Tissue glucose concentrations in the field were similar to those found in the laboratory for crabs after fighting under both normoxia (H = 0.42, p = 0.515) and hypoxia (H = 1.06, p = 0.302). Under normoxia, crabs after fighting in the field had similar glycogen concentrations to those in the laboratory (H = 0.95, p = 0.329). However, under hypoxic conditions in the field, glycogen concentrations were higher in the tissues of fought crabs compared to those fought in the laboratory (H = 5.19, p = 0.023).

3.6 *Comparison of crabs at rest with fought crabs*

There was no significant difference in L-lactate concentrations between crabs at rest and after fighting in the field (Blood H = 2.94, p = 0.086; Tissue H = 0.07, p = 0.796) but, under laboratory conditions L-lactate concentrations were higher in fought crabs than those obtained from crabs at rest (Blood H = 5.19, p = 0.023; Tissue H = 6.01, p = 0.014). In the field, under both oxygen tensions, fought crabs had similar blood glucose concentrations when compared to crabs at rest (Normoxia H = 2.94, p = 0.086; Hypoxia H = 0.01, p = 0.903) but in the laboratory, fought crabs had higher concentrations of glucose than resting crabs (Normoxia H = 5.19, p = 0.023; Hypoxia H = 6.23, p = 0.013). In the field, there was no significant difference in tissue glucose concentrations between crabs at rest and fought crabs under both oxygen tensions (Normoxia H = 0.07, p = 0.796; Hypoxia H = 0.20; p = 0.606), however, in the laboratory, fought crabs had higher concentrations of tissue glucose than crabs at rest (Normoxia H = 6.01, p = 0.014; Hypoxia H = 8.48, p = 0.004). Under laboratory conditions, glycogen concentrations were lower in fought crabs compared with crabs at rest under hypoxia (H = 5.19, p = 0.023) but were similar under normoxia (H = 1.57, p = 0.210). Under both oxygen tensions in the field, crabs at rest had similar glycogen concentrations to fought crabs (Normoxia H = 0.60, p = 0.439; Hypoxia H = 1.07, p = 0.302).

More precise information on metabolite concentrations can be found in Sneddon (1998).

4 DISCUSSION

The agonistic behavior of the shore crab in the natural environment is similar to the behavior observed in the laboratory. However, the fights in the rock pool consisted of only one aggressive interaction and no re-engagement of an opponent occurred. This agrees with field observations on *N. puber* where similar action patterns were performed by crabs but the fights were shorter in duration and consisted of only one bout (Smith 1990). In the laboratory, the fighting pairs of shore crabs are confined to a tank where they have to remain in close proximity to one another whereas, in a rock pool, crabs were able to retreat long distances from an opponent. The behavior of fighting *C. maenas* in the pool agrees with theories that agonistic behavior should not involve lengthy interactions since this may increase attraction of predators and competing conspecifics (Crowley et al. 1988, Caine 1989, Green 1990, Smith 1990). Duration may also be influenced by the relative motivation of the crabs to fight. Crabs used in laboratory experiments were starved 7 days prior to contests in order to increase motivation, however, crabs used in field observations were only starved for 24 hours. Therefore, the crabs may have been less motivated to fight for long periods over food in the field since they may have been less hungry than those fought in the laboratory. It was decided to hold them for a short time to keep conditions as natural as possible. However, it would be interesting to hold crabs 7 days prior to the field experimentation to determine if they fought for longer. Another possibility is that the value the crabs in the field place on the food items is low, since there is abundant molluscan prey available to them on the shores of Cumbrae, and so they may not be so highly motivated to fight for long over food. It is unlikely that food will be a scarce commodity for the crabs inhabiting this area and, therefore, fights probably occur when chance items of food such as carcasses of dead fauna are available, or in the breeding season, when competition occurs over receptive females (Sekkelsten 1988, Smith 1990, Abello et al. 1994, Reid et al. 1994). In future field observations, it would be interesting to place a caged female into the pool to determine if fights become longer and more intense due to the presence of this highly valued resource.

The effects of relative size in the rock pool fights under both normoxia and hypoxia, were similar to the results of laboratory studies, with approximately equal numbers of smaller and larger crabs initiating fights but more larger crabs winning (see Sneddon et al. 1997b, 1998). Crabs of different size were used in the field study since it is likely that the encounters between crabs in nature may not be between crabs of equal size. In the field study however, 86% of fights were between equally matched crabs. The results suggest that assessment is occurring during rather than prior to interactions and that assessment can only be made when individuals are close to or engage one another. Perhaps visual perception in *C. maenas* is limited since they are mostly nocturnal and can only assess visually when opponents are in very close proximity, or that assessment is tactile when the strength of an opponent is assessed in the pushing contest or during grasping. Perhaps the costs of fighting are greater for smaller crabs and so they decide to retreat from larger opponents. When comparing body and weapon size, relative weapon size exerts a stronger influence on the outcome of fights in the field, which confirms the laboratory studies (Sneddon et al. 1997b). Animals should base their strategic decisions on their perceived risk of danger (Lima & Dill 1990), and so smaller clawed crabs may detect that their opponent has larger weapons and, therefore, a greater ability to inflict injury and this may influence their decision to de-escalate. It can be concluded that claw size may be a major indicator of status and competitive ability in this species as it is in the grapsid crabs, *Aratus pisoni* and *Goniopsis cruentata* (Warner 1970).

In the laboratory, fighting resulted in greater concentrations of L-lactate and glucose in the blood and tissues of shore crabs under normoxia and hypoxia (see Sneddon et al. 1998, 1999). Glycogen concentrations were only affected under hypoxic conditions, where there was a reduction in the tissues of fought crabs. There were no such metabolic effects shown by crabs that fought in the rock pool. This may be due to the fights being shorter in the field and so, there is not a sufficiently increased energy demand caused by fighting unlike those staged under laboratory conditions, which were of much longer duration. Fighting in the laboratory resulted in enhanced anaerobic respiration and increased accumulation of L-lactate; and breakdown of glycogen into glucose causing hyperglycaemia in the blood and tissues. Another possibility why there appears to be no effects of fighting on metabolite levels in the field is that the sample sizes in the field were low and these metabolites are highly variable in freshly caught crabs (Hill 1989). In the laboratory the crabs are starved for 7 days, which reduces the variability in these metabolites between individuals (Hill 1989) and so the crabs may have similar metabolite profiles before fighting. This variability may be confounding the field data and thus, there was no difference between crabs at rest and after fighting in the pool.

Blood L-lactate and glucose concentrations were higher in resting and in fought crabs in the field than in the laboratory. This could be due to the field crabs having recently fed before capture, which is known to elevate glucose and L-lactate concentrations (Wallace 1973). Some studies have shown that hyperglycaemia and production of L-lactate is a response to stress in the shore crab (Hill 1989) and in other crustaceans (e.g. *Nephrops norvegicus*, Spicer et al. 1990) and the greater concentrations of these metabolites in the blood of field crabs may be due to the crabs not having recovered from the stress of creeling.

The field observations confirmed that the behavioral acts performed during the agonistic interactions of male *C. maenas* in the laboratory are the same as would be seen in the natural environment. However, fights are of much shorter duration in the field than those in the laboratory, and the crabs do not seem to show the same metabolic consequences of fighting as they do in the laboratory. With increased sample sizes, a similar experimental protocol as used in the laboratory studies, and lower water oxygen tensions, perhaps the metabolic effects of fighting would be more apparent in the field manipulations. However, it could be that *C. maenas* fights in the field are never as long as fights under laboratory conditions and therefore, the crabs do not incur a metabolic cost as a result of fighting.

REFERENCES

Abello, P., Warman, C.G., Reid, D.G. & Naylor, E. 1994. Chela loss in the shore crab, *Carcinus maenas* and its effect on mating success. *Mar. Biol.* 121: 247-252.

Bjordal, A. 1986. The behavior of Norway lobsters towards baited creels and size selectivity of creels and trawl. *Fisk. Dir. Skr. HavUnders.* 18: 131-137.

Caine, E.A. 1989. Caprellid amphipod behavior and predatory strikes by fish. *J. Exp. Mar. Biol. Ecol.* 126: 173-180.

Crowley, P.H., Gillett, S. & Lawton, J.H. 1988. Contests between larval damselflies: empirical steps toward a better ESS model. *Anim. Behav.* 36: 1496-1510.

Dingle, H. 1983. Strategies of agonistic behavior in crustaceans. In D.W. Dunham (ed.) *Studies in Adaptation: the behavior of the higher Crustacea*: 85-111. New York: Wiley Interscience.

Green, A.J. 1990. Determinants of dominance participation and the effects of size, weight and competition on advertisement calling in the Tungara frog, *Physalaemus pustulosus*. *Anim. Behav.* 39: 620-638.

Hill, A.D. 1989. The anaerobic metabolism of the shore crab, *Carcinus maenas*. Ph.D Thesis, University of Glasgow.

Lima, S.L. & Dill, L.M. 1990. Behavioral decisions made under the risk of predation: a review and prospectus. *Can. J. Zool.* 68: 619-640.

Reid, D.G., Abello, P., Warman, C.G. & Naylor, E. 1994. Size related mating success in the shore crab, *Carcinus maenas*. *J. Zool.* 232: 397-407.

Sekkelsten, G.M. 1988. Effect of handicapping on the mating success in the male shore crab, *Carcinus maenas*. *Oikos* 51: 131-134.

Smith, I.P. 1990. The agonistic behavior of the velvet swimming crab, *Liocarcinus puber* (L.). Ph.D. Thesis, University of Glasgow.

Sneddon, L.U., Huntingford, F.A. & Taylor, A.C. 1997a. The influence of resource value on the agonistic behavior of the shore crab, *Carcinus maenas* (L.). *Mar. Freshw. Behav. Physiol.* 30: 225-237.

Sneddon, L.U., Huntingford, F.A. & Taylor, A.C. 1997b. Weapon size versus body size as a predictor of winning fights between shore crabs, *Carcinus maenas* (L.). *Behav. Ecol. Sociobiol.* 41: 237-242.

Sneddon, L.U. 1998. The physiological effects of agonistic behavior in the shore crab, *Carcinus maenas* (L.). Ph.D Thesis, University of Glasgow.

Sneddon, L.U., Huntingford, F.A. & Taylor, A.C. 1998. Impact of an ecological factor on the costs of resource acquisition: fighting and metabolic physiology of crabs. *Functional Ecol.*, in press.

Sneddon, L.U., Taylor, A.C. & Huntingford, F.A. 1999 Metabolic consequences of agonistic behavior: crab fights in declining oxygen tensions. *Anim. Behav.*, in press.

Spicer, J.I., Hill, A.D., Taylor, A.C. & Strang, R.H.C. 1990. Effect of aerial exposure on concentrations of selected metabolites in the blood of the Norwegian lobster *Nephrops norvegicus*. *Mar. Biol.* 105: 129-135.

Wallace, J.C. (1973) Activity and metabolic rate in the shore crab, *Carcinus maenas*. *Comp. Biochem. Physiol.* 41A: 523-533.

Warner, G.F. 1970. Behavior of two species of grapsid crab during intraspecific encounters. *Behav.* 36: 9-19.

Extended parental care behavior in crustaceans –
A comparative overview

MARTIN THIEL

Smithsonian Marine Station, Fort Pierce, USA
Present address: Facultad de Ciencias del Mar, Universidad Catolica del Norte, Campus Guaya-
can, Coquimbo, Chile

ABSTRACT

Parental care behavior is reported from different crustacean taxa and environments but at present no general pattern has been recognized. The major goal of this review is to synthesize the present knowledge about parental care in crustaceans and to identify whether common traits can be identified. Crustaceans as different as synalpheid shrimp, spider crabs, crayfish, isopods, tanaids and amphipods engage in parental care but no specific morphological traits were connected with this reproductive behavior. Both iteroparous and semelparous crustaceans have been found to engage in parental care. Peracarid species that provide extended parental care for their offspring produce similar numbers of eggs as those peracarids that provide no post-marsupial care. Decapod species with extended parental care (synalpheid shrimp, some brachyuran crabs, many crayfish), produce relatively small numbers of eggs and care for individual broods for relatively long time periods. Highly motile species often carry their offspring on their body or in 'mobile homes', while sedentary species usually host their offspring in/on their dwellings (e.g. tubes and burrows). Crustaceans that carry their offspring hosted only one age cohort of juveniles on their body but those that provided for offspring in/on a dwelling could host up to three subsequent age cohorts of juveniles simultaneously. Food sharing with offspring has been observed in both motile and sedentary species. The most advanced form of parental care is found in terrestrial crustaceans where parents forage in the vicinity of an offspring cradle to which they return food items to feed their young. In general, the high diversity of crustaceans with extended parental care suggests that this reproductive behavior has evolved independently in a variety of crustacean taxa under a variety of environmental conditions. Several examples of convergent evolution of the same form of parental care support this assumption. While the potential for parental care may be high in crustaceans, the occurrence of advanced social behavior probably is limited to crustaceans with an offspring cradle as is indicated by the exclusive finding of return-foraging and multiple juvenile age cohorts among the latter species. Crustaceans offer a unique opportunity to study the evolution of extended parental care since species from a variety of both aquatic and terrestrial environments engage in this reproductive strategy.

1 INTRODUCTION

Parental care has been reported from a variety of crustaceans inhabiting environments as different as the deep sea, shallow estuaries, dry deserts, swamps or high mountains. Crustaceans care for their offspring in many different ways. In most malacostracan crustaceans, eggs are fertilized either internally or externally on the female body. Following fertilization, the eggs are incubated on the females' bodies for varying periods of time. Egg size can vary considerably among species, depending on the yolk content. The eggs of most peracarid crustaceans (amphipods, isopods and tanaids) contain large amounts of yolk, and advanced larval stages or fully developed juveniles hatch from the eggs. Decapods with large eggs often have abbreviated development, sometimes resulting in a total reduction of pelagic larval stages (Rabalais & Gore 1985). Species with large eggs have usually longer development times than closely related species with small eggs (Bueno & Rodrigues 1995). Peracarid crustaceans and decapods with abbreviated development are upon emergence of their offspring from the females' incubation chambers confronted with adult-like juveniles. It is at this stage that parents are faced with the decision to care (or not to care). Parents may continue to care for their offspring for extended time periods after these have emerged from the parental brood structure. In this review, I will primarily focus on this form of extended parental care where parents care for fully developed juveniles that in many ways look and behave like adults. It is not in all cases entirely clear, when the 'usual' parental care phase during which embryos and larvae are incubated in a bodily brood structure ends and when 'extended' parental care starts. Herein, I treat parental care that involves potentially self-subsistent juveniles or that extends substantially beyond what is found in closely related taxa as 'extended parental care'.

Extended parental care in crustaceans has not received much attention in the past. However, it may not be uncommon among crustaceans as the relatively frequent occurrence of parent-offspring associations indicates. In most cases where parents have been found to cohabit with fully developed offspring, we do not know whether parents actively look after their offspring or whether they merely tolerate them in their vicinity. However, close examination of these parent-offspring associations almost invariably revealed that crustacean parents fed, groomed or defended their offspring (Stephan 1980, Aoki & Kikuchi 1991, Coleman 1989). Therefore, I suspect that many crustaceans that tolerate their offspring will also be shown to engage in advanced forms of parental care, such as defense, grooming and feeding of juveniles. In this review, I will summarize our present knowledge about the behaviour of crustacean parents that care for their offspring, and I will point out common behavioural traits exhibited by different species.

In contrast to previous studies that focused on parental care in decapods (e.g. Hazlett 1983, Diesel 1989, 1992), isopods (Linsenmair 1984, 1989), or amphipods (Aoki 1997, Thiel 1999a), herein I compare crustacean parental care among these crustacean orders. This overview cannot achieve complete coverage of this topic, but I hope that it may serve as a good start for future studies on crustacean parental care.

2 MATERIALS AND METHODS

I searched the scientific literature for reports mentioning the occurrence of parent-offspring associations. I have carefully scanned these reports for statements qualifying these associations. Those qualifying statements could for example be an observation that parents carry their offspring on their body, or that they share their meals with them. A first survey of these reports revealed recurring parental care patterns and I have categorized these observations on the parental behaviour accordingly (see appendix). Additionally, I recorded any information on the duration of parental care, the number of eggs/offspring produced, and the sizes of offspring at leaving their parents. This review focuses on malacostracan crustaceans such as decapods, stomatopods, amphipods, isopods and tanaids but I have also included some observations on pycnogonids.

3 RESULTS AND DISCUSSION

3.1 *Forms of parental care among crustaceans*

Two forms of extended parental care are most common among crustaceans, carrying of juveniles or accommodating them in dwellings. Some amphipod, isopod and decapod parents carry their offspring for time periods of a few hours to weeks. While being carried by their parents, offspring can reach remarkable sizes (e.g. *Arcturus baffini* – Svavarsson & Davidsdottir 1995, *Serolis cornuta* – Luxmoore 1982, *Pilumnus novaezealandiae* – Wear 1967, *Paranaxia serpulifera* – Morgan 1987). In most crustaceans that carry their offspring, the juveniles remain on the parent throughout the period of parental care. However, in some crayfish species, juveniles may temporarily leave their parents to forage, but then return to reattach to the parents body (Ameyaw-Akumfi 1976, Figler et al. 1997). In *Caprella decipiens*, juveniles are not carried by their mother as in all other caprellid species with extended parental care, but herded by her (Aoki & Kikuchi 1991).

Many crustaceans provide extended parental care for their offspring in or on dwellings. Advanced larvae or fully developed juveniles may share these dwellings with their parents (Thamdrup 1935, Forbes 1973, Linsenmair and Linsenmair 1971, Richter 1978a, Johnson & Attramadal 1982, Horwitz et al. 1985, Mattson & Cedhagen 1989, Conlan & Chess 1992, Thiel et al. 1997), or parents may deposit their offspring in specific offspring cradles which they guard and manage from the outside (Diesel 1989, 1992). Extended parental care in these dwellings may last for several days, weeks up to 3 months (Lake & Newcombe 1975, Hopkins 1967, Suter 1977, Pillai & Subramoniam 1984, Thiel 1999a), and juveniles may grow to large sizes during this parental care. Dwellings of parents that host offspring may be considerably larger than those of non-parental individuals (Chess 1993, Thiel 1997). In some species, juveniles were found to start their own dwellings originating from parental dwellings (Henderson 1924, Thamdrup 1935, Menzies 1957, Forbes 1973, Brearley & Walker 1995, Thiel et al. 1997).

3.2 *Environments inhabited by crustaceans with parental care*

Crustaceans that engage in extended parental care behaviour have been reported from many different habitats. Parent-offspring associations have been reported in crustaceans from the deep sea (Wolff 1976), from arctic (Svavarsson & Davidsdottir 1995) and from tropical waters (Duffy 1996), from the open ocean (Richter 1978b), from kelp forests (Conlan & Chess 1992), shallow soft bottoms (Shillaker & Moore 1987), from freshwater ponds (Ameyaw-Akumfi 1976), bogs (Johnston & Figiel 1997), swampy fields (Horwitz et al. 1985), tropical rain forests (Cumberlidge & Sachs 1991), mountains (Ng & Tan 1995), to dry deserts (Marikovsky 1969, Linsenmair 1972, 1979) (see also appendix). A large number of crustaceans with extended parental care has been reported from shallow coastal environments which may be caused by a sampling bias allowing for the recognition of parent-offspring associations in shallow environments (Thiel et al. 1997). Further surveys that examine whether or not crustaceans engage in extended parental care are required in order to reveal whether this reproductive behaviour occurs more frequently in some habitats than in others. Many crustaceans that inhabit biotic microhabitats such as bivalves, ascidians, brachiopods, sponges, plants, and plant remains such as wood engage in extended parental care (see e.g. Vader & Beehler 1983, Messana et al. 1994, Thiel 1999b). Most likely the abundant supply of resources within these biotic microhabitats allows for the prolonged coexistence of parents and their offspring, and possibly even subsequent generations (Duffy 1996, Thiel 1999b).

3.3 *Behavior of parents during parental care*

Various parental behavior has been observed among crustaceans. Among the most commonly reported behavior is sharing of food resources that the parents have acquired. In most of the reported cases, the parents themselves feed on these food items but allow their offspring to participate in their meal (e.g. Morgan 1987, Shillaker & Moore 1987, Coleman 1989, Aoki & Kikuchi 1991). Parents that care for offspring in a dwelling may acquire food items that are transported to or are already present in the parental dwelling. Presently it is not known whether parents that provide for offspring in these dwelling exploit resources at a higher rate than individuals without offspring. The most advanced form of parental food sharing is found in terrestrial crustaceans where parents forage in the vicinity of an offspring cradle to which they return food items. In the desert isopod *Hemilepistus reaumuri*, one parent leaves the parental burrow to forage in its surrounding. In the meantime, the other parent remains home to guard the burrow and offspring (Linsenmair 1972) until the foraging parent returns with its bounty. In the tree crab *Metapaulias depressus*, the female deposits her brood in a leaf axil to which it carries prey that it conquered in the surrounding area (Diesel 1989, 1992). Despite extended parental care is found in a variety of marine crustaceans, no such return foraging has been reported from marine crustacean parents.

Grooming of small offspring has been exclusively reported for amphipod parents (Stephan 1980, Coleman 1989, Aoki & Kikuchi 1991). It is not known whether isopod or decapod parents also engage in grooming of juveniles. Self-grooming is relatively uncommon in isopods (Svavarsson & Davidsdottir 1994), and in decapods it is primarily used to keep the gills clean (Bauer 1989), but too little is known to speculate about the relationship between self-grooming and offspring grooming.

Protection of offspring from predators is one of the most important benefits of parental care. Female crayfish that care for offspring may behave more aggressive against adult conspecifics, and they are more likely to win combats than non-reproductive individuals (Figler et al. 1995). Other crustacean parents also defend their offspring against conspecific predators and competitors (Linsenmair 1987, Stephan 1980, Aoki & Kikuchi 1991). Some crustacean parents are even able to protect their offspring from larger predators such as fish and decapods (Thiel 1999c). The parental dwelling is of particular importance, and deep burrows or otherwise inaccessible dwellings appear particularly suited to protect offspring effectively against predators (Thiel 1999c).

Parents may also protect their offspring against adverse abiotic conditions, such as steep salinity or temperature variations, desiccation risk and sediment disturbance. This is probably the case for many semi-terrestrial decapods such as the potamonid crabs and crayfish, and terrestrial isopods.

3.4 *Recognition of mates and offspring*

Similar to crustaceans in which mates recognize each other (Johnson 1977, Seibt & Wickler 1979, Caldwell 1992), parents should be expected to discriminate between their own and foreign offspring. With the exception of the desert isopod *H. reaumuri*, it is not well known whether and how crustacean parents are able to recognize their own offspring. In *H. reaumuri*, parents use chemical signals to recognize their partner and their offspring. They were able to recognize their kin after being separated from them for days or even weeks (Linsenmair 1987, 1989). In the freshwater crayfish *Procambarus clarkii*, females were not able to discriminate between their own and a foreign brood (Figler et al. 1997). Females separated from their brood would readily accept juveniles other than their own if those were at a similar developmental stage as their own offspring. The authors suggest that parents do not discriminate between their own and foreign offspring because the likelihood to encounter foreign offspring would be extremely low in nature. That is, parents would recognize their offspring by association: The likelihood of foreign juveniles to enter a parental burrow is very low and therefore all juveniles in its burrow will be accepted by a parent. In the burrow-dwelling amphipod *Casco bigelowi*, the experimental addition of a foreign juvenile to a burrow with a female and its offspring caused aggressive reactions of both the female and her own juveniles towards the foreign juvenile (Thiel, pers. obs.). This observation and the studies on desert isopods by Linsenmair (1989) demonstrate that in several crustacean species which care for their offspring in a dwelling, parents discriminate between kin and foreign offspring even within their own dwelling.

3.5 *Parental investment during parental care*

'Most aspects of parental care involve substantial expenditure of energy or other resources by the parents, though the energetic costs of some aspects of parental care have received more attention than others' (Clutton-Brock 1991).

Parental investment by crustaceans into individual offspring is usually measured in terms of yolk resources. Peracarid crustaceans produce all comparatively large eggs, and all species carry their offspring during embryonic and larval development. Egg numbers of peracarid species that provide extended parental care to their offspring are in the same range as those of peracarid species without extended care (Thiel 1997, compare also Sainte-Marie 1991). For decapods, several authors have revealed that species which brood their embryos to advanced larval stages, provide large amounts of yolk to individual eggs (Rabalais & Gore 1985). These yolk reserves sustain embryonic and larval development during the egg phase. A consequence of allocating much yolk to individual eggs is that fewer eggs can be produced. Limnic and terrestrial decapods that produce large eggs have often smaller brood sizes than marine counterparts that produce comparatively small eggs (Rabalais & Gore 1985).

Relatively little attention has been given to the fact that parents not only invest yolk resources into their offspring but also time. Crustacean parents that produce large (and few) eggs also carry the developing embryos for relatively long time periods. While carrying developing offspring, female molting and the production of a second brood may be inhibited (Aoki & Kikuchi 1991, Figler et al. 1997). Thus, most crustaceans that carry their offspring do not incubate a second brood while carrying the first brood. In contrast, several crustacean species that care for their offspring in or on a dwelling produce subsequent broods while caring for older siblings (e.g. *Engaeus leptorhyncus* – Horwitz et al. 1985, *Peramphithoe stypotrupetes* – Conlan & Chess 1992, snapping shrimp – Duffy 1996, *Leptocheirus pinguis* – Thiel et al. 1997, *Dyopedos monacanthus* – Thiel 1998). Thus, crustaceans that inhabit distinct dwellings such as tubes or burrows, have the possibility to release offspring into this dwelling and incubate a subsequent brood on their body. Crustaceans that carry offspring during extended parental care, apparently are more constrained, and produce and care for only one clutch at a time. This comparison elucidates that extended parental care may not only limit the number of offspring per brood, but also the number of broods produced.

3.6 *When parental care is over*

The end of the parental care period could be initiated by parents or by their offspring. Parents may terminate care for developed offspring because either parent offspring competition increase beyond a certain limit, or simply because they require all available resources to produce a subsequent brood. Other scenarios may be that juveniles actively leave their parents (and their siblings) because intra-clutch competition has increased too much and not all juveniles can be adequately accommodated by their parents. In several crustacean species that engage in extended parental care, juveniles build their first own dwelling 'budding' off from the parental dwelling (Thalassinidea – Forbes 1973; Amphipoda – Thamdrup 1935, Thiel et al. 1997; Isopoda – Henderson 1924, Menzies 1957; Tanaidacea – Bückle-Ramirez 1965). A consequence of this behaviour can be that parental and juvenile dwellings are located right next to each other (Flach 1992) which could give rise to continuing adult-juvenile competition (see e.g. Wilson 1989).

3.7 *Maternal, paternal or bi-parental care?*

In almost all crustaceans, females are the primary care givers. They incubate the embryos after fertilization and are present upon emergence of the larvae or juveniles from the brood chambers. In some species, males remain in the female burrows for long time periods after the juveniles have emerged from the female brood structure (many boring isopods: Henderson 1924, Menzies 1957, Brearley & Walker 1995). The role of males during this bi-parental care is even less known than the role of females during parental care. In the desert isopod *Hemilepistus reaumuri*, the males apparently share all parental tasks such as foraging, burrow defense and burrow cleaning equally with the female. In other species in which the males stay with the female during parental care (semi-terrestrial crayfish, boring isopods and amphipods), the males probably also participate in burrow maintenance and defense. At present, paternal care is only known from burrow dwelling stomatopods in which males take over the task of caring for one brood while the female incubates a subsequent brood (R. Caldwell, pers. comm.). Exclusive paternal care without any assistance from the female is only known from Pycnogonids where males incubate embryos immediately after fertilization.

3.8 *Non-parental care for small juveniles*

In several crustacean species males are frequently found to 'care' for a few, small, immature individuals. For example in some isopod species among the Janiridae, males often carry immature females (Marsden 1982, Franke 1993, Kensley 1994). Isopods from the genus *Gnathia* live in harems where one male 'monopolizes' several females and larvae (see e.g. Barthel & Brandt 1995, and references therein). I consider it likely that the immature individuals 'cared' for by these males will upon closer examination prove to be females genetically unrelated to these males. Thus, these cases probably do not represent parental care, but rather early stages of mate guarding similar to what is seen in many other crustaceans. The paternal care observed in *Siphonoecetes dellavallei* (Richter 1978a) could upon closer examination be another case of early mate guarding.

3.9 *Some consequences of parental care*

Consequences of parental care may be positive (benefits, e.g. improved offspring survival or growth), or negative (costs, e.g. reduced dispersal or increasing potential for sibling competition and parasite transmission). Experimental removal of parents has demonstrated that parental protection improves offspring survival (Diesel 1992, Thiel 1997) and growth (Aoki 1997). Little is known about the direct effects of parental grooming and feeding on offspring survival or growth. In isopod species *Porcellio* and *Hemilepistus* the juveniles are fully dependent on parental feeding (Linsenmair 1989), but nothing is known in other species. Parental care may not only have effects on the offspring but also on the parents themselves. Parents that care for offspring may be susceptible themselves. Molting appears to be inhibited in offspring-carrying crustaceans, and during parental care parents may accumulate substantial loads of epibionts on their carapace. High epibiont loads have been reported from some crustaceans that engage in extended parental care (Duffy 1992, Svavarsson & Davidsdottir 1995, Thiel 1997). These epibionts could be costly to both parents and their offspring. Parents that care for offspring may also be limited in their own ability to feed or to escape from predation. As in many crustacean species with extended parental care offspring recruits into the immediate vicinity of the parents, it appears that dispersal may also be restricted in these species. A consequence of reduced dispersal in crustaceans with parental care may be the risk of continuing adult-juvenile competition (see above), and an increase in genetic differentiation on relatively small scales (km) as has been found for snapping shrimp (Duffy 1992, 1993). Schubart et al. (1998) demonstrated high speciation rates in land crabs that engage in extended parental care. Further studies investigating the consequences of parental care in crustaceans are required to better understand the evolution of this reproductive behavior.

4 OUTLOOK

The overview presented herein indicates that parental care is not uncommon among some groups of malacostracan crustaceans. Yet, it also becomes evident that very little is known about the behavior of crustacean parents, the resources they invest into individual offspring, the benefits and costs of this reproductive behavior and its evolution. One of the most fascinating facts about parental care in crustaceans that has been revealed herein is that similar forms of parental care have convergently evolved among aquatic and terrestrial crustaceans. Most previous studies on the evolution of parental care and advanced social behavior have been conducted on insect groups that are limited to the terrestrial environment. The study of crustacean social behavior offers the unique opportunity to compare species from aquatic and terrestrial environments, and therefore appears to be a rewarding enterprise for biologists interested in the evolution of social behavior.

ACKNOWLEDGEMENTS

I thank S. Dobretsov for translating the Russian references, and K.E. Linsenmair for helpful comments on the manuscript. This is contribution No. 463 of the Smithsonian Marine Station at Fort Pierce.

Species	Parent	size (mm)	structure measured	environment	care for embryo / early larvae	care for advanced larvae / juvenile	semelparous / iteroparous	simultaneous / sequential brooders	parental behavior	duration of care for embryo (d)	duration of care for juvenile (d)	juvenile recruitment aided by parent	offspring size when leaving parent (mm)	Number of broods / female life	Number of broods / year	Number of eggs / brood	Number of large offspring / brood	Source
Amphipoda																		
Phronima sedentaria (Forskål, 1775)	f	35	BL	M	cy	dw	s		fg	10+				1		200-500	100-200	Richter, G. 1978. Senckenbergiana maritima 10: 229-242
Gammarus obtusatus	f	11	BL		cy									1+		10		Sexton, E.W. 1925. J. mar. biol. Ass. UK 13: 340-401
Gammarus palustris (Bousfield, 1969)	f	7	BL	M	cy		i	sq		10	2			1+		10		Borowsky, B. 1980. J. exp. mar. Biol. Ecol. 42: 213-223
Eulimnogammarus (Melinogammarus) obtusatus	f	15	BL		cy									1+		21		Sheader, M. & F.-S. Chia, 1970. J. mar. biol. Ass. UK 50: 1079-1099
Eulimnogammarus firmarchicus	f	13.5	BL		cy									1+		4		Sheader, M. & F.-S. Chia, 1970. J. mar. biol. Ass. UK 50: 1079-1099
Neohaustorius schmitzi	f	4	BL	M	cy	dw	i	sim						1+		30-145		Croker, R.A. 1968. Crustaceana 14: 215.
Paramphithoe stizochpoetes Conlan and Chess, 1992	f	20	BL	M	cy	dw			pg					4+		4	60	Conlan, K.E. & J.R. Chess, 1992. J. Crust. biol. 12: 410-422
Haploops vallifera Stephenson 1925	f	6	BL	M	cy	dw			pg, fg	10	[0.5-7]	n	5	1+		[8]		Dauvin, J.C. 1993. Boll. Mus. civ. St. nat. Verona 20: 47-60
Lembos websteri Bate, 1856	f	4.7	BL	M	cy	cy					1+		10			8		Shillaker, R.O. & P.G. Moore. 1987. J. exp. mar. Biol. Ecol. 110: 113-132
Siphonoecetes dellavallei Stebbing, 1899	f	5	BL	M	cy	cy					1+					6		Richter, G. 1978. Natur und Museum 108: 259-266
Corophium bonnelii (Milne Edwards, 1830)	f	3.1	BL	M	cy	dw										40		Shillaker, R.O. & P.G. Moore. 1987. J. exp. mar. Biol. Ecol. 110: 113-132
Corophium volutator (Pallas, 1766)	f	7	BL	M	cy	dw						m				14		Thamdrup, H.M. 1935. Meddr. K. Danm. Fisk.- og Hav. (Ser. Fiskeri) 10: 1
Corophium arenarium Crawford, 1937	f	5	BL	B	cy	dw				1+				1+	1+	11		Crawford, G.I. 1937. J. mar. biol. Ass. UK 21: 589-630
Leptocheirus pilosus Zaddach, 1844	f	3.5	BL	M	cy	dw		sim				n	5	5	5	60		Goodhart, C.B. 1939. J. mar. biol. Ass. UK 23: 311-325
Leptocheirus pinguis (Stimpson 1853)	f	17	BL	M	cy	dw	i		pg				10	1	1	50	40	Thiel et al. 1997. J. Nat. History 31: 713-725
Casco bigelowi (Blake 1929)	f	20	BL	M	cy	dw	s											Thiel et al. 1997. J. Nat. History 31: 713-725
Maera loveni (Bruzelius)	f	20	BL	M	cy	dw			fg		10		10					Thiel et al. 1997. J. Nat. History 31: 713-725
Paraceradocus gibber Andres, 1984	f	30-80	BL	M	cy	cy						[n]		1+	1+	[3-24]		Atkinson, R.J.A., Moore, P.G. & P.J. Morgan, 1982. J. Zool. 198: 399-416
Leucothoe spinicarpa (Abildgaard, 1789)	f	10	BL	M	cy	dw					30		[3-4]			20		Coleman, Ch.O. 1989. Polar Biol. 10: 43-48
Melita glacialis Krøyer, 1842	f	8	BL	M	cy	dw		sim	gg				3	1+	1+	20		Thiel, 1998. J. Nat. Hist., in press.
Dulichia falcata (Bate 1857)	f	8.5	BL	M	cy	dw	i									20-30		Vader, W. & C.L. Beehler, 1983. Astarte 12: 57-61
Dulichia rhabdoplastis McCloskey, 1970	f	4	BL	M	cy	dw												Kannewoff, E. & W. Nicolaisen, 1973. Ophelia 10: 119-129
Dyopedos monacanthus Metzger 1875	f	8	BL	M	cy	dw	i	sim	pg, gg	[12-22]	14		2	5	5	[10-140]	[20-40]	McCloskey, L.R. 1970. Pacific Science 24: 90-98
Dyopedos porrectus (Bate, 1857)	f	6.5	BL	M	cy	dw		sim	gg	[12-24]	30		2	5	5	20-30		Mattson, S. & T. Cedhagen, 1989. J. exp. mar. Biol. Ecol. 127: 253-272
Parajassa pelagica (Leach, 1814)	f	9	BL	M	cy	dw			gg	13-24				1+	1+			Mattson, S. & T. Cedhagen, 1989. J. exp. mar. Biol. Ecol. 127: 253-272
Jassa staudei Conlan, 1990	f	6	BL	M	cy	dw												Bate, C. Spence, 1862. British Museum, London: 399 pp.
Pseudoprotella phasma (Montagu, 1804)	f	[10-14]	BL	M	cy	cy				4			4					Conlan, K.E. 1990. Can. J. Zool. 68: 2031-2075; Staude, C., pers. comm.
Caprella scaura typica Mayer, 1890	f	10	BL	M	cy	cy					7		4	1+				Harrison, R.J., 1940. J. mar. biol. Ass. UK 24: 483-493
Caprella monoceros Mayer, 1890	f	12	BL	M	cy	cy		sq			20			1+	1+	68		Lim, S.T.A. & C.G. Alexander, 1986. Mar. Behav. Physiol. 12: 217-230
Caprella decipiens Mayer, 1890	f	15	BL	M	cy	[dw]		sim			30		5			110		Aoki, M. & T. Kikuchi, 1991. Hydrobiologia 223: 229-237
Caprella subinermis Mayer, 1890	f	11	BL	M	cy	cy					4		1.8					Aoki, M. & T. Kikuchi, 1991. Hydrobiologia 223: 229-237
Aeginina longicornis Krøyer 1842	f		BL	M	cy	cy	i						3	1+	1+	35		Aoki, M. 1997. J. Crust. Biol. 17: 447-458
																		Thiel, M. 1997. J. Crust. Biol. 17: 447-458

Appendix 1. Amphipod crustaceans with extended parental care (f – female, m – male, b – both, BL – body length, CL – carapace width, M – marine, T – terrestrial, cy – carrying, dw – dwelling, I – iteroparous, s – semelparous, sim – simultaneous, sq – sequential, fg – feeding, pg – protecting, gg – grooming).

Species	Parent	size (mm)	structure measured	environment	care for embryo / early larvae	care for advanced larvae / juvenile	semelparous / iteroparous	simultaneous / sequential brooders	parental behavior	duration of care for embryo (d)	duration of care for juvenile (d)	juvenile recruitment aided by parent	offspring size when leaving parent (mm)	Number of broods / female life	Number of broods / year	Number of eggs / brood	Number of large offspring / brood	Source
Isopoda																		
Arcturus baffini (Sabine 1824)	f	35	BL	M	cy	cy							8				20	Svavarsson, J & B Davidsdottir, 1996. Polar Biology 15: 569-574
Arcturus ubari Gurjanova, 1933	f	50	BL	M	cy	cy											20+	Kussakin, O.G. 1982. Isopoda II. Nauka, St. Petersburg, 463pp
																		Zool Mus. St. Petersburg
Arcturus hastiger Richardson, 1909	f	60	BL	M	cy	cy												Gurjanova
Arcturus setosus Gurjanova, 1933	f	30	BL	M	cy	cy										70		Gurjanova
Arcturus crassispinis Richardson, 1909	f	20	BL	M	cy	cy							4					Kussakin, O.G. 1982. Isopoda II. Nauka, St. Petersburg, 463pp
Arcturus granulatus Richardson, 1909	f	25	BL	M	cy	cy												Barnes, H.G. 1980. Invertebrate Zoology
Astacilla longicornis (Sowerby, 1806)	f	110	BL	M	cy	cy	s			600			8	1		350-1000	100-200	Janssen H H & B Hoese, Polar Biol. 13: 145-149
Glyptonotus antarcticus Eights 1853	f	53	BL	M	cy		s			600			10.9			105		Luxmoore, R A, 1982. Polar Biol. 1: 3-11.
Serolis cornuta Studer	f	19	BL	M	cy		i	sq					5.5	1+		40		Luxmoore, R A, 1982. Polar Biol. 1: 3-11.
Serolis polita Pfeffer	f	44.6	BL	M	cy								6.2			232		Luxmoore, R A, 1982. Polar Biol. 1: 3-11.
Serolis pagenstecheri Pfeffer	f	11.8	BL	M	cy								3.7			30		Luxmoore, R A, 1982. Polar Biol. 1: 3-11.
Serolis septemcarinata Miers	f	6	BL	M	cy	dw										5		Wolff, T, 1976. Aquatic Botany 2: 161-174.
Echinochambeana n.sp.	b			T	cy				fg									Schneider, P. 1971. Z. Tierpsychol. 29:121-133
Hemilepistus aphganicus Bonitzky, 1958	b			T	cy	dw			fg									Linsenmair, K.E. 1972. Z. Tierpsychol. 31: 131-162
Hemilepistus reaumuri Adduin u Savigny	b			T	cy	dw	s		fg				(4-5)	1		100		Marikovsky, P.J. 1969. Zool. Zh. 48: 677-685.
Hemilepistus rhinoceros	f			T	cy				fg									Linsenmair, K.E. 1989. In Rasa, Vogel & Voland (eds): 19-47.
Porcelio albinus	f			T	cy		i	s	fg									Warburg, M R & N Cohen, 1991. J. Crust. Biol. 11: 368-374.
Schizidium liberirianum Verhoeff, 1923	f			M	cy	dw								1		80-160		Gonzalez, M & E Jaramillo, 1991. Rev. Chil. Hist. Nat. 64: 37-51
Ecdoea magellanica Cunningham, 1871	b	4	BL	M	cy	dw	[i]			~30		c		1+		[8-15]		Henderson, J.T. 1924. Contr. Canad. Biol. n ser 2, part 1 no 14: 309-325
Limnoria lignorum (Rathke)	f	5		M	cy	dw	[i]			~30		[c]		1+				Menzies, R J, 1957. Bull. Mar. Sci. Gulf Caribb. 7: 101-200
Limnoria algarum Menzies 1957	f	2	BL	M	cy	dw	[i]		pg			[c]				[1-4]		Brearley, A & D I. Walker, 1995. Aq. Bot. 52: 163-181.
Lynseia arriae Cookson & Poore 1994	b	8	BL	M	cy	dw	[i]	sim		~50	~30	c	~2.5	1+		20-50		Messana et al 1994. Evol. Ecol. Eth. special issue 3: 125-129
Sphaeroma terebrans Bate, 1866	f			M	cy													Charmantier, G. pers. comm.
Sphaeroma serratum (Fabricius, 1787)																		
Tanaidacea																		
Tanais cavolinii Milne Edwards	f	5	BL	M	cy	dw			[fg]	15	7		1.2	1+		20	15	Johnson, S B & Y.G. Attramadal, 1982. Mar. Biol. 71: 11-16
Heterotanais oerstedi Krøyer, 1842	f			M	cy	dw			[fg]			c						Buckle-Ramirz, L.F. 1965. Z. Morph. Ökol. Tiere 55: 714-782.
Nematotanais mirabilis Bird and Holdich, 1985	f	2	BL	M	cy	dw			pg			c	1.3				4+	Bird, G J & D M. Holdich, 1985. J. mar. biol. Ass. U.K. 65: 563-572.
Mirandotanais vorax Sieg 1984	f			M	cy	dw												Sieg, J. 1986. Antarctic Research Series 45
Langitanais willemoesi (Studer, 1883)	f	4	BL	M	cy	dw							<1.7			25		Shiino, S M, 1978. Sci. Rep. Shima Marineland 5: 1-122
Leptochelia dubia (Krøyer, 1842)	f	3	BL	M	cy	[dw]							1.2	3+		10		Mendoza, J A, 1982. Crustaceana 43: 225-240
Paguropseudes largoensis McSweeny, 1982															5+				Messing, C.G., 1983. J. Crust. Biol. 3: 380-408
Mysidacea																		
Heteromysis harpax Hilgendorf 1878	b	4	CL	M	cy	dw		sq		(>30)			2	5+	5+	8		Vannini, M., Innocenti, G. & R K. Ruwa, 1993. Tropical Zool. 6: 189-205

Appendix 2. Isopod, tanaid, and mysid crustaceans with extended parental care (f – female, m – male, b – both, BL – body length, CL – carapace width, M – marine, T – terrestrial, cy – carrying, dw – dwelling, I – iteroparous, s – semelparous, sim – simultaneous, sq – sequential, fg – feeding, pg – protecting, gg – grooming).

Species	Parent	size (mm)	structure measured	environment	care for embryo/early larvae	care for advanced larvae/juvenile	semelparous/iteroparous	simultaneous/sequential brooders	parental behavior	duration of care for embryo (d)	duration of care for juvenile (d)	juvenile recruitment aided by parent	offspring size when leaving parent (mm)	Number of broods/female life	Number of broods/year	Number of eggs/brood	Number of large offspring/brood	Source
Decapoda																		
Synalpheus regalis Duffy, 1996	b*	3	CL	M	cy	dw	-	sim		33	10+	n	5 (BL)	1+	1+	20-60		Duffy, J.E. 1996. Nature 381: 512-514
Callianassa kraussi Stebbing	f	10	CL	M	cy	cy & dw												Forbes, A.T. 1973. Mar Biol 22: 361-365
Callianassa tyrrhena	f	20	CL	M	cy	[dw]	-	sq										Thessalou-Legaki, M & V Kiortsis, 1997 Mar Biol 127: 435-442
Uca subcylindrica	f		CL	T	cy					40			1,3			25-1400		Rabalais, N.N & J.N Cameron, 1983 J. Crust Biol 3: 519-541
Upogebia edulis Ngoc-Ho & Chan, 1992	f	20	CL	M	cy	cy												Shy, J.-Y & T.-Y. Chan, 1996 Crustaceana 69: 175-1186
Filumnus lumpinus Bennett, 1964	f	21	CL	M	cy	cy				50	1+					60-100		Wear, R.G. 1967 N.Z. Jl mar Freshwater Res. 1: 482-535
Filumnus novaezealandiae Filhol, 1886	f	21	CL	M	cy	cy										50-250		Wear, R.G. 1967 N.Z. Jl mar Freshwater Res. 1: 482-535
Paranaxia serpulifera (Guérin)	f	59-85	CL	M	cy	cy				50	3+		6.3 (CL)			100-1000	150	Morgan, G.J. 1987 Rec. West. Aust. Mus. 13: 337-343
Metopaulias depressus Rathbun	f	13-20	CL	T	cy & dw	dw			fg		50					14-92		Diesel, R. 1989 Anim Behav. 38: 561-575
Sesarma cookei Hartnoll	f	19	CL	M	cy						50					20		Abele, L.G. & D.B. Means, 1977. Crustaceana 32: 91-93
Sesarma jarvisi Rathbun	f	15	CL	T	cy											1500		Anger, K. 1995 J. exp. mar. Biol. Ecol. 193: 119-145
Sesarma curacaoense de Man	f	10	MT	M	cy & dw	dw				50+			[6-8]			[7-18]		Diesel, R. & D. Horst, 1995. J. Crust. Biology 15: 179-195
Sesarma nodulifera de Man, 1892	f	11	CL	T	cy					45	8		2			20-40	D18	Pesta, O. 1930 Archiv für Hydrobiologie. Suppl VIII: 92-108
Sesarma verleyi Rathbun	f	8	CL	T	cy	cy					3+					25-70		Soh, C.L. 1969 J Zool 158: 357-370
Sesarma perracae Nobili, 1903	f	20	CL		cy											[8-12]		Ng, P.K.L & C.G.S Tan, 1995 Crustaceana 68: 390-395
Geosesarma notophorum Ng & Tan, 1995	f		MFB		cy					[30]	1+		7			2360		Anger, K. 1995 J. exp. mar. Biol. Ecol. 193: 119-145
Armases miersii (Rathbun)	f	27	CL	FT	cy	cy										300-500		Tyagi, A.P. 1973 Zoologica Poloniae 22: 171-176
Paratelphusa masoniana (Henderson)	f		CL	F	cy													McCann, C. 1938 J. Bombay Nat. Hist. Soc. 39: 531-542
Paratelphusa (Barytelphusa) guerini (M.-Eds.)	f	(10-35)	CL	F	cy								7			80		Pesta, O. 1930 Archiv für Hydrobiologie. Suppl VIII: 92-108
Paratelphusa (Paratelphusa) convexa (de Haan)	f	15-35	CL	F	cy	cy												Bishop, J.A. 1963 Aust. J. Mar. Freshw. Res. 14: 218-238
Paratelphusa agassizi (Rathbun)	f	34	CL	F	cy													Fernando, C.H. 1960 Ceylon J. Sci. (Bio. Sci) 3: 191-224
Paratelphusa angustifrons (A. Milne Edwards, 1869)	f	15-30	CL	F	cy					[50?]	[50?]		7					Fernando, C.H. 1960 Ceylon J. Sci. (Bio. Sci) 3: 191-224
Paratelphusa bouvieri (Rathbun, 1904)	f	23-45	CL	F	cy													Fernando, C.H. 1960 Ceylon J. Sci. (Bio. Sci) 3: 191-224
Paratelphusa ceylonensis Fernando, 1960	f	31-55	CL	FT	cy & dw	dw				41	30					240-420		Pillai, C.K & T. Subramoniam, 1984 Hydrobiologia 119: 7-14
Paratelphusa erodis (Kingsley, 1880)	f	15-35	CL	F	cy													Fernando, C.H. 1960 Ceylon J. Sci. (Bio. Sci) 3: 191-224
Paratelphusa hippocastanum (Müller, 1887)	f	(9-27)	CL	FT	cy													Fernando, C.H. 1960 Ceylon J. Sci. (Bio. Sci) 3: 191-224
Paratelphusa hydrodromous (Herbst)	f	[8-41]	CL	F	cy						1							Chandran, M.R et al. 1980 Proc. of 1st All India Symp Invert Reprod
Paratelphusa innominata Fernando, 1960	f	20-42	CL	F	cy													Fernando, C.H. 1960 Ceylon J. Sci. (Bio. Sci) 3: 191-224
Paratelphusa (Barytelphusa) jacquemontii Rathbun	f	14-50	CL	F	cy													Fernando, C.H. 1960 Ceylon J. Sci. (Bio. Sci) 3: 191-224
Paratelphusa rugosa (Kingsley, 1880)	f	23-45	CL	F	cy													Bishop, J.A. 1963 Aust. J. Mar. Freshw. Res. 14: 218-238
Paratelphusa soror (Zehntner, 1894)	f	(10-20)	CL	F	cy													Bishop, J.A. 1963 Aust. J. Mar. Freshw. Res. 14: 218-238
Paratelphusa transversa (von Martens, 1869)	f	20-25	CL	F	cy	cy												Bishop, J.A. 1963 Aust. J. Mar. Freshw. Res. 14: 218-238
Paratelphusa valentula Riek, 1951	f		FT	FT	cy								4			15-53	20-70	Koba, K. 1936 Proceedings of the Imperial Academy of Japan 12: 105-107
Geodelphusa dehaani (White, 1847)	f	44	CL	F	cy					46	1		4,5			200	20	Pace, F. Harris R.R & V. Jaccarini, 1976 J. Zool 180: 93-106
Potamon edulis (= P. fluviatile)	f	40-49	CL	F	cy								3				240	Pesta, O. 1930 Archiv für Hydrobiologie. Suppl VIII: 92-108
Potamon (Potamon) brevimarginatum (de Man, 1892)	f	20	CL	F	cy													Pesta, O. 1930 Archiv für Hydrobiologie. Suppl VIII: 92-108
Potamon (Geodelphusa) kuhli (de Man, 1892)	f	20-30	CL	T	cy								[3-5]			30-50		Vuillemin, S. 1970 Ann. de l'Université Madagascar (Sc.) 7: 245-266
Madagapotamon humberti Bott, 1955	f																	Cumberlidge N & R Sachs, 1991 Crustaceana 61: 55-68
Globonautes macropus (Rathbun, 1898)	f																	

Appendix 3a. Decapod crustaceans with extended parental care (f – female, m – male, b – both, BL – body length, CL – carapace width, M – marine, T – terrestrial, cy – carrying, dw – dwelling, I – iteroparous, s – semelparous, sim – simultaneous, sq – sequential, fg – feeding, pg – protecting, gg – grooming).

Species	Parent	size (mm)	structure measured	environment	care for embryo / early larvae	care for advanced larvae / juvenile	semelparous / iteroparous	simultaneous / sequential brooders	parental behavior	duration of care for embryo (d)	duration of care for juvenile (d)	juvenile recruitment aided by parent	offspring size when leaving parent (mm)	Number of broods / female life	Number of broods / year	Number of eggs / brood	Number of large offspring / brood	Source
Macrobrachium nipponense Shokita & Takeda, 1989	f	23-35	BL	F	cy								4.7 BL			32-81		Shokita, S., Takeda, M., Sittlert, S & T Polpakdee, J Crust Biol 11 90-1
Macrobrachium iheringi (Ortmann, 1897)	f	55-60	BL	F												55-110		Bueno, S.L. de S. & S. de A. Rodrigues, 1995 Crustaceana 68 665-686
Cambarus virilis	f			F												100+		Little, E.E. 1975 Nature 255: 400-401.
Orconectes virilis (Hagen, 1870)	f	8		F	cy	dw				7+						several 100?		Ameyaw-Akumfi, C. 1976. Ph.D thesis. Hazlett, B.A., 1983. 171-193
Orconectes propinquus (Girard)	f	8		F	cy											several 100?		Ameyaw-Akumfi, C. 1976. Ph.D thesis. Hazlett, B.A., 1983. 171-193
Orconectes sanborni	f			F	cy											100+		Little, E.E. 1975 Nature 255: 400-401.
Procambarus clarkii	f	8	CL	F	cy	dw	[i]			100+			5 CL			several 100?		Ameyaw-Akumfi, C. 1976. Ph.D thesis. Hazlett, B.A., 1983. 171-193
Procambarus hayi (Faxon)	f	36-45	CL	F	cy	dw	s	sq		25	25		3			20		Payne, J.F., 1972. Amer. Midl. Natur. 87: 25-35
Fallicambarus gordoni Fitzpatrick	f	[10-30]	CL	T	cy	[dw]		sq		120			5 CL	1	1	[50-150]		Johnston, C.E. & C. Figiel, 1997. J. Crust. Biol. 17: 687-691
Fallicambarus fodiens	f	[30-40]	CL	T	cy			sq		110	50+			1+	1	20-150		Norrocky, M.J., 1991. Am. Midl. Nat. 125: 75-86.
Paranephrops planifrons White	f	18-30	CL	F	cy	dw				150			3.4-3.8 CL		1	200-600		Hopkins, C.L., 1967. N.Z. Jl. mar. Freshwat. Res. 1: 51-58
Cherax tenuimanus (Smith)	b	40-90	CL	F	cy	dw		sim					10-15 CL		[1]	108	45-75	Horwitz, P.H.J., Richardson, A.M.M. & P.M. Cramp, 1985. Tasm. Nat. 82 1-
Engaeus leptorhyncus	f	25-35	CL	F	cy	dw	i									45-75		Suter, P.J., 1977. Aust. J. Mar. Freshwater Res. 28 85-93
Engaeus cisternarius Suter	f	25-35	CL	F	cy	dw	i	sq		120	70		4-6 CL	1+	1	30-100		Suter, P.J., 1977. Aust. J. Mar. Freshwater Res. 28 85-93
Engaeus fossor Erichson	f	20-30	CL	F	cy	dw	i	sq		120	70		4-6 CL	1+	1	30-100		Suter, P.J., 1977. Aust. J. Mar. Freshwater Res. 28 85-93
Parastacoides tasmanicus (Erichson, 1846)	f	24-36	CL	F	cy	dw	i	sq		120	90		3.5-3.8 CL	[1+?]	1	38-80		Lake, P.S & K.J. Newcombe, 1975. Aust. Zool. 18 197:214
Stomatopoda																		
Gonodactylus bredini Manning	f			M	cy & dw			sim										Dingle, H. & R.L. Caldwell, 1972. Biol. Bull. 142: 417-426
Pycnogonida																		
Boreonymphon abyssorum	m	11	BL	M	cy													Bamber, R.N., 1983. Zool. J. Linn. Soc. 77: 65-74
Paranymphon spinosum	m			M	cy													Bamber, R.N., 1983. Zool. J. Linn. Soc. 77: 65-74
Nymphon hirtipes Bell, 1853	m			M	cy	cy							1			4		Hedgepeth, J.W., 1963. J. Fish. Rec. Bd. Canada 20: 1315-1348
Nymphon robustum Bell	m			M	cy	cy												Hedgepeth, J.W., 1963. J. Fish. Rec. Bd. Canada 20: 1315-1348
Nymphon stuarti Hoek	m			M	cy													Hedgepeth, J.W., 1963. J. Fish. Rec. Bd. Canada 20: 1315-1348
Nymphonella tapetis	m			M	cy													Ohshima, 1927. Annotnes zool. Jap. 11: 257-263, cited in KING, P.E. 197
Pycnogonum litorale	m	[4-8]	BL	M	cy		i			30-100			0.15			1700-9000		Wilhelm, E., Bückmann, D. & K.H. Tomaschko, 1997. Mar. Biol. 129 595-6

Appendix 3b. Decapod and stomatopod crustaceans and pycnogonida with extended parental care (f – female, m – male, b – both, BL – body length, CL – carapace width, M – marine, T – terrestrial, cy – carrying, dw – dwelling, I – iteroparous, s – semelparous, sq – sequential, fg – feeding, pg – protecting, gg – grooming).

REFERENCES

Ameyaw-Akumfi, C. 1976. Some aspects of breeding biology of crayfish. Ph.D. thesis, 252 pp. University of Michigan.

Aoki, M. 1997. Comparative study of mother-young association in caprellid amphipods: is maternal care effective? *J. Crust. Biol.* 17: 447-458.

Aoki, M & Kikuchi, T. 1991. Two types of maternal care for juveniles observed in *Caprella monoceros* Mayer, 1890 and *Caprella decipiens* Mayer, 1890 (Amphipoda: Caprellidae). *Hydrobiologia* 223: 229-237.

Barthel, D. & Brandt, A. 1995. *Caecognathia robusta* (G.O. Sars, 1879) (Crustacea, Isopoda) in *Geodia mesotriaena* (Hentschel, 1929) (Demospongiae, Choristidae) at 75°N off NE Greenland. *Sarsia* 80: 223-228.

Bauer, R.T. 1989. Decapod crustacean grooming: Functional morphology, adaptive value, and phylogenetic significance. *Crustacean Issues* 6: 49-73.

Brearley, A. & Walker, D.I. 1995. Isopod miners in the leaves of two Western Australian Posidonia species. *Aquatic Botany* 52: 163-181.

Bückle-Ramirez, L.F. 1965, Untersuchungen über die Biologie von *Heterotanais oerstedi* Kröyer (Crustacea, Tanaidacea). *Zeitschr. Morph. Ökol. Tiere* 55: 714-782.

Bueno, S.L. de S. & de A. Rodrigues, S. 1995. Abbreviated larval development of the freshwater prawn, *Macrobrachium iheringi* (Ortmann, 1897) (Decapoda, Palaemonidae), reared in the laboratory. *Crustaceana* 68: 665-686.

Caldwell, R.L. 1992. Recognition, signaling and reduced aggression between former mates in a stomatopod. *Anim. Behav.* 44: 11-19.

Chess, J.R. 1993. Effects of the stipe-boring amphipod *Peramphithoe stypotrupetes* (Corophioidea: Amphithoidae) and grazing gastropods on the kelp *Laminaria setchellii*. *J. Crust. Biol.* 13: 638-646.

Clutton-Brock, T.H. 1991. *The evolution of parental care*. Princeton: Princeton University Press.

Coleman, C.O. 1989. Burrowing, grooming, and feeding behaviour of *Paraceradocus*, an Antarctic Amphipod Genus (Crustacea). *Polar Biol.* 10: 43-48.

Conlan, K.E. & Chess, J.R. 1992. Phylogeny and ecology of a kelp-boring amphipod, *Peramphithoe stypotrupetes*, new species (Corophioidea: Amphithoidae). *J. Crust. Biol.* 12: 410-422.

Cumberlidge, N. & Sachs, R. 1991. Ecology, distribution, and growth in *Globonautes macropus* (Rathbun, 1898), a treeliving freshwater crab from the rain forests of Liberia (Gecarcinucoidea, Gecarcinucidae). *Crustaceana* 61: 55-68.

Diesel, R. 1989. Parental care in an unusual environment: *Metopaulias depressus* (Decapoda: Grapsidae), a crab that lives in epiphytic bromeliads. *Anim. Behav.* 38: 561-575.

Diesel, R. 1992. Managing the offspring environment: brood care in the bromeliad crab, *Metopaulias depressus*. *Behav. Ecol. Sociobiol.* 30: 125-134.

Duffy, J.E. 1992. Host use pattern and demography in a guild of tropical sponge-dwelling shrimps. *Mar. Ecol. Prog. Ser.* 90: 127-138.

Duffy, J.E. 1993. Genetic population structure in two tropical sponge-dwelling shrimps that differ in dispersal potential. *Mar. Biol.* 116: 459-470.

Duffy, J.E. 1996. Resource-associated population subdivision in a symbiotic coral-reef shrimp. *Evolution* 50: 360-373.

Figler, M.H., Twum, M., Finkelstein, J.E. & Peeke, H.V.S. 1995. Maternal aggression in red swamp crayfish (*Procambarus clarkii*, Girard): the relation between reproductive status and outcome of aggressive encounters with male and female conspecifics. *Behaviour* 132: 107-125.

Figler, M.H., Blank, G.S. & Peeke, H.V.S. 1997. Maternal aggression and post-hatch care in red swamp crayfish, *Procambarus clarkii* (Girard): the influences of presence of offspring, fostering and maternal molting. *Mar. Fresh. Behav. Physiol.* 30: 173-194.

Flach, E.C. 1992. The influence of four macrozoobenthic species on the abundance of the amphipod *Corophium volutator* on tidal flats of the Wadden Sea. *Neth. J. Sea Res.* 29: 379-394.

Forbes, A.T. 1973. An unusual abbreviated larval life in the estuarine burrowing prawn *Callianassa kraussi* (Crustacea: Decapoda: Thalassinidea). *Mar. Biol.* 22: 361-365.

Franke, H.-D. 1993. Mating system of the commensal marine isopod *Jaera hopeana* (Crustacea). *Mar. Biol.* 115: 65-73.

Hazlett, B.A. 1983. Parental Behavior in Decapod Crustacea. In S. Rebach & D.W. Dunham (eds), *Studies in adaptation – the behavior of higher Crustacea*: 171-193. New York: Wiley & Sons.

Henderson, J.T. 1924. The Gribble: a study of the distribution factors and life-history of *Limnoria lignorum* at St. Andrews, N.B. *Contr. Canadian Biol.* 2, part 1 (14): 309-327.

Hopkins, C.L. 1967. Breeding in the freshwater crayfish *Paranephrops planifrons* White. *N.Z. Jl. mar. Freshwater Res.* 1: 51-58.

Horwitz, P.H.J., Richardson, A.M.M. & Cramp, P.M. 1985. Aspects of the life history of the burrowing freshwater crayfish *Engaeus leptorhyncus* at Rattrays Marsh, North East Tasmania. *Tas. Naturalist* 82: 1-5.

Johnson, S.B. & Attramadal, Y.G. 1982. Reproductive behaviour and larval development of *Tanais cavolinii* (Crustacea: Tanaidacea). *Mar. Biol.* 71: 11-16.

Johnson, V.R. 1977. Individual recognition in the banded shrimp *Stenopus hispidus* (Olivier). *Anim. Behav.* 25: 418-428.

Johnston, C.E. & Figiel, C. 1997. Microhabitat parameters and life-history characteristics of *Fallicambarus gordoni* Fitzpatrick, a crayfish associated with pitcher-plant bogs in southern Mississippi. *J. Crust. Biol.* 17: 687-691.

Kensley, B. 1994. Redescription of *Iais elongata* Sivertsen & Holthuis, 1980, from the South Atlantic Ocean (Crustacea: Isopoda: Asellota). *Proc. Biol. Soc. Wash.* 107: 274-282.

Lake, P.S. & Newcombe, K.J. 1975. Observations on the ecology of the crayfish *Parasticoides tasmanicus* (Decapoda; Parastacidae) from South-Western Tasmania. *Aust. Zool.* 18: 197-214.

Linsenmair, K.E. 1972. Die Bedeutung familienspezifischer 'Abzeichen' für den Familienzusammenhalt bei der sozialen Wüstenassel *Hemilepistus reaumuri* Audoiun u. Savigny (Crustacea, Isopoda, Oniscoidea). *Z. Tierpsychol.* 31: 131-162.

Linsenmair, K.E. 1979. Untersuchungen zur Sociobiologie der Wüstenassel *Hemilepistus reaumuri* und verwandter Isopodenarten (Isopoda, Oniscoidea): Paarbindung und Evolution der Monogamie. *Verh. dt. zool. Ges.* 72: 60-72.

Linsenmair, K.E. 1984. Comparative studies on the social behaviour of the desert isopod *Hemilepistus reaumuri* and of a *Porcellio* species. *Symp. Zool. Soc. Lond.* 53: 423-453.

Linsenmair, K.E. 1987. Kin recognition in subsocial arthropods in particular in the desert isopod *Hemilepistus reaumuri*. In D.J.C. Fletcher & C.D. Michener (eds), *Kin recognition in animals*: 121-208. New York: Wiley & Sons.

Linsenmair, K.E. 1989. Sex-specific reproductive patterns in some terrestrial isopods. In A.E. Rasa, C. Vogel, & E. Voland (eds), *The sociobiology of sexual and reproductive strategies*: 19-47. London: Chapman and Hall.

Linsenmair, K.E. & Linsenmair, Ch. 1971. Paarbildung und Paarzusammenhalt bei der monogamen Wüstenassel *Hemilepistus reaumuri* (Crustacea, Isopoda, Oniscoidea). *Z. Tierpsychol.* 29: 134-155.

Luxmoore, R.A. 1982. The reproductive biology of some serolid isopods from the Antarctic. *Polar Biol.* 1: 3-11.

Marikovsky, P.J. 1969. A contribution to the biology of *Hemilepistus rhinoceros*. *Zool. Zournal* 48: 677-685 (in Russian).

Marsden, I.D. 1982. Population biology of the commensal asellotan *Iais pubescens* (Dana) and its sphaeromatid host *Exosphaeroma obtusum* (Dana) (Isopoda). *J. Exp. Mar. Biol. Ecol.* 58: 233-257.

Mattson, S. & Cedhagen, T. 1989. Aspects of the behaviour and ecology of *Dyopedos monacanthus* (Metzger) and *D. porrectus* Bate, with comparative notes on *Dulichia tuberculata* Boeck (Crustacea: Amphipoda: Podoceridae). *J. Exp. Mar. Biol. Ecol.* 127: 253-272.

Menzies, R.J. 1957. The marine borer family Limnoridae (Crustacea, Isopoda). Part I: Northern and Central America: systematics, distribution and ecology. *Bull. Mar. Sci. Gulf. and Carib.* 7: 101-200.

Messana, G., Bartolucci, V., Mwaluma, J. & Osore, M. 1994. Preliminary observations on parental care in *Sphaeroma terebrans* Bate 1866 (Isopoda, Sphaeromatidae), a mangrove wood borer from Kenya. *Ethology Ecology & Evolution, Special Issue* 3: 125-129.

Morgan, G.J. 1987. Brooding of juveniles and observations on dispersal of young in the spider crab *Paranaxia serpulifera* (Guérin)(Decapoda, Brachyura, Majidae) from Western Australia. *Rec. West. Aust. Mus.* 13: 337-343.

Ng., P.K.L. & Tan, C.G.S. 1995. *Geosesarma notophorum* sp. nov. (Decapoda, Brachyura, Grapsidae, Sesarminae), a terrestrial crab from Sumatra, with novel brooding behaviour. *Crustaceana* 68: 390-395.

Pillai, C.K. & Subramoniam, T. 1984. Monsoon-dependent breeding in the field crab *Paratelphusa hydrodromous* (Herbst). *Hydrobiologia* 119: 7-14.

Rabalais, N.N. & Gore, R.H. 1985. Abbreviated development in decapods. *Crustacean Issues* 2: 67-126.

Richter, G. 1978a. Einige Beobachtungen zur Lebensweise des Flohkrebses *Siphonoecetes dellavallei*. *Natur und Museum* 108: 259-266.

Richter, G. 1978b. Beobachtungen zu Entwicklung und Verhalten von *Phronima sedentaria* (Forskal), (Amphipoda). *Senckenbergiana Maritima* 10: 229-242.

Sainte-Marie, B. 1991. A review of the reproductive bionomics of aquatic gammaridean amphipods: variation of life history traits with latitude, depth, salinity, and superfamily. *Hydrobiologia* 223: 189-227.

Schubart, C.D., Diesel, R. & Hedges, S.B. 1998. Rapid evolution to terrestrial life in Jamaican crabs. *Nature* 393: 363-365.

Seibt, U. & Wickler, W. 1979. The biological significance of the pair-bond in the shrimp *Hymenocera picta*. *Z. Tierpsychol.* 50: 166-179.

Shillaker R.O. & Moore, P.G. 1987. The biology of brooding in the amphipods *Lembos websteri* Bate and *Corophium bonnellii* Milne Edwards. *J. Exp. Mar. Biol. Ecol.* 110: 113-132.

Stephan, H. 1980. Lebensweise, Biologie und Ethologie eines sozial lebenden Amphipoden (*Dulichia porrecta, Dulichia monacantha* und *Dulichia falcata* – Crustacea, Malacostraca). Unpubl. Ph.D. Thesis, 326 pp. Kiel.

Suter, P.J. 1977. The biology of two species of *Engaeus* (Decapoda: Parastacidae) in Tasmania; II. Life history and larval development with particular reference to *E. cisternarius*. *Aust. J. Mar. Freshwater Res.* 28: 85-93.

Svavarsson, J. & Davidsdottir, B. 1994. Foraminifera (Protozoa) epizoites on the Arctic isopods (Crustacea) as indicators of isopod behavior? *Marine Biology* 118: 239-246.

Svavarsson, J. & Davidsdottir, B. 1995. *Cibicides* spp. (Protozoa, Foraminifera) as epizoites on the Arctic antenna-brooding *Arcturus baffini* (Crustacea, Isopoda, Valvifera) *Polar Biology* 15: 569-574.

Thamdrup, H.M. 1935. Beiträge zur Ökologie der Wattenfauna auf experimenteller Grundlage.- *Meddr. Kommn. Danm. Fisk. – Og Havunders. (Serie: Fiskeri)* 10: 1-125.

Thiel, M. 1997. Extended parental care in estuarine amphipods. Unpubl. Ph.D.-thesis, 188 pp. University of Maine, Orono.

Thiel, M. 1998. Population biology of *Dyopedos monacanthus* (Crustacea: Amphipoda) on estuarine soft-bottoms – the importance of extended parental care and pelagic movements. *Mar. Biol.* 132: 209-221.

Thiel, M. 1999a. Duration of extended parental care in marine amphipods. *J. Crust. Biol.* 19: 60-71.

Thiel, M. 1999b. Host-use and population demographics of the ascidian-dwelling amphipod *Leucothoe spinicarpa* – indication for extended parental care and advanced social behaviour. *J. Nat. Hist.* 33: 193-206.

Thiel, M. 1999c. Extended parental care in marine amphipods. II. Maternal protection of juveniles from predation. *J. Exp. Mar. Biol. Ecol.* 234: 235-253.

Thiel, M., Sampson, S. & Watling, L. 1997. Extended parental care in two endobenthic amphipods. *J. Nat. Hist.* 31: 713-725.

Vader, W. & C.L. Beehler, 1983. *Metopa glacialis* (Amphipoda, Stenothoidae) in the Barents and Beaufort Seas, and its association with the lamellibranchs Musculus niger and M. discors s.l. *Astarte* 12: 57-61

Wear, R.G. 1967. Life-history studies on New Zealand Brachyura. 1. Embryonic and post-embryonic development of *Pilumnus novaezealandiae* Filhol, 1886, and of *P. lumpinus* Bennett, 1964 (Xanthidae, Pilumninae). *N.Z. Jl. Mar. Freshwat. Res.* 1: 482-535.

Wilson, Jr. W.H. 1989. Predation and the mediation of intraspecific competition in an infaunal community in the Bay of Fundy. *J. Exp. Mar. Biol. Ecol.* 132: 221-245.

Wolff, T. 1976. Utilization of seagrass in the deep sea. *Aq. Bot.* 2: 161-174.

The swimming behavior of the marine wood borer *Limnoria quadripunctata* (Isopoda: Limnoriidae)

SUZANNE M. HENDERSON
FPRC, Buckinghamshire College, Buckinghamshire, UK

ABSTRACT

Migration is the most vulnerable period of the life cycle of limnoriids, during which adults leave their tunnels in wood to locate a new habitat. It is during this period that alternative methods of control should be concentrated to reduce the damage to timber that limnoriids cause each year. This study investigated the swimming behavior of *Limnoria quadripunctata* during migration, using observational analysis with microscopy and filming, and behavioral experiments in laminar flow chambers. Observations showed that limnoriids performed distinct swimming patterns such as circles, flips and backflips, using their setose pleopods. Swimming of limnoriids is therefore not random, as has been previously reported. Behavioral experiments indicated that the frequency of particular limnoriid swimming patterns did not alter in response to water flow or in the presence of chemical attractants. However, certain sequences of behavior patterns were repeated and since limnoriids display positive rheotaxis, an ordered swimming pathway, in response to a flow of water, may exist to aid limnoriids to detect timber.

1 INTRODUCTION

Members of the genus *Limnoria* are small isopods that belong to a destructive group of animals known as marine wood borers. Marine borers together with wood degrading microorganisms have an important ecological role in degrading wood material in the sea (Maser and Sedall 1994). Due to the increased use of timber by humans, marine borers have found a new niche in which they thrive. Recent reports of financial costs are lacking but damage due to attack by *Limnoria* were estimated as 25 million US dollars in Egypt (El Shanshoury et al. 1994) and between 50 and 200 million US dollars annually in the United States (see Menzies 1957, Ray 1959b). However, considering the economic damage that limnoriids inflict, little work has been undertaken to investigate the biology of these animals since the work of Kofoid & Miller (1927), Ray (1959a, b) and Oliver (1962) with the exception of taxonomic research (Cookson 1997) and preservative testing (Eaton & Cragg 1996). A single study on limnoriid behavior has been documented (Geyer 1982).

Prevention of limnoriid attack is predominantly by the use of preservatives and durable timbers. However, limnoriids have been reported attacking creosote (Cookson & Barnacle 1987, Cookson 1990) and copper-chrome-arsenic treated timbers (Barnacle et al. 1983, Cookson & Barnacle 1987), as well as naturally durable wood (Pitman et al. 1995; pers.observ.). As yet, effective methods of protecting timber against limnoriids are not economically viable. Biological control of limnoriids may be an option for the future, especially in view of the environmental concern over the use of wood preservatives for maritime construction (Albuquerque & Cragg 1995) and the harvesting of naturally durable timbers (Kemp 1995). The biology of limnoriids must be fully understood before such control methods could be put into practice.

Throughout most of their life cycle, limnoriids remain protected in their tunnels but during migration adults emerge to the surface of the wood and enter the water column en masse to locate fresh timber. The most vulnerable stage of the limnoriid life cycle is therefore during the period of migration. Limnoriids have not been reported in the plankton (Menzies 1961) and the swimming behavior of limnoriids during migration is unknown. Previous research on the swimming behavior on *Limnoria* in the laboratory has described swimming as erratic, with animals swimming only for short distances of approximately one meter and unable to detect wood using chemoreception (Kofoid & Miller 1927, Oliver 1962). More recently, however, Geyer (1982) stated that *Limnoria tripunctata* were attracted to timber inoculated with marine fungi. Limnoriids show positive rheotaxis, and are attracted to chemicals from conspecifics and biofilms (Henderson 1998). No further studies on the swimming behavior of limnoriids have been reported.

The swimming behavior of marine invertebrates in general can be initiated or controlled in response to various environmental factors (Crisp 1974). Generally, abiotic factors and biotic factors associated with substrates (Morse et al. 1980), prey (Atema 1995), predators or conspecifics (Hadfield 1978) must be detected by individuals before a behavioral response is initiated. The swimming behavior shown by limnoriids during migration may therefore be induced by a response to the environment that facilitates the location of new timber sources. If this is true then this knowledge may be used to prevent these pests locating wood.

Many crustaceans use specific swimming patterns to locate particular objects. Most of these behavior patterns are known as internal motor programmes that involve simple behavior patterns (Atema 1985), which maintain individuals in optimal positions. For example, many crustacean larvae respond to pressure increases to prevent them from sinking to depths (Rice 1964, Forward 1990), others display rheotaxis (Hughes 1972; Shirley & Shirley 988, Anderson & Dale 1989), or simply increase their amount of movement (Meador 1989) to aid in the location of prey or a habitat. Alternatively, some decapods use a searching strategy to actively find their prey that involves regular changes in direction and frequency of specific behavior patterns in response to chemical detection (Tierney & Atema 1988, Moore et al. 1991).

To establish the swimming capabilities of limnoriids, this study investigates their swimming behavior in detail. Observations of limnoriid swimming behavior were made to establish if movements were random or showed distinct patterns. Experiments were designed to establish if water flow or chemical cues from established attractants (Henderson 1998) altered the swimming behavior.

2 MATERIALS AND METHODS

2.1 Culture and extraction of Limnoria quadripunctata

Stock populations of *L. quadripunctata* were obtained from mixed species of driftwood collected from Portsmouth. The driftwood was kept in 25l polythene containers at 18-25°C in aerated natural sea water of 32-35‰ salinity. The water in the cultures was periodically changed with fresh sea water collected from the University of Portsmouth, that had been pumped from the Hayling Island Marine Laboratory through a gravel and sand filter. *Pinus sylvestris* sapwood blocks were fed to the cultures and were eventually colonized by limnoriids from the original timbers.

Limnoriids were extracted from their burrows with the aid of a fine (000) sable paintbrush after gently breaking away the surface layers of the wood to expose animals in their tunnels. Only individuals with a body length of over 3 mm in length were selected to ensure adults were used in experiments. After extraction, animals were transferred with the brush, to 85 mm Petri dishes with Millipore filtered sea water (0.45 µm) to allow them to recover. Individuals were examined after experiments, using a stereo microscope, to identify the species and to note any physical damage that might have occurred during extraction. Data derived from animals showing damage were discarded.

2.2 Swimming behavior observations of L. quadripunctata

Swimming of *L. quadripunctata* was observed using individual animals on a Stereo dynascope (TS4 Zoom, Vision Engineering) with video camera attachment. Animals were filmed in 85 mm diameter Petri dishes filled with sea water and recorded with a portable video recorder and monitor. Animals were tracked using the movable stage.

Observations were also made with the camera and a macro-zoom lens (18-108 mm) from the side in a perspex vessel measuring $30 \times 55 \times 55$ mm and from above in 100 mm diameter, 50 mm deep glass beakers. Illumination was provided by 60 W lamps and a matt-black background made limnoriid movements easier to view. The film was analyzed using a Panasonic editing suite that allowed sequential viewing of frames. Positions of appendages, and speeds and patterns of limnoriid swimming were recorded from the images obtained. However, the exact positions of antennae and the detailed movements of pleopods could not be resolved with this equipment.

2.3 Laminar flow experiments

The swimming behavior of *L. quadripunctata* was filmed in a simplified version of the laminar flow chamber designed by Benfield & Aldrich (1991) (Fig. 1). The flow of water was checked visually for laminarity using dye dissolved in sea water. Flow rates used were between 0.33 and 1 mls^{-1}, and comparable to rates used in flow-through experiments with other crustaceans (Shelton & Mackie 1971, Benfield & Aldrich 1991).

Individual *Limnoria* were filmed from above for a maximum of 15 min or until they disappeared from view at either end of the chamber. The video camera and a

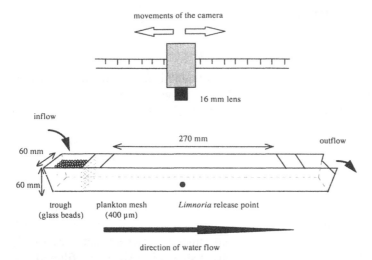

Figure 1. Laminar flow chamber constructed from PVC, based on the design of Benfield and Aldrich (1991). 'Limnoriid extract' and 'microbial' veneers were added to the glass bead trough. The total area in which limnoriids were filmed was 270 × 60 mm.

16 mm lens were attached to a horizontal pulley so that swimming *Limnoria* could be tracked along the length of the laminar flow chamber. The chamber was illuminated by two 60 W lamps positioned at each end of the chamber producing 850-1100 lux.

Swimming behavior was filmed under the following conditions: 1) Stationary sea water, 2) Flowing sea water (as near to laminar as could be achieved), 3) Flowing sea water plus 'limnoriid extract', 4) Flowing sea water plus 'microbial veneers'.

'Limnoriid extract' was prepared by crushing twenty *L. quadripunctata* in a ceramic crucible with the end of a glass rod in some filtered sea water. The crude extract was then divided into two plastic vials with mesh bases, which were immediately placed into the glass bead trough located at the upstream end of the laminar flow chamber (Fig. 1).

'Microbial veneers' were prepared by autoclaving *P. sylvestris* veneers of 0.6 mm thickness and immersing them in sea water for three weeks. Numerous microorganisms have colonized the wood surface after a period of three weeks, such as bacteria, fungi, diatoms, protozoa, and debris (Cundell & Mitchell 1977). The veneers were then cut into 10 mm strips and immersed into the glass bead trough, three to five strips were added for each experiment. Sea water was left to flow through the laminar flow chamber with the added attractant for a period of 1 to 2 min prior to placing a limnoriid within the attractant plume. Between each experiment, the experimental animal was removed and the chamber was washed out thoroughly with fresh tap water.

As no automated data collecting equipment was available, only selected components of limnoriid behavior were used. The percentage times limnoriids spent performing the following behavior patterns were calculated from the video tapes using an editing suite with minute, second and frame counter: 1) Dorsal swimming, 2) Ventral swimming, 3) Stationary on dorsal surface, 4) Stationary on ventral surface, 5) Circles, 6) Backflips, 7) Flips, and 8) Crawling. This was to determine if the frequency of limnoriid behavior characteristics was random.

The swimming behaviors were then categorized, for the purpose of statistical analysis, into swimming behavior (1, 2, 5, 6 and 7), stationary behavior (3 and 4), directional swimming behavior (5, 6 and 7) and crawling behavior (8). The total time spent performing these categorized behaviors were compared for each water flow variable using the Kruskal-Wallis one way analysis of variance (Howell 1989).

3 RESULTS

3.1 *Swimming behavior observations*

Several distinct types of swimming were observed. These were termed dorsal and ventral swimming, circles, flip, and backflips. These behaviors are described fully below.

3.1.1 *Dorsal swimming*

Observations of limnoriid swimming showed that the setose pleopods were used to propel *Limnoria* along the floor of the aquaria in short sharp bursts on their dorsal surface (Fig. 2.1). The telson was raised slightly while the first one or two peraeon somites remained in contact with the aquarium floor, while the peduncle of both pairs of antennae was positioned in front of the cephalon in the direction of swimming. Both pairs of antennae were bent at the flagellum-peduncle joint. The flagella of antenna 1 were held downwards at an angle of 45° to the horizontal and antenna 2 flagella were positioned laterally (see Fig. 2.1).

The mean dorsal swimming speed of *L. quadripunctata* was 1.9 body lengths per second (5.7 mms^{-1}, assuming a body length of 3 mm). The patterns of movements resulting from this swimming were straight lines, arcs, meanderings and large circles (three to four body lengths in diameter). Dorsal swimming continued for varying lengths of time, varying from less than one second to periods of minutes and was followed by a period of rest, grooming or a different swimming pattern.

3.1.2 Circles

Individual limnoriids often turned onto their lateral surface, frequently after dorsal swimming. By continuing to swim in this position, a tight circle was completed, measuring approximately one body length in diameter (Fig. 2.2). The latero-dorsal surface of the first two peraeon somites, and possibly the cephalon, remained in contact with the surface of the aquarium during these circles. The telson was raised slightly off the aquarium floor at an angle of between 10 and 30° to the aquarium floor. Circles were performed singly or in multiples of up to 25 revolutions. The mean time for the completion of one circle was 1.2 s. Both clockwise and anticlockwise circles were observed. Antennae were held in a similar manner to that observed during dorsal swimming. The peraeopods were relaxed, except for the seventh peraeopod held furthest from the aquarium floor which was extended slightly (Fig. 2.2). The peraeon was arched slightly dorsally between peraeon somites two to five throughout the swimming circle. The circle sequence would sometimes terminate with the individual lying on its ventral surface. This was followed by either: a short swimming burst in the ventral position before coming to rest and returning to the dorsal position; or a backflip.

3.1.3 Flip

A flip was produced by individuals beating their pleopods vigorously so the telson was raised until the *Limnoria* was vertical, with its cephalon pointing downwards. By continuously beating the pleopods, the telson was flipped over so that the individual was turned onto its ventral surface (Fig. 2.3).

3.1.4 *Backflip*

Backflips were always initiated by individuals that were on their ventral surface. Therefore a backflip was often preceded by a flip. From a ventral position, the *Limnoria* would swim forward, cephalon first and then turn vertically, before heading back down towards the floor of the aquarium, abruptly stopping when the cephalon and antennae touched the floor (Fig. 2.4).

The entire body was kept relatively rigid with very little arching of the peraeon somites. A single or a number of these backflips could be carried out in quick succes-

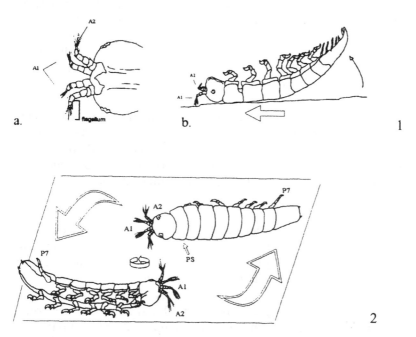

Figure 2. Illustrations showing limnoriid movements and positions of appendages during distinct swimming patterns. 1. Dorsal swimming. a) Ventral view of the cephalon showing both antennae peduncles facing anteriorly in the direction of overall limnoriid movement with the antenna 1 flagellum pointing dorsally and antenna 2 flagellum pointing laterally. b) Small arrow shows the movement of the telson due to the beating of the pleopods, the large arrow shows the general direction of movement. 2. Circle. Individuals move onto their lateral surface from either a dorsal or ventral position. They pivot on the first two pairs of peraeon somites and from a tight circle indicated by the arrows. The telson is raised off the ground at an angle between 10° and 45°. All peraeopods are held loosely except the 7th which is extended slightly.

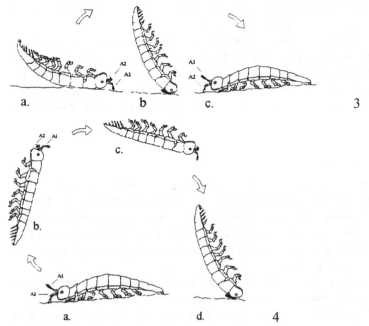

Figure 2. Continued. 3. Flip. a) The animal starts on its dorsal surface. b) By beating its pleopods the telson raises up until the animal is vertical on its cephalon. c) The animal falls down on its ventral surface. 4. Backflip. a) The animal starts the sequence on its ventral surface, often immediately after performing a flip (see 3. Flip). b) The individual beats its pleopods and starts by swimming forwards and then directly upwards. c) The animal then turns over ands heads back down to the ground. d) The individual lands on its cephalon, letting its telson drop back down to the floor, finishing on its dorsal surface. The entire loop is between 1 and 3 body lengths vertically. A1 – antenna 1, A2 – antenna 2, PS peraeon somite, P7 peraeopod 7.

sion. Backflips varied in height between one and three body lengths. The backflip would often change the direction of swimming horizontally.

3.1.5 *Ventral swimming*
Whilst in aquaria made from PVC, as opposed to polystyrene, limnoriids would frequently turn over onto their ventral surface by performing flips. In this ventral position (see Fig. 2.4a) they were able to swim along the surface of the aquarium floor by beating their pleopods and using the dactyl claws on the peraeopods to grip the surface. The uropods were in contact with the surface which supported the telson. A half crawling and half swimming motion was produced along the surface of the aquarium floor, as a sufficient grip on the surface to allow a full crawling behavior was not achieved.

3.2 *Other behavior associated with swimming*
Individual *Limnoria* could be extremely active occasionally, although the reasons for this were not clear. Circles or backflips were the most frequent behavior patterns observed at these times, with numerous repetitions of these same movements occurring.

Limnoria would also swim through the water column in the same orientation as during general dorsal swimming during these very active periods.

During periods of rest the pleopods would beat periodically without causing locomotion and grooming was frequently observed. Both antennae were groomed by the serrate setae present on the propodus-dactyl joint of the first pair of peraeopods. The uropods and telson edge were groomed by the serrate setae present on the merus and carpus of the seventh peraeopod. After grooming the peraeopods were passed over the mouthparts.

3.3 *Laminar flow experiments*

A comparison of the mean percent time that limnoriids spent stationary, swimming and crawling, when exposed to stationary water, flowing water, flowing water plus limnoriid extract and flowing water plus microbial veneers is shown in Figure 3. The greatest percentage of time was spent stationary, during all of the water flow variables, with the least percentage time spent performing directional swimming patterns (circles, backflips and flips) and crawling.

Each of the water flow variables did not alter the mean percentage of time that limnoriids spent stationary, swimming, performing directional swimming patterns and crawling (Kruskal-Wallis One-way analysis of variance, $p > 0.05$). No significant difference appeared in the mean percentage time that limnoriids spent on their ventral surface between stationary water, flowing water and flowing water with either chemical attractant added.

Certain sequences of the patterns of swimming behavior were observed to frequently be repeated. The main repeated sequence was a flip followed by a backflip. No statistical analysis was carried out for these observations, as the repeated sequences were not quantified.

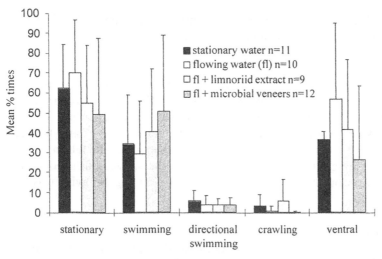

Figure 3. Bar chart showing the mean percentage times limnoriids spent performing each of the categorised behavior patterns during each of the water flow variables. Error bars indicate the standard deviations.

4 DISCUSSION

The swimming behavior of limnoriids has been previously described as irregular from initial observations in Petri dishes (Kofoid & Miller 1927, Oliver 1962). Under the more detailed investigation described here, several distinct swimming behavior patterns were observed. Circles, backflips and flip patterns resulted in a change in the overall swimming direction of limnoriids and may be particularly important during the location of wood. The alteration of swimming direction is a useful behavior used in locating prey and substrata. For example, eels remain in an attractive chemical plume by swimming perpendicular to the current when the plume is lost. This increases their chance of finding the plume again (Oliver et al. 1996). In the same way, insects fly cross wind when they lose the smell of a mating pheromone (Carde 1996), and bacteria turn more frequently in response to attractive chemicals which maintains them in the vicinity of the source (Atema 1985). Even though directional swimming patterns were not the dominant behavior during experiments with limnoriids, the performance of such distinct patterns indicates that they may be important in an overall swimming pathway, perhaps to facilitate the location of wood.

The peraeopod 7 of limnoriids was noticeably held out at an angle during the performance of circles and may be involved in balance as was suggested for the peraeopods of *Idotea* during their swimming (Alexander 1988). Throughout all swimming patterns both pairs of limnoriid antennae were held in a distinctive manner, although not streamlined. The position of the antennae may be important in sensing water flow and chemicals during the migratory period.

Limnoriids were stationary for the majority of their time during laminar flow experiments. This may be an adaptation to allow limnoriids to be swept further afield by water currents until they reach another substratum, or until they detect another environmental factor, such as an increase in hydrostatic pressure or a chemical substance which initiates a different behavior pattern. A combination of stationary periods interspersed with periods of being positively rheotactic, may maintain limnoriids in an optimal position within the water to encounter chemicals emanating from wood.

The swimming pathways of individual limnoriids were noted to vary considerably during previous research (Henderson 1998) and so limnoriids may not use an internal motor programme to aid in their location of wood. As yet, a clear searching strategy that would enable limnoriids to locate wood has not been observed. However, typical behavior patterns were performed in a certain sequence, such as the repeated flip and backflip and the circle followed by a backflip or ventral swimming. These sequences may be part of a complicated searching strategy, still to be resolved, that lead limnoriids directly to fresh sources of timbers, similar to the changes in direction and walking speeds of crayfish and lobsters (Tierney & Atema 1988, Moore et al. 1991).

Previous work by the author has shown that limnoriids are overall positively rheotactic (Henderson 1998), however, observations of limnoriids performing the various categories of swimming patterns, such as circles, backflips etc., showed that the presence of a water flow does not significantly influence the frequency of these behavior patterns. Furthermore, chemicals from the limnoriid extracts and microbial veneers did not significantly influence the frequency of the behavior patterns. This behavior is different to that of crayfish, which do alter the periods spent walking, feeding and grooming when exposed to chemicals originating from their prey (Moore et al. 1991). Both the extracts of limnoriids and microbial veneers were found to be attractive to limnoriids at close range (Henderson 1998), so limnoriids should find the same chemicals attractive at a distance. As olfaction in crustaceans requires only extremely low concentrations of chemicals in the region of 10^{-3} to 10^{-12} M (Derby 1989), the concentrations of attractants used were great enough to be detected by limnoriids.

The chemical attractants may not have been in the right context to elicit a behavioral response. For example, a chemical may not be attractive until it is present in conjunction with the correct mechanical stimuli, such as a suitable substratum (Atema 1985). Perhaps limnoriids rely predominantly on gustation and therefore require a rough substratum, such as timber, before responding to certain chemical attractants.

Limnoriids may detect and respond to limnoriid extracts and microbial veneers, but the behavioral response produced was not apparent from this analysis. Other behavioral factors that were not investigated here, such as the swimming speed, changes in direction and overall spatial and temporal distribution of the animals, may have altered in response to the chemical stimulants. It is only after precise analysis that some apparent random movements, in response to chemicals, can be interpreted as highly coordinated movements (Steele 1983). A more detailed investigation of additional swimming behavior factors of limnoriids will determine if coordinated movements of the distinct behavior patterns described in this study, exist in limnoriid swimming.

Limnoriids are photonegative (Menzies 1961, Eltringham 1966) and so the lighting used may have influenced behavior to such an extent that the varying experimental factors (water flow and chemical attractants) did not elicit any significant behavioral changes. However, if limnoriids were truly photonegative then the long periods of stationary behavior observed would not have been expected. Migration of limnoriids has been reported to occur predominantly at night (Menzies 1961), however, this seems unlikely as limnoriids can remain within the water column for days before locating new timber (Kofoid & Miller 1927, Ray 1959). Therefore, under natural conditions of migration limnoriids would be exposed to long periods of daylight.

Bourdillon (1960) showed that as light intensity increased above 1 lux, the 'trails' of limnoriids contained fewer turns until 850 lux, when a constant degree of meandering was observed. Since experiments in this study were conducted between 850 and 1100 lux, perhaps the number of flips, circles and backflips performed by limnoriids were less than they would have been under lower light intensities.

Natural swimming of limnoriids only occurs during the period of migration, which is believed to happen only once in the life cycle of individuals (Bourdillion 1960). Furthermore, migration may be triggered by environmental conditions (Eltringham & Hockley 1961) and limnoriids may only display natural swimming behavior after experiencing the migratory stimuli. Ideally experiments would use animals that had migrated naturally, but migrating limnoriids have not yet been identified in marine surveys (Menzies 1961). Experiments by the author to induce migration in the laboratory, using temperature change, have so far been unsuccessful. Migratory swimming behavior of limnoriids may not be observed until the induction of migration is possible.

Further research is required to answer the numerous questions on the biology of limnoriids. Whilst marine borers and microorganisms are the only organisms capable of degrading wood in the sea and are therefore an essential ingredient to the ecosystem, future research will focus on the biology of limnoriids to reduce the damage they cause to structures important to humans.

ACKNOWLEDGMENTS

This study was carried out during a PhD funded by Buckinghamshire College. I would like to thank Heather Daly for proof reading the manuscript.

REFERENCES

Albuquerque, R.M. & Cragg, S.M. 1995. Evaluation of impact of CCA-treated wood on the marine environment. *Proceedings of the 3rd International Wood Preservation Symposium. The Challenge – Safety and Environment, 6-7 February 1995, Cannes-Mandelieu, France*, IRG/WP95-50040, 223-236.

Alexander, D.E. 1988. Kinematics of swimming in two species of *Idotea (Isopoda, Valvifera)*. *J. Exp. Biol.* 138: 37-49.

Anderson, G. & Dale, W.E. 1989. *Probopyrus pandalicola* (Packard) (Isopoda, Epicaridea) : swimming responses of cryptoniscus larvae in water conditioned by hosts *Palaemonetes pugio* (Holthuis) (Decapoda, Palaemonidae). *J. Exp. Biol. Ecol.* 130: 9-18.

Atema, J. 1985. Chemoreception in the sea: adaptations of chemoreceptors and behavior to aquatic stimulus conditions. *Society of Experimental Biology Symposia*, 39: 387-423.

Atema, J. 1995. Chemical signals in the marine environment: dispersal, detection and temporal signal analysis. *Proceedings of the National Academy of Science USA*, 92: 62-66.

Barnacle, J.E., Cookson, L.J. & Mc Evoy, C.N. 1983. *L. quadripunctata* Holthuis – a threat to copper treated wood. *International Research Group on Wood Preservation*, IRG/WP/4100: 1-10.

Benfield, M.C. & Aldrich, D.V. 1991. A laminar-flow choice chamber for testing the responses of postlarval penaeids to olfactants. *Contrib. Mar. Science*, 32: 73-88.

Bourdillion, A. 1960. Biologie des crustacés marins xylophages. *Rectorat de la Station Marine d'Endoume* 31: 1-173.

Carde, R.T. 1996. Odor plumes and odor-mediated flight in insects. *Ciba Foundation Symposium* 200: 54-70.

Cookson, L.J. 1990. A laboratory bioassay method for testing preservatives against the marine borers *Limnoria tripunctata, L. quadripunctata* (Crustacea) and *Lyrodus pedicellatus* (Mollusca). *International Research Group on Wood Preservation*, IRG/WP4160: 1-9.

Cookson, L.J. 1997. Additions to the taxonomy of the Limnoriidae (Crustacea: Isopoda). *Memoirs of the Museum of Victoria* 56: 129-143.

Cookson, L.J. & Barnacle, J.E. 1987. The performance in Australia after ten years in the sea of single and double preservative treated timber specimens. *Material und Organismen* 22: 139-160.

Cragg, S.M., Pitman, A.S. & Henderson, S.M. 1998. The biology of marine crustaceans which bore into wood, and developments in their control. In press.

Crisp, M. 1974. Factors influencing the settlement of marine invertebrate larvae. In P.T. Grant & A.M Mackie (eds), *Chemoreception in Marine Organism*: 177-265. London: Academic Press.

Derby, C.D. 1989. Physiology of sensory neurons in morphologically identified cuticular sensilla of crustaceans. *Crustacean Issues* 6: 27-47.

Eaton, R.A. & Cragg, S.M. 1996. Evaluation of creosote fortified with synthetic pyrethroids as wood preservatives for use in the sea. Part 1: Efficacy against marine wood-boring molluscs and crustaceans. *Material und Organismen* 29: 211-229.

El Shanshoury, A.R., Mona, M.H. & Shoukr, F.A. 1994. The enumeration and characterisation of bacteria and fungi associated with marine wood-boring isopods, and the ability of these micro-organisms to digest cellulose and wood. *Mar. Biol.* 119: 321-326.

Eltringham, S.K. 1966. Environmental factors influencing the settlement, activity, activity and reproduction of the wood boring isopod *Limnoria. Material und Organismen* 1: 465-478.

Eltringham, S.K. & Hockley, A.R. 1961. Migration and reproduction of the wood-boring isopod, *Limnoria*, in Southampton water. *Limnol. Oceanogr.* 6: 467-482.

Forward, R.B. 1990. Responses of crustacean larvae to hydrostatic pressure: behavioral basis of high barokinesis. *Mar. Behavior. Phys.* 17: 223-232.

Geyer, H. 1982. The influence of wood inhabiting fungi on the food selection, feeding activity and reproduction of *Limnoria tripunctata* Menzies (Crustacea, Isopoda). *International Journal Wood Preservation* 2(2): 77-89.

Hadfield, M.G. 1978. Metamorphosis in molluscan larvae: An analysis of stimulus and response. In F.S. Chia & M.E. Rice (eds), *Settlement and metamorphosis of marine invertebrate larvae*: 165-175. New York: Elsevier.

Henderson, S.M. 1998. The detection of wood by the marine borers *Limnoria* spp. (Limnoriidae:Isopoda) during their migration. PhD thesis, Buckinghamshire college, UK.

Howell, D.C. 1989. *Fundamental statistics for the behavioral sciences*. Boston: PWS-Kent Publishing Company.

Kemp, P. 1995. *Tropical hardwoods in the maritime environment – a case for their continued use*. The Institution of Civil Engineers Maritime Board, Half day meeting, October 1995. 46 pp.

Kofoid, C.A. & Miller, R.C. 1927. Biological section. In C.L. Hill & C.A. Kofoid (eds), *Marine borers and their relation to marine construction on the Pacific coast*:1-357. Final report of the San Francisco Bay Marine Piling Committee. University of California Press.

Maser, C. & Sedall, J.R. 1994. *From the Forest to the Sea: The Ecology of Wood in Streams, Rivers, Estuaries and Oceans*. Florida: St. Lucie Press.

Meador, J.P. 1989. Chemoreception in a Lysianassid amphipod: the chemicals that initiate food searching behavior. *Mar. Behav. Phys.* 14: 65-80.

Menzies, R.J. 1957. The marine borer family Limnoriidae (Crustacea, Isopoda). Part I. Northern and Central America: systematics, distribution and ecology. *Bull. Mar. Sci. Gulf & Carib.* 7(2): 101-200.

Menzies, R.J. 1961. Suggestion of night time migration by the wood-borer *Limnoria. Oikos* 12: 170-172.

Oliver, A.C. 1962. An account of the biology of *Limnoria. J. Inst. Wood Sci.* 2(9): 32-91.

Oliver, S.J., Grasso, F.W. & Atema, J. 1996. Filament tracking and casting in American elvers *(Anguilla rostrata). Biol. Bull.* 191: 314-315.

Pitman, A.J., Cragg, S.M., & Daniel, G. 1995. The attack of naturally durable and creosote treated timbers by *Limnoria tripunctata* Menzies. *International Research Group on Wood Preservation*, IRG/WP 95-10132, 1-13.

Ray, D.L. 1959a. Nutritional physiology of *Limnoria*. In D.L. Ray (ed.) *Marine Boring & Fouling Organisms*: 46-59. Washington D.C: University of Washington Press.

Ray, D.L. 1959b. Some properties of cellulase from Limnoria. In D.L. Ray (ed.) *Marine Boring & Fouling Organisms*: 372-396. Washington D.C: University of Washington Press.

Rice, A.L. 1964. Observations on the effects of changes of hydrostatic pressure on the behavior of some marine animals. *J. Mar. Biol. Ass.* UK 44: 163-175.

Shelton, R.G.J. & Mackie, A.M. 1971. Studies on the chemical preferences of the shore crab *Carcinus maenas* L. *J. Exp. Marine Biol. Ecol.* 7: 41-49.

Shirley, S.M. Shirley, T.C. 1988. Behavior of red king crab larvae: phototaxis, geotaxis and rheotaxis. *Mar. Behavior Phys.* 13: 369-388.

Sømme, O.M. 1941. A study of the life history of the gribble *Limnoria lignorum* (Rathke) in Norway. *NYTT magasin for naturuidenskapene* 21: 145-205.

Steele, C.W. 1983. Open field exploratory behavioral toxicology. *Mar. Pollution Bull.* 14(4): 124-125.

Tierney, A.J. & Atema, J. 1988. Behavioral responses of crayfish (Orconectes virilis & O. rusticus) to chemical feeding stimulants. *J. Chem. Ecol.* 14: 123-133.

Observations and interpretations on surface swarming behavior by some marine Ostracoda and its possible correlation with their mating strategies

KENNETH G. MCKENZIE
School of Science & Technology, Charles Sturt University – Riverina, Australia

KATSUMI ABE †
Department of Geosciences, Shizuoka University, Japan

ABSTRACT

The swarming at or near the sea's surface by the planktonic halocyprid *Porroecia spinirostris* (in which it is male-associated) and the pseudoplanktonic cypridinid myodocopid *Cypridina dentata*, has been explained as due to their sex drive. We examine such rationalizations by reviewing the functional reproductive morphology and mating behavior of both species (and closely related taxa in the case of *Cypridina*); and also their spatial and temporal distributions. We conclude that swarming in surface waters for mating seems improbable with regard to the halocyprid. Rather, it may be a way for males to access bioenergetic advantages by microphagous feeding in the neuston thus accumulating reserves that maximize the probability of successful matings when they sink to levels where the species lives during the day. For *Cypridina dentata*, swarming at or near the surface certainly favors mating, both in coastal waters and offshore in nutrient-rich zones of intensive upwelling or where water masses of differing salinity and temperature meet. However, the associated male bioluminescent trails are a courtship behavior that does not continue after the onset of mating. Synchronized bioluminescence (patch displays over wide areas) by *Cypridina* and other cypridinids is a response to sudden changes in illumination, or could function as a mass light bomb to scare off patrols by fish predators.

1 INTRODUCTION

Swarming at or near the surface by zooplankton is one of the more intriguing behaviors of tiny animals. One explanation is that it may result from diel vertical migration (Angel 1979). But it can also continue for a week or more over several tens of kilometers, or occur in annual cycles such as upwellings extending over hundreds of kilometers. These are megascale phenomena (Haury et al. 1978).

We investigate surface swarming by two marine ostracodes – the planktonic halocyprid *Porroecia spinirostris* (Claus 1890), and the pseudoplanktonic cypridinid myodocopid *Cypridina dentata* (Müller 1906); these have carapace lengths of about 1.0 mm and 1.8-2 mm respectively. *P. spinirostris* is a common species worldwide (Poulsen 1973, Table 17), but we refer only to Atlantic and Mediterranean popula-

239

tions. *C. dentata* is a tropical species ranging from the northern Arabian Sea (Daniel & Jothinayagam 1979) and Indian Ocean (George & Nair 1980) to Indonesian waters (Poulsen 1962, Fig. 118) and Taiwan Straits off China (Chen 1982). *P. spinirostris* is a diel vertical migrant (Angel 1970); but *C. dentata* swarms more probably in synchronization with annual cycles which incorporate salinity and temperature variations (Chen 1982).

Current interpretations (Angel 1970, Daniel & Jothinayagam 1979) include suggestions that both species swarm to the surface for mating opportunities. Our principal objective in this paper is to evaluate critically such rationalizations.

2 SEXUAL FUNCTIONAL MORPHOLOGY AND COPULATION

2.1 *Porroecia spinirostris*

The reproductive system of males of *P. spinirostris* comprises two symmetrical testes and a penis. From each testis issues a deferent canal that unites with its pair into a single conduit which then passes into the penis. During sexual congress, numerous filamentous spermatozoids, packed closely into bundles, are evacuated from the testes into the deferent canals, then enter the penile seminal canal where they are propelled towards the penile orifice (Leveau 1965).

The penis is single, finger-shaped and situated posteriorly on the left hand side (LHS). It projects forward and has considerable relative size, being about 1/3 the carapace length of the animal. The head of the penis has two small, opposed hook-like appendages; one functions to align with the spermatheca of the female, while the other assists to hold her in place during mating (Leveau 1965).

The female of *P. spinirostris* has two symmetrical ovaries that are sheathed, like the male testes, in a conjunctive membrane. A conduit (oviduct) leads out from each; these oviducts are encased in an epithelium that provides the germinative reserves for each egg. The oviducts unite shortly before they enter the pouch of fecundation (Claus 1891) where the eggs are housed and where, prior to being covered by a protective chorion, they are fertilized by the spermatozoids.

The seminal receptacle (spermatheca) is situated posteriorly on the right side of the female. At about half its length it expands into a seminal capsule often filled with bundles of spermatozoids; from thence the canal is prolonged until it reaches the pouch of fecundation. Spermatozoids can be stored in the seminal capsule until needed (Leveau 1965).

Like all Conchoeciinae, *P. spinirostris* has a number of other sexually dimorphic characters. In particular, the 6th limbs (P2) carry long and feathered terminal setae which function probably to brush away into the seawater a hormonal mucus, which is exuded from the paired posterior dorsomedial glands of the male carapace, in order to attract nearby receptive females (Angel 1993). Other sex dimorphic characters and their function are described and illustrated in Angel (1993).

According to Claus (1891), halocyprid males and females engage in a true copulation culminating in the discharge of spermatozoids into the spermatheca, thence to lodge in the seminal capsule. Considering their sexual functional morphology, this assessment must be correct. Evidently, all Conchoeciinae copulate with the ventral margins of their carapaces juxtaposed – this is termed ventral-ventral mating in Cohen & Morin (1990, Fig. 7D; Fig. 1A). The LHS of the female can be held in position frontally by the strong RHS 'clasper' of the male A2 endopod, while the RHS of the female is aligned anteriorly by the LHS male A2 endopod and posteriorly by the two small 'hooks' of the LHS penis. The forward-projecting, finger-shaped LHS penis can then enter the RHS female orifice.

2.2 *Cypridina dentata and other cypridinids*

Unlike the case with halocyprids, all the reproductive organs of cypridinid myodocopes are paired (Cohen & Morin 1990, Table 2). From each male testis a deferent canal leads into the paired seminal vesicles and from there to the paired copulatory lobes which project forwards on either side of the posterior body. In *Cypridina*, these copulatory lobes each contain an elongate seminal canal leading to the penile orifice which is slightly turned over; a few curved spinules occur behind the penile extremity (Poulsen 1962, Fig. 121h).

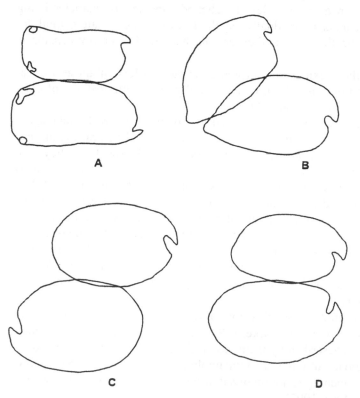

Figure 1. Mating positions in *Porroecia spinirostris*, *Cypridina dentata*, *Vargula hilgendorfii*, *Skogsbergia* sp. A) Ventral-ventral – *P. spinirostris*, B) Posterodorsal-ventral – *C. dentata* (unlikely, see text), C) Reverse ventral-ventral – *V. hilgendorfii*, D) Ventral-ventral – *Skogsbergia* sp. Figures 1A, 1B and 1D have been adapted from Angel (1993), Müller (1906) and Parker (1997) respectively.

The testes of most adult males are ripened and contain mature spermatozoids in the breeding season. Each spermatozoid is filiform and about 0.03 mm long with a helical structure. During copulation, male spermatophores containing sperm become attached to the paired depressions of the female genitalia (Cohen & Morin 1997, Fig. 4D).

In female cypridinids, the paired ovaries are positioned on either side of the midgut (Cohen & Morin 1990). The paired female genital limbs each consist of a more or less wrinkled ridge containing a rounded and often ornamented depression, a genital lobe, and the setose opening which is sited laterally of this lobe. There is also a tiny spermatophore pore near the side of the depression and a somewhat larger groove- or slit-like lateral pore located ventromedially on the surface of the genital lobe. The male spermatophore is attached to and covers the depression; it does not make contact with the setose opening. Thus, it seems unlikely that the setose opening functions as a seminal receptacle (Cohen & Morin 1997). It is hypothesized instead that ova are first discharged from the ovaries, then fertilized by sperms from the spermatophores, and finally expelled into the brooding space through the setose openings which have a similar diameter to the fertilized eggs (Cohen & Morin 1997). The very flexuous 7th limbs may assist in embryo placement (Kornicker 1981).

The spherical, oblong spermatophore containing numerous sperm and other secretion granules is enclosed in a more or less toughened capsule, its diameter being about 0.09 mm. The inclusions are gradually expelled towards the female genital depression, thereby fertilizing the eggs as they are discharged. The empty capsule is later discarded.

Cypridina has few other sexually dimorphic characters. Most of these are described and illustrated in Poulsen (1962). Other sexually dimorphic features of cyridinids are discussed by Vannier & Abe (1993).

Mating in *Cypridina dentata* has been described briefly by Daniel & Jothinayagam (1979). According to these authors, it occurs in the posterodorsal-ventral mating position, using the terminology of Cohen & Morin (1990, Fig. 7B; Fig. 1B). They note that the male spreads its valves over the posterodorsal area of the female and clasps the edges of the female valves using the endopods of its own 5th limbs (P1). It then unfolds and extends its paired penes, inserting them into the paired vaginae of the female. The female remains passive throughout copulation that lasts several minutes and is accompanied by bright luminescence from both partners. After mating, this bioluminescence becomes weaker and finally disappears (Daniel & Jothinayagam 1979). It must be added that other more recent observations of mating in *C. dentata* do not confirm that the sexes bioluminesce during copulation (Chen Ruixiang, pers. comm.; see also our Discussion).

Apart from this, the account of Daniel & Jothinayagam (1979) is unsatisfactory in several other details. We concur with Parker (1997) that the 5th limbs in *C. dentata* are too short to function as described by them; secondly, in cypridinids there is no insertion of penes into vaginae but rather a spermatophore transfer as described below; finally, the female setose opening is not involved in mating, its probable role was discussed earlier (Cohen & Morin 1997).

A more satisfying account of mating in cypridinids is given in Okada & Kato (1949) for the well-known Japanese species *Vargula hilgendorfii* (Müller 1890), originally described as a *Cypridina*. Mating was observed in a petri dish in the laboratory during daytime. According to these authors, sexual activity is initiated when the male clasps the back of the female shell, by employing the suckers (disk-like structures) and hooks on its antennules for this purpose; swimming actively the while with vigorous beating of its powerful natatory antennal exopodial setae. The female generally remains passive during this phase, occasionally also swimming actively forwards and dragging along the male. The contact, therefore, is in a reverse ventral-ventral mating position (Cohen & Morin 1990, Fig. 7C; Fig. 1C). Possibly, the powerful spinulose teeth of the P1 exopodial lobes in both sexes help to align their sex parts. During copula, which lasts more than 30 minutes, the male protrudes its copulatory lobes onto the female genital area and spermatophore transfer then follows (Okada & Kato 1949). Laboratory observations indicate that the ova are not differentiated in a copulating female which supports the hypothesis cited earlier that the eggs are fertilised after their discharge from the ovaries and oviducts. Several series of recent observations of *V. hilgendorfii* in the laboratory, including a school video tape, confirm contact between the sexes in this position, but lasting less than a second, and that the male uses its antennules to hold the female along the back (Abe and associates, unpublished).

A third mode of mating in cypridinids has been captured on video recordings by Parker (1997). 'After initial courtship the male and female ostracods, with ventral margins adjacent and anterior ends directly opposite, join their mandibular claws and furcae. Both furcae are pushed in a posterior direction until the ventral margins of the carapaces meet' (Parker 1997, Abstract). Since the genitalia are directly opposite, mating and spermatophore transfer to the female is facilitated. This ventral-ventral mating (Cohen & Morin 1990, Fig. 7D; Fig. 1D) lasts a few seconds.

3 DISTRIBUTION: SPATIAL AND TEMPORAL

3.1 *Porroecia spinirostris*

Porroecia spinirostris is a characteristic near surface warm water species of the North and South Atlantic gyres, its latitudinal distribution ranging from about 40°N to 42°S (Angel & Fasham 1975, Angel 1979). The species is also widespread throughout the Mediterranean.

Angel (1970) was the first to record predominantly male swarms of *P. spinirostris* in tropical Atlantic neuston – the male:female ratio being 800:5. His material was collected near dawn. Weikert (1972) noted that *P. spinirostris* occurred only sporadically in daytime hauls but migrated in greatest numbers into the neuston at dusk. Moguilevsky & Angel (1975) recorded surface swarms of *P. spinirostris* at two Atlantic stations, 23°48'N and 36°56'S, both hauled at dusk – the male:female ratio in their samples was 6904:17. These authors also cited a personal communication that the species swarms at the surface at sunrise and sunset. This swarming in the neuston is clearly exclusive to *P. spinirostris* because Hempel & Weikert (1972) classify only *P. spinirostris* males as facultative neuston; they regard all other planktonic halocyprids as pseudoneuston. The abundance of *P. spinirostris* in comparison with other halocyprids in the neuston is also very high, indeed it predominates in such assemblages. For the references cited above, it represented 88% of the faunule in Moguilevsky & Angel (1975), 93% in Angel (1970), and 98% in Weikert (1972).

The species is also dominant in surface and near surface waters in the Mediterranean. Off Marseilles, Leveau (1965) collected it from 300-0 m with an abundance maximum around 200-100 m that varied seasonally. Nearer inshore, he collected large numbers all year round, usually at 50 m depth, except during winter months (October-January) when it was most abundant at 15-25 m; the species also moved upwards during any temporary cooling of the water surface, as happened during a mistral in September 1963 (Leveau 1965). Similar winter results were obtained in the Adriatic by Huré (1961) who noted further that the level at which the species was most numerous dropped to around 50 m during the warmer months, the same pattern as determined by Leveau (1965).

The Italian EOCUMM-94, EOCUMM-95 and POP-95 cruises sampled in the south Tyrrhenian Sea and in the Straits of Messina, Ionian Sea during the summers of 1994 and 1995. Results respective to *P. spinirostris* for 1995 are summarized in Table 1; those for 1994 are already published (Benassi et al. 1998). The upper 100 m was sampled more closely at stations T3, T4 (Tyrrhenian) and J4 (Ionian). In all, the maximal abundances of *P. spinirostris* were at around 50 m, confirming the data of Leveau (1965) and Huré (1961). *P. spinirostris* comprised over 80% of the entire ostracode faunule from sunset through to dawn at several Tyrrhenian stations and was never less than about half the total ostracode assemblage in the upper 100 m (Table 2).

3.2 *Cypridina dentata*

Cypridina dentata is a dominant species in the upwelling waters of the Arabian Sea and off the western coast of India (George 1969, George & Nair 1980). In the northern Arabian Sea, it was collected in surface waters at numerous stations from December 1973 to May 1974 (Daniel & Jothinayagam 1979). In the 10 stations characterized by nocturnal swarms of this species, the number of specimens collected varied from about 4000 to over 100,000 per 3 minute haul, with adult males and ovigerous females occurring in about equal proportions. All other zooplankton groups were represented usually by no more than a few hundred individuals (Daniel & Jothinayagam 1979, Table 1). At 72 other Arabian Sea stations (54 night and 18 day) where it was less numerous, *C. dentata* was represented exclusively by adult males plus a few immature females. Along the Tamil Nadu coast of southeastern India, densities of 345 m^{-3} have been reported for the species (Santhakumari & Saraswarthy 1981).

Table 1. Numbers of males/females/juveniles of *Porroecia spinirostris* collected during July 1995 at stations off the Eolian Islands, southern Tyrrhenian Sea (EOCUMM 95), and in the Straits of Messina, Ionian Sea (POP 95). Maximal depth of occurrence for the species was about 700 m.

EOCUMM 95		POP 95	
> 400 m	10/54/19	> 400 m	9/41/10
400-300 m	25/68/23	400-300 m	7/19/0
300-200 m	36/85/5	300-200 m	88/66/8
200-100 m	771/693/61	200-100 m	226/195/20
100-0 m	6045/2573/66	100-0 m	1562/895/19

Table 2. Total planktonic ostracodes collected from 300-0 m during July 1995 at stations off the Eolian Islands, southern Tyrrhenian Sea (EOCUMM 95); and percentages of *P. spinirostris* (PS) in the upper 100-0 m. Stations T3 and T4, which were more closely sampled in the upper 100 m, were collected around midnight (T3A), dawn (T3B), midday (T3C), evening (T3D), midnight (T4A) and dusk (T4D) respectively, from 22-24 July 1995.

Station	T1A	T1B	T1C	T1D	T2A	T2B	T2D	T3A	T3B	T3C	T3D	T4A	T4D
300-200 m	27	30	29	28	59	30	56	40	41	66	31	152	131
200-100 m	244	288	187	227	345	1	147	367	332	306	131	616	324
100-0 m	432	736	1088	551	501	2104	1766	272	178	349	745	2601	593
% PS @ 100-0	80	82	68	73	63	50	68	57	56	77	48	87	88

Considerably fewer individuals of *C. dentata* were collected in the Malay Archipelago and Philippines waters during the 'Dana' expeditions in 1929 (Poulsen 1962) but the species still was mostly taken at or near the surface.

Offshore in southern Chinese waters, the occurrence and density of *C. dentata* show little relation to upwelling; and no specimens have been collected at depths from 500-1000 m (Chen Ruixiang, personal communication October 1997). On the other hand, along the western coast of Taiwan Straits this euryhaline warm neritic species is the dominant taxon of the ostracode zooplankton. It occurs year round, and in greatest numbers during summer and autumn until a peak in October (Chen 1982, Fig. 2), coinciding with the meeting of warm and cold water masses in the area. After November, numbers decrease and in winter months the density is only around 1 individual per 100 m^2, reflecting the movement southwards of colder coastal currents. Numbers remain low through March and April. Although regarded as a euryhaline species, its densities are markedly affected by continental runoff, typhoons and heavy rains (Chen 1982, Table 1).

3.3 *Vertical migration*

Angel (1985) noted that oceanic animals have four basic drives: to survive, feed, reproduce, and optimize their bioenergetic advantages. One of the apposite behaviors to accommodate these drives, used particularly by small species, is vertical migration which Longhurst (1976) hypothesized might be explained as avoidance of visual predation, a means to access horizontal or oblique dispersion/transport, and a way of gaining bioenergetic advantages; Angel (1985) added breeding strategies to this list. Diel vertical migration in many planktonic halocyprids, including *Porroecia spinirostris*, is well established (Angel 1970, 1979). It relates to the availability of light, food, nutrients, predation pressure and current shear (Angel 1984), in addition to seasonality and temperature (Leveau 1965).

The diel migratory activity of *Cypridina* species, especially in the Indo-Pacific, has been well studied for *Cypridina sinuosa* (Müller 1906). It is explicable as an alternation of phases of high and low activity, in response to factors such as temperature and salinity (which determine seawater density), the direction and velocity of transport, food supply, hydrostatic pressure, illumination, and the animals' specific gravity. For taxa with a high specific gravity (SG), such as *Cypridina* (SG about 1.1), these factors are most favorably combined in regions of intensive upwelling (Rudjakov 1970). Such regions include the northern Arabian Sea and off southern India, i.e. much of the oceanic range of *Cypridina dentata*. It is important to recall that diel migratory activity also features in the coastal and shallow water populations of species of *Cypridina*.

3.4 *Bioluminescence in surface waters*

That some Ostracoda bioluminesce has long been known; and it is also known that the luminescence is produced by these animals' own chemical systems, through the oxidation of luciferin in the presence of oxygen and the enzyme luciferase (Angel 1993). In some cypridinids, there is a luminescent organ in the upper lip (Saito et al. 1986). Bioluminescent halocyprids emit light from various sites in the valves (Angel 1968); but *Porroecia spinirostris* does not emit such light, only mucus as noted earlier.

The details of ostracode bioluminescent behavior have only recently been described, particularly with regard to species of the cypridinid genus *Vargula*. Irie (1953) noted that numerous *V. hilgendorfii* came to the surface during the collection period and emitted light in a helical spiral pattern. Because the ostracodes were not attracted to the bait used, the spiral emissions evidently were not related to scavenging. These observations were confirmed in September 1992 by Abe (1994) who recorded a frequency of emission of 60 per min within an area of 4 m^2, mostly associated with a spiral movement; the rather strong light lasted up to 2 seconds. The ambient characteristics were a moon age of 18-20, at high tide, in 3-4 m depth of water off Tateyama, Japan. The same author observed similar patterns at other seasons (March-May) but in much lower frequencies. Over 50 Caribbean species of *Vargula* also are known to bioluminesce, the males producing species-specific trains of light pulses lasting several seconds above reefal habitats within the hour after sunset over the whole year (Morin & Cohen 1991).

Cypridina serrata (Müller 1906), which ranges from New Guinea to Indonesia and the Philippines, emitted elliptical clouds of bright blue bioluminescence in short bursts when a flashlight was shone into the water. Collecting was done after dark between 20-23 h, and along the shoreline where the depth at high tide varied from 2-10 m. The species was not attracted by the bait used (Tsuji et al. 1970). *Cypridina dentata* was recorded as emitting dense patches of bioluminescence during nocturnal swarming to the surface at both shallow and deepwater stations in the northern Arabian Sea (Daniel & Jothinayagam 1979). The same species is known to produce bioluminescence both coastally and offshore in the western Taiwan Straits (Chen 1982). The reviewer of our first draft found that in Pittwater, off Sydney, *Cypridina* are bioluminescent but do not enter baited traps; in fact he is not aware of any scavenging species of *Cypridina* between New Guinea and Hobart, Tasmania (Andrew Parker, pers. comm.).

4 DISCUSSION

4.1 *Porroecia spinirostris*

The suggestion that this species swarms nocturnally at the surface in the Atlantic for mating (Angel 1970) seems most unlikely. The ratios of males and females captured in neuston, e.g. 800:5 (Angel 1970), 6904:17 (Moguilevsky & Angel 1975) mean both that the mature females would have no opportunity to exercise any mate choice before they were taken by a nearby male, and that for the predominance of males gene survival through mating would be impossible.

The data of Leveau (1965) from the Mediterranean further clarify the argument against mating at or near the surface by *P. spinirostris*. His results show that the greatest number of both adults and first stage juveniles occur around 50 m, i.e. below the neuston layer. This is so even given the likelihood that the species reproduces twice annually, in December-January and June-July (Leveau 1965, Figs 12, 13). The species is found in large numbers near the surface (15-25 m) either in the winter from October to January (Leveau 1965), or with a preponderance of males both day and night in July (Benassi et al. 1998; Table 1). From these data, it appears that the optimal level for mate choice by females and for male gene survival in the western Mediterranean lies below the neuston at around 50 m. Huré (1961) reported similar results from the Adriatic.

Considering the complete vertical distribution of this species in the Atlantic, it appears that male/female ratios are about equal during the day; even at night this ratio is only about 2:1, offering adequate opportunities for successful mating to both sexes (Angel 1970). Our Mediterranean data (Table 1) show similar male/female ratios, ranging from 7-11:4 in the top 100 m. Again, these favor mating success for males in the upper water column (below the neuston layer). The fact that *P. spinirostris* was the dominant halocyprid in the zooplankton (Table 2), would also tend to facilitate its male/female encounters.

If *P. spinirostris* males do not swarm at or near the surface at night in order to mate, we are left with the possibilities that they do so via some transport factor or to avoid predators or to gain bioenergetic advantages. Active or passive movement into surface waters of males almost exclusively using current shear or eddies is highly unlikely. Instead, Moguilevsky & Angel (1975, Abstract) suggest that this male behavior is to avoid predators and conclude that, 'ostracods only constitute important food items for carnivores feeding in the neuston when the latter are permanently in the neuston; the occurrence of vertically migrating carnivores does not coincide with the maximum occurrence of ostracods in the neuston'.

This conclusion seems persuasive, but does not explain why the behavior is virtually exclusive to males. There are sound reasons for emphasizing more strongly a thesis that nocturnal migration offers some bioenergetic advantages to the swarming males. *P. spinirostris*, which has a rather narrow esophagus, is an opportunist and detrital feeder on the remains of dead crustaceans (Lochhead 1968), and on such phytoplankters as centric diatoms, coccolithophores and silicoflagellates, as well as fecal pellets (Angel 1970). The swarming behavior accesses bioenergetic advantages for this species via microphagous feeding in the near surface and surface layers where such food is most abundant. This compensates both for the energy expended in swimming to the surface against the animals' negative buoyancy (Rudjakov 1970) and for any reserves lost during the day at deeper levels (around 100-50 m) when expelling hormonal mucus into the seawater to attract receptive females. The fresh reserves built up by nocturnal feeding in the neuston sustain their active or passive movement during daylight to, and mating behavior at, depths where mates are more or less readily available. The synchronization of this (exclusive among halocyprid males) swarming behavior by males of *P. spinirostris* seems to be correlated closely with the light/dark cycle (Moguilevsky & Angel 1975).

4.2 *Cypridina dentata and other cypridinids*

The likelihood that *Cypridina dentata* swarms to the surface for sex is highly probable, but this is not the whole story. Daniel & Jothinayagam (1979) noted the massive contribution to the biomass of near equal numbers of mature males and ovigerous females in the northern Arabian Sea at night stations where the species swarmed in surface waters; while at 72 other stations the species was represented by adult males and a few immature females. Clearly, the near equal sex ratios would maximize mating opportunities; and where adult males occurred with immature females this suggests that as soon as the females completed the molt to adulthood they could be fertilised with minimal delay.

On the other hand, it is unlikely that mating pairs would bioluminesce during sexual congress. Morin & Cohen (1991) have established that the species-specific luminescent trains of light pulses displayed by males of the cypridinid *Vargula* are a courtship and not a mating behavior. The prolonged bioluminescence during copulation in the laboratory noted by Daniel & Jothinayagam (1979) was most likely a response by their animals to the bright lamp used for observation. Thus, when a light was shone on the water, *Cypridina serrata* responded with a strong luminous display (Tsuji et al. 1970).

It seems probable that a female copulates only once during her life cycle and this occurs immediately after the final molting. Sperm can be stored for a considerable time and can fertilize several clumps of eggs – isolated adult females collected from nature are known to produce more than two clutches of embryos in the laboratory.

Cypridina is a member of the pseudoplankton. In shallow waters, both sexes usually live near the bottom and even within the sand layer, feeding mostly at night in these zones. By analogy with Caribbean species of *Vargula*, we speculate that shortly after dusk mature males of *Cypridina* swim to the surface where they emit species-specific patterns of light trails as courtship signals to nearby receptive females. The females approach and are mated; the pair copulates while swimming. Fertilised females return to the bottom shortly after mating while the males remain at or near the surface and continue to signal for further mating opportunities. In surface waters, couples can mate safely because this environment is relatively free of fish predators (Moguilevsky & Angel 1975).

In the deeper sea, although bottom-dwelling females could see the male light trails because there is less background light (Andrew Parker, pers. comm.) they are too distant to respond. Therefore, they migrate to the surface and join the males in regions where the general environmental factors are most favorable, e.g. in nutrient-rich upwelling areas (Rudjakov 1970) or at the meeting of cold and warm water masses (Chen 1982), and where both sexes can feed readily. Recent research indicates that these ostracodes do not luminesce while feeding (Abe et al. 1995). The luminescent displays at night are observed either as thousands of uncoordinated pinpricks of light (individual male mating signals) or, occasionally, as synchronized and spectacular patch displays over large areas of the water surface. The latter displays are often explained as response to changes in illumination, such as the sudden reappearance of the moon from clouds, but other explanations may be just as pertinent.

A video recording made recently by one of us (KA) captured a scene in which *Vargula hilgendorfii* emitted a very bright light on being spat out of the mouth of a goboid fish. Other research, including work on the *Vargula* heart and digestive system (Abe et al. 1995), also suggests that luminescence has a role in predator evasion. Thus, the patch displays over large areas may act as mass light bombs to deter occasional surface patrols by schools of the fish that might otherwise prey on these cypridinid ostracodes.

ACKNOWLEDGEMENTS

Prof Louis S. Kornicker, National Museum of Natural History: Smithsonian Institution, Washington D.C., USA; Prof Dan L. Danielopol, Limnological Institute, Austrian Academy of Sciences, Mondsee, Austria; Prof Chen Ruixiang, National Bureau of Oceanography, Xiamen, People's Republic of China; and Dr Anne C. Cohen, California Academy of Sciences, San Francisco, USA, kindly replied to our queries and provided reprints of their own and other papers. Dr Andrew Parker, The Australian Museum, Sydney, Australia reviewed an early draft of the manuscript; and very kindly presented the paper on our behalf at the 4th International Crustacean Congress, Amsterdam, in July 1998. KGM acknowledges support at the Department of Ambiental Sciences, University of Parma, Italy, and permission to publish some results of the work on *P. spinirostris*; KA acknowledges use of facilities at the Department of Geosciences, Skizuoka University; Charles Sturt University Reprographics Unit, Wagga Wagga, formatted the final diskette. Both authors acknowledge useful critiques by three referees.

Vale!

The co-author, Dr Katsumi Abe, was killed in a tragic road accident while driving home from a seminar on 23 August 1998. Beyond the high level of his scientific attainment, including mastery in the video recording of unrestricted tiny animals, his colleagues remember a sincere open personality, and a loving devotion to his family.

REFERENCES

Abe, K. 1994. *The Light of Marine Fireflies*. Tokyo: Chikuma Shobo (in Japanese).
Abe, K., Vannier, J. & Tahara, Y. 1995. Bioluminescence of *Vargula hilgendorfii* (Ostracoda, Myodocopida): its ecological significance and effects of a heart.In J. Riha (ed.), *Ostracoda and Biostratigraphy*: 11-18. Rotterdam: Balkema.
Angel, M.V. 1968. Bioluminescence in planktonic halocyprid ostracods. *J. Mar. Biol. Ass. UK* 48: 255-257.
Angel, M.V. 1970. Observations on the behavior of *Conchoecia spinirostris*. *J. Mar. Biol. Ass. UK* 50: 731-736.
Angel, M.V. 1979. Studies on Atlantic halocyprid ostracods: their vertical distribution and community structure in the central gyre region along 30°N latitude from off Africa to Bermuda. *Progr. Oceanogr.* 8: 3-124.

Angel, M.V. 1984. The diel migrations and distributions within a mesopelagic community in the north east Atlantic. 3. Planktonic ostracods, a stable component in the community. *Prog. Oceanogr.* 13: 319-351.

Angel, M.V. 1985. Vertical migrations in the oceanic realm: possible causes and probable effects. In M.A. Rankine (ed.), *Migrations, Mechanisms and Adaptive Significance. Contributions in Marine Science, Texas* 27 suppl.: 45-70.

Angel, M.V. 1993. Marine Planktonic Ostracods. In D.M. Kermack, R.S.K. Barnes & J.H. Crothers (eds), *Synopses of the British Fauna (New Series)* 48. Shrewsbury: Field Studies Council.

Angel, M.V. & Fasham, M.J.R. 1975. Analysis of the vertical and geographical distribution of the abundant species of planktonic ostracods in the North-east Atlantic. *J. Mar. Biol. Ass. UK* 55: 709-737.

Benassi, G., Ferrari, I., Rossi. V., Sei, S., Angel, M.V. & McKenzie, K.G. 1998. Planktonic ostracods off the Eolian Islands (Mediterranean Sea). In S. Crasquin-Soleau, E. Braccini & F. Lethiers (eds), *Memoir 20. What about Ostracoda!*: 3-25. Pau: Elf Ep Editions.

Chen Ruixiang 1982. The distribution of planktonic Ostracoda along the western coast of Taiwan Straits. *Acta Oceanologica Sinica* 1(2): 279-298.

Claus, C. 1891. *Die Halocypriden des Atlantisches Oceans und Mittelmeeres.* Wien: Kaiser Akademie Wissenschaftlich.

Cohen, A.C. & Morin, J.G. 1990. Patterns of reproduction in ostracodes: a review. *J. Crust. Biol.* 10: 184-211.

Cohen, A.C. & Morin, J.G. 1997. External anatomy of the female genital (eighth) limbs and the setose openings in myodocopid ostracodes (Cypridinidae). *Acta Zool. (Stockholm)* 78: 85-96.

Daniel, A. & Jothinayagam, J.T. 1979. Observations on nocturnal swarming of the planktonic ostracod *Cypridina dentata* (Müller) for mating in the northern Arabian Sea. *Bull. Zool. Surv. India* 2(10): 25-28.

George, J. 1969. A preliminary report on the distribution and abundance of planktonic ostracods in the Indian Ocean. *Bull. Nat. Inst. Sc. India* 38: 641-648.

George, J. & Nair, V.R. 1980. Planktonic ostracods of the northern Indian Ocean. *Magasagar* 13: 29-44.

Haury, L.R., McGowan, J.A. & Wiebe, P.H. 1978. Patterns and processes in the time-space scales of plankton distribution. In J.H. Steele (ed.), *Spatial Patterns in Plankton Communities*: 277-328. New York: Plenum Press.

Hempel, G. & Weikert, H. 1972. The neuston of the subtropical and boreal North eastern Atlantic Ocean. A review. *Mar. Biol.* 13: 70-88.

Huré, J. 1961. Distribution saisonnière et migration journalière verticale du zooplancton dans la région profonde de l'Adriatique. *Acta Adriatica* 9: 3-59.

Irie, H. 1953. Some ecological experiments on 'umi-botaru' (*Cypridina hilgendorfi* G.W. Müller). *Bulletin of the Faculty of Fisheries, Nagasaki University* 1: 10-13 (in Japanese, with an English abstract).

Kornicker, L.S. 1981. Revision, distribution, ecology and ontogeny of the ostracode subfamily Cyclasteropinae (Myodocopina: Cylindroleberididae). *Smiths. Contrib. Zool.* 319: 1-548.

Leveau, M. 1965. Contribution a l'étude des ostracodes et cladoceres du Golfe de Marseille. *Recueil des Travaux de la Station Marine d'Endoume, Bulletin* 37(53): 161-246.

Lochhead, J.H. 1968. The feeding and swimming of *Conchoecia* (Crustacea, Ostracoda). *Biol. Bull.* 134: 456-464.

Longhurst, A.R. 1976. Vertical migration. In D.J. Cushing & J.J. Walsh (eds), *The Ecology of the Seas*: 116-137. Oxford: Blackwell.

Moguilevsky, A. & Angel, M.V. 1975. Halocyprid ostracods in Atlantic neuston. *Mar. Biol.* 32: 295-302.

Morin, J.G. & Cohen, A.C. 1991. Bioluminescent displays, courtship, and reproduction in ostracodes. In R.T. Bauer & J.W. Martin (eds), *Crustacean Sexual Biology*: 1-16. New York: Columbia University Press.

Müller, G.W. 1906. In M. Weber (ed.), *Siboga-expeditie, uitkomsten op zoologisch, botanisch, oceanographisch en geologisch gebied verzameld in Nederlandisch Oost-Indie 1899-1900 aan boord H. M. Siboga, Monographien* 30: 1-40.

Okada, Y. & Kato, K. 1949. Studies on luminous animals from Japan. III. Preliminary report on the life history of *Cypridina hilgendorfi. Bull. Zoogeograph. Soc. Japan* 14: 21-25 (in Japanese).

Parker, A. 1997. Mating in Myodocopina (Crustacea: Ostracoda): results from video-recordings of a highly iridescent cypridinid. *J. Mar. Biol. Ass. UK* 77: 1223-1226.

Poulsen, E.M. 1962. Ostracoda-Myodocopa Part I Cypridiniformes-Cypridinidae. *Dana Reports* 57: 1-414.

Poulsen, E.M. 1973. Ostracoda-Myodocopa Part IIIB Halocypriformes-Halocypridae, Conchoeciinae. *Dana Reports* 84: 1-223.

Rudjakov, J.A. 1970. The possible causes of diel vertical migrations of planktonic animals. *Mar. Biol.* 6: 98-105.

Saito, T., Fukuda, M. & Taguchi, S. 1986. Electron microscope studies on the luminous gland of *Cypridina hilgendorfii. Zool. Sci.* 3: 391-394.

Santhakumari, V. & Saraswarthy, J. 1981. Zooplankton along the Tamil Nadu coast. *Magasagar* 14: 289-302.

Tsuji, F.I., Lynch, R.V. III & Haneda, Y. 1970. Studies on the bioluminescence of the marine ostracod crustacean *Cypridina serrata. Biol. Bull.* 139: 386-401.

Vannier, J. & Abe, K. 1993. Functional morphology and behavior of *Vargula hilgendorfii* (Ostracoda, Myodocopida) from Japan: preliminary results from video-recordings. *J. Crust. Biol.* 13: 51-76.

Weikert, H. 1972. Verteilung und Tagesperiodik des Evertebratenneuston im subtropischen Nordostatlantik während der Atlantischen Kuppenfahrten 1967 von F. S. 'Meteor'. *'Meteor' Forschungs-ergebnissen, Reihe D* 11: 29-87.

The 'eye-blink' response of mesopelagic Natantia, eyeshine patterns and the escape reaction

PETER M.J. SHELTON, EDWARD GATEN & MAGNUS L. JOHNSON
Department of Biology, University of Leicester, Leicester, UK

PETER J. HERRING
Southampton Oceanography Centre, Southampton, UK

ABSTRACT

The mechanical arrangement responsible for eye withdrawal in a variety of mesopelagic shrimps is described. The eyestalks are connected by, and articulate with, a region of stiffened cuticle that forms an interocular bar. Contraction of the antennal retractor muscles causes the bar to move backwards which results in the eyestalks swinging forwards. In the withdrawn position the compound eyes sit in antennular sockets. Tapetal reflectivity was examined in *Systellaspis debilis*, *Systellaspis cristata* and *Acanthephyra purpurea*. In *S. debilis* there is a distinct dorso-ventral gradient of increasing brightness. In the other two species upwardly directed parts of the eye are most reflective. Examination of eyeshine along the antero-posterior axis of the eye in the three species reveals a common pattern. When the eye is in the extended position eyeshine is brightest in those regions that look forwards. Eye withdrawal results in dramatic reductions in forwardly directed eyeshine. These observations are consistent with the need to minimize reflections from biolumines-cent sources during escape behavior.

1 INTRODUCTION

Reflective tapeta provide a means of increasing sensitivity in animals that are active in low light level environments. Previous studies of a number of mesopelagic deca-pod shrimps have shown that their tapeta vary in structure in different parts of the eye (Welsh & Chace 1937, 1938, Shelton et al. 1992). Two patterns of variation were identified in the latter study. First, there are marked regional changes in reflectivity along the dorso-ventral and antero-posterior axes. Second, there are distinct holes in the tapeta of some species. In addition, species from the lower mesopelagic zone have less reflective tapeta than those from the upper regions (Shelton et al. 1992). In the last study it was argued that regional variations of tapetal structure can be rationalized in terms of the highly predictable irradiance distribution in the sea (Kirk 1983) and the need for enhanced sensitivity in certain parts of the visual field. The fact that the most reflective region of the tapetum is usually downwardly directed (Shelton et al. 1992, Johnson 1998) is consistent with the fact that the upwelling irradiance is up to two orders of magnitude less bright than the ownwelling irradiance (Denton 1990). This makes objects more difficult to discriminate from the background when they are in the ventral rather than in the dorsal visual field because an object below the shrimp is seen against a dim background, whereas one above it will be silhouetted against a background 200 times brighter (Gaten et al. 1992, Shelton et al. 1992).

Similar differences in structure and reflectivity are found along the antero-posterior axis of the tapetum so that forward facing regions are more reflective than posterior facing ones (Shelton et al. 1992). Because light in the sea is distributed symmetrically about the vertical axis (Kirk 1983) and the intensity is constant at any given angle to it, the differences in reflectivity in anterior and posterior parts of the eye cannot have evolved because of differences in the light intensity in different parts of the visual field. One suggestion is that increased anterior reflectivity could be associated with maximizing sensitivity in the forward-looking part of the eye which is that region most concerned with prey capture (Shelton et al. 1992). However, because the shrimps are themselves vulnerable to predation there would also be a selective advantage in maximizi g sensitivity i. ll parts of the visual field.

Although tapeta improve the sensitivity of the eye, their reflective nature means that they can increase the shrimp's visibility to predators (Shelton 1993). Highly reflective areas on the dorsal su· ace of an aquatic animal would make it very visible against the background. This may explain why it is that the tapetum is reduced or absent from the dorsal region of the eye in species such as *Oplophorus spinosus* and *Systellaspis debilis* (Shelton et al. 1992). The need to reduce visibility to predators could explain the differences in reflectivity between anterior and posterior parts of the eye. In this paper we show that the antero-posterior gradient of reflectivity that is found in many species can be explained as an arrangement designed to minimize the visibility of the shrimp to predators during escape behavior.

2 MATERIALS AND METHODS

Material for the study was collected during cruise 122 (1995) of RRS 'Challenger' in the eastern North Atlantic within 100 nautical miles of Madeira. The animals were taken from depths down to 1000 m using the RMT 1 + 8 net system (Roe & Shale 1979) and a closing cod end (Wild et al. 1985). Shrimps were captured at night and sorted under dim red light. Prior to examination they were kept in a refrigerator at 4°C in light-proof aquaria. Eyeshine intensity was recorded and viewed in live shrimps using an Hitachi HV-620K CCTV camera mounted on a Zeiss binocular microscope, a Panasonic AG 6024 VHS videocassette recorder and an Hitachi VM 126AK video monitor. Each shrimp was mounted using a plexiglass mount that enabled it to be rotated about the pitch, roll and yaw axes. The mount was submerged in a glass aquarium containing cool seawater and the shrimp was viewed from the side using the CCTV camera and monitor. Shrimp eyes were axially illuminated using a horizontal beam from a Schott fibre optic light source placed to one side the aquarium. The light beam was redirected by 90° onto the shrimp's eye using a small mirror close to the optical axis of the binocular microscope and orientated at 45° to it. The whole apparatus was located in a darkroom maintained at 10°C. To obtain images of eyeshine from different parts of the dorso-ventral and antero-posterior axes of the

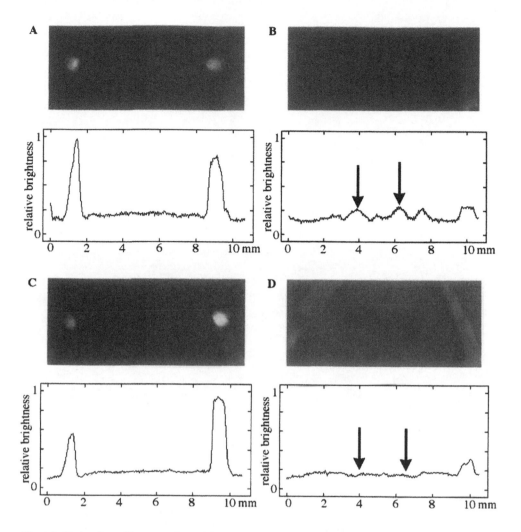

Plate 1. Examples of images of eyeshine with eyes extended and withdrawn from *Systellaspis debilis* (A, B) and *Acanthephyra purpurea* (C, D). For each image plots of eyeshine brightness are shown below; arrows mark the positions on the plots corresponding to reflection from the eyes.

eye, the shrimp was rotated in 10° steps. For the former the shrimp was orientated with its antero-posterior axis at right angles to the microscope and it was rotated in the roll axis. For the latter it was orientated with the anterior of the shrimp towards the microscope and it was rotated in the yaw axis. Images of eyeshine during 'blink' responses were recorded separately from anterior, dorsal, ventral and posterior directions during naturally occurring and induced blinks. Selected video frames were digitized to an Apple Power Macintosh computer via a Scion AG5 frame grabber. The images were analyzed using the public domain NIH Image program (developed at the US National Institutes of Health and available from the Internet at http://rsb.info.nih.gov/nih-image/). Mean eyeshine brightness was recorded in grey-scale units and, for each specimen the values at different points along the antero-posterior and dorso-ventral axes were normalized about the 90° value. Some specimens were preserved in Karnovsky's (1965) fixative for later anatomical examination. For investigating the muscles involved in the eye withdrawal response, specimens of *Palaemon serratus* were purchased from Millport Marine Laboratory.

3 RESULTS

3.1 *The eye withdrawal response*

Shrimps have a characteristic escape response in which a strong flexion of the abdomen (tail flip) results in a rapid backwards acceleration (Newland et al. 1992). We have noted that part of the escape response involves eye withdrawal in which the normally laterally directed eyestalks are swung forwards to lie next to each other parallel to the antero-posterior axis. Shrimps exhibiting this behavior have characteristic sockets in the basal segments of the antennules into which the eyes are withdrawn (Fig. 1). Examples of shrimps exhibiting this type of behavior and antennular anatomy include members of the following genera: *Systellaspis, Acanthephyra, Oplophorus, Ephyrina, Pandalus, Palaemon, Sergia* and *Sergestes*. The mechanism of the eye withdrawal response appears to involve a simple lever system. Although details of anatomy may vary, the general pattern of organization is similar from species to species. In all cases the eyes are borne on the ends of eyestalks that often taper towards the base. Below the rostrum and dorsal to the origins of the first and second antennae the two eyes are mechanically linked by a modified region of stiff anterior thoracic cuticle that has flexible folded cuticle above and below it (e.g. *Systellaspis debilis*) or anterior and posterior to it (e.g. *Sergia grandis*). In the genus *Sergia* the stiffened cuticle takes the form of an interocular bar. This is a transversely orientated tube-like structure open on one side to the haemocoel. In *Systellaspis debilis* it is supplemented by a vertically orientated stiff cuticular process that is anterior to the bar and which arises from the anterior cephalothoracic wall between the antennae. Because it is linked to the medial edges of the eyestalks, backward displacement of the bar results in the eyestalks being rotated forwards and the bases of the eyestalks being drawn back under the carapace. Using tethered specimens of *S. debilis, A. purpurea, E. figueirai* and *Sergia grandis* it was found that eye withdrawal could be elicited in a number of ways. Pushing the interocular bar posteriorly using fine forceps, or pushing the 1st and 2nd antennae backwards both resulted in the eyestalks moving to the

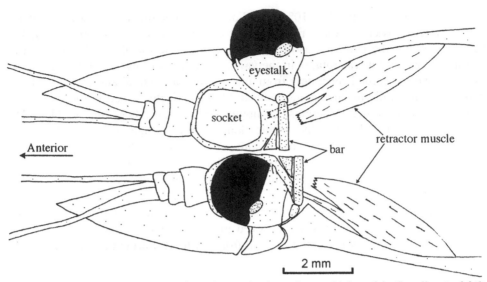

Figure 1. A schematic diagram to show the mechanism of eye withdrawal in *Systellaspis debilis*. The upper half of the diagram shows a dorsal view of an eye in the extended position; the lower half shows an eye withdrawn. The eyestalks articulate with the ends of the interocular bar. The cephalothorax has been partly cut away to show retractor muscles associated with the antennules (other muscles are not shown). On withdrawal the eyes rotate forwards to sit in the antennular sockets. Scale bar = 2mm.

withdrawn position. Reflex withdrawal of the eyestalks was found to occur naturally during the tail flip response. The latter could be elicited by gently squeezing the posterior segments of the abdomen. Reflex withdrawal of the eyestalks is brought about by prominent antennal retractor muscles associated with the first and second antennae. They arise from the dorso-lateral wall of the cephalothorax and run antero-ventrally to insert distally on the antennae. Using freshly dissected specimens of *P. serratus* it was shown that pulling on these muscles causes eye withdrawal.

3.2 *Eyeshine distribution patterns in oplophorid shrimps*

In the present study *Systellaspis debilis* was chosen as a representative of types that are found in regions of the mesopelagic zone where there is a measureable amount of downwelling light. It is a vertical migrator found at depths of up to 650 m during the daytime (Herring & Roe 1988). Its possession of well developed eyes and downwardly directed photophores (Nowel et al. 1998) is consistent with the photic environment in which it is found. Using shrimps (n = 10) with the eyestalks in the extended position, visual inspection of eyeshine along the dorso-ventral and antero-posterior axes revealed a consistent pattern with eyeshine increasing from dorsal to ventral and from posterior to anterior. A plot of eyeshine brightness from a typical example is shown in Figure 2A.

Systellaspis cristata is a non-migrant, occurs at depths of 1000 m and more (Foxton 1970), has relatively large eyes and in the adult has poorly developed photophores. Large eyes are an adaptation to increase sensitivity (Land 1981) and the need

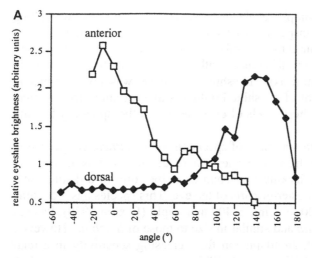

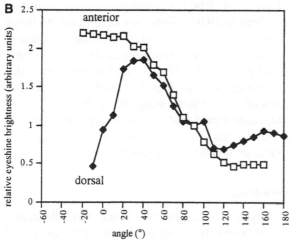

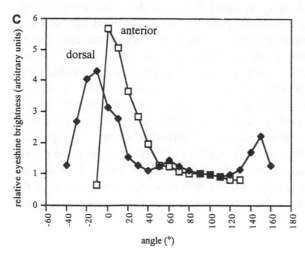

Figure 2. Plots of eyeshine brightness with angle in *Systellaspis debilis* (A), *S. cristata* (B) and *Acanthephyra purpurea* (C). In all cases eyeshine decreases from anterior to posterior. In *A.purpurea* this is due to the presence of a hole in the tapetum. In the other two species it is due to reduced reflectivity of the posterior tapetum. In *S. debilis* eyeshine is reduced in upwardly looking parts of the eye. In the other two species eyeshine brightness is maximal in upwardly looking regions.

for counterillumination using photophores decreases as the level of the downwelling light become negligible. Visual inspection of the eyeshine brightness along the antero-posterior axis (n = 8) was found to be similar to that seen in *S. debilis*. However, in *S. cristata* there were significant differences with respect to eyeshine intensity in different parts of the dorso-ventral axis. Eyeshine brightness was not maximal in ventral parts of the eye and a plot of eyeshine brightness against angle for a typical case (Fig. 2B) shows that the highest levels of eyeshine are in the upwardly looking regions (40°).

Eyeshine was also examined in a number of specimens of *Acanthephyra purpurea* (n = 15) to provide information from a species where the tapetum is incomplete (Shelton et al. 1992). The region lacking a tapetum takes the form of a distinct hole extending from the lateral to the posterior parts of the eye. This species has a daytime population maximum at about 800 m rising to 400 m at night (Foxton 1970). Its eyes are smaller than those of *S. cristata* and similar in size to those of *S debilis*. However, adjacent rhabdoms are confluent, an adaptation for increasing sensitivity in a relatively light deprived environment (Gaten et al. 1992). Consistent with its deep daytime distribution it lacks photophores (Herring 1976). With respect to eyeshine in the dorso-ventral axis it is similar to *S. cristata* in that the brightest eyeshine is found in upwardly looking parts of the eye. A plot of eyeshine brightness against angle shows a maximum at −10° and a smaller ventral peak at 150° (Fig. 2C). The antero-posterior distribution of eyeshine brightness is similar to that found in both *S. debilis* and *S. cristata* with anteriorly directed parts of the eye possessing the brightest eyeshine. The lateral to posterior hole in the tapetum means that eyeshine is greatly reduced posteriorly. During the course of the current study it was noted that *Ephyrina figueirai* has a similar tapetal hole; we presume its function to be similar.

3.3 *Changes in eyeshine during the eye blink*

Specimens of both *S. debilis* and *A. purpurea* were attached to the rotatable mount and video images of the shrimps were obtained from four directions (90° above, 90° below, 90° to one side and from directly in front). For each direction images of the eyes in the laterally extended and withdrawn positions were obtained for both species. In all cases there were changes in eyeshine brightness when the eyes moved from the extended to the withdrawn positions. When viewed from below the relatively bright eyeshine from the ventral part of the eye was particularly visible when the eyes were extended. When they were withdrawn the eyes were hidden from view by the antennal appendages. Viewed from above there was very little change in eyeshine on withdrawal but eyeshine visibly dimmed when the eyes were observed from the side. However, the most dramatic effects were observed when the shrimps were viewed from the front. On eye withdrawal in *S. debilis* light reflected from the eyes is reduced to levels comparable to those coming from the carapace (Plate 1, A and B). In *A. purpurea* the effect is even greater and the reflection from the centre of the withdrawn eye is lower than that from the cuticle adjacent to the eye (Plate 1, C and D).

4 DISCUSSION

The present results enable us to explain the significance of both the dorso-ventral and antero-posterior patterns of reflectivity in the eye. Shrimps from the mesopelagic zone live in such low light levels that mechanisms for increasing sensitivity in all parts of the visual field would seem to be advantageous. If this is the case we need to explain why certain parts of the eye have reduced or absent tapetal reflectivity. In species such as *S. debilis* the dorso-ventral pattern is consistent with the need to remain invisible to predators (Shelton et al. 1992) and the angular distribution of light within the upper mesopelagic zone (Kirk 1983). Regions of high reflectivity on the dorsal surface of an animal make it visible from above because the downwelling light reflected from it is up to 200 times brighter than the upwelling light from the surroundings (Denton 1990). Regions of high reflectivity on the ventral surface of a mesopelagic animal do not increase visibility from below in the same way because of the relative dimness of the upwelling light. In this case the higher the reflectivity of the ventral surface the less visible the animal would be against the relatively bright downwelling light. This may explain why many species of mesopelagic shrimp have a ventral tapetum that is more reflective than the dorsal. Previous investigations suggest that this is the case for a number of other species including forms like *Acanthephyra kingsleyi, A. pelagica, Oplophorus spinosus* and *Sergestes henseni* (Shelton et al. 1992), *Gennadas brevirostris, G. scutatus, G. talismani, G. valens, Parapandalus richardi, Plesionika martia, Sergia grandis* and *S. robustus* (Johnson 1998).

The present study includes two species (*Systellaspis cristata, Acanthephyra purpurea*) from the lower limits of the mesopelagic zone. In both the dorso-ventral pattern of reflectivity is different. Here upwardly directed regions of the eye were found to have relatively strong reflectivity. In other species from these depths (including *Acanthephyra gracilipes, A. stylorostratis, Bentheogennema intermedia* and *Parapasiphae sulcatifrons*) the ventral eyeshine is no brighter than the dorsal (Johnson 1998). At 1000 m the relative difference between levels of upwelling and downwelling light is negligible and the low levels of light present may mean that a highly reflective dorsal surface no longer presents such a hazard.

Although there were differences between species with respect to the dorso-ventral axis, patterns of reflectivity along the antero-posterior axis were much more consistent. In all cases anterior eyeshine was brighter than posterior eyeshine. In a previous study we suggested that this might be due to the need for increased sensitivity in those parts of the visual field associated with prey capture (Shelton et al. 1992). Since the same pattern of reflectivity is found in shrimps from different depths and photic conditions we have sought other explanations. Based on the current study, we believe that it is more likely to be due to the need to reduce reflectivity during the escape response. As part of the escape behavior many species including *S. debilis* and *A. purpurea* (and probably all oplophorids, see Herring 1985) emit clouds of bioluminescent material from the mouth during the tail flip (Herring 1976). Such light sources illuminate the eyes as the shrimp moves away backwards. Withdrawing the eyes so that the least reflective region faces the bioluminescence is of obvious adaptive significance. The present results provide an explanation for the occurrence of anterior/posterior gradients of reflectivity and for a hole in the posterior part of the tapetum in forms like *A. purpurea* and *Ephyrina figueirai*.

ACKNOWLEDGEMENTS

This work was supported by NERC grants GR9/0119A to PMJS, GR3/11212 to PMJS and PJH, and a NERC CASE award to MLJ.

REFERENCES

Denton, E. 1990. Light and vision at depths greater than 200 metres. In P.J. Herring, A.K. Campbell, M. Whitfield & L. Maddock (eds), *Light and life in the sea*: 17-148. Cambridge: Cambridge University Press.

Foxton, P. 1970. The vertical distribution of pelagic decapods (Crustacea: Natantia) collected on the SOND cruise 1965. I. The Caridea. *J. Mar. Biol. Assoc. UK* 50: 939-960.

Gaten, E., Shelton, P.M.J. & Herring P.J. 1992. Regional morphological variations in the compound eyes of certain mesopelagic shrimps in relation to their habitat. *J. Mar. Biol. Assoc. UK* 72: 61-75.

Herring, P.J. 1976. Bioluminescence in decapod Crustacea. *J. Mar. Biol. Assoc. UK* 56: 1029-1047.

Herring, P.J. 1985. Bioluminescence in the Crustacea. *J. Crust. Biol.* 5: 557-573.

Herring P.J. & Roe, H.S.J. 1988. The photoecology of pelagic oceanic decapods. *Symp. Zool. Soc. Lond.* 59: 262-290.

Johnson, M. 1998. Aspects of the visual function and adaptation of deep-sea decapods. Ph.D. thesis, University of Leicester.

Karnovsky, M.J. 1965. A formaldehyde-glutaraldehyde fixative of high osmolality for use in electron microscopy. *J. Cell. Biol.* 27: 137A-138A.

Kirk, J.T.O. 1986. *Light and photosynthesis in aquatic ecosystems.* Cambridge: Cambridge University Press.

Land, M.F. 1981. Optics and vision in invertebrates. In H. Autrum (ed.), *Handbook of sensory physiology* Vol. VII/6B: 471-492.

Newland, P. L., Neil, D.M. & Chapman, C. J. 1992. Escape swimming in the Norway lobster. *J. Crust. Biol.* 12: 342-353.

Nowel, M.S., Shelton, P.M.J. & Herring, P.J.1998. The cuticular photophores of two decapod crustaceans, *Oplophorus spinosus* and *Systellaspis debilis*. (Submitted).

Roe, H.S.J. & Shale, D.M. 1979. A new rectangular midwater trawl (RMT1+8M) and some modifications to the Institute of Oceanographic Sciences' RMT 1+8. *Mar. Biol.* 50: 283-288.

Shelton, P.M.J. 1993. Shrimp. In S.P. Parker (ed.), *McGraw-Hill yearbook of science and technology 1994*: 331-333. New York: McGraw-Hill.

Shelton, P.M.J., Gaten E. & Herring P.J. 1992. Adaptations of tapeta in the eyes of mesopelagic decapod shrimps to match the oceanic irradiance distribution. *J. Mar. Biol. Assoc. UK* 72: 77-88.

Welsh, J.H. & Chace F.A. 1937. Eyes of deep-sea crustaceans. I. Acanthephyridae. *Biol. Bull.* 72: 57-74.

Welsh, J.H. & Chace F.A. 1938. Eyes of deep-sea crustaceans. II. Sergestidae. *Biol. Bull.* 74: 364-375.

Wild, R.A., Darlington E. & Herring P.J. 1985. An acoustically controlled cod-end system for the recovery of deep-sea animals at in situ temperatures. *Deep-Sea Res.* 32: 1583-1589.

Spatial distribution of four sympatric species of hermit crabs (Decapoda, Anomura)

ALEXANDER TURRA, GIULIANO B. JACOBUCCI, FLÁVIO M.P. ARAÚJO &
FOSCA P.P. LEITE
Departamento de Zoologia, IB, Universidade Estadual de Campinas, Campinas, SP, Brasil

ABSTRACT

This study reports on the spatial distribution of four sympatric species of hermit crab in the intertidal zone of Pernambuco Islet, São Sebastião Channel, Brazil (23°49'S, 45°24'W). The samples were taken monthly, over a year, in five randomized transects perpendicular to the shore line. Each transect was subdivided in 0.25 m^2 units, which were individually sampled. *Pagurus criniticornis* was the most abundant hermit crab, occupying the lowest intertidal region and muddy substrate. *Clibanarius antillensis* occurred in higher densities than *C. sclopetarius* and *C. vittatus* and presented a contiguous distribution. On the other hand, *C. sclopetarius* and *C. vittatus* exhibited similar densities showing a regular distribution pattern. These three species of *Clibanarius* utilized different substrates (coarse sand, pebbles and rocky shore). They did not show a clear pattern of zonation, although *C. antillensis* was located predominantly near the water line. The studied clusters were typically monospecific and showed a female skewed sex ratio, except those of *P. criniticornis*. Clusters of *C. sclopetarius* and *C. antillensis* were of similar sizes, however, with a large difference in the number of individuals. Clusters of *P. criniticornis* were smaller than those of the two last-mentioned species. *Clibanarius vittatus* did not form clusters and was not included in the analysis. *Clibanarius sclopetarius* was frequently found in shells of *Chicoreus senegalensis*, while *C. antillensis* and *P. criniticornis* used mainly shells of *Cerithium atratum*. Isolated and clustered hermit crabs occupied shells of the same species, which were usually damaged. Clustered females of *C. sclopetarius* occupied less adequate shells than isolated ones. Such a difference was not verified for *C. antillensis*. The adequacy of the shells used by *C. sclopetarius* was lower than that of *C. antillensis*. The patterns of distribution and microhabitat use seemed to be influenced by hermit crab preferences, their clustering behavior, and resistance to air exposure.

1 INTRODUCTION

Small-scale spatial distribution patterns of hermit crabs, focused on the occupation of different microhabitats, were studied to analyze the relationship between crabs and

261

Table 1. Comparison of observed (N° Obs.) and expected (N° Exp.) frequencies of microhabitat utilization by the hermit crabs of Pernambuco Islet, São Sebastião, Brazil.

| Microhabitat | Description | Number of quadrats | Relative abundance of microhabitats (%) | Hermit crabs | | | | | | | |
| | | | | C. antillensis | | C. sclopetarius | | C. vittatus | | P. criniticornis | |
				N° Obs.	N° Exp.	N° Obs.	N° Exp.	N° Obs.	N° Exp.	N° Obs.	N° Exp.
S	Coarse sand	21	2.95	73	12.03	7	2.09	3	3.75	3	38.88
R	Smooth rocky shore	88	12.36	5	50.43	8	8.77	9	15.70	26	162.91
Rsd	Rocky shore with sand	46	6.46	113	26.36	0	4.59	1	8.20	50	85.14
Rcv	Rocky shore with crevices	206	28.93	144	118.04	28	20.54	59	36.74	184	381.30
Rpb	Rocky shore with pebbles	22	3.09	0	12.61	8	2.19	17	3.92	56	40.73
Rtp	Rocky shore with tide pools	10	1.40	0	5.73	0	0.99	7	1.78	7	18.45
Rht	Rocky shore with more than one characteristic	110	15.45	40	63.03	14	10.97	7	19.62	337	203.63
M	Mud	87	12.22	0	45.85	0	8.67	0	15.52	322	161.06
Mpb	Mud with pebbles	28	4.68	0	16.04	0	2.79	4	4.99	132	61.68
P	Pebbles	25	3.51	0	14.32	6	2.49	11	4.46	18	46.26
Psd	Pebbles with sand	69	9.69	33	39.54	0	6.88	9	12.31	183	127.71
Total		712		408	408	71	71	127	127	1318	1318
Statistics											
G				294.76*		42.58*		136.30*		535.12*	
df				10		10		10		10	

* significant with a critical probability of $\alpha''(\alpha/55) = 0.0009$ $(0.05/55)$, after a Bonferroni correction for non-independent data using 11 classes.

their environments (Reese 1969, Young 1978, Taylor 1981, Lowery & Nelson 1988, Leite et al. 1998), and resource partitioning among species (Grant & Ulmer 1974, Mitchell 1975, Bach et al. 1976, Kellogg 1977, Gherardi 1990, Gherardi & Nardone 1997, Rittschof et al. 1995). Different patterns of spatial distribution of hermit crabs in the intertidal region may expose them to particular degrees of desiccation and thermal stress (Reese 1969). However, shells may also play an important role in hermit crab distribution by modulating an individual's resistance to rigorous environmental conditions (Rittschof et al. 1995).

Clustering behavior is known for some species of hermit crabs (Hazlett 1966, Snyder-Conn 1980, 1981, Gherardi 1990, 1991, Gherardi & Vannini 1993), although its adaptative significance is still unknown. It is supposed that clusters may protect hermit crabs from desiccation and water currents (Reese 1969, Snyder-Conn 1981, Gherardi 1990).

Gherardi (1990, 1991) related susceptibility to desiccation of small crabs, which have higher surface/volume ratio, to their frequent occurrence in clusters. The clusters facilitate shell exchanges (Radinovsky & Henderson 1974, Hazlett 1978, 1980, Snyder-Conn 1981) by acting as meeting places where hermits can acquire relatively more adequate shells.

In order to better understand the distribution patterns of intertidal hermit crabs, we focused on the spatial distribution, microhabitat utilization and partition, and clustering behavior of four sympatric species, *Clibanarius antillensis* (Stimpson, 1858), *C. sclopetarius* (Herbst, 1776), *C. vittatus* (Bosc, 1802), and *Pagurus criniticornis* (Dana, 1852). The comparison of population structure parameters (size, sex ratio, and frequency of ovigerous females) and shell utilization patterns between clustered and scattered individuals was used to evaluate the role of clusters as places for shell exchange and reproduction for the hermit crabs studied.

2 MATERIALS AND METHODS

The study was conducted in the intertidal zone of Pernambuco Islet, São Sebastião Channel, São Paulo, Brazil (23°49'S, 45°24'W). The crabs were collected from October 1995 to September 1996 in diurnal low tide periods. Samples were taken each month in five, 0.5 m wide randomized transects perpendicular to the shore line, giving a total of 55 different transects. The length of these transects varied in function of the tidal level and substrate slope, thus depending on the irregular topography of the intertidal area. Each transect was divided into 0.25 m^2 adjacent squares sampled from the water line to the supralittoral fringe, thus covering different kinds of substrate. The number of hermit crabs and the microhabitat coverage (Table 1) were registered for each square within the transects.

Further, the size of the crabs was measured registering the shield length (distance from the basis of the shield to the top of the rostrum) under a millimetric ocular coupled with a stereo microscope. Sex and situation (scattered or clustered) of the crabs were also recorded. Crabs were considered as clustered if inactive and in physical contact with each other (groups of 5 or more individuals). Additional samples were taken to yield data on clustering structure and composition. Shells were identified, having their condition (presence of damage or encrustation) and adequacy verified.

Shell adequacy was accessed in two ways. First, through a visual index based on Abrams (1978). In this method, an index is given depending on crab retraction into the shell: 1) Crab is not visible, 2) Dactyl visible but cheliped not, 3) One cheliped visible, 4) Both chelipeds visible, 5) Both chelipeds closing shell aperture. Second, through the shell adequacy index (Vance 1972, modified by Bertness 1980).

The density of hermit crabs (individuals per $0.25 \ m^2$) was calculated for each transect (n = 55) and compared among species by ANOVA and the Tukey test (Zar 1984). To verify hermit crab zonation, all transects were divided into three equal parts representing the low, mid, and high intertidal zones. It was considered that individuals in the low intertidal zone were less susceptible to desiccation than individuals in the higher ones. Dispersion patterns through the area were assessed by the calculation of a dispersion index (Eliott 1977), which used the mean-variance ratio of the density of hermit crabs in the sampled transects. The Shannon-Wiener diversity index (H', natural logarithmic) and percent similarity (Renkonen index) were also calculated using data of density of individuals in each kind of microhabitat. The log-likelihood G test was used to test the randomness (observed versus expected densities) of microhabitat occupation by the crabs. This test was also used to evaluate the sex ratio, the proportion of ovigerous females, shell use, the visual adequacy index, shell damage and encrustation differences among clustered and scattered individuals. Shield length and shell adequacy index were compared among aggregated and isolated crabs by Student's t-test. Logarithmic transformations were performed when necessary to normalize the data. All statistics were performed at the 0.05 significance level. However, it was considered that some data obtained from squares, such as the number of hermit crabs in each microhabitat, were not independent because these squares belonged to the sorted transects. In this way, the Bonferroni correction (Sokal & Rohlf 1995) was employed to make the statistical analysis more conservative.

3 RESULTS

3.1 *Distribution*

The four hermit crab species occurred in different densities in the area (ANOVA, F = 7.573, df = 3, $p < 0.001$). *Pagurus criniticornis* (1.257 ± 1.928 individuals per $0.25 \ m^2$) was more abundant than *C. sclopetarius* (0.102 ± 0.250; Tukey $p < 0.001$) and *C. vittatus* (0.179 ± 0.336; $p = 0.001$). *Clibanarius antillensis* (0.728 ± 2.127) occurred in the same densities as *P. criniticornis*, *C. sclopetarius* and *C. vittatus* (Tukey, $p = 0.222$, $p = 0.107$ and $p = 0.194$, respectively). The density of *C. sclopetarius* was very similar to that of *C. vittatus* (Tukey, $p = 0.993$).

The dispersion index revealed two distribution patterns. *Clibanarius sclopetarius* and *C. vittatus* dispersed regularly through the area (I = 33.318, df = 54, d = –2.181, $p < 0.05$ and I = 34.074, df = 54, d = –2.089, $p < 0.05$, respectively), while *C. antillensis* and *P. criniticornis* presented a contiguous distribution pattern (I = 335.502, df = 54, d = 15.5597, $p < 0.001$ and I = 159.624, df = 54, d = 7.5335, $p < 0.001$, respectively).

Microhabitat utilization by these species did not follow availability (Table 1), showing that the crabs selected the places where they were collected. The microhabitats used by each species are presented in Figure 1. *Clibanarius antillensis* occupied mainly microhabitats associated with sand and those assigned as rocky shore with crevices. *Clibanarius sclopetarius* was found frequently in sand, rocky shore with pebbles and pebbles, *C. vittatus* in pebbles, pebbles with sand, and rocky shore with pebbles and tide pools, and *P. criniticornis* in muddy and highly complex rocky substrates.

The Renkonen index of similarity also revealed that these species occupied different kinds of substrate. The highest similarity was found between *Clibanarius sclopetarius* and *C. vittatus* (60.61%). Intermediate similarities in microhabitat utilization were registered between *C. antillensis* and *C. sclopetarius*, *C. vittatus*, and *P. criniticornis* (22.55%, 31.40% and 31.87%, respectively). The lowest values were recorded between *P. criniticornis* and both *C. sclopetarius* and *C. vittatus* (9.65% and

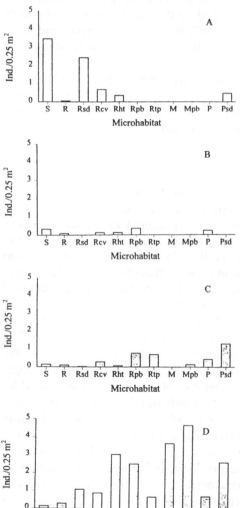

Figure 1. Comparison of the microhabitat utilization pattern of *C. antillensis* (N = 408), *C. sclopetarius* (N = 71), *C. vittatus* (N = 133) and *P. criniticornis* (N = 1318). Legend for microhabitats in Table 1.

Table 2. Comparison (ANOVA) of the density of four hermit crabs in three zones (low, mid and high intertidal) in the intertidal region of Pernambuco Islet, São Sebastião, SP, Brazil. (X, mean; SD, standard deviation)

Hermit crab	N	Low		Mid		High		F	df	$p*$
		X̄	SD	X̄	SD	X̄	SD			
C. antillensis	408	4.628	16.347	2.767	10.106	0.546	1.802	1.533	2	0.215
C. scopetarius	71	0.235	1.296	0.313	1.189	0.716	2.669	1.077	2	0.343
C. vittatus	133	0.859	1.520	0.604	1.688	0.516	2.642	0.665	2	0.516
P. criniticornis	1318	7.136	11.841	4.856	10.044	2.512	7.697	0.938	2	0.056

* non-significant difference with a critical probability of α'' $(\alpha/3) = (0.017)$, after a Bonferroni for non-independent data to compare the mean density of hermit crabs among three interidal zones.

17.59%, respectively). These results contrast with the data on diversity (Shannon-Wiener, H') and richness (R) of microhabitats used by the crabs. *Pagurus criniticornis* (H' = 1.909, R = 11) and *C. vittatus* (H' = 1.767, R = 10) used a richer and more diverse resource than *C. antillensis* (H' = 1.516, R = 6) and *C. sclopetarius* (H' = 1.616, R = 6).

Zonation was assessed by comparing densities of crabs among the three levels established in the intertidal zone (Table 2). Although the densities of the four hermit crabs did not differ among zones, a tendency for *C. antillensis* and *P. criniticornis* to occupy the lower levels of the intertidal habitat was found.

3.2 Clustering

Clustering behavior was addressed only for *C. antillensis*, *C. sclopetarius* and *P. criniticornis* because *C. vittatus* did not form clusters in this area. Data from the transects showed that *C. antillensis* presented more clustered individuals (58.10% were clustered, 226 clustered individuals in 389) than *C. sclopetarius* (20.89%, 14 in 67) and *P. criniticornis* (15.79%, 179 in 1119). Clusters of *C. antillensis* and *C. sclopetarius* presented greater size variation (n = 21 clusters, 16.80 ± 13.28 ind., range: 5 to 53 ind.; n = 4, 25.75 ± 19.62, range: 7 to 51, respectively) when compared with those of *P. criniticornis* (n = 11, 16.27 ± 7.34, range: 9 to 33).

Clustered and scattered individuals of *C. antillensis* (3.77 ± 0.93 mm and 3.85 ± 1.02 mm, respectively; t = –0.885, df = 534, ns), *C. sclopetarius* (7.81 ± 1.29 mm and 7.61 ± 1.45 mm, respectively; t = 1.024, df = 196, ns), and *P. criniticornis* (2.40 ± 0.49 mm and 2.40 ± 0.54 mm; t = 0.028, df = 478, ns) did not present differences in shield length (Fig. 2). The sex ratio (frequency of males/total number of individuals) was skewed for females in these two *Clibanarius* populations (*C. antillensis*: 0.33, $G_{[1]}$ = 65.72, $p < 0.001$; *C. sclopetarius*: 0.28, $G_{[1]}$ = 38.62, $p < 0.001$), while males were more abundant in *P. criniticornis* (0.59, $G_{[1]}$ = 14.36, $p < 0.001$) . The sex ratio differed between clustered and scattered crabs only for *C. antillensis*, which presented a higher frequency of females in clusters than isolated (Table 3). Despite these differences, clustered crabs of the *Clibanarius* species presented higher frequencies of ovigerous females than scattered ones (*C. antillensis*: 186 in 240 clustered and 83 in 125 scattered, $G_{[1]}$ = 5.10, $p < 0.05$; *C. sclopetarius*: 58 in 72 and 16 in 70, $G_{[1]}$ = 50.44, $p < 0.001$). This pattern was not verified for *P. criniticornis* (35 in 149 and 9 in 45, $G_{[1]}$ = 0.26, ns).

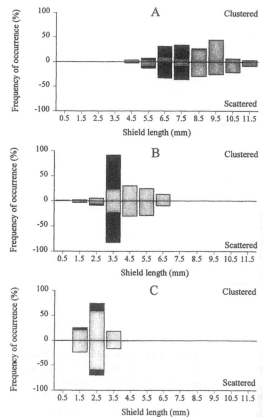

Figure 2. Shield length distribution of clustered and scattered males and females of *C. sclopetarius*, *C. antillensis*, and *P. criniticornis*. Black bars: females; hatched bars: males.

Table 3. Frequencies of clustered and scatterd males (M), females (F) and ovigerous females (F_{ov}) of *C. antillensis, C. sclopetarius* and *P. criniticornis* in the intertidal region of Pernambuco Islet, São Sebastião, SP, Brazil.

Hermit crab	Clustered				Scattered						
	M	F	F_{ov} (%)	Sex Ratio	M	F	F_{ov} (%)	Sex Ratio	G	Df	p
C. antillensis	95	240	86 (77.50)	0.28	83	125	83 (66.40)	0.40	7.68	1	< 0.01
C. sclopetarius	32	72	58 (80.55)	0.31	24	70	16 (22.86)	0.26	0.64	1	Ns
P. criniticornis	65	45	9 (20.00)	0.59	211	149	35 (23.49)	0.59	0.02	1	Ns

Pagurus criniticornis used almost exclusively shells of *Cerithium atratum*, the shell most used by *C. antillensis*. The two species of *Clibanarius* used different shell types, with *C. sclopetarius* occupying larger shells than *C. antillensis*. However, there was no evidence for a differential shell occupation between clustered and scattered individuals of these three populations (Fig. 3).

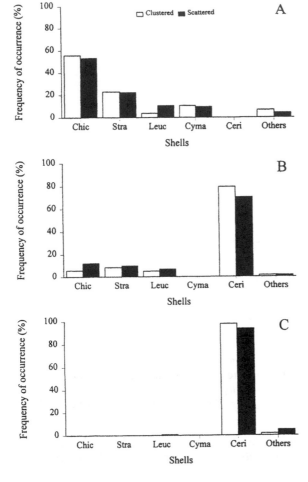

Figure 3. Shell utilization by clustered and scattered individuals of *C. sclopetarius, C. antillensis* and *P. criniticornis*. Legend: Chic, *Chicoreus* (= *Siratus*) *senegalensis* (Gmelin, 1790); Stra, *Stramonita* (= *Thais*) *haemastoma* (Linnaeus, 1757); Leuc, *Leucozonia nassa* (Gmelin, 1791); Cyma, *Cymatium partenopeum* (von Salis, 1793); Ceri, *Cerithium atratum* (Born, 1778).

The visual adequacy index for *C. sclopetarius* showed that clustered individuals used less adequate shells than scattered ones ($G_{[3]} = 16.44$, $p < 0.001$). This was not verified for *C. antillensis* ($G_{[3]} = 1.62$, ns) (Fig. 4). The log transformation of the shell adequacy index did not show any difference between clustered and scattered individuals of *C. sclopetarius* (t = -1.875, df = 169, ns) and *C. antillensis* (t = -1.077, df = 435, ns).

The proportion of used shells that were damaged was the same for *C. antillensis* (82 in 330 clustered and 47 in 196 scattered, $G_{[1]} = 0.05$, ns) and *C. sclopetarius* (20 in 103 clustered and 26 in 90 scattered, $G_{[1]} = 2.37$, ns). In contrast, shells used by clustered individuals of *P. criniticornis* presented more frequent damage than those occupied by scattered ones (54 in 78 clustered and 21 in 42 scattered, $G_{[1]} = 1.28$, $p < 0.005$). Scattered individuals of *C. sclopetarius* were found in encrusted shells more often then clustered ones (65 in 90 clustered and 40 in 103 scattered, respectively, $G_{[1]} = 22.10$, $p < 0.001$), in contrast to *C. antillensis* (124 in 330 and 64 in 196, $G_{[1]} = 1.31$, ns) and *P. criniticornis* (54 in 81 and 31 in 42, $G_{[1]} = 0.72$, ns).

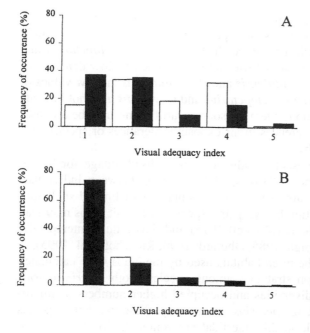

Figure 4. Comparison of the shell adequacy (visual index) between clustered and scattered individuals of *C. sclopetarius* and *C. antillensis*.

4 DISCUSSION

All four hermit crab species presented specific strategies for microhabitat utilization in the intertidal region of Pernambuco Islet. The contiguous pattern of distribution shown by *C. antillensis* and *P. criniticornis* was due to the high variance/mean ratio of the density of their individuals in sampled transects. These two species occurred in high densities, mainly in the lowest areas of the intertidal zone where they frequently occupied microhabitats associated with sand and mud. However, they also occurred in the mid and high intertidal levels occupying highly complex substrates, forming clusters only in holes and crevices. In contrast, *C. sclopetarius* and *C. vittatus* showed a regular pattern of distribution and occurred in lower densities than *C. antillensis* and *P. criniticornis*. The former species did not occupy different zones in the intertidal region, but were more frequently found in substrates associated with rocky shore and pebbles.

Microhabitats may be considered one of the dimensions of hermit crab niche and substrate use possibly has an important role in crab fitness. Percent similarity was used as an index of microhabitat utilization overlap as suggested by Abrams (1980). This index revealed that *C. sclopetarius* and *C. vittatus* overlapped strongly in the use of this resource when compared with the other species. Low values of this niche overlap index for microhabitat utilization between *P. criniticornis* and both *C. sclopetarius* and *C. vittatus* revealed that resource partitioning is taking place among them. *Clibanarius antillensis* did not show high niche overlap with the three other species, indicating that this species also partitioned microhabitats with them. Microhabitat partition among species is clearly demonstrated by the data of Figure 1.

Nevertheless, these values of niche overlap index should be analyzed carefully. A high overlap index or a low degree of resource partition does not necessarily imply that species are competing for such a resource. In the case of *C. sclopetarius* and *C. vittatus*, the high overlap in microhabitat use can be sustained by low densities of these populations in this area. For *C. antillensis* and *P. criniticornis* the low values of overlap in microhabitat use with both *C. sclopetarius* and *C. vittatus* may not indicate that competition is taking place among them. Resource partitioning may be a consequence of other factors such as differences in air exposure tolerances or specific microhabitat preferences of these crabs.

Microhabitat partition may represent an advantage or a disadvantage for hermit crabs. According to Grant & Ulmer (1974), spatial separation between individuals can be a behavioral adaptation of large specimens in situations of limited shell supply. In contrast, microhabitat partition by sympatric species of hermit crabs may lead them to a differential access to shells (Mitchell 1975) and thus reduce interspecific shell exchanges (Mitchell 1975, Spight 1985, Gherardi 1990, Rittschof et al. 1995).

The diversity and richness of the microhabitats used by the crabs were calculated also to compare spatial distribution strategies among species. Both *P. criniticornis* and *C. vittatus* presented higher diversities and occupied higher numbers of microhabitats than *C. antillensis* and *C. sclopetarius*. In this way, these two latter species did not use all substrates available in this intertidal environment. However, microhabitat utilization did not follow availability, showing that the hermit crabs are selecting the sites where they were collected.

The regularity in the distribution of *C. sclopetarius* and *C. vittatus* populations may be due to territorial behavior as suggested by Elliott (1977) for other organisms. This is possible because agonistic interactions were occasionally registered between individuals of these species in nature and in laboratory observations (A. Turra, pers. observ.). In general, agonistic behavior is common among hermit crabs in situations of intra- and interspecific competition for shells (Hazlett 1966, Bach et al. 1976, Scully 1983). However, this pattern of dispersion can be also related to the high mobility of these species (G. Jacobucci, unpubl. data), enabling them to be dispersed over a larger area than *C. antillensis* and *P. criniticornis*.

The clumped dispersion pattern shown by *C. antillensis* and *P. criniticornis* is associated with both clustering behavior and air exposure tolerances. *Clibanarius antillensis* presented a higher tendency to occur in clusters (58.10%) than *P. criniticornis* (15.79%). Although the regularity in the dispersion pattern of *C. sclopetarius*, this species also formed clusters in this area (20.89%). Clusters of *C. antillensis* and *C. sclopetarius* were collected in sites exposed to the air and in holes and crevices, while that of *P. criniticornis* were found only in highly complex areas (holes and crevices). In addition, *P. criniticornis* is less tolerant to air exposure than *C. antillensis*, which is less tolerant than *C. sclopetarius* (A. Turra, unpubl. data). These results indicate that clusters may have two adaptative implications, as also suggested by Snyder-Conn (1981). First, aggregations may be a behavioral response of *C. antillensis* and *C. sclopetarius* to a limited shell supply. It has been argued that clusters may serve as places where hermit crabs can acquire better-fitting shells, either through competitive exchange or bargaining (Hazlett 1978, 1980) or, as suggested by Snyder-Conn (1981), Gherardi (1991), and Gherardi et al. (1994), they may be considered 'shell exchange markets'. Second, clusters may function as refuges against air exposure and desiccation for both *C. antillensis* and *P. criniticornis*. Reese (1969) suggested that the clumping behavior of intertidal hermit crabs would reduce water loss during exposure at low tide. Some studies support this hypothesis (Snyder-Conn 1981, Gherardi 1990, 1991). Smaller crabs (*P. criniticornis* and *C. antillensis*) would be expected to cluster more frequently than larger ones (*C. sclopetarius* and *C. vittatus*) because the former two species have a higher surface to volume ratio and consequently a lower desiccation resistance. However, there were no differences between clustered and scattered individuals of the same species. It is being argued that such kind of clusters and the success of these species in the intertidal habitat are highly dependent on the number, size and quality of refuges, i.e., environmental complexity.

The shield length distribution (Fig. 2) reveals an unbalanced sex-ratio in the size structure of populations. The three species present a nearly equal sex-ratio in small size classes, an excess of females in intermediate classes and a dominance of males in larger classes. This is a frequent pattern in hermits (Abrams 1988, Gherardi 1991, Gherardi et al. 1994) and other decapod populations (Vannini & Gherardi 1988). In general, sex ratios were skewed for females, except for *P. criniticornis*.

Sex ratio differences between clustered and scattered individuals were registered only for *C. antillensis*, with females being more frequent within clusters. In addition, the ovigerous females of the *Clibanarius* species were also more frequent in clusters than isolated. It is supposed that clusters might facilitate the location of receptive females by males, increasing the reproductive success of both sexes (Gherardi & Vannini 1993). On the other hand, clusters could simply provide protection for ovigerous females, but this possibility was refuted by Snyder-Conn (1981).

Clustered specimens of *C. sclopetarius* wear less adequate shells than isolated individuals, as shown by the visual adequacy index. The non-significant difference in visual adequacy observed for *C. antillensis* and *P. criniticornis* may result from the high abundance of the gastropod *Cerithium atratum*, the most used shell by these species at the study site. A high shell availability would result in a greater shell choice.

The shell quality determined by the shell adequacy index (Vance 1972) did not show any difference between aggregated and isolated hermit crabs. The proportion of broken shells was the same for clustered and scattered individuals of *Clibanarius* species. Damaged shells, however, were more often in clustered *P. criniticornis* than in scattered individuals. Only clumped *C. sclopetarius* presented a different degree of encrustation relative to the isolated specimens. The opposing results of the two adequacy indices regarding *C. sclopetarius* and the significant differences in shell condition observed only for *P. criniticornis* do not lead to a generalization on the shell exchange function of clusters, at least for the species analyzed, i.e., clustered individuals did not use worse shells than isolated ones.

In a broad sense, species that present low tolerances to air exposure, as *C. antillensis* and *P. criniticornis*, display clustering behavior that is strongly related to habitat complexity. Thus, substrate choice and occupancy will be dependent on the capacity of those substrates to protect the crabs against desiccation, while clumped distributions will be associated to the patchiness of such microhabitats. On the other hand, the distribution pattern of species like *C. sclopetarius*, that are more tolerant to water loss, seems to be less related to the presence, abundance, and distribution of special kinds of microhabitats. Clustering behavior and structure in this population does not seem to be influenced by habitat complexity. This assumption may also be applied to the larger individuals of *C. antillensis*, that are more tolerant to air exposure than the smaller ones (A. Turra, unpubl. data). Thus, causes and evolutionary implications dealing with clustering behavior should be analyzed separately, for each species and, perhaps, for different size classes within one species.

ACKNOWLEDGMENTS

We would like to thank FAPESP for the fellowships to the authors G.B.J. (Proc. n^0 95/9314-8) and A.T. (Proc. n^0 93/2439-4). Thanks are also due to CNP$_q$ for the fellowship to F.P.P.L. (Proc. n^0 30337/82-5), to CEBIMar-USP and FAEP (Proc. n^0 0222/95) for logistic and financial support, respectively. We also acknowledge the valuable referees' considerations on the manuscript.

REFERENCES

Abrams, P.A. 1978. Shell selection and utilization in a terrestrial hermit crab, *Coenobita compressus*. *Oecologia* 34: 239-253.
Abrams, P.A. 1980. Resource partitioning and interspecific competition in a tropical hermit crab community. *Oecologia* 46: 365-379.
Abrams, P.A. 1988. Sexual differences in resource use in hermit crabs: consequences and causes. In G. Chelazzi and M. Vannini (eds), *Behavioral adaptations to intertidal life*: 283-296. New York: Plenum.
Bach, C.B., Hazlett, B.A. & Rittschof, D. 1976. Effects of interspecific competition on fitness of the hermit crab *Clibanarius tricolor*. *Ecology* 57: 579-586.
Bertness, M.D. 1980. Shell preference and utilization patterns in littoral hermit crabs of the Bay of Panama. *J. Exp. Mar. Biol. Ecol.* 48: 1-16.
Elliott, J.M. 1977. Some methods for the statistical analysis of samples of benthic invertebrates. *Freshw. Biol. Ass. Sc. Publ.* 25.
Gherardi, F. 1990. Competition and coexistence in two Mediterranean hermit crabs *Calcinus ornatus* (Roux) and *Clibanarius erythrops* (Latreille) (Decapoda, Anomura). *J. Exp. Mar. Biol. Ecol.* 143: 221-238.
Gherardi, F. 1991. Relative growth, population structure, and shell-utilization of the hermit crab *Clibanarius erythropus* in the Mediterranean. *Oebalia* 17: 181-196.
Gherardi, F. & Nardone, F. 1997. The question of coexistence in hermit crabs: Population ecology of a tropical intertidal assemblage. *Crustaceana* 70: 608-629.
Gherardi, F. & Vannini, M. 1993. Hermit crabs in a mangrove swamp: proximate and ultimate factors in the clustering of *Clibanarius laevimanus*. *J. Exp. Mar. Biol. Ecol.* 168: 167-187.
Gherardi, F., Zatteri, F. & Vannini, M. 1994. Hermit crabs in a mangrove swamp: the structure of *Clibanarius laevimanus* clusters. *Mar. Biol.* 121: 41-52.
Grant, W.C., Jr. & Ulmer, K.M. 1974. Shell selection and aggressive behavior in two sympatric species of hermit crabs. *Biol. Bull.* 146: 32-43.
Hazlett, B.A. 1966. Social behavior of the Paguridae and Diogenidae of Curaçao. *Stud. Fauna Curaçao* 88: 1-143.
Hazlett, B.A. 1978. Shell exchanges in hermit crabs: aggression, negotiation or both? *Anim. Behav.* 26: 1278-1279.
Hazlett, B.A. 1980. Communication and mutual resource exchange in north Florida hermit crabs. *Behav. Ecol. Sociobiol.* 6: 177-184.
Kellogg, C.W. 1977. Coexistence in a hermit crab species ensemble. *Biol. Bull.* 153: 133-144.

Leite, F.P.P., Turra, A. & Gandolfi, S.M. 1998. Hermit crabs, gastropod shells, and environmental structure: their relationship in southeastern Brazil. *J. Nat. Hist.* 32: 1599-1608.

Lowery, W.A. & Nelson, W.G. 1988. Population ecology of the hermit crab *Clibanarius vittatus* (Decapoda: Diogenidae) at Sebastian Inlet, Florida. *J. Crust. Biol.* 8: 548-556.

Mitchell, K.A. 1975. An analysis of shell occupation by two sympatric species of hermit crabs. I. Ecological factors. *Biol. Bull.* 149: 205-213.

Radinovsky, S. & Henderson, A. 1974. The shell game. *Nat. Hist.* 83: 22-29.

Reese, E.S. 1969. Behavioral adaptations of intertidal hermit crabs. *Am. Zool.* 9: 343-355.

Rittschof, D., Sarrica, J. & Rubeinstein, D. 1995. Shell dynamics and microhabitat selection by striped legged hermit crabs, *Clibanarius vittatus* (Bosc). *J. Exp. Mar. Biol. Ecol.* 192: 157-172.

Scully, E.P. 1983. The effects of shell availability on intraspecific competition in experimental populations of the hermit crab, *Pagurus longicarpus* Say. *J. Exp. Mar. Biol. Ecol.* 71: 221-236.

Snyder-Conn, E.K. 1980. Tidal clustering and dispersal of the hermit crab *Clibanarius digueti. Mar. Behav. Physiol.* 7: 135-154.

Snyder-Conn, E.K. 1981. The adaptative significance of clustering in the hermit crab *Clibanarius digueti. Mar. Behav. Physiol.* 8: 43-53.

Sokal, R.R. & Rohlf, F.J. 1995. *Biometry.* Freeman. 3rd Ed.

Spight, T.M. 1985. Why small hermit crabs have large shells? *Res. Popul. Ecol.* 27: 39-54.

Taylor, P.R. 1981. Hermit crab fitness: the effect of shell condition and behavioral adaptations on environmental resistance. *J. Exp. Mar. Biol. Ecol.* 52: 205-218.

Vance, R.R. 1972. Competition and mechanisms of coexistence in three sympatric species of intertidal hermit crabs. *Ecology* 53: 1062-1074.

Vannini, M & Gherardi, F. 1988. Studies on pebble crab, *Eriphia smithi* MacLeay 1838 (Xanthoidea, Menippidae): patterns of relative growth and population structure. *Trop. Zool.* 1: 203-216.

Young, A.M. 1978. Desiccation tolerances for three hermit crab species *Clibanarius vittatus* (Bosc), *Pagurus pollicaris* Say and *P. longicarpus* Say (Decapoda, Anomura) in the North Inlet Estuary, South Carolina, USA. *Est. Coast. Mar. Sci.* 6: 117-122.

Zar, J.H. 1984. *Biostatistical Analysis.* Prentice-Hall. 2nd Ed.

Population biology and habitat utilization of the stone crab *Menippe nodifrons* in the Ubatuba region, São Paulo, Brazil

ADILSON FRANSOZO, GIOVANA BERTINI & MICHELE O.D. CORREA
NEBECC (Group of Studies on Crustacean Biology, Ecology and Culture), Departamento de Zoologia, Instituto de Biociências – Universidade Estadual Paulista, (UNESP),São Paulo, Brazil

ABSTRACT

The aim of the current study was to evaluate a population of *Menippe nodifrons* in a reef of sabellariid worms, *Phragmatopoma lapidosa*, with regard to recruitment, population structure and sex ratio. Sampling was carried out each other month from September 1994 up to and including July 1995, on the rocky shores of the Tenório beach, São Paulo, Brazil. It resulted in 183 individuals, whose average carapace width in this biotope was 9.1 ± 5.6 m. The animals were grouped in nine size classes and the frequency distribution showed that 80% occupied the three first classes. This means that the polychaete worm colonies are of great importance for the establishment of individuals in the first juvenile stages.

1 INTRODUCTION

The Sabellariidae generally build big masses of sand tubes, and are responsible for the formation of vast reefs composed of thousands of individuals, including decapods, isopods, amphipods, etc. (Amaral 1987). The rocks supporting the formation of worm colonies have fissures and galleries that enhance the flow of breaking waters and provide shelter, protection and food to a large number of animals living in such reefs (Narchi & Rodrigues 1965).

Marine crabs, lobsters and shrimps have been the objects of various studies about resource partitioning (Navarrette & Castilha 1990, Spivak et al. 1994, Flores 1996). The occupation of refuges to avoid predation during at least the early stages of their life cycle seems to be a general habit of those crustaceans (Branch 1984, Orensanz & Galluci 1988). The interaction of decapod crustaceans with reefs formed by sabellariid worms has been studied by Gore et al. (1978), Almaça (1990), Pinheiro et al. (1997), Negreiros-Fransozo et al. (1998), and Micheletti-Flores & Negreiros-Fransozo (1999).

Despite the economical importance of *M. nodifrons* as a commercially exploited species, only a few papers are directly related to this crab. The larval development of *M. nodifrons* was described by Scotto (1979) and the juvenile development by Fransozo et al. (1988). Oshiro & Souza (1998) studied some of its reproductive aspects, obtaining individuals with a maximum size of 79.3 mm carapace width.

The present study describes the recruitment period, population structure and sex ratio, of a population of *M. nodifrons* in a sabellariid worm reef of *Phragmatopoma lapidosa*.

2 MATERIAL & METHODS

The sampling site, Tenório Beach (23°27'54''S and 45°03'30''W), is located in the Ubatuba region. Sao Paulo State, at the northern coast of Brazil.

Samples were taken every other month at low tide during one year from September 1994 onwards. Each collection in the intertidal region was composed of 20 samples of 3 kg sabellariid worm reef each. *M. nodifrons* individuals were separated, sexed and fixed in 10% formalin. Individuals whose morphological characteristics did not allow sex determination were named 'non-sexed' (ns). Carapace width (CW) was measured with a precision caliper (0.1 mm), or depending on its size, under a stereomicroscope provided with a camera lucida.

3 RESULTS

A total of 183 individuals were obtained (78 males, 82 females and 23 ns). The animals were grouped in nine size classes of 3.5 mm width. The size of the crabs varied from 2.7 to 31.7 mm and the average size was 9.1 ± 5.6 mm. The mean, maximum and minimum sizes are presented in Table 1. The size distribution of the crabs showed that almost 80% of them occupied the three first classes (Fig. 1). Approximately 78% of the total number of the 'ns' occurred in March 1995 (Fig. 2). The statistical differences related to crab size are shown in Figure 3. The sex ratio observed in juvenile *M. nodifrons* was 1:1.05 (M:F) ($\chi^2 = 0.1$; df = 1; ns).

4 DISCUSSION

Menippe nodifrons has been regularly reported as one of the various species of crustaceans present in colonies of *P. lapidosa*. We found only immature individuals of this species (≤ 31.7 mm CW) on the worm reef.

The habitats occupied by the young may differ from those occupied by the adults, probably because of factors related to cannibalism, larger refuges against predation, food availability or avoidance of competition. This has been reported for the marine molluscs *Littorina neritoides* and *Aplysia juliana*, studied by Fretter & Manly (1977) and Saver (1979) respectively. On the other hand, megalopae of other species settled in the habitats of the adults (Jensen 1989, O'Connor 1993, Micheletti-Flores 1996). This differentiated occupation of habitats during the juvenile phase of *M. nodifrons* reveals that the young need the worm colonies. The adults are found in the fissures and under the rocks in the intertidal region. The interior of the reef offers many refuges and food for the crabs during their initial development.

Table 1. *Menippe nodifons,* minimum, maximum and mean size (carapace width in mm) of non-sexed (ns) crabs, males, and females, from September 1994-July 1995.

	Variables	N	Minimum size	Maximum size	Mean ± sd
Total	Ns	23	2.7	4.4	3.7 ± 0.5
	Males	78	4.5	31.7	10.2 ± 5.4
	Females	82	4.5	28.8	9.5 ± 5.8
September 94	Ns	2	4.1	4.1	4.1 ± 0.0
	Males	14	5.9	28.1	12.6 ± 6.3
	Females	17	5.0	24.7	9.8 ± 5.2
November 94	Ns	–	–	–	–
	Males	4	7.8	22.7	11.9 ± 4.1
	Females	2	13.1	18.5	16.1 ± 1.6
January 95	Ns	–	–	–	–
	Males	1	5.0	5.0	5.0 ± 0.0
	Females	3	5.9	8.1	7.0 ± 1.1
March 95	Ns	18	2.7	4.4	3.7 ± 0.5
	Males	35	4.5	31.7	9.6 ± 5.7
	Females	40	4.5	28.8	9.2 ± 6.7
May 95	Ns	1	4.2	4.2	4.2 ± 0.0
	Males	8	5.0	12.0	8.0 ± 2.5
	Females	7	5.1	22.1	10.8 ± 5.8
July 95	Ns	2	3.1	3.3	3.2 ± 0.1
	Males	16	5.1	16.4	9.5 ± 3.3
	Females	13	5.2	17.5	9.1 ± 4.3

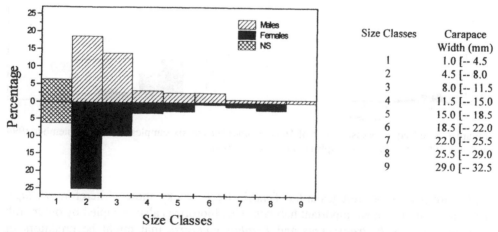

Size Classes	Carapace Width (mm)
1	1.0 [-- 4.5
2	4.5 [-- 8.0
3	8.0 [-- 11.5
4	11.5 [-- 15.0
5	15.0 [-- 18.5
6	18.5 [-- 22.0
7	22.0 [-- 25.5
8	25.5 [-- 29.0
9	29.0 [-- 32.5

Figure 1. Size frequency distribution for all male, female, and non-sexed *Menippe nodifrons.*

Other species of intertidal crabs also use shelters in the beginning of their life cycle in order to reduce the risk of predation by larger sized individuals. This is reported for the grapsid crabs *Cyrtograpsus angulatus* (Spivak et al. 1994) and *Pachygrapsus transversus* (Flores 1996), and for the xanthid crab *Dysopanopeus sayi* (Heck & Hambrook 1991).

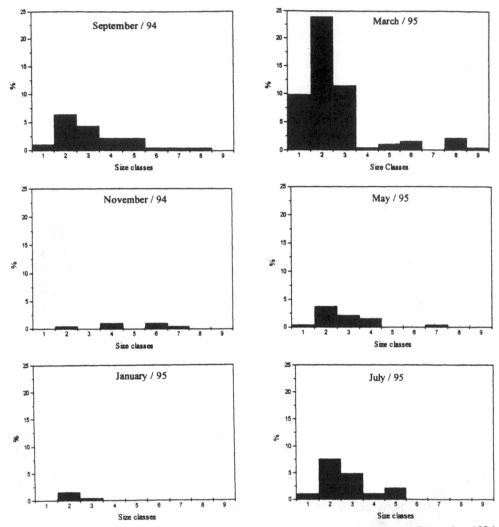

Figure 2. Size frequency distribution of *Menippe nodifrons* in six samples between September 1994 and July 1995. (See Fig. 1 for explanation of size classes)

There are few proper refuges at the Tenório Beach rocky shores. Therefore, *P. lapidosa* reefs are the most important habitats. This biotope is also occupied by other crab species, such as *P. transversus* and *Eriphia gonagra*, that might be predators of young *M. nodifrons*. Another form of predation in this biotope could be cannibalism, as observed in another crab species (Abele et al. 1986).

According to Gore et al. (1978), *M. nodifrons* feeds on the polychaete worm and on the associated cryptic fauna in large holes in the sabellariid reefs. Gore et al. (1978) reported the presence of young and adult *M. nodifrons* in *P. lapidosa* colonies in the Indian River region, Florida. We found *Menippe mercenaria* were only juveniles, whereas the adults were found mostly in the Indian River estuary. The presence of adult *M. nodifrons* in the colonies of the Indian River may be due to the formation

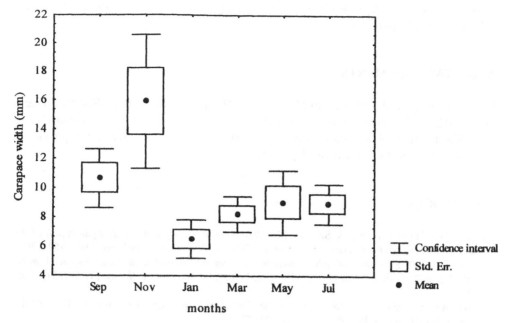

Figure 3. Mean size of *Menippe nodifrons* in six samples between September 1994 and July 1995.

of galeries and large fissures, that form a different structure than the Ubatuba colonies.

Oshiro & Souza (1998), studing *M. nodifrons* on the Manguaratiba rocky shores (Rio de Janeiro State), reported the smallest ovigerous female of 38.0 mm carapace width. As no ovigerous female was collected on the Ubatuba reef, it can be inferred that reproduction occurs outside the colony.

The size frequency distribution shows a high number of crabs in the smaller size classes, because we analyzed only one of the biotopes used by *M. nodifrons*. This also apparently holds for *Petrolisthes armatus*, in which smaller individuals were found in the *P. lapidosa* colonies while the adults lived under rocks, thus showing a remarkable habitat separation (Micheletti-Flores 1996).

As in *P. transversus*, and verified by Flores (1996), the sex ratio in *M. nodifrons* did not differ statistically during the first phases of development, proving the fundamental importance of the colony for the protection and the development of these species at the onset of ontogeny. More remarked variations may occur in the mature animals that live on the rocky surface due to the specific conditions for each sex, mainly related to the reproductive behaviors. Due to the economical importance of *M. nodifrons*, the sex ratio among the adults will change, since mainly males are captured.

The highest peak of young *M. nodifrons* (CW about 3.7 mm) occurred in March. It is supposed that hatching took place in early November, since this is consistent with data on size and age of juveniles obtained in the laboratory by Fransozo et al. (1988). Then, reproduction mainly occurs in the beginning of the spring, which is favorable in terms of temperature and availability of food.

In conclusion, *P. lapidosa* is a very important resource for the settlement and development of *M. nodifrons*.

ACKNOWLEDGEMENTS

We are grateful to the 'Fundação de Amparo à Pesquisa do Estado de São Paulo' and to the NEBECC coworkers for their help during collection. We are also thankful to Dra. Maria Lucia Negreiros Fransozo, and to anonymous reviewers, for providing valuable comments on earlier drafts of this manuscript.

REFERENCES

Abele, L.G., Campanella, P.J. & Salmon, M. 1986. Natural history and social organization of the semiterrestrial crab *Pachygrapsus transversus* (Gibbes). *J. Exp. Mar. Biol. Ecol.* 104: 153-170.

Almaça, C. 1990. Structure and interactions in the crab community inhabiting sabellariid worm colonies at Praia de Ribeira D'Ilhas (Ericeira, Portugal). *Arq. Museu Bocage, N.S.* 1(37): 505-519.

Amaral, A.C.Z. 1987. Breve caracterização de *Phragmatopoma lapidosa* Kinberg, 1867 (Polychaeta, Sabellariidae). *Rev. Bras. Zool.* 3: 471-474.

Branch, G.M. 1984. Competition between marine organisms: ecological and evolutionary implications. *Oceanogr. Mar. Bio. Anu. Rev.* 22: 429-593.

Flores, A.A.V. 1996. Biologia de *Pachygrapsus transversus* (Gibbes, 1850) (Crustacea, Brachyura, Grapsidae) na região de Ubatuba, SP. Master Science thesis. Universidade Estadual Paulista, Brazil 143 pp. (unpublished).

Fransozo, A., Negreiros-Fransozo, M.L. & Hiyodo, C.M. 1988. Développement juvénile de *Menippe nodifrons* Stimpson, 1859 (Crustacea, Decapoda, Xanthidae) au laboratoire. *Rev. Hydrobiol. Trop.* 21: 297-308.

Fretter, V. & Manly, R. 1977. The settlement and early benthic life of *Littorina neritoides* (L.) at Wembury, S. Devon. *J. molluscan Stud.* 43: 255-262.

Gore, R.H., Scotto, L.E. & Becker, L.J. 1978. Community composition, stability, and trophic partitioning in decapod crustaceans inhabiting some subtropical sabelleriid worm reefs. *Bull. Mar. Sci.* 28: 221-248.

Heck, K.L. & Hambrook, J.A. 1991. Intraspecific interactions and risk of predation for *Disopanopeus sayi* (Decapoda: Xanthidae) living on polychaete (*Filograma implexa*, Serpulidae) colonies. *Mar. Ecol.* 2: 243-250.

Jensen, G.C. 1989. Gregarious settlement by megalopae of the porcelain crabs *Petrolisthes cinctipes* (Randall) and *P. eriomerus* Stimpson. *J. Exp. Mar. Biol. Ecol.* 131: 223-231.

Micheletti-Flores, C.V. 1996. Ecologia Populacional de Porcelanídeos (Crustacea, Anomura) em Aglomerados de *Phragmatopoma lapidosa* (Polychaeta, Sabellariidae) na Praia de Paranapuã, São Vicente, SP. Master Science thesis. Universidade Estadual Paulista, Brazil. 94 pp. (unpublished).

Micheletti-Flores, C.V. & Negreiros-Fransozo, M.L. (1999). Porcellanid crabs (Crustacea, Decapoda) inhabiting sand reefs built by *Phragmatopoma lapidosa* (Polychaeta, Sabellariidae) at Paranapuã Beach, São Vicente (SP), Brazil. *Rev. Bras. Biol.* 59: 63-73.

Narchi, W. & Rodrigues, S.A. 1965. Observações ecológicas sobre *Phragmatopoma lapidosa* Kinberg. *Ciência e Cultura,* 17(2): 228.

Navarrete, S.A. & Castilla, J.C. 1990. Resource partitioning between intertidal predatory crabs: interference and refuge utilization. *J. Exp. Mar. Biol. Ecol.* 143: 101-129.

Negreiros-Fransozo, M.L., Flores, A.A.V., Reigada, A.L.D. & Nakagaki, J.M. 1998. Análise comparativa dos crustáceos decápodos de colônias de Sabellariidae em duas localidades do litoral paulista. *Anais do IV Simpósio de Ecossistemas Brasileiros* Vol. II–104: 241-220.

O'Connor, N.J. 1993. Settlement and recruitment of the fiddler crabs *Uca pugnax* and *U. pugilator* in a North Carolina, USA, salt marsh. *Mar. Ecol. Prog. Ser.* 93: 227-234.

Orensanz, J.M. & Gallucci, V.F. 1988. Comparative study of postlarval life-history schedules in four species of *Cancer* (Decapoda: Brachyura: Cancridae). *J. Crust. Biol.* 8: 187-220.

Oshiro, L.M.Y. & Souza, E.P. 1998. Aspectos reprodutivos do caranguejo guaiá, *Menippe nodifrons* (Stimpson, 1859) (Crustacea, Decapoda, Xanthidae) da Praia de Ibicuí, Mangaratiba (RJ). *Resumos do XXII Congresso Brasileiro de Zoologia*, 77 pp.

Pinheiro, M.A.A., Bertini, G., Fernandes-Góes, L.C. & Fransozo, A. 1997. Decapod crustaceans associated to sand reefs of *Phragmatopoma lapidosa* Kinberg, 1867 (Polychaeta, Sabellariidae) at Praia Grande, Ubatuba, SP, Brazil. *Nauplius* 5: 77-83.

Saver, D.J. 1979. Recruitment and juvenile survivl in the sea hare *Aphysia juliana* (Gastropoda: Opisthobranchia). *Mar. Biol.* 54: 353-361.

Scotto, L. E. 1979. Larval development of the cuban stone crab, *Menippe nodifrons* (Brachyura, Xanthidae), under laboratory conditions with notes on the status of the family Menippidae. *Fish. Bull.* 77: 359-386.

Spivak, E., Anger, K., Luppi, T., Bas, C. & Ismael, D. 1994. Distribution and habitat preferences of two grapsid crab species in Mar Chiquita Lagoon (Province of Buenos Aires, Argentina). *Helgoländer Meeresunters.* 48: 59-78.

The life history and habitat of *Allanaspides helonomus* (Syncarida: Anaspididae)

ROY SWAIN

School of Zoology, University of Tasmania, Australia

ABSTRACT

Allanaspides helonomus is a small anaspid syncarid with a restricted distribution in southwestern Tasmania. It is found almost exclusively in the ion-deficient, acidic burrow water associated with amphibious crayfish of the genus *Parastacoides*, and the small surface pools into which these burrows often open. Animals live for about 14 months, reaching a maximum size of ~14 mm. Seven size modes were identified in both sexes with no differences in modal statistics; there is no difference in size at maturity, but females outnumber males (1.6:1). Spermatophore transfer begins in early winter and continues until late spring (June-November) with a small second bout of activity in autumn. Egg release is restricted to autumn and it is estimated that hatching occurs 11-16 months later. The hatching period extends from mid-summer to early spring but peaks in early autumn. Prolonged development and hatching times, and the association with burrow water, appear to be the major safeguards against population extinction in a habitat that is subject to summer drought. Habitat characteristics were investigated through an analysis of 10 physical parameters recorded from 90 potentially suitable pools selected at random within the study area; 36 of these contained *A. helonomus*. Collections were made at a time when surface flooding was extensive and, although there was a tendency for syncarids to be found in the deeper pools, they were present in all pool types sampled, including pools without crayfish burrows. The results support the suggestions that dispersal is via surface flood waters and that the association with burrows is opportunistic rather than obligatory.

1 INTRODUCTION

Extant representatives of the Anaspididae are found only on the island of Tasmania. Species in the three genera occupy very different habitats and have very different distributions. *Anaspides tasmaniae* is the largest anaspid ($\geq$ 50 mm) with a body form designed for benthic crawling and swimming. It is widespread in mountain tarns, lakes, upland streams, and cave systems throughout central, southern and western Tasmania (Williams 1965, 1974, Swain 1983); *A. spinulae* is very similar but is restricted to Lake St Clair (Williams 1965) and nearby Clarence Lagoon in central Tasmania (Swain unpubl.). *Paranaspides lacustris* is an obligate lake dweller (≤ 24 mm), exhibiting marked lateral compression and dorsal flexure; although more widely distributed than once thought (Fulton 1982), it does not occur outside the Central Plateau. Two species of *Allanaspides* have been described (Swain et al. 1970, 1971); except for the presence of the unusual and distinctive ion transport organ (fenestra dorsalis) in the cephalothoracic segment, they are remarkably similar to *Anaspides* in body form, but considerably smaller (≤ 15 mm). In contrast to *Anaspides* and *Paranaspides*, *Allanaspides* is found in highly acidic (pH ≥ 4.0) waters; it is found in pools and seepages in the sedgelands of two adjacent valley floors in south-western Tasmania (Swain et al. 1970, 1971). Almost invariably these pools contain the openings of the extensive burrow systems of the amphibious crayfish, *Parastacoides* spp; the burrows provide continuity between surface and subsurface waters and represent the major refuge of *Allanaspides* in the summer months when the water table often falls to the underlying layer of quartz gravel (20-100 cm below surface).

Despite their interest to students of crustacean evolution, little ecological information about anaspid syncarids is available. Williams (1974) summarized the data then available, including his preliminary analysis of the life history of *Anaspides tasmaniae* (Williams 1965). Since that time a more detailed analysis of life history and population structure in *A. tasmaniae* (Swain & Reid 1983) and a brief description of the probable life history of *Paranaspides lacustris* (Fulton 1982) are the only additions to our understanding of anaspid life history. This paper presents data on population structure in *Allanaspides helonomus* and some key life history characteristics.

2 MATERIALS AND METHODS

Monthly collections of syncarids were made from January to December 1980, with the exception of June. *Allanaspides helonomus* is restricted to the Pedder Valley and Huon Plains area of south-western Tasmania and the study area (~2 hectare) was part of a small plain near Lake Pedder (146°21′E 42°57′S). Animals were collected by passing a small dip net (1.5 mm mesh) vigorously through a small pool; sorting was carried out on site and animals were preserved in 70% alcohol + glycerol.

Measurements of body length (rostral tip to telson tip) were made under a binocular microscope at appropriate magnification.

In November 1989 a preliminary analysis of habitat requirements was carried out. Ninety pools were selected at random and each sampled for syncarids for 2 min. Four physical parameters were measured for each pool: length; maximum width; maximum depth; average depth (average of 5 measurements); and a further six categorical variables were recorded: substrate cover (3 categories); substrate type (2); steepness of pool sides (4); presence of overhangs within pool (2); cover provided by surrounding vegetation (4); and the presence or absence of crayfish burrows. All observations were made on the same day, following a day of heavy rain.

3 RESULTS

3.1 *Analysis of size modes*

Separate analyses of size classes were conducted on the pooled male (N = 403) and female (N = 658) data using probability paper as described by Harding (1949) and Cassie (1954). In both sexes seven modes were identified (Table 1). It is clear that there were no substantial differences between sexes. Accordingly, data for males and females were combined to provide an overview of the life history of *Allanaspides helonomus* (Fig. 1). The smallest female observed carrying spermatophores was 7.2

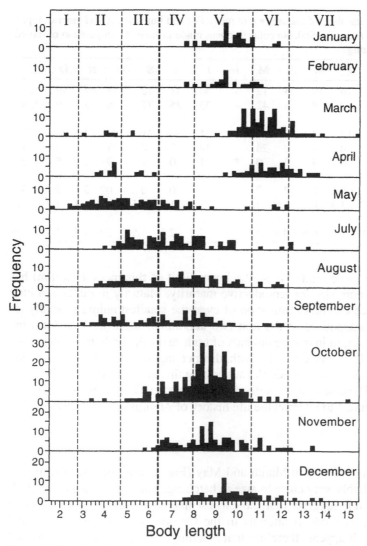

Figure 1. Size class frequency analysis of monthly collections of *Allanaspides helonomus*. Vertical lines and roman numerals mark size mode ranges identified in Table 1.

Table 1. Size class modes identified for *Allanaspides helonomus*; animals were grouped at 0.2 mm intervals for analysis.

Males Mode	Range	Mean	s.d.	N	Females Range	Mean	s.d.	N
I	≤ 2.8	–	–	2	≤ 2.8	2.6	0.81	4
II	3.0-4.6	3.8	0.38	30	3.0-4.8	4.1	0.51	64
II	4.8-6.4	5.7	0.89	36	5.0-6.4	5.7	0.59	89
IV	6.6-7.8	7.4	0.54	46	6.6-8.2	7.5	0.66	159
V	8.0-10.4	9.0	0.78	188	8.4-10.8	9.5	0.83	263
VI	10.6-12.2	11.2	0.62	76	11.0-12.4	11.5	0.48	72
VII	12.4-15.3	12.8	0.73	25	12.6-13.7	13.0	0.32	7

Table 2. Summary data for numbers of mature (● size mode IV) and immature *A. helonomus*, reproductive indicators, and sex ratios in 1980. No collection was made in June. * – copulation observed. Over-all annual mean in *italics*.

month	J	F	M	A	M	J	J	A	S	O	N	D	Σ
No. of mature females	35	16	49	25	12	–	40	41	32	148	63	39	500
No. of immature females	0	0	4	6	42	–	33	25	27	18	3	0	158
No. of mature males	12	12	54	25	11	–	21	22	21	83	44	21	326
No. of immature males	0	0	4	9	26	–	14	7	10	7	0	0	77
Females with sperma-tophores	0	0	3	0	0	*	1	0	1	9	2	0	16
Oviducal eggs visible	0	0	8	7	2	–	0	0	0	0	0	0	17
Adult sex ratio (female:male)	2.9	1.3	0.9	1.0	1.1	–	1.9	1.9	1.5	1.8	1.4	1.9	*1.5*
Total sex ratio	2.9	1.3	0.9	0.9	1.5	–	2.1	2.3	1.9	2.6	1.5	1.9	*1.6*

mm; this corresponds to size mode IV, so the lower limit of this mode (6.6 mm) has been taken as the minimum size for reproductive maturity. Maturity in males is a little harder to establish, since careful examination of even the smallest animals usually reveals the developing endopodites of the first two abdominal segments that form the copulatory style. However, in immature animals of both sexes the body tagmata are not clearly distinguished. The smallest male with distinct tagmata and fully developed copulatory structures was 7.6 mm (mode IV) suggesting that males reach maturity at the same size as females. These size limits were used to construct Table 2, which was used in conjunction with Figure 1 to interpret the life history of *Allanaspides helonomus*.

3.2 *Hatching*

Size mode I animals were found in March and May (Fig. 1) and the smallest animal found (1.6 mm) is probably very close to size at hatching. The occurrence of mode II and III animals in March suggests that the earliest hatching took place in January; likewise, the presence of mode II animals in the October sample suggests a final hatching in September. It appears therefore that hatching occurs over a very extended period (mid-summer through to early spring) with peak hatching probably occurring in early autumn (March-May).

3.3 *Breeding season*

Of the 16 animals caught with spermatophores (Table 2), five were approaching the limit of mode IV, one was borderline IV/V, 7 were in mode V, and 3 were in mode VI, indicating that females generally mate late in mode IV or in mode V. Actual copulation has been observed in June (McConnell, pers. comm); this suggests that mating starts in early winter since females from the current year do not reach maturity before then. Presumably females release spermatophores as soon as the sperm have been transferred to the spermatheca, so mating must continue until late spring (Table 2). The absence of females with spermatophores over the next few months suggests that no mating occurs during summer, but there is a short second period of mating activity in March. Thus mating, like hatching, appears to occur over a prolonged period. Most females mate in winter or spring, at around six months of age; those that are born too late to mature before the summer do not mate until early autumn.

In females with mature (fully yolked, yellow colored) eggs, the eggs are clearly visible through the ventral surface of thoracic segment 6; animals in this condition first appeared in March and were found only in the autumn months (Table 2). Of the 17 females recorded in this state, one was mode V, four were borderline mode V/VI, 10 were mode VI, and two were mode VII. Since few large animals remain in the population after April (Fig. 1), it would appear that egg release is confined to autumn and occurs over a much shorter period than hatching or mating. In any year the earliest egg laying coincides with the latest mating. Assuming that the earliest matings lead to the earliest deposition of eggs, and vice versa for late matings, maturation of eggs must take a minimum of three months (March-May, but more probably five, since a few large animals remain until July) and a maximum of eight months (June-March).

3.4 *Growth and development*

Figure 1 suggests that cohorts increase in size at a fairly constant rate until early summer (December), by which time most animals are at size mode V. Little growth occurs through the remaining summer months until March, when most animals move into mode VI and females begin laying.

Development times are difficult to estimate with confidence. However, if egg laying occurs from March-May/June and hatching takes place from January to September then two possible scenarios present themselves. One possibilitity is for a minimum of 11 months (March to January) and a maximum of 16 months (May to September in the following year; a second possibility allows a minimum development time of one month (e.g. March to April) and a maximum of eight months (May-January).

Some variation in sex ratios among the samples is apparent (Table 2) but it is clear that females consistently outnumber males. The overall sex ratio of 1.6 females for each male departs significantly different from 1:1 ($\chi^2 = 61.3$; $p < 0.001$).

3.5 *Habitat associations*

A total of 60 syncarids was collected from 36 of the 90 pools for which physical variables were recorded. Maximum depth was the only one of the 10 habitat variables measured that correlated with the presence or absence of *A. helonomus*. The association was weak (t_{88} = 2.16, *p* = 0.03) but syncarids tended to be found in deeper pools (15.7 ± 1.32 cm; cf. 12.6 ± 0.81 cm: mean ± s.e.). PCA identified three pool groupings, corresponding to well-defined, relatively deep, small pools with abundant litter and numerous parastacid burrows (these are semi-permanent pools created by many years of crayfish activity), flooded depressions, generally associated with burrows, with a peaty substrate and some litter (these also remain flooded for long periods), and shallow, poorly defined, flood pools with little substrate cover and no burrow openings (these are temporary and last for a short time only after heavy rain). *Allanaspides helonomus* were found in all three pool types. By contrast, in late summer (mid-January into February) sampling carried out in most years between 1978 and 1990 syncarids were exclusively associated with pools that were associated with crayfish burrow systems.

4 DISCUSSION

There are limited life history data available for anaspids, so extensive comparisons are not possible. It is clear that *Allanaspides helonomus* has adopted a semelparous life cycle lasting no more than about 14 months, and this is similar to that proposed for *Paranaspides lacustris* (Fulton 1982) but very different from *Anaspides tasmaniae* (Williams 1965, Swain & Reid 1983). *Anaspides tasmaniae* live up to five years and usually breed three times (Swain & Reid 1983). Size at hatching appears to follow the same order as final size: 1.6 mm in *A. helonomus* (this study); 2.0 mm in *P. lacustris* (Fulton 1982); and 2.7 mm in *A. tasmaniae* (Hickman 1937). In contrast to *A. helonomus*, which exhibits a restricted period of egg deposition in autumn, egg laying in A. *tasmaniae* extends throughout the year, with a peak in late spring (Hickman 1937). Although the difference probably relates to environmental variables such as adverse conditions faced by *A. helonomus* in summer, it is also a consequence of the difference in longevity between these species – *A. helonomus* with mature eggs simply do not survive long enough for an extended lay period.

The time of year in which hatching occurs also varies considerably in anaspids. In stream-dwelling populations of *A. tasmaniae* most hatching occurs in early winter, but continues into spring (Hickman 1937), while in alpine populations it occurs later (late winter–early spring) (Swain & Reid 1983); hatching is probably at the start of winter in *P. lacustris* (Fulton 1982) and predominantly autumnal in *A. helonomus*. Temperature may play a part in determining these differences, but only if the three genera respond differently to this cue. However, if hatching is triggered primarily by an influx of water into the habitat, they could all respond in similar fashion; autumn is the period when the sedgelands occupied by *A. helonomus* rehydrate, most cold water streams have their first spates for the year in early winter, winter rains replenish the large lakes of the Central Plateau, and alpine pools are filled by early spring snowmelt.

Restricting reproduction to a single episode, with little overlapping of generations may be fairly safe in a lacustrine species such as *Paranaspides lacustris* where habitat continuity is predictable, but is a potentially hazardous strategy in a species such as *A. helonomus* which occupies habitats that are often ephemeral during summer. The summer months represent a time of potentially critical environmental conditions in *A. helonomus* during which little growth or reproductive activity takes place. *Allanaspides helonomus* is intimately associated with crayfish burrows; it comprises a major element of the specialized burrow fauna known as the pholeteros (Lake 1977) and burrow water may be the only habitat available when no surface water is available. However, even this habitat is not guaranteed in prolonged periods of summer dryness when many crayfish burrows will be without free water for several weeks (Swain et al. 1987). While *Parastacoides* spp can survive long periods in saturated air, *Allanaspides helonomus* certainly cannot.

The vulnerability of *A. helonomus* to drought appears to be offset by the extended period over which reproductive activities occur. Most reproductive events coincide, and may well be initiated by, the arrival of autumn rains (from March onwards), after which water is abundant and surface water is regularly available for dispersal until the following summer. The only activity that appears to take place over a relatively short period is egg laying, but this is primarily a consequence of few gravid females surviving into the winter.

However, extended periods of mating and hatching will not fully protect the population if all eggs develop quickly and hatch shortly after laying. If this were to happen one exceptionally dry year could wipe out all hatched animals in summer, thus rendering a population extinct; since the distribution of the species is quite limited all populations will certainly experience the same conditions, so extinguishing one population could quite possibly terminate all. Certainly dry summers are not unusual and, since the discovery of the species in 1970, there have been several summers when attempts to collect specimens in mid-summer have been unsuccessful (unpubl. observ.). Clearly, then, rapid development of eggs is an unlikely strategy for a species with a lineage as ancient as that of the Anaspididae. It is much more likely that, as in *Anaspides tasmaniae* (Hickman 1937) development is slow and variable. Hickman (1937) demonstrated that the period between laying and hatching in *A. tasmaniae* was eight to 14 months, depending on when the eggs were laid. The above arguments suggest that, in *Allanaspides helonomus*, this period is between 11 and 16 months. Assuming that eggs can survive successfully in damp soil and debris within crayfish burrows and pools, the population would then be able to survive an extraordinarily dry summer, even if all adults died from desiccation.

The biased sex ratio in *A. helonomus* contrasts with the 1:1 ratio found in *Anaspides tasmaniae* by Swain & Reid (1983). Its significance is unclear given our poor understanding of the ecology of both of these species. Whatever mechanisms of sexual selection operate in *A. helonomus,* it is clear that they have not led to dimorphism in body size or longevity, and the extended entry of mature animals into the population and the seasonal variation in sex ratios shown in Table 2 suggest that operational sex ratios at any given time may be quite different from overall ratios.

The attempt to quantify the habitat requirements of *A. helonomus* revealed little, since syncarids were found in all pool types investigated. However, the data were gathered after heavy rains when surface flooding was extensive, and they do support qualitative conclusions reached over many years. The widespread occurrence of *A. helonomus* in surface waters having no connection with crayfish burrows suggests that the animals are able to exploit a wide range of pool habitats when water is abundant and that surface flooding provides the major opportunity for dispersal. The data further indicate that the restriction of the animals to crayfish burrows during dry periods is an opportunistic rather than an obligatory association.

ACKNOWLEDGEMENTS

I thank the Tasmanian Parks and Wildlife Service for permission to collect from the South-West World Heritage Area.

REFERENCES

Cassie, R.M. 1954. Some uses of probability paper in the analysis of size frequency distributions. *Aust. J. Mar. Freshwat. Res.* 5: 513-522.

Fulton, W.G. 1982. Notes on the distribution and life cycle of *Paranaspides lacustris* Smith (Crustacea: Syncarida). *Bull. Aust. Soc. Limnol.* 8: 23-25.

Harding, J.P. 1949. The use of probability paper for the graphical analysis of polymodal frequency distributions. *J. Mar. Biol. Assoc. UK* 28: 141-153.

Hickman, V.V. 1937. The embryology of the syncarid crustacean, *Anaspides tasmaniae. Pap. Proc. R. Soc. Tasm.* 1936: 1-35.

Lake, P.S. 1977. On the subterranean syncarids of Tasmania. *Helictite* 15: 12-17.

Swain, R., Wilson, I.S., Hickman, J.L. & Ong, J.E. 1970. *Allanaspides helonomus* gen. et sp. nov. (Crustacea: Syncarida) from Tasmania. *Rec. Queen Vict. Mus.* 35: 1-13.

Swain, R., Wilson, I.S. & Ong, J.E. 1971. A new species of *Allanaspides* (Syncarida, Anaspididae) from south-western Tasmania. *Crustaceana* 21: 196-202.

Swain, R. & Reid, C.I. 1983. Observations on the life-history and ecology of *Anaspides tasmaniae* (Thomson) (Syncarida: Anaspididae). *J. Crust. Biol.* 3: 163-172.

Swain, R., Marker, P.F. & Richardson, A.M.M. 1987. Respiratory responses to hypoxia in stream-dwelling (*Astacopsis franklinii*) and burrowing (*Parastacoides tasmanicus*) parastacid crayfish. *Comp. Biochem. Physiol.* 87A: 813-817.

Williams, W.D. 1965. Ecological notes on Tasmanian Syncarida (Crustacea: Malacostraca), with a description of a new species of *Anaspides. Int. Rev. ges. Hydrobiol.* 50: 95-126.

Williams, W.D. 1974. Freshwater Crustacea. In W.D. Williams (ed.), *Biogeography and ecology in Tasmania*: 63-112. The Hague: W. Junk.

Population structure and reproductive biology of the Camacuto shrimp, *Atya scabra* (Decapoda, Caridea, Atyidae), from São Sebastião, Brazil

RENATA GALVÃO & SÉRGIO L.S. BUENO
Departamento de Zoologia, Instituto de Biociências, Universidade de São Paulo, Brazil

ABSTRACT

A two-year field study on the population of *Atya scabra* from the Guaecá River, São Sebastião, Brazil, revealed a higher preponderance of adult females over males. No juveniles were found at the collecting site that was located approximately 3 kilometers from the river mouth. Males attain larger body size than females, but ratio between carapace length and total body size showed that the abdomen is proportionally larger in females. Based on macroscopic evaluation (color and size) of the ovaries, five distinct developmental stages were recognized. Ovigerous females were present in every sample, indicating a continuous reproductive pattern for this population. New ovarian maturation frequently occured during the egg incubation period, and revealed that a new spawning followed soon after larval hatching was completed. Eggs were numerous and small, and there was a positive correlation between fecundity and female body size.

1 INTRODUCTION

The genus *Atya* (Family Atyidae) comprises about a dozen shrimp species that typically inhabit mainland freshwater environments in tropical and sub-tropical regions of the Americas and Africa (Hobbs & Hart 1982). The Camacuto shrimp *Atya scabra* (Leach, 1815) has a wide geographic distribution in western Africa and in eastern Latin-America (Allee & Torvik 1927, Hart 1961, Hobbs & Hart 1982, Abele & Kim 1989). In Brazil, this species has been reported in the States of Pernambuco and Ceará, both in the northeastern region (Oliveira 1945, Abrunhosa & Moura 1988) and in the State of Santa Catarina, in the southern region (Hobbs & Hart 1982).

Despite the wide geographic distribution, its population is strongly determined by specific habitat requirements, like shallow rapidly flowing turbulent and oxygen-rich streams or riffles that drain towards the coast (Darnell 1956, Hobbs & Hart 1982). The river bed is usually covered with rocks and stones and provides shelter for this atyid shrimp. They are supposed to be nocturnal and to feed on detritus and organic material from the bottom (Darnell 1956).

The reproductive biology of *Atya scabra* is poorly known. A continuous reproductive cycle was suggested for the Mexican population by Villalobos (1943 cited in Hobbs & Hart 1982) and Darnell (1956). Still according to Hobbs & Hart (1982), the ovigerous females are generally found in March, May, July and August in Africa. Abrunhosa & Moura (1988) reported that eggs are numerous and small and also described the complete larval development of the species, comprising 11 planktotrophic zoeal stages. However, no relevant data on reproductive cycle of the Brazilian population of *Atya scabra* are available. The present paper, therefore, deals with a two-year field survey conducted at the Guaecá River, São Sebastião, to study the population structure and reproductive biology of *Atya scabra* from the State of São Paulo, southeastern Brazil.

2 MATERIAL AND METHODS

Regular monthly collections of *Atya scabra* were made from the Guaecá River (23°49'10"S, 45°27'30"W) in São Sebastião County, from February 1995 to January 1997 (Fig. 1). The sampling site was located at approximately 3 km from the river mouth. This section of the river resembled a mountain stream, with fast flowing, clear shallow water, and a rocky bottom with many loose stones of variable sizes. The water depth rarely exceeded 30 cm. Specimens were generally found hiding under rocks and stones. Once these rocks and stones were manually displaced, the dislodged specimens swam down the water current and could be immediately captured with a hand net.

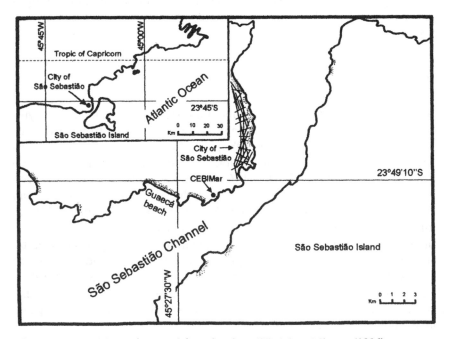

Figure 1. Map of the study area. Adapted and modified from Migotto (1996).

Specimens were transported live in habitat water and kept in the laboratory at the Centro de Biologia Marinha (CEBIMar) of the University of São Paulo, also located in São Sebastião (Fig. 1).

Presence (males) or absence (females) of the appendix masculina on the second pleopod and position of the gonopores was considered for sex determination. Specimens were measured with the aid of a caliper to the nearest of 0.1 mm. Total body length (TL) was measured from tip of rostrum to tip of telson, while carapace length (CL) was measured from post-orbital margin to mid-posterior border of the carapace. An electronic scale (Marte, model A 5000) was used to obtain net weight of shrimps to the nearest 0.1 g. Specimens were grouped according to sex, and body length in 12 size classes of 6 mm.

Macroscopic evaluation of ovarian development was done by examining the size of the ovaries through the translucent exoskeleton with the aid of a light source. Live females were dissected dorsally to register the color of the ovary.

Mean egg size was determined for both early (non-eyed condition) and late (eyed condition) embryonic stages. Measurements were taken along longest and shortest axes using a dissecting microscope (Zeiss, model Citoval 2) equipped with camera lucida. For each embryonic stage a total of 600 eggs were examined at the rate of 20 eggs each from 30 females.

The fecundity was determined based on 83 ovigerous females with eggs in the early embryonic stage only, since loss of eggs during incubation period is not uncommon in caridean shrimps (Darnell 1956, Balasundaram & Pandian 1982). The egg mass from each ovigerous females was gently removed and dried in a lab oven at 70°C. The total number of eggs for each ovigerous female was estimated on a weighed sub-sample, comprising at least 25% of the total egg mass.

3 RESULTS

A total of 1037 individuals of *Atya scabra* were collected in the Guaecá River. Monthly collections revealed that females always outnumbered the males (Fig. 2) and this factor contributed to the clear overall preponderance of females (n = 725; 70%) over males (n = 312; 30%). The average sex ratio (male:female) for the entire sample was 1:2.32, with the lowest (1:1.14) and highest (1:6.67) ratio's observed in February 1995 and July 1996 respectively (Fig. 2).

Males were generally larger than females. The average (± s.d.) total length and size range for males (n = 306) were 54.5 ± 13.5 mm and 22.7-89.3 mm respectively, while for females (n = 718) this was 46.9 ± 6.8 mm and 25.3-61.1 mm respectively. Size frequency distribution shows that size classes were more widely and more evenly distributed for males than for females (Fig. 3). The mode for males corresponded to class 7 (58.1-64.0 mm), while for females this corresponded to class 5 (46.1-52.0 mm).

No juvenile specimens were found at the collecting site. All males showed fully developed an appendix masculina, while females were either non-ovigerous showing various conditions of ovarian development, or were ovigerous. The total length of the smallest ovigerous female was 25.3 mm, which coincidentally was also the smallest female sampled. The smallest male was 22.7 mm long and already showed a fully developed appendix masculina.

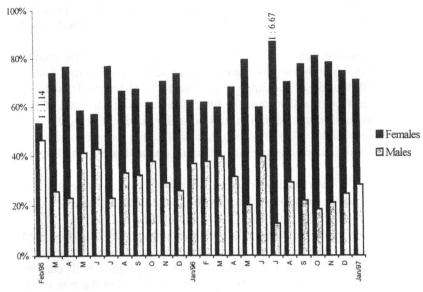

Figure 2. Monthly occurrence of males and females of *Atya scabra* in the Guaecá River, São Sebastião, Brazil. The lowest and highest value of monthly sex ratio (male:female) is indicated for February 1995 and June 1996.

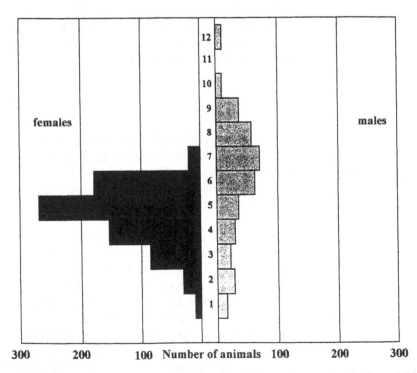

Figure 3. Size frequency distribution of male and female *Atya scabra* in Guaecá River, São Sebastião, Brazil. Limits to each size class (in mm) are 1 (22.1-28.0), 2 (28.1-34.0), 3 (34.1-40.0), 4 (40.1-46.0), 5 (46.1-52.0), 6 (52.1-58.0), 7 (58.1-64.0), 8 (64.1-70.0), 9 (70.1-76.0), 10 (76.1-82.0), 11 (82.1-88.0) and 12 (88.1-94.0).

Size of ovaries varied from barely visible to an easily distinguishable and fully developed condition, reaching as far as the fourth abdominal segment. Ovaries in live dissected adult females varied from a not developed translucent condition to ocre/brown when fully developed. Based on the size and color, five developmental stages of the ovaries could be recognized (Table 1).

Ovarian development occurred in the egg incubation period. In ovigerous females carrying early embryonic eggs, the ovarian development was found to be usually at stage 5 (spent), indicating a recent oviposition. Most berried females, with late embryonic eggs, showed ovarian development ranging from early development (stage 2) to matured condition (stage 4), indicating that a new oviposition immediately followed the completion of hatching of larvae from the previous batch.

Ovigerous females were found to be present every month, indicating a continuous reproductive pattern in this population (Fig. 4). The monthly percentage of ovigerous females was usually greater than 30% of all sampled females, with peaks at 85% and 83% in April and November 1995 respectively, and 82% in November 1996. Lower percentages of ovigerous females (minimum of 6%) were observed during late autumn (May-June) and the entire winter (June-September) in 1996.

Incubated eggs were numerous and small. The mean number of eggs was 3881 ± 1531 in the range 324-11263. Size of eggs varied with the corresponding developmental stage. At the early embryonic stage, the average size was 0.546 ± 0.039 mm across length and 0.345 ± 0.030 mm across breadth. In the late embryonic stage, eggs were slightly larger and measured 0.583 ± 0.037 mm by 0.382 ± 0.141 mm.

The ratio between carapace length and total body length, as defined by the relation $1 - (CL/TL)$, revealed that, regardless the size of the adult specimen considered, the abdomen (mean ± s.d.) was proportionally larger in females (0.6729 ± 0.0006, n = 716) than in males (0.6492 ± 0.0017, n = 307) ($t = 16.722, p < 0.001$).

Also, a positive linear correlation ($p < 0.00001$, n = 83) was revealed between number of incubated eggs and total body length (TL) of ovigerous females. This correlation is expressed by the equation $y = 201.12TL - 5722.77$ (Pearson's correlation coefficient = 0.7071).

Table 1. *Atya scabra*, color and size of ovaries in each of the five recognized developmental stages.

Developmental stage	Color	Size (extension into the tail)
1. Not developed	Translucent	Not discernible
2. Early development	Yellow	Reaching the proxmal portion of the 2nd abdominal segment
3. Intermediate	Orange	Reaching the end of the 2nd abdominal segment
4. Mature	Ocre tot brown	Reaching the end of the 3rd abdominal segment. (may even reach as far as the 4th segment)
5. Spent	Translucent with ocre or brown patches	Not discernible

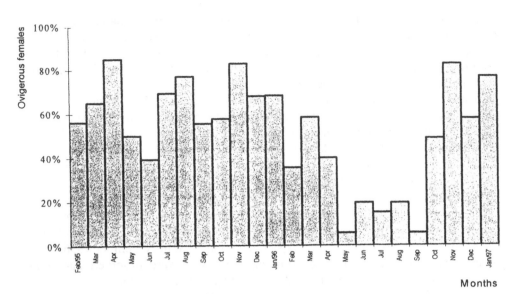

Figure 4. Percentage of ovigerous females (compared to non-ovigerous females) of *Atya scabra*, collected from the Guaecá River, São Sebastião, Brazil.

4 DISCUSSION

This paper provides the first report of *Atya scabra* in the State of São Paulo, Brazil. The preponderance of females in this atyid population contrasted sharply with the preponderance of males as observed by Villalobos (1943 *apud*, Hobbs & Hart, 1982) and Darnell (1956) for *Atya scabra* population from Mexico.

Atya scabra is a sexually dimorphic species. Apart from the usual secondary sexual characters, adult males attain larger body size, while females possess proportionally larger abdomen. The largest male in the present collection was 89.3 mm, while body size as large as 94.5 mm has been reported from Mexico by Darnell (1956). Although males are larger than females in *Atya scabra* as in some Caridean shrimps (Ruiz et al. 1996), the opposite situation has also been frequently reported (Bond & Buckup 1983, Sanz 1987, De Silva 1988a, b, Barros & Fontoura 1996b, Souza & Fontoura 1996). Larger body size in females is often regarded as an adaptation for higher egg carrying capacity in Pleocyemate decapods (Hart 1981, Berglund 1981, Guerao et al. 1994).

However, by comparing the relative size between carapace length and total body length, it became evident that the proportionately larger abdomen of females of *Atya scabra*, as compared to males, was possibly associated with egg incubation and may be regarded as an adaptation towards increased fecundity as pointed out by Darnell (1956). Such a positive correlation has also been observed in other decapods (Harris et al. 1972, Katre 1977, Bond & Buckup 1982, Barros & Fontoura 1996a).

The smallest ovigerous female was also the smallest female sampled (25.3 mm TL and 8.2 mm CL) during the present study. This size was much smaller than the sexually mature female (39.0 mm TL) as reported by Darnell (1956) for the Mexican population of *Atya scabra*. The absence of any smaller females or juveniles in the São Sebastião samples rules out any possibility of further considerations on sexual maturity in females below 25.3 mm of body size for this specific population. However, in their extensive revision on the genus *Atya*, Hobbs & Hart (1982) have reported a carapace length of 7.1 mm for the smallest ovigerous female of *Atya scabra*.

A continuous reproductive pattern is quite common in caridean shrimps that inhabit tropical regions (Lobão et al. 1978, Hart 1981, Benzie & De Silva 1988, Kumari & De Silva 1989, Graziani et al. 1993, Jalihal et al. 1994), while subtropical species usually show periodic reproductive cycles (Bond & Buckup 1982, Barros & Fontoura 1996a, Souza & Fontoura 1996). The *A. scabra* population from the river Guaecá reproduces throughout the year. Although no thorough evaluation between reproduction and environment parameters could be made in the present study, the higher occurrence of ovigerous females during late spring and summer (October-March in the southern hemisphere) might be related to warmer weather and high rain precipitation during that part of the year. Although Darnell (1956) presumed that the Mexican population of *A. scabra* reproduces throughout the year, he could not conclusively prove it because of poor sampling.

The spent condition of ovaries in females bearing early embryonic eggs indicates that the spawning is total. Further synchronous development of the ovaries and embryos is a strong evidence that a new oviposition immediately follows the completion of the hatching period. Therefore, it is reasonable to assume that the population of *A. scabra* from São Sebastião produces successive batches of eggs during its reproductive phase.

The egg size in *A. scabra* from another Brazilian population from Pernambuco (Abrunhosa & Moura 1988) was quite similar to the one presented in this paper. Darnell (1956) had reported an average length of 0.84 mm for mature (late embryonic stage) eggs from a Mexican population. This value is much higher than the average length of the eggs in the Guaecá River, irrespective of embryonic stage.

Some atyid shrimp larvae depend on salinity for successful metamorphosis (see Hunte 1975, 1979, Abele & Blum 1977, Hobbs & Hart 1982, Dudgeon 1987). As such, the estuaries play an important role in dispersal of larvae in the natal as well as other connected river systems. (Walsh & Mitchell 1995). *Atya scabra* shows a prolonged type.of larval development with dependence on salinity for successful metamorphosis (Abrunhosa & Moura 1988). The larvae hatch in freshwater but are immediately washed down to the river mouth. During the ontogenetic development of the species, new recruits probably migrate upstream and join the reproductive population by the time they become adults (Villalobos 1959). As such, the apparent absence of juveniles in the present study possibly indicates that the collecting site is within the reproduction ground of *A. scabra*.

Darnell (1956) has suggested that juveniles and large adults occupy different habitats, with the former being more at the edges of riffle. As already stated, in the Guaecá *A. scabra* individuals could be collected by displacing stones at the bottom of the river. The scarce marginal and submerged vegetation along the river edges and banks were also checked, but only a dozen or so *Atya scabra* individuals could be collected with a couple specimens of *Macrobrachium* sp. Perhaps a more intensive survey along the river course towards its mouth may be necessary to provide more information on distributional pattern in the population of *A. scabra* from São Sebastião.

ACKNOWLEDGEMENTS

The authors are grateful to CEBIMar – USP (Centro de Biologia Marinha, Universidade de São Paulo) for providing logistic and laboratory facilities that were essential for this study. Special thanks to CAPES for providing grants to one of us (RG) and to Dr Roberto M. Shimizu (IBUSP) and the Instituto de Matemática e Estatística (USP) for their assistance in the statistical analyses. We are also indebted to the Graduation Commission, Section Zoology, of the Instituto de Biociências, USP, for providing necessary laboratory facilities. Finally, we would like to express our sincere gratitude to the two anonymous referees.

REFERENCES

Abele, L.G. & Blum N. 1977. Ecological aspects of the freshwater decapod crustaceans of Perlas Archipelago, Panama. *Biotropica* 9(4): 239-252.

Abele, L.G. & Kim W. 1989. The decapod crustaceans at the Panama Canal. *Smithson. Contr. Zool.* 482: 1-50.

Abrunhosa, F.A. & Moura M.G. 1988. O completo desenvolvimento larval do camarão *Atya scabra* (Leach) (Crustacea: Decapoda: Atyidae), cultivado em laboratório. *Arqs. de Cienc. Mar.* 27: 127-146.

Allee & Torvik, 1927. Factors affecting animal distribution in a small stream of the Panama rainforest in the dry season. *J. Ecol.* 15: 66-71.

Balasundaram, C. & Pandian T. J. 1982. Egg loss during incubation in *Macrobrachium nobilli* (Henderson & Mathai). *J. Exp. Mar. Biol. Ecol.* 59: 289-299.

Barros, M.P. & Fontoura N.F. 1996a. Biologia reprodutiva de *Potimirim glabra* (Kingsley, 1878) (Crustacea, Decapoda, Atyidae), na praia da Vigia, Garopaba, Santa Catarina, Brasil. *Nauplius* 4: 1-10.

Barros, M.P. & Fontoura N.F. 1996b. Crescimento de *Potimirim glabra* (Kingsley, 1878) (Crustacea, Decapoda, Atyidae), na praia da Vigia, Garopaba, Santa Catarina, Brasil. *Nauplius* 4: 11-28

Benzie, J.A.H. & de Silva P.K. 1988. The distribution and ecology of the freshwater prawn *Caridina singhalensis* Ortmann, 1894 (Decapoda, Atyidae) endemic to Sri Lanka. *J. Trop. Ecol.* 4: 347-359.

Berglund, A. 1981. Sex dimorphism and skewed sex ratios in the prawn species *Palaemon adspersus* and *P. squila. Oikos* 36: 158-162.

Bond, G. & Buckup L. 1982. O ciclo reprodutor de *Macrobrachium borelli* (Nobili, 1896) e *Macrobrachium potiuna* (Müller, 1880) (Crustacea, Decapoda, Palaemonidae) e suas relações com a temperatura. *Revta. Bras. Biol.* 42: 437-483.

Bond, G. & Buckup L. 1983. O cultivo de *Macrobrachium borelli* (Nobili, 1896) e *Macrobrachium potiuna* (Müller, 1880) em laboratório. (Crustacea, Decapoda, Palaemonidae) *Revta. Bras. Biol.* 43: 177-190.

Darnell, R. M. 1956. Analysis of a population of the tropical freshwater shrimp, *Atya scabra* (Leach). *Amer. Midl. Nat.* 55: 131-138.

de Silva, K.H.G.M. 1888 a. Studies on Atyidae (Decapoda, Caridea) of Sri Lanka. III. Aspects of the population ecology of *Caridina simoni* Bouvier, 1904. *Crustaceana* 54: 85-103.

de Silva, K.H.G.M. 1888 b. Studies on Atyidae (Decapoda, Caridea) of Sri Lanka. IV. Some aspects of the population ecology of the endemic freshwater shrimp *Caridina pristis* Roux, 1931. *Crustaceana* 54: 225-243.

Dudgeon, D. 1987. The larval development of *Neocaridina serrata* (Stimpson) (Crustacea: Decapoda: Caridea: Atyidae) from Hong Kong. *Arch. Hydrobiol.* 110: 339-355

Graziani, C. A.; Chung K. S. & Dedonato M. 1993. Reproductive behavior and fertility of *Macrobrachium carcinus* (Decapoda, Palaemonidae) in Venezuela. *Revta. Biol. Trop.* 41: 657-665.

Guerao, G., Pérez-Baquera J. & Ribera C. 1994. Growth and reproductive biology of *Palaemon xiphias* Risso, 1816 (Decapoda: Caridea: Palaemonidae). *J. Crust. Biol.* 14: 280-288.

Harris, C., Chew K. & Prince V. 1972. Relation of egg number to carapace length of sidestripe shrimp (*Pandalopsis disper*) from Dabol Bay, Washington. *J. Fish. Res. Board Canada* 29: 464-465.

Hart, Jr. C.W. 1961. The freshwater shrimps (Atyidae and Palaemonidae) of Jamaica, W. I. *Proc. Acad. Nat. Sci. Philad.* 13: 61-80.

Hart, R.C. 1981. Population dynamics and production of the tropical freshwater shrimp *Caridina nilotica* (Decapoda: Atyidae) in the littoral of Lake Sibaya. *Freshwat. Biol.* 11: 531-547.

Hobbs, H.H., Jr. & Hart C.W. 1982. The shrimp genus *Atya* (Decapoda, Atyidae). *Smithson. Contr. Zool.* 364: 1-143.

Hunte, W. 1975. *Atya lapines* Holthuis, 1963, in Jamaica, including taxonomic notes and description of the first larval stage (Decapoda, Atyidae). *Crustaceana* 28: 65-72.

Hunte, W. 1979. The complete larval development of the freshwater shrimp *Atya innocous* (Herbst) reared in the laboratory (Decapoda, Atyidae). *Crustaceana* Suppl. 5: 231-242.

Jalihal, D.R., Almelkar G.B. & Sankalli K.N. 1994. Atyid shrimp of the genus *Caridina* H. Milne-Edwards, 1837. Potential crustacean material for experimental biology. *Crustaceana* 66: 178-183.

Katre, S. 1977. The relation between body and number of eggs in the freshwater prawn *Macrobrachium lamarrei* (H. Milne-Edwards) (Crustacea, Decapoda). *Crustaceana* 33: 17-22.

Kumari, P. & de Silva K.H.G.M. 1989. Aspects of the population ecology of a tropical freshwater atyid shrimp *Caridina fernandoi* Arud and Costa, 1962 (Crustacea, Decapoda, Caridea). *Arch. Hydrobiol.* 117: 237-253.

Lobão, V.L., Sawaya P & Santos L.E. 1978. Influência da temperatura, precipitação pluviométrica e insolação na reprodução *de Macrobrachium holthuisi* Genofre & Lobão, 1976. *Bolm. Inst. Pesca,* 5(2): 109-118.

Migotto, A.E. 1996. Benthic shallow-water hydroids (Cnidaria, Hydrozoa) of the coast of São Sebastião, Brazil, including a checklist of Brazilian hydroids. *Zool. Verh.* 306: 1-125.

Oliveira, L.P.H. 1945. Verificação da existência de *Atya scabra* Leach, camarão d'água doce da família Atyidae, Crustacea, no nordeste do Brasil. *Mems. Inst. Oswaldo Cruz.* 43: 177-190.

Ruiz, MacD. A., Peña J.C. & López Y.S. 1996. Morfometría, época reproductiva y talla comercial de *Macrobrachium americanum* (Crustacea: Palaemonidae) en Guanacaste, Costa Rica. *Revta. Biol. Trop.* 44: 127-132.

Sanz, A. 1987. Biología de *Palaemon elegans* Rathke, 1837 (Natantia: Palaemonidae) en las costas del Mediterráneo Occidental. *Inv. Pesq.* Suppl., 51: 177-187.

Souza G.D. & Fontoura N.F. 1996. Reprodução, longevidade e razão sexual de *Macrobrachium potiuna* (Müller, 1880) (Crustacea, Decapoda, Palaemonidae) no arroio Sapucaia, Município de Gravataí, Rio Grande do Sul. *Nauplius* 4: 49-60.

Villalobos, A. 1959. Contribucion al conocimiento de los Atyidae de México. II (Crustacea, Decapoda). Estudio de algunas especies del genero Potimirim (= Ortmannia), con descripción de una especie nueva en Brasil. *An. Inst. Biol. Mex.* XXX: 269-330.

Walsh, C.J. & Mitchell B.D. 1995. The freshwater shrimp *Paratya australiensis* (Kemp, 1917) (Decapoda: Atyidae) in estuaries of South-western Victoria, Australia. *Mar. Freshwat. Res.* 46: 959-965.

Population dynamics of *Dynoides* (Crustacea: Isopoda) in the Cape d'Aguilar Marine Reserve, Hong Kong

LI LI

The Swire Institute of Marine Science, Department of Ecology and Biodiversity, The University of Hong Kong, People's Republic of China

ABSTRACT

An undescribed species of *Dynoides* occurs on exposed rocky shores, inhabiting beds of the mussel *Septifer virgatus*, empty barnacle, *Tetraclita squamosa squamosa*, shells, as well as clumps of the algae *Dermonema frappierii*, *Pterocladia tenuis*, *Chaetomorpha antennina* and *Ulva fasciata* in the Cape d'Aguilar Marine Reserve, Hong Kong. The population dynamics of this species had been studied in this reserve. The maximum body length of *Dynoides* sp. was 7.9 mm for males and 5.5 mm for females. At hatching, juveniles were 0.9 mm long and 0.5 mm wide and sexually indistinguishable. Sexual characters appeared in males with the emergence of a pair of short penises or a median process projecting from the postero-dorsal margin of the pleon, and in females with the appearance of three pairs of triangular oöstegite buds at the bases of pereiopods 2-4. Recruitment occurred at three times in May, September and December. Ovigerous females were present throughout the year with peaks reaching 91% of the total complement of females in warm months and lows of 25% in cold months. Isopods grew and matured faster from spring to late autumn and attained overall smaller body lengths. They grew slower in winter though attained larger sizes. The fecundity of *Dynoides* sp. ranged from 26 to 56 eggs per brood, correlated with body length of the ovigerous females. The first juvenile molt occurred 25 days after hatching and newly released juveniles were photopositive in the laboratory and capable of swimming.

1 INTRODUCTION

The isopods from Hong Kong are poorly known, from only a few studies on the taxonomy (Markham 1980, Bruce 1980, 1986, Ma 1986, Bamber 1997) and biology (Mak et al. 1985, Dudgeon et al. 1990, Lam et al. 1991, Ma et al. 1991). Much of this work concerns terrestrial species, little being known of the marine ones. Studies on the genus *Dynoides* are relatively few and most are taxonomic (Shen 1929, Nishimura 1976, Bruce 1982, Kwon et al. 1986, Kwon 1990). Only a few authors have studied the ecology of *Dynoides* (Ong Che et al. 1992, Iwasaki 1996) in Asia.

An undescribed species of *Dynoides* occurs in the Cape d'Aguilar Marine Reserve, Hong Kong, living on intertidal rocky shores. It is commonly associated with beds of the black mussel *Septifer virgatus*, but can also be found in the empty barnacle *Tetraclita squamosa squamosa*, and is associated with the algae *Dermonema frappierii*, *Pterocladia tenuis*, *Chaetomorpha antennina* and *Ulva fasciata*, all characteristic of wave-exposed shores. The present work aims at understanding the biology and ecology of the species of *Dynoides* in the Cape d'Aguilar Marine Reserve. This is the first detailed study of the population dynamics of a species of *Dynoides*. The species is recognized by a median process projecting from the dorsal, posterior margin of the pleon of males and is at present being described and named.

2 MATERIALS AND METHODS

From April 1997 to June 1998, specimens of *Dynoides* sp. were collected monthly from the *Septifer virgatus* bed, which forms a conspicuous mid-littoral 0.5-1 m wide band, at about the level of mean high water neap tide, i.e., +1.3 m C.D. (Ong Che et al. 1992), on the western edge of the Cape d'Aguilar Marine Reserve. Five quadrats of 10 × 10 cm were randomly tossed onto the mussel bed. Within each quadrant, all mussel clumps were removed, transferred to the laboratory and frozen for 2-3 hours before sorting (Arrontes & & Anadón 1990). In the laboratory, isopods were separated from the mussels by washing them with running tap water through a 500 µm metal sieve. The isopods were separated from other organisms and preserved in 5% neutralized formalin.

Body length, i.e., from the cephalon to the telson, was measured to the nearest 0.1 mm using the micrometer scale of a dissecting binocular microscope. Males were recognized by a median process projecting from the postero-dorsal margin of the pleon (Harrison & Holdich 1984) and a pair of penes. Immature and gravid females were identified easily by three pairs of oöstegite buds at the bases of pereopods 2-4 (Harrison 1984) and a brood pouch, respectively. The internal body tissues of *post partum* females were almost empty (Holdich 1968) after releasing mancas. Sexually undetermined isopods were defined as juveniles, regardless of size.

Twenty-two gravid females were kept individually in aerated plastic containers between December 1996 and May 1997 and fed on the green alga *Ulva fasciata*. Seawater was changed every two or three days. Released juveniles (mancas) were counted, measured and the molting times recorded.

3 RESULTS

3.1 *Seasonal changes in size structure and population abundance*

The size-frequency distribution of the *Dynoides* sp. population over the study period from April 1997 to June 1998 is shown in Figure 1. From April-May, many juveniles (1-1.5 mm body length) were released as mancas and formed the first recruitment of the year. After three to four months growth, a number of gravid females occurred in the population in July and subsequently released the second recruitment in the fol-

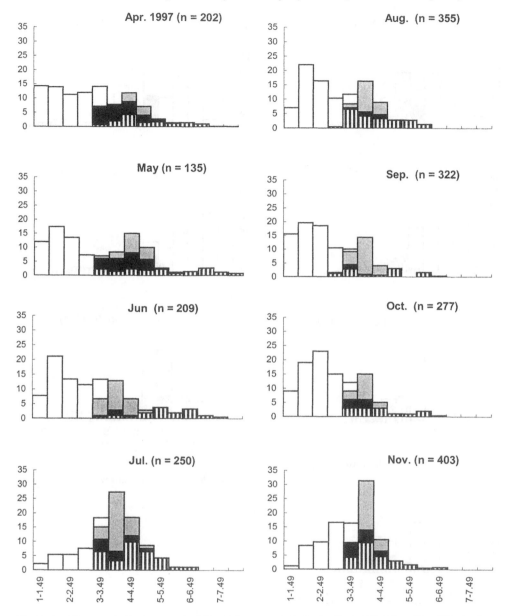

Figure 1a. Changes in the structure of *Dynoides* sp. population at Cape d'Aguilar, Hong Kong, from April to November 1997.

lowing months of August-September. The isopods matured after another three to four months in November and produced the third recruitment in December. This cohort grew subsequently, matured and overwintered to the next spring, a period lasting between five to six months. This indicates that three separate peaks in recruitment occur in each year, i.e., the spring, summer and late autumn cohorts, though juveniles can be present continuously.

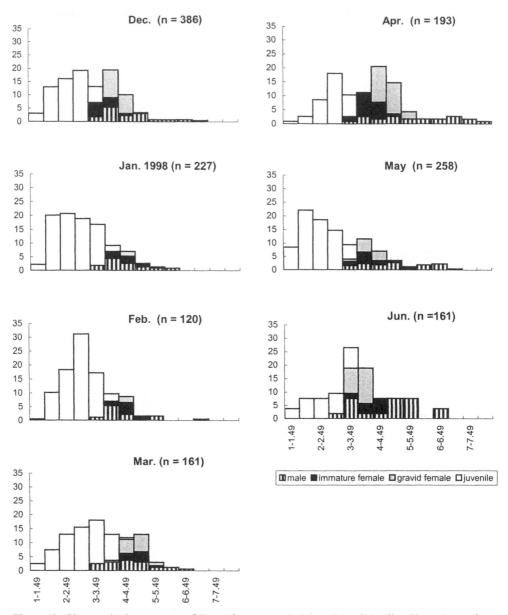

Figure 1b. Changes in the structure of *Dynoides* sp. population at Cape d'Aguilar, Hong Kong, from Ceember 1997 to June 1998.

The population of *Dynoides* sp. mostly comprised smaller individuals during the warmer months of the year, and larger ones during the cold months. From April to August, the population showed a mixed pattern, including both smaller and bigger individuals. Large animals, which had been born in the previous winter, disappeared in August. Males reached 7.9 mm in length and females up to 5.5 mm. Juveniles were constantly more abundant than adults over the course of a year.

Dynoides sp. was, thus, more abundant during the warm months from spring to early winter, with a peak density of 81 ind. 100 cm^{-2} in November. Numbers declined from winter to early spring, with the lowest abundance of 24 ind. 100 cm^{-2} being recorded in February (Fig. 2).

3.2 *Sexual activity*

Dynoides sp. was sexually active throughout the year, however, with a peak from May to December when 91% of females were ovigerous. Lower levels of such females (25%) were recorded from January to February (Fig. 3). Because isopods grew and reached maturity more quickly in the warm months, sexual characteristics could be recognized at a smaller body length, i.e., 2.7 mm in males, 3 mm in immature females and 3.1 mm in gravid females. In winter, males could be identified at 3.6 mm, immature females at 4.0 mm and gravid females at 4.5 mm body length.

3.3 *Egg production*

The eggs of *Dynoides* sp. were yellow, nearly round, and 0.3 mm in diameter. Newly hatched juveniles (mancas) were 0.9 mm long, 0.5 mm wide and sexually indistinguishable. The body is bright yellow, semi-transparent with a pair of dark compound eyes, and lacking pereiopod 7. No movement was observed other than body contractions. A week after hatching, black dots of pigment occurred on the body surface and movement commenced. Mancas tended to swim at the water surface and were distinctly photopositive. Molting occurred 25 days after hatching.

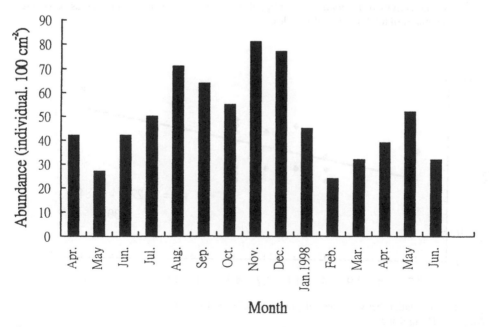

Figure 2. Changes in abundances of *Dynoides* sp. at Cape d'Anguilar, Hong Kong.

A positive relationship between the size of ovigerous females and brood size (r = 0.57, p < 0.001) was obtained. The number of brooded eggs or juveniles, significantly increased with increasing parental size, regardless of season. The brood size ranged from 26 to 56 eggs per brood (Fig. 4).

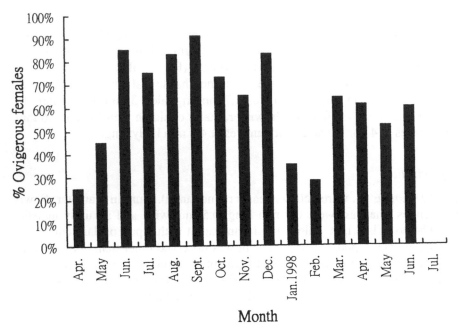

Figure 3. Sexual activity of *Dynoides* sp. at Cape d'Anguilar, Hong Kong, expressed as percentage ovigourous females of total number of females.

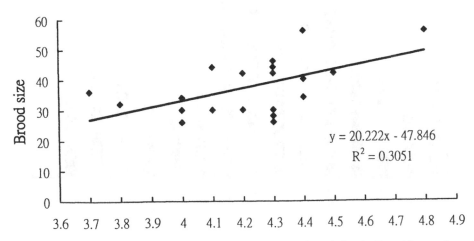

Figure 4. The relation between size of gravid females and brood size in *Dynoides* sp. from Cape d'Anguilar, Hong Kong.

4 DISCUSSION

The *Dynoides* sp. population at Cape d'Aguilar spends its life in beds of the mussel *Septifer virgatus*, unlike other isopods with different life stages in different habitats, e.g., *Dynamene bidentata*, *Dynamene magnitorata* and *Cymodoce truncata* (Holdich 1970, 1971, 1976, Arrontes & Anadón 1990). Such habitat fidelity has, however, been reported for *Idotea baltica* (Guarino et al. 1993). In the Gulf of Naples, *I. Baltica* is like *Dynoides* sp. because its habitat, thick beds of the red alga *Gracilaria* sp., is present throughout the year, providing this isopod with food and a safe shelter from predation by littoral fishes. Mussels are also available as a habitat year round at Cape d'Aguilar, forming a dense mat which supplies the isopods with food, a means of escaping desiccation, emersion and predation, protection from wave action (Hewatt 1935, Ricketts et al. 1939, Jackson 1976, Ong Che et al. 1992, Iwasaki 1996), and a permanent, stable, refuge. Gravid females occurred in all months and sexual activity was carried out year round and juveniles released continuously but obtained in greater numbers at certain times of the year.

Three recruitments occurred in each year and, thus, subsequent cohorts appeared in the population in spring, summer and late autumn. As mentioned before, the first cohort occurred in either April or May, grew and matured until July; the second cohort occurred in August or September and matured in November. Both, thus, matured after three to four months. The last, overwintering, cohort occurred in December and matured in the next April or May and, thus, needed five or six months to mature. It can be concluded that the time for maturation by *Dynoides* sp. ranged between three to six months. It is also concluded that during the warm months, isopods grew and matured faster but to a smaller size whereas they grew slower but to a larger size in the cold months. This contrasts with the results of Iwasaki (1996) on *Dynoides dentisinus*, that also occurs in *Septifer virgatus* beds in Japan, but where there are only two recruitments of mancas each year, in April and September. This may be explained by the typical sub-tropical climate of Hong Kong. Jones (1970) reported that the onset of breeding by species of *Eurydice* in Britain might be correlated with rising sea temperatures.

The life span of *Dynoides* sp. changed with the seasons. A clear longevity could not be determined for the spring cohort because of the continuous release of juveniles in the following months. The summer cohort lived for four months from September to December, and died as winter set in. This was revealed by a sharp decline in abundance after December. The third cohort overwintered and died at a large size in the following summer from January to August and thus had a life span of approximately nine months and, thus, overlapped with the spring cohort. Different life spans for different seasons have also been reported upon by Hiwatari et al. (1984) for the amphipod *Hyale barbicornis*. The abundance data for the total population of *Dynoides* sp. at Cape d'Aguilar matched the results described above, so that it was characterized by a higher number of individuals from April to December and by smaller numbers from January to March.

Many studies of isopods have shown that egg production is related to ovigerous female length (Salemaa 1979). Present data confirm this relationship, i.e., larger ovigerous females of *Dynoides* sp. had a bigger brood size.

ACKNOWLEDGEMENTS

I am grateful to my supervisor, Professor Brian Morton for his critical reading of the first drafts of this manuscript and for the comments and suggestions made by anonymous referees. I also acknowledge, with thanks, Dr N.L. Bruce for his taxonomic help with *Dynoides* sp.

REFERENCES

Arrontes, J. & Anadón, R. 1990. Seasonal variation and population dynamics of isopods inhabiting intertidal macroalgae. *Sci. Mar.* 54: 231-240.

Bamber, R.N. 1997. Peracarid Crustaceans from Cape d'Aguilar and Hong Kong, I. Mysidacea, and Isopoda: Anthuridea. In B. Morton (ed.) *The Marine Flora and Fauna of Hong Kong and Southern China IV. Proceedings of the Eighth International Marine Biological Workshop: the Marine Flora and Fauna of Hong Kong and Southern China, Hong Kong, 1995*: 78-86. Hong Kong: University Press.

Bruce, N.L. 1980. On a small collection of marine Isopoda (Crustacea) from Hong Kong. In B. Morton & C.K. Tseng (eds) *Proceedings of the First International Marine Biological Workshop: the Marine Flora and Fauna of Hong Kong and Southern China, Hong Kong, 1980*: 315-324. Hong Kong: Hong Kong University Press.

Bruce, N.L. 1982. The genus *Dynoides* Barnard, 1914 (Crustacea: Isopoda: Sphaeromatidea) from eastern Australia, with description of new species. *Mem. Queensl. Mus.* 20: 447-53.

Bruce, N.L. 1986. New records of isopod crustaceans (Flabellifera) from Hong Kong. In B. Morton (ed.) *Proceedings of the Second International Marine Biological Workshop: The Marine Flora and Fauna of Hong Kong and Southern China, Hong Kong, 1986*: 549-554. Hong Kong: Hong Kong University Press.

Dudgeon, D., Ma, H.H.T. & Lam, P.K.S. 1990. Differential palatability of leaf litter to four sympatric isopods in a Hong Kong forest. *Oecologia* 84: 398-403.

Guarino, S.M., Gambardella, C. &. de Nicola, M. 1993. Biology and population dynamics of *Idotea baltica* (Crustacea, Isopoda) in the Gulf of Naples, the Tyrrhenian Sea. *Vie Milieu* 43: 125-136.

Harrison, K. & Holdich, D.M. 1984. Hemibranchiate sphaeromatids (Crustacea: Isopoda) from Queensland, Australia, with a world-wide review of the genera discussed. *Zool. J, Linn. Soc.* 81: 275-387.

Harrison, K. 1984. The morphology of the sphaeromatid brood pouch (Crustacea: Isopoda: Sphaeromatidae). *Zool. J.Linn. Soc.* 82: 363-407.

Hewatt, W.G. 1935. Ecological succession in the *Mytilus californianus* habitat as observed in Monterey Bay, California. *Ecology* 16: 244-51

Hiwatari, T. & Kajihara, T. 1984. Population dynamics and life cycle of *Hyale barbicornis* (Amphipoda, Crustacea) in a blue mussel zone. *Mar. Ecol. Progr. Ser.* 20: 177-183.

Holdich, D.M. 1968. Reproduction, growth and bionomics of *Dynamene bidentata* (Crustacea: Isopoda). *J. Zool. London.* 156: 137-153.

Holdich, D.M. 1970. The distribution and habitat preferences of the Afro-European species of *Dynamene* (Crustacea: Isopoda). *J. Nat. Hist.* 4: 419-438.

Holdich, D.M. 1971. Changes in physiology, structure and histochemistry occurring during the life-history of the sexually dimorphic isopod *Dynamene bidentata* (Crustacea: Peracardia). *Mar. Biol.* 8: 35-47.

Holdich, D.M. 1976. A comparison of the ecology and life cycles of two species of littoral isopod. *J. Exp. Mar. Biol. Ecol.* 24: 133-149.

Iwasaki, K. 1996. Vertical distribution and life cycle of two isopod crustaceans within intertidal mussel beds *Benthos Research* 51: 75-96.

Jackson, L.F. 1976. Aspects of the intertidal ecology of the east coast of South Africa. *South African Associatio, Marine Biological Research, Investigation Report* 46: 1-72.

Jones, D.A. 1970. Population densities and breeding in *Eurydice pulchra* and *Eurydice affinis* in Britain. *J. Mar. Biol. Ass. UK* 50: 635-655.

Kwon, D.H. & Kim, H.S. 1986. *Dynoides spinipodus*, a new species of sphaeromatid isopod (Crustacea) from the south coasts of Korea. *Kor. J. Syst. Zool.* 2: 43-48.

Kwon, D.H. 1990. A systematic study on the Korean marine isopod crustaceans I. Flabellifera Part 2. Family Sphaeromatidae. *Inje University* 6(1): 151-192.

Lam, P.K.S., Dudgeon, D. & Ma, H.H.T. 1991. Ecological energetics of populations of four sympatric isopods in a Hong Kong forest. *J. Trop. Ecol.* 7: 475-490.

Ma, H.H.T. 1986. Hong Kong intertidal isopods (Isopoda: Oniscidea), with notes on the feeding and reproduction of *Armadilloniscus litoralis* Budde-Lund, 1885. In B. Morton (ed.) *Proceedings of the Second International Marine Biological Workshop: The Marine Flora and Fauna of Hong Kong and Southern China, Hong Kong, 1986*: 1023-1031. Hong Kong: Hong Kong University Press.

Ma, H.H.T., Dudgeon, D. & Lam, P.K.S. 1991. Seasonal changes in population s of three sympatric isopods in a Hong Kong forest. *J. Zool. Lond.* 224: 347-365.

Mak, P.M.S., Huang, Z.G. & Morton, B. 1985. *Sphaeroma walkeri* Stebbing (Isopoda, Sphaeromatidae) introduced into and established in Hong Kong. *Crustaceana* 49: 75-82

Markham, J.C. 1980. Bopyrid isopods parasitic on decapod crustaceans in Hong Kong and southern China. In B. Morton & C.K. Tseng (eds) *Proceedings of First International Marine Biological Workshop: The Marine Flora and Fauna of Hong Kong and Southern Chian, Hong Kong, 1980*: 325-355. Hong Kong: Hong Kong University Press.

Nishimura, S. 1976. *Dynoidella conchicola*, gen. et sp. nov. (Isopoda, Sphaeromatidae), from Japan, with a note on its association with intertidal snails. *Publ. Seto Mar. Biol. Lab.* 23 (3/5): 275-282.

Ong Che R.G. & B. Morton 1992. Structure and seasonal variations in abundance of the macroinvertebrate community associated with *Septifer virgatus* (bivalvia: Mytilidae) at Cape d'Aguilar, HongKong. *Asian Mar. Biol.* 9: 217-233.

Ricketts, E.F. & Calvin, J. 1939. *Between Pacific Tides*. Stanford, California: Stanford University Press.

Salemaa, H. 1979. Ecology of *Idotea* spp (Isopoda) in the Northern Baltic. Ophelia 18: 133-150.

Shen, C.J. 1929. Description of a new isopod, *Dynoides dentisinus* from the coast of north China. *Bull. Fan. Mem. Inst. Biol.* 1: 65-78.

Population ecology of *Callichirus major* (Crustacea: Decapoda: Thalassinidea) on a sandy beach in south-eastern Brazil

ROBERTO M. SHIMIZU & SÉRGIO DE A. RODRIGUES
Departamento de Ecologia Geral, Universidade de São Paulo, São Paulo, Brazil.

ABSTRACT

A population of the ghost shrimp *Callichirus major* was studied on Barequeçaba Beach, São Sebastião, São Paulo, a sheltered beach in southeastern Brazil. Sampling was conducted from April 1994 to December 1995. Large individuals dominated the size structure of this population. The modal class in the size frequency distribution of males was larger than that of females. Dorsal oval length (DOL) was 14 and 13 mm for males and females respectively, while the maximum DOL was very similar, 14.6 mm in males and 14.5 mm in females. Sex ratio was strongly skewed toward females (1:2.5) in adults, while was close to 1:1 in juveniles. Growth was initially rapid and maturation size (10.3 mm) was attained in both sexes within one year after hatching. In adults, growth was slower and a life span of at least four years was estimated for both sexes. The molting cycle of males was correlated with water temperature (external stimulus), while that of females was correlated with proportion of ovigerous individuals (intrinsic factor). Ovarian development intensified between late winter and early spring (August-September). This was followed by increasing proportions of ovigerous females with peaks in December, in both 1994 and 1995. The period of hatching of larvae and settlement of post-larval juveniles was estimated as December-January and January-February, respectively. Density declined from 6.1 to around 2 burrow openings per m^2, after an oil spill in May 1994, and remained unaltered until the end of the samplings. Compared to data obtained from another site (Santos Bay), the present results demonstrated that this oil spill did not affect the structure, nor the life history pattern of the population here studied. Interference of the residual oil in the sediment on the 1995 recruitment is considered.

1 INTRODUCTION

Callichirus major is currently known from sandy beaches of the Western Atlantic, ranging from North Carolina, USA (Hay & Shore 1918), up to the Island of Santa Catarina, Southern Brazil (Rodrigues 1983). This wide geographic distribution presumably represents a complex of very close species. Rodrigues (1985) detected morphological differences between specimens from Brazil and those from North Carolina. More recently, Stanton & Felder (1995) found that a population from Colombia was genetically distinct from those from the United States and Gulf of Mexico. In Brazil, *C. major* is among the most common species in the macrofauna of sheltered sandy beaches. It is the dominant species among the larger, long lived, tube or burrow dwelling components of the fauna (Rodrigues 1983, Shimizu 1991). Since the late 1980's the species has been used as fishing baits, and this activity has raised concern among local governmental agencies and non-governmental organizations. Although several biological, ecological, and palaeoecological aspects of this species were studied (see Rodrigues & Shimizu 1997 for review), little detailed information exists on its life history. The present study describes structure, growth, reproductive period, and temporal variation of density of a *C. major* population.

2 STUDY SITE

Barequeçaba beach, São Sebastião County, northern portion of the State of São Paulo coast (45°16′W, 23°49′S), is a sheltered sandy beach located about 1.5 km southward from the Marine Biology Center of the University of São Paulo (CEBIMar-USP). This beach is approximately east-west oriented, 1.2 km long, delimited by two rocky points, and is backed by a human settlement (Fig. 1). The width of the intertidal zone

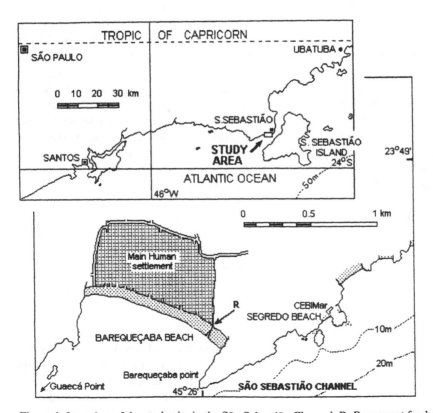

Figure 1. Location of the study site in the São Sebastião Channel. R: Permanent freshwater outlet.

is on average 90 m at spring low tides. Several freshwater outlets reach the shore, of those only the one nearest to Barequeçaba Point is permanent (Fig. 1, R). Those outlets influence interstitial water salinity, particularly in the higher reaches of the intertidal (Shimizu 1991). The sediment is composed of very well sorted to poorly sorted fine sand, the mean particle diameter ranging from 0.11 to 0.23 mm (3.22 to 2.13 ϕ). Distribution of *C. major* ranges from the infralittoral fringe down to levels exposed only during exceptional tides. Seawards, *C. major* is gradually replaced by another callianassinid, *Sergio mirim* (Rodrigues 1971).

On 15 May 1994, a severe oil spill occurred in the São Sebastião Channel, due to a rupture of the oil duct that links the local maritime oil terminal to the Cubatão refinery (central portion of the State of São Paulo coast). During the cleaning operation, 86 m^3 of oil was removed manually from the higher reaches of the intertidal (CETESB 1994).

3 MATERIAL AND METHODS

Fieldwork was conducted approximately once a month from April 1994 to December 1995 during spring low tides. For analyses of population structure and reproductive cycle approximately 100 animals were collected with a pump (Rodrigues 1966), similar to the 'yabby pump' described by Hailstone & Stephenson (1961). The specimens were preserved in 4% buffered formalin. Ovigerous females were packed individually in small plastic bags. Population density was monitored by counting burrow openings inside a 1 m^2 square wooden frame (N = 60, on each sampling date). The placement of the frame and the choice burrow opening to be explored for collection of specimens were determined with the aid of random numbers. Samplings were avoided near the upper distribution boundary of *C. major*, where densities were very low. During very low tides, levels of the beach that was not usually emerged were also avoided to prevent biases in counting burrow openings, caused by the occurrence of *S. mirim*. On each sampling date, 3 samples of water from inside *C. major* burrows, from the surrounding sediment, and from the sea were collected. Temperature (with a 0.5°C scale thermometer) and salinity (with a 1‰ scale refractometer) were measured on each sample. Beach profile was determined by the 'stake-and-horizon' method (Fox 1983).

In the laboratory, the following parameters were determined for each animal: sex, based on the position of the gonopores in juveniles, when secondary characters were not evident; dorsal oval length (DOL, in mm, Fig. 3) was measured with a 0.1 mm scale caliper. This measure was preferred to the more usual carapace length (CL), because: dorsal oval has better defined contours and provided more precise measurements; dorsal oval is more calcified and thus, more resistant to damages during collection; DOL and CL are linearly related (DOL = 0.7417CL – 0.2399; r^2 = 0.9607; N = 1137). Hardness of the exoskeleton of adult individuals was also verified. Three moderate strokes were applied with the sharp tip of a forceps on the propodus of the larger chela. When no evident damage was noted, the animal was considered to have well-calcified chela (WCC), that is, in intermolt stage. For females with well-preserved abdomens, width of the two ovaries (OW) visible through the first abdominal segment was measured. An index of ovarian development was obtained by the

ratio OW/DOL (modified from Felder & Lovett 1989). Proportion of ovigerous females was recorded for each sample. Eggs were classified according to the stages of
embryonic development described by Rodrigues (1976). Eggs in stages 1 to 4 and
stages 5 to 9 were pooled as 'uneyed eggs' and 'eyed eggs' respectively (Dworschak
1988). Eyed eggs containing identifiable zoeae (Rodrigues 1976: stage 9) was also
distinguished. The DOL's of the smallest ovigerous female (Hanekom & Erasmus
1989) and of the smallest male with the typical larger chela (Rodrigues 1985) were
assumed as the size at maturity.

Correlation between population parameters (% WCC, mean OW/DOL, % ovigerous females, % juveniles, density) and monthly means of water parameters was verified with the Spearman rank correlation coefficient. Only significant correlation will
be discussed. All statistic tests were applied according to Zar (1996). Von Bertalanfy
growth curves were adjusted on DOL frequency distributions for males and females
using the computer program The Compleat ELEFAN (Gayanilo et al. 1989). The
equation of the growth curve adjusted by the program is:

$$L_t = L\infty\left(1 - e^{-K.((t-t_o)+\frac{C}{2\pi}.sen(2\pi.(t-WP-0.5)))}\right) \qquad (1)$$

where L_t is the DOL at time t, L_∞ the asymptotic DOL, K the growth constant, C a
constant that expresses the amplitude of the growth oscillations, WP the parameter
that designates the period of the year when growth is slowest, and t_o the origin of the
growth curve. Seed values for L_∞ and K were determined as suggested by Bray and
Pauly (1986). For C, seed values were obtained according to Pauly (1987).

4 RESULTS

4.1 *Environmental parameters*

Mean salinity of the water from the burrows of *C. major* ranged from 28.0‰ to
31.3‰ during the sampling period (Fig. 2A). These values were closer to those obtained from interstitial water (25.7-31.7‰) during the period of September to December 1994 and in December 1995. Influence of sea water (31.0-35‰) was more
evident in April, May, and October 1995. Temperature varied seasonally in all three
habitats (Fig 2A), ranging from 20.0 to 28.0°C. Although temperature values of burrow water and interstitial water was higher than those of sea water in two occasions
(November 1994 and August 1995), no significant difference in temperature was detected between habitats (KW = 3.98; p = 0.14). Sea water temperature was therefore
used for correlation analyses because more monthly mean values were available. The
beach profile varied seasonally (Fig. 2B). Erosion prevailed during fall and winter
while deposition was predominant during spring and summer.

4.2 *Sex ratio and size structure*

A total of 1157 individuals of *C. major* was collected of those 330 (28.5%) were

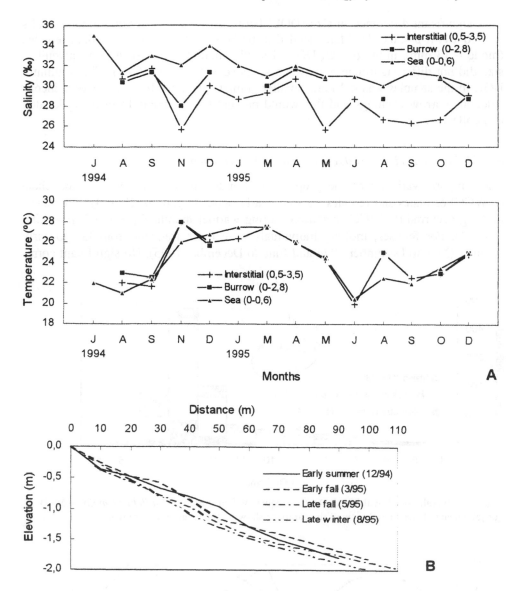

Figure 2. Environmental parameters on Barequeçaba beach during the study period. A: monthly mean salinity and temperature of interstitial water, water from *Callichirus major* burrows, and surface sea water (figures between parentheses represent ranges of standard deviation); B: most representative profiles of the beach.

males and 827 (71.5%) were females. The DOL of the smallest adult was coincident for both sexes (10.3 mm). The total male to female sex ratio of 1:2.5 differed significantly from 1:1 ($\chi^2 = 262.17$, $p < 0.001$). The more numerous adult individuals (DOL ≥ 10.3 mm) contributed for this condition (224 males, 722 females; $\chi^2 = 213.68$, $p < 0.001$). The sex ratio did not depart significantly from 1:1 among juveniles (106 males, 105 females; $\chi^2 = 0.01$, $p = 0.95$).

A significant difference in size (DOL) between sexes was verified among adults ($U = 640171$, $p < 0.001$). Males tended to be larger (mode = 14 mm) than females (mode = 13 mm), in the overall DOL class distributions (Fig. 3). Among juveniles, size did not differ significantly between sexes ($U = 5242$, $p = 0.47$). The minimum DOL in the samples was 4.2 mm. Smaller animals presumably produce very inconspicuous burrow openings and this would prevent such animals from being located and collected.

4.3 *Molting periods, reproductive cycle and recruitment*

The temporal variation of the proportion of individuals with well-calcified chela (%WCC) of males and females showed nearly opposite patterns (Fig. 4). For males, molting occurred (% WCC decreased) during warmer months ($r_s = -0.62$; $p = 0.04$; $N = 11$). For females, molting individuals were more frequent from fall to early summer (May to December 1994 and June to December 1995). No significant corre-

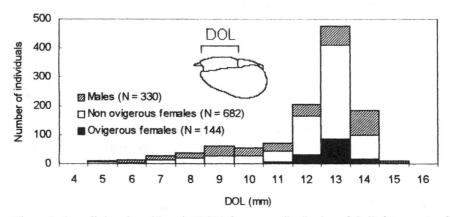

Figure 3. Overall dorsal oval length (DOL) frequency distribution of *Callichirus major* from Barequeçaba beach. Inset: lateral view of the carapace showing the length measurement.

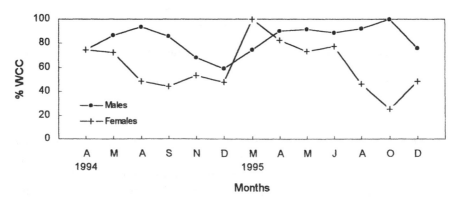

Figure 4. Temporal changes in the percentage of specimens with well calcified chela (%WCC) for males and females of *Callichirus major* from Barequeçaba beach.

lation was obtained between %WCC and water parameters. Testing against other biological events of the population, %WCC was negatively correlated with proportion of ovigerous females ($r_s = -0.71$; $p = 0.01$; N = 12).

Mean values of the index of ovarian development (OW/DOL) varied seasonally with peaks in late winter and early spring (September 1994 and August 1995, Fig. 5). Monthly mean OW/DOL were negatively correlated to water temperature ($r_s = -0.76$; $p < 0.001$; $N = 12$). Ovigerous females were progressively more frequent from August to December in both years (Fig. 5). Maximum percentages of eyed eggs were verified in December 1994 (52.3%) and in December 1995 (50.0%). It was assumed that the reproductive cycle of 1994 ceased in January 1995, when none of the 15 females collected, in a non-quantitative sampling, carried eggs. In December 1995 eggs with identifiable zoea were less frequent (12.8%) than in the previous year (22.8%) suggesting a longer reproductive period.

Recruitment was assessed by the proportion of juveniles (DOL < 10.3 mm) in the samples. Proportion of juveniles increased from August (late winter) to December (early summer) in both years, more markedly in 1994 (Fig. 6).

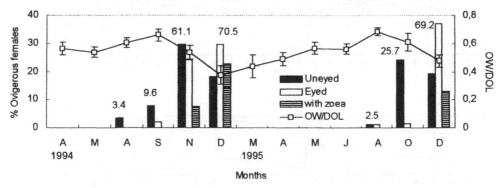

Figure 5. Monthly means of the index of ovarian development (OW/DOL) and percentage of ovigerous females in the *Callichirus major* population from Barequeçaba beach. Proportions of uneyed eggs, eyed eggs and eggs with identifiable zoeae are distinguished. Figures near the top of bars represent total percentage of ovigerous females.

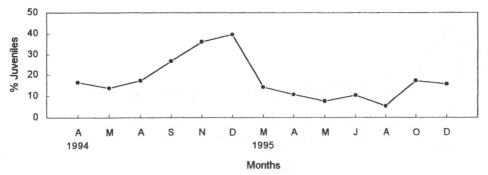

Figure 6. Temporal change in the percentage of juveniles (DOL < 10.3 mm) in the *Callichirus major* population from Barequeçaba beach.

4.4 *Growth*

The temporal sequences of size (DOL) class distribution and the fitted growth curves of males and females are shown in Figure 7. Considering that larvae hatched early in the year (January-February), juveniles were 8 to 10 months old by the time they became evident in the samples (August 1994 and October 1995). According to the growth curves, DOL at maturation (11 mm class) is attained within 12 to 14 months after hatching. In adults of both sexes, growth was slower and maximum DOL was attained around 3 years later after maturation, giving a life span of at least 4 years. Males grew more rapidly (higher growth constant, *K*) and continuously (lower seasonal oscillation of growth, *C*) than females.

4.5 *Density*

Mean density decreased markedly (Fig. 8) from 6.1 (April 1994) to 2.7 (September

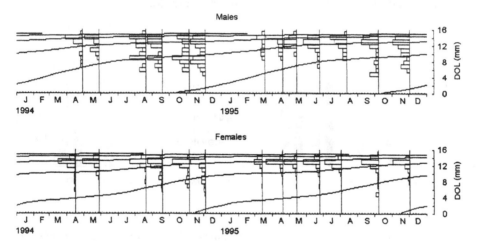

Figure 7. Growth curves for males and females of *Callichirus major* from Barequeçaba beach. Estimated parameters are: $L_\infty = 15.37$ mm, $K = 0.90$ y^{-1}, $C = 0.52$, $WP = 0.75$ for males and $L_\infty = 15.30$ mm, $K = 0.83$ y^{-1}, $C = 0.65$, $WP = 0.25$ for females.

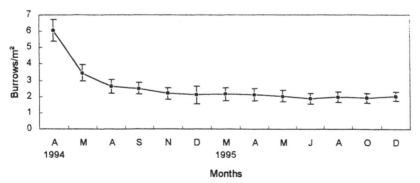

Figure 8. Monthly mean ($\pm$ 95% confidence interval) density of *Callichirus major* on Barequeçaba beach during the study period.

1994) burrows per m^2 and presented little variation on the following sampling dates (November 1994 to December 1995). This temporal variation of density was significant [$F = 16.16$, $p < 0.001$, one-way ANOVA on log ($x + 1$) transformed data]. The Tukey's multiple comparison test distinguished the April 1994 sample from a group of samples of November 1994 to December 1995. Although samples of August and September 1994 could not be distinguished with method, decrease in density is very evident after the oil spill (May 1994).

5 DISCUSSION

Sex ratio skewed towards adult females has been detected in the majority of the thalassinidean species, for both Callianassidae (Hailstone & Stephenson 1961, Devine 1966, Vaugelas et al. 1986, Felder & Lovett 1989, Pezzuto 1993, Dumbauld et al. 1996) and Upogebiidae (Tunberg 1986, Dworshak 1988, Hanekom & Baird 1992, Dumbauld et al. 1996). This pattern has been attributed to: local concentration of adult females due to migration (Hailstone & Stephenson 1961); to a differential mortality of adult males due to intraspecific agonistic interaction (Felder & Lovett 1989, Tamaki et al. 1997); and to predation on adult males that leave their burrows in search of females (Dumbauld et al. 1996). Rowden & Jones (1994) suggest that the usual sampling gears ('yabby pumps', corers) tend to favor collection of females. Although not tested for the present work, the hypothesis of mortality of males due to agonistic interaction seems to be the most adequate for *C. major*.

Males attaining larger size than females are also commonly observed in Thalassinidea (Hailstone & Stephenson 1961, Forbes 1977, Tunberg 1986, Dworschak 1988, Tamaki et al. 1996). This condition has been attributed to the amount of energy expended in reproduction, which is larger in females than in males (Hanekom & Erasmus 1989), causing a slower growth of the former, particularly during breeding periods (Forbes 1977). This is observed in the growth curves (Fig. 7). In terms of maturation time, *C. major* (1 year) is similar to other species from warm temperate waters (Forbes 1977, Felder & Lovett 1989, Tamaki et al. 1997). Among species from comparatively colder waters, maturation time is over 2 years (Buchanan 1963, Dworschak 1988, Dumbauld et al. 1996). The longevity of 4 years estimated for *C. major* is in accordance with the range of 2 to 4 years observed for thalassinidian species from warm temperate waters (Hailstone & Stephenson 1961, Forbes 1977, Felder & Lovett 1989, Hanekom & Baird, 1992, Pezutto 1993, Tamaki et al. 1997). Among these species, those that live up to 2 years are predominantly estuarine. It should be noted that longevity obtained by size class analyses may be underestimated due to the difficulty in distinguishing larger slow growing classes from each other (Hanekom & Baird 1992). The high frequency of large size classes throughout the year is another common feature among callianassid populations and is attributed to accumulation of individuals in such classes, due to the relatively slower growth of adults (Buchanan 1963, Forbes 1977, Dworschak. 1988).

Molting cycles are influenced both by intrinsic and extrinsic factors (Dworschak 1988). In the population here studied, males and females were affected differently by each of these factors. Molting before breeding activities in females was also recorded for *T. australisensis* (Hailstone & Stephenson 1961), *C. japonica* (Tamaki et al.

1996), *U. deltaura* (Tunberg 1986), and *U. africana* (Hanekom & Baird 1992), while females of *Calocaris macandreae* (Buchanan 1963), *Glypturus armatus* (Vaugelas et al. 1986) and *S. mirim* (Pezzuto 1993) molted after hatching of eggs. Ovarian growth during winter as observed for *C. major* has been attributed to increased rates of synthesis and mobilization of lipids and proteins during non-reproductive periods (Pillay & Nair 1973). Similar trends of ovarian development have been observed in *U. pusilla* (Dworschak 1988) and *L. louisianensis* (Felder & Lovett 1989).

The larval cycle of *C. major* is incompletely described. The three initial stages reared in laboratory by Rodrigues (1976) lasted longer than 20 days. Among species with larval period determined from field observations *C. japonica* is the most similar to *C. major* in terms of the temperature conditions under which larval development takes place (See Tamaki et al. 1996). Considering that larvae of *C. major* hatched in December-January, and assuming a larval period similar to that of *C. japonica* (20-25 days), the timing of post larvae settlement may be located between January and March (late summer, early fall). This corresponds to periods of increasing water temperature and lower hydrodynamic action, when deposition is the prevailing process in the study site (Fig. 2B). Seasonal breeding has been related to release of larvae or juveniles to an environment with favorable conditions for growth and development (Sastry 1983).

Recruitment was detected only 5 to 7 months after the estimated post larvae settlement period. This was attributed to selectivity of the sampling method. Annual variation in recruitment intensity has been reported for several species (Pezzuto 1993, Tamaki & Ingole 1993, Tunberg 1986, Dumbauld et al. 1996). Such variation has been related to local oceanic process affecting larval dispersal and to mortality of post larvae after settlement (Hadfield 1986). Predation, sediment disturbance, and extreme variation in temperature have been proposed as causes of juvenile mortality (Devine 1966, Tamaki & Ingole 1993). Predation on *C. major* most presumably occurs during pelagic stages, and is of interest for future studies. While no unfavorable sediment condition (extreme temperature variation, physical disturbance) was evidenced during the present study, the potential impact of the oil spill (May 1994) on the 1995 recruitment should be considered. According to Clifton et al. (1984), a layer of oil buried in the sediment remained relatively undegraded for over two years and prevented settlement of *N. californiensis* and other macroufaunal species. On Barequeçaba Beach, a layer of oil 15 cm deep in the sediment was clearly observed in May and June 1994. Seven months after the spill, residues of oil were still detectable in the sediment and also adhered to the tegument of *C. major* specimens. Thus it is reasonable to consider that the oil could have affected recruits in 1995. However, characterization of immature stages of *C. major* under natural or disturbed conditions, would benefit from further investigations on the period between hatching of larvae and settlement of juveniles.

The decrease in burrow density after the oil spill was very similar to that verified experimentally for *N. californiensis* (Clifton et al. 1984). In April 1994, prior to the spill, the density of 6.1 burrows per m^2 was comparable to those recorded for the same population from 1987 to 1989 (1.5-7.0 burrows per m^2) (Shimizu 1991). The temporal pattern in the period that followed the spill differed markedly from that obtained in 1987-1989. In the latter period, density increased from winter (July) to spring (October) and was coincident with the recruitment period verified in the present study. The lack of a similar increase in 1994 might be resultant of the reduction of burrow opening numbers caused by the spill, compensated by an increase of burrow density due to recruitment. This is plausible, considering the time interval between settlement of *C. major* post larvae and detection of the opening of their burrows (5 to 7 months) discussed earlier.

To verify possible interference of the oil spill on the life history patterns and on the structure of the population here studied, the present results were compared to those obtained from a *C. major* population (Santos Bay, State of São Paulo – Rodrigues & Shimizu, unpublished data). No marked difference was observed in terms of reproductive period, recruitment intensity (maxima: 41.6% in Santos; 39.6% in Barequeçaba), and sex ratio ($\chi^2 = 1.61$; $p = 0.20$). Thus, it is concluded that the patterns obtained here are representative for the population.

ACKNOWLEDGEMENTS

We are indebted to the Centro de Biologia Marinha da Universidade de São Paulo (CEBIMar- USP) for providing logistical support and laboratory facilities. Our sincere thanks to Drs A. Cecilia Z. Amaral, Sérgio L.S. Bueno, Sergio Rosso, and Wagner C. Valenti, members of the committee that analyzed the Doctorate thesis of the first author, from which this paper was extracted. The present results were presented by the first author at the Fourth International Crustacean Congress (Amsterdam, 1998) with financial support from FAPESP (Process No. 98/05657-6).

REFERENCES

Bray, T. & Pauly, D. 1986. *Electronic length frequency analysis – A revised and expanded user's guide to ELEFAN 0, 1 and 2*. Institut für Meereskunde an der Christian-Abrechts- Universität.

Buchanan, J.B. 1963. The biology of *Calocaris macandreae* [Crustacea: Thalassinidea]. *J. Mar. Biol. Assoc. UK*, 43: 729-747.

CETESB 1994. *Operação TEBAR V – Relatório de atendimento*. São Paulo: CESTESB.

Clifton, H.W., Kvenvolden, K.A., Rapp, J.B. 1984 Spilled oil and infaunal activity – Modification of burrowing behavior and redistribution of oil. *Mar. Environ. Res.* 11: 111-136.

Devine, C.E. 1966. Ecology of *Callianassa filholi* Milne-Edwards 1978 (Crustacea, Thalassinidea). *Trans. R. Soc. N. Z. Zool.* 8: 93-110.

Dumbauld, B.R., Armstrong, D.A., Feldman, K.L. 1996. Life-history characteristics of two sympatric thalassinidean shrimps, *Neotrypaea californiensis and Upogebia pugettensis*, with implications for oyster culture. *J. Crust. Biol.* 16(4): 689-708.

Dworschak, P.C. 1988. The Biology of *Upogebia pusilla* (Petagna) (Decapoda, Thalassinidea). III - Growth and production. *Mar. Ecol.* 9: 51-77.

Felder D.L. & Lovett, D.L. 1989. Relative growth and sexual maturation in the estuarine ghost shrimp *Callianassa louisianensis* Schmitt, 1935. *J. Crust. Biol.* 9: 540-553.

Forbes, A.T. 1977. Breeding and growth of the burrowing prawn Callianassa kraussi Stebbing (Crustacea: Decapoda: Thalassinidea). *Zool. Afr.* 12: 149-161.

Fox, W.T. 1983. *At the sea's edge*. Englewood Cliffs: Prentice-Hall.

Gayanilo, F.C., Jr, Soriano, M., Pauly, D. 1989 A draft guide to the complete ELEFAN. *ICLARM Software* 2: 1-67.

Hadfield, M.G. 1986. Settlement and recruitment of marine invertebrates: a perspective and some proposals. *Bull. Mar. Sci.* 39: 418-425.

Hailstone, T.S. & Stephenson, W. 1961. The biology of *Callianassa (Trypaea)* australiensis Dana, 1952 (Crustacea, Thalassinidea). *Univ. Qld. Pap. Dep. Zool.* 1: 259-285.

Hanekom, N. & Baird, D. 1992. Growth, production and consumption of the thalassinid prawn *Upogebia africana* (Ortmann) in the Swartkops estuary. *S. Afr. J. Zool.* 27: 130-9.

Hanekom, N. & Erasmus, T. 1989. Determinations of the reproductive output of populations of a thalassinid prawn Upogebia africana (Ortmann) in the Swartkops estuary. *S. Afr. J. Zool.* 24: 244-50.

Hay, W.P. &. Shore, C.A. 1918. The decapod crustaceans of Beaufort, N.C., and the surrounding region. *Bull. Bur. Fish. Wash.* 35: 371-475.

Pauly, D. 1987. A review of the ELEFAN system for analysis of lenght-frequency data in fish and aquatic invertebrates. *ICLARM Contribution* 232: 7-34.

Pezzuto, P.R. 1993. Ecologia populacional de *Neocallichirus mirim* (Rodrigues, 1971) (Decapoda: Callianassidae) na Praia do Cassino, RS, Brasil. M Sc. Dissertation, Fundação Universidade do Rio Grande. 124 p.

Pillay, K.K. & Nair, N.B. 1973. Observations on the biochemical changes in gonads and other organs of *Uca annulipes, Portunus pelagicus, and Metapenaeus affinis* (Decapoda, Crustacea) during the reproductive cycle. *Mar. Biol.* 18: 167-198.

Rodrigues, S. de A. 1966 Estudos sobre *Callianassa*. Sistemática, biologia e anatomia. Dr Sci Thesis, Universidade de São Paulo, 169 p.

Rodrigues, S. de A., 1971. Mud shrimps of the genus *Callianassa* Leach from the Brazilian coast (Crustacea, Decapoda). *Arq. Zool. Est. S. Paulo,* 20: 191-223.

Rodrigues, S. de A. 1976. Sobre a reprodução, embriologia e desenvolvimento larval de *Callichirus major* (Say, 1818) (Crustacea, Decapoda, Thalassinidea). *Bol. Zool. Univ. S Paulo* 1: 85-104.

Rodrigues, S. de A. 1983. Aspectos da biologia de Thalassinidea do Atlântico tropical americano. Livre Docência Thesis, Universidade de São Paulo. 174 p.

Rodrigues, S. de A. 1985. Sobre o crescimento relativo de *Callichirus major* (Say, 1818) (Crustacea, Decapoda, Thalassinidea). *Bol. Zool. Univ. S Paulo* 9: 195-211.

Rodrigues, S. de A. & Shimizu, R.M. 1997. Autoecologia de *Callichirus major* (Say, 1818) (Crustacea: Decapoda: Thalassinidea). *Oecol. Brasil.* 3: 155-170.

Rowden, A.A. & Jones, M.B. 1994. A contribution to the biology of the burrowing mud shrimp, *Callianassa subterranea* (Decapoda: Thalassinidea). *J. Mar. Biol. Assoc. UK* 74: 623-635.

Sastry, A.N. 1983. Ecological aspects of reproduction. In Vernberg, F.J. & Vernberg, W.B. (eds) *The biology of Crustacea. Vol. 8 Environmental Adaptations*: 179-270. New York: Academic Press.

Say, T. 1818. An account of the Crustacea of the United States. *J. Acad. Sci. Philad.* 1: 235-253.

Shimizu, R.M. 1991. A comunidade de macroinvertebrados da região entre marés da Praia de Barequeçaba, São Sebastião, SP. M. Sc. Dissertation, Universidade de São Paulo. 72 p.

Staton, J.L. & Felder, D.L. 1995. Genetic variation in populations of the ghost shrimp genus *Callichirus* (Crustacea: Decapoda: Thalassinoidea) in the Western Atlantic and Gulf of Mexico. *Bull. Mar. Sci.* 56: 523-536.

Tamaki, A. & Ingole, B. 1993. Distribution of juvenile and adult ghost shrimps, *Callianassa japonica* Ortmann (Thalassinidea), on an intertidal flat: Intraspecific facilitation as a possible pattern-generating factor. *J. Crustac. Biol.* 13: 175-83.

Tamaki, A., Tanoue, H., Itoh, J., Fukuda, Y. 1996. Brooding and larval developmental periods of the Callianassid ghost shrimp, *Callianassa japonica* (Decapoda: Thalassinidea). *J. Mar. Biol. Assoc. UK* 76: 675-689.

Tamaki, A., Ingole, B., Ikebe, K., Muramatsu, K. Taka, M., Tanaka, M. 1997 Life history of the ghost shrimp, *Callianassa japonica* Ortmann (Decapoda: Thalassinidea), on an intertidal sand-flat in western Kyushu, Japan. *J. Exp. Mar. Biol. Ecol.* 210: 223-250.

Tunberg, B. 1986. Studies on the population ecology of *Upogebia deltaura* (Leach) (Crustacea, Thalassinidea). *Estuar. Coast. Shelf Sci.* 22: 753-65.

Vaugelas, J. de, Delesalle, B., Monier, C. 1986. Aspects of the biology of *Callichirus armatus* (A. Milne Edwards, 1870) (Decapoda, Thalassinidea) from French Polynesia. *Crustaceana* 50: 204-216.

Zar, J.H. 1996. *Biostatistical analysis* (3rd Ed.). New Jersy: Prentice Hall.

The biology of a population of the hermit crab *Diogenes pugilator* (Decapoda, Diogenidae) in the Ligurian Sea

DANIELA PESSANI, MARCO DAPPIANO & TINA TIRELLI
Dipartimento di Biologia Animale e dell'Uomo, Università di Torino, Torino, Italy

CARLO CERRANO
Dipartimento per lo studio del territorio delle sue risorse, Università di Genova, Genova, Italy

ABSTRACT

Seventy specimens of *Diogenes pugilator* were collected monthly from a sandy bottom (Ligurian Sea, Northern Mediterranean Sea). The specimens were maintained in the laboratory for one month, in order to collect data about monthly rate of molting, population growth, and, in general, the biology of the species. Throughout the year, sex ratio was always female-biased; females were significantly smaller than males. Specimens were clustered in four size classes; on the basis of their monthly presence, it is possible to hypothesize that the life cycle of the species lasts between 1 and 1.5 year. During winter and spring, the small-sized specimens, i.e. the young ones, occupy the deepest part of the habitat; in the summer, once they become adults, they move towards the coast. Young specimens live among *Cymodocea nodosa* leaves, where they can shelter and find small-sized shells. *D. pugilator* avoids the strong hydrodynamics and muddy sand. Young specimens prefer *Hinia* sp. gastropod shells, while adults prefer *Sphaeronassa mutabilis* ones.

1 INTRODUCTION

Diogenes pugilator, the only Mediterranean representative of the genus, is strongly heterochelous and left-handed. It is a small to middle sized crab (carapace length up to 7.5 mm) and not very strikingly colored.

Ingle (1993) reports the presence of *Diogenes pugilator* from a quite wide area: the African and European Atlantic shores, the Mediterranean Sea and the Black Sea. Local restrictions to its territorial expansion are due to the bottom type, living exclusively on sandy bottoms between 0 and 40 m.

Morphometric and systematic aspects of this species were analyzed by Altes (1966) and Codreanu & Balcesco (1968).

The aim of this research is to describe *D. pugilator*'s rhythms of activity, distribution in space and time, and molting rhythms.

2 MATERIALS AND METHODS

Sampling was carried out in the eastern Ligurian Sea (Mediterranean Sea). The sampling area was located on a bottom strip (2 × 180 m) between depths of 1 and 12 m. According to Péres & Picard (1964) the bottom biocenoses consist of 'fine superficial sand' (S.F.S.) and 'fine well-classed sand' (S.F.B.C.).

Every 40 days, from June 1995 to July 1996, a scuba diver manually collected a maximum of 70 hermit crabs. In the laboratory, 54 hermit crabs were individually placed into square cells (8 × 8 cm) of a compartmented tank (pH = 8.22; T = 25°C; S = 26‰; natural lighting). Bottoms were lined with sand grains. The remaining specimens were set into the unpartitioned tank part (75 × 60 cm).

After collecting the exuviae of each specimen, the hermit crabs were put back into their original environment.

Morphological observations, sex determination, and measurement of carapace length (size) were made under a stereomicroscope with a micrometric eye-piece.

Two kinds of shell change experiments were performed: the first series lasted 24 hours and consisted in offering three gastropod shells to each crab, one similar to the occupied one and two belonging to two different species; in the second series, 15 specimens, placed into the unpartitioned tank part, were offered 30 shells; the experiment lasted 30 days.

3 RESULTS

3.1 *Population structure*

Mean males size (4.11 mm; s = 1.45, n = 98) significantly differed from that of females (3.73 mm; s = 1.02, n = 213; Student's t-test: t = 2.67, d.f. = 309, p < 0.01). Maximum female size was 6.61 mm, the maximum male size was 7.16 mm. In order to determine population structure, specimens were clustered into 4 size classes: sub-adults (1.5-3 mm; n = 74), adults I (3.1-4.5 mm; n = 151), adults II (4.6-6 mm; n = 67) and adults III (> 6 mm; n = 19). The monthly abundance of specimens belonging to the 4 size classes allows analysis of the population structure, which undergoes seasonal changes (Fig. 1).

Sub-adults make up 15% of the whole sample during summer and settle down around 30% from autumn until the next spring; adults I make up 35% of the sample from April to September (they just show a slight increase in summer) but they are over 60% in winter. On the contrary, adults II reach 45% during late summer and late spring periods then decrease to 4% in January; adults III are absent in winter and reach 15% in summer.

3.2 *Sex ratio*

The sex ratio, or number of males referred to the total number of crabs, is always in favor of females, either considering the whole sample (0.315; χ^2 = 42.5, d.f. = 1, p < 0.01) or monthly samples. While sub-adults, adults I, and adults II have values of the sex ratio approaching the general average (0.30, 0.29, and 0.24 respectively), adults III are mostly composed of males (sex ratio = 0.84) (G_{test}: G = 25.1, d.f. = 3, p < 0.001).

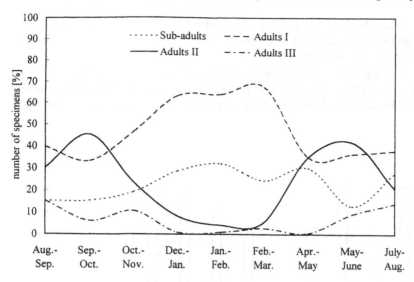

Figure 1. Monthly abundance (%) of *Diogenes pugilator* belonging to 4 size classes.

3.3 *Distribution around the habitat*

In the examined area, starting from a depth of 8 m fine sand is gradually substituted by mud. *Cymodocea nodosa*, present below 5 m depth, is replaced by *Posidonia oceanica* at about 10-12 m. *Diogenes pugilator* is frequent from 1.5 m down to 8 m, then it disappears completely while *P. oceanica* becomes present. *D. pugilator* is rare where hydrodynamics are high.

In this bathymetric range the differences in the distribution were observed. They are mainly due to season and crab size. In the range 0-1.5 m, *Diogenes pugilator* specimens were found isolated or in small groups of 4-5 animals. Adults were abundant at 3-6 m depth. Here, groups of 20-30 specimens were often observed within an area of 3-4 m^2. The specimens were at a distance of 10-20 cm from each other and reached a maximum density of 6-7 ind. m^{-2}. Specimens < 2.5 mm lived only in the deepest part of the distribution area of the species, especially where *Cymodocea nodosa* is blooming. In contrast, adults are rare there. During winter, *D. pugilator* shelters at depths a little more than 4 m, forming numerous groups of few animals close to *C. nodosa*, where hydrodynamic forces are lower.

Diogenes pugilator buries itself in a few centimeters of sand and always keeps direct contact with the overlaying water so that oxygen replacement is ensured. When not buried, *D. pugilator* strolls around.

3.4 *Molting rate*

It has been ascertained that molting rate is very small within the first 5 days and starts decreasing after 40-45 days from capture. As a consequence, exuviae collected within 35 days from capture were taken into consideration (Fig. 2). A significant and season related difference in the molting rate (G_{test}: G = 48.99, d.f. = 9, p < 0.001) was observed: the maximum rate occured in summer (46 molts out of 54 specimens in July) and decreased during the cold season (Fig. 2). A close relation between annual temperature trend and seasonal molting rate (Fig. 3) was recorded (Spearman rank correlation: t = 3.94, d.f. = 11, p < 0.01).

In order to verify a possible increase in the number of molts by females during the reproductive period, molting rate was separately analyzed for the two sexes during the whole year. The number of females exuviae is always twice or three times larger

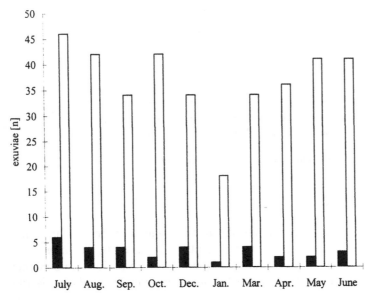

Figure 2. Molting rate of *Diogenes pugilator* within 5 (black bar) and 35 (white bar) days from sampling.

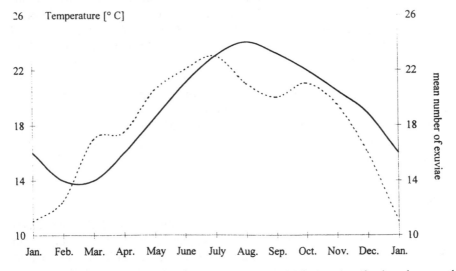

Figure 3. Relationship between annual temperature (solid line) at 4 m depth and seasonal molting rate (dashed line) of *Diogenes pugilator*.

than that of males (in agreement with sex ratio). No significant differences due to season were observed (G_{test}: G= 3.14, d.f. = 9, n.s.).

Differences in molting were also tested among different size classes. Crabs belonging to the first two classes have molting rates higher (sub-adults and adults I, respectively, 52/60 and 75/88) but not significantly different from adults II and adults III (respectively, 47/61 and 12/19) (G_{test}: G= 6.29, d.f. = 3, n.s.).

3.5 *Occupied shells*

Table 1 shows the kind of gastropod shell occupied by 137 specimens sampled between April and July 1996 and split into size classes. Shell species occupied only by one specimen were not considered; these were *Jujubinus exesperatus*, *Jujubinus striatus*, *Calliostoma conulum*, *Smaragdia viridis*, *Cerithium vulgatum*, *Naticarius hebraeus*, *Buccinum corneum*, *Columbella rustica*, and *Bela ginnania*.

The chosen shell varies with the hermit crab size. *Hinia* spp. is significantly more used by sub-adults (G_{test}: G = 27.49, d.f. = 5, p < 0.001); on the contrary, adults I as well as adults II and III prefer *Sphaeronassa mutabilis* (G_{test}, respectively, G = 73.76, d.f. = 3, p < 0.001, and G = 86.37, d.f. = 3, p < 0.001) (Table 1).

3.6 *Shell choice experiments*

Thirty–six hermit crabs, living in *Hinia incrassata*, *H. costulata*, *Payraudeautia intricata*, *Ocinebrina edwarsi*, *Sphaeronassa mutabilis*, *Neverita josephinia*, and *Murex brandaris* were offered a choice of three different shells (*H. incrassata* or *H. costulata*, *S. mutabilis*, and *N. josephinia*). The volume of the latter shells was 10% larger than that of the former ones. Twenty-four hours later, 83% of the hermit crabs had moved to a new shell. It must be remarked that, at the end of the experiment, the specimens occupying S. mutabilis shells were 31/36 (86%).

The second experiment was performed with 15 hermit crabs: 9 of them living in *Sphaeronassa mutabilis*, 3 in *Hinia incrassata*, 1 in *Neverita josephinia*, 1 in *Payraudeautia intricata* and 1 in *Ocinebrina edwardsi*. Of the 30 offered shells, 18 were S. mutabilis, while the others were equally distributed among the other 4 species of gastropods. Thirty days later, 13 adults (87%) were found inside S. mutabilis, 2 sub-adults inside H. incrassata.

Table 1. Use of different gastropod shell species by sampled specimens of *Diogenes pugilator*.

Occupied shell	Sub-adults		Adults I		Adults II		Adults III		Total
	n	%	n	%	n	%	n	%	n
Sphaeronassa mutabilis	3	10.3	33	60	36	87.8	9	75	81
Hinia spp.	17	58.7	13	23.7	–	–	–	–	30
Payraudeautia intricata	1	3.4	5	9.1	3	7.3	–	–	9
Neverita josephinia	–	–	–	–	1	2.4	3	25	4
Cyclope neritea	3	10.3	–	–	–	–	–	–	3
Jujubinus sp.	3	10.3	–	–	–	–	–	–	3
Ocinebrina edwardsi	–	–	2	3.6	1	2.4	–	–	3
Raphitoma reticulata	–	–	2	3.6	–	–	–	–	2
Tricolia pullus	2	7	–	–	–	–	–	–	2
Total	29	100	55	100	41	100	12	100	137

4 DISCUSSION

Diogenes pugilator and *D. nitidimanus* seem to be very closely related species. This allows the comparison of our results with Asakura (1984, 1991, 1992).

The life cycle of *D. pugilator* lasts 1 to 1.5 year (Fig. 1), similar as in *D. nitidimanus* (Asakura 1991). Sub-adults and adults I become more numerous during late autumn and remain abundant until February-March. In winter, the hermit crabs belonging to the mentioned classes that were born in summer, become benthonic adults and enter the adults II class during next spring. The peak in September that is shown by the adults II, corresponds to maturation of specimens born in April, when the first eggs are deposited; these crabs grow up very quickly because of the summer favorable environmental conditions. Some of them overcome their second winter and enter the adults III class in the following summer. Apparently, all of them die off in the following winter.

The maen size of *Diogenes pugilator* males was significantly larger compared to females and represented 84% of the adults III class, although they were only 31.5% of the whole population. Sex ratio fluctuates during the year but not in a statistically significant way.

In contrast to Asakura's (1992) observations, *D. pugilator* shows a single long period during which molting frequency is high but without any significant difference between the two sexes in the number of molts.

The molting rate is strictly correlated with water temperature. Therefore, the highest molt number occurred in summer when the environmental conditions (temperature, hydrodinamics, and food availability) were more favorable for the species. Consequently, *D. pugilator* can grow up very quickly in summer with an increase of molts.

In the laboratory, few hermit crabs molted twice and became sub-adults. However, during the study no significant differences were ascertained in molting rates among the 4 classes of hermit crabs.

According to Perés & Picard (1964) and Altes (1966), *D. pugilator* lives on fine sandy bottoms, which allows the animal to cover itself with sand.

In particular, this research showed differences in the distribution that are connected to size and season. In fact, young *D. pugilator* specimens were numerous on sand among *Cymodocea nodosa* during the winter and spring. In this habitat, besides weak hydrodynamic forces, sub-adults find also small suitable sized gastropod shells. Young specimens, becoming adults during summer, move towards the coast. On the contrary, adults are located in the remaining part of the seabed, most of them in groups.

Fotheringham (1975) noticed *Clibanarius vittatus* moving from the intertidal to the subtidal zone during winter; Asakura (1984) noted a summer migration of *Diogenes nitidimanus* with a reproductive purpose. The present study pointed out that *D. pugilator* moves towards deep waters when sea conditions are not good. The animals prefer to stay in groups in the deepest waters (8 m) during winter (high hydrodynamics) and spread over the bottom during the rest of the year.

Diogenes pugilator mostly occupies shells from its own habitat, similarly to *D. nitidimanus* (Asakura 1984). Only 4% of the shells used by *D. pugilator* originate from rocky bottoms. This can be explained by the combined action of two elements. The first one is biotic, related to the hermit crabs' mobility and to their mutual shell exchange. The second one is abiotic, due to sea currents, shells from either the breakwaters or the nearby coastal rocks are deposited into the sandy area.

The most commonly used shells are *Sphaeronassa mutabilis* and *Hinia* spp., which are most numerous on the sandy bottom. Since these two gastropods attain different sizes, their shells are used by hermit crabs from different size classes. Sub-adults live in *Hinia* spp. and other small gastropod shells, which are common on *Cymodocea nodosa* leaves. This could also explain the high number of young specimens in this habitat. According to Asakura (1991), the distribution of young *D. nitidimanus* specimens is bound to the availability of small sized shells. *S. mutabilis* is surely the most used gastropod shell by all adults. This preference is confirmed also by shell choice experiments performed in the laboratory.

Vance (1972) and Asakura (1984) point out that shells suitable for differently sized hermit crabs are scarce in relation to the demand. In this research no empty shells were found on the seabed and many crabs were seen inside damaged shells.

Sphaeronassa mutabilis and *Hinia* spp. have large inner volume, low number of coils, and are quite light. Although lighter and therefore weaker, they are preferred by *D. pugilator* probably because suitable for its burying habits.

ACKNOWLEDGEMENTS

The authors wish to thank the colleagues and friends Emilio Balletto and Giuseppe Rappini for the precious advises and the critical revision of the text.

REFERENCES

Altes, J. 1966. Etude comparée des variants sexuels abdominaux chez quelques pagures. *Arch. Zool. Exp. Gén.* 106: 187-377.

Asakura, A. 1984. Population ecology of the sand–dwelling hermit crab *Diogenes nitidimanus* Terao. 1. Shell utilization. *Publ. Amakusa Mar. Biol. Lab.* 7(2): 95-108.

Asakura, A. 1991. Population ecology of the sand-dwelling hermit crab *Diogenes nitidimanus* Terao. 4. Larval settlement. *Mar. Ecol. Progr. Ser.* 78: 139-146.

Asakura, A. 1992. Population ecology of the sand-dwelling hermit crab *Diogenes nitidimanus* Terao. 5. Ecological implications in the pattern of moulting. *J. Crust. Biol.* 12: 535-545.

Codreanu, R. & Balcesco, D. 1968. Etude biométrique comparée de certains caractèr dans deux populations du pagure *Diogenes pugilator* Roux de la Mer Noir et de l'Océan Atlantique. *Bull. Biol. France Belgique* 102: 369-383.

Fotheringham, N. 1975. Structure of seasonal migrations of the littoral hermit crab *Clibanarius vittatus* (Bosc.). *J. Exp. Mar. Biol. Ecol.* 18: 47-53.

Ingle, R. 1993. *Hermit crabs of the Northeastern Atlantic Ocean and the Mediterranean Sea.* London: Chapman & Hall.

Pérès, J.M. & Picard, J. 1964. Nouveau manuel de bionomie bentique de la mer Meditérranée. *Rec. Trav. Stat. Mar. d'Endoume* 31(47): 1-137.

Vance, R.R. 1972. The role of shell adequacy in behavioral interactions involving hermit crabs. *Ecology* 53: 1075-1083.

Population dynamics of *Petrochirus diogenes* (Crustacea, Anomura, Diogenidae) in the Ubatuba region, São Paulo, Brazil

GIOVANA BERTINI & ADILSON FRANSOZO

NEBECC, Departamento de Zoologia, Instituto de Biociências, Universidade Estadual Paulista, São Paulo, Brazil

ABSTRACT

This work examines the population dynamics of *Petrochirus diogenes* in the Ubatuba region (São Paulo, Brazil), focusing on size frequency distribution, sex ratio, and reproductive and recruitment period. Collections were made with two double-rig nets in the years 1993-1996. The 799 individuals obtained were separated into 14 size classes based on the length of the cephalothoracic shield. The shield size varied from 5.4 to 40.0 mm in males and from 5.7 to 32.1 mm in females. The size frequency distribution was unimodal for both sexes. Only small oscillations occurred in the sex ratio until the seventh size class, followed by preponderance of males. This suggests a standard pattern for the sex ratio in *P. diogenes*. As males were found in the largest size classes, they present a clear sexual dimorphism. This characteristic can be considered as a selective advantage, mainly during the mating processes and in the agonistic behavior. Ovigerous females were recorded in the spring and summer, indicating seasonal reproduction.

1 INTRODUCTION

Population aspects of pagurid crabs have been poorly studied despite the importance of understanding the biology of these crabs, because their populations can become totally dependent on the availability of gastropods in their habitat.

Studies of crustacean population biology usually emphasize on size frequency distribution, and growth and recruitment rates. The ratio of ovigerous to non-ovigerous females throughout the year is a common tool for the determination of the reproductive period (e.g. Fotheringham 1975, Ameyaw-Akumfi 1975, Asakura & Kikuchi 1984, Lowery & Nelson 1988, Gherardi 1991, Tunberg et al. 1994, Asakura 1995, Manjón-Cabeza & García Raso 1994, 1995, Gherardi & Cassidy 1995).

In Brazil, studies of the relation between population structure and/or shell-size were carried out for *Pagurus criniticornis, Pagurus brevidactylus, Clibanarius antillensis, Clibanarius vittatus* (all Negreiros-Fransozo et al. 1991), *Paguristes tortugae* (Negreiros-Fransozo & Fransozo 1992), *Isocheles sawayai* (Pinheiro et al. 1993), and *Calcinus tibicen* (Fransozo & Mantelatto 1998).

Recently, sexual differences in hermit crabs were observed during studies of the use and rates of shell changes (Abrams 1988), specific shell preference (Blackstone & Joslyn 1984, Elwood & Kennedy 1988), proportion of body size (Blackstone 1984), molt and growth (Blackstone 1985, Asakura 1992, 1995), and intraspecific competitive ability (Bertness 1981).

The hermit crab *P. diogenes* occurs along the western Atlantic coast from the eastern United States, Gulf of Mexico, Antilles, Guyanas, Suriname, Brazil (from Amapa to Rio Grande do Sul) and Uruguay (Rieger 1997).

The objective of the present study was the analysis of population dynamics of *Petrochirus diogenes*, in the Ubatuba region, with emphasis on the size frequency distribution, reproductive period, and sex ratio.

2 MATERIAL & METHODS

From 1993 to 1996 hermit crabs were collected in the Ubatuba region, at the north coast of the São Paulo State (between 23°20'-23°35'S and 44°50'-45°14'W) from a shrimp fishing boat equipped with two double-rig nets.

The animals were separated into four groups: males, juvenile females, adult non-ovigerous females and ovigerous females. The category 'juvenile females' was used for all females <10 mm shield length (= size of the smallest ovigerous females).

For morphometric analysis, the length of the cephalothoracic shield (LS) was measured in millimeters from the tip of the rostrum to the V-shaped groove at the posterior edge. Fifteen size classes were used of 2.5 mm width, in the range 5-42.5 mm.

For the statistical analysis of differences between the average sizes of males and females, a Student t-test was used ($p < 0.01$). Verification of significant differences in the sex ratios throughout the year, and among size classes, was done by studying the frequency of occurrence for every month of collection, and applying Goodman's test (1964, 1965) for contrasts among and within multinomials proportions, complemented by Tukey's test ($p < 0.05$).

3 RESULTS

In total, 799 individuals of *P. diogenes* were obtained, represented by 498 males, 58 juvenile females, 224 adult non-ovigerous females and 19 ovigerous females. The males had a shield length varying from 5.4 to 40.0 mm, with a mean size of 17.7 ± 6.7 mm. Female size varied from 5.7 to 32.1 mm, with an average of 14.7 ± 4.9 mm. The males were significantly larger than the females ($p < 0.01$).

The size frequency distribution of males, juveniles females, adult non-ovigerous females, and ovigerous females is shown in Figure 1. An increase in the number of females can be observed in the intervals of 7.5 to 22.5 mm, while males extended to the largest size classes. Considering the total number of collected individuals a small percentage reached sizes < 25 mm, consisting of 0.6% of the females and 9.3% of the males.

The monthly size frequency distribution of *P. diogenes* is shown in Figure 2. A recruitment in the population can be observed during almost all months of the year,

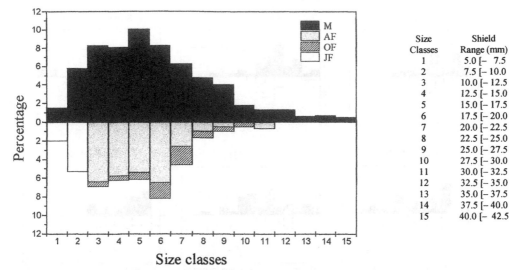

Size Classes	Shield Range (mm)
1	5.0 [– 7.5
2	7.5 [– 10.0
3	10.0 [– 12.5
4	12.5 [– 15.0
5	15.0 [– 17.5
6	17.5 [– 20.0
7	20.0 [– 22.5
8	22.5 [– 25.0
9	25.0 [– 27.5
10	27.5 [– 30.0
11	30.0 [– 32.5
12	32.5 [– 35.0
13	35.0 [– 37.5
14	37.5 [– 40.0
15	40.0 [– 42.5

Figure 1. *Petrochirus diogenes*. Overall size-frequency distribution for the total sample of hermit crabs collected during this study. (M – males; AF – adult non-ovigerous females; OF – ovigerous females; JF – juvenile females).

except for February and September, and the presence of ovigerous females from January to April and September to December. The differences in the mean size of the population among months are shown in Figure 3.

The proportion of males and females differed significantly in February, April, May, June, July, and December, when the percentage of males was higher, as shown in Table 1.

The sex ratio did not differ in the size classes 1, 2, 3, 4, and 6. In the remaining size classes the sex ratio was in favor of the males, and the last four size classes consisted of males only (Fig. 4).

The proportion of the ovigerous of potentially mature females is shown in Figure 5. The ovigerous females were most frequent from February to April and were absent from May to August.

4 DISCUSSION

In *P. diogenes*, the shield length was significantly greater in males than in females, suggesting a sexual dimorphism. Sexual dimorphism was also observed in other species of hermit crabs (Negreiros-Fransozo et al. 1991, Negreiros-Fransozo & Fransozo 1992, Tunberg et al. 1994, Asakura 1995, Fransozo & Mantelatto 1998).

In several hermit crab species, the male holds the shell of the female for a long period, until she is ready to copulate; in general this requires that the males have larger dimensions (Hazlett 1966).

Abrams (1988) mentions two benefits that males derive by being larger than females: 1) the larger reproductive effort belongs to the males, due to the ability to fertilize more than one female; 2) males with larger dimensions have greater chance of obtaining females for copulation because of intraspecific competitive fights.

Percentage

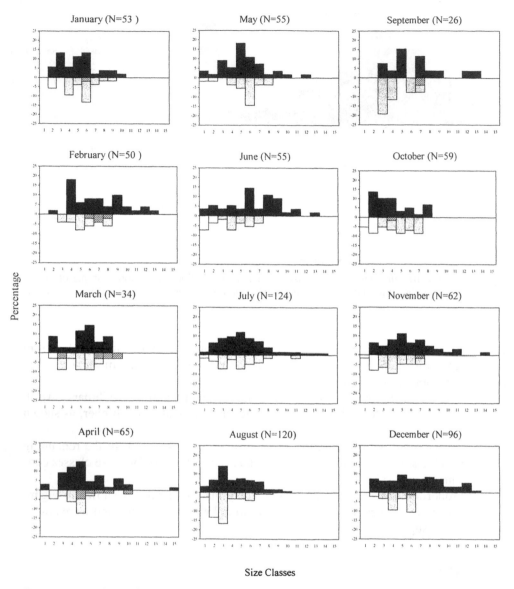

Size Classes

Figure 2. *Petrochirus diogenes*. Histograms showing frequency of individuals by months, during the study period (from 1993 to 1996). (N – number of individuals; black bars – males; white bars – juvenile females; gray bars – adult non-ovigerous females; cross-hatched bars – ovigerous females).

Williams (1984), in his survey of crustacean decapod species of the eastern coast of the United States, mentions that the maximum shield size reached by individuals of *P. diogenes* was 36 mm in males and 20 mm in females. Our animals collected in the Ubatuba region were larger in size, with males reaching 40 mm and females 32.1 mm. This difference was probably related to the small number of individuals analyzed by Williams (1984), compared to the present study.

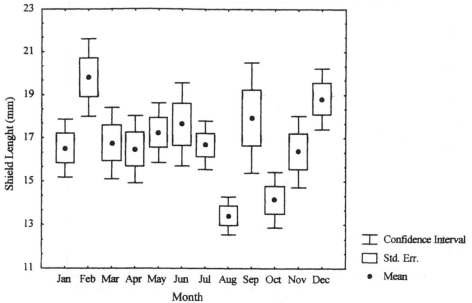

Figure 3. *Petrochirus diogenes*. Mean shield size of individuals of the population during the sampled period.

Table 1. *Petrochirus diogenes*. Total number of males and females, proportions, and sex ratios, during the months of study. Multinomial proportions analysis complemented by the Tukey test (p < 0.05). Lower case letters indicate comparison in a same row, and capital letters indicate comparison in a same line.

Month	Males	Total Females	Males	Proportions Females	Sex ratio (M:F)
January	32	21	60 aA	40 aA	1:0.67
February	34	16	68 aB	32 aA	1:0.47
March	19	15	56 aA	44 aA	1:0.79
April	42	23	65 aB	35 aA	1:0.54
May	36	19	66 aB	34 aA	1:0.52
June	37	18	67 aB	33 aA	1:0.49
July	82	42	66 aB	34 aA	1:0.52
August	66	54	55 aA	45 aA	1:0.82
September	14	12	54 aA	46 aA	1:0.85
October	30	29	51 aA	49 aA	1:0.96
November	37	25	60 aA	40 aA	1:0.67
December	69	27	72 aB	28 aA	1:0.39

The size frequency distribution of both sexes of *P. diogenes* was unimodal. A stable population structure is represented by a unimodal pattern of distribution, with a continuous recruitment and constant mortality rates, what is very common in populations of tropical decapods (Warner 1967, Ahmed & Mustaquim 1974, Hartnoll 1982). The unimodality of size frequency distributions usually characterizes a dynamic equilibrium for a certain population, and the occurrence of slight monthly variations

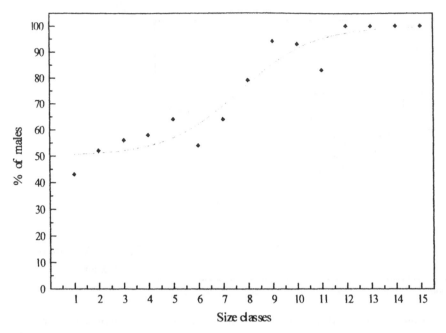

Figure 4. *Petrochirus diogenes*. Percentage of males in each size class.

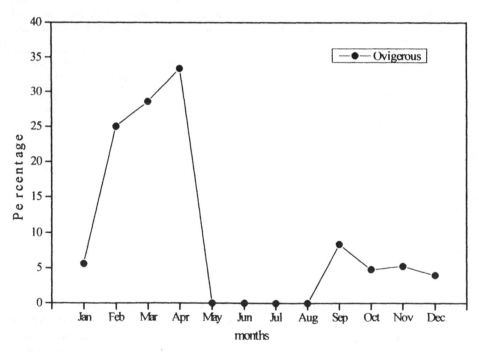

Figure 5. *Petrochirus diogenes*. Percentage of ovigerous females in the population during the study period.

could reflect recruitment pulses, growth and differential mortality rates (Díaz 1980, Díaz & Conde 1989).

The presence of ovigerous females in spring and summer revealed a reproduction peak for *P. diogenes*. These seasons were favorable for larval development mainly due to food abundance and higher temperatures. As a consequence, the recruitment of juveniles occurred with larger incidence in the fall and winter.

Notably, the separation of juveniles and adults for the females was based on the presence of one ovigerous female (LS 10 mm) in the third size class. This criterion was also adopted for three *Pagurus* species by Carlon & Ebersole (1995) and *Diogenes nitidimanus* by Asakura (1995). According to Pinheiro & Fransozo (1998) the accurate size at sexual maturity estimates should include allometric studies of secondary sexual characteres and micro- or macroscopic gonad examination since these values are often different.

According to Sastry (1983), there is much inter- and intraspecific variation in the patterns of spawning, depending on the geographic area. Generally, populations in temperate waters, spawn in a short period of time, compared to those of sub-tropical and tropical areas. In the tropics, the great majority of the species spawn for a large period of time, but peaks may occur during this period (Goodbody 1965, Ahmed & Mustaquim 1974).

P. diogenes displayed a seasonal reproduction in the study area, because ovigerous females were found mainly during two seasons (spring and summer). In a study of pagurids from the Virgin Islands, Provenzano (1961) mentioned the presence of ovigerous females of *P. diogenes* in March, and in west Florida Provenzano (1968) collected an individual in August for larval development. Those observations suggest that the reproductive period of this species, even on different latitudes, is restricted to the warmer months of the year.

Other species of the Ubatuba region, such as *Paguristes tortugae*, studied by Negreiros-Fransozo & Fransozo (1992), showed a continuous reproduction throughout the year, however, with peaks from May to November. In contrast, Fransozo & Mantelatto (1998) observed that the reproductive cycle of *Calcinus tibicen* showed a strong reproductive peak between September and May with an abrupt absence of ovigerous females during the winter months.

Seasonal variations in the reproductive activity occur in many hermit crab species. In tropical waters, a continuous reproductive period occurs with peaks in certain months of the year. Such observation were made for *Calcinus latens*, *Calcinus laevimanus* and *Clibanarius zebra* by Reese (1968), and for *Clibanarius chapini* and *Clibanarius senegalensis* by Ameyaw-Akumfi (1975). However, the reproductive cycle of hermit crab species can vary from place to place and among species, suggesting that several factors can influence their reproduction (Table 2).

According to Wenner (1972), in most mature marine crustaceans the 1:1 sex ratio does not exist. Possible explanations for this could be differential growth and mortality rates among sexes, restrictions in nutrition, behavioral differences, and reproductive migration as well as use of different habitats by each sex (Darnell 1962, Haley 1979, Díaz 1980). The sex ratio in *P. diogenes* differed significantly from the expected 1:1, and was in favor of the males. This can be explained by the fact that they are more active than females, or they use different habitats. The observed sex ratio was perhaps a real characteristic of the population, as observed for the crab *Macro-*

Table 2. Summary of reproductive patterns and peaks in the Paguridea, based on published studies.

Genus	Species	Reproductive pattern	Reproductive peak	Locality	Reference
Calcinus	*laevimanus*	Continuous	May-Aug	Kaneoho, Hawaii (21°N)	Reese (1968)
	latens	Continuous	Feb-Mar, Aug	Kaneoho, Hawaii (21°N)	Reese (1968)
	tibicen	Seasonal	Sep-May	Ubatuba, Brazil (23°S)	Fransozo & Mantelatto (1998)
	tubularis	Seasonal	Aug	Almería, Spain (37°N)	Manjón-Cabeza & García Raso (1995)
Clibanarius	*zebra*	Continuous	Feb-Apr, Aug	Hawaii, Kaneoho (21°N)	Reese (1968)
	chapini	Continuous	Jan-Oct	Ghana, Tengpobo (6°N)	Ameyaw-Akumfi (1975)
	senegalensis	Continuous	Sep – Apr	Ghana, Tengpobo (6°N)	Ameyaw-Akumfi (1975)
	vittatus	Seasonal	May-Aug	Texas, USA (29°N)	Fotheringham (1975)
	vittatus	Seasonal	Apr-Sep	Florida, USA (27°N)	Lowery & Nelson (1988)
	tricolor	Seasonal	Apr-Jun	West Indies, Barbados (13°N)	Lewis (1960)
Diogenes	*nitidimanus*	Seasonal	Apr-Nov	Amakusa, Japan (32°N)	Asakura & Kikuchi (1984)
Pagurus	*geminus*	Seasonal	Jan-Apr	Bay Sagami, Japan (35°N)	Kurata (1968)
	brevidactylus	Continuous	Apr-May	São Sebastião, Brazil (23°S)	Negreiros-Fransozo (1984)
	maclaughlinae	Continuous	Sep-Oct, Feb-Jun	Florida, USA (27°N)	Tunberg et al. (1994)
	bernhardus	Seasonal	Dec-May	South West, England (52°N)	Lancaster (1988)
Discorso-pagurus	*schmitti*	Seasonal	Jan-Apr	USA, Washington (47°N)	Gherardi & Cassidy (1995)
Cestopagurus	*timidus*	Seasonal	Mar-Oct	Almería, Spain (37°N)	Manjón-Cabeza & García Raso (1994)
Paguristes	*tortugae*	Continuous	May-Nov	Ubatuba, Brazil (23°S)	Negreiros-Fransozo & Fransozo (1992)
	erythrops	Seasonal	Feb-Apr	São Sebastião, Brazil (23°S)	Fransozo (1980)
Isochelis	*sawayaia*	Seasonal	Dec-Apr	São Sebastião, Brazil (23°S)	Negreiros-Fransozo (1980)
Petrochirus	*diogenes*	Seasonal	Feb-Apr	Ubatuba, Brazil (23°S)	Present study

pipus tuberculatus by Mori (1987). Díaz & Conde (1989) proposed that this condition could be valid for species with different growth rates or different life expectations for each sex, producing larger or older individuals for one sex. This may also apply for *P. diogenes*.

The sex ratio in the size classes for *P. diogenes* oscillated for both sexes until the seventh class (from 20 to 22.5). After this size, there was a predominance of males, suggesting the standard pattern proposed by Wenner (1972).

Studies done with other species of hermit crab revealed that there is no general rule for the sex ratio, with different patterns in species within a same genus (Table 3). An explanation for the occurrence of such deviations in certain size classes can be explained by segregation of one of the sexes, as observed by Gherardi & Cassidy (1995).

Some species show differences among sexes depending on habitat, e.g. males of *Pagurus hirsutiusculus* occupy tide pools located in the upper parts of the rocky shore, but females are dominant in certain microhabitats, without however, staying

Table 3. Classification of hermit crabs, based on sex ratio patterns proposed by Wenner (1972).

Genus	Species	References	Patterns			
			ST	RE	IN	AN
Coenobita	*compressus*	Wenner (1972)				x
Calcinus	*laevimanus*	Wenner (1972)				x
	latens	Gherardi & MacLaughlin (1994)				x
	tibicen	Fransozo & Mantelatto (1998)				x
Clibanarius	*antillensis*	Negreiros-Fransozo et al. (1991)				x
	digueti	Harvey (1988)				x
	erythropus	Gherardi (1991)				x
	laevimanus	Gherardi et al. (1994)				x
	humilis	Gherardi & MacLaughlin (1994)				x
	vittatus	Fotheringham (1975) Lowery & Nelson (1988)				x
	zebra	Wenner (1972)		x		
Diogenes	*nitidimanus*	Asakura (1995)				x
Discorsopagurus	*schmitti*	Gherardi & Cassidy (1995)	x			
Elassochirus	*tenuimanus*	Abrams (1988)				x
Paguristes	*turgidus*	Abrams (1988)				x
Pagurus	*aleuticus*	Abrams (1988)		x		
	brevidactylus	Negreiros-Fransozo et al. (1991)		x		
	beringanus	Abrams (1988)				x
	criniticornis	Negreiros-Fransozo et al. (1991)			x	
	dalli	Abrams (1988)				x
	granosimanus	Abrams (1988)				x
	hirsutiusculus	Abrams (1988)				x
	kennerlyi	Abrams (1988)				x
	samuelis	Abrams (1988)				x
	ochotensis	Abrams (1988)		x		
Dardanus	*insignis*	Fernandes-Góes (1997)	x			
Petrochirus	*diogenes*	Present study	x			

(ST- standard; RE – reverse; IN – intermediary; AN – anomalous).

out of water during the period of low tide (Abrams 1988). *Discorsopagurus schmitti*, which exclusively inhabits tubes of sabellariid polychaetes starting in the megalopa phase, does not present significant differences in sex ratio or size (Gherardi & Cassidy 1995).

Based on the present study, we can infer that the population of *P. diogenes* was stable. Nevertheless, many characteristics still need to be determined, e.g. birth and mortality rates, adults and young migration, and also larval ecology. The study of these aspects could provide a better understanding of the population dynamic of this species.

ACKNOWLEDGEMENTS

We are grateful to the Fundação de Amparo à Pesquisa do Estado de São Paulo (FAPESP), the Conselho Nacional de Desenvolvimento Científico e Tecnológico (CNPq) and the Fundação para o Desenvolvimento da UNESP (FUNDUNESP). We are also thankful to the NEBECC coworkers for their help during collection, and Dra. Maria Lucia Negreiros Fransozo who made helpful comments on earlier drafts of the manuscript.

REFERENCES

Abrams, P.A. 1988. Sexual difference in resource use in hermit crabs; consequences and causes. In G. Chelazzi and M. Vannini (eds), *Behavioural adaptations to the intertidal life*: 283-296. New York: Plenum Press.

Ahmed, M. & Mustaquim J. 1974. Population structure of four species of Porcellanid crabs (Decapoda: Anomura) occurring on the coast of Karachi. *Mar. Biol.* 26: 173-182.

Ameyaw-Akumfi, C. 1975. The breeding biology of two sympatric species of tropical intertidal hermit crabs, *Clibanarius chapini* and *C. senegalensis. Mar. Biol.* 29: 15-28.

Asakura, A. 1992. Population ecology of the sand-dwelling hermit crab, *Diogenes nitidimanus* Terao. 5. Ecological implication on the pattern of molting growth. *J. Crustacean Biol.* 12: 537-545.

Asakura, A. 1995. Sexual differences in life history and resource utilization by the hermit crab. *Ecology* 76: 2295-2313.

Asakura, A. & Kikuchi, T. 1984. Population ecology of the sand dwelling hermit crab, *Diogenes nitidimanus* Terao 2. Migration and life history. *Publ. Amakusa Mar. Biol. Lab. Kyushu Univ.* 7: 109-123.

Bertness, M.D. 1981. Interference, exploitation, and sexual components of competition in a tropical hermit crab assemblage. *J. Exp. Mar. Biol. Ecol.* 49: 189-202.

Blackstone, N.W. 1984. The effects of history on the shell preference of the hermit crab *Pagurus longicarpus* (Say). *J. Exp. Mar. Biol. Ecol.* 81: 225-234.

Blackstone, N.W. 1985. The effects of shell size and shape on growth and form in the hermit crab *Pagurus longicarpus. Biol. Bull. Mar. Biol. Lab.* 168: 75-90.

Blackstone, N.W. & Joslyn, A.R. 1984. Utilization and preference for the introduced gastropod *Littorina littorea* (L.) by the hermit crab *Pagurus longicarpus* (Say) at Guilford, Connecticut. *J. Exp. Mar. Biol. Ecol.* 80: 1-9.

Carlon, D.B. & Ebersole, J.P. 1995. Life-history variation among three temperate hermit crabs: The importance of size in reproductive strategies. *Biol. Bull.* 188: 329-337.

Darnell, R.M. 1962. Sex ratios: Aquatic animals. In P.L. Altman & D.S. Dittmer (eds), *Growth*: 439-442. Washington, D.C.: Federation of American Societies for Experimental Biology.

Díaz, H. 1980. The mole crab *Emerita talpoida* (Say): a case of changing life history pattern. *Ecol. Monogr.* 50: 437-456.

Díaz, H. & Conde, J.E. 1989. Population Dynamics and Life History of the Mangrove Crab *Aratus pisonii* (Brachyura, Grapsidae) in a Marine Environment. *Bull. Mar. Sci.* 45: 148-163.

Elwood, R.W. & Kennedy, H. 1988. Sex differences in shell preference of the hermit crab *Pagurus bernhardus* L. *I. Nat. J.* 22: 436-440.

Fernandes-Góes, L.C. 1997. Distribuição e Biologia Populacional de *Dardanus insignis* (Saussure, 1858) (Crustacea:Decapoda:Anomura) na região de Ubatuba, SP. Master Science thesis. Universidade Estadual Paulista (UNESP) 'Campus' de Botucatu, Brazil: 150 pp. [unpublished].

Fotheringham, N. 1975. Structure of seasonal migrations of the littoral hermit crab *Clibanarius vittatus* (Bosc). *J. Exp. Mar. Biol. Ecol.* 18: 47-53.

Fransozo, A. 1980. Desenvolvimento pós-embrionário de *Paguristes erythrops* Holthuis, 1959 (Decapoda, Diogenidae) e *Mithrax hispidus* (Herbst, 1790) (Decapoda, Majidae). Master Science thesis. Universidade Estadual Paulista (UNESP) 'Campus' de Rio Claro, Brazil: 89 pp. [unpublished]

Fransozo, A. & Mantelatto, F.L.M. 1998. Population structure and reproductive period of the tropical hermit crab *Calcinus tibicen* (Decapoda, Diogenidae) in the Ubatuba region, São Paulo, Brazil. *J. Crustacean. Biol.* 18: 446-452.

Giesel, J.T. 1972. Sex ratio, rate of evolution and environmental heterogeneity. *Am. Nat.* 106: 380-387.

Goodbody, H. 1965. Continuous Breeding in populations of two tropical crustaceans, *Mysidium columbiae* (Zimmer) and *Emerita portoricensis* Schmidt. *Ecology* 46: 195-197.

Goodman, L.A. 1964. Simultaneous confidence intervals for contrasts among multinomiae populations. *Annals of Mathematical Statistics* 35(2): 716-725.

Goodman, L.A. 1965. On simultaneous confidence intervals for multinomiae proporcions. *Technometric* 7(2): 247-254.

Gherardi, F. 1991. Relative growth, population structure, and shell-utilization of the hermit crab *Clibanarius erythropus* in the Mediterranean. *Oebalia* 17: 181-196.

Gherardi, F. & MacLaughlin, P.A. 1994. Shallow-water hermit crabs from Mauritius and Rodrigues Islands, with the description of a new species of *Calcinus*. *Ruffles Bull. Zool.* 42: 613-636.

Gherardi, F. & Cassidy, P.M. 1995. Life history patterns of *Discorsopagurus schmitti*, a hermit crab inhabiting Polychaete tubes. *Biol. Bull.* 188: 68-77.

Gherardi, F., Zatteri, F. & Vannini, M. 1994. Hermit crabs in a mangrove swamp: the structure of *Clibanarius laevimanus* clusters. *Mar. Biol.* 121: 41-52.

Haley, S.R. 1979. Sex ratio as a function of size in *Hippa pacifica* Dana (Crustacea, Anomura, Hippidae): a test of the sex reversal and differential growth rate hypothesis. *Am. Nat.* 113: 391-397.

Harvey, A.W. 1988. Size and sex-related aspects of ecology of the hermit crab *Clibanarius digueti* Bouvier (Decapoda: Anomura: Diogenidae). Ph. D. dissertation University of Arizona, Tucson, AZ. [unpublished].

Hartnoll, R.G. 1982. Growth. In D.E. Bliss (ed.), *The biology of Crustacea, Embriology, Morfology and Genetics*: 11-196. New York: Academic Press.

Hazlett, B.A. 1966. Social behavior of the Paguridae and Diogenidae of Curação. *Stud. Fauna Curação Other Caribb. Isl.* 23(88): 1-143.

Kurata, H. 1968. Larvae of Decapod Anomura of Arasaki, Sagami Bay, 1. *Pagurus samuelis* (Stimpson), (Paguridae). *Bull. Tokai Reg. Fish. Res. Lab.* 55: 265-269.

Lancaster, I. 1988. *Pagurus bernhardus* (L.) – An introduction to the natural history of hermit crabs. *Field Studies* 7: 189-238.

Lewis, J.B. 1960. The fauna of rocky shores of Barbados, West Indies. *Can. J. Zool.* 38: 391-435.

Lowery, W.A. & Nelson, W.G. 1988. Population ecology of the hermit crab *Clibanarius vittatus* (Decapoda: Diogenidae) at Sebastian Inlet, Florida. *J. Crustacean Biol.* 8: 548-556.

Manjón-Cabeza, M.E. & García Raso, J.E. 1994. Estructura de una población del cangrejo ermitaño *Cestopagurus timidus* (Crustacea, Decapoda, Anomura) de fondos de *Posidonia oceanica* del SE de España. *Cah. Biol. Mar.* 35: 225-236.

Manjón-Cabeza, M.E. & García Raso, J.E. 1995. Study of a population of *Calcinus tubularis* (Crustacea, Diogenidae) from a shallow *Posidonia oceanica* meadow. *Cah. Biol. Mar.* 36: 277-284.

Mori, M. 1987. Observations on reproductive biology, and diet of *Macropipus tuberculatus* (Roux) of the Ligurian Sea. *Invest. Pesq.* 51: 147-152.

Negreiros-Fransozo, M.L. 1980. Desenvolvimento pós-embrionário de *Paguristes tortugae* Schmitt, 1933 e *Isocheles sawayai* Forest & Saint Laurent, 1967 (Crustacea, Paguridae). Master Science thesis. Universidade Estadual Paulista (UNESP) 'Campus' de Rio Claro, Brazil: 73 pp. [unpublished]

Negreiros-Fransozo, M.L. 1984. Desenvolvimento pós-embrionário de *Pagurus brevidactylus* (Stimpson, 1858), *Panopeus americanus* Saussure, 1857 e *Eurypanopeus abbreviatus* (Stimpson, 1860) (Crustacea, Decapoda), em laboratório. Doctoral thesis. Universidade Estadual Paulista (UNESP) 'Campus' de Rio Claro, Brazil: 133 pp. [unpublished]

Negreiros-Fransozo, M.L. & Fransozo, A. 1992. Estrutura populacional e relação com a concha em *Paguristes tortugae* Schmitt, 1933 (Decapoda, Diogenidae), no litoral norte do estado de São Paulo, Brasil. *Naturalia* 17: 31-42.

Negreiros-Fransozo, M.L., Fransozo, A. & Hebling, N.J. 1991. Estrutura populacional e determinação do tamanho da concha em 4 espécies de ermitões (Crustacea, Decapoda, Anomura) do litoral paulista. *Biotemas* 4(2): 135-148.

Pinheiro, M.A.A. & Fransozo, A. 1998. Sexual maturity of the speckled swimming crab *Arenaeus cribrarius* (Lamarck, 1818) (Decapoda, Brachyura, Portunidae), in the Ubatuba littoral, São Paulo State, Brazil. *Crustaceana* 71: 434-452.

Pinheiro, M.A.A., Fransozo, A. & Negreiros-Fransozo, M.L. 1993. Seleção e Relação com a Concha em *Isocheles sawayai* Forest & Saint Laurent, 1967 (Crustacea, Anomura, Diogenidae). *Arq. Biol. Tecnol.* 36(4): 745-752.

Provenzano, A. J. 1961. Pagurid crabs (Decapoda Anomura) from St. John, Virgin Islands, with descriptions of three new species. *Crustaceana* 3: 151-166.

Provenzano, A.J. 1968. The complete larval development of the West Indian hermit crab *Petrochirus diogenes* (L.) (Decapoda, Diogenidae) reared in the laboratory. *Bull. Mar. Sci.* 18: 143-181.

Reese, E.S. 1968. Annual breeding seasons of three sympatric species of tropical intertidal hermit crabs, with a discussion of factors controlling breeding. *J. Exp. Mar. Biol. Ecol.* 2: 308-318.

Rieger, P.J. 1997. Os 'ermitões' (Crustacea, Decapoda, Parapaguridae, Diogenidae e Paguridae) do litoral do Brasil. *Nauplius* 5(2): 99-124.

Sastry. A.N. 1983. Ecological aspects of reproduction. In T.H. Waterman (ed.), *Biology of Crustacea. VIII Environmental adaptations*: 79-270. Academic Press.

Tunberg, B.G., Nelson, W.G. & Smith, G. 1994. Population ecology of *Pagurus maclaughlinae* García-Gómez (Decapoda: Anomura: Paguridae) in the Indian River Lagoon, Florida. *J. Crustacean Biol.* 14: 686-699.

Warner, G.F. 1967. The life history of the mangrove tree crab *Aratus pisonii*. *J. Zool.* 153: 321-335.

Wenner, A.M. 1972. Sex ratio as a function of size in marine Crustacea. *Am. Nat.* 106: 321-350.

Williams, A.B. 1984. Shrimps, lobsters, and crabs of the Atlantic coast of the eastern United States, Maine to Florida. Washington: Smithsonian Institution Press, 550 pp.

Seasonal changes (1995-1997) in the population structure of Chinese mitten crabs, *Eriocheir sinensis* (Decapoda, Brachyura, Grapsidae) in the Thames at Chelsea, London

R.S. ROBBINS, B.D. SMITH, P.S. RAINBOW & P.F. CLARK
Department of Zoology, The Natural History Museum, London, UK

ABSTRACT

Carapace widths have been measured in juvenile, male and female mitten crabs collected over a two year period (October 1995 to September 1997) from the water intake filtering screens of a power station on the Thames at Chelsea, London. Most juvenile crabs (86% of 358 collected in two years) were taken between February and September. Most male crabs (57% of 1910 collected) and most females (65% of 1385 collected) were collected between September and December. Thus juveniles were abundant at a clearly different time of the year from the males and females, indicating that different size groups of the crab population were migrating through this site at different times. Male and female crabs of the smallest size range (mode 25-30 mm carapace width) appeared at Chelsea in March and April respectively. The peak size of crabs of each sex collected increased to a maximum (40-45 mm carapace width) by the following autumn, this maximum size being maintained until the next spring. It is concluded that the large male and female crabs (mode carapace width ca 40 mm or greater) were mature crabs migrating downstream. Juvenile and small male and female crabs appearing at Chelsea in the spring were probably migrating upstream into freshwater.

1 INTRODUCTION

The Chinese mitten crab, *Eriocheir sinensis*, is an alien invader of the rivers and estuaries of Western Europe. The crab originates from China and Korea (Guo et al. 1997), and was introduced, probably in ballast water, into Germany in 1912 (Panning 1939), whence it has spread throughout northern Europe from the Baltic to the Mediterranean. It is now found from Finland (Haahtela 1963) in the north through Sweden, Russia, Poland, Germany, the Czech Republic, the Netherlands, Belgium, England (Clark et al. 1998) to France (Petit 1960, Vigneux et al. 1993).

Mitten crabs spend most of their adult life in freshwater, males and females migrating downstream to breed in estuaries (Panning 1939, Anger 1991). Larvae develop in the lower estuary and juvenile crabs gradually move upstream to complete the life cycle (Panning 1939, Anger 1991).

The Chinese mitten crab was first recorded in England by Harold (1935) when a single specimen was collected from the water intake filtering screens at Lots Road power station, at Chelsea on the Thames in London. Sightings in England were rare until the 1970s when the crab appeared to be re-established in the Thames, apparently at a low population level (Ingle & Andrews 1976, Andrews et al. 1982, Clark et al. 1998). In the 1990s, however, population levels of *E. sinensis* in the Thames have escalated (Clark et al. 1998). This population explosion has potential environmental implications, for the crabs may outcompete native crayfish, threaten other members of the British freshwater fauna and flora through their voracious omnivory, and the crabs' burrows may damage riverine earthworks (Clark et al. 1998).

This paper describes seasonal changes in the population structure of mitten crabs collected over a two year period (October 1995 to September 1997) from the filtering screens at Lots Road power station in Chelsea, London, the very screens from which the first mitten crab recorded in Britain was taken (Harold 1935). This study is part of an investigation of the population ecology (particularly the patterns of migration) of the crab in the Thames catchment, necessary to the formulation of any management plan to restrict the potential environmental damage that might be caused by this rapidly expanding population of crabs.

2 MATERIALS AND METHODS

Chinese mitten crabs were collected from the filter screens and surrounding area (the crabs will walk and hide in damp drains or piles of leaves, etc.) at Lots Road power station, Chelsea, London at irregular intervals between October 1995 and September 1997. The numbers of crabs captured on the screens varied greatly throughout the year, peaking in October, and so, population structure data were typically expressed in terms of percentages of those collected in given months. Quoted numbers of crabs collected do give an indication of the numbers available for collection, but were not strictly comparable in terms of crabs collected per unit sampling effort, affected for example by rate of flow of water filtered, time period of collection from the screens, etc. (cf. Clark et al. 1998). For this study it was essential to collect as many crabs as possible per visit, in order to increase confidence in the analysis of the population structure of the crabs present at that location in a particular month of the year. Thus living crabs were collected from drains and rubbish skips holding all debris removed from the screens, in the knowledge that such crabs had been filtered from the intake water within only a few days at most of the collection visit.

The carapace width of each crab was measured to the nearest millimeter with callipers, and the gender of the crab recorded. Those crabs which could not be recognized as male or female were recorded as juveniles. As the study progressed, it became easier with experience to allocate immature crabs to gender. Thus earlier data sets contain crabs labelled as juveniles, which would have been labelled as male or female later in the study.

3 RESULTS

The percentage division of each month's collection of crabs into males, females and juveniles is shown in Figure 1. The juveniles only made up a significant proportion of the crabs present in the months February to September, mature males and females dominating to the almost total exclusion of juveniles from September to February. Most juvenile crabs (86% of 358 collected in two years) were taken between February and September. Most male crabs (57% of 1910 collected) and most females (65% of 1385 collected) were collected between September and December. Thus juveniles were abundant at a clearly different time of the year from the mature males and females, indicating that the different size groups of the crab population were present at this site at different times.

The carapace width frequency distributions of male, female and juvenile crabs during each month in 1997 are shown in Figures 2, 3 and 4. Similar, but less complete data sets, have been compiled but are not presented for 1996 and October to December 1995. Large sexually mature male and female crabs, with mode carapace widths between 40 and 50 mm, dominated catches between October and January (Figs 2 & 3). At this time, juvenile crabs (Fig. 4) or indeed small crabs (up to 35 mm carapace width) identifiable as males (Fig. 2) or females (Fig. 3), were all but absent. Juvenile crabs started appearing in any numbers at all in March 1997 (Fig. 4) together with small crabs just recognizable as males (Fig. 2). Similarly small females appeared the next month in April 1997 (Fig. 3). Hardly any juveniles were collected in the second half of 1997 (Fig. 4), although in 1996 juvenile crabs did continue in catches in low numbers through the autumn (data not presented).

A final point to be made is that, after the appearance of small males and females in the spring, the frequency distributions indicated a size increase of the mode carapace width in each gender through the year (Figs 2 & 3).

4 DISCUSSION

There is no detailed information in the available literature on the population ecology and migration of *Eriocheir sinensis* in its native habitat in the Far East. The seasonal changes in abundances and population structure of the Chinese mitten crab at Chelsea can, therefore, be interpreted in terms of the migration pattern described by Panning (1939) for *E. sinensis* in the rivers of continental Europe. Thus the arrival of large numbers of sexually mature male and female crabs at Chelsea in September and October would represent crabs migrating downstream, their passage through Chelsea tailing off through the winter. The juveniles and the smallest identifiable males and females, particularly obvious in the spring and summer but potentially present to some degree throughout the year, would be crabs ascending the estuary to the freshwater system of the Thames.

We are taking these investigations further, sampling quantitatively through the year at different points along the estuary and comparing catches at different states of the tide in order to distinguish upstream and downstream migrations.

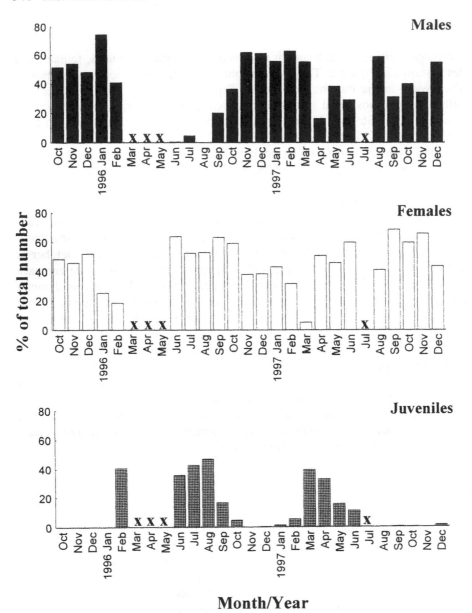

Figure 1. *Eriocheir sinensis*: the percentage division of each month's collection of crabs into males, females and juveniles from October 1995 to December 1997 at Chelsea, London. No collections were made in the months indicated by a cross.

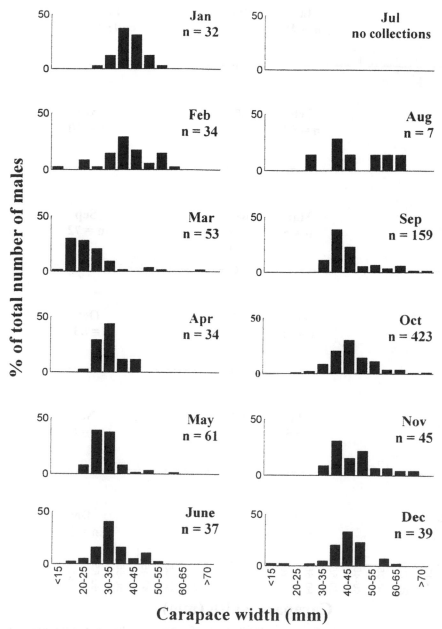

Carapace width (mm)

Figure 2. *Eriocheir sinensis*: the frequency distributions of carapace widths of male crabs collected each month in 1997 at Chelsea, London.

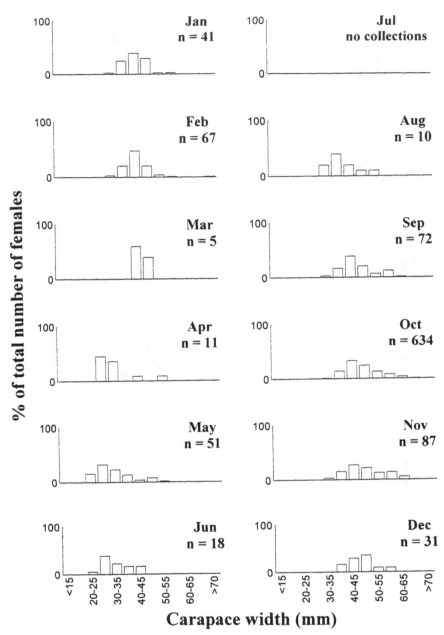

Figure 3. *Eriocheir sinensis*: the frequency distributions of carapace widths of female crabs collected each month in 1997 at Chelsea, London.

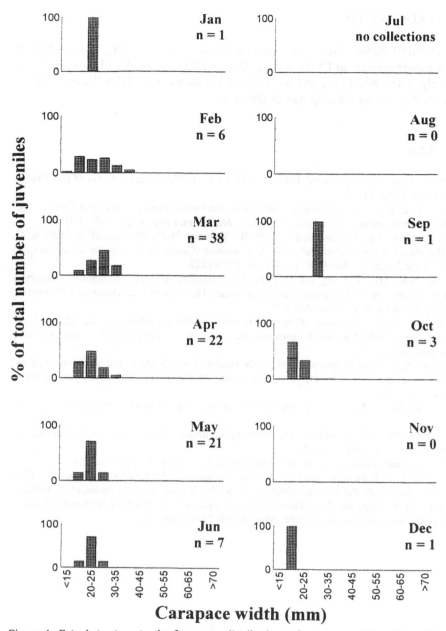

Carapace width (mm)

Figure 4. *Eriocheir sinensis*: the frequency distributions of carapace widths of juvenile crabs collected each month in 1997 at Chelsea, London.

ACKNOWLEDGEMENTS

We are grateful to Richard Bettany of London Transport for allowing us to collect at Lots Road power station, to Claire Byrne (Queen Mary & Westfield College, London) for help with collections, and to William Yeomans and Myles Thomas of the Environment Agency for funding part of this work.

REFERENCES

Andrews, M.J., Aston, K.F.A., Rickard, D.G. & Steel, J.E.C. 1982. The macrofauna of the Thames Estuary. *London Nat.* 61: 30-61.

Anger, K. 1991. Effects of temperature and salinity on the larval development of the Chinese mitten crab *Eriocheir sinensis* (Decapoda: Grapsidae). *Mar. Ecol. Prog. Ser.* 72: 103-110.

Clark, P.F., Rainbow, P.S., Robbins, R.S., Smith, B., Yeomans W.E., Thomas, M. & Dobson, G. 1998. The alien Chinese mitten crab, *Eriocheir sinensis* (Crustacea: Decapoda: Brachyura), in the Thames catchment. *J. Mar. Biol. Ass. UK* 78: 1215-1222.

Guo, J.Y., Ng, N.K., Dai, A. & Ng, P.K.L. 1997. The taxonomy of three commercially important species of mitten crabs of the genus *Eriocheir* de Haan, 1835 (Crustacea: Decapoda: Brachyura: Grapsidae). *Raffles Bull. Zool.* 45: 445-476

Haahtela, I. 1963. Some new observations and remarks on the occurrence of the Mitten Crab, *Eriocheir sinensis* Milne Edwards (Crustacea, Decapoda), in Finland. *Aquilo Societas Amicorum Naturae Oulensis* 1: 9-16.

Harold, C.H.H. 1935. Thirtieth annual report on the results of the chemical and bacteriological examination of London waters for the 12 months ending 31 December 1935. *Metropolitan Water Board London* 1-101.

Ingle, R.W. & Andrews, M.J. 1976. Chinese mitten crab reappears in Britain. *Nature*, London 263: 638.

Panning, A. 1939. The Chinese mitten crab. *Report of the Board of Regents of the Smithsonian Institution (Washington)* 3508: 361-375.

Petit, G. 1960. Le crabe chinois est parvenu en Méditerranée. *Vie et Milieu* 11: 133-136.

Vigneux, E., Keith, P. & Noël, P. (eds) 1993. *Atlas préliminaire des Crustacés Décapodes d'eau douce de France.* Coll. Patrimoines Naturels, 14, Paris: Secrétariat de la Faune et de la Flore, Laboratoire de Biologie des Invertébrés Marins et Malacologie Muséum National d'Histoire Naturelle, Conseil Supérieur de la Pêche, Ministère de l'Environment.

Seasonal occurrence of decapods in shallow waters of a subtropical area

M.L. NEGREIROS-FRANSOZO, J.M. NAKAGAKI & A.L.D. REIGADA
NEBECC, Depto. de Zoologia, IB/Caunesp, Unesp, SP, Brasil.

ABSTRACT

A comparison of the capture rate of shallow-water decapods at noon (maximum light intensity) and midnight (minimum light intensity) was accomplished in the Ubatuba Bay (23°20′, 23°35′S and 44°50′, 45°14′W), São Paulo, Brazil, over a 1-year period. Each month, three replicate samples were obtained at 3.5 m median depth during waning moon periods. Replicates consisted of bottom trawls covering a 2500 m² area. Water and sediment samples were also taken in order to monitor environmental conditions. Median values of abiotic and biotic parameters were statistically compared using non-parametric tests. Twenty-five decapod species were recorded in the sampling area. Brachyurans and penaeoideans were the most abundant decapods, comprising most of the obtained species. Differences between day and night catch rates were not significant. However, the median monthly density of crabs in diurnal samples was different, with the highest rate recorded in February. The median density of shrimps differed among months in night samples, with highest catch rates recorded in May and July. These results indicate that the catch rate of decapods in the study area was chiefly influenced by seasonal changes, instead of light conditions.

1 INTRODUCTION

Crustacean decapods are an important element in the dynamics of the littoral benthic fauna because they directly modify the environment, and control the populations of other organisms. In southeastern Brazil, most studies regarding the biology of those organisms focused on their ecological distribution, reproduction (Tommasi 1967, Forneris 1969, Abreu 1980, Pita et al. 1985, Branco et al. 1990), autoecology, and spatial and seasonal distribution (Sartor 1989, Pires-Vanin 1989, Pires 1993, Fransozo et al. 1992, Negreiros-Fransozo et al. 1992).

The constitution and geographical distribution of littoral decapod assemblages from southern and southeastern Brazil are also relatively well known. The most complete papers concerning the taxonomy, distribution and general ecology were provided by Melo (1985, 1996) and Melo et al. (1989) for the Brachyura and D'Incao (1995) for the Dendrobranchiata.

Comparisons of diurnal and nocturnal catch rates of benthic decapods, and their relation to environmental factors, were not previously made for the coast of São Paulo State. The seasonal occurrence of the more common sublittoral decapod species from the northern coast of São Paulo has been subject of recent research (Fransozo et al 1992, Santos et al, 1994, Negreiros-Fransozo & Fransozo 1995, Naka-gaki et al. 1995 and Negreiros-Fransozo & Nakagaki 1998). However, these studies have been relied on diurnal samples.

Seasonal oscillations of temperature and photoperiod exert significant influence on crustaceans in many ways. According to Bouchon (1991), long illumination periods can stimulate reproduction in *Palaemonetes*. In other crustaceans, a short illumination period can intensify their activity (Moller & Jones 1975). Additionally, many crustaceans show daily rhythms of locomotion and other activities controlled by the combination of endogenous and exogenous factors (Marsden & Dewa 1994).

In the present paper, the seasonal occurrence of brachyurans and penaeoideans in the sublittoral zone of Ubatuba Bay is investigated from diurnal and nocturnal samples obtained on a monthly basis through a 1-year period. The results are compared with studies in which only diurnal sampling was carried out.

2 MATERIAL AND METHODS

Ubatuba Bay is located in the north coast of the State of São Paulo (23°20′, 23°35′S and 44°50′, 45°14′W) (Fig. 1). According to Pires-Vanin (1989), this area is chiefly influenced by three main marine currents, i.e. the South Atlantic Central Current, the Tropical Current and the Coastal Current. Temperature, salinity and nutrient concentrations are markedly different in these water masses.

Sampling was accomplished during waning moon periods in the Ubatuba Bay. A shrimp fishery boat equipped with an otter-trawl net (15 mm of body mesh and 10 mm in the cod end with two boards of 10 kg each) was used for this purpose. Each trawl covered a 2500 m^2 area, as calculated from trawling distance (1000 m), width of the net opening (2.5 m) trawling time (30 min.) and speed (1.1 nm h^{-1}). As samples were all obtained in the same place, depth and bottom conditions were not significantly different among trawls. Special care was taken to maintain the trawling speed constant, which allowed keeping the same distance between the otter boards as was checked by SCUBA diving observations.

In every month of 1996, three replicate trawls were taken in three alternate days in each of the chosen sampling schedules, i.e. noon and midnight.

Temperature, salinity, dissolved oxygen content, sediment composition, surface brightness and transparency were monthly monitored. To obtain this, a Van Dorn bottle, a Van Veen grab, a specific rephractometer, an oxymeter (YSI 55) coupled to a thermometer, a luximeter and a Secchi disk were used. Organic matter content was determined by ash weighing. Granulometric fractions of the bottom sediment were obtained by differential sieving according to Wentworth (1922). The Kruskal-Wallis analysis of variance (Sokal & Rohlf 1979) was used to examine seasonal variation of environmental factors because these variables were not normally distributed.

All biological material was sorted, and decapods were identified, weighed, measured and sexed. Very young individuals of the genus *Callinectes* (< 25 mm carapace

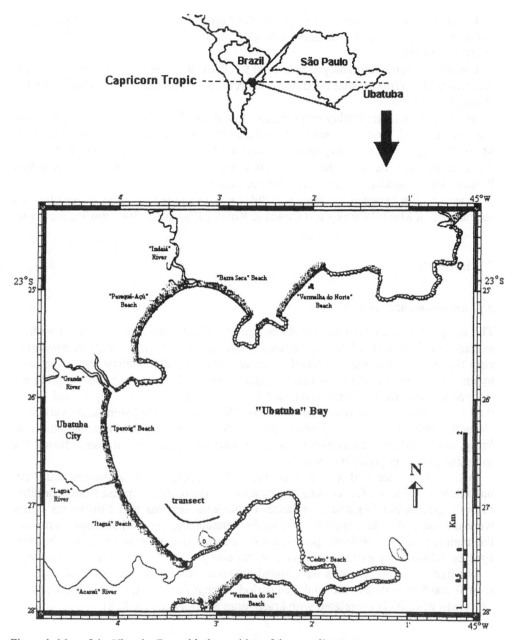

Figure 1. Map of the Ubatuba Bay with the position of the sampling transect.

width) were grouped and weighed together with adults. Identification was carried out according to Rathbun (1925, 1930, 1937), Williams (1984), Dall et al. (1990), and Christoffersen (1979). As the distinction of *Farfantepenaeus* species is based on external sexual characteristics, very young individuals of *Farfantepenaeus brasiliensis* (Latreille 1817) and *Farfantepenaeus paulensis* (Pérez-Farfante 1967) were pooled and regarded as *Farfantepenaeus* spp. Dr Gustavo A. S. Melo from the Museum of Zoology of the University of São Paulo (MUZUSP) helped in the identification of doubtful specimens.

Canonical correlation analysis was used to check if decapod abundances were associated with environmental conditions. Correlation significance was verified with a chi-square test.

In order to facilitate further comparisons, raw data of decapod density and biomass was transformed. Noon and midnight values were compared each month using the Mann-Whitney test, with hypotheses tested at the 5% significance level. Monthly variation of the values obtained for each schedule was achieved using a Kruskal-Wallis with probability values shown for each case.

Size variation between sampling schedules was examined for some of the obtained species with a Mann-Whitney test (Sokal & Rohlf 1979) at the 5% significance level.

3 RESULTS

3.1 *Environmental factors*

The sampled area showed typical characteristics of a sheltered environment with an average depth of 3.41 ± 0.61 m and bottom sediments that were mainly composed of very fine sand. Drainage of several rivers and small creeks contributed to the sedimentation of organic matter, which averaged $7.99 \pm 2.21\%$ through the sampling period. Suspended material and water visibility averaged 21.13 ± 6.7 mg l^{-1} and 1.81 ± 0.63 m respectively. However, environmental factors presented significant variations along the year ($p < 0.05$). A graphic representation of average depth, temperature, salinity, dissolved oxygen contents and light intensity along the transect during the sampling period is given in Figure 2.

A total of 25 decapod species was obtained, including 18 brachyurans and 7 penaeoideans (Table 1). Concerning the Brachyura, portunids surpassed other families in both species number and abundance. *Callinectes ornatus* and *Callinectes danae* were the most abundant species. Within the Penaeoidea, penaeid shrimps comprised the majority of all obtained species. *Xiphopenaeus kroyeri* was the most abundant species, followed by *Rimapenaeus constrictus* and juveniles of *Farfantepenaeus* spp.

No significant correlations were found between decapod occurrence and environmental conditions during the analyzed period, except for midnight samples, when salinity was negatively correlated with brachyuran abundance (Table 2).

3.2 *Brachyuran abundances*

When the number of individuals per 1000m² was compared among sampling months (Fig. 3), significant statistical differences were found ($p = 0.031$) with highest capture rates obtained in February, in the case of noon samples. However, capture rates did not differ among midnight samples ($p = 0.756$). Comparisons of capture rates between noon and midnight samples did not show significant statistical differences in none of the twelve analyzed months ($p \geq 0.05$).

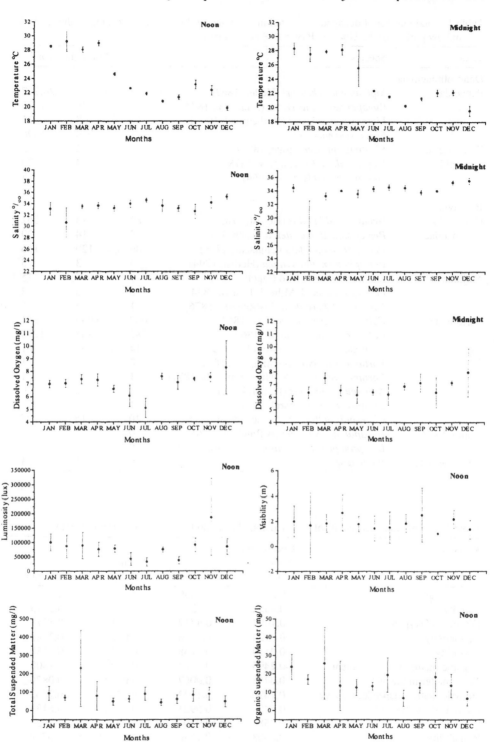

Figure 2. Ubatuba Bay. Monthly mean values and 95% confidence interval for 10 abiotic environmental factors, from January to December, 1996.

Table 1. Total number of decapods collected in noon and midnight monthly samples during 1996 in the shallow portion of the Ubatuba Bay, São Paulo, Brazil.

Taxa	Species	Noon	Midnight	Total
Dendrobranchiata				
Penaeidae	*Xiphopenaeus kroyeri* (Heller, 1862)	11,557	13,180	24,757
	Rimapenaeus constrictus (Stimpson, 1874)	319	1606	1925
	Litopenaeus schmitti (Burkenroad, 1938)	17	08	25
	Farfantepenaeus sp	612	1078	1690
Solenoceridae	*Pleoticus muelleri* (Bate, 1888)	1	2	3
Sicyonidae	*Sicyonia dorsalis* Kingsley, 1878	26	44	70
	Sicyonia typica (Boeck, 1864)	06	12	18
Brachyura				
Calappidae	*Hepatus pudibundus* (Herbst, 1785)	89	473	562
Leucosiidae	*Persephona lichtensteinii* Leach, 1817	15	14	29
	Persephona punctata (Linnaeus, 1758)	46	120	166
	Persephona mediterranea (Herbst, 1794)	0	3	3
Majidae	*Libinia ferreirae* Brito Capello, 1871	7	8	15
	Libinia spinosa H. Milne Edwards, 1834	0	3	3
	Pyromaia tuberculata (Lockington, 1876)	1	5	6
Portunidae	*Callinectes ornatus* Ordway, 1863	3047	4934	7981
	Callinectes danae (Smith, 1869)	382	888	1270
	Portunus spinimanus Latreille, 1819	14	47	61
	Portunus spinicarpus (Stimpson, 1871)	0	1	1
	Portunus ventralis (A. Milne Edwards, 1879)	1	0	1
	Arenaeus cribrarius (Lamarck, 1818)	4	6	10
	Cronius ruber (Lamarck, 1818)	1	0	1
	Charybdis helleri (A. Milne-Edwards, 1867)	10	15	25
Xanthidae	*Hexapanopeus paulensis* Rathbun, 1930	0	1	1
	Eurypanopeus abbreviatus (Stimpson, 1860)	0	1	1
Pinnotheridae	*Pinnixa* sp	0	1	1

Table 2. Correlation coefficients of decapod abundances and abiotic factors (*p < 0.001).

Variable	Coefficient			
	Noon		Midnight	
	Brachyura	Penaeoidea	Brachyura	Penaeoidea
Temperature	0.2938	−0.1450	0.1196	0.0194
Salinity	−0.4522	0.1286	−0.7411*	−0.2815
Dissolved Oxygen	0.1093	−0.4717	0.1211	−0.3771
Luminosity	0.0724	−0.3818	−0.0621	−0.3249
Visibility	0.0114	−0.1768	−0.3504	−0.2450
Organic matter of the sediment	−0.2496	0.1043	−0.0845	0.1808
Total matter in the water	−0.0755	0.0067	0.0784	−0.0876
Organic matter suspended in water	0.1490	0.1923	0.2866	0.2462
Canonical Coefficient	0.5403	0.5921	0.8355	0.6139
χ^2	10.358	12.950	35.925	14.192
Probability	0.2408	0.1136	1.82×10^{-5}*	0.0769

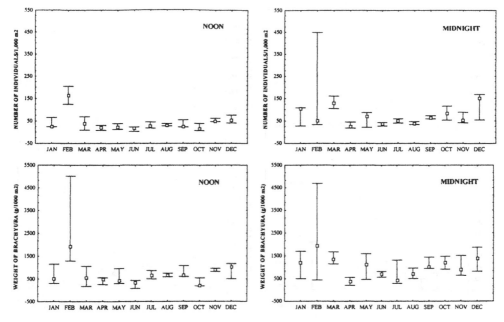

Figure 3. Brachyurans. Monthly occurrence in noon and midnight samples from Ubatuba Bay, expressed as standard number (n × 1000 m^{-2}; upper panels) and biomass (g × 1000 m^{-2}; lower panels).

3.3 *Penaeidean abundances*

Monthly numbers of individuals per 1000 m^2 did not significantly differ along the sampling period in the case of noon samples (p = 0.1878; Fig. 4). However, it significantly varied for midnight samples (p = 0.044) with the largest values obtained in May and June, which are not significantly different from each other. Comparisons between noon and midnight samples in each month revealed no statistical significant differences (p ≥ 0.05).

3.4 *Daily size variations in samples of Callinectes ornatus and Xiphopenaeus kroyeri*

In *Callinectes ornatus*, juveniles were remarkably dominant in midnight samples during January, February, April, May, June, August, September and December (p < 0.01) (Fig. 5). In the case of *Xiphopenaeus kroyeri*, between-schedules comparisons showed that juveniles were predominant (p < 0.05) in February, July, September and November in midnight samples and March, April, May and August in noon samples (Fig. 6).

4 DISCUSSION

According to Pires (1992), the benthic megafauna composition along the continental shelf of Ubatuba varies according to a bathymetric gradient and is directly affected by

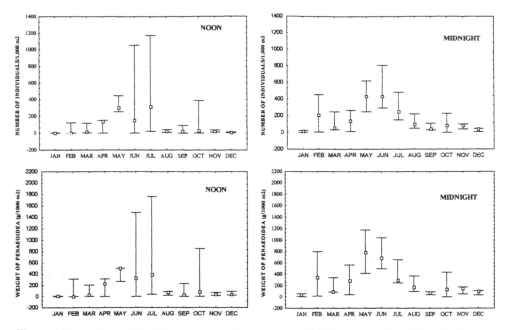

Figure 4. Penaeideans. Monthly occurrence in noon and midnight samples from Ubatuba Bay, expressed as standard number ($n \times 1000 \, m^{-2}$; upper panels) and biomass ($g \times 1000 \, m^{-2}$; lower panels).

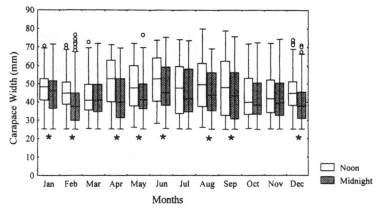

Figure 5. *Callinectes ornatus*. Monthly carapace width from noon and midnight samples. Explanation of symbols: box – median, 25 and 75 percentiles; whisker – non-outlier min. and non-outlier max.; o – outliers; * significant size difference within the same month ($p < 0.05$).

the dynamics of certain water masses operating over this region. Seasonal variations of abundance have been verified for a number of decapods in this area (Fransozo et al. 1992 Santos et al. 1994, Negreiros-Fransozo & Fransozo 1995). However, such variations are not clearly related to the fluctuation of environmental factors.

In this study, no apparent correlations were found between species abundance and variation of environmental factors, with exception of a negative correlation between

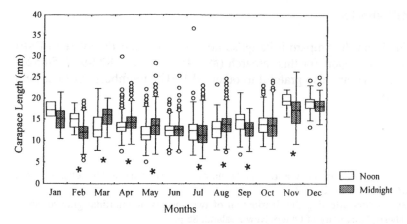

Figure 6. *Xiphopenaeus kroyeri.* Monthly carapace length from noon and midnight samples. Explanation of symbols: box – median, 25 and 75 percentiles; whisker – non-outlier min. and non-outlier max.; o – outliers; * significant size difference within the same month (p < 0.05).

abundance of brachyurans as calculated from midnight samples and salinity. Heavy rainfall in this region may cause flooding, as in February 1996, when monthly precipitation attained 638.6 mm contrasting with the annual average of 203.2 ± 154.5 mm (data from daily recordings at the North Base of the Oceanographic Institute of the University of São Paulo). Intense rain causes a considerable reduction of salinity in the Ubatuba bay and also major reworking of sediments, which results in an increased catch rate of burrowing decapods.

This investigation shows that ecological and faunistic research based on diurnal sampling is realistic, since estimates of both abundance and species richness are apparently independent of sampling schedules.

Several authors as Campbell (1967), Begg (1989), Stevens et al. (1984), Román-Contreras (1986) and Choy (1986) observed that in several crustacean species activity takes place preferably at night, but there was no evidence found for this during the present study. It is possible that local environmental characteristics are conditioning behavioral patterns of decapods, since low visibility caused by suspended material had been shown to shift their habits as a function of the 'value of protection' provided by the habitat (Laprise & Blaber 1992). Minello et al. (1985) verified that water turbidity as well as suitable burrowing substrata may either render protection against predators for *Penaeus aztecus* (= *Farfantepenaeus aztecus*). In this way, these shrimps may modify their habits depending on the nature of the occupied habitat.

A size-dependent activity pattern can be observed in *C. ornatus*, as evidenced by the prevalence of smaller individuals in night samples, and in *X. kroyeri*, in which median individual size changes among months in noon and midnight samples. Defense strategies of young individuals against predatory attacks of demersal fish can be associated with these findings. In the study region, these decapods may be largely consumed by key predators such as *Prionotus punctatus*, *Dactylopterus volitans* and several species of Chondrichthyes, which heavily prey on young brachyurans (Soares et al. 1992, Soares et al. 1993), and *Cynoscion virescens*, in which *X. kroyeri* was found to be the main food item in their stomachs (Rodrigues & Meira 1988).

ACKNOWLEDGEMENTS

We thank the 'Fundação de Amparo à Pesquisa no Estado de São Paulo' (FAPESP) for providing financial support for this research (#94/4878-9, #95/3872-9, #95/8520-3, and #98/01824-5). We are also grateful to other NEBECC members for their help during the field work.

REFERENCES

Abreu, J. 1980. Distribuição e ecologia dos Decapoda, numa área estuarina de Ubatuba, SO. *Bol. Inst. Oceanográfico São Paulo* 29: 1-3.

Beggs, S.M.C. 1989. Factors affecting the distribution of two species of intertidal grapsid crabs. Unpubl. Hons. Thesis, University of Otago, New Zealand, 86 p.

Branco, J.O., Porto-Filho, E. & Thives, A. 1990. Estrutura das populações, abundância e distribuição dentro de espécies integrantes da família Portunidae (Crustacea, Decapoda) da lagoa da Conceição e área Adjacente, Ilha de Santa Catarina, SC, Brasil. *Anais do II Simpósio de Ecossistemas da Costa Sul e Sudeste Brasileira* 2: 294-300.

Bouchon, D. 1991. Biological clock in seasonal reproductive cycle in the ditch shrimp *Palaemonetes varians* Leach. II. Ovarian state-dependent response to non-diel light-dark cycles. *J. Exp. Mar. Biol. Ecol.* 146: 13-26.

Campbell, J.A. 1967. Rhythmic spontaneous activity in an intertidal crab *Cyclograpsus lavauxi*. Unpubl. Hons. thesis University of Cantebury, 89 p.

Choy, S.C. 1986. Natural diet and feeding habitats of the crabs *Liocarcinus puber* and *L. holsatus* (Decapoda, Brachyura, Portunidae). *Mar. Ecol. Prog. Ser.* 31: 87-99.

Christoffersen, M.L. 1979. Campagne de la Calypso au large des Côtes Atlantiques de l' Amérique du Sud (1961-1962) I. 36. Decapod Crustacea: Alpheoida. In: Resultats Scientifiques de Campagnes de la Calypso. *An. Inst. Oceanogr.* 55: 297-377.

Dall, W., Hill, B.J., Rothlisberg, P.C. & Staples, D.J. 1990. The biology of the Penaeidae. In J.H.S. Blaxter & A.J. Southward (eds.) *Advances in marine biology*: 489. London: Academic Press, v. 27.

D'Incao, F. 1995. Taxonomia, padrões distribucionais e ecológicos dos Dendrobranchiata (Crustacea: Decapoda) do litoral brasileiro. 365 p. Doctoral Thesis Federal University of Paraná, Curitiba, Brazil.

Forneris, L. 1969. Fauna bentônica da baía de Flamengo, Ubatuba. Aspectos ecológicos. 215 pp. 'Livre-docência' Thesis – IB – USP, São Paulo, SP, Brasil.

Fransozo, A., Negreiros-Fransozo, M.L., Mantelatto, F.L.M., Pinheiro, M.A.A. & Santos, S. 1992. Composição e distribuição dos Brachyura (Crustacea, Decapoda) do sublitoral não consolidado na enseada da Fortaleza, Ubatuba (SP). *Rev. Brasil. Biol.* 52: 667-675.

Laprise, R. & Blaber, S.J.M. 1992. Predation by Moses perch, *Lutjanus russelli*, and blue spotted trevally, *Caranx bucculentus*, on juvenile brown tiger prawn, *Penaeus esculentus*: effects of habitat structure and time of day. *J. Fish Biol.* 40: 627-635.

Marsden, I.D. & Dewa, R.S. 1994. Diel and tidal activity patterns of the smooth shore crab *Cyclograpsus lavauxi* (Milne Edwards, 1853). *J. Roy. Soc. New Zealand* 24: 429-438.

Melo, G.A.S. 1985. Taxonomia e padrões distribucionais ecológicos dos Brachyura (Crustacea; Decapoda) do litoral sudeste do Brasil. 215 pp. Doctoral thesis – IB – USP, São Paulo, Brasil.

Melo, G.A.S. (ed.) 1996. *Manual de identificação dos Brachyura (Caranguejos e siris) do litoral brasileiro*. São Paulo: Plêiade /Fapesp, 604 p.

Melo, G.A.S., Veloso, V.G. & Oliveira, M.C.D.E. 1989. A fauna de Brachyura (Crustacea, Decapoda) do litoral do Estado do Paraná. Lista preliminar. *Nerítica*, Pontal do Sul PR 4(1/1): 1-31.

Minello, T.J., Zimmerman, R.J. & Martinez, E.X. 1985. Fish predation on juvenile brown shrimp, *Penaeus aztecus*: the effect of turbidity an substratum. *Estuaries* 8(2B): 10[A].

Moller, T.H. & Jones, D.A. 1975. Locomotory rhythms and burrowing habits of *Penaeus semisulcatus* (de Haan) and *P. monodon* (Fabricius) (Crustacea, Penaeidae). *J. Exp. Mar. Biol. Ecol.* 18: 61-77.

Nakagaki, J.M., Negreiros-Fransozo, M.L. & Fransozo, A. 1995. Composição e abundância de camarões marinhos (Crustacea, Decapoda, Penaeidea) na Enseada de Ubatuba, Ubatuba (SP), Brasil. *Arq. Biol. Tecnol.* 38: 583-591.

Negreiros-Fransozo, M.L., Reigada, A.L.D. & Fransozo, A. 1992. Braquiúros (Crustacea, Decapoda) dos sedimentos sublitorais da Praia da Enseada, Ubatuba (SP). *B. Inst. Pesca* 19: 17-22.

Negreiros-Fransozo, M.L & Fransozo, A. 1995. On the distribution of *Callinectes ornatus* Ordway, 1863 and *Callinectes danae* Smith, 1869 (Brachyura, Portunidae) in Fortaleza bay, Ubatuba (SP), Brazil. *Iheringia Ser. Zool.*, Porto Alegre. 79:13-25.

Negreiros-Fransozo, M.L & Nakagaki, J.M. 1998. Differential benthic occupation by crabs in the Ubautba bay, São Paulo, Brazil. *Journal of Shellfish Research* 17: 293-297

Pires-Vanin, A.M.S. 1989. Estrutura e dinâmica da megafauna bêntica na plataforma continental da região norte do Estado de São Paulo, Brasil. Tese de livre-docencia IO- USP, 172 pp.

Pires, A.M.S. 1992. Structure and dynamics of benthic megafauna on the continental shelf offshore of Ubatuba, southeastern Brazil. *Mar. Ecol. Prog. Ser.* 86: 63-76.

Pires, A.M.S. 1993. A macrofauna bêntica da plataforma continental ao largo de Ubatuba, São Paulo, Brasil. *Publ. Esp. Inst. Oceanogr. S. Paulo* 10: 137-158.

Pita, J.B., Rodrigues, E.S., Graça-Lopes, R. & Coelho, J.A.P. 1985. Levantamento da família Portunidae (Crustacea, Decapoda, Brachyura) no complexo baía-estuário de Santos. *B. Inst. Pesca* 12: 153-162.

Rathbun, M.J. 1925. The spider crabs of America. *Bull. US Nat. Mus.* 129: 1-613.

Rathbun, M.J. 1930. The cancroid crabs of America of the families Euryalidae, Portunidae, Atelecyclidae, Cancridae and Xanthidae. *Bull US Nat. Mus.* 152: 1-609.

Rathbun, M.J. 1937. The Oxystomatous and allied crabs of America. *Bull US Nat. Mus,* 166: 1-278.

Rodrigues, E.S. & Meira, P.T.F. 1988. Dieta alimentar de peixes presentes na pesca dirigida ao camarão sete-barbas (*Xiphopenaeus kroyeri*) na baía de Santos e Praia do Perequê, Estado de São Paulo, Brasil. *B. Inst. Pesca* 15: 135-146.

Román-Contreras, R. 1986. Análisis de la población de *Callinectes* spp (Decapoda, Portunidae) en el sector Occidental de la Laguna de Términos, Campeche, Mexico. *An. Inst. Cienc. del Mar y Limnologia Univ. Nac. Auton. Mexico,* 13: 315-322.

Santos, S., Negreiros-Fransozo, M.L. & Fransozo, A. 1994. The distribution of the swimming crab *Portunus spinimanus* Latreille, 1819 (Crustacea, Brachyura, Portunidae) in the Fortaleza bay. *Atlântica* 16: 125-141.

Sartor, S.M. 1989. Composição e distribuição dos Brachyura (Crustacea, Decapoda), no litoral norte do Estado de São Paulo. 197pp, Doctoral thesis- IO – USP, São Paulo, Brasil.

Soares, L.S.H., Rossi-Wongtschowski, C.L.B., Alvares, L.M.C., Muto, E.Y. & Gasalla, M.A. 1992. Grupos tróficos de peixes demersais da plataforma continental interna de Ubatuba, Brasil. I. Chondrichthyes. *Bolm. Inst. oceanogr.* 40: 79-85.

Soares, L.S.H., Gasalla, M.A., Rios, M.A.T., Arrasa, M.V. & Rossi-Wongtschowski, C.L.B. 1993. Grupos tróficos de onze espécies dominantes de peixes demersais da plataforma continental interna de Ubatuba, Brasil. *Publção esp. Inst. oceanogr. S. Paulo* 10: 189-198.

Sokal, R.R. & Rollf, F.J. (eds) 1979. *Biometría. Principios y metodos estadisticos en la investigación biologica.* Madrid: H. Blume Ediciones.

Stevens, B.G., Armstrong, D.A. & Hoeman, J.C. 1984. Diel activity of an estuarine population of dungeness crabs, *Cancer magister*, in relation to feeding and environmental factors. *J. Crust. Biol.* 4: 390-403.

Tommasi, L.R. 1967. Observações preliminares sobre a fauna bêntica de sedimentos moles da baía de Santos e regiões vizinhas. *Bolm. Inst. Oceanogr.* 16: 43-65.

Wenytworth, C.K. 1922. A scale of grade and class terms for clastic sediments. *J. Geol.* 30: 377-392.

Williams, A.B. (ed.) 1984. *Shrimps, lobsters and crabs of the Atlantic coast of the eastern United States, Maine to Florida.* Washington, DC: Smithsonian Institution Press.

Seasonal succession and spatial segregation of planktonic copepoda in the Zhujiang estuary in relation to temperature and salinity

C.K. WONG, P.F. TAM & Y.Y. FU
Department of Biology, The Chinese University of Hong Kong, Shatin, Hong Kong, China

Q.C. CHEN
South China Sea Institute of Oceanology, Academia Sinica, Guangzhou, China

ABSTRACT

The copepod fauna of the Zhujiang estuary was studied on 12 cruises between June 1991 and October 1992 and between January 1995 and July 1996. Forty-nine species, including 39 Calanoida, were identified. Dominant species included *Acartia spinicauda, Acartiella sinensis, Calanus sinicus, Eucalanus subcrassus, Labidocera euchaeta, Paracalanus parvus* and *Pseudodiaptomus poplesia.* The temperature and salinity ranges were specified for each dominant species. Some of the dominant species exhibited seasonal succession and spatial segregation. *C. sinicus, E. subcrassus* and *P. poplesia* were seasonal in their occurrence. *C. sinicus* and *E. subcrassus* were most abundant in oceanic waters in the outer estuary, but *C. sinicus* was most abundant in winter and *E. subcrassus* was most common in autumn. *P. poplesia* was most common in low-salinity waters in the inner estuary in spring. The other 4 copepods appeared sporadically throughout the year. *P. parvus* was restricted to the outer estuary. *A. sinensis* was most abundant in the inner estuary. *A. spinicauda* and *L. euchaeta* occurred throughout the estuary, but the spatial distribution of *A. spinicauda* in the estuary appeared to change seasonally.

1 INTRODUCTION

The Zhujiang (Fig. 1) is the largest river system in southern China. It consists of three major tributaries and has a watershed of > 230,000 km^2. During periods of maximum discharge in summer, turbid water flushing down the Zhujiang creates an estuary with an area of 8,000 km^2 at the southern coast of China.

The physical environment of the Zhujiang estuary is strongly influenced by the cold and dry NE monsoon from around December to February during winter, and by the warm and wet SW monsoon from around mid-June to early September during summer. Water temperature averages 29.0°C in July and decreases to 15°C in January. Salinity varies seasonally with the volume of river runoff. Coastal water of high salinity intrudes into the estuary during the winter, when discharge from the river is weak. Dilution begins in spring and continues through the summer. From June to September, the river runoff reaches a maximum and a plume of fresh water flows

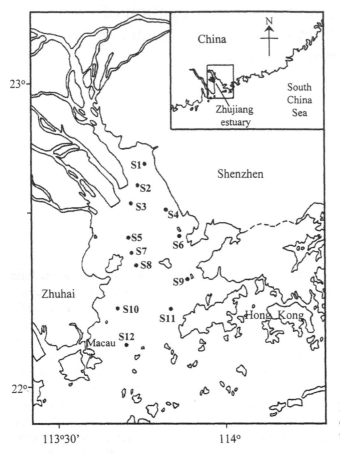

Figure 1. Map showing the location of the 12 sampling stations in the Zhujiang estuary.

over the oceanic water of the South China Sea. A horizontal salinity gradient, increasing from the inner to the outer regions of the estuary, exists in both the dry and wet seasons.

The study of the species composition and ecology of the copepod fauna in the Zhujiang estuary began in 1991. A total of 12 sampling cruises were carried out between June 1991 and July 1996. A general description of the seasonal and spatial occurrence of Copepoda in the estuary is given in Tam et al. (1998). This paper reports on the seasonal change in abundance and spatial distribution of several dominant species of calanoid copepods in the Zhujiang estuary.

2 MATERIALS AND METHODS

Zooplankton samples were made at 12 stations in the Zhujiang estuary (Fig. 1) during 12 cruises between June 1991 and July 1996. The estuary is shallow and extremely turbid. The average depth in the study area was ~ 7 m and a Secchi disk transparency of < 0.5 m was common at some of the inner stations. At each station, zooplankton was collected by a single vertical haul from ~ 1 m above the bottom to the surface with a conical net of 0.5 m mouth diameter and 667 μm mesh. The net collected mainly the adult stages of copepod species with body size > 1.0 mm, and underestimated the diversity and abundance of younger stages and smaller species. Samples were preserved immediately in 4% formaldehyde. In the laboratory, adult copepods were identified to species (whenever possible) according to Chen & Zhang (1965) and Cheng et al. (1965). At least 10% of each sample was counted and the density of each species was expressed as number of individuals per cubic meter (ind.m^{-3}). Dominant species were arbitrarily defined as species that contributed > 25% of the total copepod abundance in samples from at least one sampling cruise.

3 RESULTS AND DISCUSSION

3.1 *The dominant copepod species from the Zhujiang*

Forty-nine species of copepods are known from the Zhujiang estuary (Tam et al. 1998). The Calanoida, represented with 39 species, was the most diverse order. Seven species were considered to be dominant and are discussed below. All 7 dominant species belonged to the Order Calanoida. These were *Acartia spinicauda*, *Acartiella sinensis*, *Calanus sinicus*, *Eucalanus subcrassus*, *Labidocera euchaeta*, *Paracalanus parvus*, and *Pseudodiaptomus poplesia*. The 7 dominant species contributed for > 80% of the total copepod abundance during all 12 cruises. No Cyclopoida and Harpacticoida were found to be dominant. However, it must be noted that most of the cyclopoid and harpacticoid copepods in the samples were < 1.0 mm in body length and their abundance was likely to be underestimated by the used plankton net.

3.2 *Acartia spinicauda*

This species occurred throughout the year and showed seasonal change in spatial distribution (Fig. 2). Large numbers were found at the inner stations in January when oceanic water intruded into the estuary. A shift of the main population to the outer estuary occurred in early summer when river discharge increased. Distribution in the estuary became more even in autumn when river discharge began to recede. Dense populations of *A. spinicauda* could be found in wide range of temperature and salinity (Fig. 9). The species constituted > 60% of the total copepod abundance in January 1996. Chen (1982) reported that *A. spinicauda* is an important component of the copepod community in the Zhujiang estuary in spring. Other studies confirm that *A. spinicauda* is common in both coastal (Chen & Zhang 1965, Cheng et al. 1965, Gajbhiye et al. 1991, Yoo et al. 1991) and estuarine (Cheng 1965, Tiwari & Nair 1993) environments.

3.3 *Acartiella sinensis*

This species occurred occasionally throughout the year (Fig. 3). It reached a density of 490 ind.m^{-3} in January 1992, but was absent in January 1996. Similarly, numbers were much lower in September 1991 and October 1992 than in October 1995. *A. sinensis* occurred at temperatures ranging from 15 to 29°C (Fig. 9). Its occurrence in

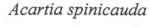

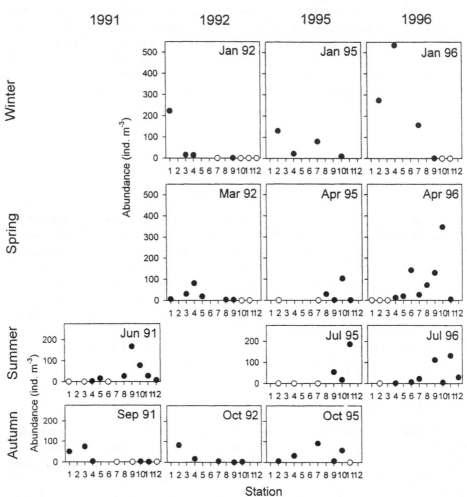

Figure 2. Seasonal abundance of *Acartia spinicauda* on 12 stations in the Zhujiang estuary during 1991, 1992, 1995 and 1996. Open circles represent the absence of the species.

the Zhujiang estuary was previously documented by Shen & Song (1979). Yanyu et al. (1991) reported that *A. sinensis* is a low salinity estuarine species. The presence of *A. sinensis* in freshwater habitats has been reported by Shen & Song (1979). In this study, dense populations were found mainly in the inner estuary where salinity was < 25‰ (Fig. 9).

3.4 *Calanus sinicus,*

Calanus sinicus was usually rare, and reached a peak in January 1992. It accounted for 43.5% of all copepods in the samples collected at that time (Fig. 4). The species

Acartiella sinensis

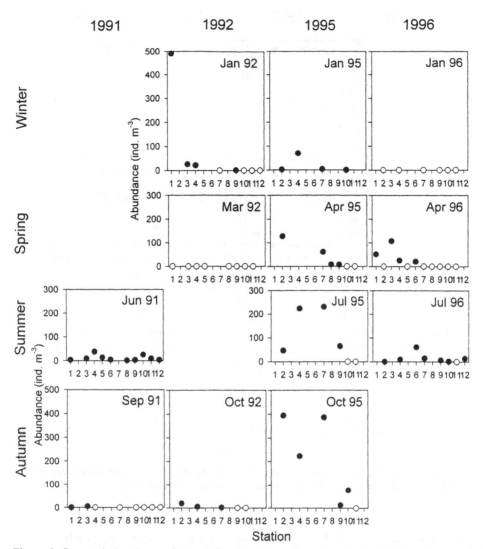

Figure 3. Seasonal abundance of *Acartiella sinensis* on 12 stations in the Zhujiang estuary during 1991, 1992, 1995 and 1996. Open circles represent the absence of the species.

exhibited strong seasonal patterns in the Zhujiang estuary. Dense populations were found only in January. The population declined during spring and summer and was absent from the estuary in autumn. *C. sinicus* showed strong preference for waters of high salinity and low temperature (Fig. 9). Individuals were found in the inner stations only during winter when oceanic waters intruded into the inner estuary. *C. sinicus* is widely distributed in the temperate waters of the East China Sea (Chen & Zhang 1965), the Inland Sea of Japan (Huang et al. 1993) and the coastal waters of Korea (Choi and Park 1993). Chen (1992) pointed out that *C. sinicus* is carried into

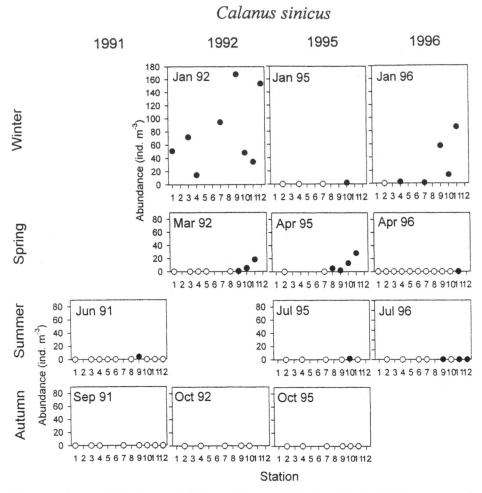

Figure 4. Seasonal abundance of *Calanus sinicus* on 12 stations in the Zhujiang estuary during 1991, 1992, 1995 and 1996. Open circles represent the absence of the species

the northern part of the South China Sea in winter by the Taiwan Current which flows along the Chinese coast from northeast to southwest. In this study, the density of *C. sinicus* was always higher in the outer than the inner regions of the Zhujiang estuary.

3.5 *Eucalanus subcrassus*

This species made up 68% of the copepod community in September 1991, but was rare or absent during most cruises (Fig. 5). *E. subcrassus* is a warm water oceanic species (Cheng et al. 1965, Yanyu et al. 1991). In the Zhujiang estuary, it was most abundant at temperatures > 25°C and salinities > 25‰ (Fig. 9). Chen (1982) supposes that *E. subcrassus* is carried into the coastal waters of the southern China every summer by ocean currents from the South China Sea. Density of *E. subcrassus* in the Zhujiang estuary was low during early summer, because the water in the estuary was

Eucalanus subcrassus

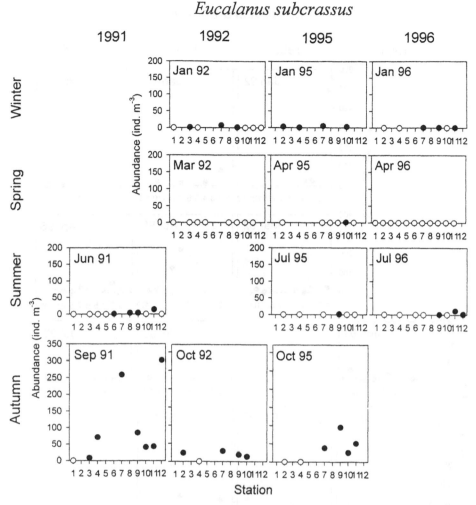

Figure 5. Seasonal abundance of *Eucalanus subcrassus* on 12 stations in the Zhujiang estuary during 1991, 1992, 1995 and 1996. Open circles represent the absence of the species.

diluted by river discharge. In autumn, river runoff receded and *E. subcrassus* was able to penetrate into the inner parts of the estuary.

3.6 *Labidocera euchaeta*

This species was one of the two that occurred throughout the sampling period (Fig. 6). No distinct seasonal pattern was observed. Densities > 400 ind.m^{-3} were recorded in January 1992 and October 1992, although numbers in most samples were low. Perennial existence of *L. euchaeta* has been reported previously from the Zhujiang estuary and a subtropical estuary in Xiamen of the Fujian Province (Lin & Li 1991, Yanyu et al. 1991). *L. euchaeta* occurred in a wide range of temperature and salinity in the Zhujiang estuary (Fig. 9).

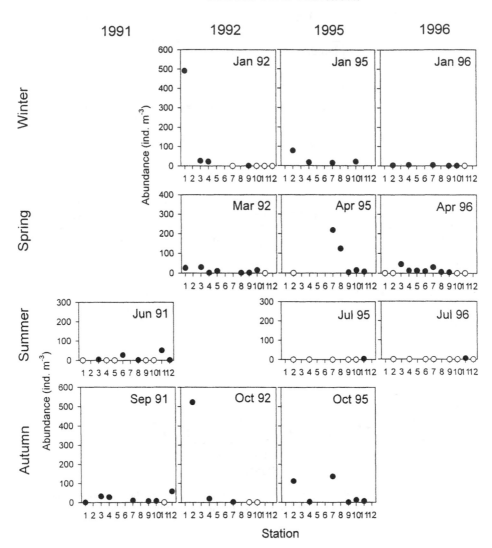

Figure 6. Seasonal abundance of *Labidocera euchaeta* on 12 stations in the Zhujiang estuary during 1991, 1992, 1995 and 1996. Open circles represent the absence of the species.

3.7 *Paracalanus parvus*

In January 1995 this species formed 27.2% of the copepod community but was relatively uncommon during the other cruises (Fig. 7). *P. parvus* occurred at temperatures ranging from 15 to 29°C, exhibited a preference for the outer parts of the estuary, and was rare at salinities < 15‰ (Fig. 9). Chen & Zhang (1965) emphasized that *P. parvus* is extremely abundant in the South China Sea (Chen & Zhang 1965). Other re-

Paracalanus parvus

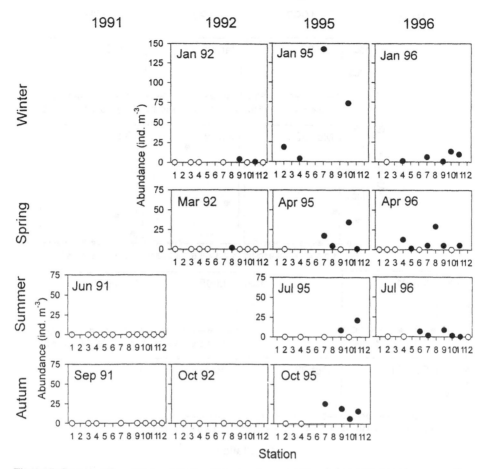

Figure 7. Seasonal abundance of *Paracalanus parvus* on 12 stations in the Zhujiang estuary during 1991, 1992, 1995 and 1996. Open circles represent the absence of the species.

ports also confirmed that *P. parvus* is common in many marine coastal areas (Chen & Zhang 1965, Cheng et al. 1965, Park et al. 1991, Choi & Park 1993, Wong et al. 1993, Dias 1994, Lokman 1994, Gomez-Gutierrez et al. 1995). Ambler et al. (1985) found that *P. parvus* was most abundant in waters of salinity > 15‰. In Australia, *P. parvus* is abundant only in the lower reaches of estuaries where salinity is high (Millar 1983). Similarly, Hulsizer (1976) reported that *P. parvus* occurred only in low numbers near the head of Narragansett Bay where salinity was low.

3.8 *Pseudodiaptomus poplesia*

This species showed strong seasonal and spatial distribution patterns (Fig. 8). It was abundant only during the spring. Its contribution to copepod community was ~ 70% in March 1992 and April 1995. During the other seasons, *P. poplesia* was either ex-

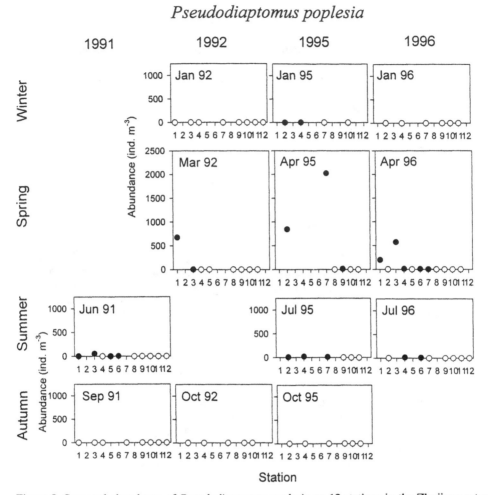

Figure 8. Seasonal abundance of *Pseudodiaptomus poplesia* on 12 stations in the Zhujiang estuary during 1991, 1992, 1995 and 1996. Open circles represent the absence of the species

tremely rare or absent. Dense populations of *P. poplesia* were found only in the inner parts of the estuary. *P. poplesia* was most common in temperatures ranging from 18 to 23°C and salinities < 12‰ (Fig. 9).

3.9 *Copepod seasonality*

Seasonal succession and spatial segregation were exhibited by the dominant copepods in the Zhujiang estuary. *Acartia spinicauda* occurred throughout the year, but its distribution in the estuary tended to change seasonally. *C. sinicus* and *E. subcrassus* were most abundant in oceanic waters in the outer estuary, but *C. sinicus* was most abundant in winter and *E. subcrassus* was most common in autumn. *P. poplesia* was restricted to low-salinity waters in the inner estuary in spring and early summer. *P. parvus, A. sinensis* and *L. euchaeta* occurred sporadically throughout the year, but

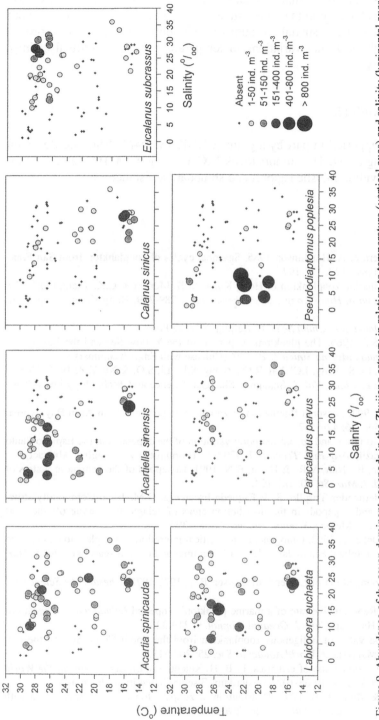

Figure 9. Abundance of the seven dominant copepods in the Zhujiang estuary in relation to temperature (vertical axes) and salinity (horizontal axes).

occupied different parts of the estuary. *P. parvus* was restricted to the outer estuary. *A. sinensis* was most abundant in the inner estuary. *L. euchaeta* was found throughout the estuary. The seasonal and spatial variations in the dominant species were probably determined by the seasonal variations in temperature and longitudinal salinity gradient in the estuary.

ACKNOWLEDGEMENTS

This research was supported in part by a grant (CUHK 185/94M) from the Research Grants Council, Hong Kong. The authors thank K.C. Cheung, L.M. Huang, Y.C. Tam and J.Q. Yin for providing logistic support and shipboard assistance.

REFERENCES

Ambler, J.W., J.E. Cloern & A. Hutchinson 1985. Seasonal cycles of zooplankton from San Francisco Bay. *Hydrobiologia* 129: 177-197.

Chen, Q.C. 1982. The marine zooplankton of Hong Kong. In B. Morton & C.K. Tseng (eds), *The marine flora and fauna of Hong Kong and southern China*: 789-799. Hong Kong: Hong Kong University Press.

Chen, Q.C. 1992. *Zooplankton of China Seas (I)*. Beijing: Science Press.

Chen, Q.C. & Zhang, S.Z. 1965. The planktonic copepods of the Yellow Sea and the East China Sea. I. Calanoida. *Studia Marina Sinica* 7: 20-129. (Chinese with English summary).

Cheng, C., Zhang, S.Z., Li, S., Fang, J.C., Lai, R.Q., Zhang, S.L., Li, S.Q. & Xu, Z.Z. 1965. *Marine planktonic copepods of China, Vol. 1*. Shanghai: Shanghai Science and Technology Press. (Chinese).

Choi, K.H. & Park, C. 1993. Seasonal fluctuation of zooplankton community in Asan Bay, Korea. *Bull. Korean Fish. Soc.* 26: 424-437.

Dias, C.D.O. 1994. Distribution and spatial-temporal variation of the copepods in the Espirito Santo Bay (Vitoria, Es, Brazil). *Arq. Biol. Tecnol.* 37: 929-949. (Portuguese with English Abstract).

Gajbhiye, S.N., Stephen, R., Nair, V.R. & Desai, B.N. 1991. Copepods of the nearshore waters of Bombay (India). *Ind. J. Mar. Sc.* 20: 187-194.

Gomez-Gutierrez, J., Hernandez-Trujillo, S. & Esqueda-Escarcega, G.M. 1995. Community structure of euphausiids and copepods in the distribution areas of pelagic fish larvae off the west coast of Baja California, Mexico. *Scientia Marina* 59: 381-390.

Huang, C., Uye, S. & Onbé, T. 1993. Ontogenetic diel vertical migration of the planktonic copepod *Calanus sinicus* in the Inland Sea of Japan III. Early summer and overall seasonal pattern. *Mar. Biol.* 117: 289-299.

Hulsizer, E.E. 1976. Zooplankton of Lower Narragansett Bay, 1972-1973. *Chesapeake Science* 17: 260-270.

Lin, S. & Li, S. 1991. Reproductive rate of a marine planktonic copepod *Labidocera euchaeta* Giesbrecht in Xiamen Harbour. *Chin. J. Oceanol. Limnol.* 9: 319-328.

Lokman, S. 1994. Food value of indigenous zooplankton from the South China Sea on the east coast of Peninsular Malaysia. *J. World Aquacult. Soc.* 25: 208-213.

Millar, C.B. 1983. The zooplankton of estuaries. In B. H. Ketchum (ed.), *Ecosystems of the World 26: Estuaries and enclosed seas*: 65-102. Amsterdam: Elsevier Scientific Publishing.

Park, C., Choi, K.H. & Moon, C.H. 1991. Distribution of zooplankton in Asan Bay, Korea with comments on vertical migration. *Bull. Korean Fish. Soc.* 24: 472-482.

Shen, C.J. & Song, D.X. 1979. Calanoida. In C.J. Shen (ed.), *Fauna Sinica. Crustacea. Freshwater Copepoda*. Peking: Science Press. (Chinese with English title page).

Tam, P.F. 1998. Distribution of marine zooplankton in coastal waters of Southern China. MPhil. Thesis. The Chinese University of Hong Kong, Hong Kong.

Tam, P.F., Wong, C.K. Chen, Q.C., Fu, Y.Y., Huang, L.M. & Yin, J.Q. 1998. Planktonic copepods of the Zhujiang estuary, 1991-1996. *J. Taiwan Mus.*, in press.

Tiwari, L.R. & Nair, V.R. 1993. Zooplankton composition in Dharamtar creek adjoining Bombay harbour. *Ind. J. Mar. Sc.* 22: 63-69.

Wong, C.K., Chan, A.L.C. & Chen, Q.C. 1993. Planktonic copepods of Tolo Harbour, Hong Kong. *Crustaceana* 64: 76-84.

Yanyu, D., Lin, J.H., Lin, M., Chen L.X. & Huang, Y.J. 1991. An ecological study of the zooplankton in western Xiamen Harbour, China. *Asian Mar. Biol.* 8: 45-56.

Yoo, K.I., Hue, H.K. & Lee, W.C. 1991. Taxonomical revision on the genus *Acartia* (Copepoda: Calanoida) in the Korean waters. *Bull. Korean Fish. Soc.* 24: 255-265. (Korean with English abstract).

The biology of the hermit crab *Calcinus tubularis* (Decapoda, Diogenidae) in nature and in the laboratory

DANIELA PESSANI, MARIA TERESA DAMIANO, GIUSEPPE MAIORANA &
TINA TIRELLI
Dipartimento di Biologia Animale e dell'Uomo, Università di Torino, Italy

ABSTRACT

Specimens of *Calcinus tubularis* were collected in different zones of the Ligurian Sea (Northern Mediterranean Sea) and maintained in the laboratory one year, in order to verify molt frequencies, growth rates, and shell preferences. Females were more numerous than males (mean sex ratio as male number/crab number = 0.36), but the mean size did not differ significantly between the sexes. Ten percent of the sampled specimens occupied calcareous tubes (belonging to vermetids and serpulids), while the others were found in gastropod shells. Fifty-eight percent of specimens (almost all > 4 mm in size) lives in *Cerithium* shells, while 19% (generally those < 4 mm-sized) lives in *Columbella rustica* shells. Out of 65 crabs that changed shell in the laboratory, 90% preferred *Cerithium*, while only 1 and 3 specimens (< 4 mm-sized) selected a calcareous tube and *C. rustica* respectively. *Calcinus tubularis* showed tendency to prolonged intermolt periods (70-80 days), with a very limited growth rate. This species does not seem to suffer from a long period of captivity, and shows a great adaptability.

1 INTRODUCTION

Calcinus tubularis (Decapoda, Diogenidae) lives in the Mediterranean Sea at depths between 0 and 40 m, generally on rocky bottoms or in *Posidonia oceanica* beds. However, according to Fenizia (1933), it prefers rocky bottoms where it can camouflage in a better way. The camouflage is often due also to the great quantity of epibionts living on its shell (Damiano et al. 1998). *Calcinus tubularis* also occupies calcareous tubes (belonging to vermetids and serpulids), which, according to Amouroux (1974), is the only shelter used by this species.

Calcinus tubularis is often considered as an uncommon species (Zariquiey Alvarez 1968) or a species limited to some areas (Kinzelbach 1990). Its geographical distribution, as well as some aspects of its biology have been studied by Gherardi (1990), Manjon Cabeza & Garcia Raso (1995), and Zibrowius (1978).

The aim of this study was to increase the knowledge of *Calcinus tubularis* with regard to sex ratio, reproductive period and fecundity, kind of the occupied shelter, relative growth, growth rates and molt frequency in the laboratory. Specimens both collected in nature and reared in the laboratory have been examined.

2 MATERIALS AND METHODS

Specimens were collected in the western Ligurian Sea (Northern Mediterranean Sea) during the reproductive season of 1995. In the laboratory, hermit crabs were individually placed in small cups with a sand lined bottom and containing about 100 ml of water (T = 20 ± 1°C; S = 36‰; photoperiod according to the season).

Although some authors prefer the shield length as size measure, the length of the carapace (from its posterior edge to the tip of the rostrum) was taken to represent a more easily measured parameter for the animal's size. These two parameters proved to be bound together by the relation:

$$\text{Size} = 1.752 \times \text{Shield length} - 0.377 \ (r = 0.983, p < 0.01, n = 21)$$

In order to evaluate growth rate, crab size and length of the larger (left) chela were measured on accidentally dead specimens or on the exuviae. Measurements were carried out with the help of a Wild M5 stereomicroscope, equipped with an ocular micrometer.

It was possible to link 75% of the sampled specimens to the shelter (gastropod shell or calcareous tube) that they occupied at the time of collection. As a shell (tube) size parameter, the diameter of the aperture was measured. The remaining 25% of specimens left their shelter during transport to the laboratory.

During shelter choice experiments, 65 hermit crabs were offered 4 different possibilities. Offered shelter volume was about 10% higher than that of the previously occupied shelter. Each hermit crab was placed into a small cup (∅ = 10 cm, height = 7 cm; water volume = 400 ml) at the same distance from each proposed shelter. After 24 hours, crabs were moved back to their original cup.

Specimens were kept in the laboratory for one year maximum. During its stay in the laboratory, each specimen moulted twice or more times. Surviving hermit crabs were finally released at the original collection site.

3 RESULTS

3.1 *Sex ratio*

Table 1 shows sampling months, the number of sampled specimens with reference to their sex and sex ratio (number of males on total amount of crabs). The ovigerous females were 26% of the ones sampled and 17% of all the specimens; the sex ratio was always in favor of females.

3.2 *Fecundity*

The mean number of eggs carried by a single female was 740 ($s = 444$, $n = 21$); the significant relationship between females size (x) and number of eggs (y) is expressed

Table 1. Numerical data relevant to *C. tubularis* specimens in the sampling period.

Month	Females [n]		Males [n]	Total [n]	Sex ratio
	Non-ovigerous	Ovigerous			
May	20	-	8	28	0.28
June	20	6	6	32	0.19
July	66	21	52	139	0.37
August	10	17	21	48	0.43
September	8	-	7	15	0.46
Total	124	44	94	262	0.36

by the equation:

$$y = 278.63x - 599.02 \; (r = 0.734, p < 0.01, \text{d.f.} = 19).$$

3.3 Size

The mean size of non-ovigerous females (4.52 mm, $s = 1.47$, $n = 124$) did not differ significantly from that of ovigerous females (4.74 mm, $s = 0.95$, $n = 44$; Student's t-test: $t = 0.93$, d.f. = 166, n.s.). As a consequence, the two categories were considered together.

There was no statistically significant difference between the mean size of females (4.58 mm, $s = 1.35$, $n = 168$) and males (4.82 mm, $s = 1.42$, $n = 94$; Student's t-test: $t = 0.33$, d.f. = 260, n.s.). Therefore males and females were considered together.

As regards hermit crabs' mean size in the months of sampling, specimens collected in May differed significantly from those collected in June as well as crabs collected in July differed from those collected in August (Table 2).

3.4 Occupied shelter

The 196 *Calcinus tubularis* specimens were found in the following shelters: *Cerithium* sp. (52.6%), *Columbella rustica* (17.4%), calcareous tubes (9.7%), *Coralliophyla* sp. (6.6%), *Buccinulum* sp. and *Gibbula* spp. (3.6%), *Hinia* sp. (1.5%), *Clanculus corallinus* and *Monodonta* sp. (1.0%), *Astraea rugosa*, *Conus mediterraneus*, *Lunatia* sp., *Mitra cornicula*, *Sphaeronassa mutabilis*, and *Turritella* sp. (0.5%).

With regard to ovigerous females, it was possible to link 20 of them to an occupied shelter, which in 14 cases was *Cerithium* sp.

Cerithium sp. and *Columbella rustica* were occupied respectively by 37 and 13 males, and 66 and 21 females, but no statistically significant difference was observed between sexes (G_{test}, respectively: $G = 1.53$, d.f. = 1, n.s. and $G = 0.04$, d.f. = 1, n.s.). Seventy-nine per cent of hermit crabs living in tubes were females.

The analyses of crab size in relation to the most often occupied shelters, demonstrated that 82.3% of the specimens occupying *Columbella rustica* and 79% of those living in tubes were of size < 4 mm (here called group 1, or 'small'), while 87.4% of those living in *Cerithium* sp. had size > 4 mm (group 2, or 'large').

Specimens living in *C. rustica* did not show any significant correlation ($r = 0.249$, d.f. = 31, n.s.) between animals' size (mean size 3.61 mm, $s = 0.59$, $n = 33$) and the

Table 2. Mean size of *Calcinus tubularis* sampled specimens in the sampling period.

Month	Mean size mm	s	n	t (*)	d.f.	p
May	3.39	1.22	28			
June	5.01	1.18	32	5.23	58	< 0.01
July	5.08	1.44	139	0.23	169	n.s.
August	4.15	0.8	48	4.27	185	< 0.01
September	4.17	0.84	15	0.1	61	n.s.

(*) t = Student's t-test; size of the specimens collected respectively in May vs those collected in June, June vs July, July vs August and August vs September.

width of the shell aperture (mean width 1.57 mm, $s = 0.33$). The opposite happened with specimens in calcareous tubes ($r = 0.707$, d.f. = 17, $p < 0.01$) (crab mean size 3.67 mm, $s = 0.71$, $n = 19$; mean aperture width 2.85 mm, $s = 0.53$) or in *Cerithium* ($r = 0.638$, d.f. = 98, $p < 0.01$) (crab mean size 5.41 mm, $s = 1.35$, $n = 100$; mean aperture width 4.37 mm, $s = 1.33$).

3.5 *Relative growth*

The relationship between size (x) and left chela length (y) showed strongly negative allometry in group 1:

$$y = 0.658 + x^{0.818} \ (r = 0.806, \text{d.f.} = 95, p < 0.01)$$

and positive allometry in group 2:

$$y = -0.656 + x^{1.187} \ (r = 0.836, \text{d.f.} = 160, p < 0.01)$$

3.6 *Growth in the laboratory*

Time between sampling of *Calcinus tubularis* specimens and their first molt lasted on average 81.8 days ($s = 52.4$, $n = 120$). Three months after sampling, 65 specimens that had not yet molted, were given the chance to change shelters; 24 of them belonged to group 1 and 41 to group 2.

Table 3 shows the kind of shelter offered and that chosen. Of 20 (out of 24) specimens belonging to group 1 that changed their shelter, 15 chose *Cerithium*. For crabs belonging to group 2, all those which changed shells (32/41), also preferred *Cerithium*, while the 9 which did not change were already living in *Cerithium*.

Ninety-two percent of the specimens that changed shelters, molted within a few days after changing; 29 of them (56%) increased their size by an average of 0.3 mm ($s = 0.19$), while the remaining showed a decrease in size (21%) or no growth (23%).

Of a further group of 65 specimens – that did not change shelters – the mean period between the first and the second molt was calculated. It lasted 115.5 days ($s = 39.9$) and was significantly different from the previous one (Student's t-test: $t = 4.53$, d.f. = 183, $p < 0.01$). Fifty-seven percent of the above mentioned specimens showed an increase in size (0.25 mm, $s = 0.21$, $n = 37$); 23% did not grow and 20% had a decrease in size. The two mean increases are not significantly different (Student's t-test: $t = 0.99$, d.f. = 64, n.s.).

Table 3. Shelter offered to 65 specimens of *Calcinus tubularis* and their choice.

Occupied shelter	[n]	s. n. c.	Tube	Columbella		Cerithium		Monodonta
				Smooth	Encrusted	Smooth	Encrusted	
Cerithium sp.*	4	1	–	–	–	3	–	–
*C. rustica**	11	1	1	2	1	6	–	–
Monodonta sp.*	2	–	–	–	–	2	–	–
Lunatia sp.*	1	–	–	–	–	1	–	–
Calcareous tube*	6	2	1	–	–	3	–	–
*A. rugosa***	1	–	–	–	–	–	1	–
Cerithium sp.**	33	9	–	–	–	12	12	–
Coralliophyla sp.**	4	–	–	–	–	3	1	–
Gibbula spp.**	1	–	–	–	–	1	–	–
*S. mutabilis***	1	–	–	–	–	–	1	–
Turritella sp.**	1	–	–	–	–	–	1	–

* = shelter occupied by hermit crabs belonging to group 1; ** = shelter occupied by hermit crabs belonging to group 2; s.c.n. = speciemens not changing shelter; s.n. = specimens changing shelter (above and at left of the tick line, shelters offered to group 1, below and at the right, shelters offered to group 2).

4 DISCUSSION

The present study allows us to point out some aspects of the biology of *Calcinus tubularis* in nature and in the laboratory.

4.1 *In nature*

During all the sampling months, the sex ratio was always female-biased as already noticed by Amouroux (1974) and Manjon Cabeza & Garcia Raso (1995). This may be explained by the increased mobility of males, which probably occupy deep waters and move towards females to mate only, as is known for *Clibanarius erythropus* (Carayon 1941). On the other hand, a biased sex ratio is common in Crustacea (Wenner 1972).

Markham (1977) found small-sized females of *Calcinus verrilli* living in *Spyroglyphus* sp. tubes, although this species usually occupies *Ceritium litteratum* shells. *Calcinus tubularis* shows a similar situation (79% of the crabs living in tubes are females). The mean size of crabs inhabiting tubes (3.69 mm, s = 0.71, n = 19) is significantly smaller than that of specimens living in shells (4.94 mm, s = 1.47, d.f. = 190; Student's t test: t = 3.648, d.f. = 194, p < 0.01). Manjon Cabeza & Garcia Raso (pers. comm.) found mostly *C. tubularis* females of small size in tubes.

Although *C. tubularis* was found in a wide range of gastropod shells, it showed a clear preference for *Cerithium* sp. and *Columbella rustica*. This may be due to both being the empty shells of these two genera more frequent in the environment and to the hermit crabs size ('small' or 'large' specimens). The narrow aperture of *C. rustica* prevents larger specimens to enter and would appear to be unsuitable even for specimens which actually occupy it.

No ovigerous females were found after August, a phenomenon in agreement with

Fenizia (1933); however, the lack of ovigerous females in the September sample may be due to the small number of specimens collected (15). The number of eggs carried by each female is in clear contrast with observations by Manjon Cabeza & Garcia Raso (1995). After replacing the parameter 'carapace length' by 'shield length' (used by the above mentioned authors), the expected mean number of eggs would be 115 instead of 740 as we counted. The egg number seems to change significantly according to size: in this research the smaller ovigerous female was 2.48 mm in size, while Spanish specimens varied between 1.02 and 1.34 mm.

According to Hartnoll (1982), the relative allometric growth (either positive or negative) of a part of the body is a sign of its major or minor functionality. The negative allometry of the left chela in the 'small' specimens of *Calcinus tubularis* – set up against the positive allometry of the same limb in the 'large' specimens – shows that in *C. tubularis* the functionality of the left chela is most probably connected with the increase in size, i.e. with crabs' age.

Ninety-five percent of the shells occupied by *Calcinus tubularis* was covered by epibionts, from about 50 to 100% of shell surface (Damiano et al. 1998). Therefore *C. tubularis* specimens may be supposed to occupy the same shells for a long period. In fact, Stachowitsch (1980) suggests that the epibiosis degree of gasteropod shells is directly proportional to the duration of their use by the hermit crabs.

4.2 *In the laboratory*

Calcinus tubularis shows long intermolt periods, which may even exceed 5 months.

Growth rate is not significantly different between the specimens that moved to a larger shelter and those that did not.

Almost 50% of the specimens did not grow after molting: this might have been connected with the reduced variations in size through out the sampling period.

A calcareous tube was chosen just once by a female. When the animal had a chance to choose freely, tubes were avoided probably because they limited the hermit crab's mobility. Moreover, also the relation between the aperture diameter of the tubes and the size of the crabs living inside them is to be remarked. The aperture is generally such as to make entry of the crab easier but, maybe, it is not easily defensible by the same crab against possible predators.

From the above discussion one may conclude that in nature *Calcinus tubularis* occupies the same shell for a long time, although it molts and therefore grows. However, the species shows a great adaptability, also underscored by its ability to survive long times in the laboratory with relatively low mortality rates (50% of the specimens over one year). Said adaptability is demonstrated also by finding in nature specimens living in glass tubes, or cohabiting in the same shell (Damiano et al. 1998).

ACKNOWLEDGEMENTS

The authors wish to thank their colleagues and friends Emilio Balletto and Giuseppe Rappini, for the precious advises and the critical revision of the text, and an anonymous and patient referee who improved the final quality of the paper.

REFERENCES

Amouroux, J.M. 1974. Sur la presence de *Calcinus ornatus* (Roux, 1830) (Crustaces Anomoures) dans la region de Banyuls-su-Mer. *Vie Milieu* 24(2)B: 421-422.

Carayon, J. 1941. Morphologie et structure de l'appareil gènital femelle chez quelques pagures. *Bull. Soc. Zool. France* 66: 95-122.

Damiano, M.T., Pessani, D., Piasco, A. & Tirelli T. 1998. Relazioni inter ed intraspecifiche nel paguro *Calcinus tubularis*: epibiosi e coabitazione. *Biol. Mar. Mediterranea* 5: 1-10.

Fenizia, G. 1933. Note biologiche sul *Calcinus ornatus* (Roux). *Boll. Soc. Nat. Napoli* 45: 129-142.

Gherardi, F. 1990. Competition and coexistence in two Mediterranean hermit crabs, *Calcinus ornatus* (Roux) and *Clibanarius erythropus* (Latreille) (Decapoda, Anomura). *J. Exp. Mar. Biol. Ecol.* 143: 221-238.

Hartnoll, R.G. 1982. Growth. In D.E. Bliss (ed.), *The Biology of Crustacea* 2: 111-196.

Kinzelbach, R. 1990. Einsiedlkrebse (Paguridea): der gartnereremit *Calcinus tubularis*. *Natur und Museum* 120(12): 393-400.

Manjon Cabeza, M.E. & Garcia Raso, J.E. 1995. Study of a population of *Calcinus tubularis* (Crustacea, Diogenidae) from a shallow *Posidonia oceanica* meadow. *Cah. Biol. Mar.* 36: 277-284.

Markham, J.C. 1977. Preliminary note on the ecology of *Calcinus verrilli*, an endemic Bermuda hermit crab occupying attached vermetid shells. *J.Zool.* 181: 131-136.

Stachowitsch, M. 1980. The epibiotic and endolithic species associated with the gastropod shell inhabited by the hermit crabs *Paguristes oculatus* and *Pagurus cuanensis*. *Mar. Ecol.* 1: 73-101.

Wenner, A.M. 1972. Sex ratio as a function of size in marine crustacea. *Am. Nat.* 106: 321-350.

Zariquiey Alvarez, R. 1968. Crustaceos Decapodos Ibericos. *Inv. Pesq.* 32: 1-510.

Zibrowius, H. 1978. Première observation du pagure *Calcinus ornatus* dans le Parc National de Port-Cros (Cote méditerranéenne de France). Repartition et bibliographie. *Travaux Scientifique du Parc national de Port-Cros* 4: 149-155.

Processes that promote decapod diversity and abundance on the upper continental slope of the southwestern Gulf of Mexico

L.A. SOTO & S. MANICKHAND-HEILEMAN
Benthic Ecology Lab., México

E. FLORES & S. LICEA
Phytoplankton Lab., México

ABSTRACT

Substrate heterogeneity across the continental shelf and particularly on the upper slope in the southwestern Gulf of Mexico is augmented by the heavy input of allochthonous and man made debris from rivers. The flux of organic materials promotes the enrichment of a bathyal crustacean megafauna community whose biomass (0.01 g C m^{-2}) is comparable to that recorded within the inner shelf. Diversity and biomass gradients are examined across the shelf and upper slope. A total of 64 species are organized into 7 faunal assemblages according to their synecological relations. The potential dietary carbon source of the bathyal forms was determined by means of carbon stable isotope and indirectly by stomach content analysis. Based on this, it is contended that a flux of organic materials is strongly pulsed by the two major climatic periods in the area and contributes to the sustenance of a detrital abyssobenthic food web.

1 INTRODUCTION

Decapod crustaceans constitute a conspicuous taxonomic assemblage in the benthic community of the southwestern Gulf of Mexico. They may contribute somewhat less than one third of the total organic carbon budget of the shelf megafaunal community, which includes an assorted number of mollusks, echinoderms, and demersal ichthyofauna. Their biomass has been estimated to fluctuate between 0.39 g C m^{-2} to 0.025 g C m^{-2} during the rainy and dry season, respectively (Soto & Escobar-Briones 1995). These authors have documented the physical processes that may regulate both seasonal and annual variability in the continental benthic production in this sector of the Gulf of Mexico, widely recognized for its important commercial fisheries and the offshore oil and gas exploitation. Some of the ecological properties seem to indicate that this subsystem of the Gulf of Mexico is fairly stable (Manickchand-Heileman et al. 1998). However, attention has been drawn to the small-scale variability in biological processes caused by physical (riverine effluents, oceanic water intrusion) or sedimentary mechanisms (sediment load) upon the benthic domain on the one hand, and the shelf's subsidy of organic materials to the adjacent deep-sea environment, on the other.

Several authors (Walsh 1983, Rowe et al. 1988, Longhurst & Pauly 1987) have postulated an imbalance between production and consumption in tropical shelf ecosystems. Presumably in these systems surface production of organic materials can be diverted to the benthic compartment on the shelf (Flint & Rabalais 1981) or exported and deposited at the adjacent continental slope.

Such deposit centers seem to play a major role in supporting benthic biomass on the upper continental slope of the SW Gulf of Mexico. Our current knowledge is practically non-existent of the meso- and microscale processes that may explain the effect upon faunal composition, shift gradients, and fluxes that sustain benthic production in the SW Gulf. This study attempts to elucidate the microscale biomass distribution, faunal depth gradients, and fluxes that support megafaunal decapod crustaceans across the shelf and upper slope of the SW Gulf of Mexico.

2 METHODS

The soft-bottom community of the continental shelf and the upper slope of the SW Gulf of Mexico was sampled during two cruises conducted onboard the R/V JUSTO SIERRA in February and September of 1997. Three perpendicular transects were sampled across the continental shelf and upper slope from depths ranging from 28 m to 530 m in the area just off the Coatzacoalcos River and the Grijalva-Usumacinta delta. Additional samples were taken in less than 200 m on the shelf in the same area (Fig. 1). A total of 19 stations were occupied at five arbitrary depth strata: inner shelf (< 60 m), middle shelf (61 to 100 m), outer shelf (101-200 m) and upper slope

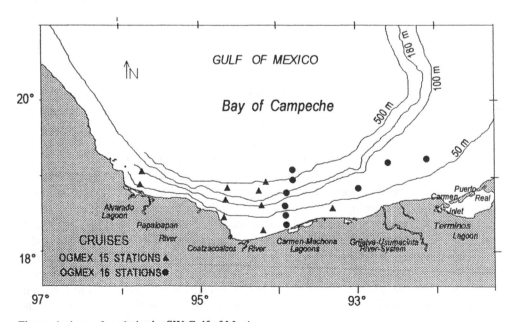

Figure. 1. Area of study in the SW Gulf of Mexico.

(201-530 m). At some of these stations we took replicate hauls, making a total of 39 sampling locations. CTD data were recorded at each station. Biological material was obtained with the use of a conventional shrimp otter-trawl with 20 and 26 m of head rope lengths and stretch mesh of 47 cm in the body and 2.5 cm in the cod end. Tows were made for 30 to 50 minutes, depending on water depth, and at an average velocity of 3 knots. Estimation of the swept area (2.2 ha) by the trawl was calculated by the equation proposed by Sparre et al. (1989); biomass values were adjusted depending upon the time spent trawling at each station. Material was identified to species level and weighed onboard. Organic carbon transformations were calculated according to Crisp (1984).

A multivariate analytical technique, i.e., the classification method Two-Way Indicator Species Analysis (TWIA, Hill 1979) implemented by the program TWINSPAN was applied to the 47 most abundant and frequent decapod crustaceans out of the 62 species identified, in order to determine faunal assemblages on the shelf and upper slope. This analysis was also employed to recognize indicator species of particular environmental conditions along the depth gradient. An indicator species that differentiates one assemblage from another is recognized by the highest biomass and frequency of occurrence in each assemblage. For this purpose 5 cut levels (0.0, 50.0, 100.0, 200.0, 500.0 kg km^2) were selected to represent biomass ranges. Additionally, species diversity at each depth stratum was estimated with the Shannon-Weaver diversity index (1963). Hutcheson's procedure (1970) for testing differences between the diversity values calculated for each cruise was also applied.

Potential organic carbon sources supporting benthic production in the area of study were studied through carbon stable isotopic analysis following the procedure described by Soto & Escobar-Briones (1995). Stomach content analysis was also performed on 8 bathyal species (*Acanthephyra purpurea, Aristaeomorpha foliacia, Benthochascon schmitti, Polycheles typhlos, Mesopenaeus robustus, Glyphocrangon aculeata, Parapenaeus longirostris*, and *Plesiopenaeus edwardsianus*) captured during OGMEX-16. Relative frequency of diatom valves was determined from a 0.5 ml aliquots of the gut content, previously cleaned of organic matter, following the procedure proposed by Hasle & Fryxel (1970) and Simonsen (1974) and later mounted in Hyrax medium. Depending on the valve concentration in each sample, counting involved from 180 to 500 valves.

3 PHYSICAL SETTING

Physical oceanography in the southwestern Gulf of Mexico is fairly well defined in terms of both water masses and circulation pattern (Emilsson et al. 1975, Monreal-Gómez & Salas de León 1990, Soto & Escobar-Briones 1995). The area of study is under the influence of a strong cyclonic gyre present in February and March whose intensity diminishes in April causing a change in the current direction. In mid year, the gyre is absent and the prevailing circulation is from east to west. From July on the cyclonic gyre is formed again and persists and intensifies toward the end of the year. Winter storms from September to February with sustained northerly winds from 34 to 63 knots promote vertical mixing of the upper water layers to a depth of 70 m, thus causing coupling conditions between the pelagic and benthic compartments. Development of the seasonal thermocline (30 to 70 m) is related to near-surface features such as intrusion of estuarine or riverine waters which form a thin, slightly cold, (26.3°C) and diluted water layer (34.7‰) that can be traced nearly 50 km off the mayor rivers in the area of study. Another feature is the presence of the shelf-slope front that may fluctuate in space and time and often occurs within the boundaries of the outer continental shelf.

Our knowledge on the mega and meso-scale biological and chemical features in this region of the Gulf is more rudimentary, particularly regarding sedimentary chemistry. The continental shelf just off the Coatzacoalcos River narrows down to about 70 km and is mostly covered by silt and clay sediments. The continental slope extends from the shelf-slope break (150-200 m) to about 3700 m in the Sigsbee Deep. A heavy load of allochthonous organic material is reflected in high sedimentary organic matter (4-6%) distributed as discontinuous bands along the inner and middle shelf (Ortega-Durán et al. 1990). Sediments at depths of 20-100 m are mostly terrigenous though in certain areas between 70 and 100 m a relic fluvial plain exist in which a significant shift in texture occurs; beyond 200 m sediments are mostly biogenic mud.

Intensive seaward transport of allochthonous materials occurs during the rainy season (June to September) when the discharge from the Coatzacoalcos and Grijalva-Usumacinta intensifies (110-6500 million m^3 $month^{-1}$) exporting to the shelf and upper slope vascular plant remains, mangrove leaves, and human related materials. This debris flow presumably accumulates on the slope but neither the precise location, nor its rate of deposition have been properly documented.

4 RESULTS

4.1 *Biomass distribution*

A total of 64 species represented by more than 2300 individuals were collected in the two cruises that were conducted across the continental shelf and the upper slope of the southwestern Gulf of Mexico under two contrasting climatic conditions. One occurs in February with strong northerly winds that causes a well-mixed water column down to about 70 m. The other occurs in September characterized by sparse rains and a moderate two layer system in the upper 36 m of the water column. In both instances, the total benthic biomass and that contributed by crustaceans follow a diminishing trend along the depth gradient (Fig. 2). Normally, the crustacean biomass represents less than one third of the total benthic organic carbon in which other macroinvertebrates and demersal fish are included. In September, the crustacean biomass along the inner shelf was slightly higher (0.02 g C m^{-2}) than the one recorded in February (0.01 g C m^{-2}). However, in both instances, at depths in excess of 100 m, there is a significant biomass decline ranging from 2 to 4 mg C m^{-2}. This decreasing pattern is interrupted at about 500 m, where peak biomass values of megafaunal crustaceans were recorded (0.01 g C m^{-2}). At this depth, the crustacean component, particularly deep-water shrimp and prawns, contribute nearly 50% to the total benthic carbon biomass.

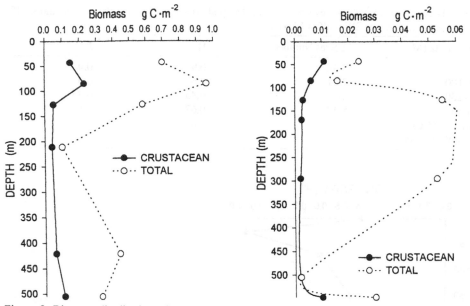

Figure 2. Biomass distribution of decapod crustaceans across the continental shelf and upper slope of the SW Gulf of Mexico.

4.2 *Faunal assemblages*

Species richness across the shelf and upper slope in the area of study varies from a maximum number of 36 species in the inner shelf strata (< 60 m) to a minimum of only 4 recorded in the limits of the outer shelf (100-200 m). On both cruises, the decapod species appear unevenly distributed, and some obvious disparities were detected between depth strata, attributed perhaps to sampling bias (Table 1). The two cruises had essentially similar species composition; the one conducted in February was slightly more diverse with a diversity index of 0.67, whereas the September cruise had an index of 0.57. Both values were statistically tested for differences prior to pooling the species data per depth strata. No significant differences were recorded between the above two indices (p < 0.50), thus allowing the analysis of the diversity gradient at four depth strata (Table 2). A declining trend in diversity was observed from the inner shelf down to the outer shelf, where values range from 0.95 to 0.45, respectively. However, this trend follows a sudden shift on the upper slope, reaching a diversity index of 0.67 because of the presence of an important number of abyssal benthic species (Fig. 3).

The multivariate analysis technique (TWINSPAN) applied in this study to 47 selected species facilitate the identification of two main assemblages, here designated as shelf (< 142 m) and upper slope (150-532 m). The addition of 15 rare species in this interpretation did not change the station groupings that comprised each assemblage. These assemblages are further divided into 7 groups (Fig. 4). The shelf assemblage includes a rather heterogeneous species complex whose bathymetric range extends from the inner shelf (24 m) to the edge of the outer shelf, arbitrarily established

Table 1. Diversity index (H') and evenness (J') of decapod crustaceans across the continental shelf and the upper slope of the SW Gulf of Mexico.

Depth strata (M)	Species Number	H'	J'
< 60	17-36	0.95	0.48
61-100	7-11	0.6	0.4
101-200	4-10	0.45	0.35
300-500	5-15	0.67 *	0.4
OGMEX -15 Cruise		0.67	
OGMEX-16 Cruise		0.57	

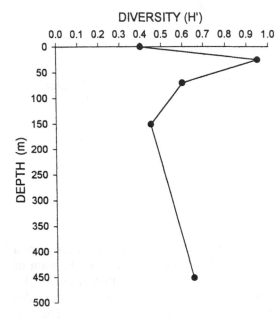

Figure. 3. Diversity index (H') and evenness (J') of decapod crustaceans across the shelf and upper slope of the SW Gulf of Mexico.

near 100 m. This diverse complex encompasses more than 36 decapod species generally referred to as the 'shrimp community' (Hildebrand 1954) because of the predominance of penaeid shrimp (0.35 g C m^{-2}; Soto 1980). Within the shelf assemblage there can be recognized four groups whose species composition is as follows.

4.2.1 *Group 1 – Inner shelf assemblage*

Most of the species included in this group were obtained during two hour replicate trawls (n = 4) made at 60 m off the Grijalva-Usumacinta deltaic system late in February of 1977. A well-mixed water column prevails at this time of the year, with a bottom temperature and salinity of 23°C and 36‰, respectively. The indicator species in this group was *Solenocera vioscai*. Other dominant species are *Calappa sulcata* and the blue crab *Callinectes sapidus*. Biomass (wt) of both the blue crab and the brown shrimp *Farfantepenaeus aztecus* attain significant values (> 200 kg km^{-2}; 100 kg m^{-2}).

Table 2. Faunal assemblages of decapod crustaceans across the continental shelf and upper slope of the SW Gulf of Mexico.

Group 1 (4 stations; 60 m)	*Callinectes sapidus* *Callinectes similis* *Farfantepenaeus aztecus* *Rimapenaeus similis* *Sicyonia dorsalis* *Solenocera vioscai*
Group 2 (3 stations; 45-60 m)	*Calappa ocellata* *Calappa sulcata* *Callinectes sapidus* *Callinectes similis* *Farfantepenaeus aztecus* *Farfantepenaeus duorarum* *Parthenope agona* *Petrochirus diogenis* *Portunus spinimanus* *Raninia louisianensis* *Raninoides loevis* *Rimapenaeus similis* *Robertsella* sp. *Sicyonia brevirostris* *Sicyonia dorsalis* *Solenocera vioscai*
Group 3 (13 stations; 28-100 m)	*Acanthocarpus alexandri* *Anasimus latus* *Calappa sulcata* *Callinectes similis* *Ethusa microphthalma* *Farfantepenaeus aztecus* *Farfantepenaeus duorarum* goneplacid sp. *Hepatus epheliticus* *Iliacantha subglobossa* pagurid sp. *Parthenope agona* *Portunus gibbesi* *Portunus rathbunae* *Portunus spinicarpus* *Portunus spinimanus* *Raninoides loevis* *Raninoides louisianensis* *Rimapenaeus constrictus* *Rimapenaeus similis* *Sicyonia brevirostris* *Sicyonia dorsalis* *Sicyonia typica* *Solenocera atlantidis*

Table 2. (Continued).

Group 4 (12 stations; 40-100 m)	*Arenarius cribarius* * *Calappa flammea* * *Calappa sulcata* *Callinectes similis* *Farfantepenaeus aztecus* *Farfantepenaeus duorarum* *Iliacantha subglobossa* pagurid sp. *Plesiopenaeus edwardsianus* * *Portunus depressifrons* * *Portunus spinicarpus* *Portunus spinimanus* *Pyromaia arachna* * *Pyromaia cuspidata* * *Rimapenaeus similis* *Sicyonia brevirostris* *Sicyonia dorsalis* *Sicyonia stimpsoni* * *Sicyonia typica* * *Stenocionops spinimana*
Group 5 (3 stations; 300-519 m)	*Benthochascon schmitti* *Glyphocrangon nobile* *Lyreidus baiirdi* *Mesopenaeus robustus* *Munida valida* *Plesiopenaeus edwardsianus* *Tomopaguropsis* sp.
Group 6 (2 stations; 500-536 m)	*Acanthephyra purpurea* *Aristaeomorpha foliacea* *Benthochascon schmitti* *Glyphocrangon aculeata* *Glyphocrangon nobile* *Leiolambrus nitidus* (?) *Mesopenaeus robustus* *Munida media* *Munida spinosa* *Munida valida* *Munidopsis erinaceous* *Munidopsis robusta* *Munidopsis spinifer* *Nephropsis aculeata* *Plesionika polyacanthoneous* *Plesiopenaeus edwardsianus* *Polycheles typhlos* *Rochinia crassa* *Scyllarus chacei* *Stenopus scutellatus* *Tomopaguropsis* sp. *Trichopeltarion nobile*

Table 2. (Continued).

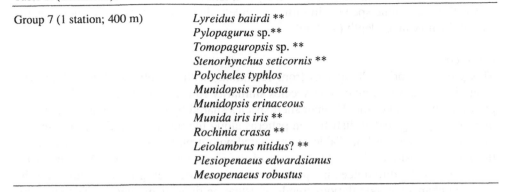

Group 7 (1 station; 400 m)	*Lyreidus baiirdi* **
	Pylopagurus sp.**
	Tomopaguropsis sp. **
	Stenorhynchus seticornis **
	Polycheles typhlos
	Munidopsis robusta
	Munidopsis erinaceous
	Munida iris iris **
	Rochinia crassa **
	Leiolambrus nitidus? **
	Plesiopenaeus edwardsianus
	Mesopenaeus robustus

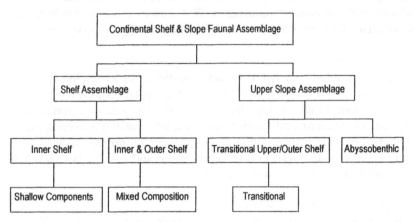

Figure 4. Faunal assemblages of decapod crustaceans distributed across the continental shelf and upper slope of the SW Gulf of Mexico.

4.2.2 *Group 2*

This assemblage is actually a subdivision of the previous group. Its depth range is slightly shallower (45-60 m) and somewhat more concentrated towards the area off the mouth of the Coatzacoalcos River. The indicator species of the rich organic sediments in this region is the brown shrimp *F. aztecus*, and *Callinectes similis* and *Calappa ocellata* are among the most abundant species.

4.2.3 *Group 3 – Mixed assemblage*

This group is represented by a diverse species assemblage that includes more than 25 inner shelf taxa, distributed from 40 to 60 m, which appear combined with some eurybathic species whose depth range extends onto the outer shelf (100-200 m). Such is the case in the species *Portunus spinicarpus, Iliacantha subglobossa,* and *Acanthocarpus alexandri.* The indicator species in this group are *Callinectes sapidus, Sicyonia dorsalis, Solenocera vioscai,* and *Farfantepenaeus aztecus.* Other predominant species are *F. duorarum* and *Portunus depressifrons.* The shallow water components of this group remain just above the seasonal thermocline under the stratified water column conditions (September) at a temperature of 24°C and a salinity of 36‰. The eurybathic species, in turn, are widely distributed across the shelf regardless of the mix-layer depth (~ 36 m).

4.2.4 *Group 4*

This group comprises 20 species from 12 stations with a depth of 40 to 100 m. The group itself shows a remarkable species similarity to the preceding assemblage. Approximately 13 out of the 21 inner shelf species shared by these two assemblages display overlapping spatial distribution patterns along the inner shelf. The 8 species (indicated with an asterisk in Table 2) not shared with group 3 are probably restricted to the muddy substrates found to the west of the Coatzacoalcos River. Another factor that explains the differences in species composition between groups 3 and 4 are the seasonal changes in populations trends of estuarine dependent species. The indicator species in this group are *Farfantepenaeus duorarum, Calappa sulcata*, and *Portunus spinicarpus*. Species with significant biomass (> 200 kg/km^{-2}) are *Sicyonia brevirostris, Farfantepenaeus duorarum, Portunus spinicarpus*, and *P. depressifrons*.

4.2.5 *Group 5 – Transitional*

Into this group were assembled only two deep-water stations, which included 7 species recorded at 300 and 500 m in the transect just off the Coatzacoalcos River. Four of these species correspond to upper slope inhabitants (*Benthochascon schmitti, Tomopaguropsis* sp., *Munida valida*, and *Lyreidus baiirdi*), whereas the remaining three constitute truly abyssal benthic shrimp normally caught at much greater depths in the Gulf of Mexico (c.f. Pequegnat 1983). Bottom temperature and salinity between 300 and 500 m range from 8° to 10°C and 34.8 to 35.4‰.

4.2.6 *Group 6 – Abyssal benthic/Upper slope*

In spite of the reduced number of stations (n = 2) included in this group, a diverse species complex of 25 bathyal and upper slope dwellers was found. Members of the families Penaeidae, Solenoceridae, Aristeidae, and Glyphocrangonidae represented roughly 80% of the total biomass recorded (0.31-0.23 g m^{-2} wt) at 500 m. Second in terms of abundance was the deep-water lobster *Nephropsis aculeata* with a biomass of 0.002 to 0.01 g m^{-2} wt. Even though the galatheids do not contribute significantly to the total biomass at this depth on the account of their reduced size, they do form an unusual species complex of six species distributed in the genera *Munida* (3 spp.) and *Munidopsis* (3 spp.). Similarly, the rare abyssal lobster *Polycheles typhlos* appears in this faunal assemblage at 400 m and again at 500 m.

The upper slope taxa included in this group are restricted to the scyllarid lobster *Scyllarus chacei*, the parthenopid crab *Leiolambrus nitidus* (?), the portunid crab *Benthochascon schmitti*, and *Tomopaguropsis* sp. The indicator species in this group is *Nephropsis aculeata*.

4.2.7 *Group 7*

This group stands alone for reasons of its species composition. It includes only one deep-water station recorded at 400 m in which were captured mostly upper slope taxa that are perhaps depth restricted (indicated with a double asterisk in Table 2). However, there are also some abyssal benthic species present at this depth, such as *Poly-*

cheles typhlos, Mesopenaeus robustus, Plesiopenaeus edwardsianus, which are more common in group 6. The indicator species is *Mesopenaeus robustus*, whose estimated biomass wt was 0.007 g m^{-2}.

4.3 *Organic carbon sources*

Carbon stable isotopic composition (δ^{13}C) of a selected group of abyssal benthic and upper slope inhabitants (Table 3) indicated highly enriched values that ranged from – 13.45‰ to –20.59‰. These (δ^{13}C values primarily indicate potential sources that support consumers in the benthic food web. They fall within the interval ascribed to allochthonous organic materials from coastal environments, such as sea grass detritus (–11.6 to –15.9‰) and estuarine phytoplankton (–16.0‰). The most depleted (δ^{13}C value of –20.59‰ corresponds to the hermit crab *Tomopaguropsis* sp., which may indicate a diet which is more dependent on marine carbon sources (> –22.0‰). In this particular case, this individual may indirectly acquire the marine carbon signal from the overlying water by consuming infaunal filter feeders that utilize particulate organic matter from an oceanic source.

Considering 1‰ of isotopic fractionation in (δ^{13}C, the values obtained for the species 2, 4, 5, and 6 (Table 3), closely approximate the (δ^{13}C estimated for estuarine phytoplankton (–16.0). Conversely, the portunid crab *Benthochascon schmitti* showed the highest enriched (13C) (–13.45‰), carrying the signal of vascular plant detritus labeled with a (δ^{13}C of –11.6‰).

4.4 *Gut content analysis*

The examination of the gut content of eight bathyal species distributed between 300 to 500 m showed different conditions of fullness. The cardiac chamber contained mostly flocculent amorphous organic material combined with silty sediment. Microscopic observations revealed the presence of phytoplankton cells with chlorophyll in centric and pennate diatoms, *Gymnodinium* spp., and unidentified phytoflagellates; an assorted group of Protozoa was also recognized. At least 69 diatom species were identified, many of them planktonic-neritic, and a few benthic. In Table 4, the subdominant diatom species have been arranged according to the frequency of occurrence and their general ecological characteristics. Only in the case of *Acanthephyra purpurea*, was the diversity of diatoms not so high as in the remaining 7 species. The feeding habits of the latter species include the ingestion of a significant number of

Table 3. δ^{13}C values of six selected abyssobenthic and upper continental decapod crustaceans from the SW gulf of Mexico.

Species	δ^{13}C (‰)	Feeding strategy
1. *Tomopaguropsis* sp.	–20.59	deposit-sifting
2. *Plesiopenaeus edwardsianus*	–16.95	opportunistic-scavenger
3. *Benthochascon schmitti*	–13.45	scavenger
4. *Nephropsis aculeata*	–16.19	scavenger-predatory
5. *Polycheles typhlos*	–17.26	scavenger
6. *Munida valida*	–17.41	deposit-sifting

Table 4. Relative frequency (%) of dominant diatom species identified in the stomach content of eight bathyal decapod crustaceans from the SW Gulf of Mexico. Subdominant species in brackets; general habitat in parentheses.

Acanthephyra purpurea (21)*	183 (m)	*Thalassiosira eccentrica* 16.3% (P); *Thalassionema nitzschioides* 13.6% (P); *Cyclotella litoralis* 11.4% (CW, E); *Thalassiosira* sp. 10.9% (P). [*Pleurosigma diverse-striatum* (EP, E); *Neodelphineis pelagica* (CW, N, ES); *Thalassiothrix longissima; Thalassiothrix heteromorpha* (O, P)].
Aristaeomorpha foliacia (37)*	500 (m)	*Thalassionema nitzschioides* 10.6% (P); *Thalassiosira eccentrica* 8.9% (P); *Nitzschia bifurcata* 6.4% (P); *Cyclotella stylorum* 6.1% (CW, E). [*Psammodictyon constrictum* (B, EP); *Neodelphineis pelagica* (CW, N, ES); *Coscinodiscus concinus* (P); *Diploneis bombus* (B, EP); *Thalassiothrix longissima* (O, P); *Synedra fulgens var. Mediterranea* (EPI)].
Benthochascon schmitti (39)*	500 (m)	*Thalassionema nitzschioides* 21.4% (P); *Thalassiosira eccentrica* 8.9% (P); *Thalassiosira* sp. 6.2% (P); *Cyclotella stylorum* 5.7% (CW, E). [*Nitzschia bifurcata* (P); *Neodelphineis pelagica* (CW, N, ES); *Cyclotella litoralis* (CW, E); *Paralia sulcata* (E, CW, ES); *Thalassionema frauenfeldii* (P)].
Polycheles typhlos (37)*	300 (m)	*Thalassionema nitzschioides* 35.8% (P); *Thalassiosira* sp. 7.5% (P); *Thalassionema frauenfeldii* 6.8% (P); *Coscinodiscus concinnus* 3.0% (P); [*Diploneis bombus* (B, EP); *Neodelphineis pelagica* (CW, N, ES); *Pseudonitzschia* sp. (P); *Thalassiothrix heteromorpha* (O, P)].
Mesopenaeus robustus (30)*	400 (m)	*Thalassionema nitzschioides* 33.0% (P); *Thalassiosira* sp. 11.7% (P); *Thalassiothrix longissima* 11.5% (O, P); *Thalassiosira eccentrica* 5.7% (P); [*Cyclotella litoralis* (CW, E); *Cyclotella striata* (CW, E, ES); *Psammodictyun constrictum* (B, EP); *Diploneis vacillans* (B, EP); *Nitzschia ganderheimiensis* (P)].
Glyphocrangon aculeata (30)*	450 (m)	*Thalassionema nitzschioides* 29.2% (P); *Thalassiosira* sp. 10.7% (P); *Thalassiosira eccentrica* 8.8% (P); *Thalassionema frauenfeldii* 7.2% (P); [*Neodelphineis pelagica* (CW, N); *Cyclotella litoralis* (CW, E); *Nitzschia* sp. (B, P); *Coscinodiscus perforatus (P)*].
Parapenaeus longirostris (31)*	445 (m)	*Thalassionema nitzschioides* 28.7% (P); *Thalassiosira* sp. 10.1% (P); *Thalassionema frauenfeldii* 9.4% (P); *Thalassiosira eccentrica* 7.1% (P); [*Thalassiothrix longissima* (O, P); *Rhizosolenia setigera* (P, ES); *Pseudonitzschia pungens* (P); *Coscinodiscus* sp. (P)].
Plesiopenaeus edwarsdianus (33)*	350 (m)	*Thalassionema nitzschioides* 30.5% (P); *Thalassiosira* sp. 10.5% (P); *Thalassionema frauenfeldii* 10.5% (P); *Thalassiosira eccentrica* 8.1% (P); [*Pseudonitzschia pungens* (P); *Rhizosolenia setigera* (P, ES); *Coscinodiscus* sp (P); *Pleurosigma diverse-striatum* (EP, E)].

* Total number of diatom species found. Abbreviations: B = benthic; CW = coastal waters; E = epipsamic; EP = epipelagic; EPI = epiphytic; ES = estuarine; N = neritic; O = oceanic ; P = planktonic.

diatoms associated with coastal and neritic environments; such is the case of the genera *Cyclotella* and *Neodelphineis*. However, there were also recognized in their gut content diatoms of an oceanic source as *Thalassiothrix longissima* and *T. heteromorpha*. Interestingly, in the 8 cases analyzed, the common planktonic diatom *Thalassionema nitzchioides* was always the most frequent in the gut content.

5 DISCUSSION

5.1 *General Issues*

Among the potential sources for organic matter in the deep sea are the coastal macroalgae and seagrass fragments (Gage & Tyler 1992). Their contribution to the deep basin of the Gulf of Mexico as particulate organic matter (POM) is likely to be associated with massive fluvial exports. The significant concentration of these allochthonous materials on the upper slope of the SW Gulf is ecologically relevant not only in terms of the energy flux that they represent to the benthic communities, but also as a factor that enhances spatial heterogeneity, as suggested by Stockton & DeLaca (1982). The downward flow of POM in the area of study may be strongly patterned by the two climatic seasons recognized in this sector of the Gulf (Soto & Escobar-Briones 1995). There is also the possibility of the flow being tied to the presence of a shelf break front that may lead to the rapid sinking of this labile material. One can not overlook the fact that the riverine nutrient enrichment may promote primary productivity (> 0.963 mg Chl a m^3; Licea 1998) in the overlying water of the shelf-slope, which eventually through passive sinking may increase the vertical flow in this region.

There is evidence that some deep sea urchins are able to metabolize macroalgae and sea grass fragments in spite of their poor nitrogen content and low decomposition rate (Suchanek et al. 1984). In this respect, the feeding guild of the natantian and reptantia decapod crustaceans captured at the upper continental shelf of the Gulf of Mexico mainly involves opportunistic scavenging and deposit sifting. Their digestive system is adapted to grind up and shred large nutritional particles. The stable carbon isotopic analysis of the muscle tissue of these bathyal inhabitants provided in this study a 'fingerprint' of their food source. The dietary carbon of the scavengers (opportunistic and predatory) mainly relies on carbon sources derived from coastal and estuarine detrital materials (^{13}C‰ –13.45 to –17.41). It may be inferred that if their carbon needs were supplied by either plankton or dissolved organic carbon, then the (^{13}C values would be more negative in the scale of –22 to –24).

The indirect evidence of stomach content analysis of 8 bathyal organisms adds support to this contention; its observation suggests the ingestion of large quantities of sedimentary organic matter that includes freshly produced estuarine phytoplankton. An earlier description of gut content in similar abyssal benthic species from the Venezuelan Basin (> 3411-5060 m) by Gore (1984, 1985) shows a remarkable similarity to our own observations. Interestingly, this author describes siliceous sponge material, large amounts of foraminiferans, and heavy fibrous plant remains presumably of terrestrial origin.

Based on the (^{13}C values, one can make some inferences as to the nature of the

benthic food web in the bathyal environment. Previous stable carbon isotopic determinations of typical inshore species (penaeid shrimp and portunid crabs) in the SW
Gulf of Mexico, showed a narrower range from –22.7 to –16.3‰ interpreted as a detrital food web with a mixed carbon source that included both estuarine and marine
input (Soto & Escobar -Briones 1995). However, in the bathyal zone of the area
studied although the food web is detrital based, the main carbon source is traced to
coastal detritus materials.

To our dismay, together with the allochthonous biogenic materials recovered from
the bathyal zone, large quantities of human generated waste were also present in
trawls made at 500 m. The deleterious effects that these pollutants may have upon the
structure and energy flow of the benthic food web have not yet been assessed. For the
past two decades, the area of study has been the convergence point of escalating industrial activity pertaining to oil and fertilizer products, and to traditional offshore
fishery. At present, because of technical constraints and high operating costs no active commercial fisheries exist in the deep water or on the edge of the continental
shelf of the Gulf of Mexico that could exploit deep-water shrimp. Included in these
groups are the genera *Mesopenaeus*, *Aristaeomorpha* and *Plesiopenaeus*, whose biomass wt may reach up to 0.01 g C m^{-2}. They are well represented in the Gulf of
Mexico (Pequegnat 1983) and in the Middle Atlantic Bight, USA (Wenner & Boesch
1979), where they reach maximum abundance at mesobathyal depths (> 600 m).

5.2 *Zonation*

Limiting depth boundaries across the continental shelf and slope has long been a de-
bated subject in benthic ecology (Pèrés 1982, Sherman et al. 1988, Emig 1997). Faunal divisions across such a depth gradient have been proposed based on physiographic or physical characteristics; at other times, arbitrary or economical divisions
have been applied. Consequently, terminology to describe faunal gradients is often
confusing. Even the use of terms like 'faunal boundary' or 'faunal zone' is considered doubtful in the examination of dynamic faunal patterns or changes in species
composition in assemblages along the continental shelf-slope coenocline (Wenner &
Boesch 1979).

The borderline between these two major environmental divisions in the SW Gulf
of Mexico depends upon shelf extension, sedimentary properties and the slope steepness. Lecuanda & Ramos-López (1998) have proposed the shelf brake (SB) as a
boundary separating the shelf from the deep realm. According to these authors, this
geomorphic feature displays in the area of study a meandering path that runs parallel
to the coast between depths of 28 and 84 m. On the other hand, the upper limit of the
slope (ULS) fluctuates from 71 down to 246 m. The area between these two boundaries is considered a transitional zone. This seems to be consistent with the criteria of
authors like Vanney & Stanley (1983) and Gage & Tyler (1992), who have emphasized the importance of the physical and biological gradients present at the shelf
edge.

The first four shelf assemblages of decapod crustaceans here described in the SW
Gulf of Mexico show overlapping patterns over much of the continental shelf, crossing the limits of the SB into the transitional zone because of the eurybathic condition
of many of their components; perhaps the only substantial changes in biomass and

species richness observed in these assemblages are attributed to seasonal variability. The transition zone represents a critical ecotone for the decapod community in which the species are depth and temperature selected. This zone serves as the point of convergence of upper slope and shelf eurybathic species. The remaining three deep-water assemblages fall within the category of the first upper slope division proposed by Hecker (1990) for the Atlantic Ocean (180-400 m). The faunal composition of the upper bathyal assemblage (Group 5) exhibits a close similarity at the generic level with that described by Gore (1984) from the Venezuela Basin, Caribbean Sea in abyssal depths (3411-5060 m).

ACKNOWLEDGEMENTS

Funds for this research were provided by the Institute of Marine Science and Limnology (ICMYL-UNAM). We extend our deep appreciation to the crew members of the RV JUSTO SIERRA and to the participating scientists in the OGMEX cruise for their invaluable support. Thanks are also extended to C. Illescas and R. Luna for data processing, P. Morales for isotopic analysis, to A. Haro for electronic instrumentation on board, and to R. Brusca and an anonymous reviewer for their comments.

REFERENCES

Crisp, D.J. 1984. Energy flow measurements. In N.A. Holme & A.D. McIntyre (eds), *Methods for the study of marine benthos*: 284-371. Norfolk, MA: Blackwell Sci. Publ.

Emig, C.C. 1997. Bathyal zones of the Mediterranean continental slope: an attempt. *Publ. Espec. Inst. Esp. Oceanogr.* 23: 23-33.

Emilsson, Y., Alatorre, M.A., Ruíz, F. & Vázquez, A. 1975. Datos oceanográficos del Golfo de México. Datos Geofísicos Serie A. Oceanografía 4. *Inst. Geofis.* UNAM: 37.

Flint, R.W. & Rabalais, N.N. 1981. Ecosystem characteristics. In Flint, R.W. & Rabalais, N.N. (eds), *Environmental studies of a marine ecosystem, South Texas outer continental shelf*: 137-156. Austin: Univ. Texas Press.

Gage, J.D. & Tyler, P.A. 1992. *Deep Sea Biology*. Cambridge: Cambridge University Press.

Gore, R.H. 1984. Abyssal lobsters, genus *Willemoesia* (Palinura, Polychelidae), from the Venezuela Basin, Caribbean Sea. *Proc. Acad. Nat. Sci. of Philadelphia* 136: 1-11.

Gore, R.H. 1985. Abyssobenthic and abyssopelagic penaeoidean shrimp (families Aristaeidae and Penaeidae) from the Venezuela Basin, Caribbean Sea. *Crustaceana* 49: 119-138.

Gore, R.H. 1985. Some rare species of abyssobenthic shrimp (families Crangonidae, Glyphocrangonidae and Nematocarcinidae) from the Venezuela Basin, Caribbean Sea (Decapoda, Caridea). *Crustaceana* 48: 269-285.

Hasle, G.R. & Fryxell G.A. 1970. Diatoms: Cleaning and mounting for light and electron microscopy. *Trans. Amer. Microsc. Soc.* 89: 469-474.

Hecker, B. 1990. Variation in megafaunal assemblages on the continental margin south of New England. *Deep Sea Res.* 37: 37-57.

Hildebrand, H.H. 1954. A study of the fauna of the brown (*Penaeus aztecus* Ives) grounds in the western Gulf of Mexico. *Publs. Inst. Mar. Sci. Univ. Tex.* 3: 229-366.

Hill, M. 1979. TWINSPAN – a FORTRAN program for arranging multivariate data in an ordered two-way table by classification of individuals and attributes. Ithaca: Cornell University.

Hutcheson, K. 1970. A test for comparing diversities based on Shannon formula. *J. Theoret. Biol.* 29: 151-154.

Licea, S. 1976. Variación estacional del fitoplancton de la Bahía de Campeche, México (1971-1972). *Symposium on progress in marine research in the Caribbean and adjacent waters.* FAO Fish Rep. 200: 253-273.

Licea, S. & Luna, R. in press. Spatio-temporal variation of phytoplankton on the continental margin in the SW Gulf of Mexico. *Rev. Soc. Mex. Hist. Nat.* 49.

Lecuanda, R. & Ramos-López, F. in press. Delimitacion de la plataforma continental mediante un criterio geomórfico. *Rev. Invest. Mar.* 19.

Longhurst, A. & Pauly, D. 1987. *Ecology of Tropical Oceans.* London: Academic Press.

Manickchand-Heileman, S., Soto, L.A. & Escobar, E.S. 1998. A preliminary trophic model of the continental shelf, southwestern Gulf of Mexico. *Estuarine, Coastal and Shelf Science* 46: 885-889.

Monreal-Gómez, M.A., & Salas de León, D.A. 1990. Simulación de la circulación de la Bahía de Campeche. *Geofis. Internac.* 29: 101-111.

Ortega-Durán, F., Ramos-López, F. & Lecuanda R. 1990. Distribución de materia orgánica en sedimentos de la plataforma y talud continental adyacente a la desembocadura de los ríos Coatzacoalcos y Tonala, Veracruz, México. Abstract. *II Congreso de Ciencias del Mar, La Habana, Cuba:* 193.

Pèrés, J.M. 1982. Ocean management. In Kinne O. (ed.), *Marine Ecology* 5(1): 9-642. Chichester: Wiley Interscience Publ.

Pequegnat, W.E. 1983. *The ecological communities of the continental slope and adjacent regimes of the northern Gulf of Mexico.* Executive Summary. US Dept. Interior. Minerals and Management Services: 40.

Rowe, G. T., Theroux, W.P., Quinby, R.W., Koschoreck, D., Whitledge, T., Falkowsky, P. & Fray, C. 1988. Benthic carbon budgets for the continental shelf south of New England. *Contr.Shelf Res.* 8: 511-527.

Sherman, K., Grosslein, M., Mountain, D., Busch, D., O'Reilley, J. & Theroux, R. 1988. The continental shelf ecosystem off the northeast coast of the United States. In D.J. Stanley & G.T. Moore (eds), *Continental shelves, Ecosystems of the World* 27: 279-338. Amsterdam: Elsevier.

Simonsen, R. 1974. The diatom plankton of the Indian Ocean expedition, R/V METEOR, 1964-1965. *Meteor Forschungsergeb. Reihe D.* 19: 107.

Sparre, P. Ursin, E. & Venema, S.C. 1996. *Introduction to tropical fish stock assessment.* Part 1. Manual. FAO Fisheries Technical Paper. 306.1: 337.

Soto, L.A. 1980. Decapod crustacean shelf fauna of the Campeche Bank: Fishery aspects and ecology. *Gulf & Caribb. Fish. Inst. Proc. 32th Sess.:* 66-81.

Soto, L.A. & Escobar-Briones E. 1995. Coupling mechanisms related to benthic production in the SW Gulf of Mexico. In A. Eleftheriou et al. (eds), *Biology and Ecology of Shallow Coastal Waters:* 233-242. Copenhagen: Olsen & Olsen.

Suchanek, T.H., Williams, S.L., Ogden, J.C. & Gill. I.P. 1984. Utilization of shallow water seagrass detritus by Caribbean deep-sea macrofauna: $\delta^{13}C$ evidence. *Deep Sea Res.* 32: 201-214.

Stockton, W.L. & DeLaca, T.E. 1982. Food falls in the deep-sea: occurrence, quality, and significance. *Deep Sea Res.* 29: 157-169.

Walsh, J.J. 1983. Death in the Sea: enigmatic phythoplankton losses. *Prog. Oceanogr.* 12: 86.

Wenner, E.L. & Boesch, D.F. 1979. Distribution patterns of epibenthic Decapoda Crustacea along the shelf-slope coenocline, Middle Atlantic Bight, USA. *Bull. Biol. Soc. Washington* 3: 106-133.

Vanney, J.R. & Stanley, D.J. 1983. The shelf break physiography: an overview. In D.J. Stanley & G.T. Moore (eds), *SEPM Special Publ.* 33: 1-24.

Background data on the decapod fauna associated with *Cymodocea nodosa* meadows in Tunisia, with observations on *Clibanarius erythropus*

KAREN J. REED & RAYMOND B. MANNING
Department of Invertebrate Zoology, National Museum of Natural History, Smithsonian Institution, Washington D.C., USA

ABSTRACT

Monthly day and night samples were taken with a pushnet on two different, shallow water *Cymodocea nodosa* meadows in the Gulf of Tunis, Tunisia, Mediterranean Sea, one at Sidi bou Said, the other at Salammbô. The samples were dominated by six caridean shrimps and one diogenid hermit crab. *Processa robusta* was the most abundant species and was present at Sidi bou Said throughout the year, in night samples only. *Palaemon serratus* was slightly more abundant than *P. elegans*, and showed its greatest abundance between May and September in both day and night samples at Salammbô. *Palaemon elegans* was most abundant at Salammbô from March to July. *Hippolyte inermis* was only slightly less abundant than *P. serratus* at Sidi bou Said. *Palaemon adspersus* was most abundant at Salammbô, where it occurred from April through June. *Hippolyte leptocerus* was found all year round at Sidi bou Said. *Clibanarius erythropus* occurred from May to January; adults dominated at day stations, juveniles at night stations.

1 INTRODUCTION

From June 1972 through June 1974 the junior author (R.B.M.) lived in Tunisia to carry out a research study on Tunisian decapod crustaceans. During this period more than 300 separate collections were made from intertidal and shallow subtidal habitats as well as from shallow sublittoral habitats. The survey included quantitative sampling with a pushnet of two similar but separate nearshore *Cymodocea nodosa* meadows in the Gulf of Tunis. Here in we provide background information on the study methods, the species collected in both quantitative and qualitative samples, and observations on temperature and salinity. In this report we present the results of our observations on *Clibanarius erythropus*, the only common shallow water hermit crab in our samples. We could find so little published information on the biology of this species that we believe our preliminary observations should be published. In three other papers in preparation we will summarize our observations and compare them with other published data on two species of *Hippolyte*, three species of *Palaemon*, and *Processa robusta*.

2 METHODS

Samples were taken at each of two sites. One site was at Salammbô (36°51′N, 10°19′E), on a flat near the Institut National Scientifique et Technique d'Océanographie et de Pêche, Salammbô, on the coast of the Gulf of Tunis at Salammbô, just north of Tunis, and the second was on the shore below Sidi bou Said (36°52′N, 10°21′E), a few kilometers north of Carthage. The Salammbô site is close to La Goulette, at the entrance to the canal from the highly polluted Lake of Tunis to the Gulf of Tunis (Fig. 1). The site at Sidi bou Said was in what was assumed to be less polluted water, primarily because of its greater distance from La Goulette.

Both sites comprised sparse *Cymodocea nodosa* Ucria meadows at a depth of less than 1 m along the shore, flanked offshore by much thicker *Posidonia oceanica* meadows. Clumps of *Caulerpa prolifera* also were interspersed with the *Cymodocea*.

Day and night samples were taken with a push net at both sites from December 1972 to January 1974, no samples were taken in November 1973. A total of 48 samples was collected. No night samples were taken in September 1973 and no day samples were taken in January 1974. Night samples were taken just after sunset.

For each sample, two stakes separated by a rope 10 m long were placed parallel to the shore near the center of each grass bed. Each side of the rope was sampled with a pushnet (Manning 1975), so that each sample was taken over 20 m of bottom. The diameter of the net mouth was 73 cm, so each sample covered 14.6 m^2. The species data taken in the quantitative samples were substantiated by data gathered from

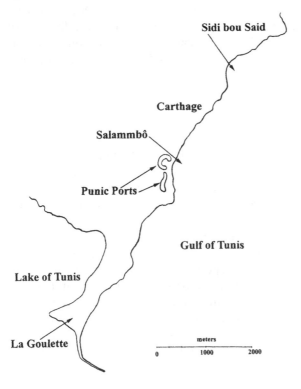

Figure 1. Sampling sites in the Gulf of Tunis.

qualitative samples which were carried out in adjacent areas on 14 different occasions at Salammbô and on 21 different ones at Sidi bou Said.

Measurements include shield length (sl), in mm, measured on the midline. Egg size is the largest diameter, in mm. Both measurements were made with an ocular micrometer in a dissecting microscope. Water temperature was measured with a thermometer; salinity was measured with a salinometer.

3 RESULTS

3.1 *Species collected*

The quantitative samples yielded more than 2200 specimens of decapods, including ten species of carideans and eight other species. Six species of caridean shrimps were dominant in our samples. A list of all species taken is given in Table 1.

3.2 *Temperature and salinity*

Observed seasonal variation in temperature, from 13-14°C in March to 27-28°C in July and September, was similar to that reported by Postel (1956) off Salammbô in 1955 and 1956 (Table 2).

3.3 *Observations on the biology of Clibanarius erythropus*

Clibanarius erythropus (n = 123) occurred from May to January (Fig. 2), with its greatest abundance from July to October; its numbers were greatly diminished in

Table 1. Species taken in quantitative and qualitative samples from Tunisian *Cymodocea* meadows.

Species	Total number	Salammbô	Sidi bou Said
Quantitative samples only:			
Processa robusta (Nouvel & Holthuis, 1957)	893	156	689
Palaemon serratus (Pennant, 1777)	486	388	98
Hippolyte inermis (Leach, 1815)	357	12	345
Palaemon elegans (Rathke, 1837)	352	266	86
Clibanarius erythropus (Latreille, 1818)	123	47	76
Palaemon adspersus (Rathke, 1837)	56	48	8
Hippolyte leptocerus (Heller, 1863)	53	7	46
All samples:			
Philocheras trispinosus (Hailstone, 1835)	35	17	18
Thoralus cranchii (Leach, 1817)	18	1	17
Philocheras fasciatus (Risso, 1816)	5	1	4
Calcinus tubularis (Linnaeus, 1767)	4	0	4
Sicyonia carinata (Brünnich, 1768)	4	1	3
Carcinus aestuarii (Nardo, 1847)	3	2	1
Macropodia czernjawskii (Brandt, 1880)	3	0	3
Athanas nitescens (Leach, 1814)	2	0	2
Liocarcinus arcuatus (Leach, 1814)	2	0	2
Pilumnus hirtellus (Linnaeus, 1761)	1	0	1
Pisa tetraodon (Pennant, 1777)	1	0	1

Table 2. Observed temperatures (°C) and salinities (‰) from December 1972 to December 1973 at Salammbô and Sidi bou Said.

		Dec	Jan	Mar	Apr	May	Jun	Jul	Aug	Sep	Oct	Dec
Salammbô	°C	14	14	13	16	21	26	28	24	28	20	14
	‰	-	39	35	34	36	36	36	38	40	38	36
Sidi bou Said	°C	16	15	14	16	19	26	27	25	27	20	15
	‰	–	37	36	37	36	35	38	39	39	36	36

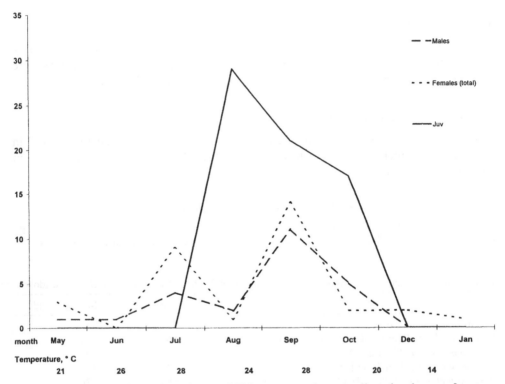

Figure 2. Monthly numbers of specimens of *Clibanarius erythropus* collected and sea surface temperature at Salammbô only, as the temperature range was similar at both stations.

May, June, November and December. It was most abundant in September (n = 46).

Males (n = 24) ranged in size from 1.4 to 3.6 mm sl; total females (n = 32) were 1.4 to 3.8 mm sl; ovigerous females (n = 5) measured 2.2 to 2.6 mm sl, and average egg size was 0.4 mm; juveniles (n = 67) measured 0.4 to 1.3 mm sl (Fig. 3).

Only five ovigerous females were taken, four in July and one in September. Ovigerous females were not the largest females in our samples.

Adults dominated at day stations (46 day vs 10 night), but juveniles were much more abundant at night than during the day (45 vs 22). Juveniles were more than seven times as abundant at Sidi bou Said than at Salammbô (59 vs 8), whereas adults were twice as abundant than juveniles (39 vs 17) at Salammbô.

Nine specimens or fewer of *C. erythropus* were taken in six qualitative samples in January, August, October and December, but 30 specimens were collected in the

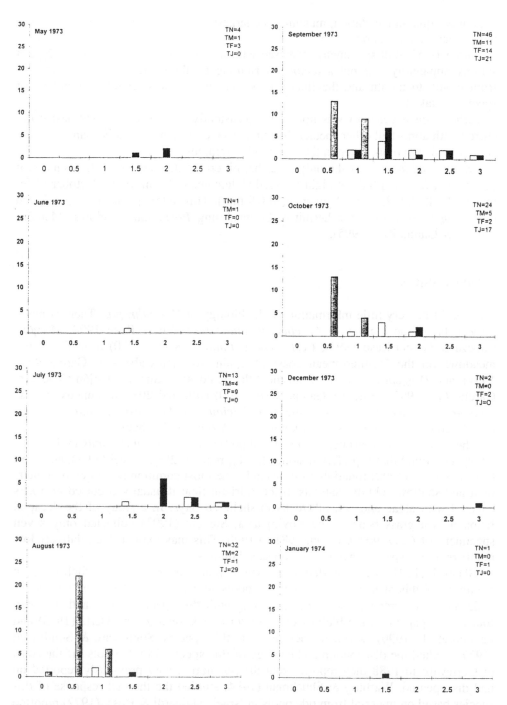

Figure 3. Length frequencies (shield lengths in mm) of *Clibanarius erythropus* collected between May 1973 and January 1974. TN, total number of specimens; TM, total number of males; TF, total number of females; TFOV, total number of ovigerous females; TJ, total number of juveniles. Clear bars = males, black bars = females, hatched bars = ovigerous females, grey bars = juveniles.

September qualitative station, matching the largest numbers taken in the same month in the quantitative stations.

Numbers of adult specimens increased with increasing temperatures (Fig. 2), but salinity apparently was not a factor contributing to abundance. Salinity fluctuated from month to month and the fluctuations were not always related to numbers of specimens taken.

Clibanarius erythropus was not taken consistently with any other species; it occurred with almost all other species. Specimens taken in our qualitative samples were similar in size to those taken in the quantitative samples.

Calcinus tubularis was the only other hermit crab collected during this survey; it occurred at Sidi bou Said at night in April (1 female, sl 2.2 mm) and October (3 females, sl 1.8, 2.2, 2.3 mm; 1 juvenile, sl 0.9 mm). This latter species has been found to be one of two dominant hermit crabs inhabiting *Posidonia* meadows (Manjón-Cabeza & García Raso 1995).

4 DISCUSSION

We could find very little information on the biology of *C. erythropus*. There is much more data on the biology of *Calcinus tubularis* (see García Raso 1990, Manjón-Cabeza & García Raso 1995), *Cestopagurus timidus* (Roux, 1830) from *Posidonia* meadows on the Mediterranean coast of Spain (Manjón-Cabeza & García Raso 1994), and *Diogenes pugilator* from the Atlantic coast of Spain (Manjón-Cabeza & García Raso 1998). Whereas García Raso (1990) collected 2084 specimens of *Cestopagurus timidus* and 1668 specimens of *Calcinus tubularis* from *Posidonia* meadows, he found no specimens of *Clibanarius erythropus* in that habitat.

Gherardi (1990) experimented on habitat preferences of *C. erythropus* in the laboratory and found that it preferred stones (35%), rocks (32%), gravel (22%), and sand (10%). Stevcic (1990) found this species to be the most common hermit crab in intertidal and shallow subtidal habitats in the Adriatic Sea; it usually occurred on rocks covered with algae and beneath rocks in sheltered bays. In a study of the decapod fauna of sea grass beds in the Rovinj area, Stevcic (1991) collected only seven specimens of *C. erythropus* from 1981 to 1990. This may explain the relatively low numbers of this species that we took on sand bottoms.

Gherardi (1991), who reported on growth, population structure, and shell usage in members of this species taken from tide pools in Sardinia, noted that 'despite its widespread occurrence, few accounts dealing with this species exist', and she cited four earlier papers on the biology of *C. erythropus* known to her. Hertz (1933) and Pessani et al. (1990) examined shell usage in this species. Southward & Southward (1977) studied the distribution and ecology of this species in tide pools off the coast of Cornwall; in 1988 they reported that the population they previously reported on had disappeared. Warburg & Shuchman (1984) studied the thermal response of this species based on material from tide pools in Israel. Mainardi & Rossi (1972) reported on social status of this species, and Pessani & Fenoglio (1994) studied its behavior. In none of these studies were samples of *C. erythropus* taken in seagrass meadows over a long time period. We have found no information on seasonal population changes in this species, so we believe our limited observations are worthy of publication.

Southward & Southward (1977) pointed out that the numbers of *C. erythropus* at one of their study sites 'fluctuated considerably'.

Southward & Southward (1988) noted that in southern England, as the ambient temperature decreased the population decreased, too, and eventually died out. Warburg & Shuchman (1984) found that the activity of *C. erythropus* eventually decreased with increasing temperature. In our samples, numbers of specimens decreased with increasing temperatures.

Gherardi (1991) reported that her samples were female biased, with females outnumbering males 2:1. In our samples, females outnumbered males with 1.3:1. The population studied by her had larger specimens (males, sl 4.85 mm; females, sl 4.38 mm); in our samples the largest male measured sl 3.6 mm, the largest female sl 3.8 mm. Our juveniles were much smaller than those reported by her (sl 0.4 mm vs sl 1.06 mm). Ingle (1993) reported that the smallest ovigerous female he examined was sl 4.1 mm and the largest specimens he examined had a sl of 8.0 mm.

Both Gherardi (1991) and Southward & Southward (1977) commented on the occurrence of aggregations of this species in small to very large groups. In our samples from grass meadows, *C. erythropus* was always widely dispersed.

Zariquiey Alvarez (1968) reported that ova measured 0.39-0.42 by 0.45-0.51 mm, similar to our observations.

In her study of the meroplankton of the Gulf of Marseille, France, Bourdillon-Casanova (1960) found ovigerous females of *C. erythropus* between May and September, with maximum abundance in June and July; our five ovigerous females in July (n = 4) and September. She noted that although ovigerous females had been reported in January, females usually carried eggs in late spring and early summer. She also found that the glaucothoe stages were abundant in the nocturnal plankton at the end of summer. All of our juveniles were collected in August, September, and October, were far more numerous at night than during the day.

Bourdon (1965) noted that ovigerous females were found from May to August at Roscoff, on the Atlantic coast of France, and Zariquiey-Alvarez (1968) reported that in Spain ovigerous females were found from May to August; García Raso (1982) found ovigerous females from June to August at Malaga on the Mediterranean coast of Spain; and Grippa (1991) collected ovigerous females in August in the Tyrhennian Sea, Italy. Lewinsohn & Holthuis (1986) reported ovigerous females in August off Cyprus. In contrast, Moncharmont (1979-1980) reported that in the Golf of Naples ovigerous females occurred all year but were most common in October. Holthuis & Gottlieb (1958) noted that ovigerous females were collected in April and June off Israel.

ACKNOWLEDGEMENTS

This study was carried out as part of a survey of Tunisian decapods with the support of the Foreign Currency Program of the Smithsonian Institution and the cooperation of the Institut National Scientifique et Technique d'Océanographie et de Pêche, Salammbô, Mr A. Azouz, Director. The support of Mr Azouz and Mme Hedia Baccar, without which this study could not have been carried out, is gratefully acknowledged. We thank Cheryl Bright, Collection Manager, Department of Invertebrate Zoology, National Museum of Natural History, for providing one of us (K.J.R.) with the uninterrupted time required to complete this study. Our colleagues Carlo Froglia and Francesca Gherardi provided us with literature which otherwise was unavailable to us, for which we are most grateful. We thank two anonymous reviewers for their suggests and corrections, which materially improved the final draft.

REFERENCES

Bourdillon-Casanova, L. 1960. Le Meroplancton du Golfe de Marseille: Les larves de Crustacés Décapodes. *Rec. Trav. Sta. Mar. Endoume* 30(18): 1-286.

Bourdon, R. 1965. Inventaire de la Faune Marine de Roscoff. Décapodes – Stomatopodes. *Éditions de la Station Biologique de Roscoff:* 1-45.

García Raso, J.E. 1982. Contribución al estudio de los Pagúridos (Crustacea Decapoda, Anomura) en el litoral submediterráneo español. *Invest. Pesquera* 46: 493-508.

García Raso, J.E. 1990. Study of a Crustacea Decapoda taxocoenosis of *Posidonia oceanica* beds from the southeast of Spain. *PSZNI, Marine Ecology* 11: 309-326.

Gherardi, F. 1990. Competition and coexistence in two Mediterranean hermit crabs, *Calcinus ornatus* (Roux) and *Clibanarius erythropus* (Latreille) (Decapoda, Anomura). *J. Exp. Mar. Biol. Ecol.* 143: 221-238.

Gherardi, F. 1991. Relative growth, population structure, and shell-utilization of the hermit crab *Clibanarius erythropus* in the Mediterranean. *Oebalia* 17: 181-196.

Grippa, G. 1991. Note sui Crostacei Decapodi dell'isola del Giglio (Arcipelago Toscano). *Atti Soc. Ital. Sci. Nat. Mus. Civ. Stor. Nat., Milan* 131 [for 1990] (24): 337-363.

Hertz, M. 1933. Über das Verhalten des Einssiedlerkrebses *Clibanarius misanthropus* gegenüber verschiedener Gehäusformen. *Z. vergl. Physiol.* 18: 597-621.

Holthuis, L.B. & Gottlieb, E. 1958. An annotated list of the decapod Crustacea of the Mediterranean coast of Israel, with an appendix listing the Decapoda of the eastern Mediterranean. *Bull. Res. Counc. Israel* 7B(1-2): 1-126.

Ingle, R. 1993. *Hermit crabs of the Northeastern Atlantic Ocean and Mediterranean Sea. An illustrated key.* London: Natural History Museum Publications and Chapman & Hall.

Lewinsohn, Ch. & Holthuis, L.B. 1986. The Crustacea Decapoda of Cyprus. *Zool. Verhand., Leiden* 230: 1-64.

Mainardi, D. & Rossi, A.C. 1972. Social status and its effects in some hermit crabs, pp. 317-326. In B. Battaglia (ed.), *Fifth European Marine Biology Symposium, Piccin Padova.*

Manjón-Cabeza, M.E. & García Raso, J.E. 1994. Estructura de una población del cangrejo ermitaño *Cestopagurus timidus* (Crustacea, Decapoda, Anomura) de fondos de *Posidonia oceanica* del SE de España. *Cah. Biol. Mar.* 35: 225-236.

Manjón-Cabeza, M.E. & García Raso, J.E. 1995. Study of a population of *Calcinus tubularis* (Crustacea, Diogenidae) from a shallow *Posidonia oceanica* meadow. *Cah. Biol. Mar.* 36: 277-284.

Manjón-Cabeza, M.E. & García Raso, J.E. 1998. Population structure and growth of the hermit crab *Diogenes pugilator* (Decapoda: Anomura: Diogenidae) from the northeastern Atlantic. *J. Crust. Biol.* 18: 753-762.

Manning, R.B. 1975. Two methods for collecting decapods in shallow water. *Crustaceana* 29: 317-319.

Moncharmont, U. 1979-80. Notizie biologiche e faunistiche sui Crostacei Decapodi del Golfo di Napoli. *Ann. Ist. Mus. Zool. Univ. Napoli* 23: 33-132.

Pessani, D. & Fenoglio, D. 1994. *Clibanarius erythropus* Crustacea, Decapoda, Diogenidae): aspetti comportamentali nelle contese intraspecifiche. *Biologia Oggi* 8: 83-90.

Pessani, D., Caltagirone, A., Palomba, I., Poncini, F. & Vetere, M. 1990. Popolazioni naturali di Clibanarius erythropus (Crustacea, Diogenidae) in relazione alla conchiglia occupata. *Oebalia* 16: 729-732.

Postel, E. 1956. Variations de la température de surface devant la Station Océanographique de Salammbô. *Bull. Sta. Océanogr. Salammbô* 53: 76-77.

Southward, A.J. & Southward, E.C. 1977. Distribution and ecology of the hermit crab *Clibanarius erythropus* in the western channel. *J. Mar. Biol. Ass. U. K.* 57: 441-452.

Southward, A.J. & Southward, E.C. 1988. Disappearance of the warm-water hermit crab *Clibanarius erythropus* from south-west Britain. *J. Mar. Biol. Ass. U. K.* 68: 409-412.

Stevcic, Z. 1990. Check-list of the Adriatic Decapod Crustacea. *Acta Adriat.* 31(1/2): 183-274.

Stevcic, Z. 1991. Decapod fauna of seagrass beds in the Rovinj area. *Acta Adriat.* 32(2): 637-653.

Warburg, M.R. & Shuchman, E. 1984. Thermal response of the hermit crab Clibanarius erythropus (Latreille) (Decapoda, Anomura). *Crustaceana* 46: 69-75.

Zariquiey Alvarez, R. 1968. Crustáceos Decápodos Ibéricos. *Invest. Pesquera* 32.

Life histories of *Triops cancriformis* and *Lepidurus apus* (Notostraca) in a group of rainpools in the Banat province in Yugoslavia

DRAGANA CVETKOVIĆ-MILIČIĆ & BRIGITA PETROV
Faculty of Biology, Institute of Zoology, Belgrade, Yugoslavia

ABSTRACT

The only place where two notostracan species *Triops cancriformis* and *Lepidurus apus* co-occur in Yugoslavia is situated in the northern Pannonian part of the country, near Melenci (Banat province). At this site *L. apus* appeared between the beginning of February and the end of April. Hatching took place when the average temperature reached 3.5°C. Sexual maturity was reached in 30 days. The average life span was about 80 days. During the reproductive season only one hatching was registered. The presence of this species at the locality was mainly dependent on temperature – it was not found at temperatures higher than 25.5°C. *T. cancriformis* was found in large bisexual populations from the beginning of March to the end of July. Hatching took place when the average temperature reached 7.5°C. Sexual maturity was reached within 20-25 days. The individual life span varied between 30 and 35 days, and that of a population between 100 and 150 days. During one season up to four hatchings may occur. The existence of this species depended on the water duration, what it is mainly independent of the temperature. In March, *Triops cancriformis* and *Lepidurus apus* co-occur, but they have never been found syntopically. They were mostly found in different periods of their life cycle: juvenile *T. cancriformis* and adult *L. apus*.

1 INTRODUCTION

Factors determining the composition of large branchiopod communities are not well known. There are opinions that the climate is one of the main governing factors (Gauthier 1928, 1930, Rzoska 1961, Belk 1977), whereas Horne (1967) emphasizes the importance of physical and chemical properties of the water and substratum. Takahashi (1976), Thiéry (1991), and King et al. (1996) emphasize that growth, reproduction, and longevity of large branchiopods are influenced by temperature, water chemistry, food, and population density. Daborn (1976) and Herbert (1982) consider that the presence of some type of segregation in feeding behavior is the most important factor enabling the co-existence of different species of large branchiopods in the community. Organisms constituting the community have different body sizes; they utilize food particles of different diameters, by which they avoid mutual competition.

Some authors consider that the occurrence of two, three, or more species of large branchiopods of one order in the same biotope is rare (Moore 1963, Huggins 1974, Mura 1985).

In Yugoslavia an area of particularly high branchiopod faunal diversity exists in the southern part of the Panonnian Plain, in the Banat province near the village Melenci. Ten species of large branchiopods (Petrov & Cvetković 1997, Cvetković 1997), among them *Triops cancriformis* and *Lepidurus apus*, are known from this site. The only co-occurrence of these two species in Yugoslavia at this site invited us to investigate their biological and ecological peculiarities and the reasons for their co-occurrence.

2 MATERIAL AND METHODS

Field studies were carried out at a site of about 0.3 hectares, situated 100 km north of Belgrade, in the vicinity of the village Melenci. The site was monitored in two-week intervals during the years 1991-1996 from the occurrence of first ponds in February until the final desiccation of the area at the end of July. The amount of water at the site and its dynamic physical and chemical properties (especially temperature and pH), as well as the vegetation and the large branchiopod assemblages (Anostraca, Notostraca, Conchostraca) were monitored. For temperature and rainfall data of the area, the official records were used.

Branchiopods were sampled with a hand net (mesh size 1 mm). The material was fixed in 70% alcohol. Sexual maturity of individuals was established, and clustering of specimens into growth classes was performed. As collecting of the material did not include sampling of larval stages, hatching time was evaluated with the aid of literature data (Marinček & Petrov 1992, Cvetković 1997).

The Fager's index of affinity (Southwood 1966) was used as a measure for the co-occurrence of the species.

3 HABITAT CHARACTERISTICS

The ponds were shallow, water depth varied from 0.10 to 0.25 m. The water was of a bicarbonate type, and pH values varied from 6 to 7.8 during the season; the physical and chemical properties indicated a certain level of pollution of agricultural origin. According to the Pantle-Buck classification, the water quality was evaluated in class III-IV. According to the Wetzel (Wetzel 1983) system, the water was between the fresh and brackish type. The salinity and conductivity were somewhat raised during the season. The amount of basic nutrients – nitrates and phosphates – indicate the eutrophic nature of the water in the ponds. The substratum was secondarily salted, with low redox potential, which indicates the presence of organic pollution by human activity.

The region experiences a continental type of climate, and a steppe (semi-arid) subtype. Within the region local meso- and microclimate differences are present. Orographic and hydrologic peculiarities, and other abiotic features create the conditions favoring the appearance of different large branchiopod associations. Some variation in the amount of water at the area between the years was found. As indicated by the vegetation (*Alopecurus pratensis, Achillea millefolium, Capsella bursa-pastoris, Seseli varium, Plantago lanceolata, Lamium purpureum, Luzula campestris,* the level of the ground water was not especially high. One-fifth of the plants belonged to the oligohaline type (*Carex divisa, Matricaria chamomilla, Festuca pseudovina, Rorippa kerneri*), because the investigated area was located on a salt-spring field.

4 RESULTS

During our investigations *Triops cancriformis* appeared almost regularly, while *Lepidurus apus* was recorded only in 1995. Thus, this was the only year when both notostracans occurred at the locality. In this year *Lepidurus apus* appeared first. It hatched at the beginning of February, when the average air temperature was 3.5°C (minimum –5°C). At the end of February juveniles were found with carapace length 0.62-1.01 cm, and a supra-anal plate from 0.03 to 0.07 cm. In the middle of March all specimens of *L. apus* were younger or adult females (Fig. 1). The maximum body size was reached at the beginning of April when individuals with carapace length of 1.34-1.82 cm, and supra-anal plate from 0.17 to 0.23 cm were collected. The abundance of the population was then decreasing. For the last time *L. apus* appeared in mid-April, when it was found in only one pool. From this moment on, *L. apus* disappeared from the site, although there was still water. The maximum air temperature at which the species was observed at the site was 25.5°C. All the time *L. apus* was present, the lowest temperatures were below zero.

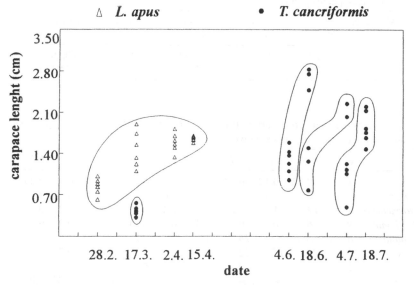

Figure 1. The time of the occurrence and carapace length of different age of *Lepidurus apus* and *Triops cancriformis*; specimens of individual hatchings grouped together.

Only females were found. Sexual maturity was reached after about one month. An individual life span and the life length of the population were about 80 days. Compared to other species recorded in 1995, *Lepidurus apus* appeared rarely (5.98%), almost equally in the single-species and in the multiple-species ponds (Table 1). This species was found in assemblages with *Branchinecta ferox*, *Chirocephalus diaphanus* and *Chirocephalus carnuntanus*. The Fager's indices of affinity with these species are shown in Table 2.

Triops cancriformis hatched at the beginning of March when average air temperature was 7.5°C (minimum –1.0°C). After the hatching, the temperature dropped below zero several times. During this month, carapace length of the juveniles was up to 0.56 cm. Because of considerably low precipitation at the end of March and the beginning of April, and in addition because of the increased temperature, all ponds inhabited by *T. cancriformis* dried out, so juveniles could not reach sexual maturity.

Raised amount of precipitation during the second decade of May caused new formation of numerous rainpools and the second hatching of *T. cancriformis* (the average air temperature was 13.7°C). Sexual maturity was reached at the beginning of June. At this time, three growth classes could be found: juveniles, younger adults and adults (Fig. 1) and 21.43% were males. Two weeks later even four growth classes were collected:

1) Juveniles from the third hatching in this season – they hatched during the first decade of June, when average temperature was 19°C, the maximum temperature reached 29°C, and the precipitation was unusually high (69 mm).

2) Younger adults from the third hatching, too.

3) Adults from the second hatching – 26.92% were males.

Table 1. Frequency of occurrence of *Triops cancriformis* and *Lepidurus apus* in all ponds, in single-species ponds, and in multiple-species ponds.

	All ponds		Single-species ponds		Multiple-species ponds	
	n	%	n	%	n	%
T. cancriformis	32	17.39	0	0.00	32	100.00
L. apus	11	5.98	5	45.46	6	54.54

Table 2. Branchiopod species that were co-occurring (number in ponds in italics) or unique (number of ponds in bold), and the Fager's affinity index between species.

	B. f.	B. s.	S. t.	C. d.	C. c.	T. c.	L. a.	E. s.	L. s.	I. b.
Branchinecta ferox	**2**	0.22	0.12	0.44	0.27	0.15	0.15	–	0.14	0.09
Branchipus schaefferi	5	**2**	0.48	0.21	–	0.61	–	–	0.73	0.59
Streptocephalus torvicornis	2	12	**1**	0.27	–	0.43	–	–	0.47	0.41
Chirocephalus diaphanus	9	6	6	**0**	0.45	0.31	0.11	–	0.27	0.24
Cirocephalus carnuntanus	8	–	–	16	**21**	0.21	0.21	–	–	–
Triops cancriformis	5	25	15	12	6	**0**	–	0.04	0.75	0.69
Lepidurus apus	2	–	–	2	6	–	**5**	–	–	–
Eoleptesheria spinosa	–	–	–	–	–	1	–	**0**	0.07	0.07
Leptestheria saetosa	5	31	17	11	–	40	–	2	**5**	0.77
Imnadia banatica	3	24	14	9	–	35	–	2	40	**4**

4) Older adults from the second hatching – 26.66% specimens were males.
In the middle of June, the population consisted of 11.11% juveniles, and 24.44% of adult specimens were males.

At the beginning of July almost all of the investigated area was covered with water. Temperatures were high, from 24 to 32°C, but less specimens could be recorded than before. They belonged to three growth classes:

1) Juveniles, whose appearance coincided with a very high amount of precipitation (73 mm) and favorable temperature conditions at the end of June (average temperature was 18.5°C). We considered that they were from the fourth hatching.

2) Younger adults from the fourth hatching too; 12.50% were males.

3) Older adults from the third hatching.

Thus, in this period, 15.38% of the population were juveniles, and 23.08% of adults were males.

T. cancriformis was present until the end of July, i.e. until the ponds dried up. At the end of the month only a few specimens (adults and older adults) were found; 33.33% were males.

The maximum air temperature at which *T. cancriformis* was observed was 33.8°C. Carapace lengths of *T. cancriformis* during the entire season varied between 0.32 cm in juveniles and 2.83 cm in adults. Sexual maturity was reached after 20-25 days, and the individual life span was 30-35 days. Life length of the population was between 100 and 150 days. During the entire season four hatchings of *Triops cancriformis* were registered. Almost one quarter of adults (24.21%) was male. Compared to other species recorded in 1995, *T. cancriformis* appeared frequently (17.39%), but only in multiple-species ponds (Table 1). It occurred in assemblages with *Branchinecta ferox*, *Branchipus schaefferi*, *Streptocephalus torvicornis*, *Chirocephalus diaphanus*, *Chirocephalus carnuntanus*, *Eoleptestheria spinosa*, *Leptestheria saetosa*, and *Imnadia banatica*, i.e. with all the species at the site except *L. apus*. The Fager's indexes of affinity, with the exception of *L. saetosa*, *I. banatica*, and *B. schaefferi*, were low (Table 2).

5 DISCUSSION

Our investigations of two species of notostracans, *Triops cancriformis* and *Lepidurus apus*, have shown some differences in the time of their occurrence and individual life span. Both species can tolerate very low temperatures at hatching as well as during the juvenile period. The presence of *L. apus* at the locality depended mainly on temperature; it appeared as a stenothermic cold-water species. *T. cancriformis* occured as a typical eurythermic species, it was present as long as pools contained water. The existence of this species depended on the duration of the water pool, irrespective of temperature.

Physiological differences and inter- and intraspecific competition might be the reasons for the striking differences in the individual life span between these similar genera, in the number of hatchings per season, and in the number of co-existing growth classes.

Lepidurus, which appeared as first during the year, had a two times longer life span than *Triops*, probably because of decreased metabolic activity and slower development at lower temperatures. On the other hand, *Triops* was present mainly in the second part of the season, when temperatures were higher; this caused an accelerated development, shortening of life span, and occurrence of several hatchings. Other reasons for this might be the limited food supplies and space competition. Cvetković (1997) reported that when only one hatching of *T. cancriformis* appears per season, the individual life span is about 80 days, however, when more than two hatchings occur it is up to 35 days.

The occurrence of several hatchings per season could be caused by heavy rains, which dilute the water and change the osmotic pressure, what stimulates a new hatching (Petrov & Cvetković 1996). The investigations of Thiéry (1997) have shown that the related species *T. granarius*, which occurs as 'peripheral' or 'edge' species, lays their cysts along the periphery of the pool in the gravel zone. Such behavior results in many cysts remaining out of water after pools incompletely dry up. This can be a reason of several hatchings in the same season found in our investigations, as heavy rains enable them to hatch when the pool is enlarged again. This was the case with *Triops,* but not with *Lepidurus,* because the temperature was too high when a new inundation happened. Eggs remained from previous seasons also could be activated by the new inundation.

Several authors give data on the co-occurrence and distribution of species of *Triops* and *Lepidurus.* Williams (1968) and Thiéry (1988) considered that *Triops* and *Lepidurus* are usually allopatric. But, there are some places where their areas overlap and species occur as sympatric, even as syntopic. Alonso (1985) reported a very interesting community of *Triops cancriformis simplex, Lepidurus apus, Cyzicus tetracerus* and *Chirocephalus diaphanus* in the La Zaida lagoon (Zaragoza) in Spain. In Austria *Triops cancriformis* was reported north of the Danube from the Lange Luss at the March River, where it was found in the same pond together with *Lepidurus apus* (Hödl & Rieder 1993), while Brtek (1976) registered both species in 0.89% of the marshes of Slovakia.

In our investigations, during the time of the co-occurrence *Triops cancriformis* and *Lepidurus apus* have been in different stages of the life cycle; in this way these ecologically similar species obviously avoid a strong interspecies competition for food and habitat niche in such small sized ponds. This was particularly emphasized at the beginning of the year, when temperatures and food resources were lower. We have registered both notostracan species at the same time. However, we have never found these two species in the same pool, which could be caused by different local conditions at the site. This points to the need of finer scale monitoring of microhabitat conditions.

ACKNOWLEDGEMENTS

We thank Mr Vjekoslav Radović for his help in writing the English version of the text. This research was supported by the National Scientific Research Found of Serbia (grant number 03E04).

REFERENCES

Alonso, M. 1985. A survey of the Spanish Euphyllopoda. *Misc. Zool.* 9: 179-208.

Belk, D. 1977. Zoogeography of the Arizona Fairy Shrimps (Crustacea; Anostraca). *Arizona Academy of Science* 12: 70-78.

Brtek, J. 1976. Anostraca, Notostraca, Conchostraca a Calanoida Slovenska (1. Cast). *Acta Rer. Natur. Mus. Nat. Slov.* 22: 19-80.

Cvetković, D.M. 1997. *Struktura i dinamika naselja Branchiopoda (Anostraca, Notostraca, Conchostraca) efemernih voda srednjeg Banata.* 132 pp. Beograd: Magistarska teza, Biološki fakultet, Univerzitet u Beogradu.

Daborn, G.R. 1976. The life history of *Branchinecta mackini* Dexter (Crustacea, Anostraca) in an agrillotrophic lake of Alberta. *Can. J. Zool.* 55: 161-168.

Gauthier, H. 1928. *Recherches sur la faune des eaux continentales de l'Algerie et de la Tunisie.* Imprim. Minerva. Algiers 419 pp.

Gauthier, H. 1930. Sur la repartition en Algerie et en Tunisie de certains Entomostraces d'eau douce Europeens (*Branchipus* et *Apus*). *Compt.-rend. Assoc. Fr. Avanc. Sci.*: 572-575.

Horne, F. 1967. Effects of physical-chemical factors on the distribution and occurrence of some southeastern Wyoming phyllopods. *Ecology* 48: 472-477.

Herbert, P.D.N. 1982. Competition in zooplancton communities. *Ann. Zool. Fenn.* 19: 349-356.

Hödl, W. & Rieder, E. 1993. *Urzeitkrebse an der March.* Orth am Donau: Distelverein.

Huggins, D.G. 1974. The sympatric occurrence of three species of Eubranchiopoda in Douglas County. *Kansas. Southw. Nat.* 20: 577-578.

King, J.L., Simovich, M.A. & Brusca, R.C. 1996. Species richness, endemism and ecology of crustacean assemblages in northern California vernal pools. *Hydrobiologia* 328: 85-116.

Marinček, M. & Petrov, B. 1992. Contribution to the knowledge of Notostraca (Branchiopoda, Crustacea) in Yugoslavia. *Bull. Mus. Hist. Nat. B* 47: 123-129.

Moore, W.G. 1963. Some interspecies relationships in Anostraca populations of certain Louisiana ponds. *Ecology* 44: 131-139.

Mura, G. 1985. Preliminary report on the sympatric occurrence of two species of fairy shrimps (Crustacea, Anostraca) in some temporary ponds from Italy. *Riv. Idrobiol.* 24: 73-82.

Pantle, R. & Buck, H. 1955: Die biologische Uberwaschung der Gewaser und die därstellung der Ergebnisse. *Gas und Wasserfach* 96, 604 pp.

Petrov, B. & Cvetković, M.D. 1996: Seasonal dynamics of *Triops cancriformis* (Bosc, 1801) (Crustacea, Branchiopoda) in the Banat province in Yugoslavia. *Tiscia* 30: 39-43.

Petrov, B. & Cvetković, D.M. 1997. Community structure of branchiopods (Anostraca, Notostraca and Conchostraca) in the Banat province in Yugoslavia. *Hydrobiologia* 359: 23-28.

Rzoska, J. 1961. Observations on tropical rainpools and general remarks on temporary waters. *Hydrobiologia* 17: 265-286.

Southwood, T.R.E. 1966. *Ecological methods.* London: Methuen & Co Ltd, 391 pp.

Takahashi, F. 1976. Pioneer life of *Triops* spp. (Notostraca: Triopsidae). *Appl. ent. Zool.* 12: 104-117.

Thiéry, A. 1988. *Triops* Schrank, 1903 et *Lepidurus* Leach, 1816, Crustaces Branchiopodes Notostraces. In Documents pour un Atlas Zoogeographique du Languedoc-Roussillon. *Universite Paul Valery, Montpellier* 33: 1-4.

Thiéry, A. 1991. Multispecies coexistence of branchiopods (Anostraca, Notostraca & Spinicaudata) in temporary ponds of Chaouia plain (western Morocco): sympatry or syntopy between usually allopatric species. *Hydrobiologia* 212: 117-136.

Thiéry, A. 1997. Horizontal distribution and abundance of cysts of several large branchiopods in temporary pool and ditch sediments. *Hydrobiologia* 359: 177-189.

Wetzel, R.G. 1983. *Limnology.* CBS College Publ. USA.

Williams, W.D. 1968. The distribution of *Triops* and *Lepidurus* (Branchiopoda) in Australia. *Crustaceana* 14: 119-126.

Differences between *Mithraculus* spp. communities in exposed and sheltered shallow-water *Thalassia* beds in Venezuela

C.A. CARMONA-SUÁREZ

Centro de Ecología. Instituto Venezolano de Investigaciones Científicas, Caracas, Venezuela

ABSTRACT

The aim of this work is to compare the community structure of spider crabs in wave exposed and sheltered *Thalassia* beds in Morrocoy National Park, Venezuela, and to determine the differences in body size of species in these areas. Two exposed (A and B) and two sheltered (C and D) shallow water *Thalassia* beds were selected in two islets in the National Park, and were sampled manually each month during one year. Three species of *Mithraculus* (Crustacea: Decapoda: Majidae) were found: *M. forceps*, *M. coryphe*, and *M. sculptus*. *Mithraculus forceps* was the most abundant species in the sheltered areas, accounting for 81% of the individuals collected at station C and 80% at station *D*. *Mithraculus sculptus* was the most abundant species at the exposed stations A and B, accounting for 59% and 68% of the total number of specimens collected at each station, respectively. Next in abundance was *M. coryphe*, with 29% (A) and 32% (B). Affinity of sampling areas was highest between the two sheltered stations C and D. Mean body size (carapace length in mm) was significantly larger in exposed areas. *M. forceps* was significantly larger in station A (12.45 mm), than in stations C (8.61 mm) and D (7.11 mm), and was absent from station B. *M. sculptus* also showed significant body size differences between stations, being larger at stations A (11.35 mm) and B (9.94 mm), than at stations C (8.28 mm) and D (7.01 mm). *M. coryphe* was significantly larger only at station A (11.15 mm), but showed no size differences between stations B (9.72 mm), C (9.28 mm) and D (9.21 mm). The appearance of *M. forceps* at both sheltered stations and at the exposed station A, may be due to the fact that these three stations were located around the same islet. Larger crabs and the greater number of berried females in the exposed stations appear mainly due to physical characteristics of the environment.

1 INTRODUCTION

Seagrass beds, especially *Thalassia testudinum*, are considered as complex ecosystems that are frequently associated with or found near coral reefs and/or mangrove forests (Hemming et al. 1994, Ogden & Gladfelter 1983, Ogden & Zieman 1977, Van Tussenbroek 1995). These highly productive ecosystems have undergone man in-

419

duced alterations (Bone et al. 1993), and suffered mass mortalities of *Thalassia testu-dinum* (Laboy 1997, Robblee et al. 1991). Little information is available on the effects of these disturbances on the structure and dynamics of the inhabitants.

Animal communities associated with *T. testudinum* were described by several authors (Heck 1977, Jiménez 1994, Vargas Maldonado & Yánez-Arancibia 1987). Carmona (1993) described the community structure of brachyuran crabs in Morrocoy National Park (northwestern Venezuela). Although this conservation site has been the target of a great deal of research (Klein & Bone 1995), little attention has been given to brachyurans living in *T. testudinum* even though they are an important component of plant ecosystems (Hines 1982, Vélez 1977, Holmquist et al. 1989). The aim of this work is: 1) To compare the community structure of spider crabs in wave exposed and sheltered *Thalassia* beds in Morrocoy National Park to obtain information concerning the distribution of different species of majids; and 2) To determine differences in body size of species in exposed and sheltered areas so as to gather basic information for future comparative studies. This work will also serve as a baseline to measure possible effects on these communities during or after drastic environmental changes.

2 MATERIALS AND METHODS

The Morrocoy National Park is located on the western coast of Venezuela between 10°47′-10°59′N, and 68°09′-68°22′W (Fig. 1). The main coastal marine ecosystems of this region are coral reefs, mangroves, and beds of the turtle seagrass *Thalassia testu-*

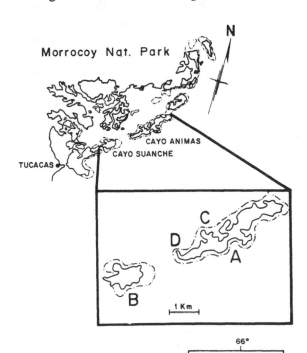

Figure 1. Morrocoy National Park (Venezuela) showing sampling sites. A and B: Wave-exposed stations. C and D: Sheltered stations.

dinum, which together with the macroalga *Halimeda* sp. are the most abundant macrophytes. For the purpose of this work, four shallow water sampling areas with *T. testudinum* were selected in two islets that border the internal part of the park (Fig. 1). Sampling areas were located on the exposed coast (A and B) and the sheltered side of one islet (C and D). Wave action in the exposed areas was strong, while almost absent from the sheltered areas. In all four sampling areas depth ranged from the 0-0.40 m. The bottom in areas A and B consisted of dead coral fragments and living *Porites* sp., and was exposed to the atmosphere during extreme low tides. The sheltered areas (C and D) were never exposed to the atmosphere, and the sediment was predominantly muddy. *Thalassia* in the sheltered sites was often found mixed with *Halimeda* sp.

Monthly sampling was conducted from October 1993 to November 1994, and each area was searched haphazardly by hand, all crabs were captured within a one hour period. Animals were identified according to the keys of Rodríguez (1980) and Wagner (1990). The carapace length was measured and the animals were sexed. Also the state of maturity and the number of berried females was determined. Salinity (ppt) was measured with a hand refractometer. Water surface temperature was taken with a mercury thermometer, and dissolved oxygen was determined with a YSI Oxygenmeter. These parameters were measured during each month, except for August and November 1994.

A R × C G-test of independence (Sokal & Rohlf 1995) was used to determine the relationship between community structures in the four sampled areas. Cluster analysis (single linkage, Euclidean distances) was used to determine the similarity in community structure between the sampled areas, using the statistical computer program Statistica (Statsoft Inc. 1993). Differences in body size (carapace length in mm) of the same species at different localities were analyzed using the t-test.

3 RESULTS

A total of 1089 crabs were collected. Three species of *Mithraculus* were found: *M. forceps* (427 individuals), *M. coryphe* (215), and *M. sculptus* (447). *M. forceps* was the most abundant in the sheltered areas: 81% (n = 162) at station C and 80% (n = 210) at station D. On the other hand, *M. sculptus* was the most abundant species at the exposed stations A and B, accounting for 59% (n = 275) and 68% (n = 112) of the total number of spider crabs at each station, respectively. These were followed in abundance by *M. coryphe*, with 29% (A) and 32% (B) (Fig. 2).

In the monthly samples, *M. forceps* appeared at station A only in July, August, and September 1994 (Fig. 3a). On the other hand, at station C, *M. sculptus* appeared in April and August 1994, and *M. coryphe* in August 1994 (Fig. 3c). At station D, *M. sculptus* appeared in February and August 1994, and *M. coryphe* in August and September 1994 (Fig. 3d).

Overall, species composition was highly dependent on the degree of exposure and type of substratum ($G_{adjust} = 664.4$; df = 6; $p < 0.001$). Species composition was independent of station in the sheltered areas C and D ($G_{adjust} = 1.6$; df = 2; $p > 0.05$; n.s.). On the contrary, species composition depended on station in the exposed areas A and B ($G_{adjust} = 34.9$; df = 2; $p < 0.001$). Cluster analysis revealed that the sheltered stations C and D were closely linked, while these had a distant association with the rest of the stations, with station B the least similar to the others (Fig. 4).

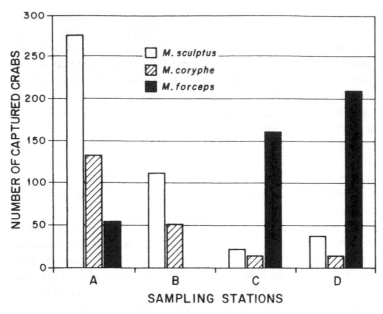

Figure 2. Total number of spider crabs collected from each sampling station from October 1993 to November 1994.

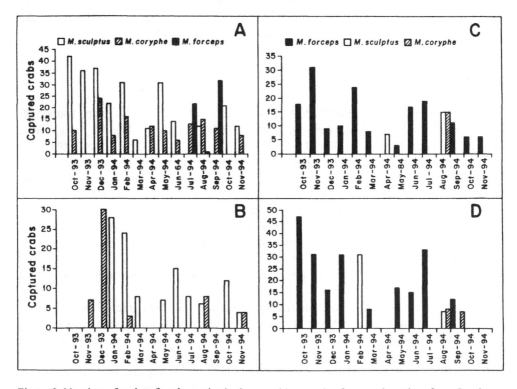

Figure 3. Number of crabs of each species in the monthly samples from each station, from October 1993 to November 1994.

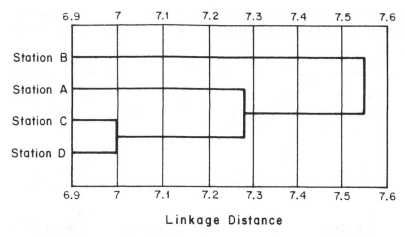

Linkage Distance

Figure 4. A graph representing affinities in species composition between the four sampling stations.

The mean, maximum, and minimum body size (carapace length in mm) for each species in each sampling area is shown is Table 1. The mean body size was significantly larger in the exposed areas (Table 2). *M. forceps* was significantly larger at station A (12.45 mm), than at stations C (8.61 mm) and D (7.11 mm). *M. forceps* was not present at station B. *M. sculptus* also showed significant body size differences between stations and was larger at stations A (11.35 mm) and B (9.94 mm) compared to stations C (8.28 mm) and D (7.01 mm). *M. coryphe* was significantly larger at station A (11.15 mm) only and showed no significant size differences between stations B (9.72 mm), C (9.28 mm) and D (9.21 mm).

The sex ratio (male:female) differed significantly from 1:1 (Table 3); males were almost always predominant over females, except for *M. coryphe* at station C, where the sex ratio was 0.7:1. Sex ratios in *M. sculptus* were dependent on the station but not in the other species. The sex ratio from pooled data was 1.5:1 for *M. coryphe*, and 1.7:1 for *M. forceps*.

At all stations, the percentage of berried females showed the following order: *M. coryphe* > *M. sculptus* > *M. forceps* (Fig. 5). Berried females of the most abundant species at exposed stations (*M. sculptus*) were present during almost all months, except for March, April, and September 1994 (Fig. 6a). In contrast, the most abundant species at the sheltered stations (*M. forceps*) had berried females only in November 1993 (at station C) and in July 1994 (at station D) (Fig. 6b).

The surface water temperature did not vary greatly during the year in any of the sampled areas, and t-tests showed no differences between the areas. For this reason, data were pooled, with a mean of 30.55°C and standard deviation of 1.5 (n = 47). The maximum temperature was registered at station A (34°C), the minimum was measured in station C (27°C). No differences in salinity were found between the sampled areas, except between stations A and C (t = –2.35; df = 22; p < 0.05). The highest values were measured at stations B, C, and D (all 34‰), and the lowest at station D (30‰). Dissolved oxygen was significantly higher at stations A and D (Table 4); however, there were no differences between these two stations (Table 5). Stations B and C had lower oxygen values, without statistical differences between these (Table 5).

Table 1. Total number (n) and body size (range and mean of carapace length in mm; standard deviation s between parentheses) of majid crabs from stations A-D in Morrocoy National Park, during 1993-1994.

Station	A	B	C	D
M. sculptus				
n	263	142	22	38
Range	2.30–19.50	4.50–17.40	5.00–14.80	4.02–11.60
Mean (s)	11.35 (3.59)	9.94 (2.86)	8.28 (2.19)	7.01 (2.24)
M. coryphe				
n	134	22	14	15
Range	4.70-16.50	7.20-12.10	8.00-11.25	4.50-13.00
Mean (s)	11.15 (2.82)	9.72 (1.23)	9.28 (1.00)	9.21 (2.76)
M. forceps				
n	55	0	161	210
Range	5.90–18.40	–	2.60–21.00	4.30–24.10
Mean (s)	12.45 (2.41)		8.61 (3.42)	7.11 (2.57)

Table 2. Difference in body size (t-test) of three majid species from stations A-D in Morrocoy National Park, during 1993-1994. t = value of t; df = degrees of freedom; p = significance; (*) significant; (**) highly significant.

	A/B	A/C	A/D	B/C	B/D	C/D
M. sculptus						
t	4.128	4.052	7.434	2.599	5.841	2.133
df	403	283	299	162	178	58
p	0.000 (**)	0.000 (**)	0.000 (**)	0.010 (*)	0.000 (**)	0.037 (*)
M. coryphe						
t	2.339	1.119	0.763	2.460	2.532	0.088
df	154	34	35	146	147	27
p	0.021 (*)	0.271 (n.s.)	0.450 (n.s.)	0.015 (*)	0.012 (*)	0.931 (n.s.)
M. forceps						
t	–	7.677	13.871	–	–	4.826
df	–	214	263	–	–	369
p	–	0.000 (**)	0.000 (**)	–	–	0.000 (**)

Table 3. Sex ratio of three majid crab species from stations A-D in Morrocoy National Park, during 1993-1994. Intraspecific difference of sex ratio tested with G-test.

	M. sculptus			*M. coryphe*			*M. forceps*		
	Males	Females	Ratio	Males	Females	Ratio	Males	Females	Ratio
A	198	77	2.6:1	86	48	1.2:1	42	13	3.2:1
B	86	56	1.5:1	12	10	1.8:1	0	0	
C	15	0	15:0	6	9	0.7:1	97	64	1.5:1
D	28	10	2.8:1	9	6	1.5:1	95	58	1.6:1
G	16.815			3.533			5.026		
Df	3			3			3		
S	P «0.001 (**)			n.s.			n.s.		

df = degrees of freedom; s = significance; (**) highly significant; n.s.: not significant.

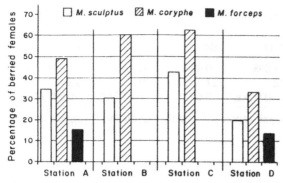

Figure 5. Percentage of berried females for each species of crab in the exposed (A and B) and sheltered (C and D) stations (October 1993 to November 1994).

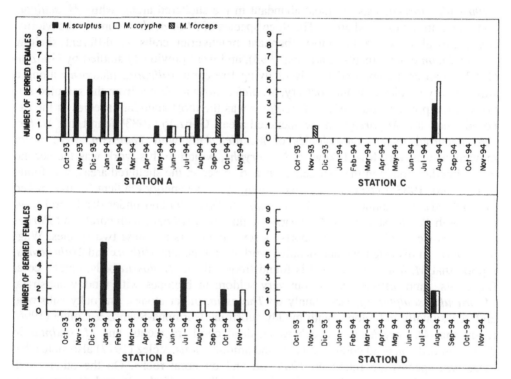

Figure 6. Berried females in the monthly samples for each species of crab in the exposed (A and B) and sheltered (C and D) stations (October 1993 to November 1994).

Table 4. Dissolved oxygen concentration (mg l^{-1}) at four stations A-D in Morrocoy National Park, between October 1993 and October 1994.

	n	mean	s.d.	min.	max.
A	12	9.50	1.93	7.0	12.0
B	11	6.91	1.30	5.0	9.0
C	12	6.67	1.37	4.0	8.0
D	12	8.25	1.21	6.0	10.0

Table 5. Differences (t-test) in dissolved oxygen concentrations between stations A-D in Morrocoy National Park, during 1993-1994.

	A/B	A/C	A/D	B/C	B/D	C/D
t	3.738	4.145	1.898	0.434	−2.556	−2.994
df	21	22	22	21	21	22
p	0.001 (**)	0.000 (**)	0.071 n.s.	0.669 n.s.	0.018 (*)	0.007 (**)

t = t-value; df = degrees of freedom; p = significance; (**) highly significant; (*) significant; n.s. not significant.

4 DISCUSSION

Mithraculus forceps was the most abundant in the sheltered areas, while *M. sculptus* dominated in the exposed areas. The three species of *Mithraculus* found in the Morrocoy National Park are the most abundant brachyuran crabs in different shallow *T. testudinum* environments (Carmona 1993), and were previously studied by Gonzalo (1983) in an area composed mainly of living fire coral *Millepora alcicornis* along a wave energy gradient. On the contrary, *M. alcicornis* was absent from the sampled stations in the present study. Also, *M. sculptus* was the most abundant species in the exposed areas, and *M. forceps* in the sheltered areas. Gonzalo (1983) found that *M. forceps* occurred mainly on colonies of *M. alcicornis* with *Thalassia*, were water movement was very low. Likewise, *M. coryphe*, which in this work was the second most abundant majid in the exposed areas and scarce in the sheltered areas, was found by Gonzalo (1983) to be rather common along the wave energy gradient (from exposed to sheltered). The abundance of *M. sculptus* on *Thalassia* and under dead coral fragments with *Porites* sp., and of *M. coryphe* mainly on *Thalassia* with corals and *Porites* sp., agrees well with published reports by several authors for these two species. Wagner (1990), in his extensive and detailed work on the genera *Mithrax* and *Mithraculus*, reports that *Mithraculus coryphe* is found frequently in *Thalassia* beds, corals, coral fragments, sand, between algae, but very seldom in biotopes with muddy substrata. *Mithraculus sculptus* appears mainly on *Thalassia*, under stones, on rocky substrata, and occasionally in biotopes with muddy substrata. Carmona-Suárez & Conde (1996) report that *M. coryphe* is found between dead coral fragments, and *M. sculptus* in *Thalassia* beds on dead coral fragments and among corals. Vélez (1977) also states that these two species are found on *Thalassia* and on coral rocks. On the other hand, *Mithraculus forceps*, which was found principally on *Thalassia* and *Halimeda* sp. growing on a muddy substratum, differed from the habitat characteristics given for this species by Carmona-Suárez & Conde (1996), Vélez (1977), Wagner (1990) and Williams (1984). These authors state that this is a spider crab found commonly on *Thalassia* growing on sandy substrata, on or under stones and dead coral, on rocky shores, and occasionally on *Thalassia* on muddy substrata. This difference may lie in the data given by the previously mentioned authors, which are mainly for inventory and taxonomic purposes, and where no information on frequency of this species is given. In the present work, a clear view of habitat preference is given for this species.

Mithraculus forceps appeared at stations A, C, and D, but was absent from station B. As the three stations are located around the same islet, migration of crabs from the sheltered stations (where *M. forceps* was most abundant) to the exposed station A is possible. This may also explain why the species composition between the exposed stations A and B is different, because at these two stations *M. sculptus* is the predominant species, followed by *M. coryphe*, but *M. forceps* was absent from station B (Fig. 2), which lies on another islet (Fig. 1). The absence of this species at this station may be due to the water current direction, which flows mainly westwards. The two islets are located on a northeastern axis, being the islet with stations A, C and D on the northern site, thus avoiding possible migration of *M. forceps* towards the southerly islet (station B). The high affinity between stations C and D (Fig. 4) was probably due to the position of these stations on the same islet and to their similar environmental conditions: *Thalassia* with *Halimeda* sp. on muddy substrata in a sheltered area.

The absence of regular population density fluctuations (Fig. 3) and the fact that berried females of any of the crab species did not appear the year round, suggest that there are no stable environmental conditions. In fact, due to the extreme shallow conditions water surface temperature can reach high values especially during extreme low tides when parts of the substrate are exposed to the atmosphere. In the sheltered areas, low water movement (or even absence of currents) can cause stagnation inducing a raise in temperature and a decrease in the dissolved oxygen values, thus causing adverse conditions in the *Thalassia* complex. This was observed in another very shallow *Thalassia* system, located on the western coast of Venezuela at 11°45'N 69°40'E. Massive mortalities of diverse marine organisms were observed in April 1991, including *Diadema antillarum*, seacucumbers, crabs from the families Calappidae and Portunidae, as well as fish belonging to the families Scorpaenidae, Acanthuridae, and the species *Abudefduf saxatilis*. This happened in a matter of hours, when wind stopped blowing and water stagnated, raising the salinity to 40‰, and the temperature to 34°C, where values normally lie around 37‰ and 27°C (Carmona, personal observations).

In the sheltered areas, less berried females were encountered, and mean specimen size was smaller than that for the same species in the exposed areas, pointing to environmental conditions as a possible source. Evidence of environmental stress causing size reduction in brachyuran populations were described by Brown & Caputti (1985), Carmona-Suárez (1992), Conde & Díaz (1992), and Conde et al. (1989). Differences in community structure may also be caused by differences in physical factors, as was determined by Holmquist et al. (1989) in a seagrass complex in Florida Bay (USA). The latter authors suggested that various physical factors play a greater role on decapod and stomatopod assemblages in different sub-environments of seagrass ecosystems (mainly *Thalassia*), than do the vegetational characteristics of the habitat.

The study of community structure of species living in seagrass meadows of *Thalassia testudinum* is important in order to understand the factors influencing changes in the species composition of assemblages occurring in different sub-environments. It is especially important to have a baseline, which will serve as a comparison for the monitoring of communities in the same area after an ecological catastrophe, as happened in Morrocoy National Park in January and December 1996 (Laboy 1997). Thus, results on the community structure of majid crabs obtained in the present work, will also serve as a baseline against which to measure changes in the future.

ACKNOWLEDGMENTS

My thanks to Sebastián Trómpiz, Omégar Céspedes and Angel López for their help during the field trips. This work was also possible due to the permission granted by INPARQUES. My deepest gratitude to Dr Jesús Eloy Conde (IVIC – Venezuela) for the thorough and critical reading of this manuscript, as well as for his valuable advice. I would like to thank also Dr Dan Rittschoff for his helpful hints, as well as Dr Richard von Sternberg for his observations on the manuscript. Most of this work was developed during my appointment as a staff professor of the Universidad Francisco de Miranda (Venezuela).

REFERENCES

Bone, D., Losada, F. & Weil, E. 1993. Orígen y efectos de la sedimentación sobre las comunidades coralinas del Parque Nacional Morrocoy, Venezuela. *Ecotrópicos* 6: 10-21.

Brown, R.S. & Caputti, N. 1985. Factors affecting the growth of undersize western rock lobster, *Panulirus cygnus* George, returned by fishermen to the sea. *Fish. Bull.* 83: 567-574.

Carmona, S.C. 1993. Estructura de la comunidad de cangrejos en campos someros de *Thalassia* en el Parque Nacional Morrocoy- Venezuela. *V Congreso Latinoamericano en Ciencias del Mar.*México. (Abstract) C-213.

Carmona-Suárez, C.A. & Conde, J. 1996. Littoral brachyuran crabs (Crustacea: Decapoda) from Falcón, Venezuela, with biogeographic and ecological remarks. *Rev. Bras. Biol.* 56: 725-747.

Carmona-Suárez, C.A. 1992. Interpopulation size variation in a tropical decorator crab, *Microphrys bicornutus* (Latreille, 1825) (Decapoda, Brachyura, Majidae). *Crustaceana* 63: 319-322.

Conde, J.E. & Díaz, H. 1992. Extension of the stunting range in ovigerous females of the mangrove crab *Aratus pisonii* (H. Milne- Edwards, 1837) (Decapoda, Brachyura, Grapsidae). *Crustaceana* 62: 319-322.

Conde, J.E., Díaz, H. & Rodríguez, G. 1989. Crecimiento reducido en el cangrejo de mangle *Aratus pisonii* (H. Milne- Edwards) (Brachyura: Grapsidae). *Acta Cient. Venez.* 40: 156-160.

Gonzalo, D. 1983. Patrones reproductivos de tres especies de *Mithrax* (Crustacea: Decapoda: Majidae) en la zona Mayorquina (Parque Nacional Morrocoy). Trabajo Especial de Grado. Universidad Central de Venezuela. Caracas.

Heck, K.L. Jr. 1977. Comparative species richness, composition, and abundance of invertebrates in Caribbean Seagrass (*Thalassia testudinum*) Meadows (Panama). *Mar. Biol.* 41: 335-348.

Hemming, M.A., Slim, F.J., Kazungu, J., Ganssen, G.M., Nieuwenhuize, J. & Kruyt, N.M. 1994. Carbon outwelling from a mangrove forest with adjacent seagrass beds and coral reefs (Gazi Bay, Kenya*). Mar. Ecol. Prog. Ser.* 106: 291-301.

Hines, A.H. 1982. Coexistence in a kelp forest: size, population dynamics, and resource partitioning in a guild of spider crabs (Brachyura, Majidae). *Ecol. Monog.* 52: 179-198.

Holmquist, J.G., Powel, G.V.N. & Sogard, S.M. 1989. Decapod and stomatopod assemblages on a system of seagrass- covered mud banks in Florida Bay. *Mar.Biol.* 100: 473-483.

Jiménez, M. 1994. Comunidad de moluscos asociados a *Thalassia testudinum* en la Ensenada de Reyes de Mochima, Edo. Sucre, Venezuela. *Bol. Inst. Oceanog. de Venezuela. Univ. de Oriente* 33: 67-76.

Klein, E. & Bone, D. 1995. Ecosistema Morrocoy: estado actual de las investigaciones y perspectivas futuras. INTECMAR. Universidad Simón Bolívar. Caracas.

Laboy, E.N. 1997. Factores ambientales que limitan la distribución y abundancia de *Isostichopus badionotus* y *Holothuria mexicana* en el Parque Nacional Morrocoy. Tesis Doctoral, IVIC, Caracas.

Ogden, J.C. & Gladfelter, E.H. 1983. Coral reefs, seagrass beds and mangroves: their interaction in the coastal zones of the Caribbean. *UNESCO Reports in Marine Science* 23: 1-139.

Ogden, J.C. & Zieman, J.C. 1977. Ecological aspects of coral reef-seagrass bed contacts in the Caribbean. *Proc. 3rd Internat. Coral Reef Symposium*: 377-382. Miami: Rosenstiel School of Marine and Atmospheric Science.

Robblee, M.B., Barber; T.R., Carlson, P.R., Durako, M.J., Fourqurean, J.W., Muelstein, L.K., Porter, D., Zieman, R.J. & Zieman, J.C. 1991. Mass mortality of the tropical sea grass *Thalassia testudinum* in Florida Bay (USA). *Mar. Ecol. Prog. Ser.* 71: 297-299.

Rodríguez, G. 1980. *Crustáceos decápodos de Venezuela*. Caracas: Instituto Venezolano de Investigaciones Científicas.

Sokal, R.R. & Rohlf, F.J. 1995. *Biometry*. 3rd. Edition. New York: Freeman.

Van Tussenbroek, B.I. 1995. *Thalassia testudinum* leaf dynamics in a Mexican Caribbean coral reef lagoon. *Mar. Biol.* 122: 33-40.

Vargas Maldonado, I. & Yánez-Arancibia, A. 1987. Estructuras de las comunidades de peces en sistemas de pasto marinos (*Thalassia testudinum*) de la Laguna de Términos, Campeche, México. *An. Inst. Cien. del Mar y Limnol. Univ. Nal Autón. Mex.* 14: 181-196.

Vélez F., M.M. 1977. Distribución y ecología de Majidae (Crustacea: Brachyura) en la región de Santa Marta, Colombia. *An. Inst. Inv. Mar. Punta Betín.* 9: 109-140.

Wagner, H.P. 1990. The genera *Mithrax* Latreille, 1818 and *Mithraculus* White, 1847 (Crustacea: Brachyura: Majidae) in the western Atlantic Ocean. *Zool. Verhand. Leiden* 264: 1-65.

Williams, A.B. 1984. *Shrimps, lobsters, and crabs of the Atlantic coast of eastern United States, Maine to Florida*. Washington: Smithsonian Institution Press.

Allocation of the portunid crab *Callinectes ornatus* (Decapoda: Brachyura) in the Ubatuba Bay, northern coast of São Paulo State, Brazil

FERNANDO LUIS MEDINA MANTELATTO

Departamento de Biologia, Faculdade de Filosofia, Ciências e Letras de Ribeirão Preto, Universidade de São Paulo (USP), Brazil

ABSTRACT

The distribution and habitat associations of the crab *Callinectes ornatus* in the Ubatuba Bay (São Paulo, Brazil) were investigated as a function of environmental factors (temperature, salinity and dissolved oxygen of the bottom water, depth, granulometric composition and organic matter of the sediment) and reproductive condition. Samples were taken monthly on three consecutive days over a one-year period (September 1995 to August 1996) using double rig trawl nets. Collections were made during daylight along eight transects of 1 km length each. The correlation of the abundance of males, non-ovigerous females, females and juveniles with the physical factors was calculated using Pearson's linear correlation. The heterogeneous spatial and seasonal distribution of the groups, particularly of ovigerous females and juvenile crabs, was associated with several factors, mainly water depth and temperature, and the texture and organic matter of the sediment. The distribution pattern shown by *C. ornatus* appeared to support the hypothesis of allocation in function of immature and mature condition, with the establishment of sites favorable to protection and egg incubation by ovigerous females and larval dispersion.

1 INTRODUCTION

Studies conducted on natural populations have shown that the individuals tend to follow a spatial distribution consisting of a clear pattern of individual clusters (Raup & Stanley 1978). In the marine environment, because of its vast extension and complex interaction of environmental factors, these patterns often become obscure. For this reason, continuous and detailed studies of species with representative abundance in small areas, as bay and inlets, may contribute to the understanding of this mechanism in marine complexes.

Although the individual or collective action of environmental factors can increase or limit the area of benthic species distribution (Pinheiro et al. 1996), these distribution patterns may differ between regions and/or species. However, intra- or interspecific factors (segregation of the sexes, competition, prey-predator relations, reproductive and molt cycles, among others) can act together with the environmental factors. This selective pressure, which changes during the life cycle of the animals, should be taken into consideration.

The Portunidae is one of the best known crab families, in function of its ecological importance, and as a promising source of food in coastal waters. Twenty-one portunid species, including *Callinectes ornatus*, have been recorded from Brazilian waters (Melo 1996). The general life history characteristics of this relatively large portunid crab of commercial value, has been documented along the western Atlantic coast. Brues (1927) mentioned the occurrence of this species in waters of 9‰ salinity; Williams (1966) and Gore (1977) compared the systematic diagnosis of this species in relation to the genus *Callinectes*; Román-Contreras (1986) and Buchanan & Stoner (1988) investigated ecological and distribution aspects in Mexico and Porto Rico respectively, and Haefner (1990a, b) reported some aspects of the natural diet and morphometric maturity of crabs in Bermuda.

The biology of *C. ornatus* from the Brazilian coast has also been investigated. Branco & Lunardon-Branco (1993a, b) reported on the growth, sexual maturity, and population aspects of the species from the Paraná State. Pita et al. (1985) reported ecological aspects in estuarine habitats of Santos along the northern coast of São Paulo. The ecology of the species was comparable to *Callinectes danae* (Negreiros-Fransozo & Fransozo 1995); size maturity, fecundity, and reproductive and molt cycles were studied by Mantelatto & Fransozo (1996, 1997, 1999, respectively). Negreiros-Fransozo et al. (1999a) studied the population structure, sex-ratio and reproductive season in the Ubatuba region. The above studies have also provided a basic understanding of the population biology and ecology of *C. ornatus*. However, patterns of habitat utilization associated with the molt and reproductive stage of individual are poorly understood

This paper describes the distribution of *C. ornatus* in the Ubatuba Bay (São Paulo State) in relation to abiotic parameters of the water and sediment, and reproductive conditions (gonadal development and maturity size). I hypothesize that the spatial and seasonal distribution and sex ratio of this species in this bay is correlated with environmental bottom conditions and/or reproductive strategy.

2 MATERIAL AND METHODS

2.1 *Study area*

The Ubatuba Bay (23°26'S-45°02'W) is located along the northern coast of Brazil. The total area of the bay is 8 km^2, and the width is 4.4 km at the entrance and decreases landward (Fig. 1). Four small rivers flow into this bay (Indaiá River, Grande River, Lagoa River and Acaraú River), which supply organic matter, and industrial effluents.

A preliminary sampling was carried out during six months to establish a sampling design for the present study. The sampling areas were chosen according to the principal sites of occurrence of *C. ornatus* in the bay, irrespective of abiotic conditions. Biological and environmental data were collected along transects of 1 km (Fig. 1). Detailed descriptions of the Ubatuba Bay in terms of physical and chemical features can be found in Mantelatto & Fransozo (1999b).

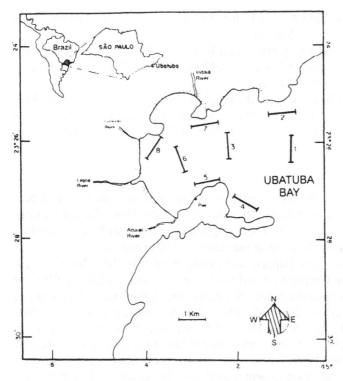

Figure 1. Ubatuba Bay with the position of the eight sampling tracks in the respective subareas.

2.2 *Site description*

– Subarea 1 is located in the mouth of the bay, and is exposed to the open sea. The current energy is high. The mean depth (MD) is 16.6 ± 0.89 m, and the organic matter content (OMC) of the sediment is 5.6 ± 5.10%. The predominant granulometric fraction (PGF) is very fine sand (75.9 ± 10.21%).

– Subarea 2 is exposed to the open sea and undergoes high wave action. It is lined by a rocky coast, which acts as a barrier to the waves, inducing strong breakers. The MD is 11.4 ± 0.62 m, and the OMC is 5.4 ± 2.91%. The PGF is very fine sand (71.6 ± 9.14%).

– Subarea 3 is located in the middle of the bay. The MD is 10.7 ± 0.72 m, and the OMC is 13.2 ± 1.47%. The PGF are very fine sand (28.0 ± 5.20%) and silt/clay (26.7 ± 8.23%).

– Subarea 4 has calm water because it is located in a sheltered part of the bay. In the coast there is an inclined small beach, situated between inclined rock shelf, with a calm wave action. The MD is 9.5 ± 1.00 meters, and the OMC is 18.5 ± 9.28%. The PGF are fine sand (36.8 ± 5.25%) and medium sand (24.5 ± 4.97%).

– Subarea 5 is located parallel to the inclined rock shelf. Residential developments are carried out along the coast, including a parking for fishing boats. The MD is 7.9 ± 1.11 m, and the OMC is 20.3 ± 6.35%. The PGF are fine sand (21.7 ± 4.86%) and medium sand (16.5 ± 2.16%).

– Subarea 6 is very similar to subarea 4. It is located in the middle of the bay, and directed towards the bay mouth in front of the Iperoig and Itagua beaches and Lagoa River. There is a littoral plain with many residencies producing domestic sewage.

The MD is 7.6 ± 0.52 m, and the OMC is 14.6 ± 2.11%. The PGF are very fine sand (23.3 ± 6.45%) and silt/clay (31.3 ± 8.64%).

– Subarea 7 is parallel to the rocky coast, and located on the left side of the estuary. The wave energy is high. The MD is 7.3 ± 0.42 m, and the OMC is 6.1 ± 2.31%. The PGF is very fine sand (70.8 ± 8.31%).

– Subarea 8 is located in front of the estuary of the Grande Ubatuba River. This site is influenced by tidal movements. The MD is 3.1 ± 0.31 m, and the OMC is 7.0 ± 1.50%. The PGF are very fine sand (54.5 ± 11.96%) and silt/clay (30.8 ± 10.91%).

2.3 *Brachyuran sampling*

Crabs were collected monthly on three consecutive days with double rig trawl nets (with a 10 mm mesh cod end) from September 1995 to August 1996. Each day, samples were made along eight transects of 1 km length during daylight hours. Trawling was at a constant speed, and the duration of the trawl was 20 minutes.

After collection, the crabs were bagged and stored frozen. In the laboratory, the material was thawed at room temperature and the sex and carapace width (CW) were determined. CW was measured to the nearest 0.1 mm between the bases of the last and penultimate lateral spines with a caliper rule. The individuals were classified into four groups: mature males (MM), mature females (MF), ovigerous females (OF) and immature males and females (IM). Maturity was defined based on criteria described by Mantelatto & Fransozo (1996, 1999a), observing the external morphology of the abdomen and the stage of gonadal development (based on color, shape, and morphology) after dissecting the crabs. According these parameters, all specimens of CW < 48.0 mm (males) and < 42.0 mm (females) were considered to be immature individuals.

The abundance (standard number of crabs) was calculated based on the total number of swimming crabs collected with three trawls on each transect per month. For a better visualization of the results, the months were grouped into seasons: spring (September, October and November), summer (December, January and February), autumn (March, April and May) and winter (June, July and August).

2.4 *Abiotic factors*

Data concerning environmental factors were obtained before trawling at three points along the trawl track in each subarea. Bottom water samples were collected with a Nansen bottle. Water temperature was measured with a reversing stem thermometer attached to the bottle. Salinity was determined with an optical refractometer. Dissolved oxygen concentration was determined according to the Winckler method proposed by Golterman & Clymo (1969), and modified by the addition of sodium azide.

At each station depth was measured using a graduated rope that was attached to the Van Veen grab sampler (0.25 m^2) that was used for sediment sampling. In the laboratory, 200 g of sediment was dried at 70°C for 72 hours, split into aliquots and organic matter and grain size were analyzed. The scale of Wenthworth (1922) was used for grain size. The percentage of organic matter was obtained by ash weight: three aliquots of 10 g each per subarea (from three stations in each subarea) per month were placed in porcelain crucibles and submitted to a temperature of 500°C over a period of three hours and weighed again.

2.5 *Statistical analysis*

Pearson's linear correlation was calculated to determine the existence of a correlation between the absolute abundance of the crabs and each environmental factor (depth, temperature, dissolved oxygen, salinity, organic matter, and size grain of the sediment). The level of significance was determined by the chi-square test.

3 RESULTS

3.1 *Seasonal distribution*

A total of 7718 specimens of *C. ornatus* were obtained during the one year sampling period: 1781 were mature males (23.08%), 1234 were mature females (15.98%), 435 were ovigerous females (5.64%) and 4268 were immature individuals (55.30%, 1749 males and 2004 females). The monthly size frequency distribution of all individuals (Fig. 2) revealed that *C. ornatus* occurred during all the collecting months, but was most abundant in February, September and August. Mature males were most abundant in February, March and September and mature females were most abundant in September, August, July, and from January to April. Ovigerous females were abundant in August, April and November, with a low occurrence in February and March. A continuous and heterogeneous recruitment of immature individuals, and also reproductive activity (based on the presence of ovigerous females and females with developed gonads) were observed during all months, with a high and low incidence in summer and winter, respectively.

3.2 *Spatial distribution*

The highest abundance of crabs occurred in subareas 6 (n = 2905), 4 (n = 1111), 8 (n = 866), and 5 (n = 833). The highest abundance of mature males was recorded in subareas 6 (n = 1632), 8 (n = 574), 5 (n = 403) and 7 (n = 385). The highest abundance of mature females was recorded in subareas 6 (n = 1153), 4 (n = 475) and 5 (n = 415). The occurrence of ovigerous females was evident in subareas 4 (n = 282) and 6 (n = 120), and were only occasional (< 2%) in other subareas. The immature individuals were highly abundant in all subareas, except subarea 4. The seasonal profile distribution in relation to subareas revealed small differences between groups (Fig. 3).

The distribution of the mean number of individuals per catch over the environmental factors is shown in Figure 4. Mature males showed the highest occurrence at moderate depths (4-8 m), moderate salinity (28-30‰), in the presence of 15-20% organic matter, and at the highest temperature (28-32°C). Mature females showed the same pattern as mature males, except in relation to temperature (16-20°C). Ovigerous females occurred at a depth of 4-12 m, and at the same temperature as mature females. The number of immature specimens tended to increase at salinities between 30 and 32‰, organic content between 10 and 20%, and depth from 4-8 m.

The significant correlation between the abundance of crabs and environmental factors can be seen in Table 1, as calculated by Pearson's correlation coefficient.

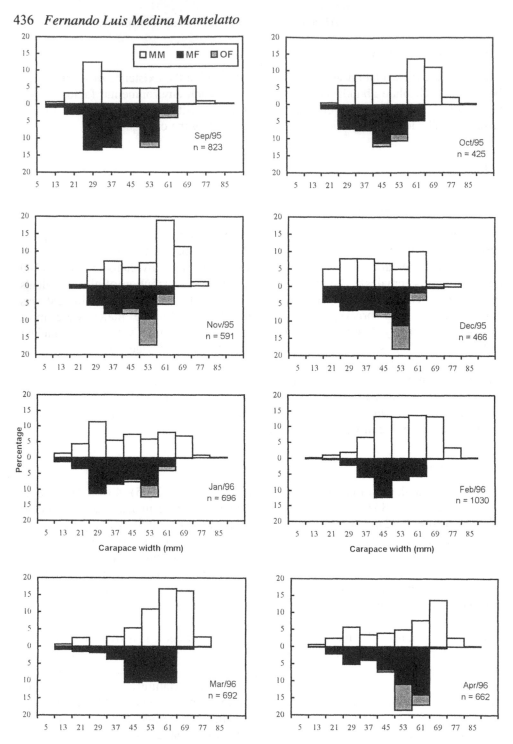

Figure 2. *Callinectes ornatus.* Monthly size frequency distribution of individuals collected in Ubatuba Bay from September 1995 to August 1996. (MM = mature males; MF = mature females; OF = ovigerous females).

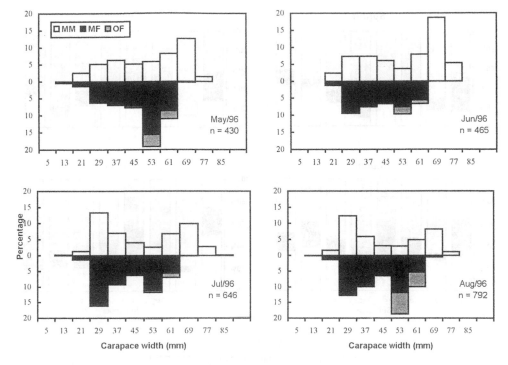

Figure 2 (continued).

Mature males, immature crabs and total number of crabs were negatively correlated with depth, with a preference for shallow water. Mature males were positively and significantly associated with higher temperature. Immature crabs and total number of specimens were correlated with very coarse sand, and ovigerous and mature females with medium and fine sand. Particularly interesting was the negative and significant correlation between very fine sand and all groups.

4 DISCUSSION

In general, all groups except for ovigerous females presented a preference for subareas 4, 5, 6, 7, and 8. This pattern existed all the year, with small oscillations. The distribution of individuals in the various subareas seem to be determined by their tolerance to a combination of factors, some of which may have a stronger effect than others, depending on the developmental stage of the individuals. The tolerance to and/or preference for a given 'limiting' factor may vary among populations at different sites, as was the case for the present species studied in Fortaleza Bay (Negreiros-Fransozo & Fransozo 1995).

Another species, *Callinectes sapidus*, has a life cycle that can be divided into an offshore and an estuarine phase, thus requiring a variety of habitats. As is the case for many decapod crustaceans, the habitat usage of blue crabs seems to be influenced by many factors. They move to specific habitats to spawn, molt and mate, to maintain

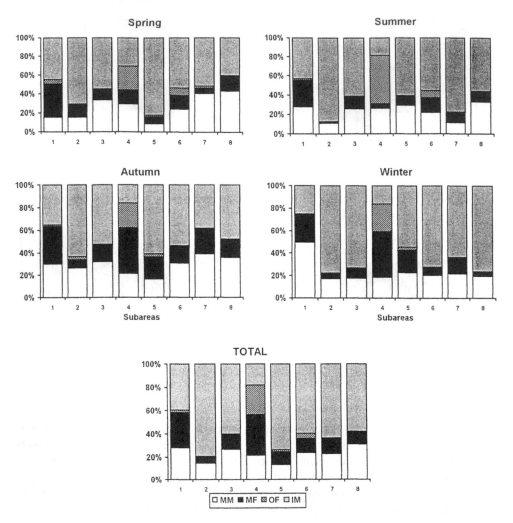

Figure 3. *Callinectes ornatus*. Seasonal and total percentage of individuals collected in the subareas of Ubatuba Bay. (MM = mature males; MF = mature females; OF = ovigerous females; IM = immature).

seasonal thermal optima, and to increase food availability or decrease the likelihood of predation (see Steele & Bert 1994 for these points).

In the Ubatuba Bay, subareas 4 and 6 can be considered as reproductive sites in function of highest occurrence of ovigerous females of *C. ornatus* during all period, suggesting that this area is favorable for embryo and possibly larval dispersion, with a year round spawning. This hypothesis is based on two evidences:

– *Substrate conditions* (texture and organic matter content): like all other portunids, *C. ornatus* incubates its eggs externally, with the sediment acting as an artificial pocket supporting and enveloping the eggs during incubation, as reported for *Arenaeus cribrarius* by Pinheiro et al. (1996). Areas with a greater supply of food on sediment load may indirectly influence the presence of the crabs, with benefit for

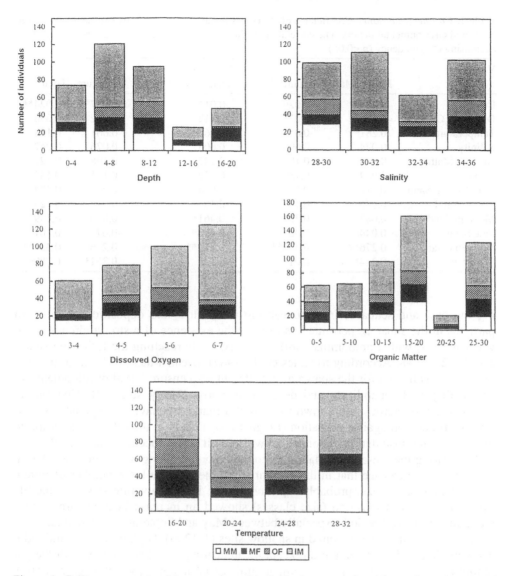

Figure 4. *Callinectes ornatus*. Mean number of individuals per trawl and by each environmental factor stratum from September 1995 to August 1996. (MM = mature males; MF = mature females; OF = ovigerous females; IM = immature).

other levels of the food chain, enriching the pelagic system and affecting the benthic system (González-Gurriarán & Olaso 1987).

– *Positioning in relation to the dynamics of the water mass*: once the crabs are located in an area protected from predation and favorable to egg incubation (subarea 4), the next step will be larval dispersion through currents that will transport them to areas favorable to development, probably on offshore sites. Field observations during the last ten years in this region (wave intensity and amplitude, buoy movement caused by waves, sites of fish nets cast by fishermen, fishing boats entering or leav-

Table 1. *Callinectes ornatus*. Coefficients of Pearson's linear correlation between the abundance of crabs and environmental factors. The correlation of the pairs of variables shows an increasing (*) or decreasing (**) tendency (p <0.05).

Variables			Coefficient		
	Mature Males	Mature Females	Ovigerous Females	Immature	Total
Depth	−0.258**	−0.029	−0.011	−0.247**	−0.024**
Dissolved Oxygen	−0.120	−0.021	0.112	0.022	−0.036
Temperature	0.281*	0.053	0.068	0.061	0.108
Salinity	−0.074	0.093	0.003	−0.076	−0.047
Organic Matter	0.049	0.109	0.142	0.108	0.127
Gravel	0.079	0.189	0.122	0.161	0.183
Very Coarse Sand	0.112	0.130	0.054	0.215*	0.203*
Coarse Sand	0.076	0.143	0.155	0.077	0.123
Medium Sand	0.082	0.350*	0.461*	−0.010	0.159
Fine Sand	0.044	0.307*	0.468*	−0.073	0.101
Very Fine Sand	−0.276**	−0.306**	−0.315**	−0.246**	−0.334**
Silt + Clay	0.344*	−0.013	−0.122	0.351*	0.289*

ing the bay, and scuba diving), concerning water flow and deposition of different particles in the subareas permitted us to infer the existence of a strong circulation of water masses, with a predominant inflow subsequently reaching the following subareas: 1→2→7→8. According to Hines et al. (1987), ocean waters of high salinity not only serve as habitat for the spawning female but also ensure larval development, increase dispersal capabilities, and decrease osmorregulatory stress. Brachyuran females, when ovigerous, are known to present a marked behavior of spatial segregation for protection against predation (Leigh 1970, Tristán et al. 1992), in addition to dislocation and installation in points that will permit larval success (Haefner 1978).

Comparing the present population to populations inhabiting other sites (see the introduction), we can see that in the present study there was a large number of immatures of very small size, probably representing the initials juvenile stages of the life cycle. The mode disposition (size classes) showed an increase in immature recruitment during studied period, especially between May and September (fall-winter).

Immature crabs were captured in shallow sites (1-12 m). High number of individuals were detected in all subareas (except 4), indicating that this is a bay with the potential for crab growth and development until sexual maturity, which occurs at 40.0-50.0 mm CW at the age of approximately 1.5 years, in the Ubatuba region (Mantelatto & Fransozo 1996). There was a variation in number of mature males and females, ovigerous females and immature captured in relation to the subareas during the seasons. This demonstrates a distribution strategy aiming at spatial accommodation in the bay, probably in function of the changes in temperature caused by currents (Castro-Filho et al. 1987), overlapping niches (Pinheiro et al. 1997), competition for food or predation, and conditions to growth. This type of behavior of habitat selection is common among the various stages of the genus *Callinectes*, including larvae and post-larvae, (McConaugha 1988), young individuals (Hines et al. 1987) and adults (Van Engel 1958, Schaffner & Diaz 1988, Lipcius & Van Engel 1990). Immature crabs preferentially occupy different complex benthic environments for foraging, growth, and protection from predators (Ryer et al. 1990, Wilson et al. 1990).

Demersal fishing methods, particularly trawling, have been shown to result in the mortality of some benthic species (Kaiser & Spencer 1995). Crustaceans respond strongly to trawling disturbance, rapidly migrating to other areas (Ramsay et al. 1996). In Ubatuba Bay trawling is the most important commercial activity and very frequent throughout the year, principally in the subareas 1, 2 and 3 because of the presence of shrimp. Exception occurs from March to May when fishing is forbidden by law. Trawling is less frequent in subarea 4 because of the low occurence of commercial shrimp species in this site (Costa 1997). This fact is probably associated to ovigerous females migration to this subarea favoring protection and larval success.

On the basis of this study and other published works about *C. ornatus*, it is possible to propose a model for the life cycle of this species in this area, that will be investigated in detail in a future study:

– The installation of mature individuals occurs in the central area of the bay. Annual reproduction is more active in summer, with ovigerous females concentrating in subarea 4, which probably represents the major center of larval hatching and dispersion.

– The settlement of immature individuals is a massive and constant process in this bay, which successfully regulates recruitment and population structure. Within approximately one and half year after settling, these young become mature and reproduce.

Obviously, much more information is needed to clarify this model, which could be confirmed by future studies on larval monitoring. However, preliminary evidence indicates a scheme similar to the life cycle model proposed for other species of this genus in similar areas (Ortiz & Gutierrez 1992), differing in relation to strong dependence of estuarine areas. This leads us to conclude that the organization of *C. ornatus* distribution results from interdependence between the phase of ontogenetic development and the conditions of the physical environment.

ACKNOWLEDGEMENTS

The author is grateful to the 'Fundação de Amparo à Pesquisa do Estado de São Paulo (FAPESP)' (Grant # 95/2833-0, 98/14539-7 and 98/03182-0) for financial support, and to the Conselho Nacional de Desenvolvimento Científico e Tecnológico for grant funding (Produtividade em Pesquisa # 300279/95-7). I thank my colleagues from the NEBECC who helped with sampling and the laboratory work.

REFERENCES

Branco, J.O. & Lunardon-Branco, M.J. 1993a. Aspectos da biologia de *Callinectes ornatus* Ordway, 1863 (Decapoda, Portunidae) da região de Matinhos, Paraná, Brasil. *Arq. Biol. Tecnol.* 36: 489-496.

Branco, J.O. & Lunardon-Branco, M.J. 1993b. Crescimento e tamanho da primeira maturação em *Callinectes ornatus* Ordway, 1863 (Decapoda, Portunidae) da região de Matinhos, Paraná, Brasil. *Arq. Biol. Tecnol.* 36: 497-503.

Brues, T.C. 1927. Occurrence of the marine crab *Callinectes ornatus*, in brackish and fresh water. *Am. Nat.* 61: 566-568.

Buchanan, B.A. & Stoner, A.W. 1988. Distributional patterns of blue crabs (*Callinectes* sp) in a Tropical Estuarine Lagoon. *Estuaries* 11: 231-239.

Castro-Filho, B.M., Miranda, L.B. & Myao, S.Y. 1987. Condições hidrográficas na plataforma continental ao largo de Ubatuba: variações sazonais e em média escala. *Bolm. Inst. Oceanogr.* 35: 135-151.

Costa, R.C. 1997. Composição e padrões distribucionais dos camarões Penaeoidea (Crustacea: Decapoda), na Enseada de Ubatuba, Ubatuba (SP). Master's Thesis, Paulista State University, Botucatu (SP), Brazil, 129p.

Golterman, H.L. & Clymo, R.S. 1969. *Methods for chemical analysis of freshwaters*. Oxford: Blackwell, Scientific Publications, 116p.

González-Gurriarán, E. & Olaso, I. 1987. Cambios espaciales y temporales de los Crustáceos Decápodos de la plataforma continental de Galicia (NW de Espanã). *Inv. Pesq.* 51: 323-341.

Gore, R.G. 1977. Studies on Decapod Crustacea from the Indian River Region of Florida. VII. A field character for rapid identification of the swimming crabs *Callinectes ornatus* Ordway, 1863 and *C. similis* Williams, 1966 (Brachyura, Portunidae). *N. Gulf Sci.* 1: 119-123.

Haefner, P.A. 1978. Seasonal aspects of the biology, distribution and relative abundance of the deep-sea red crab *Geryon quinquedens* Smith, in the vicinity of the Norfolk Canyon, western North Atlantic. *Proc. Nat. Shellfish. Assoc.* 68: 49-61.

Haefner Jr., P.A. 1990a. Morfometry and size at maturity of *Callinectes ornatus* (Brachyura, Portunidae) in Bermuda. *Bull. Mar. Sci.* 46: 274-286.

Haefner Jr., P.A. 1990b. Natural diet of *Callinectes ornatus* (Brachyura, Portunidae) in Bermuda. *J. Crustacean Biol.* 10: 236-246.

Hines, A.H., Lipcius, R.N. & Haddon, A.M. 1987. Population dynamics and habitat partitioning by size sex, and molt stage of blue crabs *Callinectes sapidus* in a subestuary of Central Chesapeake Bay. *Mar. Ecol. Progr. Ser.* 36: 55-64.

Kayser, M.J. & Spencer, B.E. 1995. Survival of by-cath from a beam trawl. *Mar. Ecol. Prog. Ser.* 126: 31-38.

Leigh, E.G. 1970. Sex-ratio and different mortality between the sexes. *Am. Nat.* 104: 205-210.

Lipcius, R.N. & Van Engel, W.A. 1990. Blue crab population dynamics in Chesapeake Bay: variantion in abundance (York River, 1972-1988) and stock-recruit functions. *Bull. Mar. Sci.* 46: 180-194.

Mantelatto, F.L.M. & Fransozo, A. 1996. Size at sexual maturity in *Callinectes ornatus* (Brachyura, Portunidae) from the Ubatuba region (SP), Brazil. *Nauplius* 4: 29-38.

Mantelatto, F.L.M. & Fransozo, A. 1997. Fecundity of the crab *Callinectes ornatus* Ordway, 1863 (Decapoda, Brachyura, Portunidae) from Ubatuba region, São Paulo, Brazil. *Crustaceana* 70: 214-226.

Mantelatto, F.L.M. & Fransozo, A. 1999a. Reproductive biology and molting cycle of the crab *Callinectes ornatus* (Crustacea, Portunidae) in the Ubatuba region, São Paulo, Brazil. *Crustaceana* 72: 63-76.

Mantelatto, F.L.M. & Fransozo, A. 1999b. Characterization of the physical and chemical parameters of Ubatuba Bay, northern coast of São Paulo State, Brazil. *Rev. Brasil. Biol.* 59: 23-31.

Mantelatto, F.L.M., Fransozo, A. & Negreiros-Fransozo, M.L. 1995. Distribuição do caranguejo *Hepatus pudibundus* (Herbst, 1785) (Crustacea, Decapoda) na Enseada da Fortaleza, Ubatuba (SP), Brasil. *Bolm. Inst. Oceanogr.* 43: 51-61.

Melo, G.A.S. 1996. *Manual de identificação dos Brachyura (caranguejos e siris) do litoral brasileiro*. São Paulo: Plêiade Ed., 603p.

McConaugha, J.R. 1988. Export and reinvaion of larvae as regulators of estuarine decapod populations. *Amer. Fish. Soc. Symp.* 3: 90-103.

Negreiros-Fransozo, M.L. & Fransozo, A. 1995. On the distributional of *Callinectes ornatus* Ordway, 1863 and *Callinectes danae* Smith, 1869 (Brachyura, Portunidae) in the Fortaleza Bay, Ubatuba (SP), Brazil. *Iheringia, Serie Zoologia* 79: 13-25.

Negreiros-Fransozo, M.L., Mantelatto, F.L.M. & Fransozo, A. 1999. Populational biology of *Callinectes ornatus* Ordway, 1863 (Decapoda, Portunidae) from Ubatuba (SP), Brazil. *Sci. Mar.* 63: 157-163.

Ortiz, C.A.S. & Gutierrez, J.G. 1992. Distribucion y abundancia de los estadios planctonicos de la jaiba *Callinectes bellicosus* (Decapoda: Portunidae), en el complejo lagunar Bahia Magdalena, B.C.S., Mexico. *Rer. Inv. Cient.* 3: 47-60.

Pinheiro, M.A.A., Fransozo, A. & Negreiros-Fransozo, M.L. 1996. Distributional patterns of *Arenaeus cribrarius* (Lamarck, 1818) (Crustacea, Portunidae) in Fortaleza Bay, Ubatuba (SP), Brazil. *Rev. Brasil. Biol.* 56: 705-716.

Pinheiro, M.A.A., Fransozo, A. & Negreiros-Fransozo, M.L. 1997. Dimensionamento e sobreposição de nichos dos portunídeos (Decapoda, Brachyura) na Enseada da Fortaleza, Ubatuba, São Paulo, Brazil. *Revta. Bras. Zool.* 14: 371-378.

Pita, J.B., Rodrigues, E.S. Graça-Lopes R. & Coelho, J.A.P. 1985b. Levantamento da família Portunidae (Crustacea, Decapoda, Brachyura) no complexo Baía-Estuário de Santos, São Paulo, Brasil. *Bol. Inst. Pesca* 12: 152-162.

Ramsay, K., Kaiser, M.J. & Hughes, R.N. 1996. Changes in hermit crab feeding patterns in response to trawling disturbance. *Mar. Ecol. Prog. Ser.* 144: 63-72.

Raup, D.M. & Stanley, S.M. 1978. Paleoecology. In *Principles of Paleontology*: 231-299. San Francisco: W.H. Freeman & Co.

Román-Contreras, R. 1986. Análisis de la población de *Callinectes* spp (Decapoda, Portunidae) en el sector Occidental de la Laguna de Terminos, Campeche, México. I. *Ann. Inst. Cien. Mar Limnol. UNAM* 13: 315-322.

Ryer, C.H., Van Montfrans, J. & Orth, R.J. 1990. Utilization of a seagrass meadow and tidal marsh creek by blue crabs, *Callinectes sapidus*. II. Spatial and temporal patterns of molting. *Bull. Mar. Sci.* 46: 95-104.

Schaffner, L.C. & Diaz, R.J. 1988. Distribution and abundance of overwintering blue crabs, *Callinectes sapidus*, in the lower Chesapeake Bay. *Estuaries* 11: 68- 2.

Steele, P. & Bert, T.M. 1994. Population ecology of the blue crab, *Callinectes sapidus* Rathbun, in a subtropical estuary: population structure, aspects of reproduction, and habitat partitioning. *Fla. Mar. Res. Pub.* 51: 1-24.

Tristán, A.L., Monteforte, M., Amador Silva, E.S. & Ramiréz, M.H. 1992. Distribución, abundância y reproducción del cangrejo rojo *Cancer johngarthi* (Decapoda, Cancridae), em Baja California Sur, Mexico. *Rev. Biol. Trop.* 40: 199-207.

Van Engel, W.A. 1958. The blue crab and its fishery in Chesapeake Bay. Part 1. Reproduction, early development, growth, and migration. *Commer. Fish. Rev.* 20: 6-17.

Wenthworth, C.H. 1922. A scale of grade and class terms for clastic sediments. *J. Geol.* 30: 377-392.

Williams, A.B. 1966. The Western Atlantic swimming crabs *Callinectes ornatus, C. danae* and a new related species (Decapoda, Portunidae). *Tulane Studies en Zoology* 13: 83-93.

Wilson, K.A., Able, K.W. & Heck Jr., K.L. 1990. Habitat use by juvenile blue crabs: a comparison among habitats in southern New Jersey. *Bull. Mar. Sci.* 46: 105-114.

Mound formation by Tanner crabs, *Chionoecetes bairdi*: Tidal phasing of larval launch pads?

BRADLEY G. STEVENS & JAN A. HAAGA
National Marine Fisheries Service, Kodiak Laboratory, Kodiak, USA

WILLIAM E. DONALDSON
Alaska Dept. of Fish and Game, Kodiak, USA

ABSTRACT

Female Tanner crabs, *Chionoecetes bairdi*, aggregate and form mounds at a deepwater (150 m) site in the spring (Stevens et al. 1994). This paper reviews observations of crab behavior, which were made each spring for 6 years (1991-1995, and 1998), via submersible, ROV, and/or video camera sled in Chiniak Bay, Alaska. Timing of mound formation was compared to water temperature, lunar cycle, tidal exchange, storm frequency, and Secchi disk depth. Mound formation was observed in 3 years (1991, 1994, and 1995) within 0-4 days of the maximum spring tide; no other environmental indicator was coincident. Crabs captured from mounds (1991, 1995) were observed releasing larvae in tanks or buckets, whereas crabs captured prior to mound formation (1995) or afterwards (1992, 1998) were not releasing larvae. Based on these data, we hypothesize that mound formation is triggered by tidal rhythms associated with the highest spring tide in April or May, coincides with larval release, and functions to improve larval dispersal by elevating spawners above the bottom sediments.

1 INTRODUCTION

The Tanner crab, *Chionoecetes bairdi*, and its congener the snow crab, *C. opilio*, together support multi-million dollar fisheries in Alaska. In April 1991 we began a study of Tanner crab reproductive behavior, using the two-person submersible *Delta* (Stevens et al. 1993). Our most surprising discovery was that female crabs formed a high density aggregation containing mounds of crabs (Stevens et al. 1994). The aggregation contained approximately 200 such mounds, spaced 1-3 m apart, with a mean of 200 crabs each, plus male crabs and mating pairs, for an estimated total of 100,000 crabs. Virtually all mounds were contained within a 2.2 ha area, at a depth of 150 m. Mounds consisted entirely of oldshell multiparous females with late stage embryos. Males and pairs in premating embrace were observed around the periphery of the aggregation. Sex ratios (M:F) varied from 1:10 overall, to 1:100 in localized areas (Stevens et al. 1994). At least seven species of crabs in the family Majidae have been shown to exhibit similar behavior (Stevens et al. 1994), including the Pacific lyre crab *Hyas lyratus* (Stevens et al. 1992), and the snow crab *C. opilio* (G. Conan, pers. comm. October 1995). Stevens et al. (1994) suggested that aggregation of Tanner crabs occurred in conjunction with mating, and that crabs in the mounds might be releasing chemical signals to attract males. Further attempts to study this phenomenon in subsequent years (1992, 1994, 1995, and 1998) were less successful due to the inability to predict timing of the event. This paper summarizes those observations, with the emphasis on timing of aggregation, and suggests a different function for the behavior.

2 METHODS

Observations were made in Chiniak Bay, Kodiak, Alaska, using the 2-person submersible *Delta*, or a remotely operated vehicle (ROV), in April and/or May each year from 1991 to 1995 (Table 1); a towed video-camera sled was used in 1998. Positions of the sub, ROV, and sled were plotted with differential GPS. Crabs observed via ROV were counted in real time, whereas those recorded on videotape (*Delta* and sled dives) were counted upon return to the lab. Crabs collected with the *Delta* were measured to the nearest mm, and reproductive and shell conditions were recorded as described in Stevens et al. (1993). Densities of crabs were difficult to estimate for several reasons. While the lengths of transects were measured, the speed of the sub, ROV or sled varied continuously throughout the observations from < 0.25 to over 1.0 knot. Therefore exact correspondence between counts and locations could only be approximated to the nearest minute, or in the case of the sled, the nearest 5 minutes. Because the aggregation was surrounded by large expanses of sea bottom virtually devoid of crabs, average numbers or densities calculated over a transect have great variance, depending on what proportion of the transect was within the aggregation area, so are not particularly meaningful. For this reason, crab densities are simply represented as the highest number of crabs per time interval for a given transect.

3 RESULTS

3.1 *1991*

Mounds were first observed on 18 April, and continued to 1 May, the last date of ob-

Table 1. Dates and tools used to locate and observe Tanner crab, *Chionoecetes bairdi*.

Year	Tool	Dates
1991	Delta sub	13 April to 1 May
1992	Delta sub	1-14 May
1993	ROV	26 April and 1 May
1994	Delta sub	17-27 April
1995	ROV	14-20 April; 5-15 May
1995	Delta sub	22 April to 4 May
1998	Sled	7, 11, 12, 13, 19, 21, 27, and 28 May
1998	Delta sub	4 June

servation. The highest spring tides (HST) occurred on 17 April (2 days after the new moon) (Fig. 1). Females recovered from the mounds and brought to the surface were placed in tanks containing seawater, and were observed releasing zoea stage larvae. Mean seawater temperature at 150 m was 3.4°C.

3.2 *1992*

Most crabs observed from 1 to 14 May had already released larvae and produced new clutches. Highest densities observed were 2.2 crabs m^{-2}. During daytime, 85% of females remained buried in the sediment, but at night, 71% were exposed and active (Stevens et al. 1994). No mounds were observed. The full moon and HST had occurred on April 16 and 18, respectively.

3.3 *1993*

Small mounds of 30-50 crabs were observed on both dates with the ROV. The full moon and HST would occur on 5 and 7 May, respectively. No temperatures were recorded.

3.4 *1994*

Crabs began to form low mounds on 27 April, coincident with the HST, and 2 days after the full moon. Temperatures were unavailable due to instrument failure.

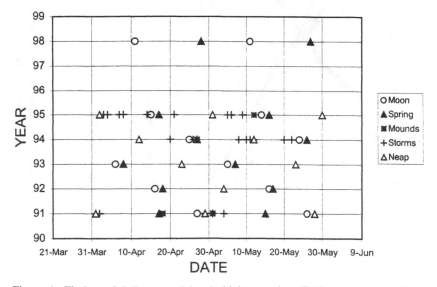

Figure 1. Timing of full moon (Moon), highest spring (Spring) and neap (Neap) tides, storm (Storm) occurrence, and observations of mound (Mound) formation by Tanner crabs (Chionoecetes bairdi), from 1991 through 1998. Mound formation was observed 1 day before (1991), coincident with (1994), and 2-4 days before (1995) the high spring tide.

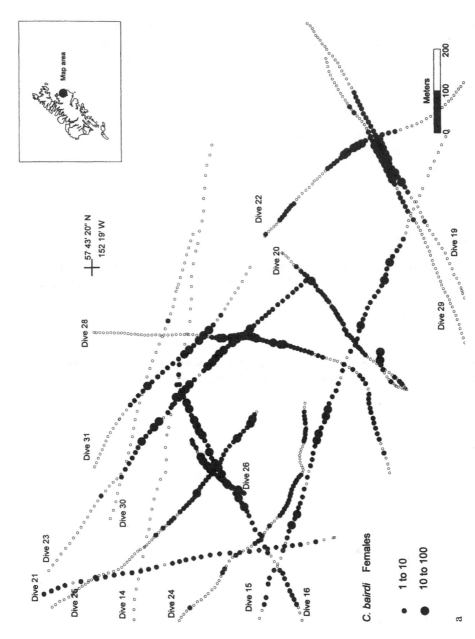

Figure 2. Numbers of female Tanner crabs observed per minute of observation by ROV in Chiniak Bay, Alaska. Virtually all crabs were buried in the mud. a. 14-20 April, 1995.

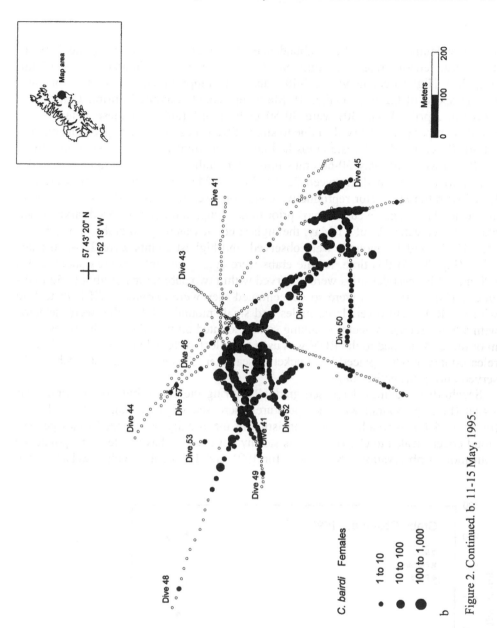

Figure 2. Continued. b. 11-15 May, 1995.

3.5 *1995*

The most detailed counts of crab abundance were obtained in 1995 using the ROV. In mid-April, female crabs were aggregated into an area approximately 1.0×0.5 km (Fig. 2A). Highest counts were 30-40 crabs min^{-1} (approximately 3-4 crabs m^{-2} at a towing speed of 0.2 m s^{-1}) (Fig. 3). Maximum counts recorded during *Delta* dives (between April 23 and 30) were 30-50 crabs min^{-1} (although these data are not strictly comparable to ROV data due to slight differences in vehicle speed and camera setup). By early May, female crabs had coalesced slightly, but highest counts (with the ROV) were still only 50-60 crabs min^{-1} (5-6 crabs m^{-2}) (Fig. 3), and all females were buried in the mud. Between 22 April and 4 May, 158 female crabs were collected with the *Delta* (for reproductive conditions, see Stevens et al. 1996); virtually all carried late-stage, eyed embryos prior to hatching, and none released larvae when placed in seawater. As of 11 May, the highest count was only 70 crabs min^{-1}, but on May 12, 13 and 14, mounds were observed, and highest counts were 225-250 crabs min^{-1} (Fig. 3). At this time, female crabs were aggregated into an area about 200 × 600 m, but highest densities were observed within two subregions of about 150 m diameter (Fig. 2B). Crabs were so concentrated they were extremely difficult to find with the ROV. However, large males and small mounds of females were detected with a high-frequency sector scanning sonar mounted on the ROV, at distances of 50 m or more. A net tied to the ROV was used to collect 10 female crabs, some of which released larvae when placed in a bucket of seawater. This was the latest hatching observed during the entire study.

Secchi disk readings decreased gradually during the study, but provided no clear signal (Fig. 4). Bottom water temperatures increased slightly, from 3.04 to 4.36°C (mean 3.64°C). A weather index consisting of mean daily wind speed times percent cloud cover (scaled to chart) indicates stormy days, which have little correspondence with mound observations (See Fig. 1 for 1991 and 1994 data). Tidal exchange (the

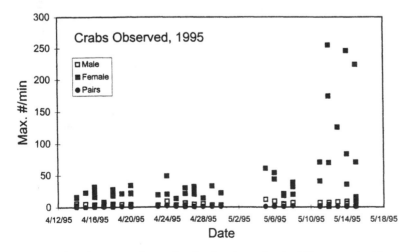

Figure 3. Maximum numbers of male, female, and paired Tanner crabs observed per minute from the Delta submersible (23-30 April) or by ROV (all other dates) on each transect in 1995. Note sudden increase on 12 May.

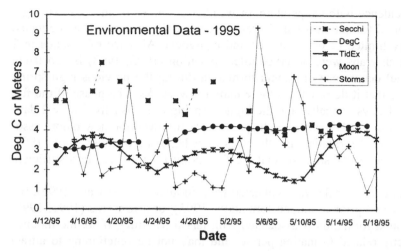

Figure 4. Environmental indices during observations of Tanner crabs in 1995. Indices include Secchi disk depth (Secchi, m), bottom water temperature (°C), tidal exchange (TidEx, m), weather index (Storms, see text), and date of full moon (Moon). Mound formation began on 12 May, 1995.

difference between highest and lowest tides of the day) and full moon are the only indices which coincide with mound formation. The full moon and HST occurred on 14 and 16 May, respectively. A previous full moon and HST had occurred on 15 and 17 April, respectively.

3.6 *1998*

The towed sled traveled faster than the *Delta* or the ROV (0.5 to 1.0 m s^{-1}), so counts were less accurate and positions more difficult to determine. Female crabs were spread over an area about 1.0×0.5 km. Counts in the highest density area were 6-12 crabs min^{-1}, equivalent to about 0.2-0.4 crabs m^{-2}. Only one grasping pair, and one group of 5-10 females was observed. Presence of exposed females suggested that hatching, mating and extrusion was completed. HST occurred on 28 April and 26 May, following the new moon, whereas the full moon occurred on 11 April and 11 May. On 4 June, 10 female crabs were collected during one dive in the *Delta*. Of these, 2 primiparous and 4 multiparous females carried newly extruded clutches. The remaining 4 females were very oldshell crabs with no external clutch, and empty or very small ovaries; they were probably senescent. Three multiparous female crabs collected by scuba from shallow water and placed in aquaria on 25 April, hatched and released larvae on 27 April (coincident with the previous HST). Mean bottom temperature was 6.1°C.

4 DISCUSSION

Although Secchi disk depths were only determined in 1995, coincidence of mounding observations with high spring tides in all 3 years observed (1991, 1994, and 1995), and lack of coincidence with temperature, or weather extremes in any year, strongly suggests that mound formation occurs in response to tidal variations. Data collected in 1993 was very limited (2 observations) and equivocal. Absence of mounds in 2 years during which crabs were observed after spawning (1992, 1998) is consistent with the (possible) occurrence of mound formation during the previous high spring tide, and indicates that it does not last more than a few weeks. The phase of the full moon advances 11 days annually, and the highest spring tides usually follow 2 days later. But in two years (1991, 1998) high spring tides followed the new moon; association between mound formation and the tides in at least one of those years (1991, and possibly 1998 as well), suggests that the full moon is probably not a sufficient signal by itself.

Release of larvae by female crabs collected from mounds (in 1991 and 1995), but not by crabs collected prior to mound formation (1995), or afterwards (1992, 1998) suggests that mound formation is associated with larval hatching. If so, the mounds may not be strictly related to mating per se, and may not be functioning to attract males, as previously suggested (Stevens et al. 1994). Instead, mounds may be a mechanism for projecting larvae up away from the silty sediments and the benthic boundary layer, with its associated turbulence and turbidity. In a flat, soft bottom with no vertical relief, climbing on the nearest crab may be the only way to position hatching embryos above the sediment. Although this explanation of crab behavior is somewhat speculative, it is the most parsimonious explanation of our observations to date.

Most crab species release larvae in synchrony with some environmental cycle. The relative importance of diel, light, lunar, and tidal influences varies with species and habitat, but virtually all crabs examined have tidal hatching rhythms (Morgan 1995). At the depths where *C. bairdi* occur, the most likely stimulus for such behavior are changes in tidal current. Starr et al. (1994) demonstrated that sinking phytoplankton could trigger release of larvae by *C. opilio* in a laboratory setting (where they were probably removed from tidal influence). Although there is an obvious adaptive advantage to such a linkage, phytoplankton blooms may not be sufficiently strong or distinct signals in the natural environment. The timing of larval release by *C. bairdi* may involve multiple signals, among which phytoplankton detritus may serve as a primer, and tidal rhythms the actual trigger. Tidally synchronized larval release probably has adaptive significance by reducing predation on the larvae (Morgan & Christy 1995). This could occur by the overwhelming of any larval predators, or by using tidal current flow to increase larval dispersion. The only potential larval predators we observed in the vicinity were euphausiids.

The coincidence between tidal exchange, mound formation by *C. bairdi*, and larval release, while supported by evidence accumulated over six years of observation, is still somewhat circumstantial, but provides a hypothesis which can be tested in the field. Testing it will require making observations with the camera sled over a 2-month period, through two complete lunar tide cycles, which we plan to do in the future.

REFERENCES

Morgan, S.G. 1995. The timing of larval release. In L. McEdward (ed.), *Ecology of Marine Invertebrate Larvae*: 157-193. Boca Raton: CRC Press.

Morgan, S.G. & Christy, J.H. 1995. Adaptive significance of the timing of larval release by crabs. *Amer. Nat.* 145: 457-479.

Starr, M., Therriault, J.-C., Conan, G.Y., Comeau, M., & Robichaud, G. 1994. Larval release in a sub-euphotic zone invertebrate triggered by sinking phytoplankton particles. *J. Plankt. Res.* 16: 1137-1147.

Stevens, B.G., Donaldson, W.E. & Haaga, J.A. 992. First report of podding behavior in the Pacific lyre crab, *Hyas lyratus*. *J. Crust. Biol.* 12: 193-195.

Stevens, B.G., Donaldson, W.E., Haaga, J.A. & Munk, J. E. 1993. Morphometry and maturity of male Tanner crabs, *Chionoecetes bairdi*, grasping pubescent and multiparous females in shallow and deepwater environments. *Can. J. Fish. Aquat. Sci.* 50: 1504-1516.

Stevens, B.G., Haaga, J.A. & Donaldson, W.E. 1994. Aggregative mating of Tanner crabs *Chionoecetes bairdi*. *Can. J. Fish. Aquat. Sci.* 51: 1273-1280.

Stevens, B.G., Haaga, J., Donaldson, W.E., & Payne, S.A. 1996. Reproductive conditions of prespawning female Tanner crabs, *Chionoecetes bairdi*, in Chiniak and Womens Bays, Kodiak, Alaska. In *High Latitude crabs: Biology, management, and economics. Alaska Sea Grant Program Report No. 96-02, University of Alaska Fairbanks*: 349-354.

3 Toxicity and physiology

The impact of the toxic strain of *Microcystis aeruginosa* on *Daphnia magna*

IRINA L. TRUBETSKOVA
Estonian Marine Institute, Tallinn, Estonia

JAMES F. HANEY
Dept. of Zoology, University of New Hampshire, Durham, USA

ABSTRACT

The effects of the toxic strain of *Microcystis aeruginosa* on mature female *Daphnia magna* were studied under controlled laboratory conditions. Daphnids cloned from a single female were grown at 1 mg C l^{-1} of *Chlorella vulgaris* until they reached the second adult instar. After production of a second clutch of eggs, some females were subjected to 1 mg C l^{-1} of *M. aeruginosa* for two subsequent instars; remaining animals served as a control at 1 mg C l^{-1} of *C. vulgaris*. *D. magna* had significantly decreased survivorship, dry body weight, and production of neonates when fed only toxic cyanobacterium. Animals lost weight, ceased reproduction and died on the sixth day of the experiment. Neonates born to mothers previously exposed for 5 days to *M. aeruginosa*, were smaller and lighter than newborns produced under the *Chlorella* treatment. Although increased mortality of *D. magna* fed *Microcystis* suggests toxicity effects, responses in growth and reproduction were similar to starvation. Possible consequences of cyanobacteria to neonate fitness are discussed.

1 INTRODUCTION

There is growing evidence of cyanobacterial blooms all over the world. Many cyanobacteria produce hepatotoxic peptides called microcystins that cause severe problems for wildlife, domestic animals, and humans (Carmichael 1981, 1988, Codd 1984, Falconer 1994, Skulberg et al. 1984). Cyanobacterial blooms can also have ecological consequences by adversely affecting different trophic levels in the aquatic ecosystems (for review see Lindholm et al. 1992, Christoffersen 1996). *Microcystis aeruginosa* is one of the most common cyanobacteria forming seasonal blooms in freshwater ecosystems. Field studies have documented seasonal changes in zooplankton community structure correlated with blooms of cyanobacteria (for review see Smith & Gilbert 1995). Laboratory studies have examined the effects of toxic cyanobacteria on zooplankton, with increasing attention to the inhibitory and toxic effects of *M. aeruginosa* on the aquatic grazer *Daphnia* (Nizan et al. 1986, Lampert 1981, 1987, Demott et al. 1991, Forsyth et al. 1992, Haney et al. 1994, 1995, Reinikainan et al. 1994, 1995, Smith & Gilbert, 1995). There are, however, some results showing that some *Daphnia* populations can persist during toxic cyanobacterial blooms (Benndorf & Henning 1989, Christoffersen et al. 1990, Jungman & Benndorf 1994, Haney, unpublished data). In our study, we attempted to answer the question: what is the response of adult *Daphnia* to a toxic strain of *M. aeruginosa* as measured by changes in their survival, body weight, reproduction, and size and weight of their progeny?

2 MATERIALS AND METHODS

Laboratory cultures of *Daphnia magna* were maintained in 1-liter glass jars filled with 0.45 µm filtered well water. They were fed exclusively with green algae *Chlorella vulgaris* grown in Bold's Basal Medium (Nichols 1973). Daphnids were kept at a density of 20 individuals l^{-1} under non-limited food conditions. *Daphnia* culture and experiments were conducted in a temperature controlled room at $20 \pm 1°C$ under a 16:8 hours light:dark photoperiod of low intensity (~ 1.5 µmol m^{-2} s^{-1}).

The unicellular microcystin-producing *Microcystis aeruginosa* strain 2385 and *C. vulgaris* were obtained from UTEX, The Culture Collection of Algae at the University of Texas, USA. Both species were grown in 0.5-liter aerated, air-filtered culture tubes at 25°C under continuous illumination (~ 20 µmol m^{-2} s^{-1}). *Microcystis* culture was maintained in ASM-1 medium (Gorham et al. 1964). *Chlorella* and *Microcystis* cultures were used in the log-phase of growth. Cells were harvested by centrifugation and then resuspended in Whatman GF/F filtered well water. The *Microcystis* concentration used (1 mg C l^{-1}) was selected to represent approximate levels during bloom conditions. *Chlorella* and *Microcystis* concentrations were determined by direct hemacytometer count, and appropriate dilutions were prepared thereafter. Average diameters ($\pm$ SD) of *Chlorella* and *Microcystis* cells were 3.8 ± 0.7 µm and 3.3 ± 0.2 µm, correspondingly. Dry weights ($\pm$ SD) of *Chlorella* and *Microcystis* were 16.9 ± 0.1 pg per cell and 7.5 ± 0.2 pg per cell, respectively. Carbon content was calculated from dry weights using a conversion factor of 0.5. *M. aeruginosa* 2385 contained 2 mg of microcystin-LR (MC-LR) g^{-1} dry weight (15×10^{-9} µg of MC-LR per cell), < 0.25 mg MC-YR and < 0.13 mg MC-RR g^{-1}.

Microcystins were analyzed with the HPLC method described by Harada et al. (1988), using the system consisted of a Hewlett-Packard Series 1050 pump operated at 1 ml min^{-1}, a HP1050 variable wavelength detector set at 238 nm, a HP3396A integrator, a Nucleosil C18, 5 micron, 150×4.6 mm, column. The mobile phase was methanol-0.05M, pH3 phosphate buffer (58:42).

To begin the test, third brood offspring of *D. magna* were collected within 12 h, randomly transferred into individual 100-ml beakers and reared at 1 mg l^{-1} of *C. vulgaris* (Fig. 1). Daily, animals were moved into clean vessels with fresh treatment media. Once they had produced the second clutch of eggs, *Daphnia* were either transferred to pure suspension of the toxic strain of *M. aeruginosa* (1 mg l^{-1}) or kept in *C. vulgaris* (1 mg l^{-1}) to serve as a control. Eight replicates were used per treatment. As decreased survivorship of animals fed cyanobacteria compared to unfed control animals is generally accepted as evidence for toxicity (for review see Smith 1993), starvation experiments were also conducted to be used as a reference for toxicity effects (Fig. 1). Animals were examined for mortality every 24 hours. When the

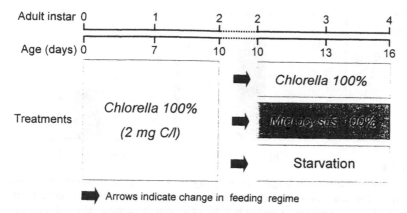

Figure 1. Experimental design for studying effects of toxic *Microcystis* on *Daphnia magna*. Dotted line indicates time of transfer of adult *D. magna* into appropriate feeding regime. See text for further details.

Daphnia reached the fourth instar, the following parameters were measured: survivorship, body length and body dry weight, number of neonates produced, neonate length and neonate dry weight. After body lengths of individual females and neonates were measured under a dissecting microscope, animals were transferred to small aluminum containers. *Daphnia* were dried overnight at 60°C, cooled in a desiccator, and weighed to the nearest 0.1 µg on a Cahn electronic microbalance. After six days, the experiment was terminated because of death of animals under *Microcystis* treatment. Treatments were tested for significant differences using *t*-test (p < 0.05).

3 RESULTS

There was no mortality of adult females of *D. magna* in the control, i.e., *Chlorella* treatment during 6-d experiment (Fig. 2). After 5 days of feeding on a pure suspension of *Microcystis*, 25% of daphnids died, and no daphnids survived after 6 days of exposure. In contrast, 50% of the *Daphnia* survived 6 days of starvation. Animals fed toxic *Microcystis*, had smaller length and lower body weight. Although body lengths of *Daphnia* kept under *Microcystis* treatment were smaller than animals in the control, the variability in their size was rather high, and their body length did not differ significantly from those under *Chlorella* treatment. Body weights of *D. magna* fed toxic *Microcystis*, were significantly lower than those fed *Chlorella*. Starved adult *Daphnia* decreased significantly both their size and body weight. There was no significant difference both in body length and body weight between *Microcystis*-fed and starved adult *Daphnia* after 6d-exposure, although under starvation animals lost more weight.

Five-day-exposure to toxic *Microcystis* caused a significant reduction in the number of neonates per clutch, as well as in neonate length and weight compared to the *Chlorella* treatment (Fig. 3). Newborns, produced under *Microcystis* treatment, were smaller and often had less developed 2nd antennae and tail spines than babies born to *Daphnia* fed *Chlorella*. Under starvation *Daphnia* produced fewer, but significantly

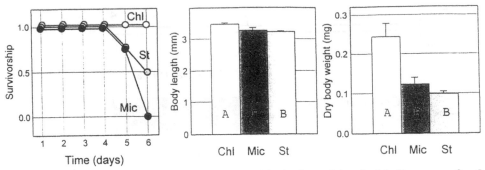

Figure 2. Changes in survivorship, body length and body dry weight of adult *D. magna* after 5-d exposure to toxic strain of *M. aeruginosa* and to starvation. Values are mean ± SE. The same letter indicates no significant difference between treatments (Chl – *Chlorella* treatment, Mic – *Microcystis* treatment, St – starvation).

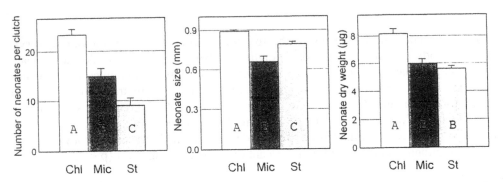

Figure 3. Changes in neonate production, neonate length and neonate dry weight of *D. magna* after maternal exposure for 5 days to toxic strain of *M. aeruginosa* and to starvation. Values are mean ± SE. The same letter indicates no significant differences between treatments (Chl – *Chlorella* treatment, Mic – *Microcystis* treatment, St – starvation).

longer neonates than those fed *Microcystis*. However, neonate weights did not differ significantly between *Microcystis* and starvation treatments.

4 DISCUSSION

There is considerable evidence that cyanobacteria have low nutritional value for *Daphnia* and may also inhibit feeding activities by mechanical and chemical means (Gliwicz & Siedlar 1980, Lampert 1987, Smith 1993, for reviews see also Lindholm et al. 1992, Smith & Gilbert 1995, Lauren-Määttä 1996). There is little known about the effects of cyanobacteria on adults and their offspring. The survivorship data could not be tested for statistical significance, but the higher mortality for daphnids fed *Microcystis* compared to complete starvation suggests *Microcystis aeruginosa* strain 2385 was toxic to adult *Daphnia magna*.

The most striking finding in our study, however, is that growth and reproductive responses of *D. magna* fed on *Microcystis* closely resemble complete starvation. For example, adult *D. magna* had the same reductions in length and weight in *Microcystis* and starved conditions, compared to feeding on *Chlorella*. Also, starved and *Microcystis*-fed daphnids produced fewer and lower weight neonates than *Chlorella*-fed animals. Food concentration appears to regulate the size of *Daphnia* eggs and neonates (Ebert 1993) and with decreasing food daphnids produce fewer, but larger eggs, to a food level where energetic constraints limit egg size (Glazier 1992, Guisande & Gliwicz 1992, Trubetskova & Lampert 1995). In fact, egg size may actually decrease as the food concentration approaches the threshold for growth (Guisand & Gliwicz 1992, Urabe 1988) or with extreme starvation (Trubetskova & Lampert 1995). We have found that the reductions in neonate weights for *Microcystis*-fed and starved mother *D. magna* can be most simply explained as a case of extreme food limitation, although an additional toxicity effect can not be excluded in the *Microcystis* condition. It is interesting that the body length of neonates fed cyanobacteria were significantly smaller than for the starved animals. If this is a repeatable response to feeding on *Microcystis*, smaller neonate carapace size might represent an adaptive advantage, such as a smaller carapace opening that reduces clogging by colonies of cyanobacteria (Gliwicz & Siedlar 1980). It is also possible that the smaller neonate length results from the toxic effects for *Microcystis* and is related to the retarded structural development noted earlier. Further research is needed to resolve the question of the effects of cyanobacteria on the fitness of *Daphnia* offspring.

ACKNOWLEGMENTS

We thank Dr Miyoshi Ikawa for doing the HPLC analysis of cyanobacterial toxins. This project has been funded by the National Research Council under the Collaboration in Basic Science and Engineering Program. The contents of this publication do not necessarily reflect the views or policies of the NRC, nor does mention of trade names, commercial products and organizations imply agreement by the NRC.

REFERENCES

Benndorf, J. & Henning, M. 1989. Daphnia and toxic blooms of *Microcystis aeruginosa* in Bautzen Reservoir (GDR). *Int. Ges. Hydrobiol.* 74: 233-248.
Carmichael, W.W. 1981. Freshwater blue-green algae (cyanobacteria) toxins – a review. In W.W. Carmichael (ed.), *The Water Environment: Algal Toxins and Health*: 1-13. Plenum Press.
Carmichael, W.W. 1988. Toxins of freshwater algae. In A.T. Tu (ed.), *Handbook of Natural Toxins*: 121-147. V 3. Dekker.
Codd, G.A. 1984. Toxins of freshwater cyanobacteria. *Microbiol Sci.* 1(2): 48-52.
Christoffersen, K., Riemann, B., Hansen, L.R., Klysner, A. & Sørensen, H.B. 1990. Qualitative importance of the microbial loop and plankton community structure in an eutrophic lake during a bloom of cyanobacteria. *Microbial Ecology* 20: 253-272.
Christoffersen, K. 1996. Ecological implications of cyanobacterial toxins in aquatic food webs. *Phycologia* 35(6): 42-50.
DeMott, W.R., Zhang, Q.-X. & Carmichael, W.W. 1991. Effects of toxic cyanobacteria and purified toxins on the survival and feeding of a copepod and three species of *Daphnia*. *Limnol. Oceanogr.* 36: 1346-1357.

Ebert, D. 1993. The trade-off between offspring size and number in *Daphnia magna*: the influence of genetic, environmental and maternal effects. *Arch. Hydrobiol. Suppl.* 90: 453-473.

Falkoner, I. 1994. Mechanism of toxicity of cyclic peptide toxins from blue-green algae. In I. Falkoner (ed.), *Algal Toxins in Seafood and Drinking Water*: 177-187. Cambridge: Academic Press.

Forsyth, D.J., Haney, J.F. & James, M.R. 1992. Direct observations of toxic effects of cyanobacterial extracellular products on *Daphnia. Hydrobiologia* 228: 151-155.

Glazier, D.S. 1992. Effects of food, genotype, and maternal size and age on offspring investment in *Daphnia magna. Ecology* 73: 910-926.

Gliwicz, Z.M. & Siedlar, E. 1980. Food size limitation and interfering with food collection in *Daphnia. Arch. Hydrobiol.* 88: 155-177.

Gorham, P.R., McLachlan, J., Hammer, U.T. & Kim, W.K. 1964. Isolation and culture of toxic strains of *Anabaena flos-aquae* (Lyngb.) de Breb. *Verh. Int. Verein. Limnol.* 15: 796-804.

Guisande, C. & Gliwicz, Z.M. 1992. Egg size and clutch size in two *Daphnia* species grown at different food levels. *J. Plankton Res.* 14: 997-1007.

Haney, J.F., Forsyth, D.J. & James, M.R. 1994. Inhibition of zooplankton filtering rates by dissolved inhibitors produced by naturally occurring cyanobacteria. *Arch. Hydrobiol.* 132: 1-13.

Haney, J.F., Sasner, J.J. & Ikawa, M. 1995. Effects of products released by *Aphanizomenon flos-aquae* and purified saxitoxin on the movements of *Daphnia carinata* feeding appendages. *Limnol. Oceanogr.* 40: 263-272.

Harada, K.-I, Matsuura, K., Suzuki, M., Watanabe, M., Oishi, H., Dahlem, A.M., Beasley, V.R. & Carmichael, W.W. 1988. Analysis and purification of toxic peptides from cyanobacteria by reversed-phase high-performance liquid chromatography. *J. Chromatogr.* 448: 275-283.

Jungmann, D. & Benndorf, J. 1994. Toxicity to *Daphnia* of a compound extracted from laboratory and natural *Microcystis* spp., and the role of microcystins. *Freshwat. Biol.* 32: 13-20.

Lampert, W. 1981. Inhibitory and toxic effects of blue-green algae on *Daphnia. Int. Rev. Gesammten Hydrobiol.* 66: 285-298.

Lampert, W. 1987. Laboratory studies on zooplankton-cyanobacteria interactions. *N. Z. J. Mar. Freshwat. Res.* 21: 483-490.

Laurén-Määttä, C. 1996. *Daphnia pulex* and exposure to toxic cyanobacteria: population and food chain experiments. Ph.D. Thesis, Turkü: Yliopisto.

Lindholm, T., Eriksson, J.E. & Reinikainen, M. 1992. Ecological effects of toxic cyanobacteria. *Environ. Toxic. Water Qual.* 7: 87-93.

Nichols, H.W. 1973. Growth media – fresh water. In J.R. Stein (ed.), *Handbook on Phycological Methods*: 7-27. Cambridge: Cambridge University Press.

Nizan, S., Dimentman, C. & Shilo, M. 1986. Acute toxic effects of the cyanobacterium *Microcystis aeruginosa* on *Daphnia magna. Limnol. Oceanogr.* 31: 497-502.

Reinikainen, M., Ketola, M. & Walls, M. 1994. Effects of the concentrations of toxic *Microcystis aeruginosa* and an alternative food on the survival of *Daphnia pulex. Limnol. Oceanogr.* 39: 424-432.

Reinikainen, M., Ketola, M. Jantunen, M. & Walls, M. 1995. Effects of *Microcystis aeruginosa* exposure and nutritional status on reproduction of *Daphnia pulex. J. Plankton Res.* 17: 431-436.

Skulberg, O.M., Codd, G.A. & Carmichael, W.W. 1984. Toxic blue-green algae. *Ambio* 13: 244-247.

Smith, A.D. (1993). Inhibitory effects of cyanobacteria on zooplankton and their implications for community structure. Ph.D. Thesis, Dartmouth College, USA: 179p.

Smith, A.D. & Gilbert, J.J. 1995. Relative susceptibilities of rotifers and cladocerans to *Microcystis aeruginosa. Arch. Hydrobiol.* 132: 309-336.

Trubetskova, I. & Lampert, W. 1995. Egg size and egg mass of *Daphnia magna*: response to food availability. *Hydrobiologia* 307: 139-145.

Urabe, J., 1988. Effect of food concentration on the net production of *Daphnia galeata*: separate assessment of growth and reproduction. *Bull. Plankton Sci. Jpn.* 35: 159-174.

Seasonal distribution and density of copepods in the sewage-polluted coastal waters of Tuticorin, India

A. SRINIVASAN & R. SANTHANAM
Fisheries College and Research Institute, Tamilnadu Veterinary and Animal Sciences University, Tuticorin, South India

ABSTRACT

The Tuticorin coastal waters of the southeast coast of India are uniquely influenced by both the northeast and southwest monsoons. This coast is presently under the threat of sewage pollution. The present annual investigation made during 1996-97 at two stations namely, sewage mixing coastal waters (station 1) and open sea free from pollution (station 2) deals with the seasonal distribution and biomass of copepod populations. While the polluted station 1 recorded a maximum number of only 12 species and a density 18,000 individuals m^{-3}, the unpolluted St. 2 recorded 24 species and 81,000 individuals m^{-3}. Interestingly, station 1 showed the abundance of cyclopoid and harpacticoid species viz. *Oithona brevicornis*, *O. rigida* and *Euterpina acutifrons*, *Longipedia coronata* and *Microsetella rosea*, and station 2 recorded more number of calanoid species viz. *Acartia erythraea*, *A. spinicauda*, *A. danae*, *Acrocalanus gracilis*, *Labidocera* spp. and *Centropages furcatus*. Such a significant variation in the distribution and biomass of copepod species in these stations could be related to the pattern of existing food organisms like diatoms etc. and hydrobiological parameters such as nutrients, dissolved oxygen, and BOD.

1 INTRODUCTION

The Tuticorin Bay, on the southeast coast of India (8°5′N; 78°6′E), is known for pearl oysters and chank fisheries resources apart from a rich and diverse finfish and shellfish resources of commercial importance. Certain localities of this coast are also fringed with mangrove vegetation. However, this coast is under the constant threat of domestic sewage, discharged approximately at the rate of 14000 m^3 per day. Investigations relating to the zooplankton of this coast were mainly by the present authors (Srinivasan et al. 1988, Srinivasan & Santhanam 1991, Santhanam & Srinivasan 1994). However, detailed investigation on the impacts of sewage pollution on the copepod potentials of this coast is wanting. Keeping this in view, the present investigation was attempted.

2 MATERIALS AND METHODS

Two stations were selected (Fig. 1), one in the vicinity of untreated sewage discharge (station 1) and the other in the open sea free from pollution (station 2). Samplings were made fortnightly at about 7 a.m. for a period of one year from March 1996 to February 1997. The physicochemical parameters such as temperature, salinity, pH, transparency of water column, dissolved oxygen, phosphate, ammonia, BOD and chlorophyll *a* of the water samples were analyzed using the standard methods of Strickland and Parsons (1972). For quantitative study, phytoplankton and zooplankton samples were collected by filtering a known volume of surface water through a conical net (0.25 m ∅) made of No. 30 bolting silk (41 µm mesh size). For qualitative study on the species composition and seasonal variation of copepods, a standard net of 1 m diameter (41 µm mesh size) was operated for 30 minutes duration. The collected plankton were preserved in 2% buffered formalin (Santhanam & Srinivasan 1994). Identification of different species of phytoplankton and copepods was made using standard keys of Santhanam et al. (1987) and Kasturirangan (1963). Biomass of adult copepods and phytoplankton in terms of density was made by direct counting of the aliquot sample taken in a Sedgwick Counting Cell. The density values were calculated by averaging three countings.

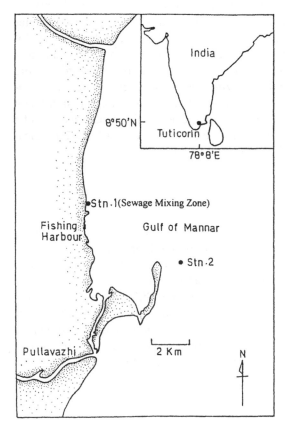

Figure 1. Map of the study areas.

3 RESULTS

3.1 *Physicochemical characteristics*

The depth of the station 1 and 2 varied from 0.9 to 3.1 m and from 3.3 to 6.3 m respectively. The surface temperature of station 1 varied from 25.5 to 31.0°C and from 26.5 to 31.6°C at station 2. The salinity showed only narrow variation as it ranged from 33.3 to 35.4‰ and 34.8 to 35.4‰ at station 1 and 2 respectively. The pH did not show marked variation as its values were from 8.4 to 9.1 (station 1) and from 8.7 to 9.0 (station 2). Owing to the discharge of sewage waters and associated turbidity, the transparency of water column was generally less in station 1 than in station 2, and Secchi disc depths ranged from 0.15 to 0.51 m and 0.45 to 0.64 m respectively. The dissolved oxygen showed drastic variations between the stations ranging from 2.4 to 5.2 ml l^{-1} (station 1) and from 4.3 to 6.2 ml l^{-1} (station 2). The values of NO_3-N, PO_4-P, NH_3-N and BOD observed in station 1 and 2 are presented in Table 1.

3.2 *Species composition of copepods*

The number of species was generally low in the sewage-polluted area (station 1) compared to unpolluted near-shore waters (station 2) and the overall number of species of

Table 1. Hydrographical parameters and phytoplankton density observed at stations 1 and 2.

Month	Stns.	O_2 (ml l^{-1})	BOD (mg l^{-1})	Salinity (ppt)	NO_3-N (μg at.l^{-1})	PO_4-P	Chl. a (mgm^{-3})	Copepods spp. (no.)	Phytoplankton density (cells m^{-3})
Mar.'96	Stn. 1	2.4	35.5	34.1	21.0	4.2	6.4	5	47000
	Stn. 2	4.3	19.68	35.0	14.5	0.2	4.5	6	36000
Apr	Stn. 1	2.8	41.4	35.0	19.4	2.4	11.4	6	29000
	Stn. 2	4.5	27.6	35.1	12.5	0.1	3.5	8	80000
May	Stn. 1	5.2	60.8	35.1	17.3	1.4	36.0	4	545000
	Stn. 2	6.1	38.15	35.4	15.1	0.4	19.5	11	340000
June	Stn. 1	4.5	52.95	35.4	9.2	1.1	24.0	5	195000
	Stn. 2	4.7	32.6	35.3	20.5	0.3	14.5	11	210000
July	Stn. 1	3.5	47.8	35.2	24.1	2.1	11.0	7	63000
	Stn. 2	5.1	29.52	35.0	26.5	0.5	18.0	9	80000
Aug	Stn. 1	4.2	39.5	34.5	29.1	9.1	7.3	4	16000
	Stn. 2	4.8	30.8	35.1	19.0	0.9	4.1	9	65000
Sep	Stn. 1	5.0	41.6	34.8	21.0	6.9	6.2	7	63000
	Stn. 2	6.2	20.8	34.8	19.2	0.5	3.6	12	36000
Oct	Stn. 1	4.8	25.4	33.3	26.5	8.7	3.0	7	11000
	Stn. 2	4.9	21.6	35.0	13.0	0.7	5.1	16	59000
Nov	Stn. 1	3.9	31.6	34.5	18.3	5.7	4.1	8	26000
	Stn. 2	4.1	22.4	35.1	19.0	1.1	4.3	14	45000
Dec	Stn. 1	3.8	29.5	35.3	15.4	4.1	2.8	6	19000
	Stn. 2	4.3	18.9	35.4	12.0	0.8	5.2	11	23000
Jan '97	Stn. 1	3.1	31.6	35.1	21.5	3.3	3.9	2	44000
	Stn. 2	5.2	25.5	35.1	11.0	0.3	3.4	7	47000
Feb	Stn. 1	4.4	36.8	34.9	16.4	3.8	4.8	5	58000
	Stn. 2	5.7	29.8	35.0	18.0	0.1	4.0	5	60000

Table 2. Checklist of copepod species in stations 1 and 2.

Species	Station 1	Station 2
Calanoids		
Acartia danae	+	+
A. erythraea	+	+
A. spinicauda	+	+
Acrocalanus gracilis	+	+
Centropages furcatus	−	+
Labidocera acuta	−	+
L. pavo	−	+
Pseudodiaptomus anandalei	+	+
Pontella danae	−	+
Temora turbinata	−	+
Tortanus gracilis	−	+
Cyclopoids		
Corycaeus danae	−	+
C. speciosus	−	+
Oithona brevicornis	+	+
O. linearis	+	+
O. rigida	+	+
Oncaea venusta	−	+
Sapphirina nigromaculata	−	+
Harpactoids		
Euterpina acutifrons	+	+
Longipedia coronata	+	+
Macrosetella gracilis	−	+
Microsetella norvegica	+	+
M. rosea	−	+
Metis jousseaumei	+	+

copepods at these stations was 12 and 24 respectively (Table 2). The number of species at any one time however, varied from 2 to 8 (station 1; Fig. 2) and from 5 to 16 (station 2; Fig. 3). Interestingly, the species composition also varied between these stations as cyclopoid species such as *Oithona brevicornis, O.rigida* and harpacticoids such as *Euterpina acutifrons, Longipedia coronata* and *Microsetella rosea* invariably made their appearance at station 1. Calanoid species such as *Acartia erythraea, A. spinicauda, A. danae, Acrocalanus gracilis, Labidocera acuta* and *L. pavo* were dominant at station 2. While the maximum number of species was recorded during November (station 1; 8 species) and during October (station 2; 16 species), the minima (station 1; 2 species and station 2; 5 species) were during January and February respectively.

3.3 Density of copepods

The density of adult copepods showed a clear-cut variation between the sewage-polluted area (station 1) and unpolluted near-shore waters (station 2), as the number varied from 6000 to 18,000 and from 14,000 to 81,000 organisms m^{-3} respectively (Fig. 4 and 5). The maxima were recorded in May 1996 at station 1 and in September 1996 at station 2, whereas the minima were recorded in August 1996 at station 1 and in October 1996 at station 2.

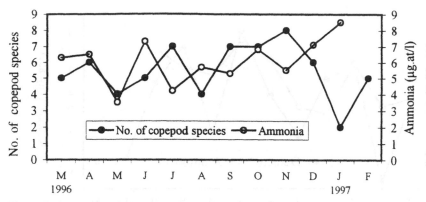

Figure 2. Seasonal variation in species composition of copepods in relation to ammonia level at station 1.

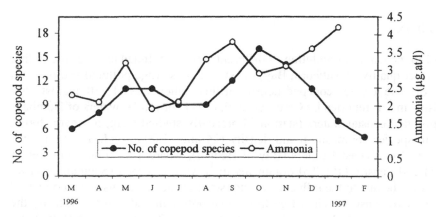

Figure 3. Seasonal variation in species composition of copepods in relation to ammonia level at station 2.

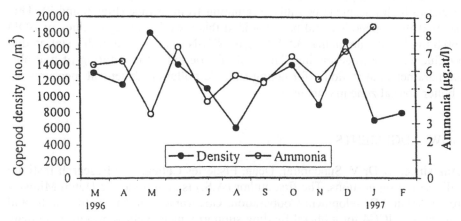

Figure 4. Seasonal variation in density of copepods in relation to ammonia level at station 1.

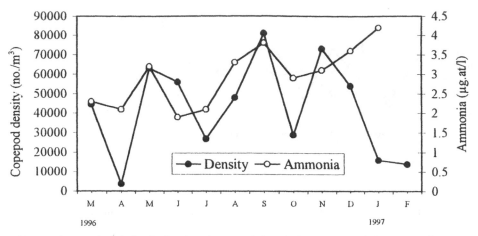

Figure 5. Seasonal variation in density of copepods in relation to ammonia level at station 2.

4 DISCUSSION

With regard to the nature of pollution of different sewage-polluted areas of the Indian coast, the water quality of Tuticorin Bay, i.e. station 1 (sewage polluted station) was not very poor. The average copepod density at this station was 11,750 against only 213 organisms m^{-3} as reported by Gajbhiye et al. (1991) in similar waters of Bombay.

The unpolluted coastal waters (station 2) presently studied were, however, found to be poor in species composition and density (24 and 81,000 organisms m^{-3}) compared to Porto Novo coastal waters (Bay of Bengal), where a maximum number of species (45) and density (106,000 organisms m^{-3}) were recorded (Santhanam et al. 1975). This may be due to the influence of industrial effluents on this coast compared to the Porto Novo coast, which is free from such pollution and is influenced by the nearby luxuriant mangrove biotope. Compared to station 2, the sewage-polluted station 1 registered low species richness (average 6) and low density of copepods (average 11,750). This may be due to its high BOD and NH_3. However, station 1, unlike station 2, registered high phytoplankton density and chlorophyll *a*, which were mainly due to high phosphate possibly originating from sewage (Figs 6 and 7). This suggests that the copepods could have avoided this area due to high values of BOD (25.4-60.8 mg l^{-1}) and ammonia (3.5-8.5 µg.at NH_3-N l^{-1}) recorded in this station due to the influence of sewage mixing. This calls for periodical monitoring of hydrobiological parameters at station 1 for taking suitable pollution abatement measures towards better coastal zone management.

ACKNOWLEDGEMENTS

We thank sincerely Dr V. Sundararaj, Dean, Fisheries College and Research Institute for facilities and suggestions. The first author (A.S.) is grateful to the Dutch Ministry of Foreign Affairs/ Development Cooperation, Government of The Netherlands, and the organisers of ICC4 for a liberal funding support which enabled him to participate and present this paper.

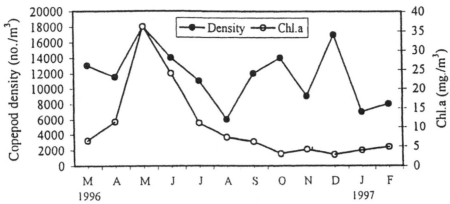

Figure 6. Relationship of copepod density with chlorophyll *a* at station 1.

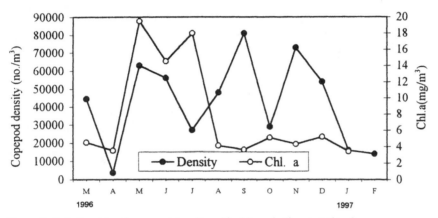

Figure 7. Relationship of copepod density with chlorophyll *a* at station 2.

REFERENCES

Gajbhiye, S.N., Stephen, R., Nair, V.R. & Desai, B.N. 1991. Copepods of the nearshore waters of Bombay. *Indian J. Mar. Sci.* 20(3): 187-194.

Kasturirangan, L.R., 1963. A key for the identification of the more common planktonic Copepoda of Indian coastal waters. Indian National Committee on Oceanic Research Publication, No.2, C.S.I.R., New Delhi. 87 p.

Santhanam, R., Krishnamurthy, K. & Sundararaj, V. 1975. Zooplankton of Porto Novo, south India. *Bull. Dept. Mar. Sci., Univ. Cochin* 7(4): 889-911.

Santhanam, R., Ramanathan, N., Venkataramanujam, K. & Jegatheesan, G. 1987. *Phytoplankton of Indian seas (An aspect of marine Botany)*. Delhi: Daya Publishing House, 134 p.

Santhanam, R. & Srinivasan, A. 1994. Impacts of sewage and thermal pollution on the water quality, plankton and fishing potentials of Tuticorin coast, S. India. In P.G. Wells & R.J. Ricketts (eds), *Proc. Intnl. Symp. Coastal Zone Canada '94* 3: 1191-1201.

Srinivasan, A., Santhanam, R. & Jegatheesan, G. 1988. Biomass and seasonal variation of planktonic tintinnids of Pullavazhi estuary, southeast coast of India. *Indian J. Mar. Sci.* 17: 131-133.

Srinivasan, A. & Santhanam, R. 1991. Tidal and seasonal variations in zooplankton of Pullavazhi brackishwater, southeast coast of India. *Indian J. Mar. Sci.* 20: 182-186.

Strickland, J.D.H. & Parsons, T.R. 1972. A practical handbook of seawater analysis, *Bull. Fish. Res. Bd. Canada* 167: pp. 310.

Biogenic amines controlling blood glucose level in the shrimp *Palaemon elegans*

SIMONETTA LORENZON
B.R.A.I.N. Center for Neuroscience, Departement of Biology, University of Trieste, Trieste, Italy & Shoreline s.ca.r.l. Area Science Park, Trieste, Italy

PAOLA PASQUAL & ENRICO A. FERRERO
B.R.A.I.N. Center for Neuroscience, Departement of Biology, University of Trieste, Trieste, Italy

ABSTRACT

Blood glucose level and its control by release of the crustacean hyperglycemic hormone (CHH) from the eyestalk is under the influence of environmental stressors. We investigated the effect of the biogenic amines serotonin (5HT) and dopamine (DA) in vivo injection into *Palaemon elegans*, whether increasing or depressing glycemia. 5-HT had a marked dose related effect in elevating blood glucose level after 2 h, and probably also on CHH circulation. No significant effect was detected on animals that had their eyestalks removed. H1 receptors seemed to be more involved in mediating 5-HT action, as cyproheptadine (CPH) was a more effective antagonist than ketanserin (H2 receptor inhibitor and also putative DA antagonist). DA injection in intact animals produced a reduction below initial levels of blood glucose within 1 h. This effect was significantly antagonized by domperidone (D2 receptor inhibitor). No significant effect of both amines occurred in animals without eyestalks. The suggestion that these actions might be mediated through Leu-enkephalin release is discussed.

1 INTRODUCTION

Toxicity induced by a pollutant in many instances is the result of interference by either the compound, or by one of its metabolites, with the biochemical events involved in the homeostatic control of a physiological process often coordinated by hormones. Dramatic changes in hormone levels would be expected to occur soon after exposure to a pollutant. Therefore, biosentinal toxicity parameters can be identified by looking for alterations in endocrine patterns (Fingerman et al. 1996).

Hyperglycemia is a typical response of aquatic animals to pollutants. Increased circulating CHH and hyperglycemia in crustaceans are reported to result from exposure to several environmental stresses. Exposure to atmospheric air induced a transitory, but large, increase in blood glucose levels in the intertidal crab *Chasmagnathus granulata* (Santos & Colares 1986). Similar effects were elicited by copper and zinc and hypoxia in *Crangon crangon* and *Carcinus maenas* (Johnson 1987), and by LPS injection in several crustacean species (Lorenzon et al. 1997). The elements Cd, Hg, and Cu induced hyperglycemia in the freshwater prawn *Macrobrachium kistenensis*

471

and the crab *Barytelphusa canicularis* (Nagabhushanam & Kulkarni 1981, Machele et al. 1989). Organic pesticides can also induce hyperglycemia in several crustaceans (Fingerman et al. 1981, Reddy et al. 1983). Reddy et al. (1996) reported that Naphtalene or Cd induced an increase in blood glucose level in *Uca pugilator*. Moreover, $CdCl_2$ induced hyperglycemia in the crayfish *Procambarus clarkii*, however, not if the eyestalks were absent. These responses suggest that Cd-induced hyperglycemia may be mediated by the eyestalk neuroendocrine structures (Reddy et al. 1994).

Neurosecretory structures in the eyestalk are the most important components of the neuroendocrine system of the stalk-eyed decapod crustaceans. Actually, the hemolymph glucose concentration is controlled by the crustacean hyperglycemic hormone (CHH) that is synthesized within the X-organ and released from the sinus gland complex in the eyestalk. In order to understand the connection between environmental stressors and their effects on blood glucose level, some knowledge of the release control mechanism is necessary. Biogenic amines have been found to mediate the release of several neurohormones from crustacean neuroendocrine tissue. 5-HT is well known as a neurotransmitter in crustaceans on several grounds, and its levels have been measured in the nervous system of various crustacean species (Elofsson et al. 1982, Laxmyr 1984, Kulkarni & Fingerman 1992). 5-HT has also been measured in the hemolymph (Livingston et al. 1980, Elofsson et al 1982), thus suggesting a possible role as a neurohormone (Rodriguez-Soza et al. 1997). Sarojini et al. (1995) found that in *P. clarkii* dopamine (DA) and Leucine-enkephalin (L-enk) inhibited the release of CHH from the sinus gland in the eyestalk. Injection of either DA or L-enk into intact crayfish resulted in lowering the blood glucose levels, while the *in vitro* experiments showed that DA or L-enk in the incubation medium suppressed the release of CHH from the eyestalk neuroendocrine tissue. Injection of DA or L-enk into eyestalkless crayfish was ineffective. The DA receptor blocker, spiperone, inhibited the hypoglycemic action of DA and was found not to affect the ability of L-enk to produce hypoglycemia. On the other hand, naloxone blocked the action of both L-enk and DA, thereby apparently allowing the release of CHH. DA and L-enk produced hypoglycemia apparently by inhibiting CHH release. These results, and the release of CHH, suggest that in the chain of neurones that calls for CHH release, terminating at the neuroendocrine cells that secrete CHH, a dopaminergic neuron precedes an enkephalinergic neuron. Luschen et al. (1993) found that 5-HT was the most potent elevator of blood glucose; DA, octopamine (OA), norephinephrine (NE), epinephrine (E), and 5-HT showed in *C. maenas* dose dependent effects and 5-HT, high doses of OA also induced a remarkable increase in glucose titer, while DA, E and NE were only weak stimulators. 5-HT and OA showed an increase of glycemia in *C. maenas* with, and without eyestalks; the catecholamines E and NE showed only weak effects on glucose levels in the crabs. Administration of DA showed a hyperglycemic effect only in intact crabs; animals without eyestalks did not respond to injection of DA; this indicated an indirect mode of action of this biogenic amine. Pre-injection of the dopaminergic antagonist trifluoperazide (TPF) markedly reduced the effect of DA on glucose titer but did not completely abolish the action of DA. Rothe et al. (1991) found that the release of CHH in the crabs *C. maenas* and *Uca pugilator* was inhibited by synthetic L-enk; the decrease in CHH release was antagonized by naloxone. In *in vivo* and *in vitro* experiments L-enk has a direct effect on the mobilization of

CHH likely from the sinus gland since there was no effect on glycogen mobilization in eyestalkless animals.

In view of 1) The difference between crayfish and crab to the reported effect of DA on the hemolymph glucose concentration, and 2) The fact that no one has yet reported on the effect of biogenic amines on glycemia of the decapod shrimp *Palaemon elegans* (Family Caridea), a standard species in environmental toxicology, the aim of this paper was to investigate the effect of 5-HT, DA and L-enk and of their antagonists, on the hemolymph glucose concentration of this shrimp.

2 MATERIALS AND METHODS

2.1 *Animal supply and maintenance*

Palaemon elegans was caught with cages in the Gulf of Trieste (Upper Adriatic Sea) and supplied by commercial fishermen. Three hundred shrimps were stocked in glass tanks of 120 l, filled with closed circuit filtered and thoroughly aerated sea water of 36‰ salinity, 16-18°C, and with natural L-D photoperiod. They were fed every second day *ad libitum* with bits of shrimp, cuttlefish or fish. Eighty-four hours before experiments, individual animals were housed in 500 cc plastic wire cages immersed in larger tanks, for individual recognition and kept unfed until the end of the experiments. Eyestalk ablation was performed under chilling anaesthesia 48-72 h before the experiments.

2.2 *Hemolymph sampling and determination of glycemia*

The hard-shelled animals, irrespective of sexes, were dry blotted and hemolymph was withdrawn from the pericardiac sinus with a sterile 1 ml syringe fitted with a 25 g needle. Animals were bled through in the same way, 50 µl hemolymph each time, at 0 h, 1 h, 2 h, 3 h, 4 h, 6 h and 24 h after injection of biogenic amines and/or of their antagonist, unless otherwise stated. Glucose content was quantified by using One Touch ® II Meter (Lifescan) and commercial kit test strips (precision strips ± 3% C.V. in the tested range). Due to the short time of the processing no anticoagulant was needed.

2.3 *Biogenic amines*

Serotonin (5-HT), dopamine hydrochloride (DA), L-enkephalin (L-enk), naloxone, ketanserin, cyproheptadine (CPH) and domperidone were obtained from Sigma Chemical Company. The amines and their antagonists were tested in doses between 10^{-9} to 10^{-6} mol per animal in a final volume of 50 µl. Each concentration was injected to groups of at least 5 individuals; where appropriate, independent replicates were carried out. The dose was administered in 50 µl saline into the pericardiac sinus. Controls, injected with saline only or just bled at the times mentioned above, were run in order to assess potential effects of saline contamination and of repetitive handling stress and blood withdrawal. Sterile saline for marine crustaceans was prepared with apyrogen distilled water and pure chemicals, reagent grade, according to Smith & Ratcliffe (1978) and autoclaved for 25 min.

3 RESULTS

3.1 *Dose dependent effects of serotonin (5-HT) on the glucose concentration of Palaemon elegans with or without eyestalks*

The dose-dependent hyperglycemic response curve for 5-HT was determined in the range 10^{-8} to 10^{-9} mol per animal over a period of 24 h (Fig. 1). On the basis of these results, the following experiments were performed on intact and eyestalkless animals at the standard concentration of 10^{-8}. Injection of 10^{-8} mol per shrimp into intact individuals causes a significant (Student's t-test, p < 0.001) increase of 7.09 ± 4.72 SD times at 2 h of the blood glucose level compared with control animals injected with saline, where the increase was 0.44 ± 0.51. In eyestalkless animals, 5-HT exerts only a small increase of glycemia of 0.89 ± 1.24 at 2h, which is not significantly different (p = 0.1) from the value 0.22 ± 0.22 of the saline control.

3.2 *Effects of dopamine (DA) and Leucine enkephalin (L-enk) on the glucose concentration in intact and eyestalkless shrimps*

As shown in Figure 2 injection of 10^{-6} mol/shrimp of DA into intact individuals resulted in a significant decrease of glucose concentration (–0.19 ± 0.27 at 1 h) as compared to that seen in the saline injected control animals (0.25 ± 0.47; p = 0.01). In contrast, injection of the same amount of DA into animals without eyestalks did not produce a significant alteration in the glucose level (p = 0.278). Injection of 10^{-6} mol

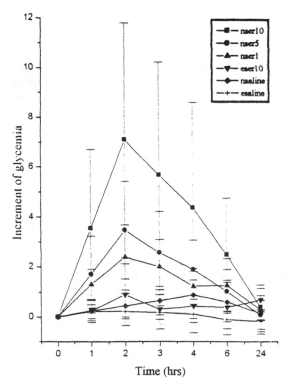

Figure 1. Time course of blood glucose levels in *P. elegans* (n = 10) injected with different concentrations of 5-HT, and with saline. Legend: *nser10*, 10^{-8}mol per animal 5-HT-injected intact animals; *nser5*, 5 × 10^{-9} mol per animal 5HT-injected intact animals; nser1, 10^{-9} mol per animal 5HT-injected intact animals; *eser10*, 10^{-8} mol per animal 5-HT-injected animals without eyestalks; *nsaline*, sterile saline-injected intact control animals; *esaline*, sterile saline-injected control animals without eyestalks. Values are expressed as means ± SD.

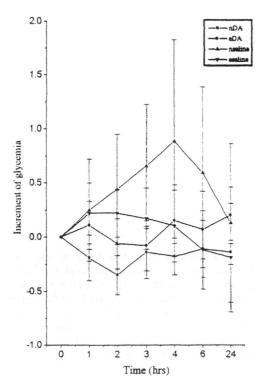

Figure 2. Time course of blood glucose levels in *P. elegans* (n = 10) injected with DA (10^{-6} mol per shrimp). Legend: *nDA*, 10^{-6} mol per animal DA-injected intact animals; *eDA*, 10^{-6} mol per animal DA-injected animals without eyestalks; *nsaline*, sterile saline-injected intact control animals; *esaline*, sterile saline-injected control animals without eyestalks. Values are expressed as means ± SD.

per shrimp of L-enk (Fig. 3) into intact animals resulted in a significant increase of blood glucose concentration (1.02 ± 0.36 at 1 h) as compared to the control animals (0.25 ± 0.47 at 1 h, p < 0.001). In individuals without eyestalks, L-enk caused a decrease of glucose concentration in the first 3h, i.e. –0.28 ± 0.25 that was significantly different from the control (p = 0.001).

3.3 *Effects of 5-HT antagonists ketanserin and cyproheptadine (CPH)*

The object of these experiments was to determine whether the hyperglycemic effect of 5-HT can be antagonized by the 5-HT receptor (H2) blocker, ketanserin 10^{-6} mol per shrimp, and by the CPH (H1 receptor blocker) 10^{-6} mol per shrimp. 5-HT produced hyperglycemia in intact animals (Fig. 4); ketanserine alone produced a significant increase in the glucose level up to 1.78 ± 0.74 at 1 h (p < 0.001 *vs* saline control). However, when 5-HT and ketanserine were coinjected, there was no significant (p = 0.07 *vs* control) effect on glycemia (maximum effect 0.76 ± 0.7 at 1 h).

Injection of 10^{-6} mol per shrimp of CPH into intact animals had no effect on blood sugar level (Fig. 5), i.e. 0.47 ± 0.27 at 2 h (p = 0.9 *vs* control). When co-injected with 5-HT, CPH was able to antagonize the effect of 5-HT. In fact there was a small and not significant (p = 0.23) increase of glycemia of 0.72 ± 0.57 at 2 h. Ketanserine and CPH were indeed able to block the strong action of 5-HT. In animals without eyestalks, injection of ketanserine or CPH did not produce a significant alteration in glucose level.

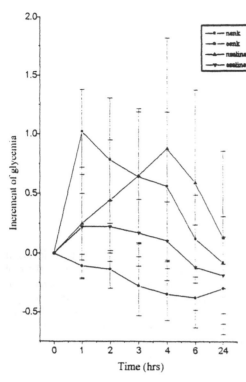

Figure 3. Time course of blood glucose levels in *P. elegans* (n = 10) injected with L-enk (10^{-6} mol per shrimp). Legend: *nenk*, 10^{-6} mol per animal L-enk-injected intact animals; *eenk*, 10^{-6} mol per animal L-enk-injected animals without eyestalks; *nsaline*, sterile saline-injected intact control animals; *esaline*, sterile saline-injected control animals without eyestalks. Values are expressed as means ± SD.

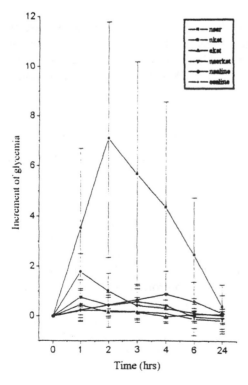

Figure 4. Time course of blood glucose levels in *P. elegans* (n = 10) injected with 5-HT (10^{-8} mol per animal) and its antagonist ketanserin (10^{-6} mol per animal). Legend: *nser*, 10^{-8} mol per animal 5-HT-injected intact animals; *nket*, 10^{-6} mol per animal ketanserin-injected intact animals; *eket*, 10^{-6} mol per animal ketanserin-injected animals without eyestalks; *nserket*, intact animal coinjected with 10^{-8} mol per animal 5-HT and 10^{-6} mol per animal ketanserin; *nsaline*, sterile saline-injected intact control animals; *esaline*, sterile saline-injected control animals without eyestalks. Values are expressed as means ± SD.

3.4 *Effects of domperidone on DA and of the competitive opioid antagonist naloxone on L-enk*

Injection of 10^{-6} mol per shrimp of domperidone (D2 receptor blocker) in intact *P. elegans* (Fig. 5) caused an increment of the glucose level by 1.76 ± 0.5 at 3 h significantly different from the saline injected control (p < 0.001). Co-injection of domperidone with DA (Fig. 6) caused an inhibition of DA action with a significant increase of glycemia to 2.05 ± 1.37 at 3 h (p = 0.008). In animals without eyestalks, domperidone caused a maximum variation of glucose level of 0.54 ± 0.26 at 3 h that was significantly different from that of the saline control animals (p = 0.007). Injection of 10^{-6} mol per shrimp of the competitive opioid antagonist naloxone (Fig. 7) caused a significant increase of glycemia (1.53 ± 0.83 at 3 h, p = 0.013 against saline control). When co-injected with L-enk, a further significant increment of blood glucose concentration up to 2.26 ± 1.42 at 2 h (p = 0.001) was obtained. In shrimps without eyestalks naloxone induced a significant hypoglycemia in the first four hours as –0.29 ± 0.23 (p < 0.001).

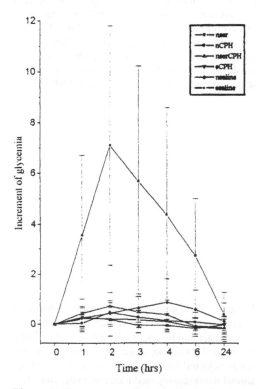

Figure 5. Time course of blood glucose levels in *P. elegans* (n = 10) injected with 5-HT (10^{-8} mol per animal) and its antagonist CPH (10^{-6} mol per animal). Legend: *nser*, 10^{-8} mol per animal 5-HT-injected intact animals; *nCPH*, 10^{-6} mol per animal CPH-injected intact animals; *eCPH*, 10^{-6} mol per animal CPH-injected animals without eyestalks; *nserCPH*, intact animal co-injected with 10^{-8} mol per animal 5-HT and 10^{-6} mol per animal CPH; *nsaline*, sterile saline-injected intact control animals; *esaline*, sterile saline-injected control animals without eyestalks. Values are expressed as means ± SD.

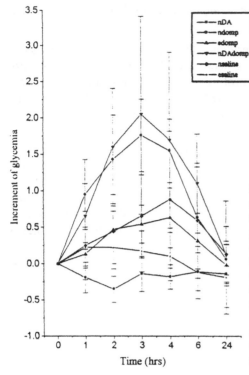

Figure 6. Time course of blood glucose levels in *P. elegans* (n = 10) injected with DA (10^{-6} mol per animal) and its antagonist domperidone (10^{-6} mol per animal). Legend: *nDA*, 10^{-6} mol per animal DA-injected intact animals; *ndomp*, 10^{-6} mol per animal domperidone-injected intact animals; *edomp*, 10^{-6} mol per animal domperidone-injected animals without eyestalks; *nDAdomp*, intact animal co-injected with 10^{-6} mol per animal DA and 10^{-6} mol per animal domperidone; *nsaline*, sterile saline-injected intact control animals; *esaline*, sterile saline-injected control animals without eyestalks. Values are expressed as means ± SD.

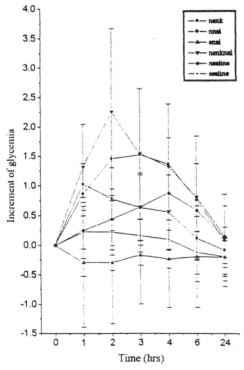

Figure 7. Time course of blood glucose levels in *P. elegans* (n = 10) injected with L-enk (10^{-6} mol per animal) and its antagonist naloxone (10^{-6} mol per animal). Legend: *nenk*, 10^{-6} mol per animal L-enk-injected intact animals; *nnal*, 10^{-6} mol per animal naloxone-injected intact animals; *enal*, 10^{-6} mol per animal naloxone-injected animals without eyestalks; *nenknal*, intact animal co-injected with 10^{-6} mol per animal L-enk and 10^{-6} mol per animal naloxone; *nsaline*, sterile saline-injected intact control animals; *esaline*, sterile saline-injected control animals without eyestalks. Values are expressed as means ± SD.

4 DISCUSSION

The results of this investigation demonstrated that biogenic amines were able to influence hemolymph glucose in the shrimp *Palaemon elegans*. 5-HT, localized in several districts of the eyestalk and of supra-oesophageal and nerve cord ganglia, was found to be a potent elevator of blood glucose levels in intact *P. elegans* as well as in *C. maenas* (Luschen et al. 1993) and *Astacus leptodactylus* (Strolenberg & Van Herp 1977). Injected antagonists ketanserin and CPH were able to inhibit the hyperglycemic action of 5-HT. In *C. maenas* both 5-HT and OA were identified as potent elevators of glucose titer in intact and animals with eyestalks removed. Both amines were able to increase glucose titer independently of the eyestalk neurosecretory system (Luschen et al. 1993). The hyperglycemic activity of 5-HT in *C. maenas* without eyestalks is in contrast with our findings for *P. elegans*, where no significant increase in blood glucose was apparent after injection of 5-HT. Injection of DA into intact *P. elegans* caused a marked hypoglycemia, while no dramatic effects were revealed in blood glucose of individuals with eyestalks removed, as reported by Sarojini et al. (1995) also in *Procambarus clarkii*. On the contrary, DA produced hyperglycemia in intact *C. maenas* but not in animals without eyestalks (Luschen et al. 1993). Injection of L-enk caused hyperglycemia within 1h in intact *P. elegans* but failed to do so in animals without eyestalks. This result is in contrast with the findings by Sarojini et al. (1995) for *P. clarkii*, and those by Rothe et al. (1991) for *C. maenas* and *U. pugilator*, where L-enk produced hypoglycemia. In the present study, the DA receptor blocker domperidone was found to inhibit the hypoglycemic effect of DA.

The specificity of action of the neuromediators was demonstrated by the corresponding inhibition of the response by co-administration of the relative antagonists. Lack of a noticeable response in animals theat had their eyestalks removed, pleas for a CHH mediated response.

REFERENCES

Elofsson, R., Laxmyr, L., Rosengren, E. & Hanson, C. 1982. Identification and quantitative measurements of biogenic amines and DOPA in the central neurons and hemolymph of the crayfish *Pacifastacus leniusculus* (Crustacea). *Comp. Biochem. Physiol.* 71C: 191-205.

Fingerman, M., Hunumante, M.M., Deshpande, U.D. & Nagabhushanam, R. 1981. Increase in the total reducing substances in the hemolymph of the freshwater crab, *Barytelphusa guerini,* produced by a pesticide (DDT) and an indolealkylamide (serotonin). *Experientia* 37: 178-189.

Fingerman, M., Devi, M., Reddy, P.S. & Katayayani, R. 1996. Impact of heavy metal exposure on the nervous system and endocrine-mediated proces in crustaceans. *Zool. Stud.* 35: 1-8.

Johnson, I. 1987. The effects of combination of heavy metals, hypoxia and salinity on oxygen consumption and carbohydrate metabolism of *Crangon crangon* (L) and *Carcinus maenas* (L.). *Ophelia* 27: 155-169.

Kulkarni, G.K. & Fingerman, M. 1992. Qualitative analysis by reverse phase high chromatography of 5-hydroxytryptamine in the central nervous system of the swamp crayfish *Procambarus clarkii. Biol. Bull. Mar. Biol. Lab. Woods Hole* 182: 341-347.

Laxmyr, L. 1984. Biogenic amines and DOPA in the central nervous system of decapod crustaceans. *Comp. Bichem. Physiol.* 77C: 139-143.

Livingstone, M.S., Harris-Warrick, R.M. & Kravitz, E.A. 1980. Serotonin and octopamine produce opposite posture in lobsters. *Science* 208: 76-79.

Lorenzon, S., Giulianini, P.G. & Ferrero, E.A. 1997. Lipopolysaccharide-induced hyperglycemia is mediated by CHH release in crustaceans. *Gen. Comp. Endocrinol.* 108: 395-405.

Luschen, W., Willing, A. & Jaros, P.P. 1993. The role of biogenic amines in the control of blood glucose level in the decapod crustacean, *Carcinus maenas. Comp. Biochem. Physiol.* 105C: 291-296.

Machele, P.R., Khan, A.K., Sarojini, R. & Nagabhushanam, R. 1989. Copper and Cadmium induced changes in blood sugar level of crab, *Barytelphusa canicularis. Uttar Pradesh J. Zool.* 9: 113-115.

Nagabhushanam, R. & Kulkarni, G.K. 1981. Freshwater palaemonid prawn, *Macrobrachium kistenensis* (Tiwari)-Effect of heavy metal pollutants. *Proc. Indian Natl. Sci. Acad.* B. 47: 380-386.

Reddy, P.S., Bhagyalaxmi, A. & Ramamurthy, R. 1983. *In vivo* acute physiological stress induced by BHC on hemolymph biochemistry of *Ozitelphusa senex senex*, the Indian rice fiddler crab. *Toxicol. Lett.* 18: 35-38.

Reddy, P.S., Devi, M., Sarojini, R., Nagabhushanam, R. & Fingerman, M. 1994. Cadmium chloride induced hyperglycemia in the red swamp crayfish *Procambarus clarkii*: Possible role of crustacean hyperglycemic hormone. *Comp. Biochem. Physiol.* 107C: 51-57.

Reddy, P.S., Katayayani, R.V. & Fingerman, M. 1996. Cadmium and Naphthalene induced hyperglycemia in the fiddler crab *Uca pugilator*: Differential modes of action on the neuroendocrine system. *Bull. Environ. Contam. Toxicol.* 56: 425-431.

Rodriguez-Soza, L., Picones, A., Rosete, G.C., Islas, S. & Arechiga, H. 1997. Localization and release of 5-hydroxytryptamine in the crayfish eyestalk. *J. exp. Biol.* 200: 3067-3077.

Rothe, H., Luschen, W., Asken, A., Willing, A. & Jaros, P.P. 1991. Purified crustacean enkephalin inhibits release of hyperglycemic hormone in the crab *Carcinus maenas. Comp. Biochem. Physiol.* 99C: 57-62.

Santos, E.A. & Colares, P.C. 1986. Blood glucose regulation in an intertidial crab, *Chasmagnathus granulata. Comp. Biochem. Physiol.* 83: 673-675.

Sarojini, R., Nagabhushanam, R. & Fingerman, M. 1995. Dopaminergic and Enkephalinergic involvement in the regulation of blood glucose in the red swamp crayfish *Procambarus clarkii. Gen. Comp. Endocrinol.* 97: 160-170.

Smith, V.J. & Ratcliffe, N. A. 1978. Host defence reactions of the shore crab, *Carcinus maenas* (L.) in vitro. *J Mar. Biol. Assoc. UK* 58: 367-379.

Strolenberg, G.E. & Van Herp, F. 1977. Mise en évidence du phénomène d'exocytose dans la glande du sinus d'*Astacus leptodactylus* (Nordmann) sous l'influence d'injections de sérotonine. *Comptes Rendus de l'Académie des Sciences,* Paris 284D: 57-59.

Dose-response to unesterified pigments of Aztec marigold, *Tagetes erecta*, in the Pacific white shrimp, *Litopenaeus vannamei*, fed with various dietary concentrations of carotenoids

JOSE LUIS ARREDONDO-FIGUEROA & E. JAIME VERNON-CARTER
DH(DCBS), AIQ(DCBI). Universidad Autónoma Metropolitana-Iztapalapa. México

JESUS T. PONCE-PALAFOX
DH(DCBS), AIQ(DCBI). Universidad Autónoma Metropolitana-Iztapalapa. México;
CIB. Universidad Autónoma del Estado de Morelos. Cuernavaca, México

ABSTRACT

This study was designed to determine the effects of various dietary unesterified Aztec marigold carotenoids concentrations on the pigmentation of Pacific white shrimp. Four pigmented diets containing the unesterified marigold extracts at concentrations of 50, 100, 200, and 350 ppm, a reference diet containing 200 ppm carophyll pink and a non-pigmented control diet were fed during 35 days to the shrimp. Abdominal muscle and exoskeleton pigmentation was influenced by the diet pigment concentration and by feeding trial duration. The degree of abdomen pigmentation achieved by the 200 ppm carophyll diet is equalled by the 200 ppm unesterified marigold diet, whereas the exoskeleton pigmentation is vastly superior for the carophyll containing diet than for any diet containing up to 350 ppm unesterified marigold pigment.

1 INTRODUCTION

Carotenoids are the main pigments of many aquatic animals (Meyers 1994). In aquaculture, astaxanthin and canthaxanthin are commonly used as pigment sources for fish and shrimp. The main fish species that were studied in this respect are rainbow trout (Lee et al. 1994), Artic charr (Hatlen et al. 1995), Atlantic salmon (Torrisen et al. 1995). Carotenoids in the Penaeidae include astaxanthin and its esters, carotenes and xanthophylls (Fisher et al. 1957). The following carotenoid pigments were found in the caparace of shrimp (Decapoda; Penaeidae): lutein, tunaxanthin, astaxanthin diester, astaxanthin monoester, and free astaxanthin. In its various forms astaxanthin was found to be the most important pigment, in both the cephalothorax and the abdomen (Carreto & Carignan 1984). Studies of shrimp pigmentation using astaxanthin from marine sources, have been described by various authors (Tanaka et al. 1976, Latscha 1990, Chien & Jeng 1992). Liao et al. (1993) and Vernon-Carter et al. (1996) have also examined pigmentation of shrimps with carotenoids from plants. Synthetic carotenoids have received considerable interest (Spinelli & Mahnken 1974, Tanaka et al. 1976, Otazu & Ceccaldi 1984, Yamada et al. 1990, Negre-Sadargues et al. 1993). It has been shown that the inclusion of pure carotenoids or crude carotenoids in diets improves pigmentation in crustaceans (Katayama et al. 1972a, b, Chien

& Jeng 1992, Okada et al. 1994). However, the feeding duration, the amount of pigment deposition, and pigment costs are important factors as to which pigment source must be included in a commercial nutrition.

The aim of this study was to determine the most effective pigment concentration of the unesterified Aztec marigold in a diet and the feeding period that is required to obtain the desired pigmentation response in the Pacific white shrimp (*Litopenaeus vannamei*).

2 MATERIALS AND METHODS

2.1 *Diet preparation*

The composition of the experimental diets is presented in Table 1. The pigment sources were carophyll pink (synthetic, 8% astaxanthin, Roche Co. Ltd.) in a concentration of 200 mg kg^{-1} (basal diet), and unesterified Aztec marigold (total carotenoids concentration = 28.6 g kg^{-1}; 0.57 g kg^{-1} β-carotene; 0.62 g kg^{-1} cryptoxanthin; 0.96 g kg^{-1} cis-luthein; 22.72 g kg^{-1} trans-luthein; 1.54 g kg^{-1} trans-zeaxanthin; 2.19 g kg^{-1} epoxy-luthein) in four concentrations 50, 100, 200 and 350 mg kg^{-1} (basal diet). Also a non-pigmented diet was used as a control. A separate experiment showed that the degradation of the pigments in these diets during storage was not significant. Moisture was determined by drying overnight at 100°C. Crude protein, crude fat, ash, and carbohydrate contents were calculated based on the official methods of analysis of the AOAC (1990).

2.2 *Feeding trials*

Assays were performed with *Litopenaeus vanname*, commonly known as the Pacific white shrimp. Three hundred and six organisms with an average weight of 10.5 ± 0.23 g were allotted randomly to 18 tanks (40 × 50 × 30 cm), each tank containing 16

Table 1. Composition of experimental diets (% as is base).

Ingredients	Pigment concentration (mg/kg)					Carophyll pink (200)
	0	50	100	200	350	
White fish meal	39.00	39.00	39.00	39.00	39.00	39.00
Squid meal	1.43	1.43	1.43	1.43	1.43	1.43
Soybean meal	13.04	13.04	13.04	13.04	13.04	13.04
Starch	29.00	27.23	25.45	21.91	16.59	26.50
Cod liver oil	4.97	4.97	4.97	4.97	4.97	4.97
Soybean oil	1.56	1.56	1.56	1.56	1.56	1.56
Vitamin premix[a]	2.00	2.00	2.00	2.00	2.00	2.00
Mineral premix[b]	2.00	2.00	2.00	2.00	2.00	2.00
Carboxymethylcellulose	3.00	3.00	3.00	3.00	3.00	3.00
Dextrin	4.00	4.00	4.00	4.00	4.00	4.00
Pigment material	0.00	1.77	3.55	7.09	12.41	2.50

[a]Vernon-Carter et al. (1996). [b]Supplemented with Fe, 21.4 mg ($FeSO_4$); Mn, 28.6 mg (MnO); Zn, 25.2 mg (ZnO); Cu, 7.2 mg (CuO); Co, 0.5 mg (Co SO_4); I, 1.0 mg (CaI_2).

shrimps, that were fed in triplicate groups. The shrimps were held in filtered seawater that was changed three times a week. Salinity was 33.0 ± 0.8 ppt, the average temperature was $28.0 \pm 0.9°C$, pH was 7.38 ± 0.21, and dissolved oxygen 6.5 mg l^{-1}. The shrimp were fed during 35 days, twice daily at 2% of body weight per day at 09:00 h and 18:00 h.

2.3 *Pigment analysis*

At the start of the trials, from each tank one organism was randomly selected and the pigment concentration was established. At day 21 and 35, eight shrimps were randomly sampled from each tank. The viscera (proventriculus, hepatopancreas, and intestine) were removed from the shrimp, while the exoskeleton and abdominal muscle were immediately stored in liquid nitrogen and subsequently analyzed for total carotenoids content (TCC). The exoskeleton and abdominal muscle of the shrimp were pooled as one sample. The extraction and determination of the carotenoid content was done in accordance to the technique described by Sommers et al. (1991).

2.4 *Statistical analysis*

Statistical analysis was carried out according to Montgomery (1984). One-way analysis of variance was used to evaluate the results. The independent variables were the five pigmented diets. Tukey's test was conducted to determine the effects among the diets.

3 RESULTS AND DISCUSSION

3.1 *Pigmentation*

The analysis of the total carotenoid concentration in the exoskeleton and muscle of the shrimps (Table 2) indicates that in any period and irrespective of the diet, the pigment content in the exoskeleton was considerably higher than in the abdominal muscle. The non-pigmented control diet showed a 2.8-fold difference in pigmentation

Table 2. Total carotenoid concentration (mg kg^{-1} of body weight) in the shrimp fed with the pigmented diets in the experiment. Values are means $\pm$ SEM. Means in the same row with different superscripts are significantly different at $p < 0.05$.

Dietary	Control	50	100	200	350	Carophyll Pink-200
Abdomen						
0	14.0 ± 0.08					
21 days	13.2 ± 0.26^a	14.9 ± 0.18^b	15.6 ± 0.17^b	18.1 ± 0.09^c	19.1 ± 0.26^d	17.5 ± 0.35^c
35 days	12.4 ± 0.07^a	13.8 ± 0.08^b	15.8 ± 0.19^c	16.6 ± 0.15^d	17.5 ± 0.27^e	16.7 ± 0.15^d
Exoskeleton						
0	40.0 ± 0.06					
21 days	39.2 ± 0.43^a	60.6 ± 0.41^b	84.3 ± 0.44^c	77.1 ± 0.21^d	86.3 ± 0.33^e	89.9 ± 0.42^f
35 days	37.6 ± 0.50^a	69.9 ± 0.62^b	86.4 ± 0.41^c	82.0 ± 0.45^d	91.0 ± 0.46^e	110.9 ± 0.30^f

exoskeleton and the abdominal muscle increased along with the increased pigment concentration from 50 to 350 ppm in the experimental diets, the latter showed a 5.2-fold pigmentation difference after 35 days of the feeding trial. However, the highest difference was exhibited by the 200 ppm carophyll reference diet (CRD) that showed a 6.6-fold difference in pigmentation between the exoskeleton and the abdominal muscle of after 35 days.

While the difference in pigmentation between the exoskeleton and the abdominal muscle remained more or less constant under the non-pigmented control diet (NPCD), for all the unesterified marigold diets (UMD) and the CRD the pigmentation difference muscle increased as the feeding time trial increased.

Furthermore, in all the experimental diets, except the 100 ppm UMD, the abdominal muscle pigmentation suffered a decrease from day 21 to day 35, while the exoskeleton pigmentation increased from day 21 to day 35. These results tend to suggest that there is a selective absorption mechanism of the dietary carotenoids, which enters in action after pigment saturation in the muscle occurs. Yamada et al. (1990) and Vernon-Carter et al. (1996) found similar tendencies, but the decrease in pigmentation was also exhibited in the exoskeleton. As no diminution in the total carotenoids concentration with time in the experimental diets was found, the explanation of tissue saturation followed by depletion is plausible. Choubert & Storebakken (1989) reported that rainbow trout fed with various concentrations of dietary carotenoids, showed a decreasing rate of accumulation of pigments, which became more pronounced with longer feeding time and increasing carotenoids concentration. If the muscle pigmentation is the dominant perception characteristic for the consumer, the feeding trials should not exceed 21 days because no perceptible gain in muscle pigmentation is obtained by extending the feeding trial to 35 days. The difference in the degree of pigmentation between the shrimp muscle and exoskeleton is accentuated, as the feeding trial is prolonged, tending to minimize the perceived degree of pigmentation achieved in the muscle.

3.2 *Abdomen pigmentation*

The total carotenoid concentration in the abdominal muscle of the shrimp for the different treatments at day 21 and 35 is shown in Table 3. At day 21, the data show that the higher pigment assimilation occurred for the 350 ppm UMD, followed by 200 ppm UMD, 200 ppm CRD, 100 ppm UMD, 50 ppm UMD, and NPCD, respectively.

Table 3. Mean weights and survival of *Litopenaeus vannamei* over 35 days. Values are means ± SEM. Means in the same row with different superscripts are significantly different (p < 0.05).

Dietary	Control	50	100	200	350	200 (Carophyll Pink)
Carotenoids						
Initial weight (g)	10.96 ± 0.47^a	10.26 ± 0.51^a	10.71 ± 0.51^a	10.68 ± 0.54^a	10.13 ± 0.50^a	10.86 ± 0.57^a
Final weight (g)	13.79 ± 0.69^a	13.16 ± 0.62^a	13.82 ± 0.71^a	13.86 ± 0.61^a	13.38 ± 0.61^b	14.18 ± 0.67^a
Weight gain (g d^{-1})	$0.081\pm0.01^*$	0.083 ± 0.02^a	$0.089\pm0.01^*$	0.091 ± 0.03^b	0.093 ± 0.03^b	0.095 ± 0.03^b
Survival (%)	76.5 ± 0.7^a	82.3 ± 0.8^a	78.4 ± 0.9^a	82.3 ± 0.7^b	$88.2\pm0.8^{b,c}$	94.1 ± 0.9^c

All the experimental diets were significantly different (p < 0.05) from the NPCD, but there was no significant difference between the 50 and 100 ppm UMD or between the 200 ppm UMD and the 200 ppm CRD. These results suggest that as far as abdominal muscle pigmentation is concerned the 200 ppm UMD is as effective as the 200 ppm CRD.

As mentioned earlier, at day 35 the pigment concentration in the abdominal muscle dropped from the values exhibited at day 21, but there existed a significant difference among all the treatments, except between 200 ppm UMD and the 200 ppm CRD.

The carotenoid concentration in the shrimp abdominal muscle for the different experimental diets found in this study using the same pigmenting sources were approximately double than those found after 28 days in Pacific white shrimp by Vernon-Carter et al. (1996).

These results indicate that the inclusion of unesterified marigold carotenoids in the basal diet has a similar effect as that of synthetic astaxanthin in the pigmentation of this particular species of shrimp. It also confirms the importance of determining the length of time and the pigment concentrations used in the feeding trials.

3.3 *Exoskeleton pigmentation*

As in the case of the abdominal muscle pigmentation, the shrimp fed with the NPCD showed a non-significant decrease in their exoskeleton pigmentation in the feeding trials. All the experimental diets showed a significant pigmentation difference with regard to the UPCD, and were significantly different from each other at day 21. The best treatment was for the CRD followed by the 350 ppm UMD, 100 ppm UMD, 200 ppm UMD, 50 ppm UMD, and the UPCD, respectively. As the feeding trial continued to 35 days, contrary to the abdominal muscle pigmentation, the degree of pigmentation in all the experimental diets increased, excepting the UPCD. There exists a significant difference in the exoskeleton pigmentation among all the diets, showing the same order from most to least as after 21 days. However, as the feeding trial continues from day 21 to day 35, the relative increase in pigmentation for the CRD was 23.3%, while for the various UMD it varied between 2.49 and 15%. These results indicate that the assimilation and deposition of astaxanthin is more effective in the exoskeleton than the unesterified marigold pigment, at whatever concentration used.

The carotenoid concentration found in the exoskeleton were higher than those found for the same pigment sources after 28 days by Vernon-Carter et al. (1996). However, in the latter study the pigments were micro-encapsulated with a biopolymer that probably hindered pigment absorption.

3.4 *Shrimp survival and weight gain*

The weight gain per day and the final weight of the shrimp after 35 days is shown in Table 3. In general terms, the shrimp exhibiting a higher weight gain and a higher survival rate, were those fed with diets of high carotenoid concentrations. In particular the CRD, 350 ppm UMD and 200 ppm UMD were significantly different among themselves and from the other experimental diets.

4   CONCLUSIONS

The Pacific white shrimp abdominal muscle can be suitably pigmented with unesterified Aztec marigold carotenoids as the pigmenting source, instead of carophyll pink. The feeding trials should not exceed 21 days as the amount of pigment deposition decreases as the feeding time trial is prolonged. However, the unesterified marigold pigment is not as effective as carophyll pink in the pigmentation of the shrimp exoeskeleton, the pigmenting difference between both sources becoming more pronounced as feeding trial time is increased.

ACKNOWLEDGEMENTS

The authors wish to acknowledge the financial support of Laboratorios Columbia S.A. de C.V., the Consejo Nacional de Ciencia y Tecnología through project 25153-B, the Universidad Autónoma Metropolitana-Iztapalapa through the Programa Multidisciplinario CBI-CBS and Secretaría de Desarrollo Agropecuario del Estado de Morelos, México.

REFERENCES

AOAC. 1990. *Official methods of analysis*, 15th Ed. Washington (D.C.): Association of Official Analytical Chemists.

Carreto, J.I. & Carignan, M.O. 1984. Pigmentos carotenoides del camarón *Artemesia longinaris* Bate (Crustacea, Decapoda, Penaeidae). *Rev. Inst. Nac. Inv. Des. Pesquero (Argentine)* 4: 5-20.

Chien, Y.H. & Jeng, S.C. 1992. Pigmentation of kuruma prawn, *Penaeus japonicus* Bate, by various pigment sources and levels and feeding regimes. *Aquaculture* 102: 333-346.

Fisher, L.R. Kon, S.K. & Thompson, S.Y. 1957. Vitamin a and carotenoids in certain invertebrates. VI. Crustacean: Penaeidea. *J. Mar. Biol. Ass. UK* 36: 501-507.

Hatlen, B., Aas, G.H., Jorgense, E.H., Storebakken, T. & Goswami, U. 1995. Pigmentation of 1, 2 and 3 year old Artic charr (*Salvelinus alpinus*) fed two different dietary astaxanthin concentrations. *Aquaculture* 138: 303-312

Katayama, T., Katama, T. & Chichester, C.O. 1972a. The biosynthesis of astaxanthin. VI. The carotenoid in the prawn, *Penaeus japonicus* Bate (Part. II) *Int. J. Biochem.* 3: 363-366.

Katayama, T., Katama, T., Shimaya, M., Deshimaru, O. & Chichester, C.O. 1972b. The biosynthesis of astaxanthin. VIII. The conversion of labelled ß-carotene-15,15' $^3H_2$ into astaxanthin in prawn, *Penaeus japonicus* Bate. *Bull. Jpn. Soc. Sci. Fish.* 38: 1171-1175.

Latscha, T. 1990. The role of astaxanthin in shrimp pigmentation. *Actes Colloq. Ifremer* 9: 319-325.

Liao, W.L., Nur, E.B., Okada, S., Matsui, T. & Yamaguchi, K. 1993. Pigmentation of culture black tiger prawn by feeding with *Spirulina* supplemented diet. *Nippon Suisan Gakkaishi Bull. Jap. Soc. Sci. Fish.* 59: 165-169.

Lee, K.H., Kang, S.J., Choe, B.D., Choi, Y.J. & Youm, M.G. 1994. Utilization of ascidian *Halocynthia roretzi* tunic.2. Optimum level of carotenoid extracts from ascidian tunica for the pigmentation of rainbow trout *Oncorhynchus mykiss. Bull. Korean Fish Soc.* 27: 240-246.

Meyers, P. S. 1994. Developments in world aquaculture, feed formulations and role of carotenoids *Pure Appl. Chem.* 66: 1069-1076.

Montgomery, D.C. 1984. *Design and analysis of experiments*. New York: John Wiley & Sons.

Negre-Sadargues, G., Castillo, R., Petit, H., Sance, S., Gomez, M.R., Milicua, G.J.C., Choubert, G. & Trilles, J.P. 1993. Utilization of synthetic carotenoids by the prawn *Penaeus japonicus* reared under laboratory conditions. *Aquaculture* 110: 151-159.

Okada, S., Nur-E-Borhan, S.A., Watabe, S. & Yamaguchi, K. 1994. Pigmentation of cultured black tiger prawn by feeding a *Spirulina* supplemented diet. *Abstracts 3rd Int. Marine Biotech. Conf. Tromsø, Norway* p. 116.

Otazu, A.M. & Ceccaldi, H.J. 1984. Influence of purified carotenoids added to compound diets on pigmentation of *Penaeus japonicus* (Crustacea, Decapoda). *Aquaculture* 36: 217-228.

Sommer, T.R., Potts, W.T. & Morrissy, N.M. 1991. Utilization of microalgal astaxanthin by rainbow trout (*Oncorhynchus mykiss*). *Aquaculture* 94: 79-88.

Spinelli, J. & Mahnken, C. 1974. Composition, processing and utilization of red crab (*Pleuroncontes planipes*) as an aquacultural feed ingredient. *J. Fish. Res. Board Can.* 31: 1025-1029.

Tanaka, Y., Matsuguchi, H. & Katayama, T. 1976. The biosynthesis of astaxanthin-XVIII. The metabolism of carotenoids in the prawn, *Penaeus japonicus* Bate. *Bull. Jpn. Soc. Sci. Fish.* 42: 197-202.

Torrisen, O.J., Christiansen, R., Struksnaes, G. & Estermann, R. 1995. Astaxanthin deposition in the flesh of Atlantic salmon, *Salmo salar* L., in relation to dietary astaxanthin concentration and feeding period. *Aquacult. Nutr.* 1: 77-84

Vernon-Carter, J., Ponce-Palafox, J.T. & Pedroza-Islas, R. 1996. Pigmentation of Pacific white shrimp (*Penaeus vannamei*) using Aztec marigold (*Tagetes erecta*) extracts as carotenoid source. *ALAN* 46(3): 243-246.

Yamada, S., Tanaka, Y., Sameshima, M. & Ito, Y. 1990. Effect of dietary astaxanthin, β-carotene and canthaxanthin on pigmentation of the prawn. *Aquaculture* 87: 323-330.

Acute toxicity and response of five size classes of *Procambarus clarkii* (Decapoda, Cambaridae) to malathion-trichlorfon mixture

ESPERANZA CANO, AGUSTINA JIMÉNEZ & MARIA ELVIRA OCETE
Laboratorio de Zoología Aplicada, Dpto. Fisiología y Biología Animal, Fac. Biología, Universidad de Sevilla, Avda. Reina Mercedes 6, 41012 Sevilla, España

ABSTRACT

In the rice fields of the Guadalquivir River marshes the farmers usually apply a mixture of malathion [S-(1,2-bis(ethoxycarbonyl)-ethyl) 0,0-dimetihyl phosphorodithioate)] (CEKUMAL 50 LE) and Trichlorfon [0,0-dimethyl-(2,2-2 trichlorohydroxiethyl)- phosphonate] (DIPTEREX 80 PS) to control the pests that are affecting the establishment of rice plants. The mean acute 96 h LC_{50} of this mixture to *Procambarus clarkii* of 2-5, 5-6.5, 6.5-8, 8-10 and > 10 cm total length, were 0.281, 2.181, 0.143, 0.743 and 0.805 respectively. These values were inferior to those observed when the products were applied separately. Crayfish that survived an exposure, manifested different post-exposure survival. We did not observe high mortality in organisms exposed to high concentration when compared with controls.

1 INTRODUCTION

Procambarus clarkii was introduced in the marshes of the Guadalquivir (Spain) in 1974 (Habsburgo-Lorena 1978, Ocete & López 1983). Since then, the crayfish population increased without control invading the whole area of rice fields and marshes. This species has become a wealthy source for an important fishing sector. In spite of the fact that the harvesting period is practically confined to the rice farming period, the area has become the first European market for this crustacean (Cano & Ocete 1994).

In marshes of the lower Guadalquivir (Seville, Spain) approximately 35,000 ha are destined for rice crops (Lliso 1993), and are especially important because they are located close to the Doñana National Park. For this reason, farmers have reduced the use of the pesticides malathion and trichlorfon. They use malathion [S-(1,2-bis (ethoxycarbonyl)-ethyl) 0,0-dimetil phosphorodithioate] to control pests that are affecting the establishment of rice plants, and a mixture of malathion and trichlorfon [0,0-dimetyl (2,2-2 trichloro hydroxiethyl)-phosphonate] to control pests affecting rice crops.

Malathion and Triclorfon are chemicals that exert their toxic effects by inhibiting the enzyme acetyl cholinesterase (AchE), causing a disruption of the central nervous system (McEwen et al. 1991). These insecticides are recommended for chironomid control during the initial phases of rice growth because they provide good initial mortality, moderate persistence, and relatively low toxicity (Stevens 1991). In fact, these insecticides are the only ones allowed for rice pest control in the lower Guadalquivir marshes.

The acute toxicity of malathion and trichlorfon to *P. clarkii* has been assessed by Cheah et al. (1980), Repetto et al. (1988), and Jiménez et al. (1998). A thorough understanding of the susceptibility of a *P. clarkii* population to a mixture of malathion and trichlorfon, requires that their acute toxicity is determined for several size classes. Distinction was made between immature (2-5 cm TL), immature-mature (5-6.5 cm TL), and mature individuals (6.5-8 cm, 8-10 cm and > 10 cm TL, this class was split according to commercial sizes in the area). We determined the 96 h LC_{50} of a mixture of malathion and trichlorfon to 5 size classes of *Procambarus clarkii*. Observations of the effects of exposure to sublethal levels of the mixture aid in the understanding of the influence of the xenobiotic substances on growth, survival and reproduction of individuals in a population.

2 MATERIALS AND METHODS

Procambarus clarkii were collected in the Guadalquivir marshes with traps ('nasa') that are used by the local fishermen. They were transported in a container (50 × 80 × 50 cm) without water to the laboratory, where they were let to be acclimated. They were stored in 10 cm deep, aged tap water that was aerated for 48 h with an aquarium aerator to remove chlorine. The water was refreshed daily, and the chemical parameters were maintained as follows: pH 7.5-8, hardness 5.5-6°d, temperature 25 ± 1°C. The photoperiod was 16:8 (light:dark).

The 96h LC_{50} values of a mixture were determined for five size classes of *Procambarus clarkii* with total lengths (TL, tip of the rostrum to the end of the telson) of 2-5 (immature), 5-6.5 (immature-mature), 6.5-8, 8-10 and ≥ 10 cm (13 cm maximum length) (all mature). The immature and immature-mature size classes were used according to Cano & Ocete (1997), who consider 5.5-6 cm length as the transit between immature and mature *P. clarkii* in the Guadalquivir marshes.

Mature crayfish (6.5-8, 8-10, and ≥ 10 cm TL) were tested in aquaria of 35 × 50 × 35 cm with 10 cm deep water. Aeration rate was 85 ml min^{-1}. Tests with crayfish of 2-5 and 5-6.5 cm TL were done in trays of 35 × 20 × 5 cm with 2 cm water without aeration. Oxygen was not a limiting factor for the crayfish.

A 24 h range-finding test was conducted to define the range of mixture dilutions to be used in the 96 h definitive test, starting from the field concentration of 1.5 l ha^{-1} of malathion plus 1.5 kg ha^{-1} of trichlorfon. The field concentration was calculated with the average water depth in a rice basin (10 cm). The concentrations in the definitive tests followed a geometric series between those concentrations in the range-finding test at which 0 and 100% crayfish mortality occurred. The five malathion + trichlorfon mixture concentrations were [1] = 1.5 µl l^{-1} + 15 × 10^{-4} g l^{-1}, [2] = 0.75 µl l^{-1} + 7.5 × 10^{-4} g l^{-1}, [3] = 0.375 µl l^{-1} + 3.75 × 10^{-4} g l^{-1}, [4] = 0.187 µl l^{-1} + 1.875 × 10^{-4} g l^{-1}, [5] = 0.094 µl l^{-1} + 0.937 × 10^{-4} g l^{-1}, and [0], a 0 mg l^{-1} control. These were assigned over 30 containers (five replicate containers per mixture concentration) for each of the five size classes. The test was repeated three times, so the total number of tests for all size classes was 15. The number of crayfish used in each toxicity test was 120.

During the test the crayfish were not fed. The dissolved oxygen concentration was

> 6 mg l^{-1}, and the temperature was 25 ± 1°C. At the start and end of the test the pH, conductivity and hardness were measured with an oxygenometer, selective ion analysis and colorimeter, respectively.

Crayfish were considered as dead if they failed to respond to antennae or leg stimuli. Dead crayfish were removed every 24 h.

A 30-day post-exposure growth and survival study was conducted using crayfish from 2-5 and 5-6.5 cm TL size classes that survived a mixture exposure in the acute toxicity tests. Crayfish were separated into groups corresponding to the respective concentration of the malathion–trichlorfon mixture. The number of crayfish used for the post-exposure test in the 2-5 and 5-6.5 cm TL size classes were 45 and 60, respectively. The crayfish were individualized in containers of 11 cm of diameter and water depth was 2.5 cm. They were fed daily with commercial food for prawns (Gallina Blanca Purina); the portions were 0.15-0.2 g per crayfish. Water was replaced every two days in order to maintain satisfactory water quality. The water temperature was maintained at 25 ± 1°C, hardness 5.5-6°d, dissolved oxygen concentration was > 6 mg l^{-1}, pH 7.5-8 and conductivity 543-590 µS. The crayfish were checked daily for molts and mortality. On day 0, 15 and 30 the increase of TL was measured to the nearest mm. Then crayfish were blotted on dry tissue paper and weighed to the nearest 0.01 g.

In the reproductive studies, 125 and 25 female crayfish were used for exposure to malathion–trichlorfon and control, respectively. Male and female crayfish of 6.5-8, 8-10, and ≥ 10 cm TL were placed together in 40 l polyethylene tanks containing 5 l water. After 24 h males were removed and the tanks with females were covered to provide a 24 h nocturnal photoperiod (Jarboe & Romaire 1991). The water was replaced every 2 days. Females were fed daily, and observed for ovoposition. The reproductive study lasted 90 days after mating.

During the acute toxicity tests, the mortality in each of the five size classes was analyzed with probit analysis to determine the 96 h LC$_{50}$ and 95% confidence limits (Finney 1971, modified by Nogueira 1995). The LC$_{50}$ was determined with the highest mixture concentration (0% mortality) and the lowest mixture concentration (100% mortality). In all other comparisons, survival and ovoposition results were sine-transformed prior to statistical evaluation. The unpaired t-test was used to determine significant differences existing in the portion of females that spawned.

Differences in total weight gain, increase of TL, and mortality of crayfish were compared using ANOVA. Statistical differences were considered significant at α ≤ 0.05.

3 RESULTS

3.1 *Toxicity tests*

During the toxicity tests water quality (Table 1) was within the range for optimal survival of crayfish (Cano 1994). Dissolved oxygen was > 6 mg l^{-1}.

The results of the toxicity tests (96 h LC$_{50}$) are shown in Table 2. The size class 5-6.5 (considered as the limit between mature and immature) was the most resistant to the malathion–trichlorfon mixture. The least tolerant size class was 6.5-8 cm, twice less tolerant than size class 2-5 cm, and 5 times smaller than size class 8-10 cm. The 96 h LC$_{50}$ for size class > 10 cm was 0.805 µl l^{-1} + 8.05 × 10^{-4} g l^{-1}.

Table 1. Mean conductivity, pH and total hardness of exposure water during 96 h LC$_{50}$ determinations with *Procambarus clarkii*, in the different concentrations (1-5 plus control C) of mixture* of malathion + trichlorfon.

Conc.	Conductivity (μS)		pH		Total hardness°d	
	0h	96h	0h	96h	0h	96h
1	254	351	7.5	8	6.16	5.5
2	264	240	7.5	8	6.72	5.5
3	245	245	7.5	8	5	5
4	251	217	7.5	8	5.5	6
5	273	272	8	8	5	6
C	619	589	8	8	5.5	5.5

* 1 = 1.5 μl l^{-1} + 15 × 10^{-4} g l^{-1}, 2 = 0.75 μl l^{-1} + 7.5 × 10^{-4} g l^{-1}, 3 = 0.375 μl l^{-1} + 3.75 × 10^{-4} g l^{-1}, 4 = 0.187 μl l^{-1} + 1.875 × 10^{-4} g l^{-1}, 5 = 0.094 μl l^{-1} + 0.937 × 10^{-4} g l^{-1} and C = 0 μl l^{-1} + 0 g l^{-1}

Table 2. Mean concentrations of malathion and trichlorfon in the 96 h LC$_{50}$ toxicity test with five size classes (TL = total length in cm) of *Procambarus clarkii*.

TL	96 h LC$_{50}$ (95%) test replicate (concentration in μl l^{-1} + g l^{-1})			
	1	2	3	Mean
2-5	0.314 + 3.14 × 10^{-4}	0.215 + 2.15 × 10^{-4}	0.307 + 3.07 × 10^{-4}	0.281 + 2.812 × 10^{-4}
5-6.5	1.956 + 19.568 × 10^{-4}	2.103 + 21.035 × 10^{-4}	2.484 + 24.843 × 10^{-4}	2.181 + 21.815 × 10^{-4}
6.5-8	0.217 + 2.17 × 10^{-4}	0.075 + 0.75 × 10^{-4}	0.138 + 1.38 × 10^{-4}	0.143 + 1.431 × 10^{-4}
8-10	1.031 + 10.31 × 10^{-4}	0.218 + 2.18 × 10^{-4}	0.98 + 9.8 × 10^{-4}	0.743 + 7.43 × 10^{-4}
> 10	1.125 + 11.25 × 10^{-4}	0.609 + 6.09 × 10^{-4}	0.681 + 6.812 × 10^{-4}	0.805 + 8.05 × 10^{-4}

In the 2-5 cm size class weight increase was detected in crayfish under concentration [4], and was significantly different in crayfish under concentration [2] and [3]. In the 5-6.5 cm size class, differences in weight increase were significant for concentration [4] and [5] (high increase) and concentration [1] and [3] (small increase).

3.2 Post-exposure growth, mortality and reproduction

The results of the 30-day post-exposure growth and survival bioassays are listed in Table 3. The TL increased from 0.05 to 0.25 cm, while the mean number of molts ranged from 0.2 to 1 molt per crayfish. Mortality among crayfish in post-exposure bioassays ranged from 0 to 70%.

In the 5-6.5 cm size class differences in mortality were not significant ($p > 0.05$) for crayfish that were exposed to different concentrations of malathion–trichlorfon, except for the ones under concentration [4] (0.187 μl l^{-1} + 1.875 × 10^{-4} g l^{-1}) that showed a significant difference with the rest. This also occurred in the 2-5 cm size class, but also in this case the difference from concentrations [0] (control) and [5] (0.094 μl l^{-1} + 0.937 × 10^{-4} g l^{-1}) was significant.

In the 2-5 cm size class, differences in length increase were not significant for the crayfish post-exposure to different malation–trichlorfon concentrations. In the 5-6.5 cm size class differences in length increase were significant for the control group and the group exposed to lower concentrations ([4] and [5]), and concentration [4] with 2 and 3.

Table 3. *Procambarus clarkii*, average (± standard deviation) length increase and weight gain, mortality, and average number of molts per crayfish in two size classes: 2-5 cm TL (upper panel) and 5-6.5 cm TL (lower panel), after 30 days following a 96 h exposure to malathion (M) + trichlorfon (T) mixtures.

Exposure concentration of M μl l^{-1} + T g l^{-1}	N	Length increase (cm)	Weight gain (g)	Mortality %	Average molt per crayfish
$1.5 + 15 \times 10^{-4}$	0	–	–	–	–
$0.75 + 7.5 \times 10^{-4}$	10	0.271 ± 0.264	0.102 ± 0.171	40	0.3
$0.375 + 3.75 \times 10^{-4}$	10	0.16 ± 0.219	0.015 ± 0.164	60	0.5
$0.187 + 1.875 \times 10^{-4}$	9	0.233 ± 0.132	0.362 ± 0.198	0	0.66
$0.094 + 0.937 \times 10^{-4}$	6	0.05 ± 0.07	0.15 ± 0.07	33.3	0.5
$0 + 0$	10	0.267 ± 0.153	0.243 ± 0.33	70	1
$1.5 + 15 \times 10^{-4}$	10	0.16 ± 0.131	-0.114 ± 0.254	50	0.2
$0.75 + 7.5 \times 10^{-4}$	10	0.086 ± 0.121	0.133 ± 0.304	30	0.2
$0.375 + 3.75 \times 10^{-4}$	10	0.075 ± 0.088	0.021 ± 0.231	20	0.3
$0.187 + 1.875 \times 10^{-4}$	10	0.25 ± 0.108	0.309 ± 0.285	0	0.2
$0.094 + 0.937 \times 10^{-4}$	10	0.20 ± 0.122	0.306 ± 0.139	50	0.6
$0 + 0$	10	0.06 ± 0.055	0.098 ± 0.068	50	0.2

3.3 *Reproduction*

Mature individuals (6.5-8, 8-10, > 10 cm) were used for reproduction assays. The results are shown in Table 4. The difference in oviposition was not significant for crayfish from the various test concentrations. Mortality was high in all classes, so there was a significant difference between the ones from lower concentrations [4] and [5] and the ones from higher concentrations [2] and [3]. Other differences were not significant. The number of molts per crayfish ranged from 0 to 0.3.

If we compare post-exposure mortality in the immature (< 5 cm), mature-immature (5-6.5 cm) and mature (> 6.5 cm) animals, the difference is significant for the mature versus the immature and immature-mature animals.

4 DISCUSSION.

The LC_{50} values found in bibliography for mature *Procambarus clarkii* individuals are 50 mg l^{-1} for malathion, and 5.15 mg l^{-1} for triclorfon (Repetto et al. 1988, Jiménez et al. 1998). These values are in all cases inferior to those obtained with the malathion-triclorfon mixture.

The increased toxicity of the mean 96h-LC_{50} value for the malathion-triclorfon mixture in the mature size classes was unexpected, because the relationship between size class response and toxicant dose among the smaller mature crayfish would suggest that larger organisms are more tolerant. Probably, mechanisms of differential uptake, absorption and excretion of insecticide vary with crayfish size and stage of sexual maturity as they do in other aquatic organisms (Guarino 1987).

These mechanisms may account for the responses observed in this study, because immature size class have a mean 96h-LC_{50} value that is inferior to the bigger size classes. However, the limit maturity size class (5-6.5 cm) had a mean 96h-LC_{50}

Table 4. *Procambarus clarkii*, mean oviposition, mortality, and average number of molts per crayfish of > 6.5 cm TL (6.5-8, 8-10, and > 10 cm TL size classes together) during 90 days following a 96 h exposure to malathion (M) + trichlorfon (T) mixtures.

Exposure concentration $M \mu l \ l^{-1} + T \ g \ l^{-1}$	N	Oviposition %	Mortality %	Average number of molt per crayfish
$0 + 0$	25	28	76	0.24
$0.094 + 0.937 \times 10^{-4}$	27	22	40.7	0
$0.187 + 1.875 \times 10^{-4}$	20	10	65	0
$0.375 + 3.75 \times 10^{-4}$	33	21.2	87.8	0.12
$0.75 + 7.5 \times 10^{-4}$	25	28	84	0.2
$1.5 + 15 \times 10^{-4}$	20	30	80	0.3

value that was highly superior ($2.181 \ \mu l \ l^{-1} + 21.815 \times 10^{-4} \ g \ l^{-1}$) to the rest of tested size classes. Anyway, further research to determine the reasons for differences in toxicant susceptibility between size classes and among stages of maturity in *P. clarkii* is necessary.

In the river marshes of the Lower Guadalquivir, the field dose of the malathion–triclorfon mixture is $1.5 \ \mu l \ l^{-1} + 15 \times 10^{-4} \ g \ l^{-1}$, which is coincident with the higher concentration used for the toxicity tests and superior to most of the mean 96h-LC_{50} values (Table 2). So, the mortality of crayfish is high and there is a decreasing in the production of *P. clarkii*, which affects the local commerce of this crustacean in the area that is the first European producer and exporter (Gaudé 1984, Cano & Ocete 1994).

Post-exposure effect of the mixture does not affect the increase in mortality, because for 2-5 cm size class the highest mortality occurs in the control population. In 5-6.5 cm size class there is the same mortality as in the control population and in the smaller and bigger ones. In mature size class there is again high mortality in concentration 0.

The mean number of molts per crayfish was higher in smaller animals and it decreased with crayfish size. Molting frequency is lower in adults (Huner & Avault, 1978). This variable, as well as weight and length increase and percentage of ovoposition are not affected by pre-exposure to different concentrations of malathion-triclorfon. It is the same as if we considered mortality because in both size classes the control population and the ones from the different concentrations show random behavior.

It is logical that in post-exposure studies the malathion–triclorfon mixture does not affect the different tested size classes, because the degradation of both insecticides occurred before the end of the toxicity tests (90 hours of mean life) (Beyers & Myers 1996, Chapman & Cole 1982).

REFERENCES

Beyers, D.W. & Myers, O.B. 1996. Use of meta-analysis to predict degradation of carbaril and malathion in freshwater for exposure assessment. *Human and Ecological Risk Assessment* 2(2): 366-380.

Cano, E. 1994. Estudios Biológicos sobre *Procambarus clarkii* Girard (Decapoda, Cambaridae) en las Marismas del Bajo Guadalquivir. Master's Thesis, Univ. Sevilla. Spain.

Cano, E. & Ocete, M.E. 1994 Estimación sobre las repercusiones socio-económicas de *Procambarus clarkii* Girard (Decapoda, Cambaridae). *Bol. San. Veg. Plagas* 20(2): 653-660.

Cano, E. & Ocete, M.E. 1997. Population Biology of red swamp crayfish, *Procambarus clarkii* (Girard, 1852) in the Guadalquivir river marshes, Spain. *Crustaceana* 70: 553-561.

Chapman, R.A. & Cole, C.M. 1982. Observations on the influence of water and soil pH on the persistence of insecticides. *J. Environ. Sci. Health* 17: 487-504.

Cheah, M., Avault Jr, J.W. & Graves, J.B. 1980. Acute toxity of selected rice pesticides to crayfish, *Procambarus clarkii. The progressive Fish-Culturist* 42(3): 169-172.

Finney, A.D. 1971. *Statistical methods in biological assay.* 2nd Edition, London: Grinffin Press.

Gaudé, P. 1984. Ecology and production of Louisiana red swamp crawfish *Procambarus clarkii* in southern Spain. *Freshwater crayfish* 6: 111-130.

Guarino, A.M. 1987. Aquatic versus mammalian toxicology: Applications of the comparative approach. *Environ Health Perspect.* 71: 17-24.

Habsburgo-Lorena, A.S. 1978. Present situation of exotic species of crayfish introduced into spanish continental waters. *Freshwater Crayfish* 4: 175-184.

Huner, J.V. & Avault Jr, J.W. 1978. Postmolt calcification in subadult red swamp crayfish, *Procambarus clarkii* (Girard) (Decapoda, Cambaridae). *Crustaceana* 34: 275-280.

Jarboe, H. & Romaire, R. 1991. Acute toxicity of permetrin to four size classes of red swamp crayfish (*Procambarus clarkii*) and observations of post-exposure effects. *Arch. Environ. Contam. Toxicol.* 20: 337-342.

Jiménez, A., Ramírez, J.L., Cano, E. & Ocete, E. 1998. Toxicidad de la azaridacctina y del triclorfon sobre especies presentes en los cultivos de arroz de las Marismas del Bajo Guadalquivir. *Bol. San. Veg.* in press.

Lliso, J. 1993. El arroz. *Agricola Vergel*, XII n°136, 181-187.

Nogueira, A. 1995. Probit Analysis. Universidade de Coimbra. Coimbra (Portugal).

McEwen, L.C., Petersen, B.E., Beyers, D.W., Howe, F.P. Adams, J.S. & Miller, C.L. 1991. Environmental monitoring and evaluation, Grasshoper Integrated Pest Management Program. In grasshopper Integrated Pest Management Project, 1991 Annual Report. USDA Animal and Plant Health Inspection Service, Boise, ID, 123-133.

Ocete, M.E. & López, S. 1983. Problemática de la introducción de *Procambarus clarkii* (Girard) (Crustacea: Decapoda) en las marismas del Guadalquivir. *Actas del I Congreso Ibérico de Entomología*, León: 515-523.

Repetto, G., Sanz, P. & Repetto, M. 1988. *In vivo* and *In vitro* of triclorfon on esterases of red creyfish *Procambarus clarkii. Bulletin of Environmental Contamination and Toxicology* 41: 597-603.

Stevens, M.M. 1991. Insecticide treatments used against a rice bloodworm, *Chironomus tepperi* Skuse (Diptera: Chironomidae): toxicity and residual effects in water. *Journal of Economic Entomology* 84: 795-800.

4 Reproduction

Comparative fecundity and reproduction in seven crab species from Mgazana, a warm temperate southern African mangrove swamp

W.D. EMMERSON
Department of Zoology, University of Transkei, Umtata, South Africa

ABSTRACT

The fecundity of four species of *Uca, U. urvillei, U. vocans, U. annulipes* and *U. gaimardi*, one species of *Macrophthalmus, M. grandidieri* and three species of sesarmids, *Neosarmatium meinerti, Sesarma catenata* and *S. eulimene* were investigated. There was a general trend of increasing egg numbers with increasing species size. For most species egg numbers increased as a power function of female crab size as females grew allometrically. Group fecundities were compared using ANCOVA and highly significant differences were found among both slopes and adjusted means. Species fecundity was then tested pairwise for differences. *N. meinerti* was found to be significantly different to the other species as it was much larger than the other species and produced more eggs. A significant difference was found between mean egg diameters with a general trend of increasing egg size with increasing species size. For all species there was a considerable scatter in egg number suggesting a possibility of multiple broods at each size class. Highly seasonal breeders such as *N. meinerti* and *U. annulipes* probably only produce one to two broods per annum, while with the other species of *Uca* and *Sesarma* the season is more extended allowing for three or four broods per season. In the lower intertidal *M. grandidieri* has a strategy of breeding all year around and probably produces around six broods per annum.

1 INTRODUCTION

There is a considerable amount of information on the feeding ecology of mangrove associated crabs including both ocypodids such as *Uca* (Icely & Jones 1978, Dye & Lasiak 1986, 1987) and sesarmids (Nielson et al. 1986, Giddins et al. 1986, Camilleri 1989, Vannini & Ruwa 1994, Slim et al. 1997). However, little research has been conducted on the breeding biology of mangrove crabs, despite their numerical and energy flow importance within the detrital food web of mangrove ecosystems (Jones 1984, Robertson 1986, Emmerson & McGwynne 1992). Most work on littoral crab reproduction has come from temperate *Spartina*-based systems (Seiple 1979, Henmi & Kanemoto 1989). While Emmerson (1994) initially investigated the seasonal breeding cycles and sex ratios of several crab species from a mangrove system, this present work focuses on another aspect of their breeding biology, namely fecundity.

Hines (1991) compared the fecundity of nine species of *Cancer* from the North Pacific and North Atlantic and found that although fecundities ranged widely due to egg size differences, female body size was the principal determinant of reproductive output. Similarly Reid & Corey (1991) also established a significant relationship between female size and fecundity in fifteen species of anomuran and brachyuran crabs. Reproductive output in Brachyura is thus dependent on allometric constraints (Hines 1982). However, the number of broods per annum is also a factor. It was thus decided to investigate whether this was also true for littoral mangrove associated crab species and to compare the fecundity and factors which may affect it for several species of crab from Mgazana, a mangrove system on the East Cape coast of South Africa.

2 MATERIAL AND METHODS

The Transkei coast is situated between the warm sub-tropical east coast of Kwa-Zulu Natal and the the warm temperate East Cape region of South Africa (Day 1981) where it forms a transition zone for marine inter-tidal flora and fauna (Kilburn & Rippey 1982). The Mgazana estuary (31°47′S, 29°25′E) lies within this zone and is characteristically rich in species, many of which are found associated within the extensive mangrove inter-tidal zone (MacNae 1963, Ward & Steinke 1982, Branch & Grindley 1979). The following species of crab were investigated: *Neosarmatium meinerti, Sesarma eulimene, Sesarma catenata, Uca annulipes, Uca gaimardi, Uca urvillei, Uca vocans* and *Macrophthalmus grandidieri*. A wide range of egg-bearing females were collected by hand during spring low tide in February 1994. This was peak breeding season when all of the above species could be found in berry (Emmerson 1994). Females were taken back to the laboratory and were identified, their maximum carapace width measured using Vernier calipers to the nearest 0.1 mm and they were subsequently fixed in 10% buffered formalin-seawater until they were processed. The eggs were then removed by teasing them from the pleopods using dissection needles. The total egg mass was then blotted on filter paper to remove excess water and weighed (mg). Five subsamples were then taken, weighed accurately to the nearest milligram and the number of eggs counted using a dissecting microscope (Hines 1991). Egg diameters were taken by pooling eggs from five females of each species, subsampling and measuring at least 150 eggs per species. This was done using a calibrated eyepiece micrometer.

The following sources were used for crab identification; Crosnier (1965) for *Sesarma* and *Macrophthalmus*, Bott (1973) for *Uca* species and Davie (1994) for *Neosarmatium*. Data were analysed using Analysis of Covariance (ANCOVA) BIOM Statistics (Sokal & Rohlf 1995) after log_{10} transformation after testing for homogeneity of variances and equality of slopes (Hines 1982) and standard Microsoft Excel Statistics.

3 RESULTS

The relationships between egg numbers and carapace widths for eight species of crab from Mgazana can be seen in Figures 1-8. In most cases egg numbers increased as a power function with increasing carapace width (*Uca annulipes, U. urvillei, U. vocans, Macrophthalmus grandidieri, Sesarma catenata* and *S. eulimene;* Figures 1, 3, 4, 5, 6, and 7, respectively), but was exponential with *U. gaimardi* (Fig. 2) and logarithmic with *Neosarmatium meinerti* (Fig. 8). This information is summarized in Table 1. In all eight species egg numbers were found to be size dependent with significant regression coefficients. A summary of the general statistics for the fecundity of the crab species is given in Table 2. There was a general trend of increasing egg numbers with the increasing size of the various species from small *S. catenata* and *U. annulipes* through to large *N. meinerti* (Table 2). This trend was linear as can be seen when correlating mean egg numbers with mean carapace widths (Fig. 9). Mean body size (CW) differed significantly among species (ANOVA, $F_{(7, 294)} = 263.175$, $p < 0.01$) as did mean egg number (ANOVA, $F_{(7, 294)} = 74.2715, p < 0.01$). All of the species overlapped in body size with the exception of the largest, *N. meinerti*, and even the smallest species *S. catenata* showed a continuum in body size of the group (Table 2: Fig. 10).

Fecundities were then compared groupwise using ANCOVA to see whether there were any differences in fecundity between species after $\log_{10}$ transformation of the data. Differences among adjusted means (F = 95.087, $p < 0.01$) and differences among slopes (F = 90.047, $p < 0.01$) were both found to be highly significantly different.

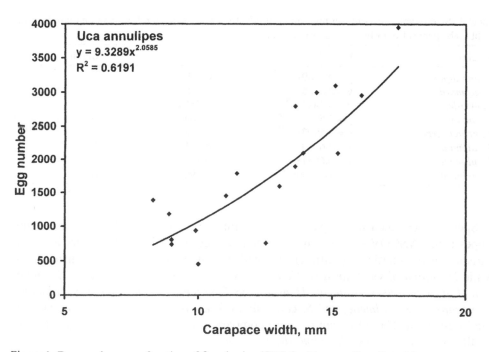

Figure 1. Egg numbers as a function of female size (CW) for *Uca annulipes* from Mgazana.

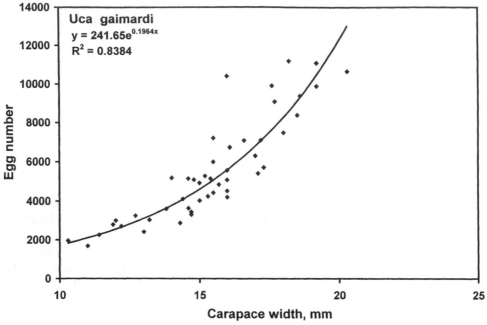

Figure 2. Egg numbers as a function of female size (CW) for *Uca gaimardi* from Mgazana.

Table 1. Size (Carapace Width in mm) and reproductive regressions (crab size vs egg number) for eight crab species from Mgazana. S, Significant (p < 0.05)

Species	Formula	R^2	Signif.
U. annulipes	$y = 9.3289x^{2.0585}$	0.6191	S
U. gaimardi	$y = 241.65e^{0.1964x}$	0.8384	S
U. urvillei	$y = 0.84x^{3.0388}$	0.7565	S
U. vocans	$y = 0.5909x^{3.338}$	0.8049	S
M. grandidieri	$y = 1.5903x^{3.1032}$	0.7724	S
S. catenata	$y = 0.6744x^{3.2431}$	0.8011	S
S. eulimene	$y = 0.6021x^{3.4623}$	0.8824	S
N. meinerti	$y = 1E - 09x^{8.6405}$	0.7907	S

Species were then tested pairwise for significance in both adjusted means and slopes using ANCOVA. The results are summarized in Table 3. Most pairwise comparisons were significantly different for both adjusted means and slope differences, while *U. gaimardi* vs *U. urvillei, U. urvillei* vs *U. vocans, U. annulipes* vs *N. meinerti, U. gaimardi* vs *S. eulimene, U. urvillei* vs *S. catenata, U. urvillei* vs *S. eulimene, U. vocans* vs *S. eulimene,* and *S. catenata* vs *S. eulimene* had one variable significantly different. The following pairwise comparisons were not significantly different at all: *U. annulipes* vs *U. gaimardi, U. gaimardi* vs *U. vocans, U. gaimardi* vs *S. catenata,* and *U. vocans* vs *S. catenata.* From these comparisons it can be clearly seen that *N. meinerti* was significantly different to the other species, being both much larger and producing more eggs. Differences between the fecundities of the other

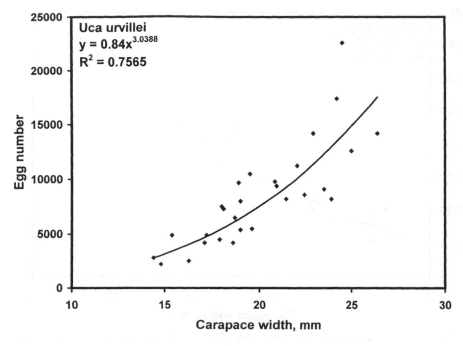

Figure 3. Egg numbers as a function of female size (CW) for *Uca urvillei* from Mgazana.

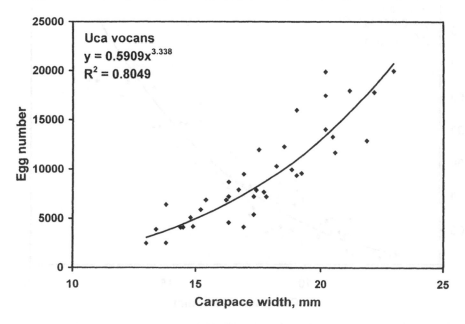

Figure 4. Egg numbers as a function of female size (CW) for *Uca vocans* from Mgazana.

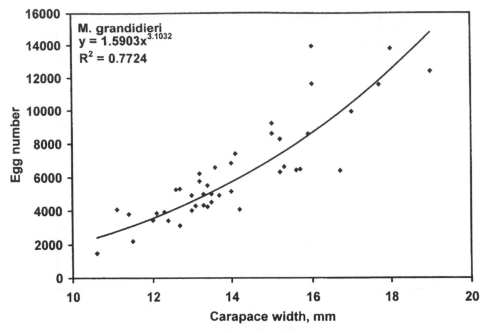

Figure 5. Egg numbers as a function of female size (CW) for *Macrophthalmus grandidieri* from Mgazana.

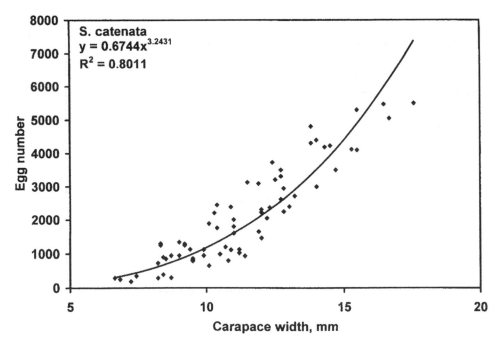

Figure 6. Egg numbers as a function of female size (CW) for *Sesarma catenata* from Mgazana.

Table 2. Statistical summary for the fecundity of eight species of crab from Mgazana.

	Freshmass, g	Carapace width, mm					Egg number					Egg diameter, μm				
	Mean	Mean	Min	Max	Rge	N	Mean	Min	Max	Rge	N	Mean	Min	Max	Rge	N
U. annulipes	0.652	12.36	8.3	17.5	9.2	18	1,834	451	3,950	3,499	18	321	288	346	58	165
U. gaimardi	0.971	15.43	10.3	20.3	10	48	5,574	1674	11,200	9,526	48	288.9	254	336	82	166
U. vocans	1.491	17.54	13	23	10	38	9,333	2494	19,980	17,486	38	304.1	254	346	92	221
U. urvillei	1.983	20.01	14.4	26.4	12	28	8,434	2205	22,590	20,385	28	234.3	192	317	125	262
U. grandidieri	0.848	14.05	10.6	19	8.4	43	6,265	1476	13,950	12,474	43	250.4	213	278	65	205
S. catenata	0.657	11.29	6.6	17.6	11	71	2,138	190	5,500	5,310	71	363.6	307	432	125	150
S. eulimene	1.285	15.73	10	21	11	30	9,568	1678	19,622	17,944	30	306.1	269	346	133	285
N. meinerti	28.488	36.27	30	43.5	13.5	26	57,730	3025	124,960	121,935	26	382.8	326	461	135	220

sesarmids, particularly *S. catenata*, and the ocypodids was not great with considerable overlap (Fig. 10).

Information on egg diameters is given in Table 2. Mean egg sizes ranged from 234 μm for *U. urvillei* to 382 μm for *N. meinerti*. *U. annulipes* eggs had the smallest range (58 μm) while *N. meinerti* had the greatest range (135 μm: Table 2). Yet despite this range, egg size did not vary significantly within species (ANOVA, $p > 0.05$). There was however a highly significant difference between the mean egg diameters of the various species (ANOVA, $F = 1414.95$, $p < 0.001$). Egg diameter was then compared with mean carapace width (Table 2). Although there was a considerable amount of scatter, with an R^2 of only 0.139 (n = 8), it was significant (ANOVA, $F = 0.967$, $p < 0.05$) with small species like *M. grandidieri* having small eggs and large species like *N. meinerti* having large eggs. There were, however, many anomalies with large species like *U. urvillei* having relatively small eggs and small species like *S. catenata* having relatively large eggs (Table 2). When mean female fresh mass was compared with mean egg diameter (Table 2) again the R^2 was low, but significant (0.338), yielding an F value of 3.062 which was significant ($p < 0.05$), despite a large scatter of points.

Size at first maturity for the eight crab species from Mgazana is given in Table 4 together with published data from elsewhere.

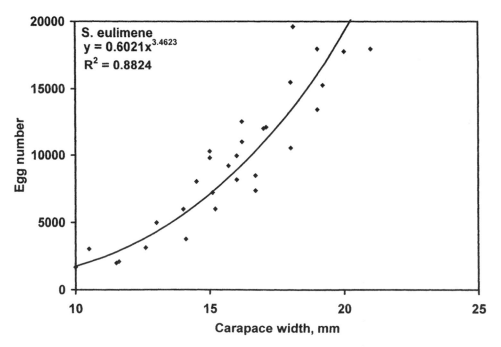

Figure 7. Egg numbers as a function of female size (CW) for *Sesarma eulimene* from Mgazana.

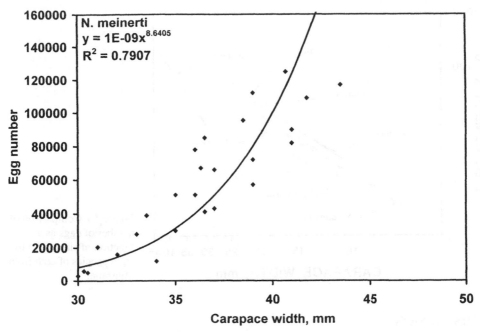

Figure 8. Egg numbers as a function of female size (CW) for *Neosarmatium meinerti* from Mgazana.

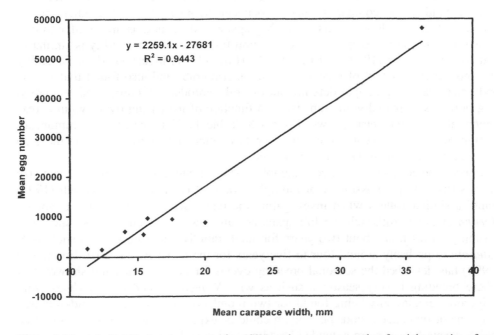

Figure 9. The relationship between mean size (CW) and mean egg number for eight species of crab from Mgazana.

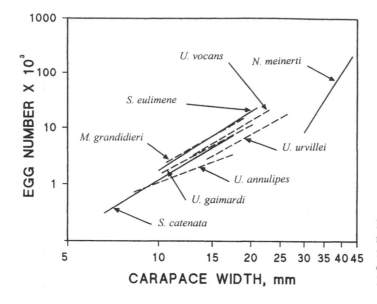

Figure 10. Comparison of number of eggs as a function of size (CW) for eight species of crab from Mgazana.

4 DISCUSSION

For all species, egg numbers increased as a function of body size. This increase was almost always by an order of magnitude between the minimum and maximum egg number per species (Table 2; Fig. 10). In the case of the largest species, *N. meinerti*, the increase was two orders of magnitude and thus showed consistent differences when fecundity was compared pairwise between *N. meinerti* and the other species (Table 3; Fig. 10). These results for both grapsids and ocypodids are similar to patterns seen for other crab species. The equation for *S. catenata* fecundity is similar to that found by Baird (1978) and Els (1982). Hines (1991) investigated the fecundity and reproductive output of nine species of cancrid crabs and also found that female body size was the principal determinant of both reproductive output and fecundity. Egg numbers increased with crab size as a function of increasing ovary volume (exponential) with allometric growth (Figs 1-8; Table 1). This pattern is common, although Stella et al. (1996) found that the best fitted model for the estuarine crab *Chasmagnathus granulatus* was logarithmic.

For all species there was a considerable scatter in egg number within female size classes suggesting a possibility of multiple broods at each size class. Hsueh (1997) found a similar pattern when investigating the reproductive output of the ocypodid *Baruna sinensis*. Although the life spans are unknown for each of these species it certainly varies from about two years for small rapidly maturing species such as *S. catenata* to probably around four to five years for the larger *N. meinerti*. Emmerson (1994) has described the seasonal breeding cycles for these species and showed that where breeding is very seasonal, such as with *N. meinerti* and *U. annulipes*, with ovigerous females only being found for two months of the year during mid-summer, only one or two large broods per annum could be expected. The other species of *Uca* and *Sesarma* have a more extended breeding period and could produce three or four broods per annum. Other species like *M. grandidieri* breed all year around and

Table 3. Pairwise ANCOVA comparisons of fecundities between crab species.

Species comparisons	Adj. means diffs., F	Signif.	Slope diffs., F	Signif.
U. annulipes vs *U. gaimardi*	42.57	S**	5.29	S*
U. annulipes vs. *U. urvillei*	3.36	NS	3.6	NS
U. annulipes vs. *U. vocans*	28.37	S**	8.01	S**
U. gaimardi vs. *U. urvillei*	29.9	S**	0.13	NS
U. gaimardi vs. *U. vocans*	2.58	NS	1.71	NS
U. urvillei vs. *U. vocans*	53.79	S**	0.49	NS
U. annulipes vs. *S. catenata*	8.44	S**	7.28	S**
U. annulipes vs. *S. eulimene*	63.93	S**	10.42	S**
U. annulipes vs. *N. meinerti*	1.03	NS	46.93	S**
U. gaimardi vs. *S. catenata*	1.7	NS	0.94	NS
U. gaimardi vs. *S. eulimene*	66.73	S**	3.44	NS
U. gaimardi vs. *N. meinerti*	16.06	S**	69.52	S**
U. urvillei vs. *S. catenata*	8.77	S**	0.2	NS
U. urvillei vs. *S. eulimene*	109.74	S**	1.08	NS
U. urvillei vs. *N. meinerti*	9.13	S**	41.23	S**
U. vocans vs. *S. catenata*	1.8	NS	0.06	NS
U. vocans vs. *S. eulimene*	34.17	S**	0.12	NS
U. vocans vs. *N. meinerti*	26.48	S**	45.52	S**
S. catenata vs. *S. eulimene*	26.5	S**	0.31	NS
S. catenata vs. *N. meinerti*	11.58	S**	46.61	S**
S. eulimene vs. *N. meinerti*	32.56	S**	42.77	S**

S** – highly significant ($p < 0.01$), S* – significant ($p < 0.05$), NS – not significant.

probably produce around six broods per annum. Mouton & Felder (1995) found similar seasonal reproduction in *Uca longisignalis* and *U. spinicarpa* from Louisiana which may reflect nutritional dependence on the annual marsh plants where they are found.

If size at first maturity is considered, some interesting comparisons can be made with published information from elsewhere (Table 4). For the *Uca*'s most species show data similar to Crane's (1975) data, except for *U. urvillei* which appeared to breed earlier at Mgazana than Pemba, probably as a result of sample size. In both the Swartkops estuary (Baird 1978, Els 1982) and Mgazana (present work) *S. catenata* is precocious, with females producing eggs very early around 6-7 mm CW, so that this species probably produces four to six broods per season similar to *S. cinereum* (Seiple 1979).

It is not known how many broods each species produces per breeding period, but work is presently in progress to determine this by collecting gravid females every three days to determine the progression of developmental stages together with ovarian development and these results will be reported separately.

In Japan, *Sesarma haematocheir* incubate eggs for about a month (Saigusa 1993), but incubation time and broods per season will generally vary with species and latitude. It is already known, though, that many estuarine crab species reproduce on a semi-lunar cycle (Paula 1989, Christy 1982, Severinghaus & Lin 1990) and that the ovaries redevelop while the female is still carrying eyed eggs. Sudha & Anilkumar (1996) for instance have shown for *Metopograpsus messor*, that spawning was im-

Table 4. Size at first maturity (smallest recorded female with eggs) for eight crab species from Mgazana. * data from Crane (1975), # data from Baird (1978) and Els (1982). CW = Carapace Width; FM = Freshmass.

Species	Mgazana CW, mm	Crab FM, mg	Elsewhere CW, mm	Locality
U. annulipes	13.4	779	11.0	Mozambique*
U. gaimardi	11.4	444	–	
U. vocans	13.0	587	13.5	Zanzibar*
U. urvillei	14.4	880	19.5	Pemba*
M. grandidieri	6.2	347	–	
S. catenata	7.0	560	6.6	Swartkops#
S. eulimene	10.0	890	–	
N. meinerti	30.0	17460	–	

mediately followed by another vitellogenic cycle which paralleled the embryogenesis of the prehatch eggs in the brood.

In some species of crab, the maximum and minimum size at maturity vary geographically and it would be interesting to make some comparisons with the species investigated. Both Baird (1978) and Els (1982) investigated the fecundity of *S. catenata* from the Swartkops estuary which is a *Spartina* dominated estuary, not mangrove as with Mgazana. However, the sizes at sexual maturity were nevertheless found to be similar, around 6 mm CW. It would be interesting to compare these fecundities with data from Inhaca Island Mozambique, which is the northernmost limit for this endemic species of *Sesarma* (Kensley 1981). Henmi (1993) found geographic variations in the reproductive traits of *Macrophthalmus banzai* from different locations, with winter breeding in one area and summer in another; one brood versus two broods, and double the egg production at the lower latitude, all of which appear to be ecologically adaptive. Comparative data for the other species is lacking.

Egg diameters were shown to vary between species, large and small, but although the ocypodids investigated tended to have smaller eggs than the grapsids, this was not consistent (Table 2) and is probably better correlated with some other variable within their respective reproductive strategies. Nevertheless, egg diameters were similar to other published values for the planktotrophic Ocypodidae and Grapsidae (Hines 1986) while *Uca* ova are usually fairly small. Although egg diameters were found to differ significantly between species, the regression coefficients were low between mean species size (CW) and egg size (diameter). All of the species investigated here had planktotrophic larvae (with all of the planktonic zoeal stages described for these sesarmid species from the region, Pereyra Lago 1987, 1989, 1993a, b), so this could possibly explain this poor correlation as the overall strategy of nearshore planktonic larval development remains the same for all the species.

Emmerson (1994) showed that with the exception of *M. grandidieri*, there was a pattern of overlapping breeding peaks for *Uca* and *Sesarma* at Mgazana, largely coinciding with summer when there is abundant food in adjacent nearshore waters for the developing zoeae. As most of these are warm, mangrove associated species near their southern limit of distribution, such a strategy would maximize survival and fitness. This is the converse of Hines' (1991) finding for the Cancridae where breeding in these cold temperate and boreal zones is restricted to a winter-spring season which coincides with the spring plankton bloom. As with the Cancridae this seasonal restriction may limit the rate of brood production, but then large, long living species like *N. meinerti*, which has a very restricted breeding period, makes up for this by producing large numbers of eggs during ideal environmental conditions to ensure good survival and subsequent recruitment back into these southeast African coast mangrove systems. *M grandidieri* on the other hand hedges its bets by producing eggs all year round to ensure some recruitment all of the time.

REFERENCES

Baird, D. 1978. Fecundity of *Sesarma catenata* Ortman (Crustacea: Grapsidae) occurring in the saltmarshes of the Swartkops estuary, Port Elizabeth. *S. Afr. J. Sci.* 74(1): 31-32.

Bott, R. 1973. Die verwandtschaftlichen Beziehungen der *Uca*-Arten (Decapoda: Ocypodidae). *Senckenberg. Biol.* 54: 315-325.

Branch, G.M. & Grindley, J.R. 1979. Ecology of southern African estuaries. Part XI. Mgazana: a mangrove estuary in Transkei. *S. Afr. J. Zool.* 14: 149-170.

Camilleri, J.C. 1989. Leaf choice by crustaceans in a mangrove forest in Queensland. *Mar. Biol.* 102: 453-459.

Christy, J.H. 1982. Adaptive significance of semi-lunar cycles of larval release in fiddler crabs (Genus *Uca*): test of an hypothesis. *Biol Bull. Woods Hole, Mass.* 163: 251-263.

Crane, J. 1975. *Fiddler crabs of the World; Ocypodidae: Genus Uca.* Princeton: Princeton University Press.

Crosnier, A. 1965. Crustacés décapodes (Grapsidae et Ocypodidae). *Faune de Madagascar* 18: 1-143.

Davie, P.J.F. 1994. Revision of *Neosarmatium* Serene and Soh (Crustacea: Brachyura: Sesarminae) with descriptions of two new species. *Mem. Queensland Mus.* 35: 35-74.

Day, J.H. 1981. *Estuarine ecology with particular reference to southern Africa.* Cape Town: Balkema.

Dye, A.H. & Lasiak, T.A. 1986. Microbenthos, meoibenthos and fiddler crabs: trophic interactions in a tropical mangrove sediment. *Mar. Ecol. Prog. Ser.* 32: 259-264.

Dye, A.H. & Lasiak, T.A. 1987. Assimilation efficiencies of fiddler crabs and deposit feeding gastropods from tropical mangrove sediments. *Comp. Biochem. Physiol.* 87A: 341-344.

Els, S. 1982. Distribution and abundance of two crab species on the Swartkops estuary saltmarshes and the energetics of the *Sesarma catenata* population. MSc Thesis, University of Port Elizabeth, 178 pp.

Emmerson, W.D. 1994. Seasonal breeding cycles and sex ratios of eight species of crabs from Mgazana, a mangrove estuary in Transkei, southern Africa. *J. Crust. Biol.* 14: 568-578.

Emmerson, W.D. & McGwynne, L.E. 1992. Feeding and assimilation of mangrove leaves by the crab *Sesarma meinerti* de Man in relation to leaf-litter production in Mgazana, a warm-temperate southern African mangrove swamp. *J. Exp. Mar. Biol. Ecol.* 157: 41-53.

Giddins, R.L., Lucas, J.S., Nielson, M.J. & Richards, G.N. 1986. Feeding ecology of the mangrove crab *Neosarmatium smithi* (Crustacea: Decapoda: Sesarmidae). *Mar. Ecol. Prog. Ser.* 33: 147-155.

Henmi, Y. 1993. Geographic variations in life-history traits of the intertidal ocypodid crab *Macrophthalmus banzai. Oecologia,* 96: 324-330.

Henmi, Y. & Kanemoto, M. 1989. Reproductive ecology of three ocypodid crabs. I. The influence of activity differences on reproductive traits. *Ecol. Res.* 4: 17-29.

Hines, A.H. 1982. Allometric constraints and variables of reproductive effort in brachyuran crabs. *Mar. Biol.* 69: 309-320.

Hines, A.H. 1986. Larval patterns in the life histories of brachyuran crabs (Crustacea, Decapoda, Brachyura). *Bull. Mar. Sci.* 39: 444-466.

Hines, A.H. 1991. Fecundity and reproductive output in nine species of *Cancer* crabs (Crustacea, Brachyura, Cancridae). *Can. J. Fish. Aquat. Sci.* 48: 267-275.

Hseuh, P-W. 1997. Reproductive output of a minute crab, *Baruna sinensis* (Decapoda: Brachyura: Ocypodidae: Camptandriinae). *Zool. Stud.* 36: 111-114.

Icely, J.D. & Jones, D.A. 1978. Factors affecting the distribution of the genus *Uca* (Crustacea: Ocypodidae) on an East African shore. *Estuar. Coast. Mar. Sci.* 6: 315-325.

Jones, D.A. 1984. Crabs in the mangal, In F.D. Por & I. Dor (eds), *Hydrobiology of the mangal*: 89-109. The Hague: Junk.

Kensley, B. 1981. On the zoogeography of southern African decapod Crustacea, with a distributional checklist of the species. *Smithson. Contrib. Zool.* (338): 1-64.

Kilburn, R. & Rippey, E. 1982. Sea shells of South Africa. Johannesburg: Macmillan.

McNae, W. 1963. Mangrove swamps in South Africa. *J. Ecol.* 51: 1-25.

Mouton, E.C. & Felder, D.L. 1995. Reproduction of the fiddler crabs *Uca longisignalis* and *Uca spinicarpa* in a Gulf of Mexico saltmarsh. *Estuaries* 18: 469-481.

Nielson, M.J., Giddins, R.L. & Richards, G.N. 1986. Effects of tannin on the palatability of mangrove leaves to the tropical sesarminid crab *Neosarmatium smithi*. *Mar. Ecol. Prog. Ser.* 34: 185-186.

Paula, J. 1989. Rhythms of larval release of decapod crustaceans in the Mira estuary. *Mar. Biol.* 100: 309-312.

Pereyra-Lago, R. 1987. The larval development of *Sesarma catenata* Ortmann (Brachyura, Grapsidae, Sesarminae) reared in the laboratory. *S. Afr. J. Zool.* 22: 200-212.

Pereyra-Lago, R. 1989. The larval development of the red mangrove crab *Sesarma meinerti* De Man (Brachyura: Grapsidae) reared in the laboratory. *S. Afr. J. Zool.* 24: 199-211.

Pereyra-Lago, R. 1993a. Larval development of *Sesarma guttatum* A. Milne Edwards (Decapoda: Brachyura: Grapsidae) reared in the laboratory, with comments on larval generic and familial characters. *J. Crust. Biol.* 13: 745-762.

Pereyra-Lago, R. 1993b. The zoeal development of *Sesarma eulimene* de Man (Decapoda, Brachura, Grapsidae), and identification of larvae of the genus *Sesarma* in South African waters. *S.A. J. Zool.* 28: 173-181.

Reid, D.M. & Corey, S. 1991. Comparative fecundity of decapod crustaceans, II. The fecundity of fifteen species of anomuran and brachyuran crabs. *Crustaceana* 61: 175-189.

Robertson, A.I. 1986. Leaf-burying crabs: their influence on energy flow and export from mixed mangrove forests (*Rhizophora* spp.) in northeastern Australia. *J. Exp. Mar. Biol. Ecol.* 102: 237-248.

Saigusa, M. 1993. Control of hatching in an estuarine terrestrial crab. 2. Exchange of a cluster of embryos between two females. *Biol. Bull. Wood Hole. Mass.* 184: 186-202.

Seiple, W. 1979. Distribution, habitat preferences and breeding periods in the crustaceans *Sesarma cinereum* and *S. reticulatum* (Brachyura: Decapoda: Grapsidae). *Mar. Biol.* 52: 77-86.

Severinghaus, L.L. & Lin, H.-C. 1990. The reproductive behaviour and mate choice of the fiddler crab (*Uca lactea lactea*) in mid-Taiwan. *Behaviour* 113: 292-308.

Sokal, R.R. & Rohlf, F.J. 1995. *Biometry*. New York: Freeman.

Slim, F.J., Hemminga, M.A., Ochieng, C., Jannink, N.T., Cocheret de la Moriniere, E. & van der Velde, G. 1997. Leaf litter removal by the snail *Terebralia palustris* (Linnaeus) and sesarmid crabs in an East African mangrove forest (Gazi Bay, Kenya). *J. Exp. Mar. Biol. Ecol.* 215: 35-48.

Stella, V.S., Lopez, L.S. & Rodriguez, E.M. 1996. Fecundity and brood biomass investment in the estuarine crab *Chasmagnathus granulatus* Dana, 1851 (Decapoda, Brachyura, Grapsidae). *Crustaceana* 69: 306-312.

Sudha, K. & Anilkumar, G. 1996. Seasonal growth and reproduction in a highly fecund brachyuran crab, *Metopograpsus messor* (Forskål) (Grapsidae). *Hydrobiologia* 319: 15-21.

Vannini, M. & Ruwa, R.K. 1994. Vertical migrations of the tree crab *Sesarma leptosoma* (Decapoda, Grapsidae). *Mar. Biol.* 118: 271-278.

Ward, C.J. & Steinke, T.D. 1982. A note on the distribution and approximate areas of mangroves in South Africa. *S. Afr. J. Bot.* 1: 51-53.

Sex in the Palaeozoic (Ostracoda, Palaeocopida)

ROGER E.L. SCHALLREUTER
Geological-Palaeontological Institute & Museum, University of Hamburg, Hamburg, Germany

KERRY M. SWANSON
Geology Department, University of Canterbury, Christchurch, New Zealand

KENNETH G. MCKENZIE
Science & Technology, Charles Sturt University-Riverina, Wagga Wagga, Australia

ABSTRACT

The demands of forming a *copula* that is both mechanically and biologically efficient has had a profound effect on the evolution of carapace architecture in the Ostracoda. Videorecordings of copulation in the straight-hinged living punciid *Manawa* show that union occurs venter to venter with valves of both sexes fully extended and about parallel to the substrate. Stability of the *copula* is likely enhanced by employment of a terminal clasper on the elongate male maxilla. It seems probable, therefore, that many extinct straight-hinged ostracodes of the Palaeozoic order Palaeocopida also mated venter to venter. Such an interpretation holds for all taxa in which the sex dimorphic features are velar structures such as antra, convex dolons and cruminae. These palaeocopid elements correspond to the frills and ventral margins of living punciids. Additionally, the marginal ornaments of many male palaeocopids likely served as recognition cues and to help align females for sex; whereas males with smooth ventral margins possibly used one or more of their limbs to engage the females securely. In those Palaeocopida females with anteroventrally loculate margins, the loculi probably functioned as egg shelters. The cavum of some other palaeocopids may well have operated as a gaseous diffusion chamber, as has been proposed for the posteroventrally loculate living podocopid *Loculicytheretta*. Palaeocopid cruminae were brood chambers.

1 INTRODUCTION

Palaeocopida are the most typical Ostracoda of the post-Cambrian Early Palaeozoic. Their size ranges from around 0.5 mm to over 5 mm and the shape from very high (length/ height less than 1.45) to very long (l/h over 2.05). In many taxa, size and shape are rather constant and instar clusters are very distinct in l/h diagrams. Outlines are postplete, amplete or preplete, and can be typical for some groups. Characteristically, the Palaeocopida are straight-hinged.

L-S sculptures are foldings of the shell and therefore are as strongly developed internally, e.g. on steinkerns, as externally. But ornament is formed externally on the shell and therefore is virtually absent on steinkerns. Lobation (L) and sulcation (S) are the characteristic beyrichicopine features. Maximally, there are 4 lobes (L1-4) and three sulci (S1-3); valves correspondingly are uni-, bi-, tri- or quadrilobate, and uni-, bi- or trisulcate. With evolution, the sulcation was reduced but the primitive quadrilobate design became expressed as ridges or by dissolution of the lobes into single nodes. The median sulcus (S2) carries the adductor muscle scars internally but these are visible only in rare instances. This S2 is the most persistent sulcus but may be reduced to only a faint depression or even disappear entirely. In front of the S2, a more or less distinct preadductorial node is developed. In some taxa the shell is thinner on this node than around it and a few authors believe therefore that it marked the position of a lateral eye. In contrast, the Binodicopa have only two lobes or nodes at the dorsal margin; and most Leiocopa are nonlobate/nonsulcate.

Specimens illustrated in Plates 1-6 originate from glacial erratic boulders (geschiebes) of Northern Central Europe and Southern Sweden. The material on Plates 1-4 and Plate 5: 1-2 is deposited at the Archiv für Geschiebekunde, Geologisches-Paläontologisches Institut der Universität Hamburg, Germany. The material on Plate 6 and Plate 5: 3-4 is deposited at the Westfälisches Museum für Naturkunde, Münster, Germany. Specimens illustrated in Plate 7 originate from the megaripple zone, 17 m, at Leigh, New Zealand. This material is deposited at the Micropalaeontological Section, New Zealand Geological Survey, Lower Hutt, New Zealand. The authors should be contacted for details on localities.

2 SEXUAL DIMORPHISM IN PALAEOCOPID OSTRACODA

The most striking characters of palaeocopids are the various types of sex dimorphic features. Both sexes occur so generally in the Silurian that Siveter (1984) has postulated that several taxa including *Kloedenia leptosoma* Martinsson, *Kloedenia wilckensiana* (Jones) and *Clavofabella reticristata* (Jones) were parthenogenetic because only female morphs are yet known.

The most important and distinctive sex dimorphic feature in female Palaeocopids is the velum, a spine-, ridge- to flange-like extension of the valve. Marginal sculptures, which can be dimorphic, are sometimes incorporated into this extension or sited close to it. Above the velum another adventral sculpture, the histium (Plate 3: 1-2), is occasionally developed.

Carapace sexual dimorphism is very common. In *Tetrada memorabilis* Neckaja, for example, males are larger and more elongate than females and are postplete whereas the females are amplete. Unlike all other ostracodes, beyrichicopine palaeocopids exhibit several special dimorphisms of the velum; these can also influence other characters of the ornament and the lobation-sulcation.

Velar dimorphism can occur anteriorly or posteriorly and is of two main kinds – antral (Plates 1-5) and cruminal (Plate 6). The crumina is a brood pouch connected with the domicilium whereas the antrum is a pouch formed outside the domicilium. Morphogenetically, cruminal velar dimorphism occurs in several lineages and therefore is polyphyletic, the crumina occupies the anteroventral and/or ventral regions of the carapace (Plate 6); posterior cruminas are unknown. In hollinomorphs the antrum is located anteroventrally (Plates 1-4). The antrum of primitiopsiomorphs, however, is restricted to the posterior part of the carapaces (Plate 5).

Plate 1. Antral dimorphism – antrum anterior and/or ventral. Females left, males/tecnomorphs right. LV = left valve; RV = right valve. All lateral exterior views. (1) *Gellensia nodoreticulata* Schallreuter, 1976; Middle Ordovician. Female LV, length 1.23 mm, antrum-forming frill (dolon) weakly convex. Tecnomorph RV, length 1.25 mm. (2) *Levisulculus* cf. *obliquus* (Steusloff, 1895); Upper Ordovician. Female RV, length 1.19 mm, antrum-forming frill strongly convex but reaching the contact plane. Tecnomorph RV, length 0.85 mm. (3) *Piretella triebeli* Schallreuter, 1964; Middle Ordovician. Anterodorsally incomplete female RV, length 1.61 mm (without posterior spines), antrum-forming frill strongly convex and developed as a 'false brood chamber'. Tecnomorph LV, length 1.90 mm. (4) *Bilobatia bidens serralobata* Schallreuter, 1976; Middle Ordovician. Female LV, length 0.96 mm, velar flange solid with peripheral spines forming a 'cage'-like structure (wehrliine antrum). Male LV, length 0.96 mm. (5) *Homeokiesowia frigida* Sarv, 1959; Middle Ordovician. Female RV, length 1.06 mm, velar flange solid and weakly convex. Male LV, length 1.07 mm.

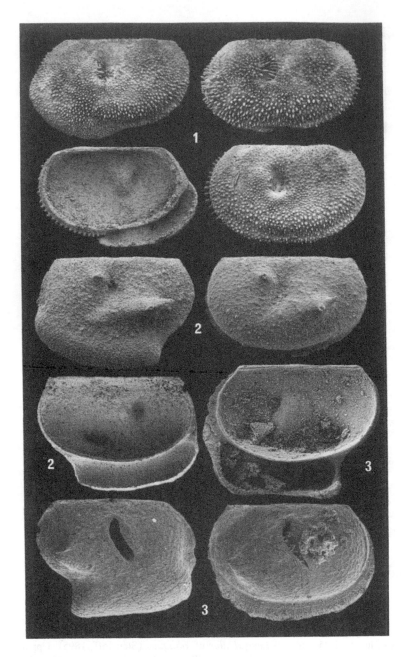

Plate 2. Antral dimorphism – females with an anterior/ ventral 'false brood chamber'. Females left, males/ tecnomorphs right. LV, RV as for Plate 1. All lateral views. (1) *Uhakiella jaanussoni* Schallreuter, 1954; Middle Ordovician. Female LV exterior, length 2.51 mm. Male LV exterior, length 2.42 mm. Female LV interior, length 2.56 mm. Juvenile (A–1) LV exterior, length 2.02 mm. (2) *Brevibolbina donbuschi* Schallreuter, 1964; Middle Ordovician. Female LV exterior, length 0.75 mm. Tecnomorph LV exterior, length 0.61 mm. Female LV interior, length 0.72 mm. (3) *Cavhithis cavi* Schallreuter, 1965; Middle Ordovician. Female RV interior, length 0.71 (including marginal flange. Female RV exterior (displaying cavum), length 0.70 mm. Tecnomorph RV exterior, length 0.61 mm.

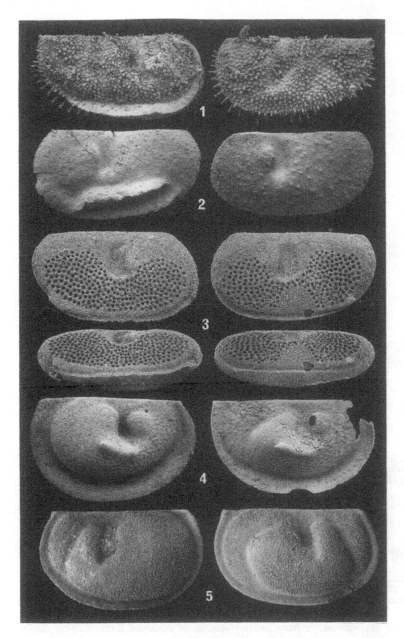

Plate 3. Antral dimorphism – antrum anterior/ventral. 1 and 2 in combination with histial dimorphism; 4 in combination with abdominal dimorphism; 5 abdominal dimorphism, velar dimorphism lost. Females left, males/tecnomorphs right. LV, RV as for Plate 1. All lateral exterior views. (1) *Kiesowia (Pseudotallinnella) scopulosa* (Sarv, 1959); Middle Ordovician. Female RV, length 1.56 mm. Tecnomorph LV, length 1.25 mm. (2) *Severella elliptica* (Steusloff, 1895); Middle Ordovician. Female LV, length 1.78 mm. Tecnomorph LV, length 1.73 mm. (3) *Scrobisylthis reticulatus* (Sarv, 1959). Middle Ordovician Female RV, length 0.62 mm. Male LV, length 0.62 mm. (4) *Bolbihithis abdominalis* Schallreuter, 1981; Upper Ordovician. Female RV, length 1.03 mm. Anterodorsally incomplete male RV, length 0.91 mm. (5) *Bolbina major* (Krause, 1892); Middle Ordovician. Female LV, length 1.44 mm. Tecnomorph RV, length 1.33 mm.

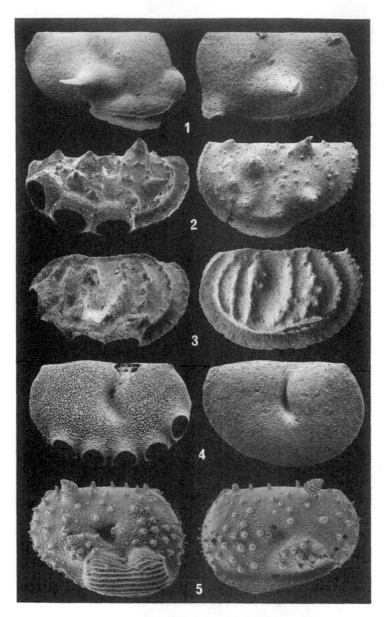

Plate 4. Loculate antral dimorphism – antrum anterior/ventral, an advanced form of antral dimorphism. Females left, males/tecnomorphs right. LV, RV as for Plate 1. All lateral exterior views. (1) *Semibolbina ordoviciana* Schallreuter, 1977; Upper Ordovician. Dolonate locular dimorphism with 3 loculi in the dolon (female velar flange). Female RV, length 0.88 mm. Tecnomorph LV, length 0.82 mm. (2) *Deefgella dajsiveteri* Schallreuter, 1981; Upper Ordovician. Admarginal loculate dimorphism with 3 loculi at the marginal surface of the valve. Female LV (slightly tilted), length 0.74 mm. Tecnomorph RV, length 0.74 mm. (3) *Tetradella separata* Sidaraviciene, 1971; Upper Ordovician. Female LV (slightly tilted) with 4 loculi, length 1.13 mm. Tecnomorph LV, length 1.01 mm. (4) *Foramenella parkis* (Neckaja, 1952); Upper Ordovician. Female RV with 5 loculi, length 0.81 mm. Tecnomorph RV, length 0.73 mm. (5) *Distobolbina tuberculata* (Henningsmoen, 1954); Upper Ordovician. Female RV, length 0.77 mm. Male LV, length 0.79 mm.

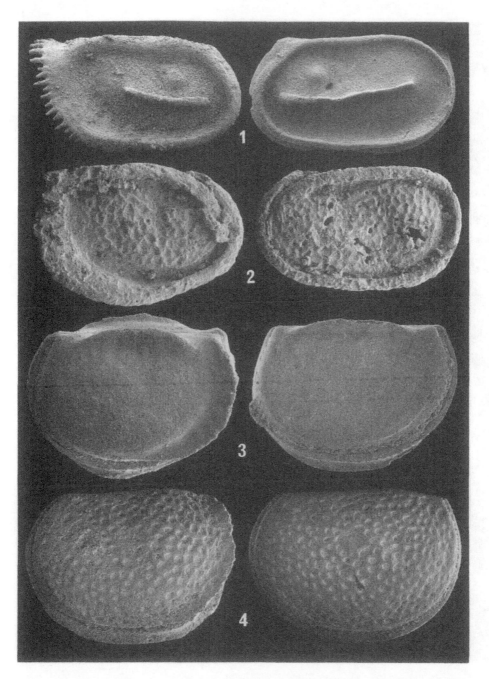

Plate 5. Antral dimorphism – antrum posterior. Females left, males right. LV, RV as for Plate 1. All lateral exterior views. (1) *Bubnoffiopsis bubnoffi* Schallreuter, 1964; Middle Ordovician. Female RV, length 0.50 mm. Male LV, length 0.49 mm. (2) *Lembitites posterovelatus* (Sarv, 1963); Middle Ordovician Female RV, length 0.51 mm. Male LV, length 0.56 mm. (3) *Pliciwemuna monasterium* Schallreuter, 1996; Upper Silurian. Female carapace (LV shown), length 0.76 mm. Male carapace (LV shown), length 0.76 mm. (4) *Macuwemuna maculata* Schallreuter, 1996; Upper Silurian. Female carapace (LV shown), length 0.87 mm. Male carapace (LV shown), length 0.88 mm.

Plate 6. Cruminal dimorphism. Females left, males/tecnomorphs right. LV, RV as for Plate 1. All lateral exterior views. (1) *Beyrichia (Beyrichia) suurikuensis* Sarv, 1968; Silurian. Female RV, length 1.92 mm. Male LV, length 1.96 mm. (2) *Hemsiella maccoyiana* (Jones, 1855); Silurian. Female LV, length 1.88 mm. Tecnomorph RV, length 1.54 mm. (3) *Berolinella praevia* Sarv, 1968; Silurian. Female LV, length 1.34 mm. Male RV, length 1.39 mm. (4) *Moierina abushikae* Schallreuter & Schaefer, 1988; Silurian. Female LV, length 0.74 mm. Tecnomorph carapace (RV shown), length 0.67 mm.

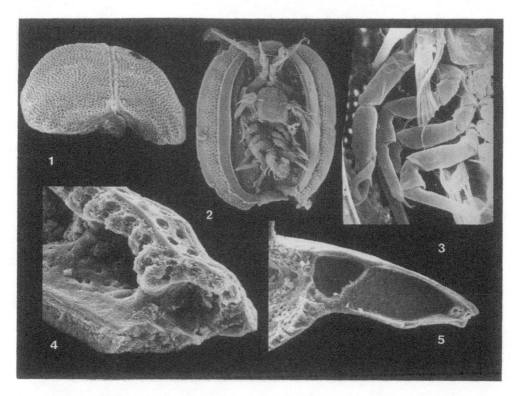

Plate 7. Anatomy of the straight-hinged living punciids *Manawa staceyi* Swanson, 1989 and *Puncia* sp. Swanson. (1) Entire carapace of *Manawa staceyi* in life position, (length 0.41 mm) × 250. (2) Ventral view of carapace and soft anatomy of female *Manawa staceyi*, (length 0.41 mm) × 210. (3) Right-hand thoracic elements of male *Manawa staceyi* showing the clasper on the elongate maxilla, × 750. (4) Internal buttressing of the right valve in *Manawa staceyi*, × 1100. (5) Internal structure of the carapace frill in *Puncia* sp., × 1000. Note closure of the cavity to the domicilium.

An antrum is developed when a more or less convex velum (dolon) forms a sausage-like extension or botulus. This may be closed in rare cases but is usually open to the outside. In the Middle Ordovician subfamily Wehrliinae the dolon developed as a series of long spines, forming a cage-like open pouch (Plate 1: 4). With evolution, the botulus became more or less distinctly divided into hemispherical compartments or loculi (Plate 4). These are formed in several lineages; thus, like cruminal dimorphism, locular dimorphism is polyphyletic. It is the most advanced form of the antrum.

Antral dimorphism evolved into abdominal dimorphism in some instances, e.g. *Bolbihithis abdominalis* Schallreuter (Plate 3: 4) and *Bolbina major* (Krause); in other cases it became rudimentary, e.g. Graviidae, or disappeared.

Egorovellid domiciliar dimorphism is restricted to the Ordovician Egorovellidae of Siberia. In this group, the anteroventral part of the domicilium of the heteromorphs is inflated and covered externally with parallel ridges.

Whereas sexual dimorphism is very common in beyrichicopines it is virtually unknown in Leiocopina, Binodicopina and Eridostraca. A lobate type of dimorphism is reported for the binodicope Neoulrichiidae but this may characterise the males only.

While the special dimorphic features are restricted to females, males also can be recognized in some cases. Thus, *Uhakiella* males feature a medioventral incisure (Plate 2: 1). In *Tetradella*, which is characterized by loculate dimorphism, the males exhibit domiciliar dimorphism via a ventrally inflated L4 (Plate 4: 3). Male characters developed commonly in the Late Palaeozoic family Hollinellidae (hollinellid dimorphism). For example, the centrodorsal node (L3) of hollinellid males is much larger than in the females which feature normal antral (open pouch) dimorphism. In the Nodellacea, males carry a special hook (hamus) anteriorly. This is hamal dimorphism – nodellacean females exhibit no special dimorphic features.

Although normally the dimorphic shell features appear with the final moult to adulthood, sometimes they can be recognized even at the A-3 stage, e.g. in *Tetradella scotti* Guber or *Oepikella tvaerensis* Thorslund. But there is no evidence that this led to premature fertility in such taxa.

3 MATING AND SEX DOMORPHIC FEATURES IN LIVING PUNCIIDAE

Punciid ostracodes are an extant group with a fossil history extending from the Mesozoic, and are regarded as Palaeozoic relics. They display affinities to extinct Eurychilinacea and Kirkbyacea and to extant Platycopida and Podocopida.

Although living specimens of the punciid *Manawa staceyi* Swanson (1989) may be coaxed into closing their valves, this extremely rare animal is normally found moving with both valves fully extended laterally. In this orientation, both valves form a shallow, dome-shaped canopy supported by the entire soft anatomy, which is exposed to and in intimate contact with the substrate (Plate 7: 1-2).

Undoubtedly, the demands of forming a *copula* in a way that is both mechanically and biologically efficient has had a profound effect on the evolution of ostracode carapace architecture. Copulation venter to venter has been observed in a number of living species (cf the orally presented paper by McKenzie & Abe, this volume). However, in the fossil record, particularly that of the Palaeozoic, the carapace geometry of many species is such that the hemipene extension required to form a *cop-

ula with valves only partly gaping would be ergonomically extreme. Similarly, the fact that most extant ostracodes have a vertically-oriented carapace with a ventromarginal valve gape of between 15-20°, was previously seen as evidence that dorsal to dorsal and posterodorsal to dorsal mating (Cohen & Morin 1990, Figs 7A, B) was more likely in Palaeozoic taxa (Kesling 1969).

Observation of the manawan gait and the soft anatomy of this animal, however, has led to development of the idea that venter to venter mating probably occurred quite commonly among Palaeozoic ostracodes because the geometric/mechanical constraints referred to above are overcome when the valves gape fully and the entire body is exposed.

Sexes in *Manawa staceyi* are easily differentiated by the presence in males of a hemipene and an elongate pediform maxilla in which the terminal segment is modified to form a clasper (Plate 7: 3). In females, the maxilla is reduced and almost vestigial but detail relating to other reproductive characters is lacking.

At the time of the original species description, mating had not been observed but the sex dimorphic adaptations in the male anatomy were seen as evidence that venter to venter copulation was likely. In 1995, a team of Japanese and New Zealand scientists working at the Leigh Marine Station, New Zealand, video-recorded venter to venter mating in *Manawa staceyi*. Analysis of these low-resolution images is still continuing. Significantly, of more than 40 live specimens examined so far none has provided evidence of brooding. Thus, punciid carapace structures such as lunettes and frills previously thought to serve as brood chambers or cavities are now seen, in the case of *Manawa*, as simple buttressing for carapace strength (Plate 7: 4) or, in *Puncia*, as possible flotation devices and/or cavities providing increased respiratory surfaces (Plate 7: 5).

4 DISCUSSION

As noted above, previous interpretation of the mating behavior of palaeocopids was influenced critically by the posterodorsal to dorsal mating position used by many freshwater podocopids (Kesling 1969). Now that more observations are available, it seems clear that at least as many podocopids mate venter to venter, e.g. *Xestoleberis* and *Loxoconcha* species (Kamiya 1988; Abe & Vannier 1993). The video-recordings of manawan mating referred to above strongly suggest that palaeocopids with velae or frills, convex dolons and cruminae also probably mated venter to venter (Cohen & Morin 1990, Figs 7C, D).

Many palaeocopid males also seem well adapted for such mating behavior. Thus, males of *Rakverella bonnemai* Opik and *Beyrichia (Altibeyrichia) kiaeri* (Henningsmoen) bear series of anteromarginal spines which could have served to help secure the females in *copula* (Kesling 1969, text Figs 2, 10). On the other hand, males of *Slependia armata* (Henningsmoen) and *Macronotella scofieldi* Ulrich possibly used one or more of their limbs to engage the female securely during mating since their convex ventral margins are smooth (Kesling 1969, text Figs 11, 22).

It is generally considered that species with closed antra lived benthically, e.g. *Uhakiella jaanussoni* Schallreuter, whereas taxa with rather unprotected and open antra might have been epibenthic or pelagic swimmers, e.g. *Foramenella* (Plate 4: 4).

As suggested by Henningsmoen (1965), the palaeocopid antrum probably served as an egg shelter. In ventrally loculate females (e.g. *Abditoloculina*), such an interpretation is strengthened by the fact that the loculi are invariably of the same size intraspecifically – even at the termini of the antrum, where on purely mechanical grounds one might expect smaller-sized chambers – as seen in *Deefgella dajsiveteri* Schallreuter (Plate 4: 2). Furthermore, these loculi may occur not only admarginally (*Deefgella*) but also in a purely dolonate position, as in *Semibolbina ordoviciana* Schallreuter (Plate 4: 1). Note, however, that loculus size does vary between species (Plate 4).

This functional purpose differs from that postulated for the living podocopid *Loculicytheretta* in which the loculi occur posteriorly on the female, not anteriorly as in female palaeocopids. In well-preserved material, the loculi in *Loculicytheretta* are covered by a chitinous membrane, that has led to the postulate that they functioned as gaseous diffusion chambers and flotation devices assisting to lift females off the bottom, thus facilitating venter to venter mating. As noted above, the frill in living *Puncia* might function in a similar manner. In palaeocopids, what may have functioned as a gaseous diffusion chamber is a sulcal sculpture, the cavum, which occurs rarely in a few unrelated genera, for example in *Cavhithis* (Plate 2: 3).

Parental brood care was first recorded in the cruminal beyrichiacean *Craspedobolbina* and subsequently confirmed by thin sections made from limestones with palaeocopid faunas (Hessland 1949). Cruminal brood pouches appear particularly well adapted for such a role (Plate 6).

Among males, the dorsally inflated L3 (in hollinellids) and the ventrally inflated L4 in *Tetradella* (Plate 4: 3), for example, may have housed large genitalia.

ACKNOWLEDGEMENTS

Responsibilities in the contribution are as follows: Plates 1-6 (Schallreuter); Plate 7 (Swanson); text (McKenzie, with input from both other co-authors); poster design (McKenzie). The authors thank Dr Andrew Parker, The Australian Museum, Sydney, for kindly presenting their poster at the 4th ICC (International Crustacean Congress), Amsterdam, July 1998. The authors thank two referees for their comments.

REFERENCES

Abe, K. & Vannier, J. M-C. 1993. The role of 5th limbs in mating behaviour of two marine podocopid ostracods, *Bicornucythere bisanensis* (Okubo 1975) and *Xestoleberis hanai* Ishizaki 1968. In K.G. McKenzie & P.J. Jones (eds), *Ostracoda in the Earth and Life Sciences*: 581-590. Rotterdam: Balkema.

Cohen, A.C. & Morin, J. 1990. Patterns of reproduction in ostracodes: a review. *Journal of Crustacean Biology* 10(2): 184-221.

Henningsmoen, G. 1965. On certain features of palaeocope ostracodes. *Geologiska Foreningens i Stockholm Forhandlingar* 86: 329-334.

Hessland, I. 1949. Investigations of the Lower Ordovician of the Siljan District, Sweden. I. Lower Ordovician ostracods of the Siljan District, Sweden. *Bulletin of the Geological Institutions of the University of Uppsala* 33(3): 97-408.

Kamiya, T. 1988. Morphological and ethological adaptations of Ostracoda in microhabitats in

Zostera beds. In T. Hanai, N. Ikeya & K. Ishizaki (eds), *Evolutionary Biology of Ostracoda*: 303-318. New York: Elsevier.

Kesling, R.V. 1969. Copulatory adaptations in ostracods Part III. Adaptations in some extinct ostracods. *Contributions from the Museum of Paleontology, University of Michigan* 22(21): 273-312.

McKenzie, K.G. & Abe, K. 2000. Swarming behaviour at or near the surface by some marine Ostracoda and observations and interpretations thereon. *Crustacean Issues* 12: 239-251.

Siveter, D.J. 1984. Habitats and modes of life of Silurian ostracodes. *Special Papers in Palaeontology* 32: 71-85.

Swanson, K.M. 1989. *Manawa staceyi* n. sp. (Punciidae, Ostracoda): soft anatomy and ontogeny. *Courier Forschungsinstitut Senckenberg* 113: 235-249.

Evolution of brachyuran mating behavior: Relation to the female molting pattern

RICHARD G. HARTNOLL

Port Erin Marine Laboratory, University of Liverpool, UK

ABSTRACT

In brachyurans mating may occur either in recently molted soft shelled females ('soft-female' mating), or in hard shelled intermolt females ('hard-female' mating). The occurrence of these two types of mating is discussed in relation to the molting pattern of females, either indeterminate (molting continues indefinitely after maturity) or determinate (molting ceases, usually at puberty). Three mating patterns are discriminated: indeterminate growth with soft-female mating; determinate growth with soft- and/or hard-female mating; indeterminate growth with hard-female mating. The costs and benefits for both males and females are discussed in relation to each of the three patterns.

1 INTRODUCTION

All aspects of reproduction have costs as well as benefits, and mating is no exception to this rule. The benefits of mating are obvious – for males there is the opportunity to pass on their genes, for females the opportunity to acquire the sperm needed to fertilize their eggs. The costs are less well defined, but several are readily apparent. Mating, and associated activities, use up time which could be otherwise deployed. In both sexes this could be feeding to build up reserves for growth and reproductive products. In males, moreover, time devoted to one female precludes mating with others for that period. Mating activities may also increase risk of predation for both sexes, by increasing visibility and restricting mobility. A well adapted mating strategy will achieve the aims of mating, whilst minimizing the risks.

A wide variety of mating behavior occurs in different brachyurans (Hartnoll 1969), as does a wide variation in molting pattern (Hartnoll 1982, 1983). The molting pattern must be considered when evaluating the adaptive value of particular mating patterns, and the evolution of the two must be closely linked. The primitive molting pattern in the Brachyura is assumed to be 'indeterminate' growth, where molting continues for a seemingly indefinite sequence following sexual maturity. This is universal in the natant and reptant decapods from which brachyurans presumably arose. This molting pattern is normally accompanied by the primitive 'soft-female' mating regime, where mating is restricted to a short period after the female has molted.

This paper will consider three aspects of this molting-mating complex. First, the adaptiveness of the 'primitive' pattern. Second, changes in mating regime when the molting pattern changes. Third, changes in mating regime without changes in the molting pattern.

2 THE ADAPTIVENESS OF THE 'PRIMITIVE' PATTERN

In the 'primitive' pattern females mate for the first time immediately after the puberty molt, and then again following each succeeding molt. The male frequently guards or carries the female for some days prior to her molt (pre-molt attendance), the female molts and they mate soon after, and the male then resumes guarding for some days following mating (post-copulatory attendance). The European edible crab, *Cancer pagurus*, provides a typical example, with pre-molt attendance averaging eight days, and post-copulatory attendance five days (Edwards 1966).

A well adapted mating strategy should be advantageous for both sexes. For females the requirements include: 1) A supply of viable sperm available at the time of ovulation, 2) A good genetic quality in the sperm-providing male, 3) Minimal disruption by mating to other activities. For males there are also requirements of an optimal mating strategy: 1) An assurance of paternity as a result of mating, 2) A limited investment of time per female.

The 'primitive' pattern assures that these various requirements are met for both sexes. This is dependent on two factors – sperm viability during storage, and sperm competition – both discussed below.

For females a supply of viable sperm is assured by mating after each molt, since brachyuran sperm will remain viable for the duration of the intermolt (they are still viable after two years in *Chionoecetes bairdi* – Paul, 1984), and sperm from one insemination will suffice to fertilize the batches of eggs produced during the following intermolt. There are normally not more than two batches in the intermolt, but sometimes three as in *Portunus sanguinolentus* (Ryan 1967). Good genetic quality in the male is a result of the female advertising her sexual receptiveness by releasing chemical pheromones into the water (e.g. Eales 1974, Gleeson 1980), which attract males so that there can be inter-male combat for females, and a superior male will displace a lesser suitor. Thus in *Necora puber* the largest male from a group of conspecifics invariably associated with and defended a pre-molt female (Choy 1986). Minimal disruption is caused by mating because the female is attractive for only a restricted period before and after the molt, and at the end of this period ceases to be of interest to males. There is the possible bonus for females in that the intermolt male protects the female during her vulnerable post-molt period. However, this apparent benefit may be limited by the increased visibility of the paired crabs.

For males, paternity can be assured by the post-copulatory guarding of the female until she ceases to be attractive to other males. Sperm competition, by which the sperm of the latest copulation are most likely to achieve fertilization (Diesel 1990, Jones 1994), ensures that earlier copulations by other males do not deprive the guarding male of paternity. Post-copulatory guarding ensures that no other male can mate with the female in that intermolt, and any eggs produced during the intermolt will be fertilized by the guarding male. The limitation of investment of time by the male follows from the limited duration of attractiveness of the female after molting, typically a matter of days – five days in *Cancer pagurus* (Edwards 1966). Once the female ceases to be attractive the male can seek out other females, with his paternity of the previous female's offspring assured.

This system operates to the apparent mutual advantage of the two sexes, but the rationale of its evolution is speculative, and deserves only a brief comment. There is a major question. Why, in so many crabs, is mating restricted to the period following molting of the female? It will be seen below that this is not the case in all crabs, and there seems no physical reason necessitating it. Perhaps in the ancestral macrurans, where spermatophores were affixed externally to the integument, a clean new integument offered substantial benefits. However, this is not relevant to the internal sperm storage situation in crabs. Possibly, since the system clearly works well to the advantage of both sexes, and changes may benefit neither, it was retained despite changes in the sperm storage mechanism.

3 CHANGES IN MATING REGIME WHEN THE MOLTING PATTERN CHANGES

The principle change in molting pattern is the replacement of indeterminate growth by determinate growth (Hartnoll 1982, 1983), where eventually molting ceases and the crabs enter a state of terminal anecdysis (Carlisle 1957). The commonest pattern is for molting to cease at the onset of sexual maturity, so that there is only a single mature instar (in females at least – in males the situation is more complex, but not relevant here).

This change has implications for the mating pattern, since the 'primitive' pattern would mean only a single brief mating window, following the molt of puberty. It is known that crabs in this terminal instar can survive for up to four years or more (Paul 1984), and produce a number of batches of eggs – eight are recorded in the terminal instar of *Pisa tetraodon* (Vernet-Cornubert 1958). The sperm received following the puberty molt would not always be sufficient and/or remain viable for long enough to enable all the egg batches to be fertilised. There is evidence that fertilisation levels decline sharply after two egg batches have been fertilized by one insemination (Paul 1984).

The mating pattern therefore has to change, and the change is for mating to occur when the female is hard-shelled in the final instar, as well as (or instead of) when she is recently molted. This scenario has been discussed in some detail (Jones & Hartnoll 1997), and once again the situation appears advantageous to both sexes. Although mating can occur when the female is hard-shelled, the female is only attractive for limited periods of a few days which precede the time of egg laying. In some species the ovaries are already mature at the time of the puberty molt, such as *Hyas coarctatus* (Hartnoll 1963) and *Chionoecetes bairdi* (Adams & Paul 1983). In these the female mates immediately after the puberty molt in the 'primitive' fashion, and the first egg batch is laid very shortly afterwards (Adams & Paul 1983, Bryant & Hartnoll 1995). She then becomes attractive again for limited periods close to the time of each subsequent ovulation. In other species the ovaries are unripe at the puberty molt, for example *Inachus dorsettensis* (Jones & Hartnoll 1997) and *Libinia emarginata* (Hinsch 1972). Here the female is not attractive at the puberty molt and does not mate. She becomes attractive only later, near the time of each ovulation, and only hard-shell mating occurs (Hinsch 1968, Jones & Hartnoll 1997).

Both patterns ensure that the requirements of the two sexes, as listed above, are again fulfilled. The female is assured of a supply of fresh sperm from competing males shortly before laying, the optimal time for successful fertilization, but with minimal interruption to her other activities. The male is assured of paternity provided that he guards the female for the relatively short time until the eggs are laid. It is not obvious which of the two patterns above is the more primitive (see discussion in Jones & Hartnoll 1997).

4 CHANGES IN MATING REGIME WITHOUT CHANGES IN THE MOLTING PATTERN

The rationale for changes in the mating pattern, when as described above the molting pattern becomes determinate, are self-evident. The changes are clearly adaptive. However, changes in the mating regime are also observed in crabs which retain the indeterminate molting pattern. Why, when the system seems well suited to the needs of both sexes? Are these changed patterns likewise adaptive?

One situation where indeterminate growth is accompanied by hard-female mating is where the habit of the species has become predominantly terrestrial or semi-terrestrial, as is found almost universally in the Ocypodidae, Gecarcinidae and Grapsidae (Salmon & Atsaides 1968, Salmon 1971, Christy 1980, Christy & Salmon 1984). Hard-female mating is essential because instead of the female attracting the male by chemical means, as in the aquatic crabs described above, the male attracts the female by visual and/or acoustic signals. This change has presumably arisen because the soft female cannot safely venture out to attract a male until the shell has hardened to provide protection and full mobility, and her chemical attractant mechanism is no longer effective in air. Thus hard-female mating becomes the norm.

How adaptive is this mating strategy? Since it is the female who chooses when to mate, she minimizes the investment of time in mating. She normally selects a mate from a number of displaying males, guaranteeing the quality of the male genes. In some cases she follows the male into his burrow – for example *Dotilla mictyroides* (Tweedie 1950, Altevogt 1957), *Ocypode saratan* (Linsenmair 1967) and *Uca* spp. (Crane 1958). She mates there, and may remain there until she lays shortly afterwards, though information on the latter aspect is often limited. For the male this both guarantees paternity, and limits the investment of time in each female. Again this pattern appears to be adaptive for both sexes. However, in other species such as *Uca vocans* mating occurs on the surface, the female is immediately released, and may mate with another male shortly after (Christy & Salmon 1984). The adaptiveness of mating strategies in *Uca* is discussed at some length by Christy & Salmon (1984).

However, there are also fully aquatic crabs with indeterminate growth which have nevertheless switched from soft-female to hard-female mating. This has been recorded in two portunids. In *Thalamita sima* extensive field and laboratory studies indicated that only hard-female mating occurred (Norman 1996). Mating was restricted to females without eggs, or with late stage eggs. However, there was no post-copulatory guarding observed (compare the spider crabs discussed above), and females readily re-mated. Rather unusually females produced up to seven broods within an intermolt. The second example is *Thalamita prymna,* which is also restricted to hard-female mating, again without post-copulatory guarding (Norman et al. 1997). Mating occurred in all females except those bearing very early eggs, but mated females generally did not readily re-mate. Only a few broods were produced within an intermolt.

Hard-female mating has also been observed in a number of xanthid crabs. It was recorded in *Lophopanopeus bellus* (Knudsen 1960, 1964) and *Paraxanthias taylori* (Knudsen 1960), and females may mate several times before laying. In *Neopanope sayi* mating occurs while the female is hard, and may occur when the female is ovigerous (Swartz 1976). A female may mate several times between successive ovulations (Swartz 1978). The same author also records hard-female mating in *Eurypanopeus depressus* and *Panopeus herbstii,* but without any details. In *Pilumnus hirtellus* females mate when hard, normally shortly before laying (Bourdon 1962). There is also hard-female mating in *Xantho incisus* (Bourdon 1962). Most of these accounts are short on relevant details, but there are several common factors. Courtship and mating are both quite brief, and there are no records of post-copulatory guarding. Females may mate several times between ovulations, presumably with different males. In some species females may mate while ovigerous, or shortly before they lay.

Various suggestions have been proposed as to why some aquatic indeterminate crabs have shifted to hard-female mating. Swartz (1976) suggests that it might be a response to living in very high density populations with consequent high levels of intra-specific competition. Norman (1996) considers that the ability to minimize predation risks by rapid mating in the hard shelled condition is a major consideration. There is currently not enough evidence for a sufficient variety of species to draw any firm conclusions. It is certain that hard-female mating in aquatic indeterminate crabs is much commoner that is so far known, and only detailed studies from a larger range of species will help resolve this question.

5 CONCLUSIONS

There has been a detailed discussion of the adaptive value of different mating strategies in males by Christy (1987). In this paper I have attempted to analyze the adaptive value to the two sexes, considering mating strategy as a process of co-evolution in both sexes. The presumption has been that if a strategy is not mutually beneficial, then it will not persist. The analysis has concentrated primarily upon the effect of changes in growth pattern from indeterminate to determinate molting. Here the accompanying changes in mating strategy appear both logical and adaptive. However changes in mating strategy in crabs retaining indeterminate growth have been harder to account for, especially in those retaining at the same time an aquatic habit. Only further comprehensive studies will clarify this uncertainty.

REFERENCES

Adams, A.E. & Paul, A.L. 1983. Male parent size, sperm storage and egg production in the crab *Chionoecetes bairdi. Int. J. Invert. Reprod.* 6: 181-187.

Altevogt, R. 1957. Beiträge zur Biologie und Ethologie von *Dotilla blandfordi* Alcock und *Dotilla myctiroides* (Milne-Edwards) (Crustacea Decapoda). *Zeitschr. Morph. Ökol. Tiere* 46: 369-388.

Bourdon, R. 1962. Observations préliminaires sur la ponte des Xanthidae. *Bull. Soc. Lorraine Sci.* 2: 3-28.

Bryant, A.D. & Hartnoll, R.G. 1995. Reproductive investment in two spider crabs with different breeding strategies. *J. Exp. Mar. Biol. Ecol.* 188: 261-275.

Carlisle, D.B. 1957. On the hormonal inhibition of molting in decapod Crustacea. 2. The terminal anecdysis in crabs. *J. Mar. Biol. Ass. UK* 36: 291-307.

Choy, S.C. 1986. Ecological studies on *Liocarcinus puber* (L.) and *L. holsatus* (Fabricius) (Crustacea, Brachyura, Portunidae) around the Gower Peninsula, South Wales. Ph.D. thesis, University of Wales, Swansea.

Christy, J.H. 1980. The mating system of the sand fiddler crab, *Uca pugilator*. Ph.D. thesis, Cornell University, New York.

Christy, J.H. 1987. Competitive mating, mate choice and mating associations of brachyuran crabs. *Bull. Mar. Sci.* 41: 177-191.

Christy J.H. & Salmon, M. 1984. Ecology and evolution of mating systems of fiddler crabs (genus *Uca*). *Biol. Revs. Cambridge Phil. Soc.* 59: 483-509.

Crane, J. 1957. Basic patterns of display in fiddler crabs (Ocypodidae, genus *Uca*). *Zoologica, New York* 42: 69-82.

Diesel, R. 1990. Sperm competition and reproductive success in the decapod *Inachus phalangium* (Majidae): a male ghost crab seals off rival's sperm. *J. Zool.* 220: 213-227.

Eales, A.J. 1974. Sex pheromone in the shore crab, *Carcinus maenas*, and the site of its release from females. *Mar. Behav. Physiol.* 2: 345-355.

Edwards, E. 1966. Mating behavior in the European edible crab (*Cancer pagurus* L.). *Crustaceans* 16: 161-181.

Gleeson, R.A. 1980. Pheromone communications in the reproductive behavior of the blue crab, *Callinectes sapidus. Mar. Behav. Physiol.* 7: 119-134.

Hartnoll, R.G. 1963. The biology of Manx spider crabs. *Proc. Zool. Soc. London* 141: 423-496.

Hartnoll, R.G. 1969. Mating in the Brachyura. *Crustaceana* 16: 161-181.

Hartnoll, R.G. 1982. Growth. In D.E. Bliss & L.G. Abele (eds), *The Biology of Crustacea, vol. 2, Embryology, Morphology and Genetics*: 111-196. New York: Academic Press.

Hartnoll, R.G. 1983. Strategies of crustacean growth. *Mem. Aust. Mus.* 18: 121-131.

Hinsch, G.W. 1968. Reproductive behavior in the spider crab, *Libinia emarginata* (L.). *Biol. Bull.* 135: 273-278.

Hinsch, G.W. 1972. Some factors controlling reproduction in the spider crab, *Libinia emarginata. Biol. Bull.* 143: 358-366.

Jones, D.R. 1994. Aspects of the reproductive biology of the crabs *Inachus dorsettensis* (Majidae) and *Carcinus maenas* (Portunidae). Ph.D. thesis, University of Liverpool, 261 pp.

Jones, D.R. & Hartnoll, R.G. 1997. Mate selection and mating behavior in spider crabs. *Estuar. Cstl Shelf Sci.* 44: 185-193.

Knudsen, J.W. 1960. Reproduction, life history and larval ecology of the California Xanthidae, the pebble crabs. *Pacif. Sci.* 14: 3-17.

Knudsen, J.W. 1964. Observations of the reproductive cycles and ecology of the common Brachyura and crablike Anomura of Puget Sound. *Pacif. Sci.* 18: 3-33.

Linsenmair, K.E. 1967. Konstruktion und Signalfunction der Sandpyramide der Reiterkrabbe *Ocypode saratan* Forsk. (Decapoda, Brachyura, Ocypodidae). *Zeitschr. Tierpsychol.* 24: 403-456.

Norman, C.P. 1996. Reproductive biology and evidence for hard-female mating in the brachyuran crab *Thalamita sima* (Portunidae). *J. Crust. Biol.* 16: 656-662.

Norman, C.P., Hirano, Y.J. & Miyazaki, T. 1997. Hard-female mating in the brachyuran crab *Thalamita prymna* (Portunidae). *Crust. Res.* 26: 62-69.

Paul, A.J. 1984. Mating frequency and viability of stored sperm in the tanner crab *Chionoecetes bairdi* (Decapoda, Majidae). *J. Crust. Biol.* 4: 375-381.

Ryan, E.P. 1967. Structure and function of the reproductive system of the crab *Portunus sanguinolentus* (Herbst) (Brachyura: Portunidae). 2. The female system. *Proc. Symp. Crust.*, Ernakulam, India, 522-544.

Salmon, M. 1971. Signal characteristics and acoustic detection by fiddler crabs, *Uca rapax* and *Uca pugilator*. *Physiol. Zool.* 44: 210-224.

Salmon, M. & Atsaides, S.P. 1968. Visual and acoustic signalling during courtship by fiddler crabs (genus *Uca*). *Amer. Zool.* 8: 623-639.

Swartz, R.C. 1976. Agonistic and sexual behavior of the xanthid crab, *Neopanope sayi. Chesapeake Sci.* 17: 24-34.

Swartz, R.C. 1978. Reproductive and molt cycles in the xanthid crab *Neopanope sayi* (Smith, 1869). *Crustaceana* 34: 15-32.

Tweedie, M.W.F. 1950. Notes on grapsoid crabs from the Raffles Museum, 2. On the habits of three ocypodid crabs. *Bull. Raffles. Mus.* 23: 317-324.

Vernet-Cornubert, G. 1958. Biologie générale de *Pisa tetraodon. Bull. Inst. Océanogr. Monaco* 1113: 1-52.

Fecundity and reproduction period of the red mangrove crab *Goniopsis cruentata* (Brachyura, Grapsidae), São Paulo state, Brazil

VALTER JOSÉ COBO
Dept. de Biologia, Universidade de Taubaté – UNITAU, Taubate, Brazil

ADILSON FRANSOZO
Dept. de Zoologia, I.B. UNESP 'Campus' de Botucatu, Brazil

ABSTRACT

A study was made of the reproductive biology of *G. cruentata* with emphasis on its breeding period and fecundity. Monthly collections were performed from January 1995 to December 1996 in a mangrove area on the coast of the São Paulo State (23°29′24″S and 45°10′12″W). Mature gonads were observed throughout the analyzed period. The overall mean fecundity observed for this species was 57,235 ± 35,235 varying from 12,249 to 169,400. The presence of mature gonads during the whole period suggests continuous reproduction, which is common for tropical crabs. The great variation of fecundity may be related to multiple spawning in this species resulting from a single copulation. Fecundity in *G. cruentata* is among the highest within the Grapsidae, suggesting a high reproductive potential for this species.

1 INTRODUCTION

Reproduction is certainly the main mechanism to guarantee species continuity, and it is also an important regulatory process in determining the size of natural populations. According to Hartnoll & Gould (1988) the large variety of reproduction patterns in Crustacea is a result of adaptive processes determined by evolutionary pressures to improve survival of the offspring. This may explain the great variation observed in brachyuran life histories when different geographical distributions are considered.

In spite of many articles on brachyuran reproduction, including those on the Grapsidae (Abelle 1977, Kyomo 1986, Diesel & Horst 1995, Tschusida & Watanabe 1997), little is known regarding the biology of the *Goniopsis cruentata*. Hartnoll (1965) provided some general information in his paper concerning grapsid crabs of Jamaica, Schöne, (1968) on its behavior, Zanders (1978) on its osmoregulatory physiology, and Cobo & Fransozo (1998) on its relative growth.

G. cruentata is a semi-terrestrial representative crab of the Grapsidae and a common inhabitant of estuarine areas, mainly those covered by mangrove swamps. The species is remarkably agile and fast. According to Melo (1996) this species is distributed along the western Atlantic from Bermuda to Santa Catarina, Brazil, and in the eastern Atlantic from Senegal to Angola. The aim of our research was to study the different aspects related to reproduction of the *G. cruentata* with emphasis on the breeding period and fecundity.

2 MATERIAL AND METHODS

Individuals of *G. cruentata* were collected monthly from January 1995 to December 1996 in an estuarine area north coast of São Paulo State, Brazil, (23°29′24″S and 45°10′12″W) (Fig. 1). The animals were captured by hand, establishing a catch effort of three collectors during one hour.

After collection, the animals were sexed, carapace width (CW) was measured using a Vernier caliper, and their sexual maturity was determined based on the macroscopic analysis of the gonads. Fecundity was analyzed using the egg batches of 33 females. In this procedure, each egg batch was homogeneously suspended in seawater, and five 1-ml sub-samples were removed for egg counting under a stereomicroscope. The average value obtained was then extrapolated for the whole suspension to estimate the total number of eggs. Macroscopic analysis of gonad and the identification of development stages were based on Haefner (1976), Johnson (1980), Choy (1988), and Abelló (1989a/b) (Table 1).

3 RESULTS

Some 542 animals, 282 males and 260 females, were collected in 1995, while 628, 291 males and 337 females were caught in 1996. The frequency distribution of

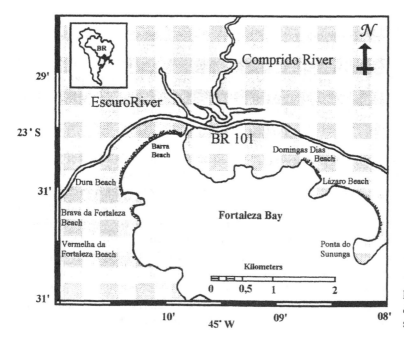

Figure 1. *G. cruentata* – Map of the study area.

gonad developmental stages in females during the study period indicates the presence during the entire period of both mature individuals, as well as crabs with developing gonads. Otherwise, immature crabs were sometimes absent (Fig. 2). The size class including crabs between 20 to 24 mm CW marked the transition between the immature and mature stages for the sampled period (Fig. 3A, B).

Ovigerous females were recorded during most of this period. However, they were scarce between May and September. During 1995, the smallest ovigerous female captured measured 25.1 mm and the largest was 41.4 mm CW. In 1996, the same corresponding values were 25.7 and 41.2 mm of CW (Fig. 4A, B).

An average fecundity of 57,235 ± 35,235 was found, varying between 12,249 and 169,400. Mean size of females for which fecundity was estimated measured 33.55 mm ± 4.01 CW. Fecundity variation relative to carapace width was fitted into a power function given by the equation $y = .084x^{3,8}$ and yielded a correlation coefficient of $r = 0.61$ (standard error = 0.56, standard deviation = 0.75, and $F > 2.25 \times 10^{-7}$). The scatter plot representing this correlation is shown in Figure 5.

Table 1. *G. cruentata* Description of the developmental stages for male and female gonads (modified from Haefner 1976, Abelló 1989).

Gonad Stages	Characteristics	
	Male	Female
Immature – IM	observable only with magnification $> 40 \times$	observable only with magnification $40 \times$
Developing – D	filamentous vas deferens, colorless	filamentous ovaries, colorless
Mature – M	testicles + vas deferens, about 1/4-1/2 hepatopancreas, white	ovary from 1/2-larger than hepatopancreas, orange-dark brown

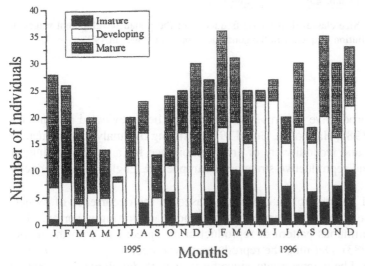

Figure 2. *G. cruentata* – Monthly distribution frequency of the gonad development stages.

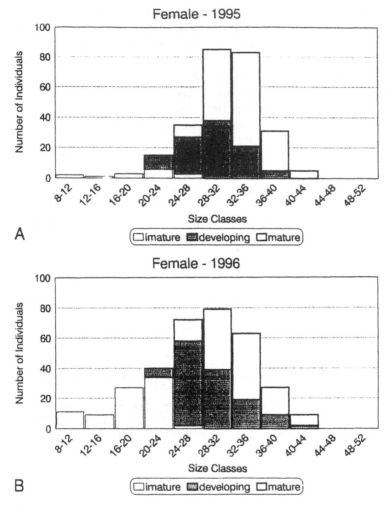

Figure 3. *G. cruentata.* A. Size classes distribution frequency of the gonad development stages in 1995. B. Size classes distribution frequency of the gonad development stages in 1996.

4 DISCUSSION

The diversity of the reproductive patterns in Crustacea has been reviewed in many respects (Hartnoll 1982, 1985, Hartnoll & Gould 1988). Among the analyzed parameters, size at sexual maturity and fecundity are of great importance for an understanding of population dynamics and reproductive strategies (Campbell & Eagles 1983, Kennelly & Watkins 1994).

The breeding period in Brazil of *G. cruentata,* based on the occurrence of ovigerous females, shows a peak in the warmest months of the year, starting in spring. Most species shown a discontinuous pattern of reproduction in agreement with the model proposed by Sastry (1983). Defining the reproduction type occurring in *G. cruentata* as 'discontinuous' would be inappropriate, since females with developing and mature gonads were present in large numbers throughout the study period. The occurrence of

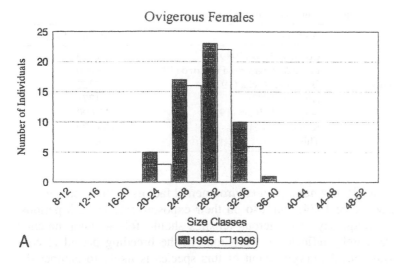

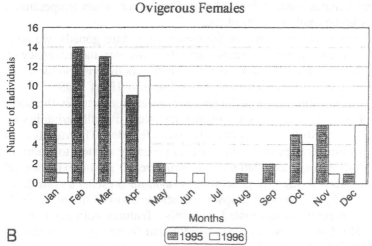

Figure 4. *G. cruentata.*
A. Size classes distri-
bution frequency of
ovigerous females.
B. Monthly distribu-
tion frequency of
ovigerous females.

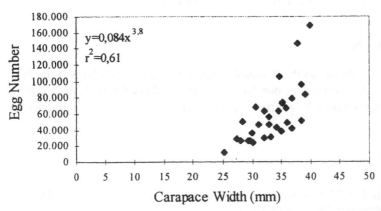

Figure 5. *G. cruentata.*
Relation between
carapace width and
number of eggs.

Table 2. Comparison of fecundity among species of the Grapsidae.

Species	Authors	Number of eggs
Aratus pisonii	Conde & Diaz (1989)	16,379
Aratus pisonii	Leme & Negreiros-Fransozo (1998)	15,000
Pachygrapsus transversus	Ogawa & Rocha (1976)	9222
Pachygrapsus maurus	Almaça (1987)	1771
Armases cinerum	Seiple & Salmon (1987)	15,000
Sesarma intermedia	Kyomo (1986)	100,000
Goniopsis cruentata	This work	57,235

ovigerous females depends on a number of environmental factors such as temperature and salinity (Giese & Pierce 1974) and also on their exposure rate in such periods. Considering this, the frequency of ovigerous females should follow environmental dynamics, which are thereby reflected in variations of the breeding period in consecutive years. In addition, the cryptic habit of this species is likely to hamper the catch rate of ovigerous females, mainly from May to September, when temperatures reach minimum values in the southern hemisphere.

Based on the degree of maturation and the frequency of mature gonads, we concluded that *G. cruentata* shows continuous sexual activity. However, the frequency of ovigerous females is influenced by favorable environmental conditions conducive to larval survival, as noted in other species (Kyomo 1986, Diesel & Horst 1995).

As most Brachyura, *G. cruentata* exhibited a positive correlation between the body size and the number of eggs. Expressed as a standard deviation, overall variability of fecundity also concerns variability within differently sized females. The variation in the number off eggs within females of similar size is probably related to multiple spawning originated from a single copulation. This could result in the extrusion of consecutive batches of different sizes. In general, *G. cruentata* presents a considerably higher fecundity if compared to other grapsid species, as showed in the Table 2. High fecundity values are probably associated with other features defining r- or k-strategies (Hartnoll 1985). For *G. cruentata*, we believe that fecundity values should be related to a high reproductive output because of the high abundance of this species in mangrove areas. Other studies must be undertaken in order to complete our understanding about reproduction in *G. cruentata*, such as those concerning its larval and juvenile development and the appearance of secondary characters.

ACKNOWLEDGEMENTS

We wish to express our thanks to the 'Conselho Nacional de Desenvolvimento a Pesquisa – CNPq'. We also grateful to our NEBECC co-workers for their help during collection and Dr Maria Lucia Negreiros-Fransozo who made helpful comments on the manuscript.

REFERENCES

Abele, L.G. & Means, D.B. 1977. *Sesarma jarvisi* and *Sesarma cookei*: montane, terrestrial grapsid crabs in Jamaica (Decapoda). *Crustaceana* 32: 91-93.

Abelló, P. 1989. Reproductive biology of *Macropipus tuberculatus* (Roux, 1830) (Brachyura: Portunidae) in the northwestern Mediterranesn. *Ophelia* 30: 47-53.

Almaça, C. 1987. Egg number and size in *Pachyugrapsus maurus* (Lucas, 1846), from Praia da Laginha (Faial, Azores islands). *Investigaciones Pesqueras* 51: 157-163.

Campbell, A. & Eagles, M.D. 1983. Size at maturity and fecundity of rock crabs *Cancer irroratus*, from the Bay of Fundy and southwestern Nova Scotia. *Fish. Bull.* 81: 357-362.

Choy, S.C. 1988. Reproductive biology of *Liocarcinus puber* and *L. holsatus* (Decapoda, Brachyura, Portunidae) from the Gower Peninsula, South Wales. *Mar. Biol.* 9: 227-241.

Cobo, V.J. & Fransozo, A. 1998. Relative growth of *Goniopsis cruentata* (Crustacea, Brachyura, Grapsidae), on the Ubatuba region, São Paulo, Brazil. *Iheringia, Série Zoologia* 84: 21-28.

Conde, J.E & Diaz, H. 1989. The mangrove tree crab *Aratus pisonii* in a tropical estuarine lagoon. *Estuarine. Coastal and Shelf Science* 28: 639-650.

Diesel, R.S. & Horst, D. 1995. Breeding in a snail shell: ecology and biology of the Jamaican montane crab *Sesarma jarvisi* (Decapoda, Grapsidae). *J. Crust. Biol.* 15: 179-195.

Giese, A.C. & Pearce, J.S. 1974. Introduction: General principles. In A.C. Giese & J.S. Pearce (eds), *Reproduction of Marine Invertebrates, vol. 1*: 1-49. New York: Academic Press.

Haefner, P.A. 1976. Distribution, reproduction and molting of the rock crab *Cancer irroratus* in the Mid-Atlantic Bight. *J. Nat. Hist.* 10: 377-397.

Hartnoll, R.G. 1965. Notes on the marine grapsid crabs of Jamaica. *Proc. Linn. Soc. Lond.* 176: 377-397.

Hartnoll, R.G. 1982. Growth. In L.G. Abele (ed.), *The Biology of Crustacea. vol. 2*: 111-196. New York: Academic Press.

Hartnoll, R.G. 1985. Growth, sexual maturity and reproductive output. *Crust. Issues* 3: 101-128.

Hartnoll, R.G. & Gould, P. 1988. Brachyuran life history strategies and the optimization of egg production. In A.A. Fincham and P.S. Rainbow (eds), *Aspects of Decapod Crustacean Biology*: 1-9. Symposia of the Zoological Society of London.

Johnson, P.T. 1980. *Histology the blue crab Callinectes sapidus: a model for Decapoda*. New York: Praeger Scientific Publishing.

Kennelly, S.T. & Watkins, D. 1994. Fecundity and reproduction period and their relationship to catch rates of spanner crab, *Ranina ranina*, off the coast of Australia. *J. Crust. Biol.* 14: 146-150.

Kyomo, J. 1986. Reproductive activities in the sesarmid crab *Sesarma intermedia* in the coastal and estuarine habitats of Hakata, Japan. *Mar. Biol.* 91: 319-329.

Leme, M.H.A. & Negreiros Fransozo, M.L. 1998. Fecundity of *Aratus pisonii* (Decapoda, Grapsidae) in Ubatuba region, State of São Paulo, Brazil. *Iheringia, Série Zoologia* 84: 73-77.

Melo, G.A.S. 1996. *Manual de Identificação dos Brachyura (Caranguejos e siris) do Litoral Brasileiro*. São Paulo: Ed. Plêiade.

Ogawa, E.F. & Rocha, C.A.S. 1976. Sobre a fecundidade de crustáceos decápodos marinhos do Estado do Ceará, Brasil. *Arquivos de Ciência Marinha* 16: 101-104.

Sastry, A. N. 1983. Ecological aspects of reproduction. In F.J. Vernberg and W.B. Vernberg (eds), *The Biology of Crustacea, Vol 8*: 179-254. New York: Academic Press.

Schöne, H. 1968. Agonistic and sexual display in aquatic and semi terrestrial brachyuran crabs. *Amer. Zool.* 8: 641-654.

Seiple, W.H. & Salmon, M. 1987. Reproductive, growth and life-history contrasts between two species of grapsid crabs, *Sesarma cinereum* and *S. reticulatum*. *Mar. Biol.* 94: 1-6.

Skinner, D.G. & Hill, B.J. 1987. Feeding and reproductive behavior and their effect on catchability of the spanner crab *Ranina ranina*. *Mar. Biol.* 94: 211-218.

Tsuchida, S. & Watanabe, S. 1997. Growth and reproduction of the grapsid crab *Plagusia dentipes* (Decapoda: Brachyura). *J. Crust. Biol.* 17: 90-97.

Zanders, I.P. 1978. Ionic regulation in the mangrove crab *Goniopsis cruentata*. *Comp. Biochem. Physiol.* 60: 293-302.

Density dependent regulation of the reproduction cycle of *Moina macrocopa (Cladocera)*

YEGOR S. ZADEREEV & VLADIMIR G. GUBANOV
Institute of Biophysics, Krasnoyarsk, Russia

ABSTRACT

The allelopathic effect of population density on gametogenesis induction of *Moina macrocopa* (Crustacea: Cladocera) was investigated at varying food concentration, photoperiod, and temperature. The presence of the non-trophic effect of population density on gametogenesis induction was demonstrated. The experiments with single females have shown that if the photoperiod, temperature, and food concentration are favorable for parthenogenesis, the crowding water has the effect of stimulating females into changing the reproduction mode. This effect was enhanced by using a higher density of the population to achieve crowding. The change of reproduction mode occurs in the experiments with populations of *M. macrocopa* with the achievement of food availability favorable for the parthenogenesis of single females. Chemical interactions between animals are the most obvious explanation for the obtained results.

1 INTRODUCTION

The reproductive cycle of many species of *Cladocera* can be described as an alternation of parthenogenesis and gametogenesis. Gametogenesis allows for the survival of a population by means of the production of resting eggs that resist unfavorable environments. The change of reproduction mode usually precedes the adverse conditions and is induced by the environment. The main factors that have been mentioned to stimulate gametogenesis are photoperiod, temperature, food availability, and population density (Alekseev 1990, Carvalho & Haghes 1983, Korpelainen 1989, Stross 1987). Even though the phenomenology of gametogenetic induction has been investigated for many years, the questions of the role and character of the effect of population density on the reproduction cycle of Crustacea are not clearly answered (Gilyarov 1991). The population density effect can become manifest either indirectly, through changes in trophic conditions, or through chemical or behavioral signals.

Our study has been designed to separate and evaluate different effects of population density on gametogenetic induction of *M. macrocopa* at varying food concentration, photoperiod, and temperature.

2 MATERIAL AND METHOD

Moina macrocopa is a typical representative of *Cladocera* and an inhabitant of temporary ponds. The effect of population density on gametogenesis was investigated with individual females and populations of *M. macrocopa*.

The effect of population density in the experiments with single females was simulated by the use of filtered water from a dense culture of *Moina*. The intensity of the crowding factor was calculated as total wet weight of the animals per volume of the medium. In order to produce water with a different crowding factor (100, 200, and 300 µg), a different number of females were added to the culture volume and allowed to grow one day in the medium. In a number of experiments, the water was filtered trough a 0.85-0.95 µm membrane filter.

Experiments with single females were conducted in beakers containing 20 ml of the crowded water or control medium (aged tapwater). Crowded water or control medium were enriched with a suspension of green algae *Chlorella vulgaris* to make a tested algae concentration (50, 100, 200, 300, 400, or $500*10^3$ cells per ml). Beakers were inoculated with one juvenile female. Inoculants were taken from a culture of *Moina*, grown under the favorable for parthenogenesis nutritional conditions. Beakers were placed in chambers with controlled temperature and photoperiod. The medium in the beakers was renewed daily. The formation of ephippial eggs was the criterion of gametogenesis.

A flow-through system was used in the population experiments, and the volume of chambers was 180 ml. The chambers were refreshed by a rate of 1200 ml per day with new medium containing tested *Chlorella* concentration. Aged tap water was used as the culture medium. Flow-through chambers were placed in the climate chambers with a temperature 26°C and photo-period 16 hours light and 8 hours dark. The population experiments were conducted at five food concentrations – 100, 200, 400, 800 and $1600*10^3$ cells per ml.

Population experiments started with single parthenogenetic female with the eggs of her first or second brood. All individuals in a population were counted and measured daily. Size of the population, and age and sex structures were determined on the basis of these measurements. The population experiments were continued until appearance of gametogenetic females.

3 RESULTS

The presence of the non-trophic effect of population density on gametogenesis induction was demonstrated both in experiments with single females and populations of *M. macrocopa*.

3.1 *Experiments with single females*

With an optimal temperature (26°C) and photoperiod (16l:8d) for parthenogenesis, the dependence of gametogenesis efficiency (as percentage of females that change a mode of reproduction) from particular food concentration and population density used for simulation of crowding is shown in Figure 1. In clean water, females changed repro-

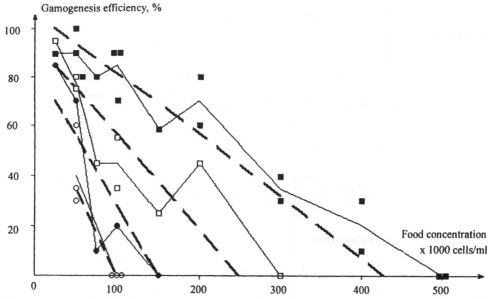

Figure 1. Effect of population density and food concentration on gametogenesis efficiency. Experimental data (–), regression (- – -), Population density: O – control, ● – 100 µg/ml, □ – 200 µg/ml, ■ – 300 µg/ml.

duction mode at food concentration 50,000 cells per ml. The food concentration 100,000 cells per ml is favorable for parthenogenesis in clean water. However, crowding has the effect of stimulating single females into changing the reproduction mode with a food concentration favorable for parthenogenesis in clean water. This effect was enhanced by using a high population density to simulate crowding. The increase of food concentration reduces efficiency of population density effect. The number of ephippial females is reduced to zero with increasing food concentration even with a population density of 300 µg per. However, the higher the population density used for the preparation of crowding experiments, the higher the food concentration is needed to prevent gametogenesis induction. The effect of population density and food concentration on gametogenesis efficiency is described by the following regression equation:

$$y = 72 + 0.13x_1 - 0.7x_2 + 0.0015x_1x_2,$$

where y = gametogenesis efficiency (%), x_1 = population density (µg/ml), x_2 = food concentration (thousand cells/ml).

3.2 *Population experiments*

For all tested food concentrations, the main phases of population growth that precede gametogenesis induction can be described as follows. The 'initial' female reproduces by parthenogenesis, and her offspring (first generation) become mature and also start parthenogenetic reproduction. Females and males are produced in the first brood of the first generation (second generation). Females from this second generation start the formation of ephippial eggs (Fig. 2).

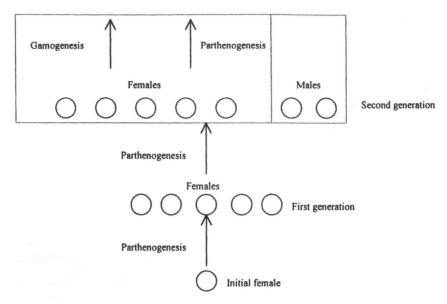

Figure 2. Schematic description of population development until gametogenesis.

Despite such a constant population cycle, the change of population parameters at the moment of gametogenesis induction has been observed with the increase in food concentration (Table 1). The number of parthenogenetic females in a population increased from 15-20 with a food concentration of 100,000 cells/ml to 185 with a food concentration 1,600,000 cells/ml. The increase in food concentration also resulted in an increase of biomass and food concentration in flow-through chambers. However, the most significant in our case is the increase of food availability. The experiments with single females have demonstrated that the change of reproduction mode in the controls starts with a food availability of 1.5 mg of food per 1 mg of biomass. With the food availability increased to twice that amount, females reproduce by parthenogenesis. In the population experiments, the food availability during the formation of ephippial eggs was less than 1.5 mg of food only with food concentrations 100,000 and 200,000 cells per ml. With food concentrations 400,000, 800,000 and 1,600,000 cells per ml, the food availability during the development of second generation, which changed the mode of reproduction, were 3.2, 3.7, and 4.4 mg of food, respectively (Fig. 3). Therefore, the change of reproduction mode occurs in the experimental populations of *M. macrocopa* despite favorable food availability for the parthenogenesis of single females.

4 DISCUSSION

Chemical interactions between animals are the most reasonable explanation for the results obtained.

The experiments with single females were based on the assumption that the concentration of metabolites in the crowded water is proportional to the population den-

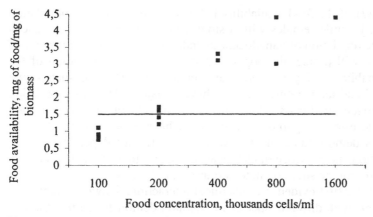

Figure 3. Effect of initial food concentration on the availability of food at the moment of gameto-genesis induction in laboratory populations of *M. macrocopa*. Solid line = level of food availability that stimulates gametogenesis in the experiments with single females.

Table 1. Characteristics of batch culture of *M. macrocopa* at the moment of gametogenesis induction.

Algae concentration, cells/ml	Food concentration in flow-through chamber, Thousand cells/ml	Number of parthenogenetic females	Population biomass, mg	Food availability, mg of food/biomass of population
100000	20, 30, 25, 30	4, 30, 2, 16	20, 21, 16, 14	0.8, 0.8, 0.9, 1.1
200000	40, 50, 60, 50	32, 42, 33, 29	20, 17, 17, 22	1.2, 1.7, 1.7, 1.4
400000	90, 110	37, 53	22, 22	3.3, 3.1
800000	160, 110	139, 136	41, 29	3.0, 4.4
1600000	185	147	57	4.4

sity used for preparation of crowded conditions and time of exposure to the population. Such an approach has significant limitations. First, the metabolism of an animal depends on different factors (temperature, nutritional conditions, etc.). Second, the concentration of metabolites in the medium is determined not only by the rate of excretion but also by the rate of degradation. The critical population density that switches the reproduction mode in the experiments with single females was 100 µg per ml. Due to the aforementioned limitations, this value could be either under or overestimated. Applying the same approach to the results of the population experiments, we have revealed that at the moment of detection of the first gametogenetic females, the population density was equal to 20-30 µg/ml, with food concentrations 400 and 800 10^3 cells per ml and 50 µg/ml with food concentration 1600 10^3 cells per ml. Taking into consideration that the food availability was higher than the limiting quantity for parthenogenesis it is reasonable to suggest that the possible stimulus for gametogenesis induction is the effect of chemical substances released by animals. Another possible cause for gametogenesis induction in the population experiments is tactile communications between animals.

It is known whether successful reproduction of *Cladocera* is possible only if the food availability exceeds certain limiting level (Manuilova 1964) that is specific for species (Gilyarov 1987). If the food availability is less than this level, the population will consist of small, juvenile females with a slowed down growth rate (Chmeleva 1989). In this case, the regulation of gametogenesis induction requires the presence of regulating signals that will protect the population from sudden deterioration of the environment. For inhabitants of permanent water bodies the effect of such signal factors as photoperiod and temperature seems to have a major role in the regulation of induction of gametogenesis (Stross 1987). This is not the case in temporary aquatic ecosystems, where the effect of photoperiod does not have ecological justification (d'Abramo 1984). As demonstrated in our experiments, density dependent factors like nutritional condition and presence of chemical substances that are released by the population also can control the situation in temporary water bodies.

Our results are not the first evidence of chemical interactions between planktonic organisms. Experiments with different species of *Daphnia* demonstrated that the water soluble factor released by different predators has significant effect on behavior of tested animals (Dodson 1988), the size of females and theirs offspring (Dodson & Havel 1988), and clutch size (Engelmayer 1995). In addition, the water soluble factor causes the formation of defensive structures (Schwartz 1991). It was shown that different species of *Daphnia* altered their fecundity under the effect of water from mass cultures (Burns 1995, Seitz 1984). Several authors reported a reduced filtering rate of *Daphnia* in the presence of crowding (Helgen 1987, Matveev 1993). It has been demonstrated with regard to gametogenesis induction that the chemical signal released by a fish predator can induce the formation of resting eggs in *Daphnia* (Slusarczyk 1995). The list of examples and complexity of the scheme could be enlarged substantially. Thus, change in reproduction strategy during the lifetime of *M. macrocopa* can be considered as new evidence of direct chemical interactions between planktonic animals.

Chemical communications between planktonic organisms in nature is often argued (Gilyarov 1987). The claim is that the population's densities, used in laboratories, are never achieved in nature. The results of our experiments have demonstrated that density dependent control of gametogenesis induction is possible with considerably low population densities. This is supported by the results of Hobaek & Larsson (1990), who demonstrated an increase in the proportion of males in the offspring of tested females in response to their own metabolites. This means that species of *Cladocera* can be extremely sensitive to the effect of metabolites. Chemical interactions between and within species are probably the preferable way of communication in aquatic ecosystems. Water allows for efficient and quick distribution of chemical substances. However, it also keeps chemical substances within a limited area (water body). In this case, chemical communications require lower amounts of chemical substances in comparison with air, where the distribution of the chemicals is unlimited. Also, the chemical composition of the aquatic medium could be altered by different factors and, in this case, can be a good indicator of the environmental quality.

REFERENCES

Alekseev, V.R. 1990. *Diapausa rakoobraznih: ecologo-fiziologicheskie aspekti.* [Diapause in Crustacean: Ecological-Physiological aspects]. Moscow: Nauka.

Burns, C.W. 1995. Effects of crowding and different food levels on growth and reproductive investment of Daphnia. *Oecologia* 101: 234-244.

Carvalho, G.R. & Haghes, R.N. 1983. The effect of food availability, female culture density and photoperiod on ephippia production in *Daphnia magna* St. (Crustacea, Cladocera). *Freshwater Biol.* 13: 37-46.

Chmeleva N.N. 1988. *Zakonomernosti razmnojeniya rakoobraznich* [Regularities of the reproduction of Crustacea]. Minsk: Nauka i texnika.

d'Abramo, L.R. 1980. Ingestion rate decrease as the stimulus for sexuality in populations of *Moina macrocopa*. *Limnol. Oceanogr.* 25: 422-429.

Dodson, S.I. 1988. The ecological role of chemical stimuli for the zooplankton: Predator-avoidance behaviour in Daphnia. *Limnol. Oceanogr.* 33: 1431-1439.

Dodson, S.I. & Havel, J.E. 1988. Indirect prey effect: Some morphological and life history responses of *Daphnia pulex* exposed to *Notonecta undulata*. *Limnol. Oceanogr.* 33: 1274-1285.

Engelmayer, A. 1995. Effects of predator-released chemicals on some life history parameters of *Daphnia pulex*. *Hydrobiologia* 307: 203-206.

Gilyarov, A.M. 1987. *Dinamika chislennosti presnovodnix planktonnix rakoobraznich.* [Population dynamic of freshwater zooplankton]. Moscow: Nauka.

Gilyarov, A.M. 1991. Chto mi znaem o diapause rakoobraznich? [What do we know about the diapause in Crustacea]. *Gidrobiologicheskii jurnal.* 27(5): 101-103.

Helgen, J.C. 1987. Feeding rate inhibition in crowded *Daphnia pulex*. *Hydrobiologia* 154: 113-119.

Hobaek, A. & Larsson, P. 1990. Sex determination in *Daphnia magna*. *Ecology* 7: 2255-2268.

Korpelainen, H. 1989. Sex ratio of the cyclic parthenogen *Daphnia magna* in a variable environment. *Z. zool. Syst. Evolut.-forsch.* 27: 310-316.

Manuilova, E.F. 1964. *Vetvistousie rachki (Cladocera) fauni SSSR* [Cladocera of USSR fauna]. Moscow: Nauka.

Matveev, V. 1993. Investigation of allelopathic effects of Daphnia. *Freshwater Biol.* 29: 99-105.

Seitz A. 1984. Are there allelopathic interactions in zooplankton? Laboratory experiments with *Daphnia*. *Oecologia* 62: 94-96.

Schwartz, S.S. 1991. Predator-induced alteration in Daphnia morphology. *J. Plancton Res.* 13: 1151-1161.

Slusarczyk, M. 1995. Predator-induced diapause in Daphnia. *Ecology* 76: 1008-1013.

Stross, R.G. 1987. Photoperiodism and phased growth in Daphnia populations: co-actions in perspective. *Mem. Ist. ital. idrobiol. 'Dott.M.Marchi.'.* 45: 413-437.

Pattern of reproduction in *Sicyonia disdorsalis* (Decapoda: Penaeoidea: Sicyoniidae)

TERESA JERÍ
Alfred Wegener Institute for Polar- and Marine Research, Bremerhaven, Germany

ABSTRACT

The temporal pattern of reproduction of the penaeoid shrimp, *Sicyonia disdorsalis*, is described from monthly population samples (November 1995 through November 1996) from shallow waters in a tropical locality in northern Peru. The reproductive condition of females was assessed from the degree of ovarian maturity and the presence or absence of sperm masses in the seminal receptacles. Adult females with mature ovaries occurred in samples from May to November with maximal numbers in May, August, and November. Insemination followed the same pattern as ovarian maturity. Sexual maturity of males was deduced from the fusion of the gonopodal endopods (petasma). Mature males were present throughout the year, whereas a small proportion of juveniles was found from May to November. Recruitment intensity was described for both female and male juveniles in terms of their percentage in monthly population samples. Recruitment occurred throughout the year with higher incidence in summer and autumn months.

1 INTRODUCTION

Penaeoid prawns contribute an important part to commercial coastal fisheries in many parts of the world, particularly in tropical and sub-tropical regions (Garcia & le Reste 1981, Dall et al. 1990). From the 39 benthic species recorded for the eastern Pacific coast from Mexico to north Peru, 35 of them are commercially important (Hendrickx 1996).

Sicyonia disdorsalis, a non-commercial species, occurs as a by-catch in commercial prawn trawl fisheries off northern Peru. To date biological studies on this species in this area are virtually absent. Most of the recently obtained information on seasonal patterns of breeding of sicyoniids was assessed in temperate areas off California (Anderson et al. 1985) and in Puerto Rico (Bauer 1992b, Bauer & Rivera Vega 1992). Females of this genus have a closed thelycum, which permits storage of simple sperm masses deposited by males during mating (Bauer 1991, Bauer 1994). Mating takes place in newly molted females with immature ovaries. Afterwards vitellogenesis and the final maturation of oöcytes are completed (Yano 1995).

543

Therefore all these studies on sicyoniid reproduction were based on oöcyte characteristics, using the presence of yolk and cortical crypts in oöcytes as indicators of imminent spawning of females. Anderson et al. (1985) showed that females of *S. ingentis* had multiple spawning occurring throughout the summer, while a high percentage of adult females of *S. parri* and *S. laevigata* displayed a year-round spawning activity (Bauer & Rivera Vega 1992). The breeding patterns reported in other species of *Sicyonia* are compared to those found for *S. disdorsalis* of northern Peru in this study. The objective was to see whether continous or seasonal reproduction patterns depend on the latitudinal distribution of the species and to provide biological information on reproductive patterns in *S. disdorsalis* by: 1) observing the annual variation of molting activity, ovarian maturity, and insemination in females, and of gonopodal fusion in males; and 2) comparing these results with those previously reported for this genus from other areas.

2 MATERIAL AND METHODS

Monthly samples of *S. disdorsalis* were obtained with a 10 mm mesh Agassiz trawl from November 1995 to November 1996 in coastal waters off northern Peru (03°38'S) (Fig. 1). The sampling area was the same area where the commercial prawn trawling fishery takes place.

All individuals were sexed and their molt condition determined touching the carapace of the individuals (either soft or hard cuticle). A characteristic of individuals with soft cuticle is the absence of coloration in comparison with individuals with hard cuticle. The size of each individual was measured as carapace length (CL), the linear

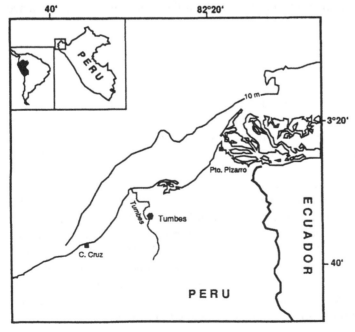

Figure 1. Study area of the coast of Tumbes, northern Peru, showing the trawling area between Caleta Cruz and Puerto Pizarro.

distance from the posterior edge of the orbit to the posterior mid-dorsal edge of the carapace. Total wet weight (TW) of all individuals from a trawl was recorded. The complete ovaries (GW) were removed from a random sub-sample of females and weighed separately to the nearest 0.01 g.

Reproductive conditions of females were determined from the degree of ovarian maturity. Ovary maturation stages were determined based on macroscopic criteria, such as color and volume occupied by the gonads (Bauer & Rivera Vega 1992, Bauer & Lin 1994), and histological examination considering size and development of oöcytes, size of the nucleus, granulations in the cytoplasm, presence of follicle cells, and degree of vitellogenesis (Yano 1988). Five stages were distinguished.

– *Stage 1*. Undeveloped, very thin, and transparent ovary with short, paired lobes along the hindgut. Oöcytes are small and spherical with clear cytoplasm and large nucleus with basophilic chromatin.

– *Stage 2*. Early development, thicker but colorless ovary cords above and around the hindgut. Oöcytes with cytoplasm weakly basophilic, follicle cells on the outer surface of the oöcyte.

– *Stage 3*. Late development, larger and thicker ovary, which becomes yellow. Oöcytes with strongly basophilic vesicles in the cytoplasm, nucleus with peripheral circular nucleoli, follicle cells.

– *Stage 4*. Pre-maturation, ovary increases in size, fills most of the cephalothorax, varies in color from dark yellow to orange. Oöcytes present cytoplasm with acidophilic yolk granules and cortical crypts, nucleus begins to migrate to the periphery decreasing in size, follicle cells are still observed.

– *Stage 5*. Maturation, ovary fills the whole cephalothorax and the abdomen along the hindgut, colored from dark yellow to orange. In the oöcytes peripheral end of cortical crypts and the cytoplasmic membrane are coated by the vitellin envelope, nucleus shrinks and migrates towards the cytoplasmic membrane after ovulation, follicle cells disappear from around the oöcyte.

Insemination was considered as another indicator of female breeding activity. The area of the ventral cephalothorax with the seminal receptacles of females was dissected, and the presence or absence of sperm masses inside the seminal receptacles was observed with a stereomicroscope. Males were checked for presence or absence of fusion of gonopods as an indicator of male sexual maturity.

Measurements of surface water temperature were taken during each sampling episode. Pairwise rank correlation coefficients were calculated in order to correlate water temperature with percentage of adult females with mature ovaries.

3 RESULTS

Only 11 samples were obtained during the year, January and February 1996 samplings were missed due to mechanical failure of the vessel. A total of 1087 females and 877 males were collected over the year and measured for size-frequency distributions. Observations on reproductive condition were done on 279 females and on all males collected.

3.1   *Ovary weight – carapace length relationship*

Ovary weight was negligible (< 0.52 mg) in females ≤ 7.0 mm CL. In larger females, it was highly variable in all size classes, but generally increased with carapace length. A maximum ovary weight of 490 mg was recorded for a female of 23.9 mm CL (Fig. 2). The relationship between these variables can be described by the equation:

$$GW = 1.0^{-5} \cdot CL^{5.6194}$$

$$(n = 279, r^2 = 0.769, p < 0.0001)$$

3.2   *Molting*

Molting activity showed the same pattern for females and males throughout the year with two major peaks in May and September. Minor peaks were observed in December 95 and November 96. In July, no females and only two male specimens were captured (Fig. 3). Both females and males showed the highest molting frequency at similar carapace lengths between 6 to 12 mm CL (females), and 5 to 12 mm CL (males).

3.3   *Ovarian development*

Ovaries could be discerned in individuals ≥ 7 mm CL. Therefore, all smaller females were classed as undeveloped (stage 1).

Females with immature ovaries were found at higher frequency in the first months of sampling (November 1995 to April 1996). This pattern was also observed for males with unfused gonopods. Stage 2 females occurred regularly in all months. The same was observed for females with mature ovaries at stage 5, with the exception of November 1995, April 1996 (with only one mature female) and September 1996 (Fig. 4). Maximal numbers of males with fused gonopods were found from May to November 96 (Fig. 5).

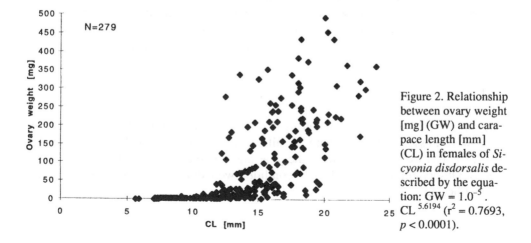

Figure 2. Relationship between ovary weight [mg] (GW) and carapace length [mm] (CL) in females of *Sicyonia disdorsalis* described by the equation: $GW = 1.0^{-5} \cdot CL^{5.6194}$ ($r^2 = 0.7693$, $p < 0.0001$).

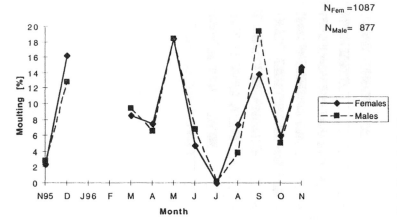

$N_{Fem} = 1087$

$N_{Male} = 877$

Figure 3. Percentage of molting activity of females and males of *Sicyonia disdorsalis* (based on the frequency of soft animals).

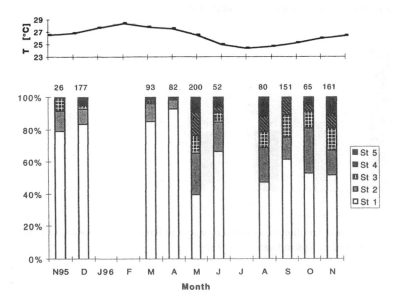

Figure 4. Temporal pattern of ovarian maturity in *Sicyonia disdorsalis*. Numbers on the top of individual bars indicate the sample size. Figure on the top shows surface water temperature T [°C] variations in the sampling area.

3.4 *Insemination*

Some 78.5% of all 279 analyzed females of *S. disdorsalis* had sperm masses inside the seminal receptacles. Insemination in females from stage 1 was negligible (5.5%) and increased in mature females to a maximum of 40.6% at stage 2 (Fig. 6).

Insemination in adult females followed a pattern similar to that found for ovarian maturity (Fig. 7). Inseminated females were found in all monthly samples with peaks occurring in May and August 1996, while the lowest proportion of inseminated females was found in April and September 1996, when females lacked ripe ovaries.

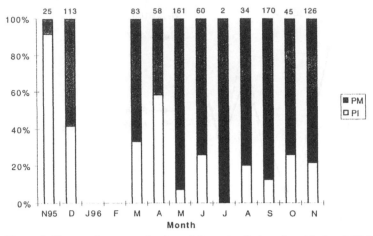

Figure 5. Temporal pattern of males of *Sicyonia disdorsalis* with fused (PM) and unfused gonopods (PI). Numbers on the top of individual bars indicate sample size.

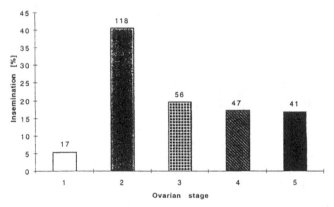

Figure 6. Incidence of insemination for all ovarian maturity stages in females of *Sicyonia disdorsalis*. Numbers on the top of individual bars indicate sample size.

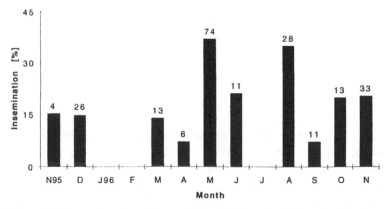

Figure 7. Temporal pattern of insemination in *Sicyonia disdorsalis*. Numbers on the top of individual bars indicate number of inseminated females.

3.5 *Correlation between temperature and breeding*

Surface water temperature varied between 24.4 and 25.0°C in the winter months (June to September) and between 26.8 and 28.5°C in summer (December to March). There was a significant correlation ($p < 0.05$, r = −0.578) between temperature and percentage of females with mature ovaries (stages 4 and 5).

4 DISCUSSION

4.1 *Molting*

Monthly molting activity for *S. disdorsalis* females and males displayed patterns varying with ovarian stage and season. Low winter molting frequencies from June to August were followed by higher monthly rates in autumn and spring. The bulk of females with soft cuticle was in developmental stage 2, at which the highest percentage of insemination was observed. Probably the high incidence and longer duration of this stage make these females more available for mating and insemination by one or more mature males.

From laboratory observations in other species of *Sicyonia* Bauer (1992a) reported repetitive copulation and polyandry of newly molted females. He suggested that the polyandry of females is an adaptive advantage for the offspring, as it may increase genetic diversity.

4.2 *Ovarian maturity*

Spawning, estimated by the presence of females with mature ovaries at stages 4 and 5, took place from May to November 1996. Obviously there is a marked seasonality of ovarian maturity and insemination from the last months of autumn to spring, although neither mature nor inseminated females were found in September. Anderson et al. (1985) presented similar results for *S. ingentis* on the coast off California, indicating a spawning season from June to October. For *S. parri* and *S. laevigata*, Bauer & Rivera Vega (1992) reported a continuous year-round reproduction. Extended reproductive seasons might be expected in tropical areas where high and relatively constant water temperatures prevail (Sastry 1983, Longhurst & Pauly 1987).

However, other factors such as precipitation, salinity, and river discharge should be considered in the area of this study, which are important for the spatial and temporal distribution of juveniles and mature specimens of another penaeid species (Jerí, 1991). The frequent occurrence (40.8%) of early development stage females probably relates to the extended duration of this stage. By contrast, females at stages 4 and 5 were less frequent, indicating a short time span between prematuration and maturation stages, during which spawning takes place. For *S. ingentis*, Anderson et al. (1984) reported a time span of approximately 90 hours between prematuration and spawning in the laboratory. If this time is applied to *S. disdorsalis*, each of these fourth or fifth stages would be indicative of spawning activity. Both stages could be used to provide temporal and spatial information on the spawning activity in a stock.

4.3   *Insemination*

The low percentage of insemination in April and September 1996 seemed to be related to the absence of mature females (stages 4 and 5) during these months. The reduction of insemination in September 1996 could be caused by a peak of molting, after which the mating plug and stored sperm are lost. As in other closed thelycum sicyoniids and some penaeoids, seminal receptacles are lined with cuticle and thus stored sperm is cast off when the female molts. Therefore, mating and insemination must take place again after a molt in order to replenish the store of sperm (Bauer 1991, Bauer 1992a, Bauer & Lin 1993, Yano 1995).

The high frequency of inseminated females at stages 2 and 3 reflects the reproductive mode of closed thelycum species, in which the sperm masses transfer takes place often at previtellogenic stages of the female. After several weeks, ovaries reach final maturation and eggs are released directly into the seawater where they are fertilized externally (Yano 1995). The low proportion of large inseminated females (> 20 mm CL) would be explained by the size differences between sexes (Lmax = 23.9 mm CL for females and 17.7 mm CL for males). Obviously mating selection by smaller males favors smaller females, as has been shown in other species of penaeid (Courtney et al. 1995, Courtney & Masel 1997).

4.4   *Correlation between temperature and breeding*

Several studies on penaeoid reproduction have indicated that variation of water temperature is an important factor influencing spawning (Dall et al. 1990, Bauer & Rivera Vega 1992, Crocos & van der Velde 1995). In this study, mass occurrence of females with mature ovaries followed increasing water temperatures from August (winter) to November (spring). The negative correlation can be explained by the major peaks of spawning females in the last months of autumn (May and June) and August, in which last year's recruits probably reached maturity together with the current mature females, as result of a extended reproductive season from spring to summer. An aspect, that reflects the long spawning time of the adults are the numerous juveniles of both females and males in summer months until the beginning of autumn.

Bauer (1992b) reported a continuous breeding and an episodic recruitment in two sicyoniids of tropical regions. In my study, the presence of mature females and males in the mentioned months was principally related with rainy months and the consequently increase of the Tumbes River discharges followed by a decrease of salinity, which probably has a clearer effect than the temperature on the reproduction of this species.

In this one-year-study my aim was to obtain a first overview of the temporal distribution of breeding activity of *S. disdorsalis* including its reproductive aspects. A exact description of the synchronous molting and breeding activity deduced from sexual maturity, ovarian development and insemination of females can not be given at the present stage of this study. The complex life history of penaeoids and the high variability of this tropical system make longer periods of sampling necessary, in order to establish a model and to compare it with species of the same taxa at different sites and latitudes.

ACKNOWLEDGEMENTS

This study was funded by the Alfred Wegener Institut. I thank Juan Cabanillas and the fishermen of Caleta Grau for assisting with the fieldwork, and Mrs Elisabeth Schwarz of the ELMI Center, University Rostock for her assistance with histological treatments. Comments and suggestions of anonymous reviewers of the manuscripts are gratefully acknowledged. This is contribution No. 1524 of the Alfred Wegener Institute.

REFERENCES

Anderson, S.L., Chang E.S., & Clark, W.H. 1984. Timing of postvitellogenic ovarian changes in the ridgeback prawn *Sicyionia ingentis* (Penaeidae) determined by ovarian biopsy. *Aquaculture, Amsterdam* 42: 257-271.

Anderson, S.L., Clark, W.H., & Chang E.S. 1985. Multiple spawning and molt synchrony in a free spawning shrimp (*Sicyonia ingentis*: Penaeoidea). *Biol. Bull.* 168: 377-394.

Bauer, R.T. 1991. Sperm transfer and storage structures in Penaeoid shrimps: A functional and phylogenetic perspective. In R.T. Bauer & J.W. Martin (eds), *Crustacean Sexual Biology*: 183-207. New York: Columbia University Press.

Bauer, R.T. 1992a. Repetitive copulation and variable success of insemination in the marine shrimp *Sicyonia dorsalis* (Decapoda: Penaeoidea). *J. Crust. Biol.* 12: 153-160.

Bauer, R.T. 1992b. Testing generalizations about latitudinal variation in reproduction and recruitment patterns with sicyoniid and caridean shrimp species. *Invert. Reprod. Dev.* 22: 193-202.

Bauer, R.T. 1994. Usage of the terms *Thelycum* and *Spermatheca* in the reproductive morphology of the decapoda, with special reference to the Penaeoidea. *J. Crust. Biol.* 14: 715-721.

Bauer, R.T.. & Rivera Vega, L.W. 1992. Pattern of reproduction and recruitment in two sicyoniid shrimp species (Decapoda: Penaeoidea) from a tropical sea grass habitat. *J. Exp. Mar. Biol. Ecol.* 161: 223-240.

Bauer, R.T. & Jun Min Lin 1993. Spermatophores and plug substance of the marine shrimp *Trachypenaeus similis* (Crustacea: Decapoda: Penaeidae): Formation in the male reproductive tract and disposition in the inseminated female. *Biol. Bull.* 185: 174-185.

Courtney, A.J. & Masel, J.M. 1997. Spawning stock dynamics of two penaeid prawns, *Metapenaeus bennettae* and *Penaeus esculentus*, in Moreton Bay, Queensland, Australia. *Mar. Ecol. Prog. Ser.* 148: 37-47.

Courtney, A.J., Montgomery, S.S., Die, D.J., Andrew, N.L., Cosgrove, M.G. & Blount, C. 1995. Maturation in the female eastern king prawn *Penaeus plebejus* from coastal waters of eastern Australia, and consideration for quantifying egg production in penaeid prawns. *Mar. Biol.* 122: 547-556.

Crocos, P.J. & van der Velde, T.D. 1995. Seasonal, spatial and interannual variability in the reproductive dynamics of the grooved tiger prawn *Penaeus semisulcatus* in Albatros Bay, Gulf of Carpentaria, Australia: the concept of effective spawning. *Mar. Biol.* 122: 557-570.

Dall, W., Hill, B.J., Rothlisberg, P.C. & Sharples, D.J. 1990. The Biology of the Penaeidae. *Adv. Mar. Biol.* 27: 1-489.

Garcia, S. & le Reste, L. 1981. Life cycles, dynamics, exploitation and management of coastal penaeid shrimp stocks. *FAO Fish. Tech. Pap.* 203: 1-215.

Hendrickx, M.E. 1996. *Los camarones penaeoidea bentónicos (Crustacea: Decapoda: Dendrobranchiata) del Pacífico mexicano. Comisión Nacional para el Conocimiento y Uso de la Diversidad (CONABIO), Instituto de Ciencias del Mar y Limnología*. Mexico: Universidad Nacional Autónoma de México.

Jerí, T. 1991. *Estudio de la pesquería y algunos aspectos biológicos del langostino (Fam. Penaeidae) frente a Tumbes, Perú. Tesis Licenciado en Biología*. Lima: Universidad Ricardo Palma.

Longhurst, A.R. & Pauly, D. 1987. *Ecology of Tropical Oceans.* New York: Academic Press.

Sastry, A.N. 1983. Ecological aspects of reproduction. In J.J. Vernberg & W.B. Vernberg (eds), *The Biology of Crustacea, vol. 8*: 179-270. New York: Academic Press.

Yano, I. 1988. Oöcyte development in the kuruma prawn *Penaeus japonicus. Mar. Biol.* 99: 547-553.

Yano, I. 1995. Final oöcyte maturation, spawning and mating in penaeid shrimp. *J. Exp. Mar. Biol. Ecol.* 193: 113-118.

On the reproduction of *Nephrops norvegicus* in the southern Adriatic Sea: sex ratio, maturity lengths and potential fecundity

NICOLA UNGARO, CHIARA ALESSANDRA MARANO, ROBERTA MARSAN
& ANNA MARIA PASTORELLI
Laboratorio di Biologia Marina, Bari, Italy

ABSTRACT

Reproductive aspects of Norway lobster were studied from ten seasonal trawl surveys (1991-1996) carried out in the southern Adriatic Sea. Species carapace lengths ranged from 8 to 66 mm (males from 12 to 66 mm; females from 8 to 60 mm). The overall sex ratio value (all surveys) was 1:1; on the other hand the same ratio value per carapace length was significantly lower than expected 1:1 in smaller sizes (CL < 20 mm) and significantly higher in larger sizes (CL > 40 mm). This last result was probably linked to different growth by sex. Female carapace length at first maturity (50%) was 27.5 mm (the smallest mature specimen measured 20 mm CL) while the potential fecundity ranged from 105 to 3716 eggs (carapace length range = 22-50 mm); analysis of relative propodus growth gave a value of 40 mm CL for male maturity. Environmental features could affect the above mentioned estimates, as other Authors suggested it. The reported information about reproductive biology could be useful for the management of the Norway lobster fishery in the study area, although methods of study could affect the results.

1 INTRODUCTION

Nephrops norvegicus represents one of the most valuable, demersal, fishery resources in the central Mediterranean and particularly in the Adriatic Sea (Stamatopoulos 1995). The species represents more than 30-40% of the total crustacean catch in the southern Adriatic (Pastorelli et al. 1996).

Scientific information about the species biology in the above mentioned area is scarce (Marano et al. 1998), in spite of the extensive literature from other geographic zones (see below). Herein, some aspects about Norway lobster reproduction were analyzed in order to improve the knowledge of species biology in the investigated area.

2 MATERIALS AND METHODS

Biological data were collected during ten trawl surveys (1991-1996) carried out in the southwestern Adriatic Sea (39°40'-42°30' lat. N; 15°20'-19°10' long. E). Specifically data came from 353 hauls – 151 'spring' hauls ('91, '92, '93, '94, '95 surveys), 202 'autumn' hauls ('91, '92, '94, '95, '96 surveys). The sampling gear was an 'Italian' trawl net (cod-end stretched mesh = 36 mm) while the sampling design was random stratified (five bathymetric strata: 10-50 m, 51-100 m, 101-200 m, 201-450 m, 451-800 m) (AA.VV. 1993, AA.VV. 1996). The sampling periods coincide with the referenced reproductive season for the species in the Mediterranean Sea (from spring to autumn) (Falciai & Minervini 1992).

The collected specimens were analyzed using the following measurements (mm) taken by a vernier calliper (Fig. 1): 1. Carapace length per sex (from eye socket to mid-posterior margin of carapace; measure code = CL, mm); 2. Length of right (crusher) and left (cutter) male propodus (first pair of pereiopods; the males with both the chelipedes were measured only, excluding anomalous handedness and regeneration episodes. Measure code = PL); 3. Width of right (crusher) and left (cutter) male propodus (measure code = PW); 4. Thickness of right (crusher) and left (cutter) male propodus (measure code = PT); 5. Sexual maturity stage for female specimens by external observation of the gonads following a modified version of the color method by Farmer (1975) (Immature = white – cream; mature = green; ovigerous = external eggs); 6. Total number of external eggs per ovigerous female (alcohol preserved sub samples; the newly laid eggs were counted only).

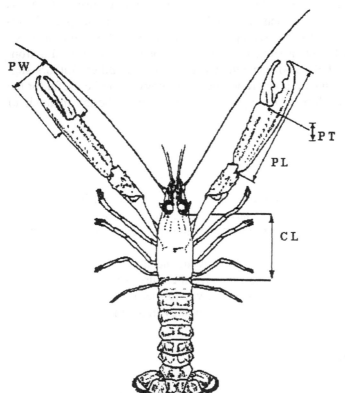

Figure 1. *Nephrops norvegicus*: biometric measures (mm). CL= Carapace Length; PL = Propodus Length; PT = Propodus Thickness; PW = Propodus Width.

The raw data were processed to obtain the following results: 1. Length frequency distributions per sex and surveys; 2. Sex ratio values per survey and per carapace length interval (2 mm) (statistically tested by means of χ^2 test with respect to the expected 1:1 ratio); 3. Female sexual maturity percentages per carapace length; 4. Female length at first maturity (50%) by fitting maturity curve (logistic model); 5. Eggs number *versus* carapace length; 6. Relative growth of propodus by using regression analysis (STATISTICA software package) in order to compute sexual maturity length for male specimens. Specimen dissection was not undertaken and the male maturity stages cannot be identified by external observation of the gonads (Hartnoll 1978). This kind of analysis was also utilized by other authors for the same species in different geographic areas (Farmer 1975, Sardá et al. 1981, Biagi et al. 1990a). In this paper, data were processed by means of 'piecewise regression analysis,' but other methods are also available (Somerton 1980).

The mentioned methods (different approaches for male and female) were used in order to minimize the research time and costs, rather than using the more accurate method of histological analysis.

3 RESULTS

A total of 9946 individuals were caught during ten seasonal surveys in the southwestern Adriatic Sea, from thirty-five to 730 m depths; the species was found in 158 sampling hauls (44.8% of samples) (Fig. 2).

The length frequency distributions per survey are shown in Figure 3. The whole stock carapace lengths range from 8 to 66 mm (mostly between 22 and 36 mm) while male specimens range from 12 to 66 mm (mean CL = 28.8 mm ± 7.9) and female specimens from 8 to 60 mm (mean CL = 25.6 mm ± 5.6).

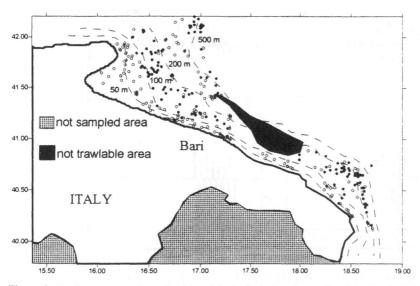

Figure 2. *Nephrops norvegicus*. Positive (black circles) and negative (empty circles) samples in the investigated area.

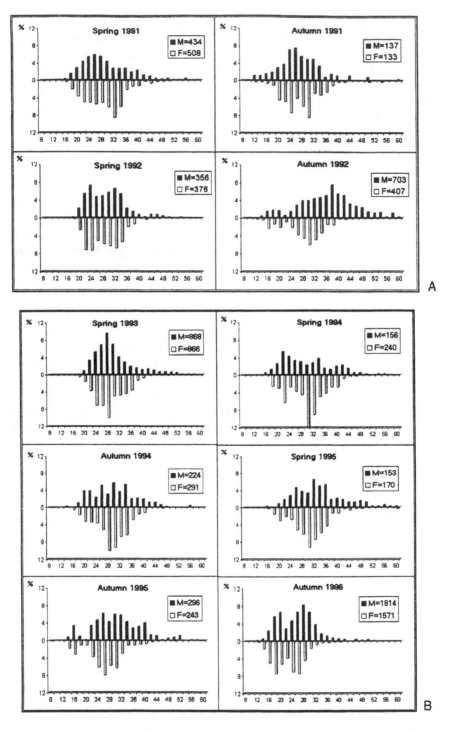

Figure 3. *Nephrops norvegicus*. A. Carapace length frequency distributions (CL, mm) in the southern Adriatic Sea (1991-1992 surveys) (M = male, F = female); B. Carapace length frequency distributions (CL, mm) in South Adriatic Sea (1993-1996 surveys) (M = male, F = female).

Sex ratio (m:f) values per survey range from 0.75 (spring 1994 survey) to 1.73 (autumn 1992 survey). The same values were significantly biased (expected value = 1; χ^2 test, $P < 0.05$) towards the males in the surveys autumn 1992 and autumn 1996, while the ratios were biased towards the females in the surveys spring 1994 and autumn 1994. The sex ratio value referring to the total number of collected specimens (all surveys) was 1:1. Moreover sex-ratio values per carapace length (2 mm interval) were computed, referring to the total number of collected specimens (all surveys) and to the pooled spring surveys (4127 specimens) and autumn ones (5819 specimens). Figure 4 (all surveys) highlights the predominance of females in smaller sizes and of males in larger sizes, and similar results are found in spring and autumn length distributions (Fig. 5).

Female maturity stages results are shown in Figure 6; the smallest mature specimen measured 20 mm CL. Maturity curve fit (females with green ovaries and ovigerous ones) is in Figure 7, while back-calculation of length at first maturity (50%)

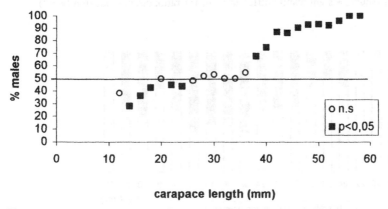

Figure 4. *Nephrops norvegicus.* Sex ratio (% of males) per carapace length (total number of collected specimens) (square = significantly different from 1:1 ratio; circle = not significant).

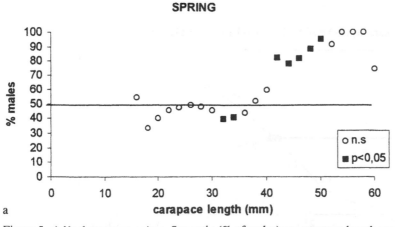

Figure 5. a) *Nephrops norvegicus.* Sex ratio (% of males) per carapace length per spring pooled survey (square = significantly different from 1:1 ratio; circle = not significant).

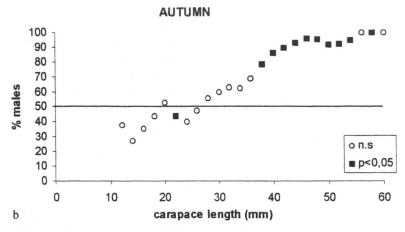

b   **carapace length (mm)**

Figure 5. Continued. b) *Nephrops norvegicus*. Sex ratio (% of males) per carapace length per autumn pooled survey (square = significantly different from 1:1 ratio; circle = not significant).

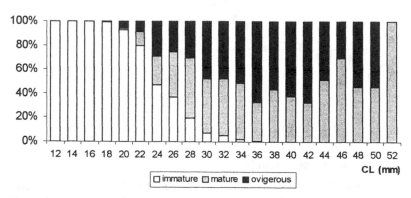

Figure 6. *Nephrops norvegicus*. Maturity percentage per carapace length (females).

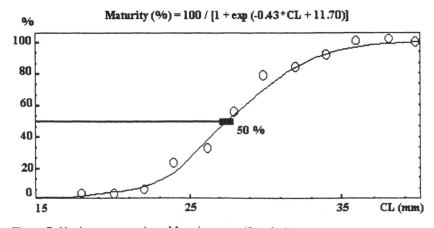

Figure 7. *Nephrops norvegicus*. Maturity curve (females).

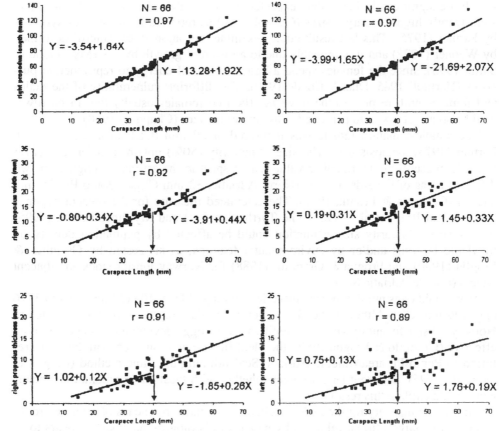

Figure 8. *Nephrops norvegicus*. Right (crusher) and left (cutter) chelipedes biometric relationships (→ break point).

by means of logistic equation parameters, gave the value of 27.5 mm (CL). The estimated value was 31.8 mm CL if we consider the ovigerous females only.

Potential fecundity (newly laid eggs total number per carapace length) was measured for a sub-sample (from 22 to 50 mm CL) of well preserved ovigerous females (no. 47). The same fecundity ranged from 105 (22 mm CL) to 3716 eggs (49 mm CL) (mean fecundity referred to the mentioned length range = 975 eggs ± 675).

Male propodus relative growth, computed by means of piecewise regression, gave the same 'breakpoint' value at 40 mm (CL) (smaller sizes = immature specimens; larger sizes = mature specimens) for all the variables (carapace length *versus* propodus length, propodus width and propodus thickness) (Fig. 8).

4 DISCUSSION AND CONCLUSIONS

The overall (total number of specimens) sex-ratio value was 1:1, very close to the results from other Mediterranean areas such as in Greek waters (Mytilinou et al. 1995) and the Ligurian Sea (Orsi Relini & Relini 1985). Sex ratio values per carapace length are significantly lower than expected 1:1 in smaller sizes (CL < 20 mm) and significantly higher in larger sizes (CL > 40 mm) as reported in the species synopsis by Farmer (1975). This last result could resemble the 'anomalous pattern' described by Wenner (1972) and it is probably due to a different growth by sex (Mytilineou & Sardá 1995), since the females spend a large amount of energy to reproductive purposes (Hartnoll 1985, Díaz & Conde 1989). The differing vulnerability of the males and females with respect to the trawl net (berried females usually live in burrows) could also affect sex ratio per length class in larger sizes (Chapman 1980).

The smallest mature female was measured at 20 mm (CL), in agreement with Farmer (1975) synopsis data. The size at maturity (50%) obtained to females (27.5 mm CL) presents an intermediate value with respect to the estimated lengths for two different areas of the Adriatic Sea (North Adriatic, 32 mm $CL_{50\%}$; Pomo Pit, 26 mm $CL_{50\%}$) (Gramitto & Froglia 1980). The calculated measure for the present paper is smaller than Biagi et al. (1990b) (the Tyrrhenian Sea) and Mytilineou et al. (1995) (Greek seas). Maturity size estimation could be affected by population density on trawlable bottoms and/or by environmental features, as suggested by Gramitto & Froglia (1980) and Froglia & Gramitto (1988) for Norway lobster stock in adjacent waters (middle Adriatic Sea).

Newly laid eggs number per carapace length ranged from 105 (22 mm CL) to 3716 eggs (49 mm CL). Although the above mentioned data represent rough estimates of Norway lobster fecundity and the identification of eggs development stages is more effective (Gramitto & Froglia 1980), the identified range is smaller than the result referred to Pomo Pit area (middle Adriatic Sea) utilizing the same method (Froglia & Gramitto 1995). Different length ranges together with local environmental characteristics could explain this result.

The size at maturity for the male specimens, estimated by means of relative propodus growth analysis (to be validated by means of histological analysis), appears to be larger than Farmer (1975) Norway lobster synopsis data for Atlantic population. However, it is quite similar to data from Sardá et al. (1981) and Biagi et al. (1990c) for other Mediterranean areas. This last result could confirm the influence of environmental parameters (mostly water temperature) in differential growth and maturity of Norway lobster between the Atlantic and Mediterranean populations (Froglia & Gramitto 1988).

The reported results concerning reproductive biology, although they could be improved upon, could be useful for the management of the Norway lobster fishery in the study area. Concerning the regulation of the fishery remember that the species minimum legal size in Italian and Mediterranean waters is 20 mm carapace length.

ACKNOWLEDGEMENTS

We wish to thank anonymous reviewers for valuable criticism and helpful suggestions.

REFERENCES

AA.VV. 1993. Valutazione delle Risorse Demersali nell'Adriatico meridionale dal Promontorio del Gargano al Capo d'Otranto. Relazione finale triennio '90-'93. Rome: Min. Mar. Mer.

AA.VV. 1996. Valutazione delle Risorse Demersali nell'Adriatico meridionale dal Promontorio del Gargano al Capo d'Otranto. Relazione finale triennio '94-'96. Rome: M.R.A.A.F.

Biagi, F., de Ranieri, S. & Mori, M. 1990a. Relative growth of the Crusher Propodite of Nephrops norvegicus (L.) in the Northern Tyrrhenian Sea. Rapp. Comm. Int. Mer Médit. 32: 33.

Biagi, F., de Ranieri, S. & Mori, M. 1990b. Reproductive biology of the females of Nephrops norvegicus in the Northern Tyrrhenian Sea. Rapp. Comm. Int. Mer Médit. 32: 33.

Biagi, F., de Ranieri, S. & Mori, M. 1990c. Size at sexual maturity for males of Nephrops norvegicus (L.) in the Northern Tyrrhenian Sea. Rapp. Comm. Int. Mer Médit. 32: 32.

Chapman, C.J. 1980. Ecology of juvenile and adult Nephrops. In Cobb J.S. & Phillips B.F. (ed.), *The biology and management of lobster* (2): 143-178. New York: Academic Press.

Díaz, H. & Conde, J.E. 1989. Population dynamics and life history of the Mangrove crab Aratus Pisonii (Brachyura, Grapsidae) in a marine environment. *Bull. Mar. Sci.* 45: 148-163.

Falciai, L. & Minervini, R. 1992. Guida dei Crostacei decapodi d'Europa. Padova: Franco Muzzio.

Farmer, A.S.D. 1975. Synopsis of biological data on the Norway lobster. Nephrops norvegicus (Linnaeus, 1758). FAO Fish. Synop. 112: 1-97.

Froglia, C. & Gramitto, M.E. 1988. An estimate of growth and mortality parameters for Norway lobster Nephrops norvegicus in the central Adriatic Sea. FAO Fish. Rep. 394: 189-203.

Froglia, C. & Gramitto, M.E. 1995. Crustacea Decapoda Assemblage of the western Pomo Pit. II – Reproduction. Rapp. Comm. Int. Mer. Medit., 34: 1-30.

Gramitto, M.E. & Froglia C. 1980. Osservazioni sul potenziale riproduttivo dello scampo Nephrops norvegicus (L.) in Adriatico. *Mem. Biol. Marina e Oceanog.* suppl. 10: 213-218.

Hartnoll, R.G. 1978. The determination of relative growth in crustacea. *Crustaceana* 34: 281-293.

Hartnoll, R.G. 1985. Growth, sexual maturity and reproductive output. *Crust. Issues* 3: 101-129.

Mytilineou, C.H., Fourtouni, A. & Papaconstantinou, C. 1995. Preliminary study on the biology of Norway lobster Nephrops norvegicus, in the Gulf of Chalkidiki (Greece). Rapp. Comm. Int. Mer. Medit. 34: 1-38.

Mytilineou, C.H. & Sardá F. 1995. Age and growth of Nephrops norvegicus in the Catalan Sea, using length-frequency analysis. Fish. Res. 23: 283-299.

Marano, G., Marsan, R., Pastorelli, A.M. & Vaccarella, R. 1998. Areale di distribuzione e pesca dello scampo, Nephrops norvegicus (L.), nelle acque del Basso Adriatico. *Biol. Mar. Med.* 5: 284-292.

Orsi Relini, L. & Relini, G. 1985. Notes on the distribution, reproductive biology and fecundity of Nephrops norvegicus in the Ligurian Sea. FAO Fish. Rep. 336: 107-111.

Pastorelli, A.M., Vaccarella, R., Marano, G., & Ungaro, N. 1996. I crostacei dei fondi strascicabili del Basso Adriatico. *Nova Thalassia* 12: 27-35.

Sardá, F., Miralles, L.M. & Palomera, I. 1981. Morfometrìa de Nephrops norvegicus (L.) del mar catalàn (NE. de España). *Inv. Pesq.* 45: 279-290.

Somerton, D.A. 1980. A computer technique for estimating the size of sexual maturity in crabs. *Can. J. Fish. Aquat. Sci.* 37: 1488-1494.

Stamatopoulos, C. 1995. Trends in catches and landings. Mediterranean and Black Sea fisheries: 1972-1992. FAO Fish. Circ. 855.4 Suppl.: 1-53.

Wenner, A.M. 1972. Sex ratio as a function of size in marine Crustacea. *Amer. Nat.* 106: 321-350.

5 Larvae

Taxonomy of solenocerid larvae and distribution of larval phases of *Pleoticus muelleri* (Bate, 1888) (Decapoda: Solenoceridae) on the southern Brazilian coast

DANILO CALAZANS

Departamento de Oceanografia – FURG, Rio Grande, Brazil

ABSTRACT

Larval phases of three solenocerids, *Pleoticus muelleri, Solenocera necopina* and (?) *Mesopenaeus tropicalis* were taken from quantitative oblique samples between August 1982 and 1983 in the southern Brazilian coast. Morphological features of larval stages of these species were observed and compared. Abundance of *P. muelleri* were analysed and results showed that this species has an annual stock of mature adults spawning surrounding the Patos Lagoon mouth. A high abundance of 147,7 protozoea larvae per 100 m^3 (67% of total larvae capture) during autumn indicates this season as the most important to larval hatch. Despite the number of protozoea larvae during winter, the low number of megalopae denotes high mortality during this season. Spring and summer have low density of protozoeae but also low mortality suggesting these two seasons as more important than winter to recruitment.

1 INTRODUCTION

Solenoceridae is represented by the genera *Gordonella, Haliporus, Solenocera, Pleoticus, Mesopenaeus, Haliporoides, Cryptopenaeus, Hadropenaeus* and *Hymenopenaeus*. Three species occur in southern Brazilian coastal waters: Pleoticus muelleri (Bate 1888), *Solenocera necopina* Burkenroad, 1939 and *Mesopenaeus tropicalis* (Bouvier 1905) (Pérez-Farfante & Kensley 1997). *Pleoticus muelleri* is present from 5 to 600 m usually below 100 m and is the most commercially important shrimp in Argentinean waters (Wyngaard & Bertuche 1982) with average annual landings of 11,500 t during the 1981-1988 period (Boschi 1989). *Solenocera necopina* has been reported by Pérez-Farfante and Bullis (1973) to the southern Brazilian coast at depths between 90 and 550 m usually below 180 m. *Mesopenaeus tropicalis* is distributed at depths between 30 and 915 m, usually below 190 m (Pérez-Farfante 1977).

Very few contributions about larval taxonomy of solenocerids are known from plankton samples. Iorio et al. (1990) described the complete larval development of *Pleoticus muelleri* under laboratory condition. Heldt (1955) described *Solenocera membranacea* larval stages from plankton samples taken in the Mediterranean.

Heegaard (1966) described, under the name *S. muelleri*, two mysis stages of *Solenocera* from plankton samples taken offshore from the central Brazilian coast. There is no description for *M. tropicalis* larvae. The few studies on larval distribution of *P. muelleri* are reported at Desterro (Santa Catarina State, Brazil) (Müller 1863); offshore Mar del Plata, Argentina (Boschi & Scelzo 1969; Boschi 1989) and Baia Blanca, Argentine (Mallo & Celvellini 1988).

This paper describes and compares aspects of protozoea and mysis phases of these 3 species and megalopa of *P. muelleri* taken from plankton samples off southern Brazil. Abundance and distribution of *P. muelleri* are also presented and discussed.

2 MATERIAL AND METHODS

The study was conducted between 30 miles south and 30 miles north of the mouth of Patos Lagoon in southern Brazil. Zooplankton samples were taken during 4 seasonal surveys, using R/V 'Atlântico Sul', with oblique hauls from one meter above the bottom to surface with a conical 330 µm mesh size net, 60 cm mouth diameter and 250 cm length, with flowmeter centrally positioned in the mouth. Further details about the study area and sampling can be found in Calazans (1994).

All solenocerid larvae were sorted and removed from the zooplankton samples using a Nikon MZ 10 microscope. Drawings were made using an Olympus BH-2 microscope with Nomarski interference contrast and camera lucida. At least 10 specimens of each larval stages were observed.

Larval abundance of *P. muelleri* (x) was standardized as number per 100 m^3 of filtered water. Since number of larvae may vary between individual tows from absent to extremely abundant, a geometric progression of base 3 was used to construct maps of distribution to allow rapid interpretation of the abundance categories (Calazans 1994). Homeocedasticity of larval abundance was obtained by the transformation log (x+1) (Sokal & Rohlf 1981) and three-way analysis of variance (ANOVA) was used to investigate the effect of 1. Season; 2. Depth and 3. Position of sampling point to the north or south of the lagoon's mouth. Correlation analysis was used to assess the effect of temperature and salinity on the abundance of the larval phases.

3 RESULTS

3.1 *General larval morphology of solenocerids*

All described solenocerids have 5-6 nauplii, 3 protozoeae and only 2 mysis stages, and are more abbreviated in comparison with other penaeoideans. Gross morphological features common to all solenocerids are: 1. Spines and/or setae on surface of carapace; 2. Spines and/or processes on edge of carapace; 3. Stout spines and short setae on abdominal somites; and, 4. Dorsal organs present.

3.2 *External morphological features*

Pleoticus muelleri: The antennal endopod of the protozoea phase is 2-segmented

(Fig. 1D). First segment with 2 setae on middle and 3 setae on distal inner margin; second segment with 5 terminal setae.

– *Protozoea I* (Fig. 1A). Total length = 0.87 mm. Anterior portion of carapace with pair of frontal spines. Posterior portion of carapace with dorsal process bearing 2 short spines on edge. Caudal furca narrow, bearing 7 + 7 setae.

– *Protozoea II* (Fig. 1B). Total length = 1.72 mm. Rostrum smooth, ventrally curved. Carapace with supraorbital spine well developed on anterior region. Laterally carapace with antennal, pterygostomial, latero-branchial spines and branchial process and posteriorly with postbranchial and epi-intestinal processes. Caudal furca narrow, bearing 7 + 7 setae.

– *Protozoea III* (Fig. 1C). Total length = 2.72 mm. Posterior portion of carapace with a postbranchial spine smooth and one epi-intestinal process centrally. First 5 abdominal somites with lateral and dorsal spines shorter than somites. Sixth somite bearing small lateral and ventro-lateral spines.

– *Mysis I* (Fig. 2A) Total length = 4.93 mm. Rostrum smooth almost 3 times longer than eyestalk. Posterior portion of carapace with smooth postbranchial, epi-intestinal and laterobranchial spines. First 5 abdominal somites with dorsal spine almost the same length of somite. Sixth somite with dorsal spine bigger than previous somites. Telson (Fig. 2H) fork shaped with spinules irregularly arranged around the edge, bearing 1 + 2 lateral spines on outer margin, 4 pairs of setae on inner margin and 2 rows of paired setae present on dorsal surface.

– *Mysis II* (Fig. 2D). Total length = 7.15 mm. Rostrum almost 5 times longer than eyestalk.

– *Megalopa* (Fig. 2G). Total length = 6.67 mm. With remarkable changes in comparison with previous phase. Rostrum shorter than in mysis phase, almost same size as eyestalk and bearing 8 spines. Anterior and posterior dorsal organs absent. Lateral and posterolateral margins with row of small setae. First 5 abdominal somites without dorsal spines. Telson triangular in shape with pointed end.

Solenocera necopina: The antennal endopod of the protozoea phase is 2-segmented (Fig. 1G). First segment with 2 setae on middle and 2 setae on distal inner margin; second segment with 5 terminal setae.

– *Protozoea II* (Fig. 1E): Total length = 2.51 mm. Rostrum spinulate, curved downwards, of same size as eyestalk. Laterally carapace with antennal, pterygostomial and branchial processes and posteriorly with postbranchial and epi-intestinal processes. Group of spines present between latero-branchial and postbranchial processes. Caudal furca wide bearing 7 + 7 spines.

– *Protozoea III* (Fig. 1F): Total length = 4.54 mm. Posterior portion of carapace with postbranchial spine with serrules and two epi-intestinal processes centrally. First 4 abdominal somites with dorsal spine of same length as somite and more prominent on the fifth somite. Sixth somite bearing median size lateral and ventro-lateral spines. Caudal furca wider than in protozoea II with 8 + 8 spines.

– *Mysis I* (Fig. 2B). Total length = 8.16 mm. Rostrum spinulate, almost 3 times longer than eyestalk. Posterior portion of carapace with laterobranchial, postbranchial and epi-intestinal processes. Fifth and sixth abdominal somites with pair of long pappose setae on posterior dorsal portion. Telson (Fig. 2I) fork shaped with spinules irregularly arranged around the edge, and bearing 1 + 1 + 1 lateral spines on outer margin, 4 setae on inner margin and 3 pairs of setae present on dorsal surface.

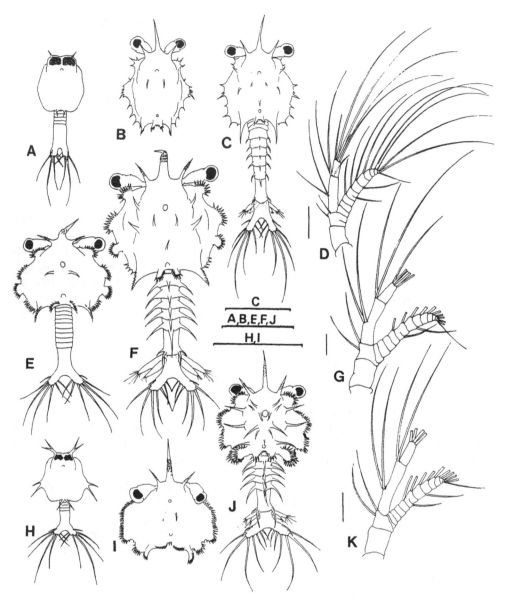

Figure 1. *Pleoticus muelleri*: A-C, protozoea I-III, D, antenna; *Solenocera necopina*: E-F, protozoea II-III, G, antenna; (?)*Mesopenaeus tropicalis*: H-J, protozoea I-III, K, antenna. Scales: A-C, E-F, H-J = 0,5 mm; D, G, K = 0,2 mm.

– *Mysis II* (Fig. 2E): Total length = 12.28 mm. Lateral and dorsal regions of carapace bearing considerable number of small spines. Fourth abdominal somite bearing pair of long pappose setae on dorsal portion in addition to those on fifth and sixth somites. Dorsal spine on second and third somites as hook-like structures pointed upwards. Telson still fork shaped bearing 10 pairs of setae on dorsal surface.

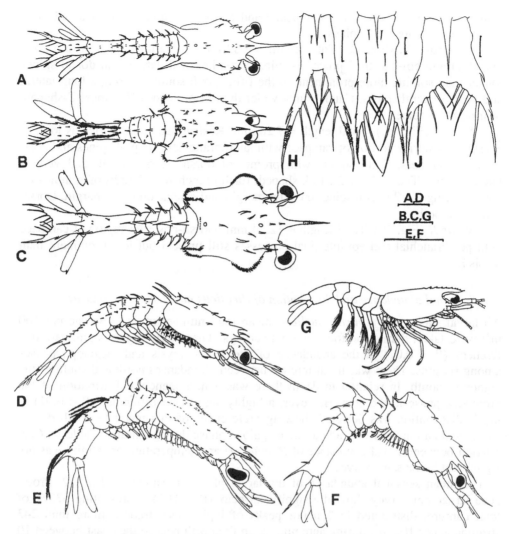

Figure 2. *Pleoticus muelleri*: A, mysis I, D, mysis II, G, megalopa, H, Telson; *Solenocera necopina*: B, mysis I, E, mysis II, I, Telson; (?) *Mesopenaeus tropicalis*: C, mysis I, F, mysis II, J, Telson. Scale: A-G = 1,0 mm; H-J = 0,2 mm

Mesopenaeus tropicalis: The antennal endopod of protozoea phase 2-segmented (Fig. 1K). First segment with 2 setae on middle and 2 setae on distal inner margin; second segment with 4 terminal setae.

– *Protozoea I* (Fig. 1H): Total length = 0.99 mm. Anterior portion with pair of bifurcated frontal spines. Posterior portion of carapace with dorsal process bearing 2 long spines on edge. Caudal furca very wide, bearing 7 + 7 setae.

– *Protozoea II* (Fig. 1I): Total length = 2.30 mm. Rostrum spinulate, curved downwards, longer than eyestalk. Carapace with spinulated supraorbital spine well developed on anterior region. Edge of carapace serrulated from anterior to posterior portions. Posteriorly carapace has postbranchial and epi-intestinal processes; group of spines also present between postbranchial and epi-intestinal processes. Caudal furca very wide, bearing 7 + 7 spines.

– *Protozoea III* (Fig. 1J): Total length = 3.82 mm. Posterior portion with a post-branchial and epi-intertinal processes. First 5 somites with lateral and dorsal spines longer than somite, more prominent on the fifth. Sixth somite bearing a long lateral and ventro-lateral spines. Caudal furca wider than in protozoea II, almost T-shaped, bearing 8 + 8 setae.

– *Mysis I* (Fig. 2C). Total length = 7.26 mm. Rostrum spinulate, twice as long as eyestalk. Posterior portion of carapace with laterobranchial group of spines and post-branchial process. Sixth somite with prominent dorsal, dorso-lateral, and ventro-lateral spines. Telson (Fig. 2J) fork-shaped. Each branch with 2 rows of spinules on edge, bearing 1 + 2 lateral spines on outer margin and 4 setae on inner margin. Dorsal surface smooth.

– *Mysis II* (Fig. 2F). Total length = 12.24 mm. Dorso-posterior region of carapace with postbranchial and epi-intestinal processes still present but less prominent than mysis I.

3.3 *Seasonal abundance and distribution of Pleoticus muelleri larval phases*

All larval stages of *Pleoticus muelleri* (mean total number of 218.4 larvae per 100 m^3) were taken during the four seasonal surveys. There were highly significant differences (P < 0.01) in the abundance of protozoea, mysis and megalopa phases among seasons. There was no difference between abundances north and south of the lagoon's mouth. In relation to depth there was a homogeneous distribution of the protozoea phase in the area. However, a highly significant difference (P < 0.01) in mysis distribution was detected showing preferences of this phase to depth during spring and autumn seasons, and to the megalopa phase during autumn (Table 1). Correlation between larval abundance of *P. muelleri* and temperature or salinity was not significant in the same survey.

The mean seasonal abundance of the larval phases is shown in Table 2. Proto-zoea larvae per survey showed a significant peak (P < 0.01), representing 67.6% of total capture, distributed in 2 major peaks of high concentration (more than 243 specimens per 100 m^3) during autumn, one in the north near to the coast between 10 and 20 m and another in the south located between 30 and 60 m (Fig. 3). Winter with 9.6% of total capture and spring with 5.9% represents also an important concentration of protozoea (Table 2) with small peaks near to the coast but with more homogeneous distributions. During summer a small number of protozoeae were present between 10 and 30 m located in the north (Fig. 3).

Table 1. Summary of the results of ANOVA on the effect of season, depth and position of sampling point to the north or south of the lagoon mouth on larval abundance (transformed data) of *Pleoticus muelleri*.

Phase	Season	Depth	Position
Protozoea	*	n.s.	n.s.
Mysis	*	*	n.s.
Megalopa	*	*	n.s.

* = P < 0.01; n.s. = not significant.

Table 2. Seasonal abundance (mean number per 100 m^3) and percentage of the total species catch of *Pleoticus muelleri* larvae.

	Winter		Spring		Summer		Autumn	
	n	%	n	%	n	%	n	%
Protozoea	21.0	9.6	12.9	5.9	0.6	0.3	147.7	67.6
Mysis	1.2	0.6	11.1	5.1	0.7	0.3	20.0	9.2
Megalopa	0.04	0.0	0.6	0.3	0.4	0.2	2.2	1.0
Total	22.2	10.2	25	11.3	1.7	0.8	169.9	77.8

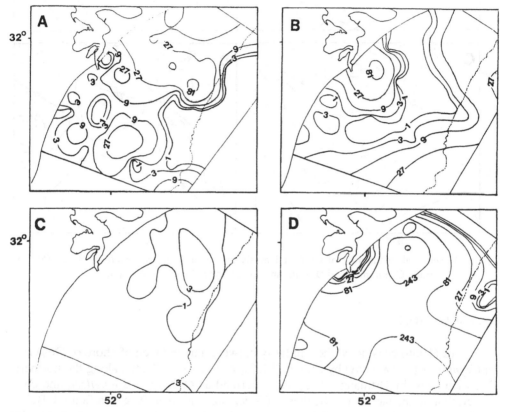

Figure 3. Seasonal distribution of combined protozoea larvae of *Pleoticus muelleri*, Bate 1888. A, Winter; B, Spring; C, Summer; and D, Autumn. Dotted line indicates 50 m depth.

The mysis phase is distributed throughout the region with preferences to depth between 30 and 60 m mostly in spring and autumn. During winter in the north the distribution is also between these depths and in the south close to the coast (Fig. 4). The abundance of megalopae (Table 2) suggests high disappearance rates in all seasons. A small number of megalopa is distributed homogeneously in the southern area during autumn with an abundance peak between 20 and 60 m. During spring they are offshore but no peak is shown. In summer the distribution is located in the northern area close to the coast, and in the winter a few megalopae are found in the south very close to the coast below 10 m depth (Fig. 5).

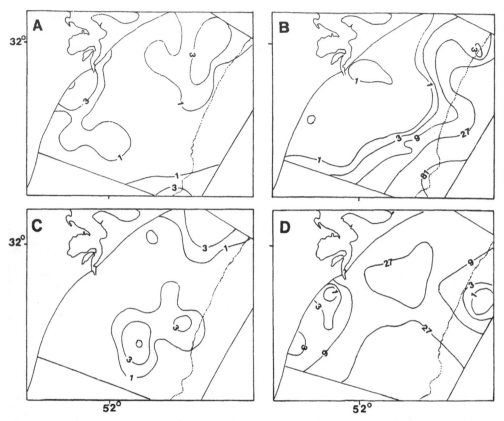

Figure 4. Seasonal distribution of combined mysis larvae of *Pleoticus muelleri*, Bate 1888. A, Winter; B, Spring; C, Summer; and D, Autumn. Dotted line indicates 50 m depth.

4 DISCUSSION

Heegaard (1966) considered the similarity between larvae taken offshore of Desterro, as *Cryptopus* (= *Pleoticus*) by Müller (1863), and larvae collected along the northern Brazilian coast by Ortmann (1893) and described as *Opisthocaris muelleri*, and concluded that all the larvae may be those of *Solenocera vioscai*. Pérez-Farfante & Bullis (1973), studying the western Atlantic distribution range of *Solenocera* adults, on the other hand, reported difficulties in accepting Heegaard's suggestion because *S. vioscai* is distributed from the North Atlantic to the southern part of the Gulf of Mexico. In fact three species of *Solenocera* have been reported along the Brazilian coast: *S. geijskesi* to the northern region, *S. atlantidis* in the central region and *S. necopina* in the southern region. Based on this, it is considered here that the species described as *S. muelleri* by Heegaard (1966) is in fact *S. atlantidis*, but its original name will be conserved until further confirmation. Together with *Pleoticus muelleri* and *Solenocera necopina*, *Mesopenaeus tropicalis* is the other solenocerid reported in southern Brazilian coastal waters. For this reason it is possible to deduce that the other solenocerid larvae found in this study belong to this species. Nevertheless, some doubt must remain and the larvae are described as (?) *Mesopenaeus tropicalis*.

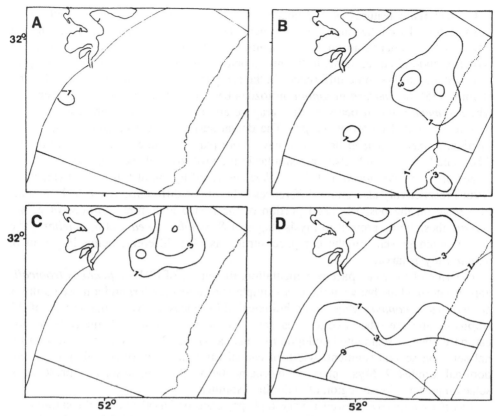

Figure 5. Seasonal distribution of megalopae of *Pleoticus muelleri*, Bate 1888. A, Winter; B, Spring; C, Summer; and D, Autumn. Dotted line indicates 50 m depth.

Pleoticus and other genera of the family Solenoceridae have an abbreviated development with only 2 mysis stages, but earlier and later specimens of the same stage show variation in total length and degrees of pleopodal development. These differences in *P. muelleri* are so pronounced that Scelzo & Boschi (1975) recognized 4 mysis stages during the development of laboratory reared specimens. Close observations of structure and setal formulae of other appendages, such as mouthparts, showed no major changes apart from increase in size of the 2 stages. There are no previous reports of the larvae of *Solenocera necopina* being reared in the laboratory. Nevertheless, from descriptions of the larval development of other species of *Solenocera*, 5-6 nauplii, 3 protozoeae, and 2 mysis stages may be expected to represent complete larval development. There are also no recorded descriptions of larvae of *Mesopenaeus tropicalis* reared in the laboratory, but between 5 and 6 naupliar stages, 3 protozoeae, and 2 mysis stages are expected.

Solenocerid larvae are distinctive in comparison to other penaeoidean larvae, by being larger, with a well ornamented body. The larvae of *Pleoticus muelleri* have the following main morphological features: protozoea phase with carapace edge bearing few spines, 2-segmented endopod of antenna with 2, 3 setae on first and 5 terminal setae on second segment; and, a mysis phase with a long rostrum and smooth post-branchial spine.

Characteristic features of *Solenocera necopina* are: protozoa phase with processes on the edge of the carapace, 2-segmented endopod of antenna with 2, 2 setae on first and 5 terminal setae on second segment, and a long setulose pair of setae on posterior abdominal segments in the mysis phase. The present description of the protozoea stages shows some differences in the carapace margin from those described by Heldt (1955). *Solenocera necopina* protozoea bear processes in four distinct regions, while *S. membranacea* bears on the carapace an anterior marginal process and continuous serrulated posterior margin. The dorsal surface of the carapace is also distinctive. *S. membranacea* protozoea have more pairs of spines than in *S. necopina*. Abdominal somites and telson are similar in the protozoea of the two species. Mysis stages of *S. membranacea* and *S. muelleri* described by Heldt (1955) and Heegaard (1966) respectively also show differences. The major difference is the lack of setulose setae on the postero-dorsal portion of the first 3 abdominal somites of *S. necopina*, these being present in mysis stages of *S. membranacea* and *S. muelleri*. The dorsal organ present on the carapace surface is also different, being larger in *S. muelleri* mysis larvae.

In terms of their carapace ornamentation, the larvae of (?)*Mesopenaeus tropicalis* appear intermediate between the more simple *Pleoticus muelleri* and a more complex *Solenocera necopina*. Major morphological differences between the larvae of *M. tropicalis* and the other two species are: 1) Protozoea phase with many spines on carapace edge, 2-segmented endopod of antenna with 2, 2 setae on first and 4 terminal setae on second segment, width of caudal furca and size of dorsal spine of abdominal somites; 2) Mysis phase with ornamentation of carapace surface; presence of setae on postero-dorsal region of abdominal somites.

Pleoticus muelleri showed a very high presence of larval phases throughout the year in shallow waters surrounding the Patos Lagoon mouth, with higher abundances during winter and mainly autumn indicating a great spawning activity of this species during these seasons. The low abundance of protozoea larvae during summer, when mean temperature is 24.5°C (23.1°C-25.6°C) may indicate that low spawning activity in this period was related to high temperature. On the other hand a very abundant number of protozoeae during autumn seems to indicate that the minimum (18.8°C) and maximum (20.8°C) temperatures observed during this season are appropriate for the spawning activity of this species. The spawning activity does denote the presence of an annual adult stock in the region. This stock can support an artisanal fishery, mainly during autumn and winter, when the higher number of larval stages suggests a higher number of adults in the area. The high abundance of protozoeae but low abundance of megalopae during winter and the similar abundance of megalopae during spring and summer, despite low abundance of protozoeae, suggest that the concentration of adults is less important to total recruitment during winter than in summer and spring.

ACKNOWLEDGEMENTS

I wish to thank Dr José H. Muelbert for his comments on the manuscript and Conselho Nacional de Desenvolvimento Científico e Tecnológico – CNPq. (Proc. N. 20,0527/87.8).

REFERENCES

Bate, C.S. 1888. Report on the Crustacea Macrura collected by HMS Challenger during the years 1873-76. In: *Report on the Scientific Results of the Voyage of H.M.S. Challenger 1873-76. Zoology* 24; 2nd Part, Plates I-CL.

Boschi, E.E. 1989. Biologia pesquera del langostino del litoral patagônico de Argentina (*Pleoticus muelleri*). *INIDEP, Ser. Contrib.* 646: 1-71.

Boschi, E.E. & Scelzo, M.A. 1969. Nuevas campanas exploratorias camaroneras en el litoral Argentino, 1967-68. Con referencia al plancton de la region. *FAO., Proy.Des.Pesq., Ser.Inf.Tecn.* 16: 1-31.

Calazans, D. 1994. Morphology, abundance and distribution of larval phases of two sergestids in the southern Brazilian coast. *Nauplius* 2: 75-86.

Heegaard, P. 1966. Larvae of decapod Crustacea. The oceanic penaeids. *Solenocera-Ceratasis-Cerataspides. Dana Rep.* 67: 1-147.

Iorio, M.I., Scelzo, M.A. & Boschi, E.E. 1990. Desarrollo larval y postlarval del langostino *Pleoticus muelleri* Bate, 1888 (Crustacea, Decapoda, Solenoceridae). *Sci. Mar.* 54: 329-341.

Iwai, M. 1973. Pesquisa e estudo biológico dos camarões de valor comercial. In: Relatório sobre a segunda pesquisa oceanográfica e pesqueira do Atlântico Sul entre Tôrres e Maldonado (Lat. 29°S-35°S). Programa Rio Grande do Sul – II. *Publ. esp. Inst. oceanogr.* 3: 501-534.

Mallo, J. & Celvellini, P. 1988. Distribution and abundance of larvae and postlarvae of *Artemesia longinaris, Pleoticus muelleri* and *Peisos petrunkevitchi* (Crustacea: Decapoda: Penaeidae) in the coastal waters of Blanca Bay, Argentina. *J. Aqua. Trop.* 3: 1-9.

Müller, F. 1863. Die Verwandlung der Garneelen. *Arch. f. Naturgesch.* 29: 8-23.

Pérez-Farfante, I. 1977. American solenocerid shrimp of the genera *Hymenopenaeus, Haliporoides, Pleoticus, Hadropenaeus* new genus, and *Mesopenaeus* new genus. *Fish. Bull.* 75: 261-346.

Pérez-Farfante, I. & Bullis, H.R. 1973. Western Atlantic shrimps of the genus *Solenocera* with description of a new species (Crustacea: Decapoda: Penaeidae). *Smithsoniam Contrib. Zool.* 153: 1-33.

Pérez-Farfante, I. & Kensley, B. 1997. Penaeoid and sergestid shrimps and prawns of the world – Key and diagnoses for the families and genera. *Mem. Mus. nat Hist. Nat.* 175: 1-233.

Scelzo, M.A. & Boschi E.E. 1975. Cultivo del langostino *Hymenopenaeus muelleri* (Crustacea, Decapoda, Penaeidae). *Physis* 35: 37-45.

Sokal, R.R. & Rohlf, F.J. 1981. *Biometry. The principles and practice of statistics in biology research.* 2nd ed., New York: W.H. Freeman and Co.

Williams, A.B., 1984. *Shrimps lobsters and crabs of the Atlantic coast of the eastern United States, Maine to Florida.* Washington: Smithsonian Institution Press.

Wyngaard, J.G. & Bertuche, D.A. 1982. Algunos aspectos de la biologa pesquera del langostino (*Pleoticus muelleri*) de la Bahìa Blanca y un análisis del desenbarco comercial en el período 1955-1979. *Rev. Invest. Des. Pesq.* 3: 59-76.

Development of a crangonid shrimp *Philocheras parvirostris* from Karwar, west coast of India, in the laboratory

K. JAGADISHA
Department of Zoology, Vijaya College, Karnataka State, India

SHAKUNTALA SHENOY & K.N. SANKOLLI
College of Fisheries, Shirgaon, Maharashtra State, India

ABSTRACT

We describe the life history of the sand shrimp *Philocheras parvirostris* comprising 3 zoeal and a post-larval stage reared in the laboratory. Though the larvae show general crangonid features, the segmented antennal scale has not been reported so far in any other species. The Crangonidae is a complex family with systematics of some genera not yet satisfactorily understood. The larval evidence elucidates the taxonomic problems of separating of the genera *Philocheras* and *Pontophilus*, the heterogeneity of the genus *Philocheras*, and the basic diversity of known crangonid larvae into two groups. We comment on different developmental patterns observed in the family, and egg mass attachment of the present species is also described. General features of *Philocheras* larvae are given based on the so far known larvae of the genus.

1 INTRODUCTION

While studying the intertidal caridean fauna of Karwar, a single species of sand shrimp belonging to the family Crangonidae, *Philocheras parvirostris*, was collected from Baburwada sandy bay, Ankola, about 30 km south of Karwar port (west coast of India). This shrimp has several interesting features such as feigning death upon disturbance and peculiar egg-mass attachment. The berried females were 13-16 mm (total length) bearing 250-270 eggs that are ivory colored, oval and 0.40-0.45 × 0.50-0.55 mm in diameter. The larvae were reared in the laboratory in saline conditions from 32-34 ppt and the temperature ranging from 27-29°C, using *Artemia* nauplii as food. The larvae passed through 3 zoeal stages before post-larva, and though sand was provided as substratum to simulate the natural conditions, wherein burying behavior started immediately, post-larvae died after 2-3 days without molting further.

The systematics of the family Crangonidae, particularly from the tropical region, are far from satisfactory (Chace 1984, Holthuis 1993) and importance of larval evidence in systematics is now well accepted. A good deal of larval information is available on Crangonidae mainly from the Atlantic region like the North Sea, the Norwegian Sea, the Mediterranean Sea, the Adriatic Sea, and Japan. However, that of the Indian Ocean region is restricted to only Makarov (1975), who deals with larval material from the International Indian Ocean Expedition (IIOE) plankton collections. Some of the larval identification of Makarov (1975) is doubtful as discussed later in this paper. However, as far as the littoral species of the Indian waters are concerned, larvae reared in laboratory are known only in two species, viz. *Pontocaris pennata* (cf. Sankolli & Shenoy 1976) and the present species *Philocheras parvirostris*. In the present paper, description of first stage and post-larva is given in detail while for remaining stages, only changes over the preceding stage are given. Also some of the taxonomic problems of the family elucidated by larval evidence are discussed.

2 DESCRIPTION OF LARVAL STAGES

2.1 *First Zoea*

(Fig. 1, 1-15): Total length (TL)= 1.75 mm; duration: 70-78 hrs. Eyes sessile. Rostrum small, smooth, slightly bent downwards, reaching distal 1/3rd of antennular peduncle. Carapace with only a pterygostomial spine followed by 4 sharp anterolateral denticles. Antennule (A1) peduncle long; exopod with 3 aesthetascs + 2 setae; endopod a broad-based sparsely plumose seta. Antenna (A2) peduncle with a large distal spine; scale (exopod) typical, 2-segmented, distal segment small bearing 2 tiny setae, outer margin with 2 long plumose setae and inner with 9; endopod spiniform, more than half scale in length, with swollen base and fine spine at tip. Mandibles (Mdbs) slightly asymmetrical, minute teeth on incisor and molar processes; no palp. First maxilla (Mx1) coxal endite with 6 strong bristles, basal with 4 broad, serrated teeth; palp (endopod) 2-segmented, distal segment with 3 thick plumose and proximal with 2 setae. Second maxilla (Mx2) both coxal and basal endites bilobed with apical setae; palp unsegmented but 4-lobed with 2, 2, 1 and 2 setae distalwards; scaphognathite with 5 marginal setae on outer and fine hairs on inner margins. First maxillipede (Mxpd1) endopod little more than half of exopod, 4-segmented, setation distalwards 3, 1, 2, 3 ; exopod unsegmented, with 3 apical + 2 subapical natatory setae as in 2nd and 3rd Mxpds; basis slightly bilobed , large, with 5 and 8-9 setae. Second maxillipede (Mxpd2) endopod 4-segmented, shorter than exopod, setation distalwards 3, 1, 2 and 4 + 1 outer, 8 setae on basis. Third maxillipede (Mxpd3) endopod 5-segmented, subequal to exopod, setation distalwards 2, 1, 0, 2 and 3 + 1 outer, and 3 short setae on basis. Pereiopods (Prpds) all 5 pairs small, uniramous buds. Abdomen (Abd) with 5 somites, only 5th with a pair of posterolateral spines; no pleopod (Plpd) or uropod (Urpd) buds. Telson (Tel) typical, elongatedly triangular, median notch on posterior margin spinulose; 3rd, 4th, and 6th processes longer than others. Chromatophores: Larvae brownish black to the naked eye, brownish-black, densely stellate chromatophores distributed all over the body throughout the zoeal stages.

2.2 *Second Zoea*

(Fig. 1, 16-30): TL = 2.15 mm; duration = 46-52 hrs. Eyes stalked. Rostrum straight, tip acute reaching upto half of A1 peduncle. Carapace anterolateral denticles larger extending to half of margin. A1 peduncle with 2 groups of small setae. A2 scale distal segmentation lost, 11 long plumose setae; endopod enlarged overreaching scale. Mdb teeth more developed. Mx1 basal endite with 7 serrated teeth, coxal bris-

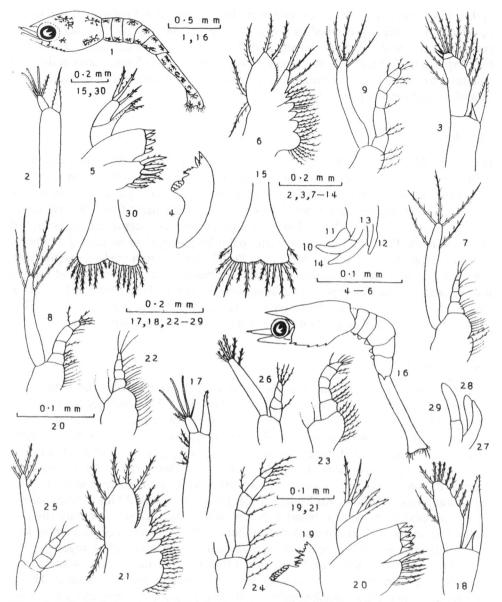

Figure 1. *Philocheras parvirostris*. First Zoea (1-15); Second Zoea (16-30). 1 & 16 – Entire larva; 2 & 17 – A1; 3 & 18 – A2; 4 & 19 – Mdb; 5 & 20 – Mx1; 6 & 21 – Mx2; 7 & 22 – Mxpd1; 8 & 23 – Mxpd2; 9 & 24 – Mxpd3; 10 & 25 – Prpd1; 11 & 26 – Prpd2; 12 & 27 – Prpd3; 13 & 28 – Prpd4; 14 & 29 – Prpd5; 15 & 30 – Tel.

tles longer and thinner. Mx2 scaphognathite with 7 setae. Mxpd1 endopod setation increased to 3, 2, 2 and 3; basis setation increased; exopod with 4 apical + 2 subapical natatory setae on all Mxpds. Mxpd2 endopod 5-segmented now, setation distalwards 3, 2, 0, 2, 5 + 1 outer. Mxpd3 endopod setation distalwards increased to 2, 1, 1 + 1 outer, 2, 4 + 1 outer. Prpds: First 2 pairs biramous, 1st endopod 5-segmented, setation

1, 0, 1, 0 + 1 outer, 1 + 1 outer; exopod much longer, with 3 + 2 natatory apical setae, tip of endopod slightly notched as a precursor of subchela; Prpd2 smaller, with 4-segmented endopod, terminal segment with 4 long and basal with 1 setae, exopod in both Prpds with 3 apical + 2 subapical natatory setae; Prpds 3-5 uniramous buds. Abd somites 1 to 5 with small, uniramous Plpd buds. Tel process formula 8 + 8 with addition of a small median pair. No Urpds.

2.3 *Third Zoea*

(Fig. 2, 31-47): TL = 2.30 mm; duration = 92-98 hrs. A2 scale with 14 plumose setae + small outer spine, endopod unequally 2-segmented, elongated, more than 1.5 times longer than scale, tip pointed. Mx2 scaphognathite with a small posterior lobe, and 12 setae. Mxpd1 endopod setation not changed but last 2 segments fused. Mxpd2 endopod setation 2 + 1, 0 + 1, 0 + 1, 2 and 5 + 1 outer, an outer seta added on 1st to 3rd segments each. Prpd1 stumpy, subchelate, palm large, equal to remaining segments together; exopod with 4 + 1 natatory setae. Prpd2 smallest, non-chelate, exopod with 4 natatory setae. Prpds 3-5 uniramous, long, segmented but non setose. Abd 6th somite separated from Tel; Plpd buds still uniramous. Tel median notch much reduced, process formula 7 + 7, 1st process spine–like having shifted laterally. Urpds biramous, shorter than Tel, exopod functional with 7 setae, endopod small bud.

2.4 *Post-larva*

(Fig. 2, 48-64): TL = 2.45 mm. The postlarvae were provided with sandy substratum wherein they buried readily but died after 2-3 days. Feigning death behavior had not yet started. Eyes set much closer, adult-like. Rostrum reduced, small, triangular, reaching posterior margin of cornea. Pterygostomial spine continues but denticles following it lost; antennal, hepatic or branchiostegal spines of adult not yet developed. Most appendages with adult–like features but less setose. A1 peduncle 3-segmented, basal segment twice length of others with small, sharp middoral spine, stylocerite broad; exopod with 2 unequal segments, distal with 2 long aesthetascs + 2 setae; endopod unsegmented, broad with few small setae. A2 scale broadly rounded with 17 setae and a short outer spine, flagellum of 30 segments, peduncle spine lost. Mdb totally changed from zoeal to an elongated molar process with 4 large, terminal teeth. Mx1 endites thinner, coxal with only 3 thin setae, distal 4 teeth + 4 setae, palp unsegmented with single terminal seta. Mx2 both endites reduced to small knobs, palp reduced, nonsetose; scaphognathite with distinct posterior lobe and 20-21 marginal setae. Mxpd1 endopod rod-shaped, unsegmented with 3 small setae; exopod broad at basal part with 3-4 small outer setae, flagellar distal part inbent with 4 apical + 1 subapical setae. Mxpd2 endopod incurved, broad, dactylus diagonally articulated to propodus and with 4-5 strong bristles at tip + a row of outer setae; exopod long, flagellar, 2-segmented with 4 apical + 1 subapical setae. Mxpd3 endopod 3-segmented, basal segment longer than remaining two together, last 2 segments setose; exopod reduced to about half endopod and with 4-5 apical setae. Prpd1 subchelate, propodus ending in a sharp spine, dactylus about 2/3rd propodus, carpus short, merus long with 1 outer distal spine, basis and ischium short. Prpd2 smallest, chelate, uniramous, fingers short, claw–like terminal spines. Prpds 3-5 uniramous, segmented,

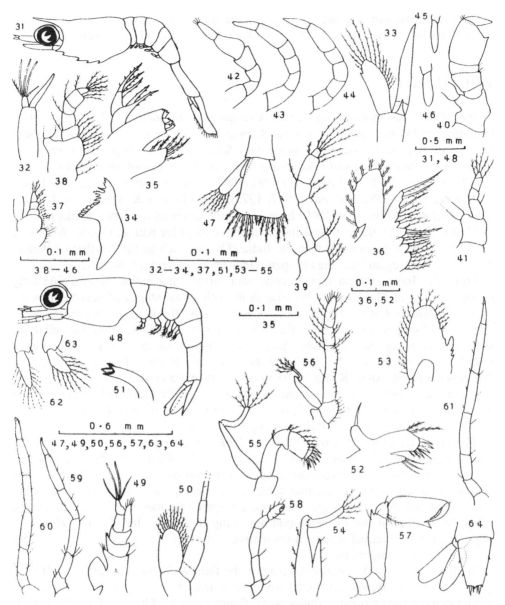

Figure 2. *Philocheras parvirostris*. Third Zoea (31-47); Postlarva (48-64). 31 & 48 – Entire larva; 32 & 49 A1; 33 & 50 – A2; 34 & 5 – Mdb; 35 & 52 – Mx1; 36 & 53 – Mx2; 37 & 54 – Mxpd1; 38 & 55 – Mxpd2; 39 & 56 – Mxpd3; 40 & 57 – Prpd1; 41 & 58 – Prpd2; 42 & 59 – Prpd3; 43 & 60 – Prpd4; 44 & 61 – Prpd5; 45 & 62 – Plpd1; 46 & 63 – Plpd2; 47 & 64 – Tel + Urpd.

5th longest, few setae on all three pairs. Abd with 6 somites, all smooth, zoeal spine on 5th lost. Plpds functional, setose but still uniramous, projecting laterally characteristically like in adult, endopod and appendix interna totally absent. Tel almost of adult shape, little shorter than 6th Abd somite; posterior margin with a short median point and 2 pairs of spines – a small outer and a large inner + a median pair of plumose setae, 2 pairs of tiny lateral spines not yet shifted dorsally. Urpds overreaching Tel, narrow, elongated, both rami functional, exopod with small outer spine.

3 DISCUSSION

Chace (1984) has revised the family Crangonidae to include 17 genera, while Holthuis (1993) has listed 20 genera. Larvae are known in almost half of them namely in *Crangon, Argis, Sclerocrangon, Sabinea, Notocrangon, Paracrangon, Philocheras, Pontocaris, and Pontophilus* by the contributions of authors like Gurney (1942), Kurian (1956), Williamson (1960, 1967), Kurata (1964), Tesmer & Broad (1964), Squires (1965), Makarov (1967, 1975), and Villamer & Brusca (1988). Gurney (1942) and Williamson (1960) have listed and reviewed the larval works prior to 1942 and, therefore, those are not included here except for Kurian (1956). All above works are from regions other than the Indian Ocean, such as the Northern Atlantic, North Sea, Norwegian Sea, Sea of Japan, Mediterranean, and the Adriatic.

From the Indian Ocean region, crangonid larval information is available only through works of Makarov (1975) who dealt with larval stages of seven series assigned by him as *Pontocaris* sp., *Philocheras* spp. A, B and C, *Pontophilus* spp. A and B, and Crangonidae n 1, from the IIOE plankton collection. From the Indian inshore waters, except for three laboratory reared stages described for *Pontocaris pennata* (cf. Sankolli & Shenoy 1976) and the entire life history of the present species *Philocheras parvirostris* (Kemp 1916) no other larvae are known.

The parent material of the present species was initially identified as *Pontophilus parvirostris* following Kemp (1916) but in light of Chace (1984) and Holthuis (1993), it has been herein assigned to *Philocheras parvirostris*.

The larval features of the present material *Philocheras parvirostris* broadly agree with the crangonid characters listed by Gurney (1942), Kurata (1964), and Makarov (1967) in possessing a pterygostomial spine followed by anterolateral denticles and absence of supraorbital or antennal spines on the carapace; smooth rostrum; spiniform broad endopod of antennule and antenna, spines on abdominal somites, and the subchelate nature of the 1st pereiopod appearing in late the larval stage. However, segmentation of antennal scale is a distinctive feature of the present species, not observed so far in Crangonidae.

With regard to the developmental pattern, the family Crangonidae exhibits all three types. There is the typical prolonged type with at least 5 zoeal + postlarval stages observed in the genera *Crangon, Philocheras, Pontophilus* (cf. Gurney 1942, Williamson 1960, Makarov 1967, 1975, Villamar & Brusca 1988), *Pontocaris* (cf. Sankolli & Shenoy 1976). The abbreviated type is seen in *Argis* (cf. Squires 1965) comprising 2 zoeal + postlarval stages to completely abbreviated type hatching on pleopods of mother with only 1 stage as in deep water and cold water forms of *Sclerocrangon* (cf Williamson, 1960, Makarov 1967). There is a partially abbreviated type passing through 3 zoeal stages + postlarva observed in *Sabinea* (cf. Gurney 1942, Williamson 1960, Makarov 1967). The present species *Philocheras parvirostris* shows partially abbreviated development with 3 zoeal + a postlarval stage unlike the known larval forms of the genus showing prolonged type (Williamson 1960, Kurian 1956, Makarov 1975) with at least 5 zoeal stages. Such abbreviation may possibly be assigned to reproductive strategy in response to the specialized habitat of the present species which is the surfy, sandy inter-tidal zone with pounding wave action. The family as a whole occupies a wide range of habitats from estuarine, inshore to deep waters in boreal, temperate through tropical belts, and developmental patterns appear to have helped in attaining the success in the species diversity.

Based on adult taxonomy, the genus *Philocheras* continues to be synonymized with *Pontophilus* despite the differences between the two, as reported by Chace (1984). Larval study was, therefore, carried out to find out whether it could elucidate the taxonomic problem.

Comparison of the known larvae of *Pontophilus* species (Gurney 1942, Kurian 1956, Williamson 1960) and *Philocheras* species (Kurian 1956, Williamson 1960, Makarov 1975) shows that they distinctly differ from each other and reveals the following.

1. Characters of the species of *Pontophilus*, viz. *P. spinosus* and *P. norvegicus*, like large size of larvae from 5.5 mm to 16.0 mm total length, long rostrum reaching tip of antennular peduncle, presence of large median spine on 3rd abdominal somite reaching end of 4th somite, 1st pereiopod with setose exopod in I stage, deeply in-cised / forked telson, are distinct from those of *Philocheras* spp. (cf. Williamson 1960) viz. *P. faciatus, P. sculptus, P. echinulatus, P. trispinosus*, and the present spe-cies *P. parvirostris*, which have smaller size ranging from 1.3 to 5.0 mm, shorter ros-trum, 1st pereiopod exopod not setose in stage I, absence of large median spine on 3rd abdominal somite and a typically non-forked, triangular telson with a shallow median notch. These larval differences adequately merit separation of the two genera.

2. Absence or presence of exopod on 1st pereiopod is given as a diagnostic feature of *Philocheras* and *Pontophilus*, respectively, by Chace (1984) and Holthuis (1993) based on adult taxonomy. The present species shows absence of this exopod even in the postlarval stage thus indicating its inclusion in *Philocheras* and separating it from *Pontophilus* even on larval evidence.

Based on so far known larvae, characters of *Philocheras* can be summarized as follows.

Development prolonged with 5 or more zoeal stages or partially abbreviated with 3 zoeal stages before postlarva; rostrum generally shorter than antennular peduncle; denticles on ventral margin of carapace mostly present (absent in *P. faciatus*); anten-nal scale unsegmented except in *P. parvirostris* (where it is 2-segmented in stage I); endopod of Maxilla I unsegmented or 2-segmented; first two pairs of pereiopods bi-ramous but exopod of first not setose in stage I; abdominal somites with paired spines on 3rd to 5th, or only on 5th somites (sometimes absent as in *P. trispinosus*); pleo-pods mostly biramous, rarely uniramous (*P. parvirostris*); telson typical, triangular with shallow median notch, process formula 7 + 7, in stage I.

In light of the above features, larvae from the Indian Ocean plankton described by Makarov (1975) can be assigned as follows. His *Philocheras* spp. A, B, C appear to belong to the genus *Philocheras*. His *Pontophilus* spp. A & B do not belong to that genus since the telson is not deeply incised and there is no large median spine on the 3rd abdominal segment. The larva described as Crangonidae n 1 has dorsal spines on the carapace described so far only in *Notocrangon antarcticus* (cf. Gurney 1942) but the latter is an Antarctic species with only 3 zoeal stages, a large telson, long antennal flagellum and only the first pereiopod biramous, 5th abdominal segment with long postero-lateral spines. On the other hand Makarov (1975) describes zoea V with 2 pairs of biramous pereiopods, abdominal somites 3 to 5 with paired spines and telson resembling the triangular type, thus similar to *Philocheras* features, though dorsal spines of carapace remain unexplained. Regarding his *Pontocaris* sp. A, it lacks the long antennular flagellum characteristic of known *Pontocaris* larvae. Also the curved hook-like spine on 2nd abdominal somite is not observed in any known crangonid larvae-thus this species does not seem to belong to any genera in which larvae are described. Since Makarov (1975) states that these larvae are most common and abundant in his collection and are from inshore waters, it would be worthwhile to study the adults and rear the larvae from the investigated regions as the species apparently belongs to a genus other than *Philocheras, Pontophilus* or *Pontocaris*.

Within the genus *Philocheras*, larvae show distinct differences (see Kurian 1956, Williamson 1960) – the present species differs from remaining mainly in number of developmental stages and also in segmented antennal scale and uniramous pleopods. These differences also reflect the heterogeneity of the genus as observed by Chace (1984) based on adults. In fact, Kemp (1916) had proposed 5 distinct groups for his *Pontophilus* species (some of which are now assignable to *Philocheras*) based on pleopod characters. The pleopods are uniramous in the larvae of present species, while they are biramous in other *Philocheras* and *Pontophilus* larvae. Thus development of pleopods through juveniles would be also interesting to observe. His *Pontophilus parvirostris* belonged to 'group V' (most advanced) and *Pontophilus spinosus* and *P. norvegicus* to 'group I' (most primitive) – such separation is also supported by present larval evidence as already suggested by Williamson (1982).

Lastly, based on an important, consistent larval feature, i.e. number of biramous pereiopods, all the known crangonid larvae can be divided into two basic groups. The 'group I' with only first pair of pereiopod biramous, comprises the genera *Crangon, Sabinea, Argis, Notocrangon,* and *Sclerocrangon,* and 'group II' with first two pairs of pereiopods biramous covering *Philocheras, Pontophilus, Pontocaris,* and *Paracrangon*. The implications of this on adult taxonomy need to be studied since the complexity of the family is not yet fully resolved.

Along with different patterns of development, attachment of eggs in adults also is an important reproductive adaptation dictated by the environment or habitat. Egg attachment in the present species is very interesting as described below: Eggs forming a single mass, are attached to the ovigerous setae of only bases of pleopods, while both rami are free, also helped by the absence of an appendix interna. Eggs are held in a single mass by what appears to be a thin mucilagenous substance or sheath (?), which remains intact even on preservation. Also, the abdomen is ventrally grooved forming a kind of brood pouch between the pleural projections of either side, wherein the egg mass is safely lodged, both rami of the pleopods laterally projecting outside from the brood pouch for free movement of the pleopods.

ACKNOWLEDGEMENTS

The authors gratefully acknowledge the following: The Dutch Ministries of Foreign Affairs, and Development Cooperation for financial support for SS to participate in the ICC IV, Associate Dean, College of Fisheries, Ratnagiri and the authorities of the Konkan Agricultural University for facilities.

REFERENCES

Chace, F.A. Jr. 1984. The caridean shrimps (Crustacea: Decapoda) of the Albatross Philippines Expedition, 1907-1910, part 2: Families Glyphocrangonidae and Crangonidae. *Smithsonian Contributions to Zoology* 397: 1-63.

Gurney, R. 1942. Larvae of decapod Crustacea. *Ray Soc. Publs.* 129: 1-306.

Holthuis, L.B. 1993. *The recent genera of the caridean and stenopodidean shrimps (Crustacea, Decapoda) with an appendix on the Amphionidacea.* Leiden: Nationaal Natuurhistorisch Museum.

Kemp, S. 1916. Notes on Crustacea Decapoda in the Indian Museum, VI: Indian Crangonidae. *Rec. Indian Mus* 12: 355-384.

Kurata, H. 1964. Larvae of Decapoda Crustacea of Hokkaido. H. Crangonidae. *Bull. Hokkaido Reg. Fish. Res. Lab.* 28: 35-50.

Kurian, C.V. 1956. Larvae of decapod Crustacea from Adriatic Sea . *Acta Adriatica* 6: 45-48.

Makarov, R.R. 1967. Larvae of the shrimps and crabs of the west Kamchatkan shelf and their distribution (translation of Russian book): 80-124. Boston Spa, Yorkshire: National Lending Library for Science and Technology.

Makarov, R.R. 1975. Larvae of Decapod Crustacea of the International Indian Ocean Expedition collections – Family Crangonidae (Decapoda: Caridea). *Indian J. Mar. Sci.* 4: 68-76.

Sankolli, K.N. & Shenoy, S. 1976. Laboratory behaviour of a crangonid shrimp *Pontocaris pennata* Bate and its first three larval stages. *J. Mar. Biol. Assc. India* 18: 62-70.

Squires, H.J. 1965. Larvae and megalopa of *Argis dentata* (Crustacea, Decapoda) from Ungava Bay. *J. Fish. Res. Bd. Canada* 22: 69-82.

Tesmer, C.A. & Broad, A.C. 1964. The larval development of *Crangon septemspinosa* (Say). *Ohio J. Sci.* 64: 239-250.

Villamar, D.F. & Brusca, G.J. 1988. Variation in larval development of *Crangon nigricauda* (Decapoda , Caridea) with notes on larval morphology and behavior. *J. Crust. Biol.* 8: 410-419.

Williamson, D.I. 1960. Crustacea, Decapoda: larvae family Crangonidae, Stenopidae. *Fishes Ident. Zooplankton* 90: 1-5.

Williamson, D.I. 1967. On a collection of planktonic Decapoda and Stomatopoda (Crustacea) from the Mediterranean coast of Israel. *Fish. Res. Sta. Haifa Bull.* 45: 31-64.

Williamson, D.I. 1982. Larval morphology and diversity. In: L.G. Abele (ed.) *Biology of Crustacea* (2): 43-110. New York: Academic Press.

Larval development of the Indian atyid shrimp *Caridina kempi* reared in the laboratory

D.R. JALIHAL[†], SHAKUNTALA SHENOY & K.N. SANKOLLI
College of Fisheries, Ratnagiri, Maharashtra State, India

† 9 July 1999

ABSTRACT

The atyid shrimp *Caridina kempi* passes through 6 zoeal + 1 postlarval stages during its metamorphosis in fresh water. Some larvae pass through 1 or 2 pre-postlarval stages before the actual post-larval stage. Most of the adult features appear at the third or fourth instars. Successful completion of its larval development even in 20 ppt sea water provides experimental evidence for the marine ancestry of atyids.

1 INTRODUCTION

The Indian inland atyid shrimp *Caridina kempi* constitutes about 3% of total prawn catch in the Dharwad area of Karnataka State. It occurs both in permanent as well as seasonal ponds almost throughout the year and compensates for its small size (about 30 mm) by large biomass. As such, it was deemed necessary to study its larval biology so that we might have an idea as to what led to its successful establishment in different water bodies, as well as to describe its life history.

2 MATERIAL AND METHODS

Caridina kempi bears brownish and elliptical eggs having dimensions of 0.33-0.40 × 0.58-0.65 mm. Berried females (20-30 mm) carry 240 to 630 eggs. After hatching, the larvae were reared individually in 10 ml vials containing tap water while checking once a day for molts or dead larvae. A few drops of green water from an anoxic pond were added as feed. The larvae passed through 6 zoeal stages before the post-larva at the room temperature of 26-28°C.

At each stage, at least 10 larvae were preserved for further studies in the preservative suggested by Thakur (1960). Before preservation, the larvae were narcotized with dilute clove oil to keep the appendages in extended condition. Larvae were dissected in diluted glycerine using entomological needles under an Olympus stereoscopic binocular. Drawings were made with the help of a camera lucida attached to a Labo

587

Optik IS 3886 monocular microscope. Larval size was measured from tip of rostrum to tip of telson. Only average size of each larval stage is indicated.

3 DESCRIPTION OF LARVAL STAGES

3.1 *First Zoea*

(Fig. 1, 1-14): Total length = 1.80 mm; duration = 1 day. Rostrum short, smooth, slightly extending beyond sessile eyes and reaching middle of antennular peduncle. Carapace with a prominent pterygostomial and a minute infraorbital spines. Antennule (A1): Peduncle unsegmented; inner ramus with a long plumose seta, outer short and with 2 aesthetascs + 1 seta. Antenna (A2): Peduncle short, with a sharp and serrated spine at base of unsegmented endopod; scale with 3 anterior segments and 2 plumose setae along outer margin; endopod about half as long as scale and terminally with a long plumose + 1 small plain seta. Mandibles (Mdb): Slightly asymmetrical; incisor with 2 or 3 teeth and molar with small tubercles. First maxilla (Mx1): Coxal and basal endites with 4 and 2 spines, respectively; palp slightly bifid at tip with 2 + 2 setae; exopod with 2 plumose setae. Second maxilla (Mx2): Endites bilobed and with 7, 2, 3 and 3 setae, respectively; endopod with 3 + 2 + 1 + 3 setae; scaphognathite with 5 plumose setae, posteriormost being stouter. First maxillipede (Mxpd 1): Coxa with 2 and basis with 8 plumose setae; endopod about half exopod, 4-segmented and with 3, 1, 1 and 3 + 1 setae; exopod with 4 natatory setae. Second maxillipede (Mxpd 2): Coxa with 2 and basis with 6 plumose setae; endopod 4-segmented, about 2/3 rd of exopod and with 3, 1, 2 and 3 + 1 setae; exopod with 5 natatory setae. Third maxillipede (Mxpd 3): Basis with 3 setae; endopod 4-segmented with 2, 1, 2 and 3 + 1 setae; exopod with 6 natatory setae. Pereiopods (Prpds): Only first 3 pairs present as small buds. Abdomen (Abd) 5-segmented; telson (Tel) broadly triangular with process formula 7 + 7, all processes plumose but first 2 on inner margin only. Chromatophores: Violet-brown on orbits, base of Tel, at dorsal junction of Abd and thorax, dorsolaterally on 3rd, 4th and 5th Abd segments and ventrally at base of A1 peduncle, orbits and sides of oral region; orange-red at base of Tel and midventrally on 2nd, 3rd and 4th Abd segments.

3.2 *Second zoea*

(Fig. 1, 15-24): Total length = 2.00 mm; duration = 2 days. Eyes stalked. A1 peduncle 3-segmented. Mbds with 1 or 2 bristles between molar and incisor processes. Mx more setose. Mxpd endopods 5-segmented; 2nd segment of Mxpd3 with 1 outer plumose seta. Buds of first 3 Prpds increased in size. Tel process formula 8 + 8; 2nd process plumose on both margins. Uropod (Urpd) seen as bud within Tel cuticle.

3.3 *Third zoea*

(Fig. 1, 25-34): Total length = 2.65 mm; duration = 2 days. A1 basal segment with a bristle in place of future stylocerite; inner ramus separated. A2 scale with 2 distal segments. Mx1 exopod with its setae totally lost. Mx2 scaphognathite with a small

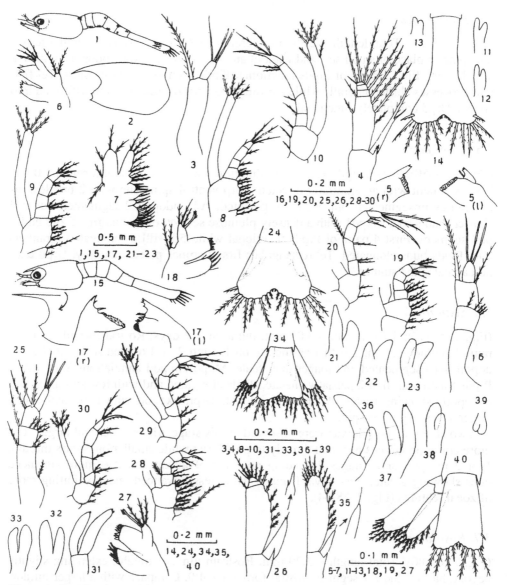

Figure 1. *Caridina kempi*, First zoea (1-14); Second zoea (15-24); Third zoea (25-34); Fourth zoea (35-40). 1 & 15: Entire larva; 2: Rostrum + carapace; 3, 16 & 25: A1; 4, 26 & 35: A2; 5 & 17: Mbd (r & l indicate right and left); 6, 18 & 27: Mx1; 7: Mx2; 8: Mxpd1; 9, 19 & 28: Mxpd2; 10 & 20: Mxpd3; 11, 21 & 29: Prpd1; 12, 22 & 30: Prpd2; 13, 23, 31 & 36: Prpd3; 32 & 37: Prpd4; 33 & 38: Prpd5; 39: Plpd; 14, 24, 34 & 40: Tel/Tel + Urpd/s.

proximal lobe. Mxpd2 basal long seta now prominent. First 2 Prpds functional with 5-segmented endopods; last 3 pairs represented as elongated biramous buds. Abd with traces of uniramous pleopod (Plpd) buds; 6th segment separated from telson. Tel process formula 8 + 8, 1st being transformed into a spine. Urpd exopods functional but endopods still narrow elongated buds.

3.4 *Fourth zoea*

(Fig. 1, 35-40): Total length = 2.80 mm; duration = 2-3 days. A2 scale with a terminal spine in place of small seta and without any outer setae. Mxpd endopods with 7-8 annulations. First 3 pairs of Prpds functional. Plpds seen as biramous buds. Tel first pair processes shifted laterally while 2 posterior outermost processes transformed into spines. Urpd completely functional.

3.5 *Fifth zoea*

(Fig. 2, 41-50): Total length = 2.85 mm; duration = 2-3 days. Rostral formula 1/0. A1 basal segment with an anteriorly directed sharp ventral spine and with 2 bristles; inner ramus unsegmented but outer 2-segmented. A2 endopod 2-segmented but still shorter than scale. Mxpd1 with a densely plumose seta on outer margin of 2nd endopod segment. First 4 pairs of Prpds functional while 5th still as elongated biramous bud. Plpds buds elongated. Tel rectangular; first 2 spines placed laterally. Urpd exopod apical spine stouter.

3.6 *Sixth zoea*

(Fig. 2, 51-57): Total length = 3.00 mm; duration = 3 days. Rostral formula 2/0. A1 rami 2-segmented; outer with 1 aesthetasc on 1st and 2 or 3 on 2nd segment. A2 endopod 8-10 segmented and longer than scale. Mxpd exopod annulations 8-10. All 5 Prpds functional; first 2 partially chelate. Plpds elongated and with few stumpy setae; endopods showing traces of appendix interna. Urpd exopod with subapical and a larger accessory inner spine.

Two size groups of larvae were observed at this stage, viz. smaller and larger. The more common larger ones molt directly to post-larva while smaller ones pass through 1 or 2 pre-postlarval stages depending on stage of advancement – more advanced ones showing greater post-larval features while less advanced ones exhibiting more of zoeal features (Fig. 2, 58-64).

3.7 *Post-larva*

(Fig. 2, 65-80): Total length = 3.20 mm. Rostrum almost straight, extending slightly beyond half of 2nd A1 segment; rostral formula 2-4/0. Carapace with a larger infraorbital/antennal and a minute pterygostomial spines. A1 outer ramus 3-segmented, with 2 aesthetascs each on 2nd and 3rd segments. A2 peduncle 4-segmented; flagellum long and whip-like; terminal spine of scale overreaching lamellar apex. Mbds & Mxs almost adult-like. Mxpds as in adults, but less developed and less setose. Prpd chelipedes fully functional with fingers carrying typical atyid brushes of setae; endopod segments of last 3 pereiopods with 1, 2, 2, 2 and 2 spines, respectively. Plpds completely functional; appendix interna absent on 1st but well developed on last 4 pleopods and with 2-4 hooks. Tel posterior margin convex and narrower than base; first 2 spines slightly dorsal (submarginal) in position. Urpd exopod terminating in a larger subapical tooth with a movable accessory spine on traces of future uropod diaeresis.

Most of adult features like lower teeth on rostrum, rudimentary (or even absence of) pterygostomial spine, lengthening of A1 flagella, extension of A2 lamella beyond

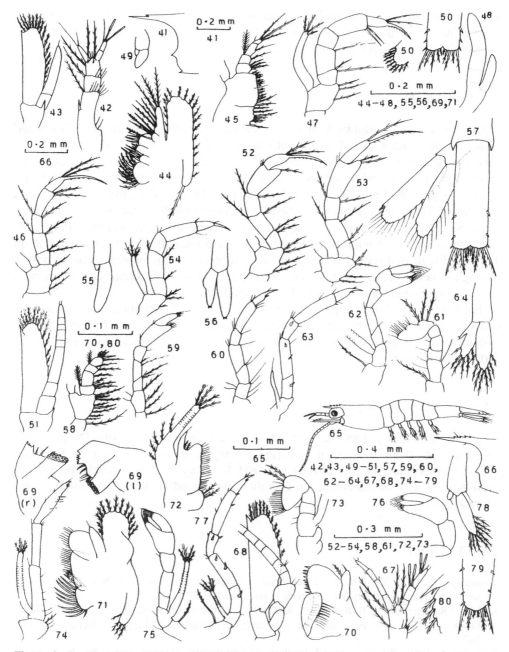

Figure 2. *Caridina kempi* Fifth zoea (41-50); Sixth zoea (51-57); Pre-postlarva I (58-60); Pre-postlarva II (61-64); Post-larva (65-80). 41 & 66: Rostrum + carapace; 42 & 67: A1; 43, 51 & 68: A2; 65: Entire larva; 69: Mbd (r & l indicate right and left); 70: Mx1; 44 & 71: Mx2; 45 & 72: Mxpd1; 58, 61 & 73: Mxpd2; 74: Mxpd3; 46, 52, 59, 62 & 75: Prpd1; 53 & 76: Prpd2; 47: Prpd4; 48, 54, 60, 63 & 77: Prpd5; 49, 55 & 78: Plpd1; 56 & 64: Plpd2; 50, 57, 79 & 80: Tel & Urpds.

outer spine, hairs on mandibles, reduction of Mx1 palp and posterior setae of scaphognathite, Prpd dactylar spines, Urpd diaeresis spinules etc appear / take place later at the third or fourth juvenile instars.

4 DISCUSSION

The genus *Caridina* exhibits 3 types of larval developmental patterns, viz. Prolonged Type, Partially Abbreviated Type, and Completely Abbreviated Type (Jalihal et al. 1994). By passing through 6 larval stages, *C. kempi* exhibits the Prolonged Type of development. The only other species of the genus wherein the prolonged type of development is published are *C. weberi* (cf. Chinnayya 1974) and *C. pseudogracilirostris* (cf. Pillai 1975) – both from India. The former completes its metamorphosis in purely fresh water while the latter essentially requires saline water. *C. weberi* differs from the present species in possessing 9 zoeal stages and total absence of exopod setae on first maxilla in first zoea. On the other hand, although *C. pseudogracilirostris* passes through 6 larval stages, its first zoea differs from that of *C. kempi* in possessing a longer rostrum and a total absence of pereiopod buds.

Even though *C. kempi* does not need any salinity for its metamorphosis, it can successfully complete its larval development in sea water up to 20 ppt. This incidentally provides experimental evidence for the marine ancestry of atyids, as postulated by Ortmann (1902).

ACKNOWLEDGEMENTS

The authors sincerely thank the authorities of the Karnatak University, Dharwad and the Associate Dean, College of Fisheries, Ratnagiri, for providing all the facilities. The generous financial help provided to DRJ & SS by the Dutch Ministries of Foreign Affairs and Development Co-operation for attending ICC-4 is gratefully acknowledged.

REFERENCES

Chinnayya, B. 1974. The embryonic and larval development of *Caridina weberi* De Man in the laboratory (Decapoda, Atyidae). *Broteria (Cienc. nat.)* 63(70): 119-134.
Jalihal, D.R., Almelkar, G.B. & Sankolli, K.N. 994. Atyid shrimps of the genus *Caridina* H. Milne Edwards, 1837 – potential crustacean material for experimental biology. *Crustaceana* 66: 178-183.
Ortmann, A. 1902. The geographical distribution of freshwater decapods and its bearing upon ancient geography. *Proc. Amer. Phil. Soc.* 41: 267-400.
Pillai, N.N. 1975. Larval development of *Caridina pseudogracilirostris* reared in the laboratory. *J. Mar. Biol. Ass. India* 17: 2-17.
Thakur, M.K. 1960. A new technique for preserving prawn larvae. *Curr. Sci.* 29(4): 138.

Prolonged larval development of an inland palaemonid prawn *Macrobrachium walvanensis* from India

G.B. ALMELKAR
'Ashram' Saraswatpur, Karnataka State, India

SHAKUNTALA SHENOY, D.R. JALIHAL & K.N. SANKOLLI
College of Fisheries, Shirgaon, Maharashtra State, India

ABSTRACT

The inland palaemonid prawn, *Macrobrachium walvanensis*, from India, requires at least some traces of salinity (i.e. 1.60 ppt) for successful completion of its larval development with 8 zoeal stages in 16 to 17 days. When reared in pure fresh water, besides an extended duration of 21 days, there is heavy mortality of larvae but the surviving larvae pass through 9 zoeal stages before the post-larva.

1 INTRODUCTION

Macrobrachium walvanensis is an inland freshwater prawn that inhabits fish farms at Lonavala in Maharashtra state of India. Despite possessing greenish eggs like many other impounded forms of *Macrobrachium*, its egg size of 0.70-0.87 × 0.82-1.02 mm is considerably smaller as compared to others. It was, therefore, decided to study its larval development in order to have a better understanding of this specialized feature.

2 MATERIAL AND METHODS

Berried females were collected at the fish farm of Tata Hydro-Electric Company Limited, near the Walvan dam in Lonavala. The eggs numbered 125 to 300 per female. Initial life history studies in fresh water, with *Artemia* nauplii as feed, resulted in heavy mortality of larvae although a few of them managed to reach the post-larval stage on the 21st day after passing through 9 zoeal stages. Therefore, they were specially reared in various grades of diluted seawater. It was observed that in salinity grades ranging from 1.60 to 13.20 ppt all the larvae could metamorphose into post-larvae, without any mortality, in 16 to 17 days after passing through only 8 zoeal stages. The latter is, therefore, considered as its normal developmental pattern and as such the 8 larval stages have been described hereunder, with a brief note on the 9th zoeal stage obtained in freshwater rearing. All appendages are described in first zoea and post-larva, while in remaining stages only changes are given.

3 DESCRIPTION OF LARVAL STAGES

3.1 *First Zoea*

(Fig. 1, 1-15): Total length (TL) = 3.30 to 3.50 mm; duration (Drn) = 1 day. Rostrum smooth, pointed, subequal to sessile eyes and with small papilla-like hump at base. Carapace with only small pterygostomial spine. Antennule (A1): Peduncle unsegmented, sometimes with distal setae; inner flagellum represented by long plumose seta; outer short with 1 bristle + 3 aesthetascs and one seta terminally. Antenna (A2): Peduncle unsegmented, bearing short spine at endopod base; scale with 5 distal segments and with 1 Or 2 denticles + 2 small plain setae on outer margin; endopod unsegmented, about half length of scale and with long plumose seta + small spine terminally. Mandibles (Mdb): Slightly asymmetrical; incisor and molar parts not clearly differentiated and with 2 prominent sharp teeth and 4 Or 5 small denticles, respectively. First maxilla (Mx1): Coxal and basal endites with 1 tiny tooth + 3 stumpy setae and 4 teeth, respectively; palp slightly bifid. Second maxilla (Mx2): Both coxal and basal endites bilobed, each lobe bearing 2 denticles; endopod with 1 and scaphognathite with 7 plumose setae. First maxillipede (Mxpd1): Basis with 2 setae; endopod small, bearing 3 plain terminal setae; exopod with 4 natatory setae. Second maxillipede (Mxpd2): Basis with 1 seta; endopod 3-segmented with 0, 1, and 3 setae; exopod with 4 + 1 natatory setae. Third maxillipede (Mxpd3): Basis with 1 seta; endopod 4-segmented with 1, 0, 2 and 1 serrated + 1 plain setae; exopod with 4 + 1 natatory setae. Pereiopods (Prpds): All 5 pairs represented as elongated buds – first 4 biramous but last one uniramous and longer; endopods of first 2 pairs showing traces of 4 faint segments. Abdomen (Abd): With 5 smooth segments. Telson (Tel): Triangular with 7 + 7 process formula, all processes plumose but first 2 along inner margin only; posterior margin with 5 or 6 minute hair-like setae between processes, excepting first 3. Chromatophores: Dorsally pair of orange-red stellate and smaller pair of diffused blue behind orbit on diffused orange-yellow patch. Ventrally a pair of orange-yellow patch at mandibular region.

3.2 *Second Zoea*

(Fig. 1, 16-30): TL = 3.30 to 3.50 mm; Drn = 2 days. Eyes stalked. Rostrum extending up to 1st segment of A1 peduncle. Carapace with supraorbital and pterygostomial spines. A1 peduncle 3-segmented and setose, basal with 2 bristles at future stylocerite region; inner ramus small bud with short apical seta. A2 scale 4-segmented distally and without denticles; endopod 3-segmented. Mdb with more teeth. Mx s and Mxpds more setose. All Prpds segmented excepting 4th; first 3 pairs 4-segmented, last segment terminated by 1 large plumose + 1 small plain setae; exopod with 5 natatory setae; fifth Prpd 5-segmented, with few setae and 1 long terminal seta. Tel process formula 8 + 8.

3.3 *Third Zoea*

(Fig. 1, 31-35): TL = 3.50 to 3.60 mm; Drn = 2 days. A1 basal segment with anterior ventral spine. A2 scale without distal segments or outer setae; peduncle 2-segmented. Endopod of 4th Prpd feebly segmented but exopod non-functional. Tel separated from Abd; 1st pair of processes transformed into spines. Uropods (Urpd): Exopods functional but endopods still buds.

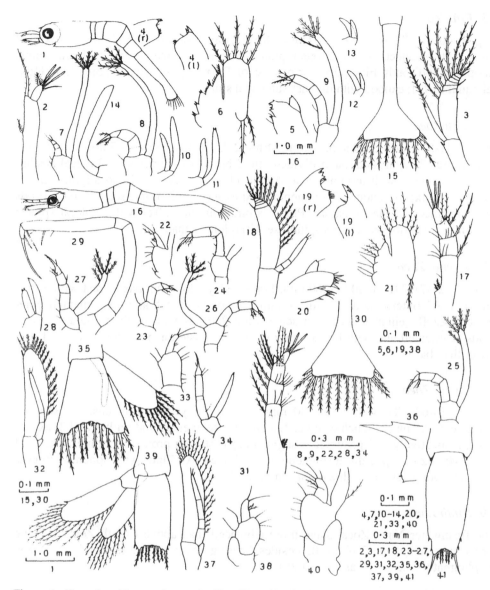

Figure 1. *Macrobrachium walvanensis*. First Zoea (1-15); Second Zoea (16-30); Third Zoea (31-35); Fourth Zoea (36-39); Fifth Zoea (40-41). 1 & 16: Entire larva; 36: Carapace; 2, 17 & 31: A1; 3, 18, 32 & 37: A2; 4 & 19: Mdb (r & l indicate right and left); 5 & 20: Mx1; 6 & 21: Mx2; 7, 22, 33, 38 & 40: Mxpd1; 8 & 23: Mxpd2; 9 & 24: Mxpd3; 10 & 25: Prpd1; 11 & 26: Prpd2; 12 & 27: Prpd3; 13, 28 & 34: Prpd4; 14 & 29: Prpd5; 15, 30, 35, 39 & 41: Tel / Tel + Urpd.

3.4 *Fourth Zoea*

(Fig. 1, 36-39): TL = 4.00 to 4.30 mm; Drn = 2 days. Rostrum with 1 dorsal tooth. Carapace with antennal spine. A2 scale with outer spine. Basis of Mxpd1 with epipod. Prpd4 fully functional. Tel rectangular with 2 pairs of lateral denticles and 5 + 5 posterior processes. Urpd completely functional; exopod with slender subapical spine.

3.5 *Fifth Zoea*

(Fig. 1, 40-41): TL = 4.50 to 4.70 mm; Drn = 2 days. Mxpd1 epipod bilobed; exopod basal part expanded and with outer plumose seta. Abd with 5 pairs of uniramous pleopod (Plpd) buds. First posterior process of Tel non-plumose and spine-like. Urpd subequal to Tel; exopod with distinct subapical spine.

3.6 *Sixth Zoea*

(Fig. 2, 41-51): TL = 5.70 to 5.90 mm; Drn = 2 days. Rostrum with 2 dorsal teeth. A1 inner ramus 2-segmented; outer divided into 2 branches – inner with 3 apical aesthetascs while outer with terminal seta. A2 flagellum 5-segmented and slightly longer than scale. Mxpd1 endopod 2-segmented and exopod with 3 or 4 setae. First 2 Prpds showing traces of chelate nature. Plpds as biramous buds. Tel with 3 lateral spines; outermost posterior process non-plumose. Urpd longer than Tel.

3.7 *Seventh Zoea*

(Fig. 2, 52-57): TL = 6.30 to 6.60 mm; Drn = 2 days. Rostrum basal crest with 3 dorsal teeth. A1 outer flagellum with 2-segmented outer branch and unsegmented inner branch. A2 flagellum longer than scale. First 2 Prpds with distinct chelate nature. Plpds segmented and with few stumpy setae. Tel posterior margin convex and narrower than base.

3.8 *Eighth Zoea*

(Fig. 2, 58-63): TL = 7.20 to 7.30 mm; Drn = 3 days. Rostrum basal crest more prominent. A1 inner branch of outer flagellum 2-segmented and with 1 and 3 aesthetascs, respectively. A2 peduncle 4-segmented; first with prominent spine at base of scale. Mxpd2 endopod distal segments more dilated. All Plpds with appendix interna buds excepting 1st. Tel with 2 dorsal spines.

3.9 *Ninth Zoea*

The freshwater Ninth Zoeal stage more advanced than above Eighth Zoea in being larger in size (about 7.70 mm), besides bearing more rostral teeth (4 or 5) and scaphognathite setae (30 as against 25).

3.10 *Post-larva*

(Fig. 2, 64-81): TL = 9.10 to 9.50 mm. Rostral crest prominent; formula 6-8/0-1 (1 postorbital). Carapace without supra-orbital spine; pterygostomial shifted slightly upwards. A1 basal segment with pointed stylocerite and antero-lateral tooth; both rami flagellar – inner branch of outer 3-segmented, bearing 2 and 3 aesthetascs on last 2 segments, respectively. A2 peduncle spine shifted towards outer margin. Mdb incisor part with 4 sharp teeth; palp still absent. Mxs adult-like but less setose. Mxpds almost adult-like with epipods and podobranchs. All Prpds fully functional with 5-segmented endopods as in adult, but first 4 pairs still with exopod buds. Plpds fully functional. Tel posteriorly with a median point + 3 to 5 plumose processes, flanked by 2 pairs of unequal spines. Urpd exopod with subapical and longer accessory spines.

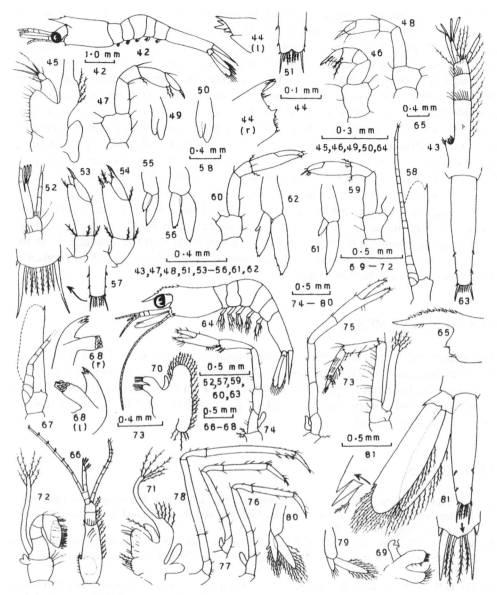

Figure 2. *Macrobrachium walvanensis*. Sixth Zoea (42-51); Seventh Zoea (52-57); Eighth Zoea (58-63); Postlarva (64-81). 42 & 64: Entire larva; 65: Carapace; 43, 52 & 66: A1; 58 & 67: A2; 44 & 68: Mdb (r & l indicate right and left); 69: Mx1; 70: Mx2; 45 & 71: Mxpd1; 46 & 72: Mxpd2; 73: Mxpd3; 47, 53, 59 & 74: Prpd1; 48, 54, 60 & 75: Prpd2; 76: Prpd3; 77: Prpd4; 78: Prpd5; 49, 55, 61 & 79: Plpd1; 50, 56, 62 & 80: Plpd2; 51, 57, 63 & 81: Tel / Tel + Urpd.

4 DISCUSSION

Macrobrachium walvanensis fits into the Group B of Type I development, i.e. Prolonged or Normal Type of larval developmental pattern as described by Jalihal et al. (1993). The only other known species of *Macrobrachium* that belong to this group are *M. lanceifrons* (cf. Rasalan et al. 1969) from the Philippines, *M. niloticum* (cf. Williamson 1972) from West Africa, and *M. lanchesteri* (cf. Chong & Khoo 1988, Jalihal 1992) from Singapore. Among these species, only the first zoea is known in *M. lanchesteri*. In *M. niloticum*, the larval stages have been described from plankton. Although laboratory development in *M. lanceifrons* was studied, there is no proper description or illustrations of larval stages. As such, only the first zoea comparison is possible between *M. walvanensis* and the above species. The first zoea of *M. walvanensis* stands out in possessing 5 pairs of pereiopods, postrostral hump, scaphognathite with 7 setae and 3–segmented endopod of the first maxillipede. The remaining species possess the only 4 pairs of the pereiopods, non-humped rostrum, scaphognathite with 5 setae and unsegmented endopod of the first maxillipede. Thus, *M. walvanensis* shows advanced features over the 3 remaining species.

Although essentially a land-locked form, *M. walvanensis* has prolonged development as is seen in salinity-dependent coastal species. It may, therefore, be inferred that the salinity required for its successful metamorphosis might be available from the land salts brought in by terrestrial drains during usual heavy rainfall in monsoon. This, however, needs further study.

ACKNOWLEDGEMENTS

The authors wish to express their sincere thanks to the Associate Dean, College of Fisheries and the authorities of Konkan Agricultural University, Dapoli, for all the facilities. The financial support extended by the Dutch Ministries of Foreign Affairs and Development Cooperation to SS & DRJ for attending the ICC-4 at Amsterdam is gratefully acknowledged.

REFERENCES

Chong, S.S.C. & Khoo, H.W. 1988. The identity of *Macrobrachium lanchesteri* (De Man 1911) (Decapoda, Palaemonidae) from Peninsular Malaysia and Singapore and description of its first zoea. *Crustaceana* 54: 196-206.

Jalihal, D.R. 1992. Erroneous records of the Malayan riceland prawn *Macrobrachium lanchesteri* (De Man 1911) from India – larval evidence (Decapoda, Caridea, Palaemonidae). *Crustaceana* 62: 101-105.

Jalihal, D.R., Sankolli, K.N. & Shenoy, S. 1993. Evolution of larval developmental patterns and process of freshwaterization in the prawn genus *Macrobrachium* Bate, 1868 (Decapoda, Palaemonidae). *Crustaceana* 65: 365-376.

Rasalan, S.B., Delmendo, M.N. & Reyes, T.G. 1969. Some observations on the biology of the freshwater prawn *Macrobrachium lanceifrons* (Dana), with notes on the fishery. FAO Fish. Rep. 57(3): 923-933 .

Williamson, D.I. 1972. Larval development in a marine and a freshwater species of *Macrobrachium* (Decapoda, Palaemonidae). *Crustaceana* 23: 282-298.

Enrichment of the nauplii of two *Artemia* species with docosahexaenoic acid (DHA, 22:6n–3)

HAN KYUNG-MIN, INGE GEURDEN & PATRICK SORGELOOS
Laboratory of Aquaculture and Artemia Reference Center, University of Gent, Gent, Belgium

ABSTRACT

Freshly-hatched *Artemia* nauplii from two different strains (*Artemia franciscana*, Great Salt Lake, USA and *Artemia tibetiana* in Lagkor Co Lake, Tibet, PR China) were enriched with 22:6n–3 (docosahexaenoic acid, DHA) by using a lipid emulsion containing 95% of DHA ethyl esters (% total fatty acids). Temperature and salinity during the 24 h enrichment and subsequent 24 h starvation were 28°C and 34 g l^{-1}, which are the standard conditions for *Artemia franciscana*. The initial DHA values after hatching were 0 and 0.2 mg g^{-1} for *A. franciscana* and the *A. tibetiana*, respectively. Initial 20:5n–3 (eicosapentaenoic acid, EPA) values were exceptionally high in the Tibet nauplii (44.6 mg g^{-1} DW as compared to 8.6 mg g^{-1} DW in the *A. franciscana*). To our knowledge this is the highest EPA level ever reported in freshly hatched *Artemia* nauplii. During the first enrichment period (0-12 h), the DHA content increased to 22.2 and only to 3.4 mg g^{-1} DW in the *A. franciscana* and *A. tibetiana*, respectively. The low DHA incorporation in the latter strain might be due to a retardation in developmental stage of the nauplii, possibly related to sub-optimal rearing conditions for the Tibet species. After the 24 h enrichment the respective DHA values were 26.2 and 20.2 mg g^{-1} DW. During the subsequent starvation (24-48h), the DHA levels in both species decreased to 8.9 and 3.8 mg g^{-1} DW. This is accordance with previous findings, showing an important and rapid catabolism of DHA in *Artemia* spp.

1 INTRODUCTION

Artemia nauplii are common live prey organisms used in marine fish and shrimp larviculture. However, freshly-hatched *Artemia* nauplii do not fulfill the nutritional requirements of most of the larval marine species because of their very low n–3 highly unsaturated fatty acids (HUFA), especially docosahexaenoic acid (DHA) (Watanabe & Kiron 1994). Numerous enrichment techniques have been developed to improve the nutritional value of *Artemia* nauplii using emulsified lipids (Watanabe et al. 1983, Léger et al. 1991). Enrichment with DHA in *Artemia* nauplii has been demonstrated to improve survival, growth, and pigmentation of marine fish larvae (Watanabe 1993, Reitan et al. 1994).

The metabolism and catabolism of DHA during enrichment and subsequent starvation has recently been compared in two species, i.e., *Artemia franciscana* (Utah, USA) and *Artemia sinica* (Inner Mongolia) (Triantaphyllidis et al. 1995, Evjemo et al. 1997). There was no difference in the increase of the DHA content in both species during enrichment. However, in contrast to the strong DHA catabolism observed during starvation of *A. franciscana*, DHA levels in the Chinese strain remained fairly constant. The present study examines the metabolic fate of the major n–3 and n–6 HUFA in two different *Artemia* species by enriching them with an emulsified and highly purified DHA concentrate, followed by a starvation period.

2 MATERIALS AND METHODS

2.1 *Cyst hatching and enrichment*

Two different *Artemia* strains, *Artemia franciscana* (Great Salt Lake, USA) and *Artemia tibetiana* (a new species described by Abatzopoulos et al. 1998 from Lagkor Co Lake, Tibet, PR China), were used in this experiment. The cysts (4 g l^{-1}) were disinfected with a hyphochlorine solution of 200 ppt for 20 minutes before hatching. After washing with tap water to remove the hyphochlorite, the cysts were incubated in filtered and disinfected seawater (34 g l^{-1} salinity) at 28°C under continuous aeration and light for hatching. After hatching, the nauplii were separated from the cyst shells and transferred to 2 liter cylindroconical glass tubes at 28°C and a density of 200 ind ml^{-1} with continuous aeration from the bottom of the cone using an additional airstone to keep oxygen levels above 5-6 mg l^{-1} for enrichment. Freshly hatched *Artemia* nauplii were enriched with 22:6n–3 (DHA) by using a lipid emulsion containing 95% of DHA ethyl esters (% total fatty acids). The emulsion (0.2 g l^{-1}) was added at the beginning of enrichment (t = 0h) and after 12h (t = 12h). After 24h enrichment (t = 24h), surviving nauplii were transferred into new glass tubes at a density of 100 ind. ml^{-1} and at 28°C for a subsequent starvation of 24h (s = 24h). Samples from the triplicate tubes were taken at t_0, t_6, t_{12}, t_{24}, and s_{24} for lipid analysis.

2.2 *Fatty acid analysis*

The *Artemia* nauplii were analyzed by a direct transmethylation method according to Lepage & Roy (1984). The sample amount was 250 mg wet weight and the internal standard was eicosadienoic acid (20:2n–6). Fatty acid methyl esters (FAME) were separated by gas chromatography (Chrompack CP 9001) with a very polar column (BPX 70). The carrier gas was hydrogen and the detection mode FID. The 'Maestro' (Chrompack) program was used to integrate the chromatogram.

3 RESULTS

The fatty acid composition of the freshly-hatched nauplii of both *Artemia* strains are shown in Table 1. The initial levels of DHA after hatching were 0 and 0.2 mg g^{-1} DW in the *Artemia franciscana* and *A. tibetiana*, respectively. The initial 20:5n–3 (eicosapentaenoic acid, EPA) value was exceptionally high in the Tibet strain (44.7 mg g^{-1} DW) as compared to the value of *A. franciscana* (8.6 mg g^{-1} DW).

During the first enrichment period (t = 12h), the DHA content increased to 22.2 mg g^{-1} DW in *A. franciscana* and only up to 3.4 mg g^{-1} DW in *A. tibetiana* species (Table 2). After 24h enrichment, the DHA content in both strains were elevated (26.2 and 20.2 mg g^{-1} DW, respectively) as well as the EPA content also increased slightly from 8.4 to 14.7 mg g^{-1} DW only showing in *A. franciscana* (Fig. 1). During the subsequent starvation (24-48h), the DHA levels decreased drastically in both species to 8.9 and 3.8 mg g^{-1} DW, respectively (Table 2).

Table 1. Fatty acid composition (mg g^{-1} dry weight) of freshly-hatched (t = 0h) *Artemia franciscana* (Great Salt Lake, USA) and *A. tibetiana* (Lagkor Co Lake, Tibet, PR China). Nd: not detected; Tr: < 0.1 mg g–1 dry weight [*] ≥ 20:3n–3.

	GSL	Tibet
14:0	2.0	2.7
14:1n–5	1.8	1.5
15:0	0.5	0.6
16:0	20.6	23.6
16:1n–7	7.9	2.4
17:0	0.8	1.0
17:1n–7	2.1	Nd
18:0	6.9	9.0
18:1n–9	34.9	40.4
18:1n–7	13.6	30.9
18:2n–6	10.1	6.2
18:3n–6	0.5	1.9
18:3n–3	39.4	Tr
18:4n–3	4.8	4.3
19:1n–9	0.5	Tr
20:1n–9	Nd	0.7
20:4n–6	1.7	1.8
20:3n–3	0.4	Tr
20:4n–3	0.6	0.6
20:5n–3	8.4	44.7
21:5n–3	0.2	0.6
22:6n–3	Nd	0.2
Total n–3 HUFA [*]	9.7	50.3
Total FAME	163.8	245.2

Table 2. Contents of 22:6n–3, DHA, 20:5n–3, EPA, and 20:4n–6, ARA (mg g^{-1} dry weight) of nauplii of *Artemia franciscana* (Great Salt Lake, USA) and *A. tibetiana* (Lagkor Co Lake, Tibet, PR China) sampled at subsequent intervals. t0h: freshly-hatched, t6h, t12h, and t24h: enriched and s24h starved (28°C). Data represent means (n=3), except for the Tibet strain (n=2).

	Artemia franciscana			*Artemia tibetiana*		
time (h)	DHA	EPA	ARA	DHA	EPA	ARA
t0h	0	8.4	1.7	0.2	44.7	1.8
t6h	10.7	9.2	1.6	0.5	47.1	1.9
t12h	22.2	11.3	1.6	1.9	45.2	1.8
t24h	26.3	14.7	1.7	20.0	41.6	1.7
s24h	8.9	13.8	1.6	3.8	23.3	1.0

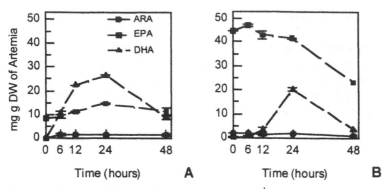

Figure 1. Change of fatty acid composition (mg g^{-1} dry weight) of *Artemia franciscana* (A) and *Artemia tibetiana* (B) during enrichment with DHA (t = 0-24h) and subsequent starvation (t=24-48h).

4 DISCUSSION AND CONCLUSIONS

In this study, both strains reached more or less the same DHA content during the enrichment and both rapidly catabolized DHA during starvation. EPA levels were more stable than DHA during starvation, in accordance with previous findings in *A. franciscana* (Triantaphyllidis et al. 1995, Evjemo et al. 1997). The former value represents, to our knowledge, the highest EPA level ever reported in freshly-hatched *Artemia* nauplii. The fatty acid composition of *Artemia* is determined by its natural habitat condition (Lavens et al. 1989). Therefore, the salt lake from which the Tibet cysts were collected probably contains specific microalgal species (diatoms, phytoplankton) with very high EPA levels.

The low DHA incorporation in the Tibet strain during the first enrichment period might be due to a slower development of the nauplii, possibly related to suboptimal culture conditions (e.g. temperature, salinity) for the Tibet species. During the enrichment, increasing the content of EPA which suggests the bioconversion of DHA to EPA in *A. franciscana*, not found in the Tibet strain (Fig. 1).

Future study is necessary to enrich with other highly-purified lipid emulsions rich in EPA and ARA (arachidonic acid, 20:4n–6) in order to examine the metabolic fate of these HUFA in *Artemia* nauplii.

REFERENCES

Abatzopoulos, T.J., Zhang, B. & Sorgeloos, P. 1998. *Artemia tibetiana*: preliminary characterization of a new *Artemia* species found in Tibet (PR China). International Study on *Artemia*. LIX. *Intl. J. Salt Lake Research* 7: 41-44.

Evjemo, J.O., Coutteau, P., Olsen, Y. & Sorgeloos, P. 1997. The stability of DHA in two *Artemia* species following enrichment and subsequent starvation. *Aquaculture* 155: 135-148.

Lavens, P., Léger, P. & Sorgeloos, P. 1989. Manipulation of the fatty acid profile in *Artemia* offspring using a controlled production unit. In N. de Pauw, H. Ackefors & N. Wilkins (ed.), *Aquaculture a biotechnology in progress*: 731-739. Bredene, Belgium: European Aquaculture Society.

Léger, P. & Sorgeloos, P. 1991. Optimized feeding regimes in shrimp hatcheries. In: A.W. Fast & L.J. Lester (eds), *Culture of Marine Shrimp: Principle and Practices*: 225-244. New York: Elsevier.

Lepage, G. & Roy, C.C. 1984. Improved recovery of fatty acids through direct transesterification without prior extraction or purification. *J. Lip.Res.* 25: 1391-1396.

Reitan, K.I., Rainuzzo, J.R., & Olsen, Y. 1994. Influence of lipid composition of live feed on growth, survival and pigmentation of turbot larvae. *Aquaculture Int.* 2: 33- 48.

Triantaphyllidis, G.V., Coutteau, P. & Sorgeloos, P. 1995. The stability of n–3 highly unsaturated fatty acids in various *Artemia* populations following enrichment and subsequent starvation. In P. Lavens, E. Jaspers & I. Roelants (eds), *LARVI '95. Fish and Shellfish Symposium*: 149-153. European Aquaculture Society Special Publication No. 24, Ghent, Belgium.

Watanabe, T. 1993. Importance of docosahexaenoic acid in marine larval fish. *J. World Aquacult. Soc.* 24: 152-161

Watanabe, T. & Kiron, V. 1994. Prospects in larval fish dietetics. *Aquaculture* 124: 223-251.

Watanabe, T., Tamiya, T., Oka, A., Hirata, M., Kitajima, C. & Fujita, S. 1983. Improvement of dietary value of live foods for fish larvae by feeding them on ω3 highly unsaturated fatty acids and fat-soluble vitamins. *Bull. Jpn. Soc. Sci. Fish.* 49: 471-479.

The effect of salinity on larval development of the spider crab *Libinia spinosa* (Brachyura, Majidae) reared in the laboratory

CLARA MARÍA HEREU & DANILO CALAZANS
Departamento de Oceanografia-FURG-Cx, Rio Grande, Brasil

ABSTRACT

Ovigerous females of *Libinia spinosa* were collected in the oceanic region adjacent to the Patos Lagoon and were maintained in the laboratory at 20°C and 30 PSU until spawning. After hatching larvae were transferred to compartmented plastic boxes of approximately 25 ml with filtered sea water of different salinities (25, 30 and 35 PSU) and kept under photoperiod of 12 h light: 12 h dark. Water was changed every other day, and larvae were checked daily to assess molt and death rates. Larvae were fed *Artemia* sp. nauplii newly hatched. Survival to megalopa was higher at salinities of 30 and 35 than at salinity 25. Although larvae reached megalopal phases in all three salinities mortality was 100% during this phase. Salinity had no effect on duration of zoea stage I whereas development of zoea II was significantly delayed in the lowest salinity tested. The results obtained were compared to other species of the same genus which were maintained under similar conditions. Nevertheless, there is a difference between the larval development length reported in another study for this species, where the mean length of larval stages was longer.

1 INTRODUCTION

Salinity is one of the factors that may influence larval development in aquatic ecosystems. Salinity effects have been reported for larvae of several decapod species. A change in salinity can modify developmental pathways and behavior of decapod larvae (Criales & Anger 1986, Forward 1989). The minimum change in salinity necessary to induce behavioral response can vary ontogenetically within a species. The response to differences in salinities will depend on the salinity regime experienced by individuals in their natural habitat (Costlow & Bookhout 1962, Sastry 1983, Foward 1989, Anger et al. 1998).

The present study deals with the larval development and the effect of salinity in the spider crab, *Libinia spinosa*, under laboratory conditions. This species is distributed from Brazil (Espirito Santo State), to Argentina (Santa Cruz State) (Souza 1994). Ovigerous females are found throughout the year along the southern Brazilian coast and larval hatching occurs in all seasons, with a larval peak in spring and summer

(Hereu 1999). Larvae may be exposed to reduced salinities with higher influence of continental water discharges from the Patos Lagoon and La Plata River estuaries during winter and spring months.

2 MATERIAL AND METHODS

A female of *Libinia spinosa* carrying eggs in advanced stages of embryonic development was collected in the oceanic region adjacent to the mouth of Patos Lagoon estuary. It was maintained at laboratory temperature (20°C) and local salinity (30 PSU) until larval hatching.

Four days later, newly hatched larvae in first zoeal stage were transferred to compartmented plastic boxes of 25 ml capacity with filtered sea water. Experiment was conducted with larvae in three salinities: 25, 30 and 35. Larvae tested at 25 and 35 were kept previously in intermediate salinities for a couple of hours until exposed to the final salinity. Larvae were fed *ad libitum* on *Artemia* sp. nauplii newly hatched. Checking for molts and deaths was done every day and change of water every other day. It was maintained at a 12L:12D regime and a constant temperature of 20°C during the experiment.

A one way ANOVA was employed to test treatment effects and when significant differences were detected, a Tukey's test of comparison of means complemented the analysis. Survival data to megalopa the phase were compared applying a Chi-square test (Sokal & Rohlf 1981).

3 RESULTS

The effect of salinity on survival of larval stages of *Libinia spinosa* is shown Table 1. Mortality was significantly higher at salinity 25 (p = 0.00), where 47.8% molted to zoea II and only 6.8% molted to a megalopa stage. Higher survival was obtained in the highest salinities tested: 75.9% of zoea I molted to zoea II and 51.7% attained mega-

Table 1. *Libinia spinosa* survival and development times of larval stages in three salinities tested. n_0, initial number of larvae; $X \pm SD$, mean duration and standard deviation (in days); x_m e x_M, minimum and maximal duration of larval stage; n, number of larvae that reached next stage; % S, survival percentage from initial number of larvae. ZI, zoea I; ZII, zoea II; M, megalopa. Same letters means no differences in mean duration of larval stages (Tukey Test, $\propto$ = 0.05).

Stage	N_0	25	30	35
		88	87	88
Z I	$X \pm SD$	$5.3 \pm 0.5\,a$	$5.1 \pm 0.4\,a$	$5.3 \pm 0.6\,a$
	x_m-x_M	5-7	5-6	5-7
	N	42	66	66
	% S	47.8	75.9	75
Z II	$X \pm SD$	$6.0 \pm 0.6\,b$	$4.9 \pm 0.5\,a$	$5.1 \pm 0.5\,a$
	x_m-x_M	5-7	4-6	4-7
	n%	6	45	57
	S	6.8	51.7	64.8

lopa in salinity 30 and in salinity 35, 75% molted to zoea II and 64.8% molted to megalopa. The megalopae lived for a longer period in salinities of 30 and 35, but in salinity of 25 they survived only for 2 days.

Duration of development did not show differences within the salinities tested during first zoeal stage (p = 0.08). Mean duration of zoea I was 5.3, 5.1, and 5.3 days in salinities 25, 30, and 35, respectively. Nevertheless, developmental time of zoea II in the lowest salinity tested was significantly longer (p = 0.00) than in salinities 30 and 35, with mean duration of 6.0, 4.9 and 5.1 days, respectively (Table 1).

4 DISCUSSION

It was verified that the effect of salinity on development of larvae of *Libinia spinosa* was that survival in salinities 30 and 35 were higher than in salinity 25. At this salinity survival was higher in zoea stage I but decreased abruptly in zoea stage II and, although megalopa stage was reached, larvae survived only for two days. In other species with larvae developing in coastal or oceanic waters, development is not affected in the lower range of higher salinities, whereas larvae from estuarine or brackish waters show a more euryhaline response and tolerate a wider range of salinities. In natural habitats, larvae of *Libinia spinosa* are found in coastal areas where surface salinities can be lowered with higher runoff of continental waters from the Patos Lagoon estuary. A migratory pattern in the water column during larval development may be an important way to allow the presence of larvae in an area with such characteristics. Nevertheless, larval abundance decreases when higher runoff of continental waters and lower temperatures occur, mainly in winter (Hereu 1999).

Total mortality of megalopae in all three salinities may indicate deficiencies in culture conditions to this stage, such as quality of food source, the absence of an adequate substrate, or chemical stimuli to allow megalopae molt to first crab stage. For megalopae of *Libinia emarginata* better survival rates were obtained when fed on a diet based on *Artemia* nauplii with a supplemental food source such as rotifers and algae (Bigford 1978). Anger et al. (1989) studying development of *Libinia ferreirae* reported an extremely high mortality at metamorphosis that continued in the juveniles, and they concluded that causes could be inadequate food quality and the absence of a particular substrate. Harms & Anger (1990) observed that for the megalopa stage of the spider crab *Hyas araneus* survival was strongly affected by nutritional conditions.

Salinity effect on development duration was not so strong. Development of the zoeal phase lasts 10-11 days, and larvae would stay in the water column for about 20 days until metamorphosis to crab I. Boschi & Scelzo (1968) reported for the same species an average duration of 8-10 days for each zoeal stage and 20-30 days required to reach first crab. The longer period for larval development observed by the authors may be typical for this species that develops in colder waters at higher latitudes.

Duration of development also varies within the genus *Libinia*, being in general shorter for *L. spinosa* than that usually reported. *Libinia erinacea* required 14 days or 9 days to reach first crab stage at 20° and 25°C, respectively (Yang 1967 in Johns & Lang 1977). For *L. emarginata,* the duration of pelagic phases until the first crab stage reported was 14 days in rearing conditions of 20° and 25°C and 30 of salinity (Jhons & Lang 1977). Bigford (1978) reported for the same species a period of about 10 days until the megalopa, and about 16-19 days to reach first crab stage in the same conditions of salinity and temperature and varying food concentrations. Anger et al. (1989) found that larvae of *L. ferreirae* required about 15 days to complete larval development at 25°C, whereas *L. dubia* required only 9 days (Sandifer & Van Engel 1971). This shortness of developmental time in this species may have been in part due to the higher temperature of culture (25.5°-28.5°C).

REFERENCES

Anger, K., Harms, J., Montú, M. & Bakker, C. 1989. Growth and respiration during the larval development of a tropical spider crab, *Libinia ferreirae* (Decapoda: Brachyura). *Mar. Ecol. Prog. Ser.* 54: 43-50.

Anger, K., Spivak, E. & Luppi, T. 1998. Effects of reduced salinities on development and bioenergetics of early larval shore crab, *Carcinus maenas. J. Exp. Mar. Biol. Ecol.* 220: 287-304.

Bigford, T.E. 1978. Effect of several diets on survival development time, and growth of laboratory-reared spider crab, *Libinia emarginata*, larvae. *Fish. Bull.* 76: 59-64.

Boschi, E. & Scelzo, M.A. 1968. Larval development of the spider crab *Libinia spinosa* H. Milne Edwards, reared in the laboratory (Brachyura, Majidae). *Crustaceana*, Suppl. 2: 170-180.

Criales, M.M. & Anger, K. 1986. Experimental studies on the larval development of the shrimps *Crangon crangon* and *C. allmanni. Helgoländer Meeresunters* 40: 241-265.

Costlow, J.R. & Bookhout, C.G. 1962. The larval development of *Hepatus epheliticus* (L.) under laboratory conditions. *J. Elisha Mitchel Sci. Soc.* 78: 113-125.

Foward, R.B. 1989. Behavioural response of crustacean larvae to rates of salinity change. *Biol. Bull.* 176: 229-238.

Harms, J. & Anger, K. 1990. Effects of nutrition (herbivore vs. carnivore) on energy budget of a brachyuran megalopa. *Thermochimica Acta* 172: 229-240.

Hereu, C.M. 1999. Aspetos biológicos e ecológicos durante o período planctônico de *Libinia spinosa* H. Milne Edwards 1834 (Decapoda, Brachyura, Majidae) no sul do Brasil. Rio Grande, 141p. These (Mestrado em Oceanografia Biológica, Fundação Universidade do Rio Grande).

Johns, D.M. & Lang, W.H. 1977. Larval development of the spider crab *Libinia emarginata* (Majidae). *Fish. Bull. US* 75: 831-841.

Sandifer, P. & Van Engel, W. 1971. Larval development of the spider crab *Libinia dubia* H. Milne Edwards (Brachyura, Majidae, Pisinae) reared in laboratory culture. *Chesapeake Sci.* 12: 18-25.

Sastry, A.N. 1983. Pelagic larval ecology and development. In D.E. Bliss (ed.), *The Biology of Crustacea*. Vol. 7: 213-282. New York: Academic Press.

Sokal, R.R. & Rohlf, F.J. 1981. *Biometry. The principles and practice of statistics in biological research.* San Francisco: W.H. Freeman.

Souza, J.A.F. 1994. Distribuição dos Brachyura (Crustacea: Decapoda) da plataforma rio-grandina (Rio Grande do Sul). Recife, 131p. These (Mestrado em Oceanografia Biológica, Universidade Federal de Pernambuco).

Day time larval release by decapod crustaceans in southeastern Thailand

PEDRO RAPOSO, R. RIBEIRO & A.M.V.M. SOARES
Instituto do Ambiente e Vida, Departamento de Zoologia, Universidade de Coimbra, Portugal

F. GONÇALVES
Departamento de Biologia da Universidade de Aveiro, Campus Universitário de Santiago, Aveiro, Portugal

ABSTRACT

Decapod crustaceans release larvae rhythmically in relation to lunar, light-dark, and tidal cycles. Those rhythms related to lunar phase are usually semilunar with larval release mainly occurring at the time of the largest amplitude nocturnal ebb tides, which usually corresponds to spring tides at the new and full moon. Semilunar, tidal, and diel timing of larval release by decapod crustaceans in southeastern Thailand were studied to confirm a preliminary sampling, which indicates that in this region larval release occurs during the largest-amplitude high tides, either during night or day. For this purpose, the first larva's stages in the water column were collected in the ebb tide during a complete lunar cycle. Densities were determined mainly for the Ocypodidae and Grapsidae, which represent more than 70% of all brachyuran larvae present.

1 INTRODUCTION

Decapod crustaceans exhibit three general reproductive patterns: 1. Spawning in coastal waters with migration of larvae or juveniles to estuaries (e.g. *Scylla serrata*), 2. Spawning and retention of larvae within the estuary (e.g. Xanthidae), and 3. Spawning within the estuary with the export of larvae to coastal waters and subsequent return of post-larvae or juveniles (e.g. *Uca* spp.). Thus, the larvae can be retained within the estuary, advected to the adjacent continental shelf, or expelled and widely distributed across the shelf until settlement (e.g. Christy 1982, Morgan 1987, Epifanio 1988, McConaugha 1988, Christy 1989, Morgan & Christy 1994).

Many decapod crustaceans have a true larval release connected to cycles in environmental factors. For many estuarine species, females synchronously release larvae near the time of nocturnal high slack water (e.g. De Vries et al. 1983, Provenzano et al. 1983, Christy 1986, De Vries & Forward 1989, Paula 1989, De Vries et al. 1994, Queiroga et al. 1994, Gonçalves et al. 1995). With few exceptions, larval release occurs approximately every two weeks when the ebb tides are between sunset and midnight (local time). Early evening ebb tides typically occur near the time of spring tides. However, Christy (1986) (for the eastern Pacific) and Paula (1989) and Gonçalves (1992) (for the Portuguese coast) found that intertidal crabs release their larvae over the neap tides, which are crepuscular in those areas. Paula (1989) concludes that the amplitude of tides does not seem to be important for successful larval development and any crepuscular high tide should be suitable for larval release. Thus, the timing of larval release by crabs tracks the amplitude of the ebb tides (usually the nocturnal ebb tides) not a particular lunar phase.

According to Morgan & Christy (1995), the timing of larvae release ultimately depend on whether adults or larvae are most vulnerable to predation by visual predators. The safest time for intertidal crabs to release would be at night. Species that live high in the intertidal zone should release near highest tides because only then will they be immersed.

The main purpose of this study was to determine lunar and diel influence in larval release of decapod crustaceans in the SE of Thailand.

2 MATERIAL AND METHODS

2.1 *Study area*

The present study was conducted on the Andaman Sea coastline of southern Thailand (9°50′N, 98°35′E) in the estuary of the Ngao River, surrounded by mangrove forests, and numerous smaller creeks (Fig. 1). The Ngao River is part of a larger mangrove system covering the extensive delta of the Kra Buri River, which forms the border between Thailand and Burma. A dry season from November to June and a rainy season from July to October control the physical and chemical characteristics of this estuary. The tidal regime along the Andaman Sea coast is predominantly semidiurnal and the mean tidal range at the mouth of the River Ngao measures 2.5 meters. Spring tides have an average range of 3.3 meters and neap tides an average range of 1.0 meters (Itthipatachai et al. 1991). The largest amplitude tides follow the full and new moon and peak during the night. At the first and last quarters the largest amplitude ebbs occurred during the day.

2.2 *Sampling*

Larvae release of the most abundant decapods was studied by sampling the plankton at a fixed station during ebb tides and beginning 1h after high tide. Three samples were taken each nocturnal and diurnal ebb tide during two situations: full moon and first quarter in February/March 1996.

Decapod larvae were sampled with a conical plankton net (40 cm in diameter, 150 cm long and 335 μm mesh size) equipped with a Hidrobios Kiel flowmeter. Tows were made horizontally at a depth of 1 m, at a speed of about 1 ms^{-1}. Samples were fixed and preserved in 5% formaldehyde. At the laboratory, samples were sub-sampled with a Folsom plankton splitter, and the fraction counted had at least 500 larvae. Larvae were counted and identified under a binocular microscope.

2.3 *Statistical procedures*

Analysis of variance (ANOVA) was used to test the significance of the effects of

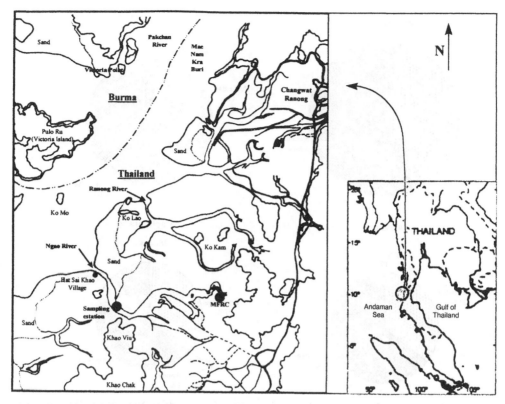

Figure 1. Location of study site in the Andaman Sea, Thailand (9°50′N, 98°35′E).

time of day (day/night) and phase of moon (Full Moon/First Quarter) upon the density of the larvae. The data were transformed by the equation $x = \log(x+1)$ to be in agreement with the assumptions of this analysis (Zar 1984).

3 RESULTS

The first stage zoeae of natantians account for 59% of the total, Brachyura 37%, and Anomura 4% (Fig. 2). In the first group, Sergestidae were the most abundant (84%). Within Brachyura, the Ocypodidae (*Uca* spp.) (37%) and Grapsidae (36%) predominated. The Diogenidae (64%) and Porcellanidae (35%) dominated the Anomura.

Stage I of the Sergestidae were significantly more abundant during first quarter than new moon (F = 23.977, df = 1, P = 0.001), suggesting that these larvae were released during diurnal and nocturnal ebb tides, preferentially during the First Quarter (Fig. 3). In this case, larvae were more abundant during diurnal ebb tide, but we did not find significant differences between day and night. Another pattern of larval abundance was observed in species of Caridea where there was no significant effect of lunar phase (Fig. 3). Larvae were significantly more abundant during night ebb tides than ebb day tides (F = 6.049, df = 1, P = 0.03). Thus, within natantians we find two different patterns of larval release.

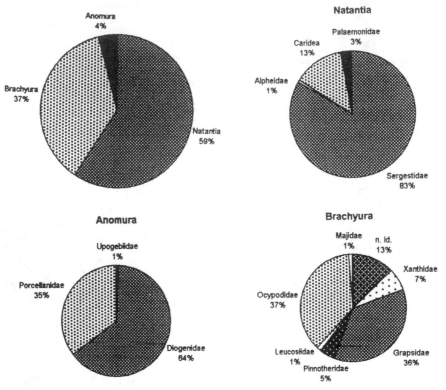

Figure 2. Percentage of zoeae stage I of main groups of decapods in Ngao River (SE Thailand), during new moon and first quarter ebb tide.

Stage I of the Grapsidae showed greater abundance during new moon, day ebb tide. The overall pattern of abundance suggests that spawning in this group occurred preferably during New Moon (F = 135.76, df = 1, P < 0.001) and during daytime (F = 40.34, df = 1, P < 0.001) (Fig. 3). Ocypodids (*Uca* spp.) showed reproductive activity during both lunar phases, but more larvae were released during new moon (F = 33.37, df = 1, P < 0.001). Despite the abundance peak observed during daytime (new moon), we do not find differences in larval abundance between night ebb tide and day ebb tide (F = 0.021, df = 1, P = 0.89) (Fig. 3).

Analysis of variance showed a significant effect of lunar (F = 47.14, df = 1, P < 0.001) and diel (F = 9.23, df = 1, P = 0.01) phases on abundance of zoeae stage I in the Diogenidae, which were more abundant during new moon and daytime (Fig. 3).

4 DISCUSSION

In the present study, the larvae of the main groups that were released in the mangrove system may be exported to the ocean, as referred by other authors (e.g. Epifanio & Dittel 1984, Dittel & Epifanio 1990, Dittel et al. 1991, Morgan & Christy 1995, Wooldridge & Loubser 1996). On the other hand, this study suggests that larval release of species inhabiting the high intertidal mangrove system occurs during large

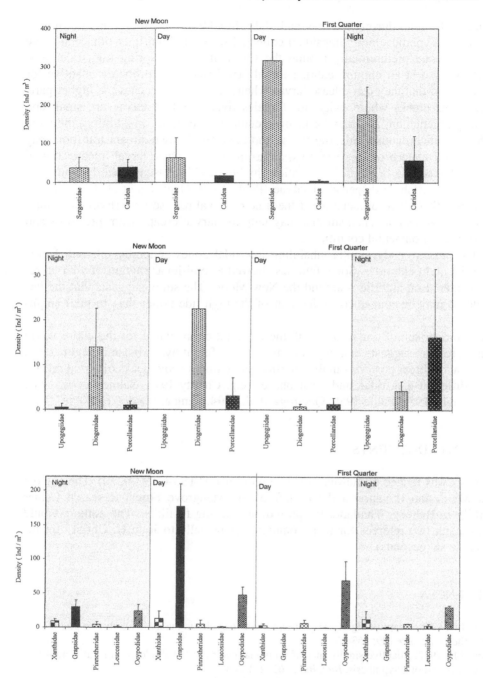

Figure 3. Densities of zoeae stage I for each taxon during the study period in Ngao River (SE Thailand).

amplitude ebb tides during both day and night. In addition, we found that all ovigerous crabs (Ocypodidae and Grapsidae) collected from the high inter-tidal mangrove
system spawned preferentially around the times of diurnal spring high tide (new
moon) (Raposo et al. unpubl. data). Diurnal larval release is probably adaptive for
these crabs so that they can release larvae whenever the tide reaches the higher parts
of mangrove creeks where ovigerous females live. The first larvae are small and
without pigmentation, and may be inconspicuous to predators. In addition, the large
number of synchronously released larvae and their rapid down-stream transport may
increase their chance of successful export to the ocean. Thus, larval release around
high tide linked to a larval tidal migration rhythm of ebb-phased upward swimming,
may enhance offshore dispersal of larvae (Paula 1989, Queiroga et al. 1997, Rodriguez et al. 1997). This association of the time of larval release with rhythmic behavior could serve as a mechanism that may help the larvae escape from predators and
unsuitable environmental conditions.

In the present study, adults inhabiting intertidal zones generally released larvae
during the night ebb tides. Some families showed a semi-lunar rhythm of a larval release, centered on high tides around the New Moon. The study suggests that the larval release may be connected to the hour of the high tide rather than to tidal amplitude.

The comparison of our results with the information available for the entire world
on larval release suggests that populations under different hydrodynamic regimes and
that live at different positions in the intertidal zone might show peaks of larval release
during different diel, tidal, and lunar phases (e.g. Christy 1978, Salmon et al. 1986,
McConaugha 1988, Paula 1989, Queiroga et al. 1994, Zeng & Naylor 1997).

ACKNOWLEDGEMENTS

This work was funded by European Union (contract n° TS3-CT93-0230). The authors
thank Mr Sopon Havanon and the staff of the Mangrove Forest Research Center
(MFRC), in Ranong, Thailand, for providing working facilities. The authors would
like to thank two referees for their comments (especially to John H. Christy for his
invaluable suggestions).

REFERENCES

Christy, J.H. 1978. Adaptive significance of reproductive cycles in the fiddler crab *Uca pugilator*: a
    hypothesis. *Science* 199: 453-455.
Christy, J.H. 1982. Adaptive significance of semilunar cycles of larval release in fiddler crabs (ge-
    nus *Uca*): test of a hypothesis. *Biol. Bull.* 163: 251-263.
Christy, J.H. 1986. Timing of larval release by intertidal crabs on an exposed shore. *Bulletin of Ma-
    rine Science* 39: 176-191.
Christy, J.H. 1989. Rapid development of megalopae of the fiddler crab *Uca pugilator* reared over
    sediment: implications for models of larval recruitment. *Mar. Ecol. Prog. Series* 57: 259-265.
Dittel, A.I. & Epifanio, C.E. 1990. Seasonal and tidal abundance of crab larvae in a tropical man-
    grove system, Gulf of Nicoya, Costa Rica. *Mar. Ecol. Prog. Series* 65: 25-34.
Dittel, A.I. & Epifanio, C.E. & Lizano, O. 1991. Flux of crab larvae in a mangrove creek in the Gulf
    of Nicoya, Costa Rica. *Estuarine, Coastal and Shelf Science* 32: 129-140.

Epifanio, C.E. 1988. Transport of invertebrate larvae between estuaries and the continental shelf. *Amer. Fish. Soc. Symp.* 3: 104-114.

Epifanio, C.E. & Dittel, A.I. 1984. Seasonal abundance of brachyuran crab larvae in a tropical estuary: Gulf of Nicoya, Costa Rica, Central America. *Estuaries* 7(48): 501-505.

Goncalves, F. 1992. Zooplâncton e Ecologia Larvar de Crustáceos Decápodes no estuário do Mondego. Ph.D. thesis at Departamento de Zoologia da Universidade de Coimbra, 351 pp.

Goncalves, F., Ribeiro, R., Soares, A.M.V.M. 1995. *Rhithropanopeus harrisii* (Gould): an American crab in the Mondego River (Portugal). *J. Crust. Biol.* 15: 756-762.

Itthipatachali, L., Kjerfve, B., Rakkhiew, S., Siripong, A., Srisangtong, D., Tangjaitrong, S., Wattahakorn, G. & Wolanski, E. 1991. Oceanography and hydrology studies. In D.J. Macintosh, S. Aksornkoae, M. Vannucci, C. Field, B. Clough, B. Kjerfve, N. Paphavasit & G. Wattayakorn (eds), *The integrated multidisciplinary survey and research programme of the Ranong mangrove ecosystem*: 8-34. Bangkok: Funny Publishing Limited Partnership.

Morgan, S.G. 1987. Adaptive significance of hatching rhythms and dispersal patterns of estuarine crab larvae: avoidance of physiological stress by larval export. *J. Experim. Mar. Biol. Ecol.* 113: 71-78.

Morgan, S.G. & Christy, J.H. 1994. Plasticity, constraint, and optimality in reproductive timing. *Ecol.* 75: 2185-2203.

Morgan, S.G. & Christy, J.H. 1995. Adaptive significance of timing of larval release by crabs. *Amer. Natur.* 145: 457-479.

McConaugha, J.R. 1988. Export and reinvasion of larvae as regulators of estuarine decapod populations. *Amer. Fish. Soc. Symp.* 3: 90-103.

Paula, J. 1989. Rhythms of larval release of decapod crustaceans in the Mira estuary, Portugal. *Mar. Biol.* 100: 309-312.

Provenzano, A.J., McConaugha, J.R., Phillips, K.B., Johnson, D.F. & Clark, J. 1983. Vertical distribution of first stage larvae of the blue crab *Callinectes sapidus* at the mouth of the Chesapeake Bay. *Estuarine Coastal and Shelf Science* 16: 489-499.

Queiroga, H., Costlow, J.D. & Moreira, M.H. 1994. Larval abundance patterns of *Carcinus maenas* (Decapoda, Brachyura) in Canal de Mira (Ria de Aveiro, Portugal). *Mar. Ecol. Prog. Series* 111: 63-72.

Queiroga, H., Costlow, J.D. & Moreira, M.H. 1997. Vertical migration of the crab *Carcinus maenas* first zoea in an estuary – implications for tidal stream transport. *Mar. Ecol. Prog. Series* 149: 121-132.

Rodriguez, A., Drake, P. & Arias, A.M. 1997. Reproductive periods and larval abundance patterns of the crabs *Panopeus africanus* and *Uca tangeri* in a shallow inlet (SW Spain). *Mar. Ecol. Prog. Series* 149: 133-142.

Slamon, M., Seiple, W.H. & Morgan, S.G. 1986. Hatching rhythms of fiddler crabs and associated species at Beaufort, North Carolina. *J. Crust. Biol.* 6: 24-36.

Vries, M.C. de & Forward, R.B. 1989. Rhythms of larval release of the sublittoral crab *Neopanope sayi* and the supralittoral crab *Sesarma cinereum* (Decapoda: Brachyura). *Mar. Biol.* 100: 241-248.

Vries, M.C. de, Epifanio, C.E. & Dittel, A.I. 1983. Lunar rhythms in the egg hatching of the subtidal crustacean, *Callinectes arcuatus* Ordway (Decapoda: Brachyura). *Estuarine, Coastal and Shelf Science* 17: 717-724.

Vries, M.C. de, Tankersley, R.A., Forward, R.B., Kirby-Smith, W.W. & Luettich, R.A. 1994. Abundance of estuarine crab larvae is associated with tidal hydrological variables. *Mar. Biol.* 118: 403-413.

Wooldridge, T.H. & Loubser, H. 1996. Larval release rhythms and tidal exchange in the estuarine mudprawn, *Upogebia africana. Hydrobiol.* 337: 113-121.

Zar, J.H. 1984. *Biostatistical analysis*. New Jersey: Prentice-Hall.

Zeng, C.S. & Naylor, E. 1997. Rhythms of larval release in the shore crab *Carcinus maenas* (Decapoda, Brachyura). *J. Mar. Biol. Assc. U.K.* 77: 451-461.

6 Fisheries and aquaculture

Current practice and future research trends in salmon aquaculture in Scotland

GORDON H. RAE
Scottish Salmon Growers Association, Scotland

ABSTRACT

The Scottish Salmon Farming Industry produced 100,000 tons of Atlantic salmon in 1997. While the value of this tonnage might once have been as much as £600 million, salmon is now widely available and this is reflected in the present day value of about £230 million ex-farm. In global terms, Scottish salmon fills the quality niche market and it is supported in this by 'The Product Certification Scheme for Scottish Quality Farmed Salmon', the scheme behind the Tartan Quality Mark, which is promoted by the Scottish Salmon Board. The scheme is run and monitored by Food Certification (Scotland) Ltd. which is accredited to the European Standard, EN 45011. In order to guarantee supplies of high quality salmon for the scheme, SSGA has been conducting a robust research program for over a decade. The mechanism for this is a Technical Committee, which comprises farmers and technical experts. Smaller Working Groups discuss specific issues, contract research and monitor progress. Two major health problems will be used to illustrate how a common approach can yield results. Furunculosis is a bacterial disease that can cause serious mortality to farmed salmon. It is endemic in Scotland and wild fish carry the bacterium. Improved management strategies and efficacious vaccines have been developed and the disease is presently under control. Sea lice (*Lepeophtheirus salmonis* and *Caligus elongatus*) are crustacean parasites of salmon and numbers can build up to epizootic proportions causing stress and mortality of salmon. Control of sea lice is the number one priority for salmon farmers today and the SSGA research strategy is designed to reduce the impact of these parasites on farms. Many of the projects have now been completed but significant bureaucratic hurdles remain to be overcome before the results can be translated into real improvements on salmon farms.

1 INTRODUCTION

The Atlantic Salmon (*Salmo salar*, Linnaeus) is known as the king of fish and has traditionally been a high priced delicacy. Salmon farming started in Scotland in the early 1970s and by the 1980s was producing significant tonnage. The estimated production for 1997 is 100,000 metric tons with a farm gate value of £230 million and a retail value of £500 million. Atlantic salmon is also produced by other countries both North and South of the equator and the main competition comes from Norwegian and Chilean products. This competition in the market place has resulted in a fall in price and, where salmon once commanded £6,000 per ton, it now sells at £2,300 per ton ex-farm and has become available to everyone through the retail chain stores.

The Scottish Salmon Growers Association (SSGA) was formed in 1982 with the aim of marketing farmed salmon for its members. Since then, it has become established as the lead industry body for representation of salmon farming in Scotland. In addition to marketing, its remit includes political representation and technical issues. Funds are raised on a yearly basis from subscriptions and a voluntary levy on each ton of fish produced by members. The Technical Department receives a proportion of the money generated to spend on issues which our members decide require attention.

Scotland is the third largest Atlantic salmon producing country. Scottish salmon is perceived to be the best and fills the quality niche market. In this respect, it is supported by 'The Product Certification Scheme for Scottish Quality Farmed Salmon', the scheme behind the Tartan Quality Mark which is promoted by the Scottish Salmon Board. The scheme is run and monitored by Food Certification (Scotland) Ltd. which is accredited to the European Standard, EN 45011. This means that independent inspectors monitor farming practices according to the standards set in the Quality Manual. These standards have been set after many years of research and development at a cost of about £20 million. The process is dynamic and adapts to incorporate the best standards of fish welfare, quality and hygiene.

In order to guarantee supplies of high quality salmon for the scheme, SSGA has been conducting a robust research program for over a decade. The mechanism for this is a Technical Committee that comprises farmers and technical experts. Smaller Working Groups discuss specific issues, agree research contracts and monitor progress through the services of a Research and Development Manager. Two major health problems will be used to illustrate how a common approach can yield results.

2 FURUNCULOSIS

In the middle of the 1980s the fledgling industry was booming but a number of production problems were becoming more persistent with the result that fish welfare was compromised and survival was poor. The worst of these was a bacterial disease called furunculosis, caused by *Aeromonas salmonicida*. Salmon farmers had lived with it for some years but resistance to the available antimicrobials had developed and the volume of fish for harvest was being affected.

In order to combat the problem, SSGA agreed to increase its levy and significant funds were generated. A research program was devised to concentrate on three main approaches. These were to improve management practices, to develop new medicines and to find an efficacious vaccine.

Good farm management practices are crucial to the control of infectious diseases and systems of daily dead fish removal were developed and adopted. Once this practice was introduced it greatly helped during outbreaks of the disease. Also, a study undertaken by the Institute of Aquaculture (IOA), University of Stirling into the epidemiology of the disease confirmed that fallowing of a farm followed by restocking with healthy smolts can have significant beneficial effects.

However, despite the best management practices, furunculosis still appeared and the search for new efficacious medicines was extended. SSGA contracted the IOA to act as a focus for pharmaceutical companies wishing to enter the market and both amoxycillin and florphenicol were identified by this screening project. The contractors also worked on the application of licensed products for best on-farm efficacy.

The search for a vaccine received the most backing. The *A. salmonicida* bacterium is well adapted to coping with host defenses and it had proved to be very difficult to make an efficacious vaccine against it. There were a number of institutes and pharmaceutical companies working in the field and SSGA set out to meet them all in order to assess where best to spend its members' money. Contracts were agreed with scientists in Ireland, Norway and England but the main thrust of the program took place at the Scottish Office Marine Laboratory in Aberdeen in conjunction with the IOA and also at Glasgow University.

When the Stirling/Aberdeen work came to fruition, a consortium of funding bodies was formed and SSGA was asked to find a manufacturer to take the product to license. In teaming up with Aquaculture Vaccines Limited, SSGA was able to organize large field trials on its members' farms and these proved that the vaccine was indeed efficacious. The authorities were subsequently able to grant a product license and the vaccine became commercially available through the 'Furovac' range. There are now a number of vaccine products and there is good competition in the market place. SSGA is no longer involved directly in this field but, because injection of fish is labour intensive, is still looking at better methods of delivery. Over ninety per cent of fish are now vaccinated before they are delivered to sea and the impact of furunculosis is now much reduced.

3 SEA LICE

Sea Lice (*Lepeophtheirus salmonis* and *Caligus elongatus*) are natural parasites of Atlantic salmon and were always around exacerbating the problems of furunculosis control but once that had been minimized they became the prime welfare issue. One medicine based on dichlorvos and called 'Aquagard' was authorized as a bath treatment but there were problems with administration and, due to constant use, in some areas resistant populations of lice had developed.

Recognizing this problem, SSGA had been contracting IOA to screen potential products against lice since 1989. However, in 1993, the program was expanded and a strategy was developed. New projects were contracted under the headings, Novel Treatments, Biological Research, and Medicinal Treatments.

3.1 *Novel Treatments*

Novel treatments included the use of wrasse as cleaner fish, the development of light lures and the use of ultrasound.

3.1.1 Wrasse

The use of cleaner wrasse was developed in Norway and SSGA carried out their own proving trial before advising members to try them out. Although not applicable to all farms their use has now become established as an aid to control of lice in Scotland. Various species of wrasse are caught around the Scottish coasts and are introduced to the salmon smolt pens in the summer. Goldsinny wrasse are favored and they soon learn to pick lice off the young salmon and this can mean that treatments are not necessary. There is, however, a problem with overwintering of wrasse and this means that new supplies need to be caught every year.

3.1.2 *Light Lures*

Sea lice are positively phototactic and, in 1993, this behavior was exploited by a Glasgow based engineer and inventor. With sponsorship from SSGA, he studied the reaction of nauplii and copepodid larvae to certain light sources. Through his company, he then engineered a prototype 'light lure' which, when tested in tanks of sea water, attracted all stages of lice. Trials at fish farms proved to be more problematical. Lice were attracted to the lures when they were deployed in and around salmon farms, but the numbers caught were very small. In order to quantify the effect of the lures, a field study was undertaken on behalf of the SSGA by IOA scientists, and the resulting report concluded that there were no significant differences in the numbers of lice obtained from control and experimental samples.

3.1.3 *Ultrasound*

SSGA has also sponsored research into the effects of ultrasound on lice and salmon. Farmers would welcome a control method based on this type of technology because it is believed that the bureaucratic constraints put on them regarding medicines would not affect progress. The work is being carried out under secrecy and cannot be reported further at this time.

3.2 *Biological Research*

3.2.1 *Reproductive biology and semiochemicals (pheromones).*

An understanding of the reproductive biology of sea lice is important when considering epidemiological aspects of infestations on salmon farms. SSGA sponsored a studentship at the University of Aberdeen to look at both structural and behavioral aspects of reproduction in *L. salmonis*. An interesting finding was that, during mate searching, males showed behaviors that suggested that they were reacting to chemical signals from females. It was decided to follow this line of research and a separate study was undertaken to confirm the existence and role of semiochemicals in the biology of *L. salmonis*.

Many pests are now managed by the use of semiochemicals and the SSGA was sufficiently impressed with the potential to sponsor a CASE studentship to continue the work. This work has now been completed and a paper on the work was given at the Sea Louse Workshop that is part of this conference.

A successful application for funds to continue this line of research was made to the UK Aquaculture LINK Initiative and the research project that is designed to find ways of applying the results of the previous studies is now underway.

3.2.2 *Epidemiology and management options*
The concepts of Rotational Farming and Fallow Periods for salmon farms were developed in Scotland in the late 1970s in order to combat the effects of the bacterial disease, furunculosis. The objective was to break the cycle of disease by fallowing a farm, disinfecting the hardware, and re-stocking with healthy smolts. This worked well as an aid to combating furunculosis and the resulting single year class sites also showed a benefit in terms of sea louse numbers.

Further proof of the efficacy of rotational farming in lowering louse numbers was generated through an SSGA sponsored study at IOA. Infestation levels of sea lice on single and multi-year class sites were monitored for a period of 20 months. The results demonstrated that single year class sites had a slower build up of lice and less treatments were required when compared with smolts stocked on multi-year class sites.

Today, fallowing is used where possible and farmers have joined together in certain areas under Management Agreements in order to make the best use of this technology.

3.3 *Medicinal Treatments*

Since 1989, SSGA has sponsored work to identify new candidates for licensing. The main thrust of this research has been at IOA through the funding of a research assistant to screen potential products.

3.3.1 *Azamethiphos*
The first medicine to be identified by the screening process was the organophosphate, azamethiphos (Novartis). The licensed product is known as 'Salmosan', but, although it has been available in Norway since 1993, it has only just been made available to Scottish salmon farmers on a limited basis. This is due to the different approval systems in place in the two countries. Although there is cross resistance in lice between dichlorvos and azamethiphos, Salmosan is much more 'fish-friendly' and will be a useful addition to any Integrated Pest Management Strategy (IPMS) that might be devised.

3.3.2 *Cypermethrin*
The project at IOA also identified the potential for pyrethroids (synthetic pyrethrins) to kill lice and the product 'Excis' (Grampian Pharmaceuticals), which is based on cypermethrin, was first evaluated through this work. 'Excis' has been widely available in Norway since 1995, but again the different approval system in the EU and UK accounts for the fact that it is not yet widely available to Scottish salmon farmers. The product has a good safety margin for fish and is effective against all stages of lice. It will therefore be a very good addition to any IPMS that is devised.

3.3.3 *In-feed Treatments*
Three in-feed treatments are currently undergoing trials on Scottish salmon farms. Two of these are the Insect Growth Regulators (IGRs), teflubenzuron and diflubenzuron, and these are already widely available in Norway having been developed by two major salmon feed companies. The other is a novel avermectin that is effective against all stages.

3.3.4 *The search for a Vaccine*

The successful use of vaccines to control bacterial diseases like vibriosis and furunculosis has encouraged farmers to invest in the development of a vaccine against sea lice. Research is following a line of enquiry using a biotechnological approach to finding 'hidden antigens'. This method was successful in the development of a vaccine against cattle ticks and aims to produce antibodies in the fish that attack the gut cells in the louse. The project was mainly funded by EU grants and was coordinated by Dr Munro of the Scottish Office Agriculture Environment and Fisheries Department (SOAEFD) at the Marine Laboratory, Aberdeen. This is a long term line of research and, while the EC project reached its objectives in terms of potential antigens tested, none was found to be consistently protective. EU money is difficult to obtain and funding for this project now lies with the Aquaculture LINK Initiative and the collaborators are the UK Government, SSGA and a vaccine manufacturing company.

3.4 *Sea Lice Conclusions*

Over the last ten years, in order to develop an IPMS for sea lice, SSGA has spent £1,700,000 of its members' money on research. The results of much of the work have been published in the scientific press while more papers are being prepared and will appear in due course. In terms of new medicines, most of the science has been done and it is up to the European Medicines Evaluation Agency (EMEA), Veterinary Medicines Directorate (VMD) and the Scottish Environment Protection Agency (SEPA) to decide whether these new medicines can be used in Scotland. When a number of different active ingredients become licensed as medicines for fish, SSGA intends to conserve sensitivity by developing a strategy for use based on published experiences of insect pest control.

In the meantime, farmers monitor lice levels on a regular basis by taking samples of fish from the pens and counting the number of parasites. The population structure of the lice is analyzed and treatments are agreed with the veterinarian involved. The number of treatments is both minimized, to preserve efficacy, and synchronized to make sure the maximum number of lice is killed. Two treatments may be required to kill off those immature lice that were attached by a filament when the first dose was given. However, while every effort is made to use the best technique and to administer the correct dose, the difficulties inherent in bath treatments can mean that therapy is required more frequently than farmers would like, resulting in further disturbance to routine husbandry with subsequent loss of growth. The expected authorization of efficacious new medicines cannot come too soon and must be welcomed by all interested parties.

4 EUROPEAN UNION FUNDING – THE AIR AND FAIR PROGRAMS

While the European Union might be seen as a good source of funding for collaborative research into aquacultural problems, in practice, this has not proved to be the case. Scottish salmon farmers have benefited from one project on the development of a vaccine against sea lice but the project was very ambitious and has not resulted in a vaccine being submitted for license. In general, industry finds that the AIR and FAIR programs are administered by bureaucrats for academics and there is no money for 'real issues'. The application procedure is also very complex and when the qualifying criteria cannot be met, industry tends to look elsewhere for the required funds.

5 FUTURE RESEARCH – COLLABORATIVE STUDIES – THE LINK AQUACULTURE INITIATIVE.

The majority of the research described above was funded by the Industry with some assistance from Highland and Islands Enterprise, a Government Agency and the Crown Estates Commissioners, who act as landlords as regards seabed leases. In the United Kingdom, this type of collaboration has given way to participation in a new 'Aquaculture LINK Initiative' which started in 1996. The program will run for five years in the first instance and it is worth £8 million. The allocation of the money is in the hands of a Program Management Committee (PMC) which comprises experts from both the aquaculture industry and academic institutions. Advertisements are placed in the popular scientific press inviting consortia to bid for the funds on a project by project basis. Outline proposals are submitted and the panel decides which should be worked up into full proposals.

A full proposal has to show that it will address a constraint to the development of aquaculture, is scientifically sound and is 50% supported by Industry. Applications are peer reviewed and then discussed by the PMC which awards the grant. SSGA has been prominent in making successful applications for the funds and to date has been a successful partner in 13 projects that have a total value of £2.5 million at a cost of around £0.5 million. This apparent disparity on the 50/50 expectation is due to the contribution of other industrial partners such as the trout industry and the pharmaceutical companies. While it may be seen that by participating in the LINK program the SSGA research program has lost some of its focus, it has gained in terms of value for money and it is also able to attend to some problems that are lower down the priority list.

This method of funding projects has been well received and, as margins tighten and money levied from salmon farmers becomes harder to find, is likely to be favored in the future.

Selective effects of fishing on the population dynamics of crustaceans

GLEN S. JAMIESON
Science Branch, Fisheries and Oceans Canada, Pacific Biological Station, Canada

ABSTRACT

To fishery scientists, selectivity generally means the effectiveness with which fishing gear captures and retains animals of variable size. However, with full exploitation today of most benthic crustacean resources, concern about another aspect of selection, the consequences of selective harvest on species' genetics, population dynamics and ecology, is increasing. The results of intentional selection for specific traits in terrestrial species are well known, being reflected in diverse varieties of farmed plants and animals, and pet species such as dogs and cats. Significant evolution can occur in 3-10 generations. Selective wild harvest of a species is not intentionally directed towards modification of any specific attribute, but this does not mean that there are no resultant physiological or behavioral responses, some of which may be genetically based. Most harvest practices in crustaceans are size selective to ensure that individuals theoretically have opportunity to breed before they can be captured. If all individuals in a population have equal reproductive opportunity, then apart from decreasing overall population fecundity, effects from selective harvest may be minimal. However, because selective harvest alters the size structure of the breeding population by removing the largest individuals, the relative reproductive contribution of smaller individuals may be increased. If inadvertent genetic selection for earlier sexual maturity or smaller adult size is occurring because of the relative timings of molting and fishery openings, or if the maximum size attained by an individual is effected in a non-genetic manner by intraspecific factors, then continued fishing in a proscribed historical manner may be eroding the potential yield which could be obtained from a population. Despite difficulties in evaluating possible genetic change in a wild population, it is useful to consider the intensity of selective harvests, the manner in which they are conducted, possible consequences, and whether mitigative management measures may be required.

1 INTRODUCTION

To fishery scientists, selectivity generally means the effectiveness with which fishing gear captures and retains animals of variable size. However, Policansky (1993a,b) has argued that since fishing mortality is often very high and nonrandom with respect to

627

at least partly heritable life history traits, it is likely that fishing can cause selection, or evolution, in specific characteristics. Policansky (1993a) discussed the degree to which such evolution might occur and how to detect and measure it but concluded the actions of other factors affecting phenotypes was likely to make detection and measurement of induced genetic change difficult. For evolution to occur there must be heritable variation in the traits being considered and fishing must cause differential reproduction of genotypes.

With full exploitation today of most benthic crustacean resources, the implications of fisheries selection, i.e. the consequences of selective harvest on a species' genetics, population dynamics and ecology, need to be considered. The selective capture of individuals with specific characteristics, usually larger size, is a common management goal. The literature on gear selectivity is too extensive to be considered here in any detail, as the effects of gear selection are often complex. Although size-selectivity is an obvious result of nonrandom fishing, other traits, though, can be selected for by fishing practices. Some of these other traits may be related to size, such as locomotion rate. Larger individuals, for example, generally move over larger distances than do smaller individuals, and so would be more likely to encounter stationary fishing gear (see Rudstam et al. 1984), such as traps, in a given time interval. Selection could thus directly be for individuals with less mobility.

Similar selection for other, non-size related behaviors, such as greater crypsis, may also be inadvertently selected for. A documented example is in intensive abalone (*Haliotis* sp.) harvest areas (whether by a fishery or by sea otter (*Enhydra lutris*) predation), where individual abalone which survive longest are those that tend to remain inaccessible in crevices or under rocks (Breen 1986). The extent to which this specific behavior is genetically determined and how it applies to crustaceans, which tend to roam further than abalone, is unknown, but animal behavior often has a genetic component.

Strong nonrandom selection as a result of mortality, whether from predation or fishing, can be assumed to influence the evolution of all organisms, including crustaceans. This has been described to a limited extent in the literature for crustaceans, with natural selective predation of crustaceans reported to influence evolution in a variety of traits (e.g. shore avoidance: Gliwicz & Rykowska 1992; hatching rhythms and dispersion: Anger et al. 1994; reproductive success: Wellborn 1994, 1995; phototactic behavior: De Meester 1996; adult size: Sparkes 1996). Most of these examples involve relatively short-lived, planktonic prey species, and there is limited documentation on the effects of possible selection on larger, commercially exploited crustaceans that typically have meroplanktonic larvae. Effects arising from the consequences of selective harvest may be manifested through a species' genetics, but even if a genetic effect cannot be proven, there may be resultant consequences in terms of a species' population dynamics and ecology. Here I will consider the consequences of selective harvest from both the perspective of the harvested crustacean and the predator, often human, causing the mortality.

2 FISHERY EFFECTS ON SPECIES CHARACTERISTICS

While there is a tendency to consider species as stable entities from the narrow perspective of our few centuries of published literature, the reality is that individual species are simply the current state in a never-ending evolutionary process of life adapting to the environment around it. In any population of a species, there is considerable range of genetic diversity for many traits, and some genetically determined traits will under certain conditions positively or negatively effect the reproductive output of a specific genotype. If a selective pressure is intense enough or sustained long enough on such traits, then an influenced, genetically based trait may become sufficiently dominant in the population to then characterize our perception of the species.

The rate at which this happens is determined by both the heritability of the trait, and the differential reproductive rates of different genotypes that occur in the population. In animal husbandry, which includes aquaculture, only individuals with the most desired traits are often permitted to breed, and genotype selection may be quite rapid. Significant evolution can occur in 3-10 generations (Falconer 1989, p.33). Selective wild harvest of a species is not intentionally directed towards modification of any specific attribute, but this does not mean that there are no resultant physiological or behavioral responses within individuals, some of which may be genetically based. One of the best documented examples of this happens to involve pandalid shrimp. Jensen (1965, 1967) first suggested that the age timing of sex reversal in the protandrous hermaphrodite *Pandalus borealis* was related to exploitation level. Ghiselin (1969) subsequently suggested that natural selection favors sex reversal over dioecism when an individual's reproductive success as a male or female is related to size or sex, and is different for each sex. Jensen's suggestion was supported by Charnov (1979, 1981), who noted that formal models allow predictions to be made as to the age of sex change which would be favored by natural selection, based on different levels of reproductive success by sex. Charnov (1981) found that the predicted age of sex reversal was most sensitive to adult mortality rate (Z), not growth rate. Higher mortality rate selects for a shorter time as a male, and ultimately for some first breeders to become females (= primary females). This was because although each individual must spend some time as a male, the male phase is expressed as what the average individual does, meaning that under extreme selective pressure, some individuals may never reproductively function as adult males. Fishery-induced changes in the population sex-age structure of *Pandalus jordani* have been similarly reported (Hannah & Jones 1991).

It is interesting, though, to note that this response has only been noted in pandaliid species that are fished by trawls, i.e. gear which does not strongly select by adult size. Pandaliid populations, e.g. the spot prawn, *Pandalus platyceros*, in the northeast Pacific, which are fished by size selective traps having escape rings that efficiently allow the release of smaller, mostly male, adults below a minimum size limit, show little if any presence of primary females even under intense fishing pressure (J. Boutillier, Dept. Fish. Oceans, Nanaimo, B.C., Canada, pers. comm.). In this case, male mortality with size selective traps is not as intense as male mortality apparently is with less selective trawls. There is thus insufficient selection to noticeably indicate earlier sex reversal.

Jamieson et al. (1998) considered a number of ways in which the effects of fishing could selectively affect a species' population dynamics and/or reproductive potential.

2.1   Removal of larger individuals

2.1.1   Decrease in mean adult size

Although few in total abundance, the larger individuals typically contain more muscle or meat and have larger reproductive potential. Fishery management, which tries to balance growth with mortality, based on optimal yield, or egg/recruit, favors exploitation of larger individuals. In fisheries where individual size can be cost-effectively selected for, such as most crustacean trap fisheries, the largest individuals are typically preferred, and often command the highest prices. However, the largest individual is a relative term, and the average absolute size can decrease as a fishery evolves. Using Dungeness crab (*Cancer magister*) as an example, in relatively virgin populations, crab averaging around 200 mm carapace width (CW, notch-to-notch) have been reported, either anecdotally (T. Butler, Dept. Fish. and Oceans, Nanaimo, B.C., pers. comm.) or in the literature (Merritt 1985), to characterize fisheries. The largest Dungeness crab landed was in excess of 230 mm CW. In some locations, this may in part reflect habitat conditions, but large size has also been suggested to result from different rates of growth per molt (Merritt 1985), at least by a portion of the population. When a population has been heavily exploited for more than a few decades, though, few crabs survive to reach these larger sizes, and mean harvested size becomes relatively close to the minimum legal size limit. If growth potential is at least partially genetically determined, then minimizing the reproductive opportunities of larger individuals may have selective consequences. Larger individuals often produce disproportionately more larvae than smaller individuals, and their removal from the population likely accentuates the relative reproductive contribution from slower-growing individuals.

2.1.2   Decrease in mating opportunity of larger adults

The potential problem here is unique to arthropods, and particularly crustaceans, which are the only members of this Phylum routinely harvested by humans for food. All other animals show continuous growth, which means that individual animals gradually increase in size through a species' normal size range, and perform the activities typically associated with that particular size, such as reproduction if mature. Crustaceans, on the other hand, show discontinuous growth and by sudden size increases during molting, individuals leapfrog through the size range of that species. This means that a relatively large crustacean may not have previously had opportunity to reproduce, even though it is quite larger than other mature individuals in the population. Minimal size selection thresholds, while particularly effective with crustaceans which because of their shells, can be accurately measured and released with minimal harm if sublegal, are thus not selective with respect to what life history activities that individual may have previously experienced. Use of a size limit alone, for example, if set within an instar's size range rather than between instars, cannot necessarily ensure that only individuals are caught which may have reproduced.

With Pacific Dungeness crab (*Cancer magister*), only males are fished, and fisheries only typically commence as soon as exoskeletons harden up after the major molting period to legal size. Most male crab usually molt a few months prior to periods of major female molting, which allows them to have the hard shells required to mate successfully with a soft-shell female. Regular removal by fisheries of the largest, presumably fastest growing, individuals in a year class before such individuals have opportunity to mate has the potential to exert strong selective pressure.

In a possible documentation of resulting effects, Jamieson et al. (1998) described for two British Columbian fisheries potential consequences of having only the smallest adults do most of the mating. In the populations studied, the largest 50% of individuals in the each recruiting year class (the minimum size limit of 165 mm CW generally equally splits recruiting male year classes, which also happen to be the first, sexually mature male instar) of males is mostly exploited in the time interval between male and female moultings, as the fishery can be very intense (F = 5.1-6.9, Smith & Jamieson 1989). This leaves most mating to be effected by the remaining males in each year class, and perhaps as a result of increased mating activity, most of these sublegal size males do not molt the following year, and indeed never seem to subsequently molt to a larger, and legal, size. With existing fishing regulations, not only do the smaller, presumably slower growing males thus do most of the mating, but fishers may be forgoing a significant percentage of the potential yield obtainable from each year class. With no unexploited crab populations on the coast, it is uncertain whether removal of larger males has resulted in reduced probability of a smaller male molting, but regardless, the situation is primed to maximize the opportunity for genetic selection to occur.

A variation on this theme occurs with the Caribbean stone crab (*Menippe mercenaria*), which supports a unique fishery in which only crabs' claws are harvested. The minimum legal claw size is 70 mm propodus length which in South Carolina, at least, is attained with the larger, crusher claw at average male and female sizes of 84 and 90 mm CW, respectively (Caldwell 1992). The smaller pincer claw only becomes harvestable in male crab > 90 mm CW, and few female crabs ever have harvestable pincer claws (Restrepo 1992). Declawed crabs must be returned to the ocean, in the hope that they will regenerate their massive claws, which can each weigh up to 250 g; claw regeneration accounts for 1-10% of the annual landing (Savage et al. 1975; Ehrhardt & Restrepo 1989). Up to 50% of harvested stone crab may survive after release, dependent on fishing and handling practices (Davis et al. 1979). However, although some declawed crab may survive, the effect of declawing on an individual's reproductive success has not been studied. Declawing may decrease the ability of that individual to reproduce, either because it inhibits feeding, or it affects the ability of males in either agonistic mating interactions with other males or their ability to clasp females successfully. From a selective perspective, declawing may greatly reduce an individual crab's fitness, subjecting the population to the same selective pressure as would have resulted from its complete retention when it was first captured.

2.1.3 *Exploitation below the size of sexual maturity*

A more extreme example of the above is with populations which are exploited often as immatures, such as near shore populations of the American lobster, *Homarus americanus*, in the Bay of Fundy (Campbell & Robinson 1983), or the spider crab, *Maja squinado*, in Spain (ICES 1996). These populations apparently persist because of immigration of either meroplanktonic larvae or juveniles from refugia, such as deeper-water populations, which may be exploited at a size above that of sexual maturity. Not only is total population fecundity reduced, but where such populations are a source, rather than a sink, population in terms of recruitment, again a smaller size at sexual maturity would be strongly selected for.

Fisheries may also selectively affect the seasonal timing of successful mating, either directly or inadvertently, by their timing and intensity of removal of specific cohorts of individuals. In eastern Canada, mating in snow crab (*Chionoecetes opilio*) occurs at the end of winter or the beginning of spring and while the ocean is still covered with ice, i.e. before fishing can commence in the northeastern part of the Gulf of St. Lawrence. However, ice cover disappears earlier in the major crab fisheries of the remaining parts of the Gulf. There, spring fisheries take competent breeder males of commercial size (size limit = 95 mm CW) before they mate that year, although because of the complexity of snow crab reproductive behavior, some of these males may have had a chance to mate previously as adolescents or morphologically immature males. In years with high exploitation rate, most successful mating in these later exploited populations may thus be accomplished by sub-legal adult or adolescent males, which can mate as small as 40 mm CW, and so fishing might be inadvertently selecting for earlier, smaller size of morphological maturity, or terminal molt. Creation of 'pigmy males' by fisheries selection against larger size males has been actively discussed in the past. The subject has been of primary importance since a male terminal molt and its possible size-genetic component were accepted (see Conan & Comeau (1986) and Jamieson & McKone (1988)), but research to date has never proved, nor disproved this hypothesis.

2.2 *Removal of mainly smaller individuals*

Fishery preference for small individuals within a population is practiced when the monetary return from a harvested individual is determined either by some quality aspect other than its absolute size or weight, or the population is so heavily fished that, in what might be a multi-year class fishery, each year's new recruits are the primary target. Concerning the former, an example is the harvest of immature peeler blue crab (*Callinectes sapidus*) for the soft shell market (Oesterling & Moore 1995). The soft crab fishery is smaller than the hard shell fishery, but more valuable per unit weight (in 1995, soft crabs were $US 5.93 kg^{-1}; hard crabs were $US 0.86 kg^{-1} (Oesterling 1995)). In 1995, over 5 million soft-shell crab were produced from peeler crab; there is no minimum size limit for peeler (either sex) or hard-shell female crab, although there is for fishing soft shell crab (90 mm CW, spine tip to spine tip) and hard-shell male crab (127 mm CW). Harvested peeler and soft-shell crab have not usually had prior opportunity to reproduce, and the effect of this collective removal remains unknown on species' dynamics. It is relatively small as a percentage of landed weight (1.5-3.3%, Oesterling & Moore 1995), but because soft-shell crabs are smaller, they comprise a greater percentage of landed individuals.

Roe is a targeted product is some fish (caviar and herring roe) and invertebrate (uni, from sea urchins) species. However, while crustacean roe is often a desirable gastronomic dish, e.g. lobster coral roe from the Chinese mitten crab (*Eriocheir sinensis*) and roe from the Asian blue crab (*Portunus trituberculatus*) (I. Yeon & S.Y. Hong, Korea, pers. comm.), most crustacean roe consumed appears to be an ancillary item and not the target of directed fishing. There is thus unlikely to be any selective effect on crustacean populations associated with this product.

2.3 *Change in local carrying capacity*

As described above, fished Caribbean stone crabs are often released alive after their largest claw(s) are removed. This may directly decrease their future ability to be successful in mating, but it also has implications with respect to habitat carrying capacity, and the latter can affect overall population fecundity. Experimental increase in the availability of refuges has been argued either to increase population size, or to simply attract animals, with no real population size consequences. It has been usually assumed that size classes of animals at significant predation risk, such as smaller individuals, are the most likely to be shelter limited, but as Beck (1995), in his study of stone crab, concluded, this may mostly apply to larger individuals in some species. Using artificial shelters of different sizes, he found that the largest shelters were occupied in significantly greater proportions than smaller shelters. Small crabs were not excluded from shelter by inter-specific competition and seemed always able to find shelter by burrowing in the substrate. No crabs > 60 mm CW were observed to burrow, but use of shelter also had no effect on survival of > 60 mm, hard shell crab, which seemed to have few major predators. However, shelters strongly attracted gravid females and molting individuals, and shelter had a strong effect on the fecundity of females. Females that had shelters produced egg masses twice as frequently as those did without shelters; female stone crab can lay up to 13 egg masses in an intermolt period (Cheung 1969). Their egg masses, each containing 100,000-500,000 eggs, are held by females for 7-10 days; after hatching an egg mass, another can be spawned within 2 days (Beck 1995). Shelter also reduces post-ecdysial vulnerability (e.g. Beck & Connor 1992), and the availability of shelter has been hypothesized as an important external cue for molting in at least some crustaceans (Addison 1986). Lost appendages in crustaceans can only be regenerated through molting (Bliss 1960, Warner 1977), and loss of limbs stimulates onset of the next molt. Environmental stimuli also affect molting, and in the brachyuran *Gecarcinus lateralis* at least, light inhibits pre-molt limb generation, which only proceeds in darkness. Male claw removal should increase an individual's molting frequency, which increases competition for shelters and potentially limits shelters for smaller, gravid females. Decreased mating opportunity by handicapped, large crabs may favor the mating success of smaller, slower growing males, and overall population fecundity may be decreased by fewer spawnings of females.

3 CONDITIONS FAVORABLE TO RAPID GENETIC SELECTION

3.1 *General issues*

Species are most likely to experience significant genetic selection as a result of fishery harvest if their larval production and dispersal are relatively limited, and/or if fishing limits the reproductive contribution of a particular cohort in the population. Limited population fecundity and dispersal likely means the gene pool is relatively small, which in turn is likely to have resulted in greater specialization or adaptation to local conditions. Changed reproductive capacity from fishing can arise from a number of causes. These include greater exposure to predation from habitat disruption; regular, almost complete, exploitation of a portion of the adults before reproduction;

food chain path modification as a result of exploitation of other species; or modified natural mortality rate on a specific genotype in the population. Factors which might minimize genetic selection are: if only one sex is predominantely exploited, such as with many crabs, as the genetic contribution of females may most influence growth and/or fecundity; and if the habitat occupied by the population is a sink in terms of recruitment within the metapopulation.

3.2 Relevant factors

3.2.1 Heritable population variation

Policansky (1993a) described the limited number of examples in fish where phenotypic traits have been shown to be influenced by genotypic variation, and there have been a few more examples since then (e.g. Baumgartner 1995). There have also been similar investigations of invertebrates (e.g. Kirby et al. 1994; Ream 1996; Bruno & Edmunds 1997), and even of crustaceans (e.g. Ebert et al. 1993a, Clancy 1995, De Meester 1996), but there appears to have been no such study with harvested crustaceans.

However, there has been evaluation of the implications of phenotypic plasticity (Bradshaw 1965). Pollock (1995) differentiated between phentoypic responses by populations of spiny lobsters to changing environmental variables (including those caused by humans) and those presumably more persistent, measurable differences among populations which have a genetic origin. Changes in rates of growth, mortality and reproduction can all be responses of phenotypic plasticity. Animal density, for example, can significantly affect both growth (Pollock 1993; Breen & Booth 1989) and survival rates (Breen & Booth 1989). In fishes, growth rate is a major force behind life-history variation (Metcalfe 1993), and age at sexual maturation is related to growth rate. Faster growing individuals tend to mature at an earlier age, but larger sizes make them more vulnerable to size-selective fishing, particularly when there is a minimum size threshold.

Growth rate is in turn influenced by food availability, or at least by the energy reserves in an animal. Food level not only affects phenotype but may also affect the heritability of traits, i.e. the relative importance of heredity in determining phenotypic values. However, a character can be hereditary in the sense of being determined by the genotype (the broad sense) or in the sense of being transmitted from parents to offspring (the narrow sense), and the extent to which these are both true may differ (Falconer 1981). Table 1 lists heritabilities in the broad sense for the only crustacean

Table 1: Heritabilities in the broad sense of characters at specific food levels (Ebert et al. 1993 a) and their plasticity across food levels (Ebert et al. 1993b) in *Daphnia magna* for selected characteristics.

Trait	Heritability (%)	Heritability of plasticity (%)
Adult length (7th instar) (high food level)	81.0	~60.0
Adult length (7th instar) (low food level)	36.0	~60.0
Age-at-maturity (high food level)	22.6	35.0
Age-at-maturity (low food level)	40.2	35.0
Clutch size (high food level)	30.0	~20.0
Clutch size (low food level)	15.0	~20.0

data (Ebert et al. 1993a, 1993b) I am aware of; heritabilities in the narrow sense were generally lower.

Increased food supply, perhaps as a result of reduced recruitment, increases growth, and since the mean number of molts to maturity is species-specific in most crustaceans (Hartnoll 1985), increases mean size at maturity. This in turn would likely normally increase lifetime fecundity by females, but as Pollock (1995) observed this compensatory response can be offset by an intensive fishery directed at sexually mature adults. Increased density should tend to act in the opposite direction, and reduce length at maturity.

3.2.2 *Distinct metapopulations*

Physical parameters such as water temperature also affect molt frequency in crustaceans, and the latter influences generation time. However, most exploited crustacean species have meroplanktonic larvae, which are typically relatively widely dispersed by water currents. In order for genetically distinct sub-populations to develop, some isolating mechanism is required. Pollock (1995) documents how *Panulirus* species apparently evolved in response to physical oceanographic parameters that isolated populations, and Jamieson & Phillips (1993) showed that even within a time period as short as 13,000 years, i.e. since the last ice age in British Columbia, significant life history features can evolve. Dungeness crab (*Cancer magister*) have about a 4 month larval period, with larvae potentially moving great distances from their hatching location (Jamieson et al. 1989, McConnaughey et al. 1995). Jamieson & Phillips (1993) studied inner (Strait of Georgia) and outer coast Dungeness crab megalopae to determine how the inner coast population could persist, since opportunity for crab larvae retention in, or emigration into, waters only connected to the outer coast by the 25 km wide, 100 km long Juan de Fuca Strait seemed limited. They found that diel vertical migratory behavior differed substantially between the two stocks, with inner and outer coast megalopae migrating about 160 and 25 m, respectively; both groups of megalopae were located at the surface at night. Because of a large freshwater flow into inner waters, Juan de Fuca Strait has estuarine circulation characteristics. The top 100 m of Juan de Fuca Strait mostly flows seawards, while the bottom 100 m flows inwards, and at a latitude of 49°N, day length when megalopae are present is about 16 h. This means that outer coast megalopae are always in outward flowing surface water, while for two-thirds of the day, inner coast megalopae are in inward flowing water. This difference in spatial distribution largely isolates the two crab populations, and because the deeper water is only 7-9°C, in contrast to 12-16°C outer coast surface waters, also separates peak periods of larval settlement between the two populations by 2-3 months. The inner coast was glaciated during the last ice age, and since inner coast crab are unique in this behavior, presumed to now be genetically determined, this study illustrates the potential for relatively rapid genetic selection to occur naturally.

4 POTENTIAL FOR GENETIC SELECTION OR DETECTION OF EVOLUTION

Most crustacean fisheries have only been really intense over the entire range of a species for the past half century or so at most, much shorter that the thousands of years suggested for the Dungeness crab evolution described above. However, this latter evolution rate may have been initially quite rapid, and then have slowed down in more recent years. What time frames are thus required to effect genetic change, and is it practical to try and separate genetic change from possible phenotypic plasticity? A major difficulty here is the limited duration of documented data – only decades for life history traits, and a few thousand years at most for hard structure remains found in the middens of indigenous peoples. The fossil record is not informative over the relatively short time frames involved, i.e. a few thousand years or less. As Policansky (1993a) noted, the difficulty in disentangling genetic from phenotypic based changes is probably responsible for the general lack of attention paid to this issue by both fisheries researchers and managers, despite the topic being repeatedly brought up in the literature. Confusion arises from compensatory effects (described above); Baranov's 'fishing-up' effect (Baronov 1918 – in longer lived species, old fish are subject to repeated fishing for a longer time than younger ones causing the average growth rate of remaining fish to be higher than actually for the population as a whole since growth rates of younger fish are higher than for older fish); possible migrations into or out of the study area; and environmental changes, such as natural cycles or even global warming, not directly related to fishing (Policansky 1993a). Further confusion arises from changing fishing practices over time, due to technological developments and changing market demand, and changes in the quality and quantity of research data available for time series analyses. Data collected for a particular reason may not be particularly appropriate to investigate a different factor such as possible genetic change in a species' characteristics over time.

Nevertheless, while reasons to either defer study or fail to adopt appropriate management actions might be understandable, they are not really justifiable. Arguments that selective fishing can cause evolution go back to at least 1957 (Miller 1957, see Policansky 1993a for a review). With the declared goal of most fisheries management authorities to lean towards the side of conservation, fisheries should be managed, even in the absence of clear scientific advice, to minimize the likelihood of their causing species' selection (Nelson & Soulé 1987). Clear evidence of the effects of fishing selection may best be evaluated through experimental fisheries management and/or simulation of species population dynamics (Policansky 1993a). However, I suggest that fisheries managers not even wait for these evaluations before adopting preventive measures, since there is considerable evidence in the literature that heritable variation exists and that fishing can cause differential reproduction of genotypes. Law & Grey (1989) note that ignorance of understanding does not affect the progress of a species' evolution, and argue that its possible effects, both biological and economic, not be minimized or overlooked. What, then, can managers of crustacean fisheries do immediately to minimize potential negative evolutionary consequences of fishing practices, which can potentially have serious economic consequences to the fishers involved. The reality is that fisheries resource managers mostly manage people, i.e. fishermen exploiting the resource, not the resource *per se*, which from a recruitment perspective at least, is in most cases largely determined by natural environmental factors. Since fishing is the cause of the potential for evolution being considered here, managers do then have a real opportunity to ensure selection is not occurring, or is at least minimized.

5 PROPOSED PREVENTATIVE, OR MITIGATIVE, MANAGEMENT ACTIONS TO MINIMIZE SELECTION

The are several instances where phenotypic or genotypic selection is likely already happening.

5.1 *Significant fishery removal*

Significant harvesting, including poaching, remove individuals below the age/size of sexual maturity.

5.2 *Fishing mortality*

Few animals survive beyond the age of recruitment, which is low enough that many, if not most, fished animals have not mated before removal. Consequently, a specific cohort, e.g., smallest mature males, in the population subsequently does most of the mating.

5.3 *Increased mortality of cohorts having specific traits*

Cohort mortality, which might be independent of size, can have, for example, specific behavioral or morphological traits that make them vulnerable to the fishing gear or indirect fishing mortality, and thus decrease their lifetime reproductive contribution. There are several examples. 1. Large females can be captured in male-only crab fisheries (small females can escape through the escape ports), and their subsequent increased mortality occurs through rough handling in the discard process. 2. There can be an incidental capture of by-catch shrimp species in shrimp trawling, which might favor selection for other habitat preferences. Increased mortality may thus arise from exploitation of another species entirely, i.e. the impacted crustacean species doesn't have to be the fishery targeted species.

5.4 *Distinct stocks with limited larval dispersal*

Although not currently exploited (there was a limited fishery in the 1970-80s), a relic population of golden king crab, *Lithodes aequispina*, occurs in an isolated, silled, deep British Columbian fjord (Sloan 1985), geographically distant from open ocean populations now found off the Aleutian Peninsula. There is no source of larval recruitment for this isolated population other than its own larvae production.

5.5 *Significant disruption of habitat by fishing*

Habitat disturbance is likely to be associated with population effects only when habitats are significantly disrupted, such as with hydraulic dredging gear or bottom drags or trawls. Relatively small, isolated populations are again potentially most vulnerable.

5.6 *Significant trophic food chain disruption*

This is significant when it affects the crustacean species being considered. I know of no examples where trophic chain disruption has negatively affected a crustacean, but a release from possible selective pressure may be currently happening because of the collapse of northern cod (*Gadus morhua*) populations from overfishing off New-foundland. Cod was a significant crab predator, and a significant increase in *Chiono-ecetes opilio* abundance is presently occurring, with subsequent crab fishery expansion (D. Taylor, Dept. Fisheries and Oceans, Newfoundland, pers. comm.).

6 PREVENTATIVE MANAGEMENT

Preventative management actions to minimize opportunity for selective modification of a species' life history characteristics, whether the result of genetic change or phenotypic plasticity, which should exist, or be implemented, in every crustacean fishery include the following.

6.1 *Equal opportunity to reproduce*

Fishing practices should ensure that every exploited individual has had an appropriate opportunity to reproduce before it is subject to capture. This may mean increasing the size of the minimum legal size limit, or, if the recruiting size class is the first sexually mature instar, delaying onset of the fishery until after the mating and/or egg hatching period(s). This will at least ensure that every genotype in the population reproduces.

6.2 *Marine protected areas*

Much of the selective effects of fishing might be ameliorated if managers ensure that the entire population of the exploited species is not subject to fishing. Marine protected areas are increasingly becoming recognized (Orensanz & Jamieson 1998) in temperate waters as a potentially important management tool, both to enhance recruitment and to conserve genetic diversity. Many so called protected areas effectively protect little. They are often established by a level of government that has no mandate to control fishing (e.g. the municipal or provincial), when control of fishing is a federal mandate. However, if strict preservation, or no-take, refugia are established at a scale and spatial distribution that is meaningful for effective population conservation, then this approach can go a long to mitigating the effect of a less-than-perfect fishing process.

6.3 *Education*

To ensure the continuance of effective management measures, whether there be a modified fishing process or establishment of protected areas, the public and fishers need to be educated as to the need for this conservation, and their buy-in into the process maintained. Education is an ongoing process, as not only is there turnover among fishermen, there is turnover in the general human population of an area. The general public, and in particular children, need to be reminded regularly about the potential for cryptic change that may be occurring in organisms around us because of perturbation by humans of natural systems.

REFERENCES

Addison, J.T. 1986. Density-dependent mortality and the relationship between size composition and fishing effort in lobster populations. *Can. J. Fish. Aquat Sci.* 43: 2360-2367.

Anger, K, Spivak, E. , Bas, C., Ismael, D. & Luppi, T. 1994. Hatching rhythms and dispersion of decapod crustacean larvae in a brackish coastal lagoon in Argentina. *Helgol. Meeresunters.* 48: 445-466.

Baronov, F.I. 1918. On the question of the biological basis of fisheries. *Nauchnyi issledovatelskii ikhtiologicheskii Institut Isvestia* 1: 81-128 (in Russian; translated by W.E. Ricker, mimeograph, 1945).

Baumgartner, J.V. 1995. Phenotypic, genetic and environmental integration of morphology in a stream population of the threespine stickleback, *Gasterosteus aculeatus. Can. J. Fish. Aquat. Sci.* 52: 1307-1317.

Beck, M.W. 1995. Size-specific shelter limitation in stone crabs: a test of the demographic bottle-neck hypothesis. *Ecology* 76: 968-980.

Beck, M.W. & Connor, E.F. 1992. Factors affecting the reproductive success of the spider crab *Misumenoides formosipes*: the covariance between juvenile and adult traits. *Oecologia* 92: 287-295.

Bliss, D.E. 1960. Autotomy and regeneration. In T.H. Warner (ed.), *The Physiology of Crustacea. Vol. 1. Metabolism and Growth*: 561-590. New York: Academic Press.

Bradshaw, A.D. 1965. Evolutionary significance of phenotypic plasticity in plants. *Adv. Genet.* 13: 115-155.

Breen, P.A 1986. Management of the British Columbia fishery for northern abalone (*Haliotis kamschatkana*) In G.S. Jamieson & N. Bourne (eds), *North Pacific Workshop on Stock Assessment and Management of Invertebrates*: 300-312. Can. Spec. Publ. Fish. Aquat. Sci. 92.

Breen, P.A & Booth, J.D. 1989. Puerulus and juvenile abundance in the rock lobster *Jasus edwardsii* at Stewart Island, New Zealand. *New Zealand J. Mar. Freshwater Res.* 23: 519-523.

Bruno, J.F. & Edmunds, P.J. 1997. Clonal variation for phenotypic plasticity in the coral *Madracis mirabilis. Ecology* 78: 2177-2190.

Caldwell, M.A. 1992. Aspects of the biology of the stone crab, *Menippe mercenaria* (Say), from South Carolina, with comments on the South Carolina stone crab fishery. In T.M. Bert (ed.), *Proceedings of a Symposium on Stone Crab (Genus* Menippe*) Biology and Fisheries*: 99-107. *Fla. Mar. Res. Publ.* 50.

Campbell, A. & Robinson, D.G. 1983. Reproductive potential of three American lobster (*Homarus americanus*) stocks in the Canadian Maritimes. *Can. J. Fish. Aquat. Sci.* 40: 1958-1967.

Charnov, E.L. 1979. Natural selection and sex change in pandalid shrimp: test of a life-history theory. *Amer. Nat.* 113: 715-734.

Charnov, E.L. 1981. Sex reversal in *Pandalus borealis*: effect of a shrimp fishery. *Mar. Biol. Letters* 2: 53-57.

Cheung, T.S. 1969. The environmental and hormonal control of growth and reproduction in the adult female stone crab, *Menippe mercenaria* (Say). *Biol. Bull.* 136: 327-346.

Clancy, N. 1995. Environmental and population-specific contributions to growth rate variation in the marine amphipod *Jassa marmorata. Benthic Ecology meeting, New Brunswick, New Jersey, March 17-19, 1995. Abstract*: p. 23.

Conan, G.Y. & Comeau, M. 1986. Functional maturity of male snow crab (*Chionoecetes opilio*). *Can. J. Fish. Aquat. Sci.* 43: 1710-1719.

Davis, G.E., Baughman, D.S., Chapman, J.D., MacArthur, D. & Pierce, A.C. 1979. Mortality associated with declawing stone crabs, *Menippe mercenaria. Rep. SFRC T-552, Natl. park Serv.*: 23. Homestead, South Fla. Res. Center.

De Meester, L 1996. Evolutionary potential and local genetic differentiation in a phenotypically plastic trait of a cyclical parthenogen, *Daphnia magna. Evolution* 50: 1293-1298.

Ebert, D., Yampolsky, L. & Stearns, S.C. 1993a. Genetics and life history in *Daphnia magna*. 2. Heritabilities at two food levels. *Heredity* 70: 335-343.

Ebert, D., Yampolsky, L. & van Noordwijk, A.J. 1993b. Genetics and life history in *Daphnia magna*. 2. Phenotypic plasticity. *Heredity* 70: 344-352.

Ehrhardt, N.M. & Restropo, V.R. 1989. The Florida stone crab fishery: a reusable resource? In Caddy, J.F. (ed.), *Marine Invertebrate Fisheries: Their Assessment and management*: 225-240. New York: Wiley.

Falconer, D.S. 1981. *Introduction to Quantitative Genetics*. New York: Longman.

Ghiselin, M.T. 1969. The evolution of hermaphroditism among animals. *Q. Rev. Biol.* 44: 189-208.

Gliwicz, Z.M. & Rykowska, A. 1992. 'Shore avoidance' in zooplankton: a predator-induced behaviour or predator-induced mortality? *J. Plankton Res.* 14: 1331-1342.

Hannah, R.W. & Jones, S.A. 1991. Fishery-induced changes in the population structure of pink shrimp *Pandalus jordani*. *Fish. Bull, US* 89: 41-52.

Hartnoll, R.G. 1985. Growth, sexual maturity and reproductive output. *Crustacean Issues* 3: 101-128.

ICES, 1996. *Report of the Study Group on Majid Crabs, La Coruna, 20-23 November 1995*. ICES C.M. 1996/K:1, 20 pp.

Jamieson, G.S. & McKone, W.D. 1988. Proceedings of the International Workshop on Snow Crab Biology, December 8-10, 1987, Montreal, Quebec. *Can. MS Rep. Fish. Aquat. Sci.* 2005: 145 pp.

Jamieson, G.S. & Phillips, A. 1993. Megalopal spatial distribution and stock separation in Dungeness crab. *Can. J. Fish. Aquat. Sci.* 50: 416-429.

Jamieson, G.S., Phillips, A. & Hugget, S. 1989. Effects of ocean variability on the abundance of Dungeness crab larvae. *Can. Spec. Publ. Fish. Aquat. Sci.* 108: 305-325.

Jamieson, G.S., Phillips, A. & Smith, B.D. 1998. Implications of selective harvests in Dungeness crab (*Cancer magister*) fisheries. *Can. Spec. Publ. Fish. Aquat. Sci.* 125: 309-321.

Jensen, A.J.C. 1965. *Pandalus borealis* in the Skagerrak (length, growth and changes in the stock and fishery yield). Rapp. et Procès-Verbaux des Réunions, Conseil Int. Explor. Mer 156: 109-111.

Jensen, A.J.C. 1967. The *Pandalus borealis* in the North Sea and Skagerrak. *Mar. Biol. Assoc. India, Proc. Symp. on Crustacea, Part* 4: 1317-1319.

Kirby, R.R., Bayne, B.L. & Berry, R.J. 1994. Phenotypic variation along a cline in allozyme and karyotype frequencies, and its relationship with habitat, in the dog-whelk *Nucella lapillus* L. *Biol. J. Linn. Soc.* 53: 255-275.

Law, R. & Grey, D.R. 1989. Evolution of yields from populations with age-specific cropping. *Evol. Ecol.* 3: 343-359.

McConnaughey, R.A., Armstrong, D.A. & Hickey, B.M. 1995. Dungeness crab (*Cancer magister*) recruitment variability and Ekman transport of larvae. *ICES Mar. Sci. Symp.* 199: 167-174.

Merritt, M.F. 1985. The Lower Cook Inlet Dungeness crab fishery from 1964-1983. *Univ. Alaska, Alaska Sea Grant Rep.* 85-3: 85-95.

Metcalfe, N.B. 1993. Behavioural causes and consequences of life history variation in fish. *Mar. Behav. Physiol.* 23: 205-217.

Miller, R.R. 1957. Have the genetic patterns of fishes been altered by introductions or selective fishing. *J. Fish. Res. Bd. Canada* 14: 797-806.

Nelson, K. & Soulé, M. 1987. Genetical conservation of exploited fishes. In N. Ryman and F. Utter (eds), *Population Genetics and Fishery Management*: 3435-368. Seattle: University of Washington Press.

Oesterling, M. 1995. The soft crab fishery. *Virginia Mar. Resource Bull., Virginia Sea Grant Prog.* 27: 13-14.

Oesterling, M. & Moore, N. 1995. Characterization of the Virginia soft crab/peeler industry. *Virginia Sea Grant Adv. Prog.* VSG-95-11: 23 pp.

Orensanz , J. M. & Jamieson, G. S. 1998. The assessment and management of spatially structured stocks: an overview of the North Pacific Symposium on Invertebrate Stock Assessment and Management. *Can. Spec. Publ. Fish. Aquat. Sci.* 125: 441-459.

Policansky, D. 1993a. Fishing as a cause of evolution in fishes. In T.K. Stokes, J.M. McGlade & R. Law (eds), *The Exploitation of Evolving Resources*: 2-18. Heidelberg: Springer-Verlag.

Policansky, D. 1993b. Evolution and management of exploited fish populations. In Proceedings of the Int. Symp. on Management Strategies for Exploited Fish Populations. *Alaska Sea Grant Rep.* 93-02: 651-664.

Pollock, D.E. 1995. Notes on phenotypic and genotypic variability in lobsters. *Crustaceana* 68: 193-202.

Ream, R.A. 1996. Variation in morphology and larval development of Puget Sound *Dendraster excentricus*. 24th Annual Benthic Ecology Meeting, Columbia, South Carolina, March 7-10, 1996. Abstract: 68.

Restrepo, V.R. 1992. A mortality model for a population in which harvested individuals do not necessarily die: the stone crab. *Fish. Bull., US* 90: 412-416.

Rudstam, L.G., Magnuson, J.J. &.Tonn, W.M 1984. Size selectivity of passive fishing gear: a correction for encounter probability applied to gill nets. *Can. J. Fish. Aquat. Sci.* 41: 1252-1255.

Savage, T., Sullivan, J.R. & Kalman, C.E. 1975. An analysis of stone crab (*Menippe mercenaria*) landings on Florida's west coast, with a brief synopsis of the fishery. *Fla. Dep. Nat. Res. Lab. Publ.* 13: 1-37.

Sloan, N.A. 1984. Incidence and effects of parasitism by the rhizocephalan barnacle, *Briarosaccus callosus* Boschma, in the golden king crab, *Lithodes aequispina* Benedict, from deep fjords in northern British Columbia. *J. Exp. Mar. Biol. Ecol.* 84: 111-131.

Sloan, N.A. 1985. Life history characteristics of fjord-dwelling golden king crabs *Lithodes aequispina*. *Mar. Ecol. Prog. Ser.* 22: 219-228.

Smith, B.D. & Jamieson, G.S. 1989. Exploitation and mortality of male Dungeness crab (*Cancer magister*) near Tofino, British Columbia. *Can. J. Fish. Aquat. Sci.* 46: 1609-1614.

Sparkes, T.C. 1996. Effects of predation risk on population variation in adult size in a stream-dwelling isopod. *Oecologia* 106: 85-92.

Warner, G.F. 1977. *The Biology of Crabs*. London: Elek Science.

Wellborn, G.A. 1994. Size-based predation and prey life histories: a comparative study of freshwater amphipod populations. *Ecology* 75: 21004-2117.

Wellborn, G.A. 1995. Determinants of reproductive success in freshwater amphipod species that experience different mortality regimes. *Anim. Behav.* 50: 353-363.

Diel variations in decapod catch rate and size of captured individuals in a subtropical area of Brazil

M.L. NEGREIROS-FRANSOZO, A.L.D. REIGADA & J.M. NAKAGAKI
NEBECC (Group of studies on crustacean biology, ecology and culture), Depto. de Zoologia, IB, Unesp, Botucatu, São Paulo, Brasil

ABSTRACT

Diel variations in decapod crustaceans catch rate, as well as variations in size of sampled individuals, were investigated in a sublittoral portion of Ubatuba Bay (23°20', 23°35'S and 44°50', 45°14'W) in order to detect differential patterns of occurrence. Three replicate trawls, each enclosing a 2,500 m^2 area, were performed at a median depth of 3.5 m during the waning moon period, in 3 consecutive summer and winter months. Trawls were conducted at dawn, noon, dusk and midnight. Hydrological and substratum features were monitored. Penaeoideans did not show a significant diel catch rate variation during the sampling periods, but the catch rate of brachyurans was highest at dusk and midnight during winter ($p < 0.01$). Fixing diel variation, catch rates of both brachyurans and penaeoideans are subjected to significant seasonal differences ($p < 0.05$). The largest specimens of *Callinectes ornatus, Xiphopenaeus kroyeri, Rimapenaeus constrictus* and *Farfantepenaeus* spp were found at twilight during summer. Differences on size of captured individuals mainly in samples of portunids and penaeids taken during the course of the day evidences that significant daily movements take place. This confirms that activity alterations depend on characteristics of daily schedules and on environmental demands of studied species.

1 INTRODUCTION

Decapods constitute an important element in the dynamics of the benthic fauna of coastal areas because they modify the physical environment and may control populations of other organisms.

The current knowledge of the general biology of the benthic macrofauna shows that its components are subjected to different environmental factors that can act isolated or simultaneously. However, one or another factor may determine certain behaviors at a larger scale. Subrahmanyam (1976), who studied the activity rhythms of *Penaeus duorarum*, provided an example illustrating how this influence works. This shrimp exhibits night activity, which is specially pronounced at nocturnal high tides. When this shrimp was maintained in captivity after one week, the tidal component disappeared but the nocturnal one persisted.

Light intensity can be one of the factors influencing the occurrence of benthic organisms in shallow waters. Campbell (1967) and Begg (1989) found that the crab *Cyclograpsus lavauxi* shows maximum activity during dark periods and reduced activity in the presence of light. Coles (1979), who studied penaeidean larval and post-larval forms, mentioned different responses to light in three species. In his study on the decapod crustaceans in the Lagoon of Términos (Mexico), Román-Contreras (1986) found that these organisms are more abundant at night. Choy (1986) verified that, although feeding during the day when submerged, *Liocarcinus puber* feeds mainly at night. Stevens et al. (1982, 1984) verified that decapod crustaceans are known to exhibit strong diel behavioral variations. They also found an increase in the density of *Cancer magister* in certain areas during twilight periods, which is, according to the authors, related to feeding since crangonid shrimps, their main prey, are abundant in these periods. In spite of that evidence, there is some controversy in the literature (Lawton 1987, Wassenberg & Hill 1989, Fernandez & González-Gurriarán 1991) concerning activity timing.

Biological rhythms provide adaptive advantages for several animal and plant species in different biological processes, e.g. reproduction, feeding, predation avoidance, and migratory behavior. These rhythms can be directly adjusted by environmental cues and may involve five different components known to modulate behavioral periodicity: circadian, circatidal, circasemilunar, circalunar and circannual (de Coursey, 1983). According to Palmer (1974), circadian rhythms are those enclosing cycles of about one day in length. This stands for solar-day rhythms persisting in constant conditions, which may slightly deviate from the 24 hour period displayed in nature.

As an attempt to contribute for a better understanding of daily movements in these organisms, replicate samplings during the course of the day were performed. This was done to verify if there are differences in the day/night catch rate of benthic decapods (brachyurans and penaeideans) and if the size of sampled organisms is variant on a diel basis.

2 MATERIALS AND METHODS

Samples were conducted in a shallow portion (3.5 ± 0.5 m) at Ubatuba Bay (SP) (23°20′, 23°35′S; 44°50′, 45°14′W) (Fig. 1) with a shrimp fishery boat provided with an otter-trawl (2.5 m mouth width; 15 mm mesh in the body net and 10 mm mesh in the cod end) accompanied by two boards of 10 kg each. Trawls were accomplished along a 1 km transect during 30 minutes at a constant speed (1.1 nauts/hour). Sampled area in each trawl was approximately 2,500 m².

Field work was carried out during summer and winter of 1996 at waning moon periods (mean level 0.66 m) to minimize the effect of tides in the results obtained. In each season, replicate samples were taken in three alternate days over three consecutive months. For each day of collection, 4 scheduled trawls were performed over a 24 hour period as follows: a first trawl beginning at dawn (± 6:00 h); a second trawl at noon (± 12:00 h), a third trawl at dusk (± 18:00 h) and a fourth one beginning at midnight (± 24:00 h). For each trawl a set of abiotic factors was measured. Water samples were obtained with a van Dorn Bottle, from which temperature, salinity, and dissolved oxygen contents were recorded using a thermometer, a specific refractometer, and an oxymeter (YSI 55), respectively. A luxmeter (Panlux eletronic 2) and a Secchi

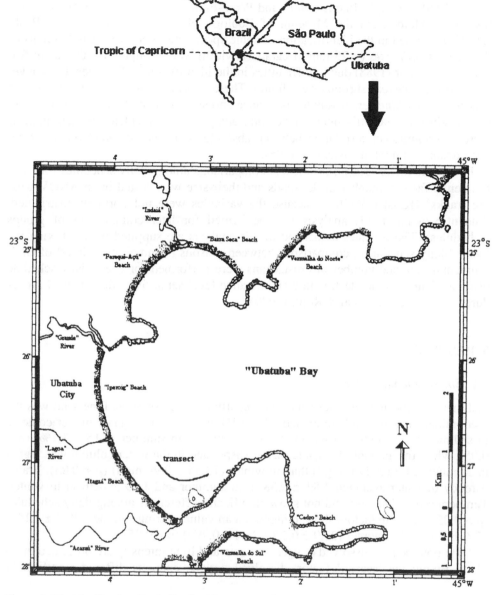

Figure 1. Ubatuba Bay locality indicating the transect site.

disk were used to measure light intensity at the water surface and water transparency. Sediment samples were obtained with the aid of a van Veen sediment grab and then analyzed at the laboratory. Organic matter content was obtained by ash weighing, and granulometric fractions of sediment were obtained by differential sieving, based on the American scale (Wentworth 1922).

All crustaceans were first separated, from which brachyurans and penaeideans were sorted out and counted. Each decapod was measured with a caliper to the nearest 0.01 mm and its sex recorded. Decapod identification was based on Rathbun (1925, 1930, 1937), Williams (1984) and Pérez-Farfante & Kensley (1997). Dr Gustavo A.S. Melo, from the Museum of Zoology of the University of São Paulo (MZUSP) helped in the identification of some specimens. Very young individuals of the genus *Callinectes* (< 25 mm of carapace width) were grouped together as in Buchanan & Stoner (1988) due to difficulties in the identification of species. Only juvenile *Farfantepenaeus* shrimps were found. These specimens were pooled together as proceded in *Callinectes*. Absolute abundance and size of most abundant species were then analyzed. Pearson's linear correlation analysis was calculated to determine if there is a significant correlation between absolute abundance of brachyurans or penaeoideans and light intensity ($p < 0.05$).

Fixing the factor 'season', statistical comparisons among daily schedules for environmental factors, number of decapods and their size were tested in an ANOVA using 'ranks' (Kruskal-Wallis), because the variables were not normally distributed. Student-Newman-Keuls analysis was performed for later comparisons of groups (schedules). The p-values associated to this analysis are supplied below. Fixing the factor 'daily schedules', comparisons between seasons (summer and winter) of environmental data and number of crustaceans were performed for every daily schedule using a Mann-Whitney test, with a significance level set at 5%. All statistical procedures were based on Sokal & Rohlf (1979).

3 RESULTS

3.1 *Environmental factors*

Sediment composition did not vary significantly between seasons. Fine sand was the predominant fraction and comprised 54 to 61% of the total. Organic matter contents in sediments was higher in winter (8.43 ± 1.48%) than summer (6.15 ± 3.1%) ($p < 0.05$), while the amount of organic matter suspended in the water column was higher in summer (21.85 ± 12.6 mg/l) than in winter (13.34 ± 9.94 mg/l) ($p < 0.05$). Transparency of water averaged 1.83 ± 0.58 m in summer and 1.59 ± 0.56 m in winter. Temperature and salinity did not show significant differences among daily schedules ($p > 0.05$), but both factors showed higher mean values during summer (28.2 ± 1.1°C; 32.6 ± 3.4‰) than winter (21.9 ± 0.5°C; 34.4 ± 0.7‰) ($p < 0.05$).

Data obtained on dissolved oxygen contents and light intensity are presented in table I. Fixing the 'season' factor, dissolved oxygen contents differed significantly

Table 1. Results of dissolved oxygen and light intensity (mean ± standard deviation) obtained in Ubatuba Bay for different time schedules.

	Dawn	Noon	Dusk	Midnight
Dissolved oxygen (mg/l)				
Summer	6.31 ± 0.70	7.14 ± 0.44	7.62 ± 0.86	6.58 ± 0.84
Winter	5.96 ± 1.05	6.26 ± 1.35	6.94 ± 1.25	6.51 ± 0.70
Light intensity (Lux)				
Summer	111.76 ± 282.53	91111.11 ± 45979.37	25.74 ± 57.82	0
Winter	302.29 ± 441.62	49722.22 ± 27796.08	0.45 ± 1.36	0

between every daily schedule (p < 0.05) during summer, with the highest values occurring at dusk, followed by noon. During winter, values of dissolved oxygen contents at noon and midnight were not significantly different, but dusk and dawn measurement significantly differed (p < 0.05). Fixing the 'daily schedule' factor, between-season comparisons of dissolved oxygen contents among scheduled trawls do not show statistical significant differences (p > 0.05).

Disregarding seasonal variations, among-schedules comparisons revealed that light intensity varied significantly (p < 0.05) during summer. This was not the case for dusk and midnight during winter, but in remaining comparisons significant differences were observed (p < 0.05). Excluding daily-schedules variation, light intensity at dawn and noon were significantly different between summer and winter (p < 0.05), which was not the case for dusk and midnight (p > 0.05).

3.2 Decapod crustaceans

3.2.1 Species involved

Thirty-two species of decapod crustaceans were captured, including 19 brachyurans, 7 penaeideans and 6 anomurans. Portunids outnumbered other brachyurans in both species number and abundance, *Callinectes ornatus* and *Callinectes danae* being the most abundant species. Brachyuran abundance was negatively correlated with light intensity (r = –0.343; p < 0.05) during winter. No significant correlation was found for summer data.

Penaeid shrimps prevailed within the Penaeoidea, *Xiphopenaeus kroyeri* being the most abundant species followed by *Rimapenaeus constrictus* and juvenile *Farfantepenaeus*. Other penaeoids were much less abundant and their occurrence considered accidental. Penaeoid abundance was not correlated with light intensity in any season analyzed.

A list of decapod species with total number of captured individuals along the study period is presented in table II.

3.2.2 Comparison of catch rates of decapods among daily schedules, disregarding seasonal variation

Total catch of brachyurans and penaeideans is shown in Figures 2 and 3. No statistically significant differences were found during summer for brachyuran (p = 0.455) and penaeidean (p = 0.392) abundance. During winter, no differences were found in the abundance of penaeideans (p = 0.972), but in brachyurans there was a statistically significant difference (p = 0.0019) between dusk and midnight. Among brachyurans (Fig. 4), catch rates of *Callinectes ornatus* were higher at dusk (p = 0.0314), and captures of *Callinectes danae* and *Hepatus pudibundus* were more expressive at dusk and midnight (p = 0.0006 and p = 0.0169, respectively).

3.2.3 Comparison of catch rates of decapods between seasons, disregarding diel variations

For penaeideans, number of captured specimens per trawl did not differ between summer and winter samples (p ≥ 0.05). In the case of brachyurans, seasonal differences were only significant for dawn and midnight samples (p < 0.05).

Table 2. List of decapod species and total number of specimens captured at Ubatuba Bay during 1996. *Farfantepenaeus* spp (= juveniles of *F. brasiliensis* and *F. paulensis*. **Callinectes* spp (juveniles of *C. ornatus* and *C. danae*).

Decapod species	Summer	Winter
Penaeoidea		
Xiphopenaeus kroyeri	3214	31202
Rimapenaeus constrictus	240	843
Litopenaeus schmitti	0	20
Farfantepenaeus spp*	1841	77
Sicyonia dorsalis	3	13
Sicyonia typica	10	3
Brachyura		
Hepatus pudibundus	85	140
Persephona lichtensteinii	2	9
Persephona punctata	38	77
Persephona mediterranea	1	0
Libinia ferreirae	1	16
Pyromaia tuberculata	0	3
Callinectes ornatus	8422	2198
Callinectes danae	1068	543
Callinectes spp**	257	87
Portunus spinimanus	36	2
Portunus spinicarpus	0	1
Portunus ventralis	2	1
Arenaeus cribrarius	10	0
Cronius ruber	1	0
Charybdis helleri	38	4
Eurypanopeus abbreviatus	1	0
Hexapanopeus schmitti	1	1
Hexapanopeus paulensis	0	1
Eucratopsis crassimanus	0	1
Pinnixa sp	1	0
Anomura		
Dardanus insignis	0	1
Petrochirus diogenes	2	1
Loxopagurus loxochelis	0	4
Pagurus exilis	0	1
Porcellana sayana	3	2
Minyocerus angustus	0	1

3.2.4 *Size analysis of sampled decapods*

In *Callinectes ornatus* (Fig. 5), median size of trawled specimens was highest at dawn ($p < 0.0001$) during summer, and highest at dawn and midnight ($p < 0.0001$) during winter. In *Callinectes danae* (Fig. 6), highest median size was obtained at noon ($p < 0.0001$) during summer, but no significant differences were observed in winter ($p = 0.193$). For *Hepatus pudibundus* (Fig. 7), no statistical differences of size were found in summer ($p = 0.217$) nor in winter ($p = 0.209$).

Median size of *Xiphopenaeus kroyeri* (Fig. 8) was highest in dusk samples ($p < 0.05$) during both seasons. A same trend was also verified in *Rimapenaeus constrictus*

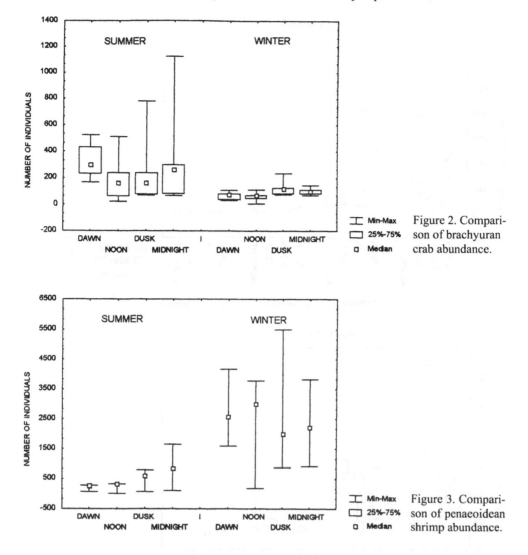

Figure 2. Comparison of brachyuran crab abundance.

Figure 3. Comparison of penaeoidean shrimp abundance.

(Fig. 9) during summer, but during winter no significant differences (p > 0.05) were found in this species. In the case of *Farfantepenaeus* spp (Fig. 10), highest median sizes were observed in dawn samples (p < 0.05) in both seasons.

4 DISCUSSION

Ubatuba Bay is located on the north coast of São Paulo State. The marine coastal environment in this region is mainly influenced by three marine currents: South Atlantic Central Water (SACW), Tropical Water (TW) and Coastal Water (CW) (Pires-Vanin 1989). According to the same author, these water masses present very different characteristics. Temperature of SACW is typically lower than 15°C; its salinity is about 35‰ and nutrients contents are high. Temperature of TW is usually lower than 24°C,

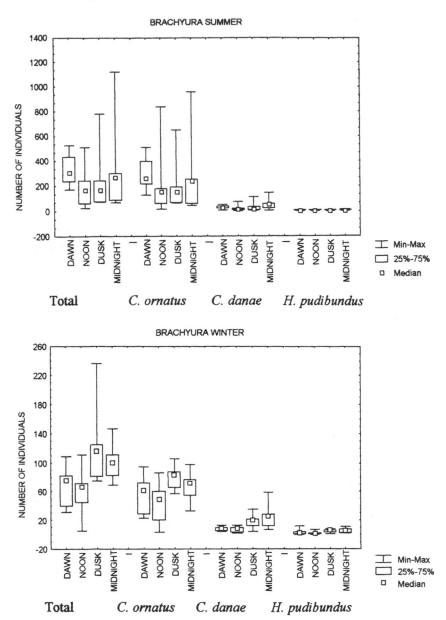

Figure 4. Comparison of abundance among brachyuran species

its salinity varies from 36 to 37‰, and nutrients contents are low. Temperature of CW is about 24°C and salinity conditions range from 34.3 to 35‰.

Water temperature during summer is constantly higher at all daily schedules, showing that upwelling of SACW occurring in summer in this region did not influenced the sampling site. Instead, seasonal climatic changes seem to have more direct effect. Reduced rainfall and consequent decrease of freshwater input in the bay explain the higher salinity recorded in winter. Dissolved oxygen contents were similar

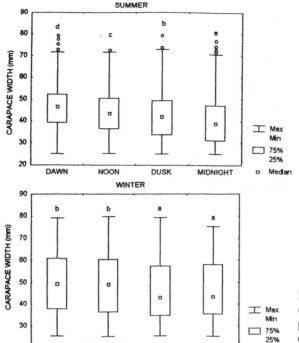

Figure 5. Size comparison in *Callinectes ornatus* specimens (box plots with the same letter above do not differ statistically; 25% lower quartile and 75% upper quartile).

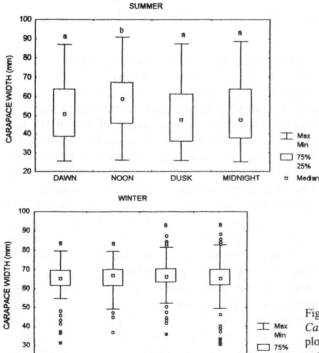

Figure 6. Size comparison in *Callinectes danae* specimens (box plots with the same letter above do not differ statistically; 25% lower quartile and 75% upper quartile).

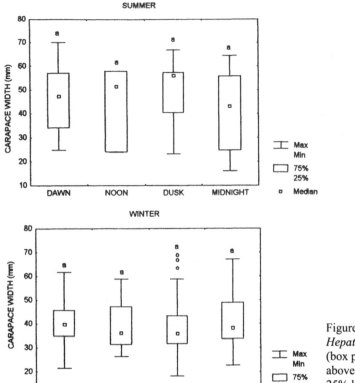

Figure 7. Size comparison in *Hepatus pudibundus* specimens (box plots with the same letter above do not differ statistically; 25% lower quartile and 75% upper quartile).

between seasons, which is an indication that trawlings were actually performed at the very same area and primary productivity is similar in both seasons.

Organic debris, mainly of vegetable origin, discharged during intense rainfall in summer explains the high values of suspended organic matter recorded at that time. The Acaraú and Lagoa rivers, which drain the sewerage system of part of Ubatuba City, may contribute to increase the suspended organic contents in the water column during summer because of tourist activity. Otherwise, organic matter in the sediment increases in winter, when rainfall and other sources of organic drainage into the bay decrease. By that time, major processes of sedimentation have probably occurred.

The factors showing more intense diel variations, which were observed in both seasons, were light intensity and dissolved oxygen content. Light intensity was higher during dawn and noon during summer, due to climatic characteristics of the season. Dissolved oxygen content increased during the day, as a consequence of a gradual increase of light facilitating the photosynthetic activity of organisms.

For any given system, including this study area, timing of predation risk in benthic organisms needs to be experimentally tested for different prey species, before attributing diel patterns of activity and habitat use to predator avoidance mechanisms. Sogard & Able (1994) stated that diel variability of predation risk varies for different prey species and depends on species composition, abundance of potential predators and alternative prey.

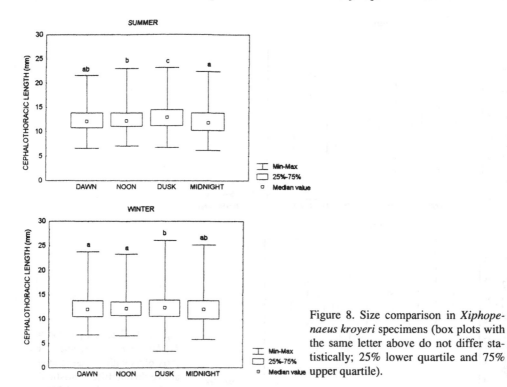

Figure 8. Size comparison in *Xiphopenaeus kroyeri* specimens (box plots with the same letter above do not differ statistically; 25% lower quartile and 75% upper quartile).

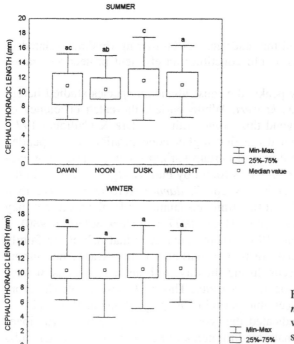

Figure 9. Size comparison in *Rimapenaeus constrictus* specimens (box plots with the same letter above do not differ statistically; 25% lower quartile and 75% upper quartile).

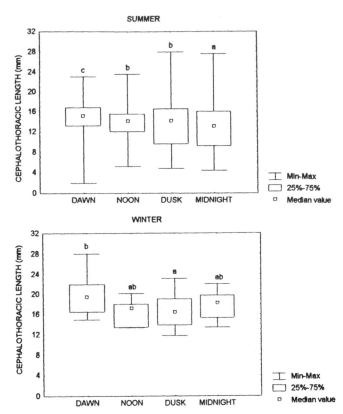

Figure 10. Size comparison in *Farfantepenaeus* spp specimens (box plots with the same letter above do not differ statistically; 25% lower quartile and 75% upper quartile).

Moller & Jones (1975) suggested that endogenous burrowing rhythms, related to photonegative response in *Penaeus*, could constitute an efficient protection against visual predators in nature.

In spite of lacking clear activity peaks, the most abundant species studied herein present burrowing habits, except *X. kroyeri*. While buried, the catch efficiency of these organisms may decrease. Beyond this, some authors (Clark & Caillouet 1975, Carothers & Chittenden 1985) pointed out that higher concentrations of suspended material could modify behavioral patterns of certain shrimp species. Burrowing habits may be dependent on shrimp size or age as found by Funs & Ogreen (1966). In the present study, larger shrimps and crabs (except *C. danae*) were more active than smaller ones during twilight periods. At that time, remaining under the sediment may prevent or minimize predation as asserted by Heck (1977) and Heck & Orth (1980).

According to Soares & Apelbaum (1994), there is evidence that fish of the family Triglidae feed mainly during the day, resting in darkness hours. The same authors verified that *Prionotus punctatus* feeds during the day, with an activity peak at dusk when it consumes prey displaying nocturnal activity. This triglid fish would therefore take some advantage in a more susceptible period of its prey. Braga & Braga (1987) and Soares & Apelbaum (1994) indicated that young swimming crabs (Portunidae) and shrimps (Penaeidea) were the main feeding items of *P. punctatus* in the Ubatuba area. However, their collections were accomplished in deeper areas. It is possible that specimens living in shallower grounds present different feeding habits, since food availability and predator size are likely to vary as well.

Petti (1990) found that *C. ornatus* feeds mainly during the day on winter, which coincides with highest densities of its main prey species, the penaeid *Xiphopenaeus kroyeri*. The abundance of its main feeding resource may explain the activity pattern of that crab. Similarly, Freire et al. (1991) found that feeding activity in *Liocarcinus depurator* is highest in areas where potential food items are present.

Differences on size of captured individuals mainly in samples of portunids and penaeids taken during the course of the day evidences that significant daily movements take place. This confirms that activity alterations depend on characteristics of daily schedules and on environmental demands of species studied.

Foraging migrations would also provide an explanation for diel variations in benthic occupation. Trophic relationships and their connection to daily foraging migrations are aspects that deserve more detailed investigation.

REFERENCES

Begg, S.M.C. 1989. Factors affecting the distribution of two species of intertidal grapsid crabs. Unpubl. Hons. Thesis Univ. Otago, New Zealand, 86 p.

Braga, F.M.S. & Braga, M.A.A.S. 1987. Estudo do hábito alimentar de *Prionotus punctatus* (Bloch, 1797) (Teleostei, Triglidae) na região da Ilha Anchieta, Estado de São Paulo, Brasil. *Rev. Brasil. Biol.* 47: 31-36.

Buchanan, B.A. & Stoner, A.W. 1988. Distribution patterns of blue crabs (*Callinectes* sp.) in a tropical estuarine lagoon. *Estuaries* 11: 231-239.

Campbell, J.A. 1967. Rhythmic spontaneous activity in an intertidal crab *Cyclograpsus lavauxi*. Unpubl. Hons. thesis Univ. Canterbury, 89 p.

Carothers, P.E. & Chittenden, M.E., Jr. 1985. Relationships between trawl catch and tow duration for penaeid shrimp. *Trans. Amer. Fish. Soc.* 114: 851-856.

Choy, S.C. 1986. Natural diet and feeding habitats of the crabs *Liocarcinus puber* and *L. holsatus* (Decapoda, Brachyura, Portunidae). *Mar. Ecol. Prog. Ser.* 31: 87-99.

Clark, S.H. & Caillouet, C.W. 1975. Diel fluctuations in catches of juvenile brown and white shrimp in a Texas estuarine canal. *Contrib. Mar. Sci.* 19: 119-124.

Coles, R.G. 1979. Catch size and behavior of pre-adults of three species of penaeid prawn as influenced by tidal current direction, trawl alignment, and day and night periods. *J. Exp. Mar. Biol. Ecol.* 38: 247-260.

de Coursey, P. 1983. Biological timing. In D. Bliss (ed.), *The biology of Crustacea*, Vol. 7: 107-162, New York: Academic Press.

Fernandez, L. & Gonzalez-Gurriaran, E. 1991. Population biology of *Liocarcinus depurator* (Brachyura: Portunidae) in mussel ralf culture areas in the ría Arousa (Galicia, NW Spain). *J. Mar. Biol.* 71: 375-390.

Freire, E., Muiño, R. & Gonzalez-Gurriaran, E. 1991. Life cycle of *Liocarcinus arcuatus* (Brachyura, Portunidae) in the ría de Arousa (Galicia, NW Spain): role of the beach and mussel ralf culture areas. *Mar. Ecol.* 12(3): 193-210.

Fuss, C.M. & Ogreen, L.H. 1966. Factors affecting activity and burrowing habits of the pink shrimp, *Penaeus duorarum* Burkenroad. *Biol. Bull. Mar. Biol. Lab., Woods Hole* 130: 170-191.

Heck, K.L. 1977. Comparative species niches, composition and abundance of invertebrates in caribbean seagrass (*Thalassia testudinum*) meadows (Panamá). *Mar. Biol.* 41: 335-345.

Heck, K.L. & Orth, R.J. 1980. Structural components of elgrass (*Z. marina*) meadows in the lower Chesapeake Bay. Decapod Crustacea. *Estuaries* 3: 289-295.

Lawton, P. 1987. Diel activity and foragging behavior of juvenile American lobsters *Homarus americanus*. *Can. J. Fish. Aquatic. Sci.* 44: 1195-1205.

Moller, T.H. & Jones, D.A. 1975. Locomotory rhythms and burrowing habits of *Penaeus semisulcatus* (de Haan) and *P. monodon* (Fabricius) (Crustacea: Penaeidae). *J. Exp. Mar. Biol. Ecol.* 18: 61-77.

Palmer, J.D. 1974. *Biological clocks in marine organisms.* New York: John Wiley.

Perez-Farfante, I. & Kensley, B. (eds) 1997. *Penaeoid and Sergestoid shrimps and prawns of the world. Keys and diagnoses for the families and genera.* Mémoires du Muséum National d'Histoire Naturelle, Paris, France. Tome 175, p. 233.

Petti, M.A.V. 1990. Hábitos alimentares dos crustáceos decápodos braquiúros e seu papel na rede trófica do infralitoral de Ubatuba (Litoral norte do Estado de São Paulo, Brasil). Thesis IO, USP, São Paulo, 150 p.

Pires-Vanin, A.M.S. 1989. Estrutura e dinâmica da megafauna bêntica na plataforma continental da região norte do Estado de São Paulo, Brasil. Thesis IO – USP, 172 pp.

Rathbun, M.J. 1925. The spider crabs of America. *U.S. Nat. Mus Bull.* 129: 1-613.

Rathbun, M.J. 1930. The cancroid crabs of America of the families Euryalidae, Portunidae, Atelecyclidae, Cancridae and Xanthidae. *U.S. Nat. Mus. Bull.* 152: 1-609.

Rathbun, M.J. 1937. The Oxystomatous and allied crabs of America. *U.S. Nat. Mus. Bull.* 166: 1-278.

Roman-Contreras, R. 1986. Análisis de la población de *Callinectes* spp (Decapoda, Portunidae) en el sector Occidental de la Lagurna de Términos, Campeche, Mexico. *An. Inst. Cienc. del Mar y Limnologia Univ. Nac. Auton. Mexico* 13(1): 3154-322.

Soares, L.S.H. & Apelbaum, R. 1994. Atividade alimentar diária da cabrinha *Prionotus punctatus* (Teleostei: Triglidae) do litoral de Ubatuba, Brasil. *Bolm. Inst. Oceanogr.* 42: 85-98.

Sogard, S.M. & Able, K.W. 1984. Diel variation in immigration of fishes and decapod crustaceans to artificial seagrass habitat. *Estuaries* 17: 622-630.

Sokal, R.R. & Rohlf, F.J. (eds) 1979. *Biometria. Princípios y métodos estadísticos en la investigación biológica.* Madrid: H. Blume Ediciones.

Stevens, B.G., Armstrong, D.A & Cusimano, R. 1982. Feeding habits of the Dungeness crab *Cancer magister* as determined by the index of relative importance. *Mar. Biol.* 72: 135-145.

Stevens, B.G., Armstrong, D.A & Hoeman, J.C. 1984. Diel activity of an estuarine population of dungeness crabs, *Cancer magister*, in relation to feeding and environmental factors. *J. Crust. Biol.* 4: 390-403.

Subrahmanyan, C.B. 1976. Tidal and diurnal rhythms of locomotory activity and oxygen consumption in the pink shrimp, *Penaeus duorarum. Contrib. Mar. Sci.* 20: 123-132.

Wassemberg, T.J. & Hill, B.J. 1989. Diets of four decapod crustaceans (*Linuparus trigosus*, *Metanephrops andamancius*, *M. australiensis* and *M. boschmai*) from the continental shelf around Australia. *Mar. Biol.* 103: 161-167.

Wentworth, C.K. 1922. A scale of grade and class terms for clastic sediments. *J. Geol.* 30: 377-392.

Williams, A.B. (ed.) 1984. *Shrimps, lobsters and crabs of the Atlantic coast of the eastern United States, Maine to Florida.* Washington: Smithsonian Institution Press.

Concepts of reproductive value and reproductive potential and their utility for stock assessment and management: spiny lobster *Palinurus elephas* from the coast of Portugal as a model

PAULA SERAFIM & ANTÓNIO. S.T. AUBYN
Dep. Matemática, Instituto Superior de Agronomia, Lisbon, Portugal

MARGARIDA CASTRO
CCMAR, Universidade do Algarve, Faro, Portugal

ABSTRACT

The objective of this study was to investigate the usefulness of the reproductive value and reproductive potential concepts (Fisher 1930, MacArthur 1960) as tools for the management of crustacean stocks; in particular for defining minimum landing sizes. The formulas were adapted to size structured populations using the equations of length-based cohort analysis (Jones 1984) for expressing growth and mortality. The spiny lobster *Palinurus elephas* of the southwest coast of Portugal was used as a model. The reproductive potential of the population, which is the sum of the reproductive values of all the size classes present, was evaluated for different levels of fishing mortality and minimum landing sizes. A value of minimum reproductive potential was chosen to represent a threshold level below which a sustainable exploitation of the resource can not be guaranteed. Curves of reproductive potential as a function of minimum landing size, each one representing a different fishing mortality rate, were compared with this threshold level in order to make decisions about the appropriate minimum landing size and fishing effort. The results indicate that with increasing fishing mortality, larger minimum landing sizes are required in order for the stock to remain within biologically safe limits. On the other hand, at low levels of exploitation minimum landing sizes are not necessary.

1 INTRODUCTION

The reproductive value is the expected average future contribution of a female in terms of the reproduction of the species during the remainder of its life. The concept of reproductive value was first used by Fisher (1930) to indicate the degree of reproductive success of a female during its lifetime. MacArthur (1960) applied this concept to population dynamics and ecology and introduced the concept of age class reproductive value. In this case, the reproductive value is the expected contribution of all females of a give age class to the reproduction of a species and is a function of age-specific fecundity and age-specific survival rates. Several authors have applied this concept to exploited fish populations (Kanciruck 1980, Kanciruk & Herrnkind

1976, Jensen 1981, Ware 1985, Saila & Erzini 1989). A review of the use of these concepts in fisheries is presented by Katsukawa (1997).

The summation of the reproductive value of all age classes (all females) present in the population is called the reproductive potential or spawning potential and represents the expected future contribution in terms of spawning of the stock (Katsukawa 1997).

In this work, the equations expressing reproductive value and reproductive potential were modified to describe these concepts in length structured populations. This is of particular importance for crustacean populations because individual age estimation is not possible. The spiny lobster, *Palinurus elephas*, also known as *Palinurus vulgaris*, was used as a model.

This species is an important resource of the southwest coast of Portugal. It is caught by tangle nets at depths below 100 meters and the only management measure applied to this stock is a minimum landing size (Galhardo 1994). Minimum landing sizes are effective management measures for this type of resource since survival of released undersized animals can be assumed to be very high.

The objective of this study was to evaluate the applicability and the utility of the reproductive value and reproductive potential concepts in the management of this resource, particularly for the definition of minimum landing sizes.

2 MATERIALS AND METHODS

The approach used in this work consisted in simulating an exploited lobster population and verifying the theoretical reproductive potential under different fishing pressures and minimum landing sizes.

The simulated populations were based on a simple model with the following assumptions: 1. Only females are considered; 2. The number of females in a cohort evolves according to a decaying exponential; 3. Individual growth follows a von Bertalanffy growth curve and is considered to be the same for all individuals; 4. Instantaneous annual natural mortality rate is constant throughout the life of the cohort; 5. Instantaneous annual fishing mortality rate is constant throughout the exploited phase; 6. Size-at-first capture equals the minimum landing size; and 7. Only the exploitable phase is considered. Recruitment (R) to this phase is considered constant. R is considered to be the number of females at the beginning of the first class (age or length) considered.

Considering this simple model, the evolution of a cohort will represent the age (size) structure of a population at a given time. The equations for a cohort can therefore be used to simulate the structure of the population. The simulations were done using an Excel spreadsheet (1995-96 Microsoft Corporation).

The reproductive value and reproductive potential were expressed using notation in stock assessment to express numbers of a cohort instead of the more common life history table notation. Considering an age structured population, the reproductive value of a female at age t can be expressed approximately by:

$$RV_t = \frac{1}{N_t} \sum_t^{t_{max}} (\overline{N}_t \cdot MAT_t \cdot FEC_t) \tag{1}$$

Where RV_t: Reproductive value of a female belonging to age class t; N_t: Number of females at the beginning of interval t; N_t: Average number of females for interval t, equal to $(N_t+N_{t+1})/2$, where N_{t+1} is simultaneously the number at the end of interval t and at the beginning of interval t+1; MAT_t: Percentage of mature females for age class t; FEC_t: Average fecundity for age class t.

Since the reproductive value concept is age specific and corresponds to individual females, the reproductive potential of the population will be the summation of the reproductive values of all females present:

$$RP = \sum_0^{t_{max}} RV_t \cdot \overline{N}_t \qquad (2)$$

Where RP: Reproductive potential of the population, and all other symbols are the same as in Equation 1.

Equations 1 and 2 can be modified to express the reproductive value and reproductive potential of length class j. In this situation, the length intervals j will be considered to correspond to unequal time intervals representing the time it takes to grow from the length at the lower boundary ($L1_j$) to the upper boundary ($L2_j$) of the length class. The age-at-length is obtained by inverting of the von Bertalanffy growth curve. This methodology is used in length-based cohort analysis (Jones 1984). Equations 1 and 2 can therefore be rewritten to represent length classes in the following way:

$$RV_j = \frac{1}{N1_j} \sum_j^{j_{max}} (\overline{N}_j \cdot MAT_j \cdot FEC_j) \qquad (3)$$

Where RV_j: Reproductive value of a female belonging to length class j; $N1_j$: Number of females at the beginning of interval j; N_j: Average number of females for length interval j, equal to $(N1_j+N2_j)/2$, where $N2_j=N1_{j+1}$; MAT_j: Percentage of mature females for length class j; FEC_f: Average fecundity for length class j, and

$$RP = \sum_0^{j_{max}} RV_j \cdot \overline{N}_j \qquad (4)$$

Where RP: Reproductive potential of the population, and all other symbols are the same as in Equation 3.

According to the assumption stated previously and considering the correspondence between age and length, the number of individuals at the beginning (N1j) and at the end (N2j) of length class j are related by the following equation:

$$N2_j = N1_j \cdot e^{-(M+F)\Delta_j} \qquad (5)$$

Where $N2_j$: Number of females at the end of length class j ($L1_j$); $N1_j$: Number of females at the beginning; F: Instantaneous annual fishing mortality rate; M: Instantaneous annual natural mortality rate; Δ_j: Time it takes to grow from $L1_j$ to $L2_j$. Δ_j is equal to the difference between $t2_j$ and $t1_j$, the ages corresponding to $L2_j$ and $L1_j$ respectively. The average number of individuals for length class j (N_j) is considered to be the arithmetic average between $N1_j$ and $N2_j$.

Data on the biology of the model species was taken from Galhardo (1994). The von Bertalanffy growth parameters were obtained by non-linear least-squares fitting

Table 1. Example of simulated data for F = 0.5 and minimum landing size, MLS = 80 mm. The different columns represent the following: L_j – Center of length class (standard length) in millimetres; AGE – Age: $t1_j$ – age corresponding to lower limit of length class j, $t2_j$ – age corresponding to upper limit of length class j, Δ_j – time to grow from the beginning to the end of length class j and equal to $t1_j$-$t2_j$; L_∞, K and t_0 – Parameters for the von Bertalanffy growth equation, used for age determination (parameters were the same for all simulations); F – Instantaneous annual fishing mortality rate; M- Instantaneous annual natural mortality rate; F_j – Instantaneous fishing mortality rate adjusted for the length interval, and corresponding to the annual rate defined (F) multiplied by Δ_j; F_j is considered zero below minimum landing size; M_j – Instantaneous natural mortality rate adjusted the same way as for F; R – Recruitment in numbers of females at the beginning of the exploitable phase of the lifecycle (carapace length larger than 40 mm); $N1_j$ – Number of individuals at the beginning of length class j; $N2_j$ – Number of individuals at the end of length class j; N_j – Mean number for length class j, equal to ($N1_j + N2_j)/2$; MAT_j – percentage of mature females in the length class j; FEC_j – Average number of eggs produced by a female in length class; R_j – Individual contribution of length class j for the reproductive value, equal to N_j · MAT_j · FEC_j; RV_j – Reproductive value for the length class j; RP– Reproductive potential.

L_j (mm)	Age $t1_j$	$t2_j$	Δt_j	F_j	M_j	$N1_j$	$N2_j$	N_j	MAT_j	FEC_j	R_j	RV_j	L_j contribution for RP
		K= 0.184					R= 1000						
		L_{inf}= 160					F= 0.5						
		T_0= 0					M= 0.2						
42.5	1.56	1.79	0.23	0.00	0.05	1000	955	977	0	0	0	8.24E-04	8.05E+07
47.5	1.79	2.04	0.24	0.00	0.05	955	910	932	0	0	0	8.63E-04	8.04E+07
52.5	2.04	2.29	0.25	0.00	0.05	910	865	887	0	0	0	9.05E-04	8.03E+07
57.5	2.29	2.55	0.26	0.00	0.05	865	820	843	0	0	0	8.52E-04	8.02E+07
62.5	2.55	2.83	0.27	0.00	0.06	820	776	798	0	0	0	1.00E-05	8.01E+07
67.5	2.83	3.13	0.8	0.00	0.06	776	731	754	0	0	0	1.06E-05	8.00E+07
72.5	3.13	3.44	0.29	0.00	0.06	731	687	709	0	0	0	1.13E-05	7.99E+07
77.5	3.44	3.77	0.31	0.00	0.07	687	644	665	0.074	7.12E+03	3.51E+05	1.20E-05	7.98E+07
82.5	3.77	4.12	0.33	0.18	0.07	644	503	574	0.133	2.52E+04	1.92E+06	1.27E-05	7.31E+07
87.5	4.12	4.49	0.37	0.19	0.07	503	387	445	0.227	4.33E+04	4.38E+06	1.59E-05	7.09E+07
92.5	4.49	4.90	0.40	0.20	0.08	387	292	340	0360	6.15E+04	7.52E+06	1.96E-05	6.64E+07
97.5	4.90	5.33	0.44	0.22	0.09	292	215	254	0519	7.96E+04	1.05E+07	2.33E-05	5.92E+07
102.5	5.33	5.80	0.47	0.24	0.09	215	155	185	0.674	9.77E+04	1.22E+07	2.68E-05	4.96E+07
107.5	5.80	6.32	0.52	0.26	0.10	155	108	131	0.798	1.16E+05	1.21E+07	2.94E-05	3.86E+07
112.5	6.32	6.89	0.57	0.29	0.11	108	72	90	0.883	1.34E+05	1.06E+07	3.10E-05	2.79E+07
117.5	6.89	7.53	0.64	0.32	0.13	72	46	59	0.936	1.52E+05	8.40E+07	3.16E-05	1.87E+07
122.5	7.53	8.26	0.73	0.36	0.15	46	28	37	0.965	1.70E+05	6.06E+06	3.12E-05	1.15E+07
127.5	8.26	9.10	0.84	0.42	0.17	28	15	22	0.982	1.88E+05	3.99E+06	3.00E-05	6.47E+06
132.5	9.10	10.09	0.99	0.50	0.20	15	8	12	0.990	2.06E+05	2.36E+06	2.81E-05	3.25E+06
137.5	10.09	11.30	1.21	0.61	0.24	8	3	6	0.995	2.24E+05	1.23E+06	2.56E-05	1.41E+06
142.5	11.30	12.86	1.56	0.78	0.31	3	1	3	0.997	2.43E+05	5.32E+05	2.25E-05	4.95E+05
147.5	12.86	15.07	2.20	1.10	0.44	1	0	2	0.999	2.61E+05	1.74E+05	1.90E-05	1.27E+05
152.5	15.07	18.84	3.77	1.88	0.75	0	0	1	0.999	2.79E+05	3.52E+04	1.49E-05	1.89E+04

RP = 1.07E+09

of estimated length at relative age data from Galhardo (1994). The estimated parameters were $L_\infty = 160.28$ mm and $K = 0.185$. The parameter t_0 was set to zero since its value does not affect the calculation of Δ_j.

Fecundity was measured by counting the number of external eggs in ovigerous females. The number of eggs was found to be a linear function of length (Galhardo 1994):

$$FEC_j = 3623 \, L_j - 273663 \tag{6}$$

Where FEC_j – Average number of external eggs produced by a female of length j (in millimeters); L_j – Center of length class j. The proportion of mature females for each length class was also based on Galhardo (1994), and was assumed to follow a logistic curve:

$$MAT_j = 1/(1 + e^{12.6 - 0.13 \, L_j}) \tag{7}$$

Where; MAT_j = proportion of mature females for length class j (in millimeters), and all other symbols are the same as in Equation 6.

The simulations involved only the exploitable phase of the female population. The smallest size considered was 40 mm of carapace length, corresponding to the smallest size caught in the area that provided biological information for this work (Galhardo 1994). The number of recruited females to the exploitable phase was set at 1000 for all simulations.

An example of a simulated population is presented in Table 1. Whenever a given minimum landing size (size at first capture) was considered, the value of F (annual instantaneous fishing mortality rate) was set to zero up to that point, and kept constant after that. Table 1 illustrates the example with F equal to 0.5 and a minimum landing size of 75 mm.

3 RESULTS

Figure 1 shows the curves of reproductive potential of the stock as a function of minimum landing size. Each curve corresponds to a different fishing mortality rate. The maximum reproductive potential is obtained in a situation of no fishing, and represents an asymptotic value for the species.

In order to apply this concept in fisheries management one has to define a minimal acceptable value for the reproductive potential. This level, referred to here as 'biological safety level', was in this case arbitrarily set to be 1/2 of the maximum reproductive potential. Any combination of F and minimum landing size that produces a reproductive potential below this level is considered negative for the sustainable exploitation of the resource.

These results show that the minimum landing size has to increase with higher fishing levels if the reproductive potential of the species is to be maintained within 'biological safety' levels. Low fishing rates (such as F = 0.1 in Fig. 1) do not require the establishment of a minimum landing size. With increasing fishing mortality, the reproductive potential decreases and larger minimum landing sizes are required to maintain reproductive potential at or above the biological safety level (Fig. 1). As can be seen in Figure 1, fishing mortality rates from 0.5 to 2.0 require minimum landing sizes varying from approximately 115 to 120 mm in order for the stock to stay within the biological safety level.

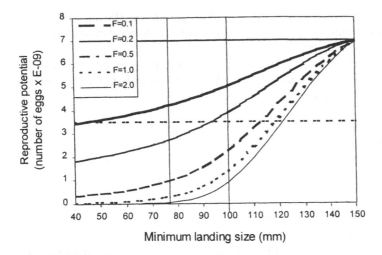

Figure 1. Reproductive potential as a function of minimum landing size for varying annual fishing mortality rates (F) values. The horizontal lines represent the maximum egg production obtained in a situation of no fishing (solid line) and the egg production considered to be the minimum for biologically sustainable exploitation of the resource (dotted line), called 'biological safety level', set at 50% of maximum egg production. The vertical lines indicate the present minimum landing size (77.5 mm) and the size at 50% maturity (100 mm).

4 DISCUSSION

Minimum landing sizes are usually established on the basis of the maturation curve (Equation 6), and are set around the size at 50% maturity. In this species, this corresponds roughly to 100 mm carapace length. If reproductive value is to be used in making this decision, the minimum landing size would have to be calculated for each exploitation pattern. In the fishery used as an example in this work, the legal minimum landing size is 20 cm of total length that corresponds roughly to 77.5 mm of carapace length.

If we consider the threshold for biological safety to be reasonable (aim at an egg production half of the maximum), a minimum landing size based on the maturation curve would only be acceptable for fishing efforts below 0.5. This fishing rate corresponds roughly to a 40% fishing mortality per year for the exploited phase. In the absence of information on effective fishing effort, and with the knowledge that intensive fishing is applied to this resource, this could be considered a minimum estimate for F, making the present minimum landing size inadequate for this resource. A similar conclusion was reached by Kanciruk & Herrnkind (1976), for *Panulirus argus* in the Bahamas. The results of this study supports Mac Arthur's theory (Mac Arthur 1960) that harvesting should be scaled in reproductive value units.

In a situation such as the one studied here, where F is not known, simulations can be a useful tool for understanding the effect of different exploitation patterns on the resource. This study suggests that for this species, it may be more beneficial to fish all length classes at a lower rate than to impose measures to protect females below maturity as proved by Jensen (1981), Saila & Erzini (1989), and Katsukawa (1997).

These authors agree that reproductive value may be the best descriptor of the effect of fishing mortality on population reproduction, justifying management decisions based on the knowledge of a population's reproductive value distribution.

The concepts discussed here can also be useful for trawl fisheries, where the selectivity of the gear can be linked to the age of first capture.

ACKNOWLEDGEMENTS

We would like to thank to Dr Karim Erzini for his critical review of the manuscript.

REFERENCES

Fisher, R.A. 1930. *The Genetical Theory of Natural Selection.* Oxford: Oxford University Press.
Galhardo, A.C. 1994. Biologia e pesca da lagosta da costa Sudoeste. MS thesis, University of the Algarve.
Herrnkind, W.F. 1980. Spiny lobsters, patterns of movement. In J.S. Cobb & B.F. Phillips (eds), *The biology and management of lobsters,* Vol 1: 345-407. London: Academic Press.
Herrnkind, W.F. & Kanciruk, P. 1976. Autumnal reproduction in *Panulirus argus* at Bimini, Bahamas. *Bull. Mar. Sci.* 26: 417-432.
Jensen, A.L. 1981. Population regulation in lake whitefish, *Coregonus clupeaformis* (Mitchill). J. *Fish Biol.* 19: 557-573.
Jones, R. 1984. Assessing the long term effects of changes in exploitation pattern using length composition data. FAO Fisheries Technical Paper No. 256.
Katsukawa, T. 1997. Points of view Introduction of spawning potential: improvement in the threshold management theory. *Rev. Fish Biol. Fisheries* 7: 285-289.
MacArthur, R.H. 1960. On the relation between reproductive value and optimal predation. *Proc. Natl. Acad. Sci. USA* 46: 143-145.
Saila, S.B. & Erzini, K. 1989. A comparison of the relationship between optimal harvesting strategies and reproductive values in four marine species with different life history characteristic. In: R.B. Pollnac & M.T. Morrissey (eds) *Aspects of Small-Scale Fisheries Development:* 1-21. International Centre for Marine Resource Development, University of Rhode Island.
SAS Institute Inc. 1988. *SAS/STAT^{TM} User's Guide,* Release 6.03 Edition. Cary, NC: SAS Institute Inc.
Ware, D.M. 1985. Life history characteristics, reproductive value and resilience of Pacific herring (*Clupea harengus pallasi*). *Can. J. Aquatic Sci.* 42: 127-137.

Fishery and tagging of *Palinurus elephas* in Sardinian seas

E. SECCI, D. CUCCU, M.C. FOLLESA & A. CAU
Dipartimento di Biologia Animale ed Ecologia, Università degli Studi di Cagliari, Cagliari, Italy

ABSTRACT

Lobster fishing was developed gradually in the last decades of the nineteenth century, reaching as much as 25% of the small local artisan fishing boats. Over and above this an indefinite number of fishermen from mainland Italy seasonally practice lobster fishing in Sardinian waters. At present there are more than 254 lobster fishing boats from the whole island, 71% of them fish along the west coast and 29% along the east; some 587 fishermen practice this trade. The area available for lobster fishing is about 10,011 sq. km and the bathymetric range is between 50-200 m. Because of its geomorphology, most of the sea bottom is inaccessible to lobster fishing, especially along the eastern coast where the continental shelf is narrower. Considering the depths, between 50-200 m, fishing pressure on the eastern coast is less than on the western. An analysis of the landings and density indices in two sample areas between 1983 and 1997 showed that the production of *Palinurus elephas* is now rising slowly after the decrease recorded between 1970 and 1980. Using catch and tagging data, studies on the fishing and biology of the common spiny lobster were carried out between 1994 and 1998 in two sampling areas in the center of western and northern Sardinia (central western Mediterranean Sea). The modal classes of the catches of lobsters from the two sample areas were, respectively, 20.8 and 20.5 cm total length, while the percentages below the minimum landing size of 24 cm total length were, respectively, 80.9 and 82.2%. A sample of 1066 lobsters were tagged and released in an area of about 400 ha where all fishing was restricted. Though still in an initial phase, this tagging experiment has suggested that lobsters move by a monthly average of 1,024 km. Mean total length increments of 11.1% have been observed after recapture.

1 INTRODUCTION

The development of *Palinurus elephas* fishing in Sardinia dates back to the last decades of the past century, when the number of dedicated fishing boats increased significantly (Santucci 1928). Fishing occurred between 15 and 100 m of depth normally on hard bottoms that were rarely mixed with sand. The boats were usually 7 m long and were equipped with tanks in which to place the lobsters. The most widely used fishing gear was the lobster pot, which varied according to the fishing area, while the trammel net was less frequently used. The number of lobster pots per boat was usually between 50 and 100. Lobster fishing was not allowed between January and April (according to a National Act prohibiting lobster fishing within 6 miles of the coast), while the greatest number of catches were made between July and October (Santucci 1928, Manunta 1955).

A decrease in the number of catches was observed from the 1950s onwards. This was probably due to uncontrolled fishing for lobsters in the previous thirty years (Manunta 1955, Cottiglia et al. 1976). The continuous increase in fishing pressure had been partly due to the abandonment of other fishing activities, the conversion of equipment, the increase in the number of lobster pots per fishing boat, and more efficient fishing systems. This is correlated to a constant decrease in the number of catches per fishing unit and to a decrease in the mean size of caught specimens.

The decreasing tendency in catches continued, but it was compensated by a constant increase in unit prices (Secci et al. 1995). This led to more restrictive standards being set by the regional administration in order to limit the fishing effort (time restrictions), and to protect both the reproducers and the recruits (season and minimum catch size restrictions). The decree established that fishing and selling of lobsters was prohibited from 1 September to 28 February (Decree of the President of the Republic 02.10.68 no.1639 art. 132) and set a minimum catch size limit of 24 cm of total length (DPR 02.10.68 no. 1639 art. 88). Since 1991 the period in which fishing for lobsters is prohibited has been extended by 45 days (DPR 22.07.91 n. 25).

Considering the great social and economic importance of lobster fishing in Sardinia, it is urgent to extend our knowledge of lobster biology and ecology by capture-recapture studies and implementing experimental restocking programs. This calls for a new management policy. The lobster restocking experiment that has been operating since 1997 provides indications of how to increase the lobster stock in the Sardinian seas. Similar surveys on *P. elephas* have been carried out in the past in Corsica and Cornwall (Marin 1981-1985, Campillo & Amadei 1978, Hepper 1977).

This paper gives a general picture of lobster fishing in Sardinia and provides a first approach to a method for tagging and release in a protected area aimed at assessing the growth and mortality rates as well as the movement displacement of lobsters.

2 MATERIALS AND METHODS

The following parameters were assessed on each lobster catch: 1. Carapace length measured along the median line from the top of the rostrum to the back edge of the carapace (expressed in mm); 2. Total length (TL) from the infraorbital spine up to the posterior edge of the telson along the median line (expressed in cm); 3. Total weight; 4. Sex; and 5. Any sign of a recent molt. The measurements were made following Hepper (1966).

Furthermore, lobster fishing in Sardinia has been studied by several means. First, a census was made of all the fishing boats in all the fishing ports of the island (gross tonnage, GT; horse power, HP); the number of lobster fishermen; and of the type and characteristics of fishing gear used. Second, a calculation was made of fishing pressure expressed as square kilometer per GT. More precisely the fishing pressure exerted by all fishing boats registered for the 50-100 m bathymetry was calculated, while in the 100-200 m bathymetry only fishing boats of a GT greater than 3 tons were considered. Third, an analysis was performed of the landing from two sampling areas in central-western and northern Sardinia from 1983 to 1997 aimed at studying fishing in time before and after the enforcement of legal restrictions. Fourth, an assessment was made of the catches. A few fishing campaigns were carried out in two sample areas with the aim at calculating density indices (catches in grams per length of net per day at sea).

We carried out several tagging and restocking experiments. A restocking area was chosen along the central-western coast of Sardinia for its geomorphological and bionomic characteristics. It extends over an area of about 400 hectares at a depth of between 50 and 100 m and characterized by coastal detritus, pre-coralligenous and coralligenous formations (Fig. 1). We recorded the biometric and sex parameters of all specimens caught in the area and of a sample of lobster of a TL smaller than 24 cm caught in neighboring areas. These lobsters were tagged and released in the restocking area. Tagging was made with plastic T-BAR tags. The tags were placed respectively between the carapace and the first abdominal segment in specimens with a TL of less than 15 cm, and between the first and second abdominal segments in specimens with a TL of more than 15 cm (Campillo & Amadei 1978).

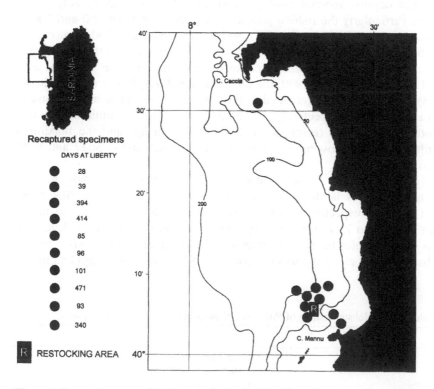

Figure 1. Restocking area of *Palinurus elephas* and location of recaptured specimens.

3 RESULTS

Lobster fishing involves over 25% of the local artisanal small fishing boats. More than 254 boats operate in Sardinian waters, involving about 587 fishermen. However, lobster fishing is evolving continuously due to the variability and rarefaction of the catches. In the different fishing ports, the number of boats fluctuates as well as their characteristics. At present 71% of the boats operate on the western coast and only 29% on the eastern. The most commonly used fishing boats have a GT of between 7 and 10 tons. This is true of both coasts (the found values are 55% on the eastern coast and 43% on the western). Almost all fishing boats use different kinds of fishing gears either contemporaneously or in succession, because lobster fishing is restricted to less than five months a year, considering regional provisions and meteorological conditions at sea.

In line with the meteorological conditions and depending on the fishing ports, trammel net fishermen stay at sea for 24 to 48 hours. Trammel nets are lowered at depths of between 50 and 200 m. Considering this depth range, the potential area of exploitation for lobster fishing would be 10,011 square km (Table 1), while the area actually used for lobster fishing is far smaller on account the shape of the sea bottom at these depths. Moreover the two coasts of Sardinia are remarkably different since the continental shelf along the western coast is more extensive. Consequently, a larger area in terms of square km can be exploited on the western coast than on the eastern, both in the 50-100 m and in the 100-200 m depth range.

Considering all depths involved, fishing pressure on the eastern coast is less than on the western. Particularly the fishing pressure was greatest between 50 and 100 m along the western coast (26.16 sq. km/GT), and between 100 and 200 m along the eastern (33.36 sq. km/GT) (Fig. 2). Upon analysis, the landings from the two sample areas in the central western and northern coasts of Sardinia, it can be seen that after an abrupt fall between 1985 and 1990, the production of *P.elephas* is now gradually rising (Fig. 3). The peak values recorded in the central-western area in 1984 and in the northern area in 1985 (respectively, 10,494 and 1033 kg), the minimum landing values in both areas in 1990 (respectively, 2404 and 3508 kg) and the maximum values recorded in 1992 following regional provisions (respectively 3627 and 4689 kg) are significant.

The density indices (Table 2) based on samplings from both areas show a slight increase in the amount of landed lobsters. These vary from 5.8 g and 70.8 g per length of net per day respectively in the central-western area in 1994 and in the northern area in 1995, to 75.7 and 118.4 g per length of net per day in 1997. In both studied areas captures refer to individuals below the minimum landing size. In fact 80.9% of lobsters caught along the western coast and 82.2% of those caught along the

Table 1. Suitable areas for lobster fishing of *Palinurus elephas* in the seas around Sardinia.

Depths	Areas
50-100 m	4066 sq. km.
100-200 m	5945 sq. km.
50-200 m	10,011 sq. km.

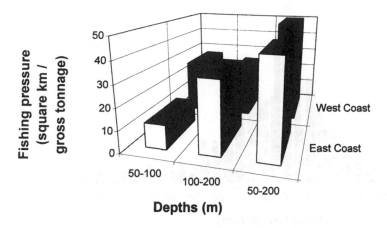

Figure 2. Lobster fishing pressure along the western and eastern coasts of Sardinia (square kilometers/gross tonnage).

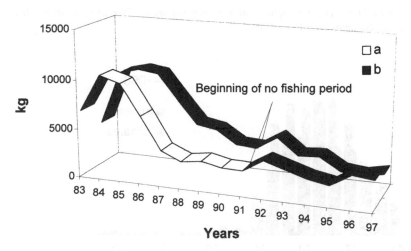

Figure 3. Lobsters landed in two sampling areas of the central-western (a) and northern (b) coasts of Sardinia.

Table 2. *Palinurus elephas* density indices from 1994 to 1997 in the central-western and northern areas of Sardinia.

Years	Central-western area (gr/net/day)	Northern area (grams/net/day)
1994	5.8	–
1995	56.67	70.86
1996	52.13	28.37
1997	75.77	118.44

northern coast had a TL less than 24 cm (Fig. 4). The catch range was wide in both samplings. The smallest lobsters measured respectively 7.5 and 8.7 cm TL in the central-western and northern areas. The largest had a TL of 35.8 and 33.0 cm respectively. The modal classes were respectively 21 and 20 cm of TL.

An experimental restocking program started in 1996 and became operative in 1997 in an area where fishing is prohibited by regional provisions. A total of 1066 tagged lobsters, precisely 579 males and 487 females, were released in the area (Fig. 5). Ten

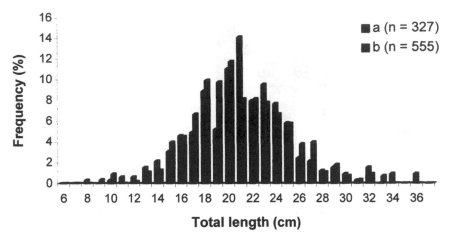

Figure 4. Specimens of *Palinurus elephas* captured in two sampling areas of the central-western (a) and northern (b) coasts of Sardinia from 1994 to 1997.

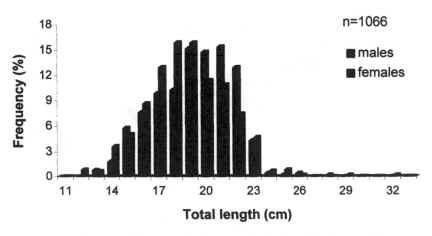

Figure 5. Specimens of *Palinurus elephas* tagged and released in the restocking area.

of these (7 males and 3 females) were recaptured after 28 to 414 days at liberty (Table 3). Using cartographic maps it appears that, if lobsters traveled in a straight line, the maximum distance traveled was about 4,118 km in 414 days at liberty (Fig. 1). The smallest distance traveled, on the other hand, was 0.549 km after 101 days at liberty. It is calculated therefore that the lobsters traveled at a mean 1,025 km per month. However, one specimen was recaptured after 340 days at liberty some 50 km from where it was released.

Four lobsters out of seven (3 males and 1 female) showed a mean increase of TL of 11.1% (more precisely the minimum increase was 6.9%, while the maximum increase was 16.7%).

Table 3. Recaptures of tagged *Palinurus elephas* (TL = total length). *exceptional recapture.

Recapture	Sex	TL increment (%)	km/month	Days at Liberty
1	M	–	4,157	28
2	M	–	2,815	39
3	M	–	33	394
4	M	16.7	302	414
5	M	–	323	85
6	F	–	286	96
7	M	7.8	163	101
8	M	13.0	157	471
8	F	6.9	785	93
means	–	11.1	1,025	–
10*	F		4,412	340

4 DISCUSSION

Lobster fishing has evolved remarkably since the 1920s, especially in regards to the fishing gear used and in areas fished. This is due both to stock impoverishment at the shallower depths, and to the use of larger fishing boats equipped with ultrasonic echolocation and loran devices for the location of the most suitable sea banks and bottoms.

The areas where lobsters can be readily fished are fewer than shown on the maps. This is because the bottoms in these areas are soft and not a good habitat for lobsters, and because the large coral formations would very soon damage the nets and make them impossible to recover. This has led to the establishment of areas of biological protection for the species.

In spite of the fact that *P. elephas* fishing is gradually increasing, a comparison with the capture data reported in the literature shows that the past critical situation was due to an excess in the fishing effort. The samples collected in both areas show that compared to the 1970s the modal class of landed lobsters became gradually smaller, and decreased from 25 cm TL (Cottiglia et al. 1976), respectively, to 21 and 20 cm TL (present paper). Moreover, the results showed significant shifts especially when compared to the literature data for specimens of *P. elephas*. Hepper (1977) observed a mean speed of 0.005 km per week. The observations carried out so far are insufficient to define growth constants. However, from the data in our possession it is possible to hypothesize that the size increase seems larger than in the literature (Marin 1981, Campillo & Amadei 1978).

REFERENCES

Campillo, A., & Amadei, J. 1978. Premieres données biologiques sur la langouste de Corse, *Palinurus elephas* Fabricius. *Rev. Trav. Inst. Peches Marit.* 42: 347-373.
Cottiglia, M., Ionta, G., Masala, B., Serra, E. & Tagliasacchi Masala, M.L. 1976. La pesca di *Palinurus elephas* Fabr. in Sardegna. *Ecologia, etologia, produzione e sforzo di pesca. La programmazione in Sardegna* 59: 1-14.

Hepper, B.T. 1966. Mesurement of carapace length and total length in the crawfish, *Palinurus vulgaris. J. Cons. Perm. Int. Explor. Mer.* 30: 316-323.

Hepper, B.T. 1977. The fishery for crawfish, *Palinurus elephas,* off the coast of Cornwall. *J. Mar. Biol. Ass. UK* 57: 925-941.

Manunta, C. 1955. Sulla biologia sessuale di *Palinurus vulgaris* Latreille. La pesca delle aragoste nella baia di Alghero negli ultimi decenni. *Symposia Genetica* V: 1.

Marin, J. 1981. Données préliminaires sur la croissance de la langouste rouge, *Palinurus elephas* Fabr. en Corse. Rapp. Comm. *Int. Mer Médit.* 27: 173-174.

Marin, J. 1985. Etude de la croissance des crustacés à partir des données de marquages – recaptures. Application à la langouste rouge de Corse, *Palinurus elephas* Fabricius. *Conseil int. pour l'explor. de la Mer C.M.* 26: 1-17.

Santucci, R. 1928. La pesca dell'aragosta in Sardegna. *R. Comitato Talassografico Italiano – Memoria* 136: 1-23.

Secci, E., Addis, P. Stefani, M. & Cau, A. 1995. La pesca di *Palinurus elephas* (Fabricius 1787) nei mari circostanti la Sardegna. *Biol. Mar. Medit.* 2: 455-457.

A mass stranding, or 'walk out' of west coast rock lobster, *Jasus lalandii*, in Elands Bay, South Africa: Causes, results, and implications

A.C. COCKCROFT, D.S. SCHOEMAN, G.C. PITCHER, G.W. BAILEY & D.L. VAN ZYL
Marine and Coastal Management, Cape Town, South Africa

ABSTRACT

During 14 March-7 May, 1997, South Africa experienced its worst ever low-oxygen induced rock lobster mortality. About 2000 tons (compared to an annual TAC of 1,680 tons for the 1996/97 season) of *Jasus lalandii* were stranded in and around Elands Bay, an important fishing area on the west coast of South Africa. The persistent accumulation and eventual decay of a dense dinoflagellate bloom (most likely as a result of nutrient depletion) and concomitant depletion of oxygen, as a result of aerobic bacterial activity, was the cause of the low oxygen levels that precipitated the sequence of stranding events. A massive stranding of some 1,500 tons in Elands Bay during 5-8 April (spring tides) was preceded by two large strandings and followed by three less severe events. This sequence followed a general north-south progression. Females constituted the bulk (about 80%) of the stranded lobster up to early April, with males increasing in importance during the latter events. The proportion of stranded lobster larger than the commercial minimum size limit (75 mm CL) increased from < 20%, during 14 March-25 April, to around 30% for the last two stranding events. In total, some 318 tons (16% of total lobster stranded during the sequence of events) were returned to the sea alive in areas unaffected by low oxygen. Trap and diving surveys conducted after the strandings illustrated the success of this operation. These surveys, together with decreased commercial catch per unit effort in the area, indicated that lobster distribution, but not abundance, recovered within a year of the strandings. Decreased lobster growth rates were also noted in the season following the strandings. The long-term implications of this substantial loss (5-10% of total spawning biomass) to the resource as a whole are difficult to assess. Since the recruitment dynamics of this species are poorly understood at present, and recruitment to the fished population requires in excess of six years, the full impact of this event may only be manifested in the future.

1 INTRODUCTION

West Coast rock lobster *Jasus lalandii* are distributed from about 23°S, just north of Walvis Bay in Namibia, to about 28°E, near East London on the east coast of South

Table 1. A summary of estimated rock lobster and other faunal mortaly on the South African west and south east coasts during 1962 - April 1998. The values cited are based on the best estimates of Sea Fisheries inspectors, rock lobster industry members and research scientists. * Approximate scaling: Small to moderate = up to 10 tons; severe = up to 30 tons. 1 and 2 considered the same stranding event.

Date	Region	Associated dinoflagellate	Rock lobster stranding	Other species affected	References
1962	False Bay	Alexandrium poly-gramma	No	About 100 t of fish	Grindley & Taylor (1964)
12/66-01/67	Elands Bay	Gonyaulax grindleyi	No	Beach clams and black mussels	Grindley & Nel (1970)
02/67	False Bay	Noctiluca miliarus	No	A small mortality of fish	Horstman (1981)
12/67	Elands Bay to north of Olifants river	A. catenella	Yes – small to moderate*	Beach clams, black mussels and fish	Horstman (1981)
03/68	Elands Bay	Not recorded	Yes – severe*	Not recorded	Newman & Pollock (1974a)
12/68	Elands Bay Lamberts Bay	A. catenella	No	Beach clams and black mussels	Grindley & Sapeika (1969)
11/73	Dwarskersbos	Prorocentrum micans	No	Beach clams	Horstman (1981)
09/74	Elands Bay	G. grindleyi	No	Beach clams and black mussels	Horstman (1981)
08/76-09/76	False Bay	Gymnodinium sp.	No	Inshore fish	Brown et al. (1979)
04/78	St Helena Bay	Mesodinium rubrum	Yes –small to moderate*	Many fish species	Horstman (1981)
05/78	Saldanha Bay	A. catenella	No	Beach clams and black mussels	Popkiss et al. (1979)
03/80	Elands Bay	A. catenella	No	Beach clams and black mussels	Horstman (1981)
03/86	Elands Bay	Not recorded	Yes – 27 t	Not recorded	Pollock & Bailey (1986)
03/89	Elands Bay	Not recorded	Yes – 18 t	Not recorded	Pollock pers. Comm.
1989	Betty's Bay	Gymnodinium sp.	No	30 t of abalone (Haliotus midae)	Horstman et al. (1991)
02/93	Lamberts Bay northwards	Ceratium lineatum P. micans	Yes – small to moderate*	Mussels, crabs and fish	Fisheries Inspectorate pers. comm.
02/94[1]	Elands Bay	P. micans, C. furca	Yes – 5 t	Not recorded	Matthews & Pitcher (1996)
03/94[1]	Dwarskersbos	P. micans, C. furca	Yes – 3 t	Mussels and some fish species	Fisheries Inspectorate pers. comm.
03/94[1]	St Helena Bay	P. micans, C. furca	Yes – 60 t. These were washed ashore dead.	Many invertebrate species and about 1500 t of fish, comprising about 50 species	Matthews & Pitcher (1996)
03/97-04/97[2]	Elands Bay	C. furca, C. lineatum	Yes – 1925 t	Some mussels, crabs and fish	This study
05/97[2]	Dwarskersbos	C. furca, C. lineatum	Yes – 30 t	Mussels and some fish species	This study
04/98	Dwarskersbos	C. furca, C. lineatum	Yes – 20-30 t	Many fish species	This study

Africa. Commercial densities are, however, only encountered along the west coast, from about 25°S in Namibia to Cape Hangklip (34°23'16"S) in South Africa (Pollock 1986). Commercial exploitation of this resource in South Africa started in the late nineteenth century and expanded rapidly. The fishery, which is divided into zones comprising two areas, is currently valued in excess of about R150 million (c. $30 million) per annum, providing employment for more than 4,000 people.

The western and southern coasts of South Africa are characterized by seasonally modulated, wind-induced upwelling; spatial and temporal variability in the oxygen content of bottom water (Bailey 1991, Bailey et al. 1985, Pollock & Shannon 1987); the regular occurrence of red tides (Pitcher et al. 1995); and sporadic faunal mass mortalities (see Table 1). Many of these mass mortalities may be considered direct or secondary consequences of dinoflagellate blooms (red tides). The mechanisms of dinoflagellate bloom development in the region of the upwelling front, and their transport and accumulation inshore is well documented (Pitcher et al. 1995, Pitcher et al., in press). Red tide toxicity has been shown to be the direct cause of both mussel (Grindley & Nel 1970, Popkiss et al. 1979, Horstman 1981) and abalone mortalities (Horstman et al. 1991). Mass mortalities resulting from the secondary effects of red tides include gill clogging in fish (Grindley & Taylor 1964, Brown et al. 1979) and anoxia resulting from the effects of biomass accumulation (Horstman 1981, Matthews & Pitcher 1996). However, mass mortalities on the South African west coast are not always linked to red tide activity. The advection of oxygen-deficient water from a poleward flowing undercurrent (De Decker 1970) into the near shore region was considered the cause of the rock lobster strandings or 'walkouts' reported by Newman & Pollock (1974a) and Pollock & Bailey (1986).

The responses of *J. lalandii* populations to low oxygen conditions are well documented and include commercial catch rate declines (Newman & Pollock 1971, Bailey et al. 1985, Beyers & Wilke 1990), reduced growth rates (Pollock & Beyers 1981, Pollock & Shannon 1987, Beyers et al. 1994), smaller size at sexual maturity and reduced fecundity (Beyers & Goosen 1987), decreased consumption rates (Beyers et al. 1994), restricted distribution (Newman & Pollock 1971, Bailey et al. 1985, Pollock & Shannon 1987, Beyers & Wilke 1990), and mass mortalities (Newman & Pollock 1974a, Pollock & Bailey 1986, Matthews & Pitcher 1996). Behavioral and physiological responses to reduced oxygen levels have also been recorded in other lobster species, including *Panulirus cygnus* (Chittleborough 1975, Phillips et al. 1980) and *Homarus americanus* (McMahon & Wilkens 1975, Sinderman & Swanson 1979).

This paper documents the most severe rock lobster stranding ever recorded along the South African coast (some 2,000 tons stranded over a sequence of six events), which occurred at Elands Bay, an important lobster fishing area on the west coast, from March to May 1997. The magnitude of this event is placed in perspective when considering that the Total Allowable Catch (TAC) for the entire South African lobster resource was 1,680 tons for the 1996/97 season. The causes, results and implications are examined and discussed.

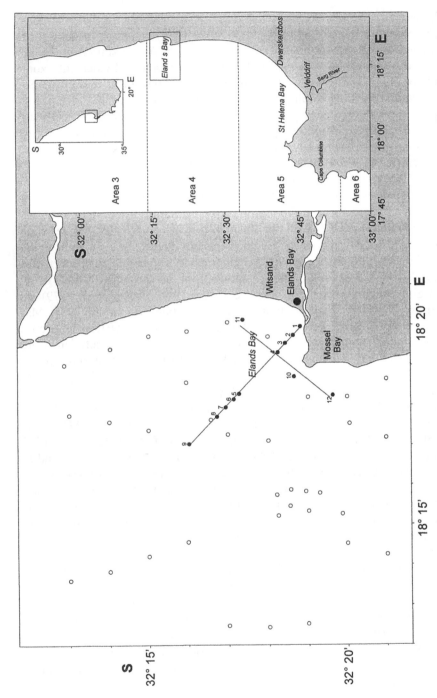

Figure 1. Study area indicating the sites of lobster strandings in and around Elands Bay and the positions of sampling stations monitored on a long-term basis (solid dots and cross). The additional sites sampled during the study period are represented by the open dots.

2 MATERIAL AND METHODS

2.1 *Baseline information*

Baseline physical, chemical, and biological data were obtained from routine annual surveys conducted in the Elands Bay area (Fig. 1) during March/April from 1992 to 1996. Research and/or commercial fishing vessels deploying both standard commercial gear (100 × 60 mm mesh traps) and small (60 × 60 mm) mesh traps were used to sample the lobster population. Traps were deployed overnight and soak times ranged between 16 and 24 hours. Additional information on benthic community structure and lobster abundance was obtained using SCUBA (10 minute random collections by two divers), benthic sampling, and remote video imaging at selected sites. Oxygen, salinity, and temperature profiles during these surveys were constructed using data from a Grant Water Quality Logger. Bottle-derived Winkler titrations were used to determine oxygen concentration at discrete depths.

2.2 *Wind, temperature and dinoflagellate cell densities: January-May 1997*

Wind observations were obtained from a meteorological station situated at Cape Columbine (Fig. 1). Daily water samples for phytoplankton analysis collected in Elands Bay were fixed in buffered formalin and enumerated by the Utermohl method (Hasle 1978). Water temperature was logged hourly by means of a sensor and data logger deployed from a fish factory jetty in Elands Bay at approximately 2 m deep.

2.3 *Surveys conducted during the strandings: March-May 1997*

Two surveys were conducted in Elands Bay during the stranding event. The first, conducted on 8 April, involved a detailed water quality survey. Temperature and in situ chlorophyll fluorescence profiles of the water column were conducted by means of a Chelsea Instruments Aquapack. Chlorophyll *a* samples were analyzed by fluorometric analysis, as detailed by Parsons et al. (1984), and were used to calibrate the in situ fluorescence profiles. Oxygen profiles were conducted by means of a Seabird CTD and oxygen probe. Additional oxygen measurements were obtained by Winkler titration of samples collected at discrete depths by means of NIO bottle samples. The second, conducted during 20-25 April, consisted of both trap (60 mm mesh) and diving samples. A hydrological survey conducted in the Dwarskersbos area (Fig. 1, the site of the last stranding) on 7 May 1997 included the analysis of hydrogen sulphide, as described in Bailey (1991).

The size composition and sex ratios of stranded rock lobster were obtained from random samples collected at the various stranding locations. Lobster carapace length (CL) was measured (to nearest mm) using Vernier callipers.

2.4 *Additional information*

Follow-up surveys in the area were conducted in July and September 1997 (traps only) and again in April 1998 (trap, diving and oxygen profiles). Fishery performance in the area was calculated from routine commercial catch monitoring, and lobster growth rates were determined from routine tagging studies (Cockcroft & Goosen 1995, Goosen & Cockcroft 1995).

3 RESULTS

During the first half of 1997, periods of reduced southerly winds at Cape Columbine corresponded with higher water temperatures and increased dinoflagellate abundance in Elands Bay (Fig. 2). The first red tides of the 1996-97 upwelling season were observed during December 1996 and were dominated by the dinoflagellate *Prorocentrum micans*. On 30 December this species attained concentrations exceeding 29 × 10^6 cells.l^{-1} After dissipation of this bloom on 3 January 1997, no further red tides were observed in Elands Bay until the 6 March 1997. On this occasion the bloom was dominated by the dinoflagellates *Ceratium furca* and *C. lineatum*. Although cell concentrations fluctuated, this bloom essentially persisted for a period of two months during which total dinoflagellate cell concentration typically exceeded 1 × 10^6 cells.l^{-1} and occasionally exceeded 25 × 10^6 cells.l^{-1} (Fig. 2). The bloom was finally dispersed in early May.

The subsequent sinking and decay of the bloom (probably as a result of nutrient depletion) and concomitant depletion of oxygen as a result of aerobic bacterial activity is clearly illustrated by the water quality survey conducted on 8 April (Fig. 3).

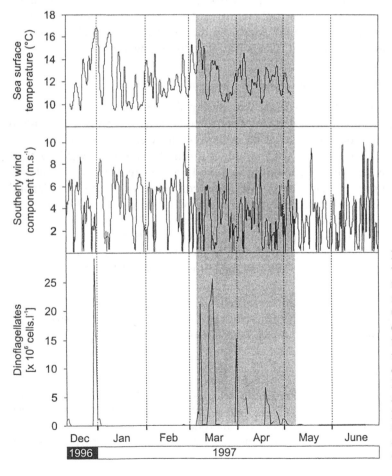

Figure 2. Equator wind speed measured by the anomometer at Cape Columbine, concomitant sea surface temperature and dinoflagellate cell counts in Elands Bay. Wind speed and sea surface temperature have been smoothed. Shaded area denotes period of strandings.

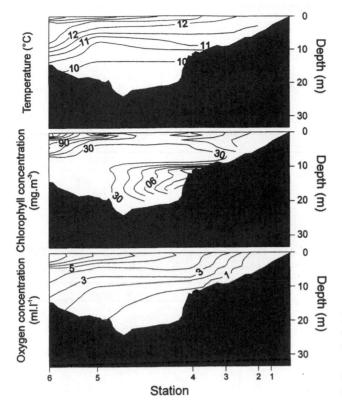

Figure 3. Isopleths of the hydrological components measured in Elands Bay on 8 April 1997. Station numbers refer to positions on the transect represented in Figure 1.

Very low oxygen levels (< 1 ml.l^{-1}) inshore, compared to the 'normal' levels (Fig. 4), were associated with very high bottom chlorophyll *a* concentrations (> 100 mg.m^{-3}). The continuation of extremely low oxygen levels (< 1 ml.l^{-1}) throughout the water column to 40 m depth was confirmed during the survey conducted in Elands Bay during late April 1997 (Fig. 4). Extremely low oxygen levels in the water column were also detected during the final stranding at Dwarskersbos on 7 May. The presence of hydrogen sulphide at levels of up to 200 μM.l^{-1} in the water column were also detected during that survey.

Nearly 2,000 tons of rock lobster were stranded in the greater Elands Bay area in three distinct phases (Table 2). The massive 'walkout' (5-8 April, the period of spring tides), during which around 1500 tons of lobster were stranded, was preceded by two large events and followed by three less severe incidences. The locations of the strandings indicate a general north to south progression of the phenomenon over time, with the last stranding at Dwarskersbos on 7 May 1997. The sex ratios and size composition of stranded lobsters revealed clear changes during the course of the event (Table 2). Females constituted about 80% of the lobsters sampled in Elands Bay and Mossel Bay during the strandings that occurred up to early April (including the major event) but comprised only 48%-55% in those areas during subsequent events. Males constituted 63% of the lobsters in the final stranding at Dwarskersbos. The proportion of stranded lobster larger than commercial minimum size limit (75 mm CL) increased from less than 20% during 14 March-25 April to around 30% during the last two stranding events.

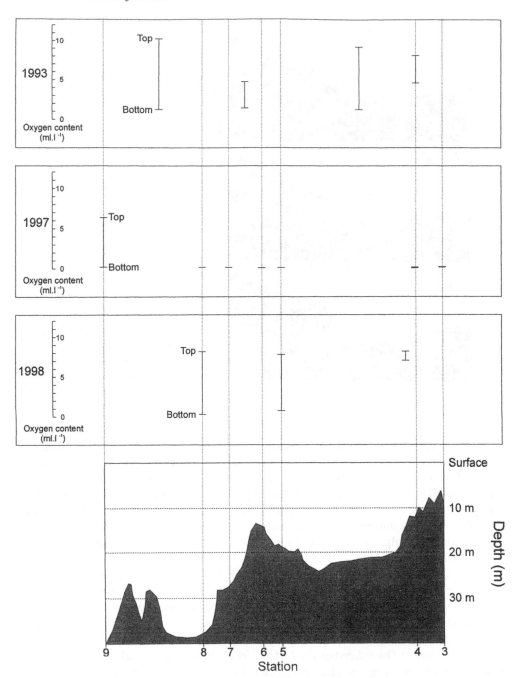

Figure 4. Oxygen content of surface and bottom waters measured during the surveys conducted in Elands Bay during April 1993, April 1997 and April 1998. Upper and lower bars represent surface and bottom measurements, respectively. Station numbers refer to positions on the transect represented in Figure 1.

Table 2. Detailed summary of rock lobster stranding event in the Elands Bay region during March-May 1997. All tonnages are estimated whole mass (derived from various sources including Sea Fisheries Inspectors, Industry representatives and the authors).

Date	Locality	Lobster stranded (tonnes)	Lobster returned (tonnes)	Sex ratio (♂:♀)	% legal size (> 75 mm CL)	n	Mean CL (mm)
14-24 March	Mussel Bay	75	35	20:80	16	420	68.5
23-30 March	Elands Bay	150	15	21:79	16	420	66.9
5-8 April	Elands Bay	1500	200*	18:82	11	418	65.9
5-8 April	Mussel Bay	20					
5-8 April	Witsands	50		31:69	17	210	67.6
24-25 April	Elands Bay	30	20	45:55	14	523	66.1
30 April	Mussel Bay	100	30	52:48	33	420	72.1
7 May	Dwarskersbos	30	18	63:37	30	467	70.2

*In addition, some 300 tonnes whole mass were processed by Industry (275 tons to fishmeal tailed and packed).

Lobster distribution and abundance in Elands Bay during autumn (Fig. 5), based on surveys conducted from 1992 to 1996, are characterized by substantial quantities of lobster on the reef at Stations 5, 6, 10 and 11. Lobster were moderately to sparsely distributed at the remaining sites, generally mirroring oxygen levels. Large numbers of juveniles and females at Stations 1 and 11, respectively, were recorded during some, but not all of the surveys. This contrasts markedly with the results obtained from the diving and trap survey conducted during 20-25 April 1997 that revealed the complete absence of lobster at all sites sampled, except for the area immediately behind the breaker line in extremely shallow water in Elands Bay and Mossel Bay where lobster were found concentrated in large numbers (Fig. 5). This accumulation at Elands Bay extended along some 2 km of shoreline, about 20 m wide, at an average density of 20 lobsters.m^{-2}. Assuming a mean mass of 0.22 kg.lobster^{-1} (calculated from diver-collected and stranded length-frequency samples), this projected to roughly 175 tonnes of lobster. A stranding of about 30 tons was recorded in the same area on the following day.

The survey conducted in July 1997 (using 100 mm mesh traps) showed an absence of lobster at the sites normally associated with high densities. However, a dense aggregation of lobster was observed in an area coinciding with the site of live lobster release (about 50 tonnes, Table 2) during the stranding events in March (Fig. 5). The trap survey in September 1997 (using 60 mm mesh traps), although not directly comparable to that conducted in July, revealed a similar lobster distribution, with some evidence of dispersion into surrounding areas (Fig. 5). The trap (60 mm mesh) and diving survey conducted in April 1998 (Fig. 5) indicated a lobster distribution approaching that described for a 'normal,' i.e. non-stranding, year. The corresponding bottom oxygen concentrations in Elands Bay during that period were high (Fig. 4). No major differences in sessile benthic community structure were noted during the diving surveys conducted before and after the stranding events.

As a result of the strandings, a Total Allowable Catch (TAC) of 152 tons was allocated to Zone B (the fishing zone encompassing the stranding event) for 1997/98, a reduction of 61% compared to the previous fishing season (Table 3). More than 99%

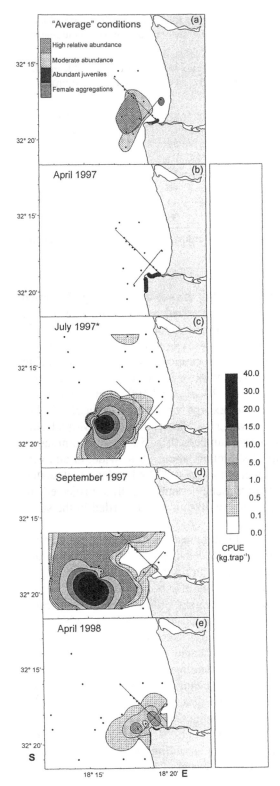

Figure 5. Interpolated representations of (a) rock lobster distribution and abundance (using both trap and diving results) during routine surveys conducted from 1992-1996 and (b-e) trap catch-per-unit-effort during the surveys conducted during the study period.

Table 3. A summary of the catch statistics and growth rates measured in Zone B for the 1992/1993–1997/1998 commercial fishing seasons. The units for the hoopnet and trap catch-per-unit effort are kg/rowboat/day and kg/trap/day, respectively.

Season	TAC Allocated Global (tons)	TAC Allocated Zone B (%)	Rock lobster landed in Zone B Total (tons)	Rock lobster landed in Zone B Area 3 (%)	Rock lobster landed in Zone B Area 4 (%)	Catch statistics in Zone B Caught by hoopnets	Catch statistics in Zone B CPUE hoopnets	Catch statistics in Zone B CPUE traps	Annual growth increment (mm CL) 70-80 mm CL size class	Annual growth increment (mm CL) 80-90 mm CL size class
92/93	2200	21	541	39	61	40%	43	7	2.5	2.1
93/94	2200	22	555	1	99	55%	57	15	–	–
94/95	2000	25	502	14	86	47%	27	6	3.0	2.2
95/96	1520	24	373	4	96	41%	45	10	2.7	1.2
96/97	1675	24	386	4	96	50%	56	8	3.6	3.9
97/98	1920	7	152	< 1	> 99	97%	44	6	2.2	2.1

of the TAC was caught in Area 4 (the sub-area of the lobster stranding), mostly with hoop nets (97%). Substantial declines (ca 20%) in hoop net and trap catch-per-unit effort (CPUE) were recorded in the 1997/98 season (Table 3). Similarly, somatic growth rates (determined from routine tagging) decreased by 39% and 46% in the 70-79 mm and 80-89 mm CL size classes, respectively, over the same time period.

4 DISCUSSION

Lobster strandings, or 'walkouts', on the South African west coast are the result of lobster moving into shallow waters in an attempt to avoid low oxygen conditions. This is coupled with an accumulation in extremely shallow water during high tides and their subsequent inability to retreat fast enough to avoid exposure during ebb tides (Newman & Pollock 1974a, Pollock & Bailey 1986, this study). Rock lobster tend to avoid dissolved oxygen values lower than 2 ml.l^{-1} by moving to shallower, better-oxygenated depths (Newman & Pollock 1971, Pollock & Beyers 1981, Pollock 1982, Bailey et al. 1985), with females being less tolerant of low oxygen water than males (Pollock & Bailey 1986). This oxygen concentration is similar to that at which significant differences in growth increment, ingestion, intermolt period and molt-stress-related mortality occurred in juvenile aquarium-held lobster (Beyers et al. 1994). Recent field observations (Cockcroft, unpubl. data) have, however, revealed that lobsters showed no signs of directed away movement when exposed to low (ca 1-1.5 ml.l^{-1}) oxygen concentrations for short periods (6-24 hours). Therefore, although an oxygen level of 2 ml.l^{-1} may be considered a threshold level, the duration of exposure to low oxygen levels is a critical component affecting lobster avoidance behavior. Earlier attempts to identify the exact point at which *J. lalandii* is affected by oxygen deficiency were inconclusive (Bailey et al. 1985).

The sequence of the six strandings between 14 March and 7 May 1997 should be considered collectively as a single major event rather than as a set of discrete or separate responses to low oxygen water. The similarities between that event and those reported in 1968 (Newman & Pollock 1974a) and 1986 (Pollock & Bailey 1986) include general locality, the accumulation of lobster in shallow waters prior to stranding and that the most serious strandings occurred during spring tides. However, the magnitude or severity, and the origin of the low oxygen water responsible for the strandings, are major differences between these earlier strandings and the 1997 event.

The development of and spatial and temporal variation in oxygen-deficient bottom water along the southern African coast is well documented (Bailey et al. 1985, Pollock & Shannon 1987, Bailey 1991). Newman & Pollock (1971) showed that the incursion of oxygen-poor water, which occurs along the entire western coast of South Africa (de Decker 1970), causes the inshore/offshore migration cycle of lobsters in the Elands Bay area. The movement of this low oxygen bottom water close to the shore under calm conditions was postulated as the cause of the 1968 and 1986 strandings (Newman & Pollock 1974a, Pollock & Bailey 1986). These authors made no mention of red tide activity during the strandings that they described. In contrast, the sinking and decay of the dense dinoflagellate bloom (most likely as a result of nutrient depletion), and concomitant depletion of oxygen as a result of aerobic bacterial activity, caused the low oxygen levels recorded in the shallow water during the sequence of stranding events reported here. The mechanism of dinoflagellate bloom development in the region of the upwelling front and the transport and accumulation of these blooms inshore has been described by Pitcher et al. (1995), Pitcher et al. (in press). A mass mortality of lobster and fish (ca 60 tons and 1500 tons, respectively) at St Helena Bay during March 1994 as a result of suffocation and/or hydrogen sulfide poisoning was also attributed to the entrapment and decay of a dense dinoflagellate dominated red tide (Matthews & Pitcher 1996).

It is possible that the severity and magnitude (about 30 times greater than any previously recorded walkout in South African waters) of the overall sequence of events indicates a combination or coincidence of causative factors. The quiescent conditions, which favor the shoreward transport of frontal blooms and the formation of inshore red tides (Pitcher et al. 1995), also favors the incursion of low oxygen bottom water into shallow depths (Newman & Pollock 1971). Although the initial inshore movement of lobster in the present study may have been in response to the low oxygen bottom water, the major causative factor of the strandings is considered to be the oxygen depletion associated with bloom decay.

Although the underlying cause of the strandings is fairly clear, the reasons for the severity or magnitude of the massive 1500 ton stranding in Elands Bay on the 5-8 April warrants further discussion. Elands Bay has been the site of numerous lobster strandings (see Table 1). It is postulated that the orientation of the bay, the bottom topography and the normal inshore/offshore lobster migration all contribute to the concentration and trapping of lobsters inside the bay, making this area a focal point for low oxygen induced lobster strandings on the west coast. The prolonged duration of the low oxygen event, which prevented the dispersal of lobsters back to deeper water, during March/early April, as well as a possible 'herding' of lobster into Elands Bay from farther north, are factors considered here to have contributed to the massive stranding during the spring tides of 5-8 April. Under quiescent conditions, poleward surface flow is responsible for the southward propagation of red tides (Pitcher & Boyd 1996, Pitcher et al. in press). The general north to south sequence of the strandings and the poor lobster catches in the fishing region (Area 3) immediately north of the affected area in the season following the stranding (Cockcroft, unpubl. data), supports this postulation.

The calm conditions during the sequence of strandings ensured that the vast majority of lobster were alive when first beached. Subsequent exposure to the sun, desiccation by wind, crushing as a result of the formation of huge piles of lobsters by wave action, and trampling and removal of lobsters by people all contributed to the massive mortality in Elands Bay during 5-8 April. Some 300 tons (whole mass) of lobsters were processed (275 tons to fishmeal and 20 tons tailed and packed) during that period and about 200 tons were returned to the sea in areas unaffected by low oxygen water. Improved crowd control and the smaller nature of the subsequent strandings reduced the mortality and increased the percentage of lobsters returned to the sea alive during these events. In total, some 318 tons (16% of total lobster stranded during the sequence of events) were returned to the sea alive.

It is likely that the normal distributional dynamics of *J. lalandii* in Elands Bay contributed to the demographic characteristics observed during the period of strandings. Juvenile lobsters (< 50 mm CL) generally occur mainly on shallow reefs and females, being more sensitive to low oxygen than males, are restricted to habitats shallower than about 15 m (Pollock 1982). Therefore, the anomalously high (on average, the population in this area is comprised of 70% females (Newman & Pollock 1971)) incidence of small and female lobsters observed in the initial strandings reported here, and those of March 1968 (Newman & Pollock 1974a) and March 1986 (Pollock & Bailey 1986) can be explained. Subsequently, as conditions worsened, male lobsters could have been forced landwards from the deeper waters (Pollock & Bailey 1986), resulting in their higher representation in strandings towards the end of the period. The preponderance of males in strandings south of Elands Bay was probably exacerbated because of the more male dominated populations in these areas (Newman & Pollock 1971, Pollock 1982).

The trap and diving surveys conducted after the stranding events highlight some important issues. The good survival rate of lobsters in the vicinity of the area where they had been returned to the sea justifies the attempts by the authorities to rescue as many of the lobsters as possible. The subsequent dispersal of the lobsters and the return to a more or less 'normal' distribution within a year indicates that at least the lobster distribution, if not abundance, was restored fairly rapidly. Nevertheless, even in terms of abundance, the recovery of the lobster population in the Elands Bay area has been fairly rapid compared to that of an offshore reef area, or blinder, in the St Helena Bay area after the black tide event of 1994. This is probably because the sessile benthic community structure around Elands Bay was not impacted to any large degree by the low oxygen event, while that of St Helena Bay was decimated as a result of hydrogen sulfide poisoning (Matthews & Pitcher 1996). The benthic reef community structure of the latter area was altered from sponge to mussel dominated (Cockcroft, unpubl. data). Although the presence of hydrogen sulfide was detected in the Dwarskersbos area during the latter part of the final lobster stranding in May 1997, it was not considered an important factor contributing to lobster mortality during the present study. That most of the lobsters were alive when stranded (as opposed to being washed up dead as in the 1994 black tide event), supports this conclusion. It has been demonstrated that whelks can achieve and maintain dominance in disturbed inshore benthic assemblages of the west coast by effectively excluding lobsters (Barkai & McQuaid 1988). However, concerns that this may occur in Elands Bay have, to date, proved unfounded.

The partial recovery of the lobster resource around Elands Bay, indicated by surveys conducted after the strandings, was mirrored by the commercial catches in the area when the reduced TAC for the 1997/98 season was landed, albeit at a lower CPUE. Similarly, a marked drop in local CPUE during the fishing season that followed the 1968 stranding was reported by Newman & Pollock (1971). Factors such as physical damage, i.e. appendage loss (G.G. Newman, unpublished data), and unfavorable nutritional regimes (Newman & Pollock 1974b, Pollock 1982), especially during the reserve accumulation phase of the molt cycle (Cockcroft 1997), are known to influence somatic growth rates. Most of the lobsters returned to the sea experienced some degree of damage (usually the loss of one or more appendages) during the stranding, collection, transportation and release process. This, coupled with reduced feeding during the critical reserve accumulation phase could account for the reduction in measured growth rates in the area.

While the short term implications of strandings in the immediate area are evident and relatively easy to monitor, the overall implications of this substantial loss of spawning biomass to the resource as a whole (possibly 5-10% of total spawner biomass) is more difficult to assess. Since the recruitment dynamics of *J. lalandii* are poorly understood at present, and recruitment to the fished population typically requiring in excess of six years (Pollock 1986), the full implications of this event may only be manifested in the future. However, the recent improvement in indices of resource performance and a management strategy incorporating stock rebuilding may diminish the potential repercussions.

Examination of the frequency of lobster strandings or mass mortality (Table 1) reveals that the 1960s, 1970s and 1980s were characterized by one or two such events per decade. The 1990s, however, has experienced four recorded mass strandings to date (including that at Dwarskersbos in 1998). However, more importantly, is that the three events that have occurred within the past 5 years have included by far the two most serious strandings (ca 60 tonnes in 1994 and 2000 in 1997) ever recorded. They have been very closely linked to the decay of massive dinoflagellate dominated red tides. A continuation of this trend of increasing frequency and severity of lobster strandings would not only be devastating to the rock lobster industry, and especially the small fishing communities on the South African west coast, but it would have a serious impact on the ecology of the region.

The magnitude and diversity of the activities and participants involved in the various aspects related to the sequence of stranding events reported here make the calculation of an overall financial cost extremely difficult. However, many millions of Rands are estimated to have been spent and/or lost as a direct result of this natural disaster.

A contingency plan, instituted by Sea Fisheries management authorities after the 1997 event, to cope with the type of situation described here, has as its main objective the return of as many live lobsters to the sea in as good a condition as possible. The implementation of this plan resulted in an improved overall management of the stranding at Dwarskersbos in 1998 and the successful return of a large percentage of lobster stranded. This plan, coupled to the monitoring of red tide activity and a proposed early warning system to detect the presence of low oxygen water in Elands Bay, will hopefully assist in the management of future lobster strandings in South Africa.

ACKNOWLEDGEMENTS

The technical staff of Marine and Coastal management are gratefully acknowledged for their assistance. Drs. D.E. Pollock, S.C. Pillar, and A.I.L. Payne are thanked for their constructive comments on the manuscript.

REFERENCES

Bailey, G.W. 1991. Organic carbon flux and development of oxygen deficiency on the modern Benguela continental south of 22°S: spatial and temporal variability. In R.V. Tyson & T.H. Pearson (eds), *Modern and Ancient Shelf Anoxia*: 171-183. *Geological Society Special Publication* 58.

Bailey, G.W., Beyers, C.J. De B. & Lipschitz, S.R. 1985. Seasonal variation of oxygen deficiency in waters off southern South West Africa in 1975 and 1976 and its relation to the catchability and distribution of the Cape rock lobster *Jasus lalandii*. *S. Afr. J. Mar. Sci.* 3: 197-214.

Barkai, A. & McQuaid, C. 1988. Predator-prey role reversal in a marine benthic ecosystem. *Science* 242: 62-64.

Beyers, C.J. De B., & Goosen, P.C. 1987. Variations in fecundity and size at sexual maturity of female lobster *Jasus lalandii* in the southern Benguela ecosystem. In A.I.L. Payne, J.A. Gulland & K.H. Brink (eds), *The Benguela and Comparable Ecosystems*. *S. Afr. J. Mar. Sci.* 5: 513-521.

Beyers, C.J. De B., & Wilke, C.G. 1990. The biology, availability and exploitation of rock lobster, *Jasus lalandii* off South West Africa/Namibia, 1970-1980. *Investl. Rep. Sea Fish. Res. Inst. S. Afr.* 133: 56 pp.

Beyers, C.J. De B., Wilke, C.G. & Goosen, P.C. 1994. The effects of oxygen deficiency in growth, intermoult period, mortality and ingestion rates of aquarium-held juvenile rock lobster *Jasus lalandii*. *S. Afr. J. Mar. Sci.* 14: 79-87.

Brown, P.C., Hutchings, L. & Horstman, D.A. 1979. A red-water outbreak and associated fish mortality at Gordon's Bay near Cape Town. *Fish. Bull. S. Afr.* 11: 46-52

Chittleborough, R.G. 1975. Environmental factors affecting growth and survival of juvenile western rock lobsters *Panulirus longipes* (Milne-Edwards). *Aust. J. Mar. Freshwat. Res.* 26: 177-196.

Cockcroft, A.C. 1997. Biochemical composition as a growth predictor in male west coast rock lobster (*Jasus lalandii*). *Mar. Freshwat. Res.* 48: 845-856.

Cockcroft, A.C. & Goosen, P.C. 1995. Shrinkage at moulting in the rock lobster *Jasus lalandii* and associated changes in reproductive parameters. *S. Afr. J. mar. Sci.* 16: 195-203.

Decker, A.H.B. de 1970. Notes on an oxygen-depleted subsurface current along the west coast of South Africa. *Investl. Rep. Div. Sea Fish. S. Afr.* 84: 1-24.

Goosen, P.C. & Cockcroft, A.C. 1995. Mean annual growth increments for male West Coast rock lobster *Jasus lalandii*, 1969-1993. *S. Afr. J. Mar. Sci.* 16: 377-386.

Grindley, J.R. & Nel, E.A. 1970. Red water and mussel poisoning at Elands Bay, December 1966. *Fish. Bull. S. Afr.* 6: 36-55.

Grindley, J.R. & Sapeika, N. 1969. The cause of mussel poisoning in South Africa. *S. Afr. Med. J.* 43: 275-279.

Grindley, J.R. & Taylor, F.J.R. 1964. Red water and marine fauna mortality near Cape Town. *Trans. Roy. Soc. S. Afr.* 37: 111-130.

Hasle, G.R. 1978. The inverted-microscope method. In A. Sournia (ed.), *Phytoplankton Manual*: 88-96. Paris: UNESCO.

Horstman, D.A. 1981. Reported red-water outbreaks and their effects on fauna of the west and south coasts of South Africa, 1959-1980. *Fish. Bull., S. Afr.* 15: 71-88.

Horstman, D.A., McGibbon, S., Pitcher, G.C., Calder, D., Hutchings, L. & Williams, P. 1991. Red tides in False Bay, with particular reference to recent blooms of *Gymnodinium* sp. *Trans. Roy. Soc. S. Afr.* 47: 611-628.

McMahon, B.R. &. Wilkens, J.L. 1975. Respiratory and circulatory responses to hypoxia in the lobster *Homarus americanus*. *J. Expl Biol.* 62: 637-655.

Matthews, S.G. & Pitcher, G.C. 1996. Worst recorded marine mortality on the South African coast. In T. Yasumoto, Y. Oshima & Y. Fukuyo (eds), *Harmful and toxic algal blooms*: 89-92. Paris: Intergovernmental Oceanographic Commission of UNESCO.

Newman, G.G. &. Pollock, D.E. 1971. Biology and migration of rock lobster *Jasus lalandii* and their effect on availability at Elands Bay, South Africa. *Investl Rept, Div. Sea Fish., Repub. S. Afr.* 94: 1-24.

Newman, G.G. &. Pollock, D.E. 1974a. A mass stranding of rock lobsters *Jasus lalandii* (H. Milne-Edwards, 1837) at Elands Bay, South Africa (Decapoda, Palinuridae). *Crustaceana* 26: 1-5.

Newman, G.G. & Pollock, D.E. 1974b. Growth of the rock lobster *Jasus lalandii* and its relationship to benthos. *Mar. BioL.* 24: 339-346.

Parsons, T.R., Moita,Y. & Lalli, C.M. 1984. *A Manual of Chemical and Biological Methods for Seawater Analysis*. Oxford: Pergamon Press.

Phillips, B. F., Cobb, J.S. & George, R.W. 1980. General biology. In J.S. Cobb & B. F. Phillips (eds), *The Biology and Management of Lobsters*: 1-82. New York: Academic Press.

Pitcher, G.C. & Boyd A.J. 1996. Across-shelf and alongshore dinoflagellate distributions and the mechanisms of red tide formation within the southern Benguela upwelling system. In T. Yasumoto, Y. Oshima & Y. Fukuyo (eds), *Harmful and toxic algal blooms*: 243-246. Paris: Intergovernmental Oceanographic Commission of UNESCO.

Pitcher, G.C., Agenbach, J., Calder, J., Horstman, D., Jury, M. & J. Tauntonne-Clark. 1995. Red tides in relation to the meteorology of the southern Benguela upwelling system. In T. Yasumoto, Y. Oshima & Y. Fukuyo (eds), *Harmful and toxic algal blooms*: 657-662. Paris: Intergovernmental Oceanographic Commission of UNESCO.

Pitcher, G.C., Boyd, A.J., Horstman, D.A. & Mitchell-Innes, B.A. (In press). Subsurface dinoflagellate populations, frontal blooms and the formation of red tide in the southern Benguela upwelling system. *Mar. Ecol. Prog. Ser.*

Pollock, D.E. 1982. The fishery and population dynamics of west coast rock lobster related to the environment in the Lamberts Bay and Port Nolloth areas. *Investl Rept, Div. Sea Fish. Inst., S. Afr.* 124: 1-57.

Pollock, D.E. 1986. Review of the fishery for and biology of the Cape rock lobster *Jasus lalandii* with notes on larval recruitment. *Can. J. Fish. Aquat. Sci.* 43: 2107-2117.

Pollock, D.E. & Bailey G.W. 1986. Rock lobster stranding at Elands Bay, March 1986. *S. Afr. Ship. News and Fish. Indust. Rev.* June: 31.

Pollock, D.E. & Beyers, C.J. De B. 1981. Environment, distribution and growth rates of west coast rock-lobster *Jasus lalandii* (H. Milne Edwards). *Trans.R. Soc. S. Afr.* 44: 379-400.

Pollock, D.E.& Shannon, L.V. 1987. Response of rock-lobster populations in the Benguela ecosystem to environmental change – a hypothesis. *S. Afr. J. Mar. Sci.* 5: 887-899.

Popkiss, M.E.E., Horstman, D.A. & Harpur, D. 1979. Paralytic shellfish poisoning: a report on 17 cases in Cape Town. *S. Afr. Med. J.* 55: 1017-1023.

Sinderman, C.J. & Swanson, R.L. 1979. Historical and regional perspective. In R.L. Swanson & C.J. Sinderman (eds) *Oxygen Depletion and Associated Benthic Mortalities in New York Bight, 1976:* 1-16. *US Department of Commerce, NOAA Professional Paper* 11.

Laboratory estimates of the daily ration of the Norway lobster *Nephrops norvegicus* from the southern coast of Portugal

M. CRISTO & P. ENCARNAÇÃO
Universidade do Algarve, UCTRA, Campus de Gambelas, Faro, Portugal

ABSTRACT

Daily ration and preference for different types of food were studied in *Nephrops norvegicus* by means of simple laboratory experiments. Lobsters were collected off the south coast of Portugal at depths of 600 m in 1997 and were held in captivity individually for a period of several weeks at 14°C of water temperature in a closed circuit of sea water with a salinity of 37‰, for adaptation. Two groups of 10 lobsters were fed three different types of food alternate: a. Fish – *Engraulis encrassicolus*; b. Polychaeta – *Diopatra neapolitana* or *Marphisa sanguinea*; c. Crustacea – *Palaemonetes varians*. Daily ration was estimated at 0.0077 g dried weight (DW) per g body weight per day, for the group fed every 24 h, and 0.0087 g DW per g body weight per day for the second group, fed every 48 h. These can therefore be considered as maximum values for food intake per day. Considering the amount of food ingested daily as a measure of preference of different types of food, we can also conclude that Norway lobsters prefer crustaceans to fish, and these again to polychaetes.

1 INTRODUCTION

Several complementary aspects must be considered in any feeding ecology study. Although the description of the preys of a certain species, based on the analysis of stomach contents, is the basis for any feeding study, these qualitative studies should be complemented with quantitative studies, such as determination of daily ration and of gastric evacuation rate. The latter rates can be determined by means of field experiments or laboratory experiments.

Over the past decades, numerous studies have focused on daily rations estimates in fish, either through field experiments (Brodeur & Pearcy 1987, Amundsen & Klemetsen 1988, Tudela & Palomera 1995, Pakhomov et al. 1996) or laboratory experiments (Walh & Stein 1991). In crustaceans, however, comparatively few experiments on this subject were carried out in the field (Pakhomov & Perissinotto 1996, Perissinotto & Pakhomov 1996, Maynou & Cartes 1997) or in the laboratory (Sardá & Valladares 1990). All the results derived from these studies have direct application in the field of trophic ecology, for example the determination of predation pressure on prey species, sustainable yields of different habitats, impact of environmental modifications, or even for testing bioenergetic models in different species (Héroux & Magnan 1996).

The present work arises as a complement of a feeding study on *Nephrops norvegicus* off the south coast of Portugal (Cristo 1998), and is part of a broader study that has as objective the estimation of gastric evacuation rates and daily ration in the laboratory and in the field.

2   MATERIAL AND METHODS

The methods emplyed differ from the work of Sardá & Valladares (1990), since the specimens were kept individually, making possible an accurate knowledge of the food ingested. Also in our case food was given in excess, to make sure the full satiation of the individuals.

2.1   *Experimental conditions*

The specimens of *Nephrops norvegicus* used in the present experiment were collected in 1997 as a single sample in the winter by means of a trawler at depths of approximately 600 m off the southern coast of Portugal. The lobsters were transported without water in humid conditions in insulated plastic containers and acclimated in a closed-sea water circuit. They were held in captivity individually for a period of several weeks for adaptation. During this period they were fed with food pellets. The physico-chemical conditions of the water through out the experiment were 14°C water temperature, 37‰ salinity, 8 pH, and stabilized levels of nitrites and ammonia (< 0.2 mg/l) (Encarnação et al. 1998). The light regime corresponded to total darkness to simulate the light regime observed at the depth at which the lobsters were caught. All the specimens used fell within the carapace length classes of 30 and 40 mm in order to eliminate possible variations due to size. All individuals were at intermolt stage C (Sardá 1983). Only lobsters in good physical condition were used in the experiments.

2.2   *Description of the experiments*

An 8 days period in which the individuals were deprived of food preceded the experiment in order to guarantee the total emptiness of the digestive tract. Two sets of ten individuals were used. Each group was composed of males and females (2 berried/3 not berried), in a proportion of 1:1. Each individual was held in its own plastic box so that the food supplied could only be eaten by itself and the remainder could not be lost. All the plastic boxes were immersed in flowing sea water in a 200 l tank. (Encarnação et al. 1998). The three types of food supplied were: a. Fish: *Engraulis encrassicolus*; b. Polychaeta: *Diopatra neapolitana* or *Marphisa sanguinea*; c. Crustacea: *Palaemonetes varians*. The first group was fed every day and the second group every second day. Only one type of food per day was given.

2.2.1   *Experiment 1*
The fresh food supplied was weighed individually for the first group of lobsters. The food was given in excess (approximately 2 g) so that the lobster could eat at libidum.

The first day's meal was composed of fish, the second of Polychaeta, the third of Crustacea, the fourth fish, and so on until the fourteenth day, which was the last day of the experiment. Every day, before supplying new food, the uneaten remains were individually collected and dried at 60°C until there was no further lost of weight, usually 48 hours, in order to obtain dry weight.

2.2.2 *Experiment 2*

For the second group of lobsters the supplying of food was done the same way as before (experiment 1), but the lobsters were only fed every second day, although the remains were removed the day after feeding. The remains were also weighed dry. This experiment also ran for 14 days.

2.3 *Data processing*

Since the presentation of data in fresh weights could lead to significant errors, due to the fact that the tissues not eaten absorb water; the data are presented in dry weight. Regressions of fresh weight against dry weight were established for the three types of food supplied.

3 RESULTS

The results of the regressions of wet weight (WW) against dry weight (DW), are:
– Fish (*Engraulis encrassicolus*): $DW = 0.0015 + 0.2871WW$, $r = 0.9919$, $n = 30$;
– Polychaeta (*Diopatra neapolitana*): $DW = -0.0043 + 0.227WW$, $r = 0.9835$, $n = 30$
or *Marphisa sanguinea*: $DW = 0.0329 + 0.1752WW$. $r = 0.8919$, $n = 30$;
– Crustacea (*Palaemonetes varians*): $DW = 0.0013 + 0.2679WW$, $r = 0.9897$, $n = 30$.

3.1 *Experiment 1*

During the total duration of the experiment the average amount of fresh food supplied was 2.15 g, which is equivalent to an average of 0.55 g in dry weigh. The average amount of food actually eaten was 0.25 g. Table 1 shows the actual food ingested in dry weight (g) per g of lobster body weight for the three categories of food.

3.2 *Experiment 2*

In this second experiment, where the lobsters were deprived of food in alternate days, the results do not differ greatly from the results in experiment 1. Table 2 presents the results of the amount of food ingested in dry weight (g) per g of lobster body weight, for the three categories of food, every second day.

The total amount of food ingested in each experiment was compared using a Mann-Whitney test (Conover 1980). The differences where not significant ($p > 0.05$).

To compare the amount of food ingested by species, the Kruskal-Wallis test was used (Freund 1979). Three classes of food were compared, fish, Crustacea, and Polychaeta. The test was significant ($p < 0.05$). The consumption of Crustacea was higher than fish, and fish higher than Polychaeta.

Table 1. Average values of dry-weight ingested (DW Ing.) in g per g of lobster body weight (BW), from experiment 1.

Date	Type of food	DW Ing. (g)/g of BW
22-10-1997	Fish	0.0132
25-10-1997		0.0058
28-10-1997		0.0069
31-10-1997		0.0069
03-11-1997		0.0056
Average	Fish	0.0077
24-10-1997	Crustacea	0.0114
27-10-1997		0.0112
30-10-1997		0.0147
02-11-1997		0.0124
Average	Crustacea	0.0115
23-10-1997	Polychaeta	0.0077
26-10-1997		0.0045
29-10-1997		0.0019
01-11-1997		0.0017
04-11-1997		0.0044
Average	Polychaeta	0.0040
Total Average		0.0077

Table 2. Average values of dry weight ingested (DW Ing.) in g per g of lobster body weight (BW), from experiment 2.

Date	Type of food	DW Ing. (g)/g of BW
22-10-1997	Fish	0.0107
28-10-1997		0.0105
03-11-1997		0.0061
Average	Fish	0.0091
24-10-1997	Crustacea	0.0107
30-10-1997		0.0138
Average	Crustacea	0.0123
26-10-1997	Polychaeta	0.0053
01-11-1997		0.0037
Average	Polychaeta	0.0045
Total Average		0.0086

4 DISCUSSION AND CONCLUSIONS

There are very few studies concerning daily ration or maximum daily ration either in the field or in the laboratory, for decapod crustaceans (Sardá & Valladares 1990, Maynou & Cartes 1997).

From the results obtained with experiment 1 and 2, which revealed that the amount of food ingested is independent of the quantity available, we can assume that the daily ration for this species is also the maximum daily ration that an individual can ingest any time (considering the source of food is not limited). In contrast to fish, lobster stomachs do not expand to any great extent. Therefore lobsters are more easily satiated. It should be noted that this maximum amount of food ingested did not seem to be affected by the period of starvation to which they were submitted every other day.

The greater preference for crustaceans, followed by fish and polychaets, are consistent with the results from field studies (Gual-Fau & Gallardo-Cabello 1988, Sardá & Valladares 1990, Mytilineou et al. 1992, Cristo & Cartes 1998).

The results obtained in this particular study are within the range of other published daily ration estimates (Table 3). Our results are quite similar to those obtained by Sardá & Valladares (1990) for this species, and are of the same order of magnitude of the values calculated for *Aristeus antennatus* (Maynou & Cartes 1997) both for field observations and for the daily ration estimates. These two species are decapods of the upper slope, with *N. norvegicus* being a less mobile and burrow related species and *A. antennatus* a more active benthic crustacean one. Maynou & Cartes (1997) consider *A. antennatus* an atypical species among deep water decapods having probably one of the highest daily rations within the benthic bathyal decapod communities, since it is always characterized by high stomach fullness (Cartes 1994). Maynou & Cartes (1997) also consider the values of daily ration obtained for *N. norvegicus* by Sardá & Valladares (1990) as overestimated. However, these results were not unexpected since in contrast to what has been reported in the literature for *N. norvegicus* in other areas (Lagardère 1977, Mytilineou et al. 1992) in our study area very low proportions of empty stomachs were found (Cristo & Cartes 1998). On the other hand in our experiments, despite the continuous presence of food, individuals did not eat more and there were always leftovers of all three types of food to remove. So in this particular case, although the experiments took place under conditions that were very different from those in nature, we believe that the daily ration was mainly dependent on the volume of the stomach and could be considered as the maximum food intake for *Nephrops norvegicus*.

Our results are comparable to the daily ration estimates of the brook charr in an oligotrophic lake (T = 17.8 to 19°C) in Canada (Héroux & Magnan 1996) or to those obtained by Tudela & Palomera (1995) for the European anchovy (T > 20°C). However, they are one or two orders of magnitude lower than those obtained for an Ant-

Table 3. Comparison of daily-ration in % BW Wet = (g Wet Weight/g Body Weight Wet) x 100 and % BW Dry = (g Dry Weight/g Body Weight Wet) × 100 of *Nephrops norvegicus* with daily-rations obtained for other species. Sources: (1) Sardá & Valladares 1990, (2) Maynou & Cartes 1997, (3) Recalculated from Pakhomov & Perissinotto 1996, (4) Héroux & Magnan 1996, and (5) Tudela & Palomera 1995.

	% BW Wet	% BW Dry	Source
Nephrops norvegicus (minimum – maximum values)	2.65-4.11	0.40-1.23	This study
N. norvegicus (average values)		0.77-0.86	This study
N. norvegicus (maximum food consumption)	2.5	0.86	(1)
Aristeus antennatus (actual ration)	2.588		(2)
A. antennatus (range of daily ration models)	1.666-2.315	0.130-0.223	(2)
Themisto gaudichaudi (field estimates)		4.5	(3)
T. gaudichaudi (laboratory experiments)		5.2-13.4	(3)
Salvelinus fontinalis (field estimates)		0.62	(4)
Engraulis encrasicolus (field estimates)	3.70-3.92		(5)

arctic amphipod (T < 10°C). In spite of the different temperatures at which these estimates were obtained, we might expect that the Norway lobster, an opportunistic scavenger species characterized by a burrowing live style, might have a lower daily ration in relative terms to body weight than species lower in the food chain, but a higher daily rations than predators higher in the food chain (Maynou & Cartes 1997).

In conclusion, the experiments revealed that an average lobster of 40g can eat approximately 0.3 g of dry food, (0.304 g for experiment 1 and 0.344 g for experiment 2), which corresponds to approximately 1g of fresh food a day. This daily ration seems not to be greatly affected by food deprivation in alternate days. We consider that these kinds of studies are important but have to be accompanied by field evaluations such as those which we are presently undergoing.

REFERENCES

Amundsen, P.A. & Klemetsen, A. 1988. Diet, gastric evacuation rates and food consumption in a stunted population of Arctic charr *Salvelinus alpinus* L., in Takvatn, northern Norway. *J. Fish. Biol.* 33: 697-709.

Brodeur, R.D. & Pearcy, W.G. 1987. Diel feeding chronology, gastric evacuation and estimated daily ration of juvenile coho salmon, *Oncorhynchus kisutch* (Walbaum), in the coastal marine environment. *J. Fish.Biol.* 31: 465-477.

Cartes, J.E. 1994. Influence of depth and season on the diet of the deep-water aristeid *Aristeus antennatus* along the continental slope (400 to 2300 m) in the Catalan Sea (western Mediterranean). *Mar. Biol.* 120: 639-648.

Conover, W.J. 1980. *Practical Nonparametric Statistics*. New York: John Wiley.

Cristo, M. 1998. Feeding ecology of *Nephrops norvegicus* (Decapoda: Nephropidae). *J. Nat. Hist.* 32: 1493-1498

Cristo, M. & Cartes J.E. 1998. A comparative study of the feeding ecology of *Nephrops norvegicus* (Decapoda: Nephropidae) in the bathyal Mediterranean and the adjacent Atlantic. *Scientia Marina* 62 (suppl. 1): 81-90.

Encarnação, P., Reis, J. & Castro, M. 1998. A closed sea water system for experimental trials with the Norway lobster, *Nephrops norvegicus*. *Proceedings and Abstracts. Fourth Int. Crustacean Congress.* Amsterdam, July 1998.

Freund, J.E. 1979. *Modern Elementary Statistics* New York: Prentice-Hall.

Gual-Frau, A. & Gallardo-Cabello, M. 1988. Análisis de la frecuencia y habitos alimenticios de la 'cigala', *Nephrops norvegicus* (Linneo, 1758) en el Mediterraneo Occidental (Crustacea: Nephropsidae). *An. Inst. Cienc. Del Mar Y Limnol. Univ. Nal. Autón. México* 15: 151-166.

Héroux, D. & Magnan, P. 1996. In situ determination of daily ration in fish: review and field evaluation. *Environmental Biology of Fishes* 46: 61-74.

Lagardère, J.P. 1977. Recherche sur la distribution verticale et sur l'alimentation des crustacés décapodes benthiques de la pente continentale du Golfe de Gascogne. *Bull. Cent. Etud. Rech. sci, Biarritz* 11: 367-440.

Maynou, F. & Cartes, J.E. 1997. Field estimation of daily ration in deep-sea shrimp *Aristeus antennatus* (Crustacea: Decapoda) in the western Mediterranean. *Mar. Ecol. Prog. Ser.* 153: 191-196.

Mytilineou, C., Fourtouni, A. & Papaconstantinou, C. 1992. Stomach content analysis of Norway Lobster *Nephrops norvegicus*, in the North Aegean Sea (Greece). *Rapp. Comm. Int. Mer Médit.* 33: 46.

Pakhomov, E.A., Perissinotto, R. & Mcquaid, C.D. 1996. Prey composition and daily rations of myctophid fishes in the Southern Ocean. *Mar. Ecol. Prog. Ser.* (134): 1-14.

Pakhomov, E.A. & Perissinotto, R. 1996. Trophodynamics of the hyperiid amphipod *Themisto gaudichaudi* in the South Georgia region during late austral summer. *Mar. Ecol. Prog. Ser.* (134): 91-100.

Perissinotto, R. & Pakhomov, E.A. 1996. Gut evacuation rates and pigment destruction in the Antarctic krill *Euphausia superba*. *Mar. Biol.* 125: 47-54.

Sardá, F. 1983. Determinación de los estados de intermuda en *Nephrops norvegicus* (L.), mediante la observación de los pleópodos. *Inv. Pesq.* 47: 95-112.

Sardá, F. & Valladares, F.J. 1990. Gastric evacuation of different foods by *Nephrops norvegicus* (Crustacea: Decapoda) and estimation of soft tissue ingested, maximum food intake and cannibalism in captivity. *Mar. Biol.* 104: 25-30.

Tudela, S. & Palomera, I. 1995. Diel feeding intensity and ration in the anchovy *Engraulis encrasicolus* in the northwest Mediterranean Sea during the spawning period. *Mar. Ecol. Prog Ser.* 129: 55-61.

Walh, D.H. & Stein, R.A. 1991. Food consumption and growth of three esocids: field tests of a bioenergetic model. *Trans. Amer. Fish. Soc.* 120: 230-246.

Close water system for experimental trials with the Norway lobster (*Nephrops norvegicus*)

PEDRO ENCARNAÇÃO, JOÃO REIS, MARGARIDA CASTRO
Centro de Ciências do Mar (CCMAR), Universidade do Algarve, Portugal

ABSTRACT

A closed water system was built with the objective of running several dietary and growth monitoring experiments with the Norway lobster (*Nephrops norvegicus*). The experimental lab was set up in a commercial container that had been previously thermally insulated. It consists of a closed water circuit composed of a 500 l elevated reservoir, 12 (200 l) plastic tanks, 1 (200 l) settling tank, 1 (500 l) bio-filter, 1 (500 l) refrigerating tank, and 2 (500 l) ground reservoirs. The system is also equipped with a foam separator that pumps the water from and to the settling tank. Each 200 l tank is divided into 16 (23 × 13 × 15 cm) individual compartments with a PVC frame. The compartment's bottom is made 15 cm above the tank bottom with a plastic net over a plastic grid in order to improve removal of all organic material (feces and uneaten food) from the individual cell. For the dietary trials 5-L plastic bottle bottoms replaced the PVC frame compartments. The water is pumped from the ground reservoirs to the elevated reservoir by a water pump controlled by a level device control and distributed by gravity to the trial tanks and after to the settlement tank, foam separator, bio-filter, refrigerator, and ground reservoir again. Water monitoring revealed a good performance by the system, with low and stabilized levels of nitrites and ammonia (< 0.2 mg/l) and constant pH of 8. In general, the *Nephrops* showed a good adaptation to the captivity conditions; mortality records normally occurred in the first week after introduction in the tanks, and probably were due to post-catch stress. After that, mortality accounted for 5 % in a total of 208 specimens during a period of 20 months in which 4 females spawned in the tanks and 129 molted once, 42 twice, and 1 four times.

1 INTRODUCTION

The Norway lobster, *Nephrops norvegicus*, represents a very important fishery resource in the southern region of Portugal. Due to its importance the Marine Biology Department of the University of Algarve places particular emphasis on research on this species. Thus, a closed sea water system was designed and built so that the lobsters could be reared independently with the objective of running several dietary and growth monitoring experiments.

2 MATERIAL AND METHODS

The laboratory set-up started with the conversion of a commercial transport container, to an experimental lab. The first step was the thermal insulation of the container, in order to keep constant low water temperatures (12-14°C) during the trial period. Thus, a layer of WALLMATE® was added to the inside wall of the container and after covered with plywood boards.

We built a closed water circulation system of 12 × 200-l tanks divided into 16 individual containers to hold the *Nephrops*. This closed circuit was composed, of a 500 l elevated reservoir, 12 (200 l) plastic tanks, 1 (200 L) settling tank, 1 (500 l) bio-filter, 1 (500 L) refrigerating tank, and 2 (500 L) ground reservoirs (Fig. 1). The water is pumped in this system from the ground reservoirs to the elevated one by a water pump (Espa-Squiper 30 M, nonofasic version 220/240 V; 50/60 Hz; 0.25 kW, 2900 r.p.m. with a capacity of 265 l/min), controlled by a level control device. After being pumped to the reservoir the water is distributed by gravity to the trial tanks and after to the settlement tank, bio-filter, refrigerator, and ground reservoir again. Primary filtration was accomplished by using a net bag covered with glass wool inside the settling tank at the mouth of the outlet pipe. The system is also equipped with an ATK (500 l) foam separator functioning with air injected by a high-speed motor (45 W). The water is pumped from the settling tank to the foam separator by a pump (Iwaki MD 40F, 65 W, 45-50 l/min) and afterwards to the settling tank. The bio-filter was made with 0.5 m3 of URSAN®, and oxygenation (using an air pump, 200 mbar) was added to improve nutrient remineralization. Activated charcoal inside stockings was introduced in one of the ground reservoirs, and oyster shells were also introduced in one of the ground reservoirs to act as a buffer and to keep the pH around 8. The refrigerator used was a thermostatically controlled (set for ± 1°C) immersion copper tube (covered with polyethylene) unit powered by a 2.2 kW compressor.

Each 200 l tank (93 × 54 × 50 cm) is divided into 16 (23 × 13 × 15 cm) individual compartments, with a PVC frame 1 mm thick (B, Fig. 2). The compartments were located 15 cm above the tank bottom, using a plastic grid (∅ 2 cm) over PVC supports. This plastic grid was then covered with a plastic net (∅ 1 cm) and a small amount of

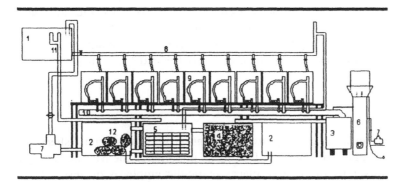

Figure 1. Representation of the closed water system. 1. Elevated reservoir; 2. Ground reservoirs; 3. Settling tank; 4. Bio-filter; 5. Refrigerating tank; 6. Foam separator; 7. Water pump; 8. Main distribution pipe; 9. Trial tank; 10. Drain pipe; 11. Security overflow pipe; 12. Oyster shell bags.

Plate 1. Upper figure. General overview of the closed water system. Middle figure. Individual compartments for growth trials. Lower figure. Individual compartments for dietary trials.

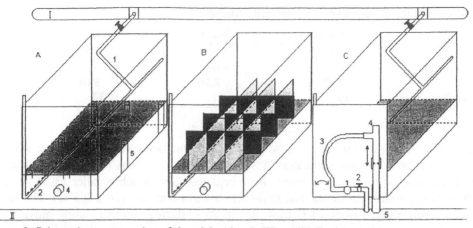

Figure 2. Schematic representation of the trial tanks. A. Water distribution (1. Inlet system; 2. Bottom cleaning system; 3. Fake bottom (grid mesh); 4. Outlet; 5. Grid mesh supporter. B. Individual compartments representation. C. Outlet system (1. Outlet; 2. Drainage tap; 3. Flexible pipe; 4. Water level control pipe; 5. Main drain pipe

gravel over the top (Plate 1, upper figure). This system was designed to reduce water deterioration allowing a better removal of all organic material (feces, uneaten food) from the individual tanks. To prevent that individuals might escape from its one compartment, a plastic net covers all individual tanks (Plate 1, middle figure). For running dietary trials the PVC frame compartments were replaced by 5 l plastic bottle bottoms of approximately the same size of the PVC compartments (Plate 1, lower figure).

The *Nephrops* collected in the Algarve waters (ranging from 25-55 mm carapace length) were accommodated in the individual compartments of the trial tanks, from October 1996 to December 1998. During this period they were fed with, experimental pellets and fresh food. The fresh food used included shrimp (*Palaemonetes varians*), clams (*Cerastoderma edule*), and fish (*Atherina boyeri*). The feeding regime used was ± 1 g of experimental diets for 5 days, then one day of fresh food, then one day of starvation. Water renewal was done every 3 to 4 month with natural seawater and accounted for 10% of total water volume. Water monitoring was performed every week, to check water quality regarding nitrite, ammonia and pH. The analyses were performed after the feeding period, using for that purpose the kits Aquamerk®8024 Ammonium-test, Aquaquant®14424 nitrite, and the pH meter (Crison pH 2001), according to procedures outlined in the handbooks.

3 RESULTS AND DISCUSSION

In terms of general layout, we can say that the implemented system is according to the extensive recommendations made by Beard & McGregor (1991), in terms of storage and care of live lobsters. All the analyses performed revealed a good performance by the system, with low and stabilized levels of nitrites, ammonia (< 0.2 mg/l) and constant pH of 8 (Table 1), that may reflect the efficiency of the foam separator and/or the biological filter. Boghen & Castell (1979) in a system designed for juvenile

Table 1. Results of water monitoring during the trial period.

	Min. (mg/l)	Max (mg/l)	Normal values
Nitrites	0.1	0.6	0.2
Ammonia	0.1	< 0.6	0.2
PH	7.87	8.16	8.04
Sal. %o	36	40	37

lobsters (*Homarus americanus*) also got low and constant levels of ammonia without the existence of a foam separator. However, their system was much smaller (360 l) and the size of the 160 specimens held in it was also much smaller (30-400 mg). Water temperature inside the test tanks was always constant (14 ± 0.5°C). The room temperature ranged from 14°C in winter to 25°C during summer, while outside the container ranged from 8 to 34°C reflecting the effectiveness of the thermal insulation.

The only problem registered in the system was the appearance of some bacteria, probably chitinolytic, that could be responsible for the appearance of some shell necroses on the *Nephrops* that disappeared after molting, also reported by Sarda & Arté (1982). These bacteria accumulated mostly in the tank bottom with the feces and uneaten food and had to be removed periodically. Thus, the introduction of UV light in the system appears to be of great importance to improve the system, controlling the occurrence of these bacteria and algal fouling as reported by Boghen & Castell (1979).

In terms of running costs, energy represented the main expense, around 54 kW/day.

In general, *Nephrops* showed good adaptation to the captivity conditions; mortality normally occurred in the first week after introduction in the tanks, and probably was due to post-catch stress. After that, mortality was 6% in a total of 208 specimens, during a period of 20 months, lower than the < 2% a month recorded by Sarda & Arté (1982) in their system, or the stabilized 1 to 2 % over a period of weeks referred by Beard & McGregor (1991). During the captivity period we where able to get 4 females to spawn in the tanks and 129 individuals to molt once, 42 twice, and 1 four times.

4 CONCLUSIONS

The implemented system showed a good performance and was found to be adequate for keeping specimens of Norway lobster in captivity for a wide range of experimental trials, such as studies of effects of diets, molting, reproduction, or metabolism. The system is simple to set up in any wet laboratory, easy to maintain, and readily adapted for other crustacean species when individual data is needed.

REFERENCES

Beard, T. W. & McGregor D. 1991. *Storage and care of live lobsters*. MAFF – Directorate of Fisheries Research. Laboratory Leaflet Number 66. Lowestoft.

Boghen, A. D. & Castell, J. D. 1979. A recirculating system for small-scale experimental work on juvenile lobsters (*Homarus americanus*). *Aquaculture* 18: 383-387.

Sardà, F. & Arté, P. 1982. Note sur le maintien en captivitie de la langoustine *Nephrops norvegicus* (L.). *Quad. Lab. Tecnol. Pesca, Ancona*. 3: 147-150.

Catchability of the lobster (*Homarus americanus*): Late spring versus autumn

M. JOHN TREMBLAY

Invertebrate Fisheries Division, Science Branch, Fisheries and Oceans Canada, Bedford Institute of Oceanography, Dartmouth, Canada

ABSTRACT

While there are several estimates of effective fishing area (EFA, measured in m^2 per trap) for *Homarus americanus* in autumn, there are none for late spring, when most of the Canadian fisheries are active. In 1997, a study of size- and sex-specific EFA's was initiated in Lobster Bay, off southwest Nova Scotia. The 45,000 m^2 study site was characterized by gravel, sand, mud, cobble, isolated boulders and kelp patches at depths of 7-19 m. Catchability was estimated in June and in September by enumerating lobsters on belt transects using SCUBA, and then trapping with research and commercial traps. The density of lobsters > 50 mm carapace length (CL) in September ($0.14\ m^{-2}$) was about double that of June, probably due to movement of lobsters into the study area coupled with molting of lobsters into the larger sizes. Research traps were more effective at capturing lobsters < 70 mm CL while commercial traps were better at catching legal sizes (> 80 mm CL). Strong seasonal differences in EFA were apparent. In June, females > 80 mm CL were more than 3 times as catchable as in September, and more catchable than males of the same size. In September large males were more catchable than large females. If these seasonal differences in catchability are generally applicable, there are several important implications. First, size composition data from spring trap samples used for mortality estimates would underestimate female mortality. Second, abundance indices based on spring trap samples would be inflated relative to other times of the year. Third, conservation of large reproductive females would be enhanced by reducing fishing effort in spring.

1 INTRODUCTION

The catch rate of fishing gear is a function of animal abundance and catchability, where catchability (q) can be defined as the probability of an animal being captured by a randomly applied unit of effort (Paloheimo 1963). In the case of lobster fisheries the unit of effort is the baited trap, and the catchability is affected by a host of factors related to lobster biology (e.g. molting), social behavior, the environment (e.g. temperature, currents), mechanical design of traps, and fishing strategy (Caddy 1979, Krouse 1989, Miller 1990). Estimates of q for lobsters are of interest primarily because of their potential application in stock assessments. Catch rate from the commercial fishery is often used as an index of stock abundance (Skud 1979, Fogarty 1995) and its usefulness depends on constant catchability from season to season. In addition, size composition from trap catches is often used to estimate total mortality, and thus size-specific catchability should be accounted for.

Catch (*C*), the number of lobsters per trap, is related to the number of lobsters on the bottom (*N*), catchability (*q*), and effort (*f*, number of trap hauls) as:

$$C_t = N_t\left(1 - e^{-qf_t}\right) \text{ (Paloheimo 1963)}$$

If *q* is small and time intervals are short, catch is estimated as (Morgan 1974, Ricker 1975):

$$C_t = N_t q_t f_t$$

If densities (*D* in n per unit area) are substituted for abundance then:

$$q_t = \frac{C_t}{f_t D_t}$$

In this formulation *q* has units of area per trap, and can be thought of as the effective fishing area (EFA) (Miller 1975). The EFA is the theoretical area from which lobsters would be removed if all had a probability of capture equal to 1. It can be used to translate catch rates into population density estimates.

The other concept, related to measuring how a trap influences the target animal, is the area of attraction. The area of attraction of the trap is larger than the EFA, since it circumscribes the area within which at least some lobsters can detect the bait. It embodies the fact of decreasing probability of capture with increasing distance from the trap.

There have been several estimates of EFA for *Homarus americanus* in recent years; all have been in September (Miller 1995a, Tremblay et al. 1998b). For three areas in coastal Nova Scotia and one in the southern Gulf of St. Lawrence, large males have been much more catchable (larger EFAs) than females at this time of year. Application of these EFAs to a modeled population suggests that females should outnumber males in the catch in the long run but this is not observed in any fisheries, where sex ratios are usually about 1:1 (Miller 1995a). In addition, other approaches for estimating catchability (tag returns) during commercial fishing do not support higher catchability of males than females in spring (Tremblay et al. 1998a). The catchability of the Western rock lobster varies seasonally and is correlated with water temperature, salinity and the percentage of lobsters in pre-molt condition (Morgan 1974). Here we provide the first EFA estimates for *Homarus* in a productive nearshore lobster area in two seasons, June and September. Lobster Bay (Fig. 1) is part of the fishing area off southwest Nova Scotia, which has high landings per unit area compared to other areas of coastal Nova Scotia (Hudon 1994).

2   METHODS

2.1   *Study Area*

One of the requirements for the study area was that it be searchable for lobsters by

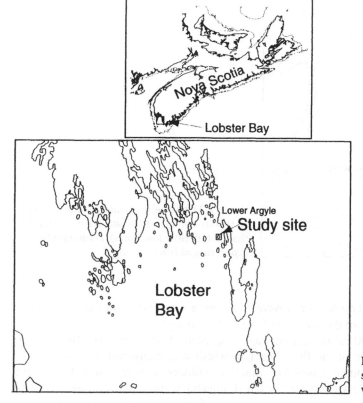

Figure 1. Location of study site in Lobster Bay, with inset of Nova Scotia.

divers. This eliminated areas with dense kelp cover or areas with large immovable boulders or stacked smaller boulders. Several locations were surveyed prior to selection of a site off Gooseberry Island (Fig. 2). Depth was 7-19 m with a tidal amplitude of approximately 3 m. Bottom habitat was characterized by gravel, sand, cobbles, a few isolated boulders (most < 50 cm diameter) and low density kelp patches. Cobbles and boulders were usually set in the substrate and were not stacked. The study site was 300 m long and 150 m wide, delineated by a 300 m groundline anchored at both ends, with lines to surface buoys every 25 m. For trap setting, additional marker buoys were placed perpendicular to the groundline.

The spring diving and trapping was completed from June 12-21, when bottom temperatures were 11.0-11.8°C. The autumn study took place from Sept. 9-17, when bottom temperatures were 15.2-16.9°C.

2.2 *Lobster density estimates*

Lobsters were censused within the study area using belt transects. Each transect line was 150 m of biodegradable twine which was laid across the study area. The location of the transects were randomly determined from a possible 60 positions spaced 5 m apart. If randomization resulted in 2 transects being only 5 m apart, another position

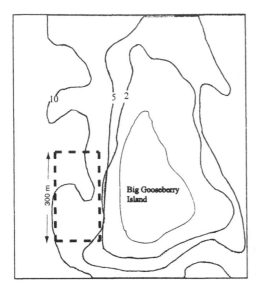

Figure 2. Details of study sites where dive transects were conducted. Depths are in meters at chart datum (low low water). During study, depths ranged from 7-19 m.

was randomly chosen. Transect lines were laid from a small boat, the start position of which was estimated from the position of the surface buoys.

For each transect 300 m² of bottom was searched by two divers counting lobsters within 1 m of each side of the line. The 1 m width was measured from the diver's chest to the end of his outstretched arm. Smaller boulders were overturned; any cavities under larger boulders (> about 50 cm diameter) were searched visually or by careful hand probing. Lobsters in the open were often aware of the divers approach and typically faced the diver with claws outstretched in an aggressive manner. The lobster either remained stationary, or backed slowly away. There were relatively few rapid escape reactions prior to the diver attempting to grasp the lobster. Lobsters were caught if possible, sexed, measured with a gauge and assigned to 10 mm size groups. If lobsters could not be captured their size group was estimated. Divers focused on lobsters > 50 mm CL. The transect lines were marked at 25 m intervals, and the habitat type was noted by divers every 25-50 m. Data were recorded on underwater slates.

A total of 10 divers participated in the 2 surveys; 9 had previous experience at searching for lobsters.

2.3 Trapping Survey

Lobsters were trapped in the study area within 5 days of completing the dive surveys in June and September. Traps were of plastic coated wire mesh with 3.8 cm mesh openings. Two trap types were used: research and commercial. The research traps were single parlor types (see Miller 1995b for a description of trap design) of dimensions 91 cm long by 53 cm wide by 36 cm high with entrance ring diameters of 12.7 cm and no escape gaps. The commercial traps were double parlor types with dimensions 120 cm long by 53 cm wide by 36 cm high with entrance ring diameters of 15.2 cm. By regulation, the commercial traps had slots 15 cm long by 4.4 cm high to allow escapement of lobsters below the minimum legal size of 81 mm CL. Both trap types were baited with about one pound of frozen mackerel, placed in wire mesh bait boxes.

For trap placement, the 300 m by 150 m study area was divided into 24 blocks of 50 m by 38 m. The long dimension was parallel to the long axis of the island, in the direction of the tidal currents. Twelve traps of each type were randomly allocated to the 24 blocks. In both June and September the traps were set on day 1, hauled and re-assigned random positions on day 2, and hauled again on day 3. Soak times were 22-26 hours.

To assess whether some lobsters were captured on both days, all lobsters captured on day 1 were banded. To obtain a measure of exchange at the survey area boundary, 5 traps each were set 50-100 m beyond the perimeter of the study area on both days.

3 RESULTS

3.1 *Lobster counts and density estimates*

3.1.1 *June*

A total of 385 lobsters were counted on 15 belt transects from June 12-18 (Table 1). Of the 385, 15% (57) could not be sexed, because the lobsters escaped. Of these 57, 15 lobsters could not be measured but were estimated as 50-80 mm CL. All 57 lobsters were proportionally assigned to a sex and size based on the proportion within the appropriate 10 mm size grouping. Mean lobster densities (no. m^{-2}) ranged from 0.004 (females > 90 mm CL) to 0.014 (males 60-70 mm CL) (Fig. 3). Non-overlapping confidence intervals indicated a significant drop in density from 70-80 mm CL to the larger sizes. Although the mean density of males was higher than females, the confidence intervals always overlapped.

Table 1. Counts of lobsters on 15 belt transects in Lobster Bay, June 12-18, 1997. Shown are raw counts and aggregated counts, corrected for unknown sex or precise size. CL = carapace length, M = male, F = female, F-O = female ovigerous, Unk = unknown sex.

| CL (mm) | Counts | | | | | Counts corrected | | | |
	M	F	F-O	Unk	Total	CL (mm)	M	F	Total
< 50	28	18	0	13	59	< 50	35.9	23.1	59.0
50-60	25	18	0	10	53	50-60	32.4	23.3	55.8
60-70	57	33	0	8	98	60-70	65.8	38.1	103.8
70-80	52	47	0	8	107	70-80	59.6	53.8	113.4
80-90	21	7	0	2	30	80-90	22.5	7.5	30.0
90-100	5	1	1	1	8	90-130	12.7	6.3	19.0
100-110	4	0	0	0	4	130-160	4.0	0.0	4.0
110-120	2	1	0	0	3	Total	232.8	152.2	385.0
120-130	1	0	3	0	4				
130-160	4	0	0	0	4				
50- 80				15	15				
Total	199	125	4	57	385				

3.1.2 September

A total of 738 lobsters were counted on 16 belt transects from September 9-10 (Table 2). Of the 738, 10% (71) could not be sexed, and 11 of these were estimated as 50-80 mm CL. These unknowns were apportioned as for the June data. Mean lobster densities were higher in September than in June for all size classes, but confidence intervals were non-overlapping with the comparable June intervals only for the larger sizes (> 80 mm CL) (Fig. 3). Based on shell hardness, about 5% of the lobsters on the bottom had molted within the previous 1-2 weeks.

3.2 Trapping surveys

3.2.1 June

In 24 trap hauls of each trap type within the grid from June 19-21, a total of 195 lob-

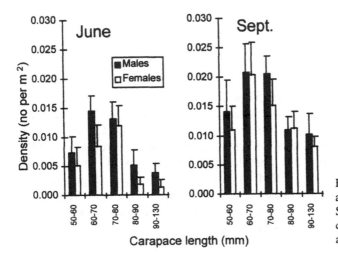

Figure 3. Lobster density in study area in June and September 1997. Shown are means and upper 95% confidence limits for each group and sex.

Table 2. Counts of lobsters on 16 belt transects in Lobster Bay, September 9-10, 1997. Shown are raw counts and aggregated counts, corrected for unknown sex or precise size. CL = carapace length, M = Male, F = female, F-O = female ovigerous, Unk = Unknown sex.

| CL (mm) | Counts | | | | | Counts corrected | | | |
	M	F	F-O	Unk	Total	CL (mm)	M	F	Total
< 50	24	25	0	11	60	< 50	29.4	30.6	60.0
50-60	62	47	0	11	120	50-60	69.8	52.9	122.6
60-70	95	92	0	9	196	60-70	101.9	98.7	200.5
70-80	90	67	0	14	171	70-80	100.2	74.6	174.8
80-90	49	33	0	6	88	80-90	52.6	35.4	88.0
90-100	31	17	0	7	55	90-130	47.8	41.2	89.0
100-110	4	5	1	2	12	130-160	3.0	0.0	3.0
110-120	5	11	1	0	17	Total	404.7	333.3	738.0
120-130	3	2	0	0	5				
130-160	3	0	0	0	3				
50- 80				11	11				
Total	366	299	2	71	738				

sters were captured in the research traps, and 197 in the commercial traps. The number of lobsters trapped on the 2 days was about the same (207 on day 1, 185 on day 2). The research traps caught relatively more small pre-recruits (< 70 mm CL); the commercial traps more recruits (Fig. 4). To test whether there was a significant interaction between size group and trap type, a contingency table was used. Chi-square tests (d.f. = 4) were significant for both days (p < 0.01).

Of a total of 256 lobsters banded on day 1 (including 49 from outside the grid), 6 (2%) were returned on day 2 (Table 3). All had been tagged within the grid and 5/6 were recaptured within the grid, indicating minimal movement.

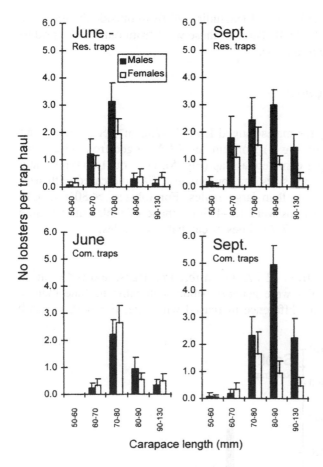

Figure 4. Catch rate of lobsters in research and commercial traps during June and September. Shown are means and upper 95% confidence limits for each size group and sex.

Table 3. Number and location of lobsters (inside or outside of grid) tagged on day one of trapping, and recaptured on day 2 of trapping. Trap types combined. 'In-In' indicates lobster(s) was tagged inside the grid and recaptured inside the grid.

Season	Number tagged		Number of recaptures by location			
	Inside grid	Outside grid	In-In	In-Out	Out-In	Out-Out
June	207	49	5	1	0	0
September	265	79	10	1	3	3

3.2.2 September

Substantially more lobsters were captured in the traps from September 15-17 compared to June (Fig. 4). In 23 trap hauls of each trap type within the grid (one haul of each trap type was excluded because the lid was unsecured or bait missing), 288 lobsters were captured in research traps, 306 in commercial traps. 1.5% of the lobsters in the traps were 'buckle-shelled', indicating molting within the previous few weeks.

As in June the number of lobsters caught on the 2 days was similar (265 on day 1, 329 on day 2). Again research traps caught more small pre-recruits than the commercial traps while the commercial traps captured more legal sized animals. As in June chi-square tests for interaction between size group and trap were significant for both days (p < 0.001).

Of a total of 344 lobsters banded on day 1 (including 79 from outside the grid), 17 (5%) were recaptured on day 2 (Table 3). Three of these went from outside the grid to inside, and one went from inside to outside.

3.3 Estimates of effective fishing area

3.3.1 June

Estimates of catchability (EFA) for males ranged from 11-238 m^2 per research trap and 0-188 m^2 per commercial trap (Fig. 5). For females, EFA ranged from 33-250 m^2 per research trap and 0-375 m^2 per commercial trap (Fig. 5). For both trap types male EFAs were maximum at intermediate sizes, but for females EFAs increased with size. Large females were more catchable than large males. For lobsters > 80 mm CL, EFAs for females in commercial traps were 1.6-4 times those of males of the same size; in research traps females were 3.7-7.6 times as catchable as males.

3.3.2 September

EFA estimates for males ranged from 12-272 m^2 in research traps, and 6-450 m^2 in commercial traps (Fig. 5). Females were generally more catchable in June than in September in both trap types; this difference increased with size. For both research

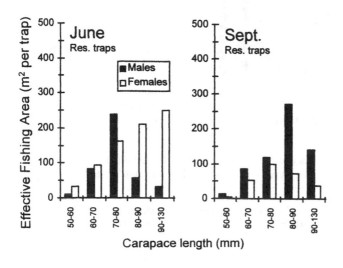

Figure 5. a) Effective fishing areas of research traps in June and September.

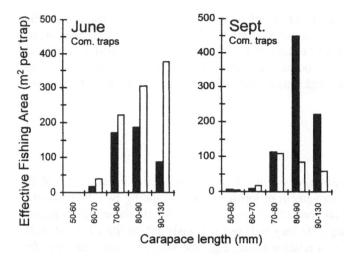

Figure 5. Continued. b) Effective fishing areas of commercial traps in June and September.

and commercial traps, females > 80 mm CL were 3 times more catchable in June. Males > 80 mm CL showed the reverse trend – more catchable in September than in June. For the research traps the difference was about 4-fold; for commercial traps, EFA's in September were about 2.5 times those of June.

4 DISCUSSION

4.1 *Density estimates*

The lower density estimates for the 50-60 mm CL lobsters in June and September are likely due to under sampling. These smaller lobsters were more likely to be missed during short periods of reduced visibility when cobbles were overturned. The sharp drop in density between 70-80 mm CL and 80-90 mm CL in June was likely due to removals of lobsters greater than 80 mm CL by the fishery (minimum size of 81 mm CL), which ended on May 31.

The increase in density between June and September is probably a result of: 1) Lobsters molting into sizes greater than 60 mm CL (and possibly becoming more apparent to divers), and 2) Immigration into the study area. Summer occupation of Nova Scotian shallow waters by *Homarus americanus* in summer is documented elsewhere (Roddick & Miller 1992).

While variability in counting ability between divers probably contributed to the variation in numbers within surveys, it does not explain the differences between June and September. Five of the divers did over half the transects in June and over 80% of the September transects; each diver showed an increase in the number of lobsters counted over this period by factors ranging from 1.9-2.4. A learning effect between June and September is unlikely since most of these divers were experienced in conducting lobster counts along transects prior to the June study.

4.2 *Research vs. commercial traps*

The commercial traps used here are clearly more efficient at capturing legal size lobsters and reducing the catch of lobsters below the minimum legal size. The higher catch of legal size lobsters in commercial traps could be due to a reduction in trap saturation (Miller 1990) – either because of their larger size, or because fewer prerecruits would remain in the traps because of the escape slots. Both traps showed the same trends with size, gender and season, indicating that in a relative sense the two trap types behave similarly.

4.3 *Effective fishing area*

4.3.1 *Methods*

The EFA estimates are sensitive to a number of factors including accuracy of the density estimates by diving, movement of lobsters due to trapping two days in a row in the same area, and potential competition among traps. While these factors require more study, the evidence suggests they will not seriously affect the EFA estimates presented here, particularly in a relative sense, e.g., the size, sex, and season differences.

The density estimates may be low for the smallest size class, which would inflate the EFA estimates. For lobsters > 60 mm CL however, we likely counted a high percentage of the lobsters on the transects, since the bottom was searchable and experienced divers did the counts. The potential aggregating effect of trapping on day 1 cannot be fully evaluated, but the low recapture rate of lobsters on day 2 (less than 3%) suggests only a small fraction of the population was affected by the trapping.

A potentially important factor affecting the absolute value of EFA estimates is the distance between traps. There are no data here to evaluate the area of influence of the traps, nor are there data from other studies of *Homarus*. In the present study traps were separated by 50 m in the along-current direction and 38 m in the cross current direction. How far might a lobster travel to a trap? Jernakoff & Phillips (1988) reported that the greatest distance traveled by a tagged rock lobster to a trap was 120 m. Although it seems unlikely that this lobster actually sensed the bait at this distance, more study on this topic is needed. Data on daily movements of lobsters from ultrasonic tracking may provide some answers (Tremblay, unpubl.).

4.3.2 *September estimates of EFA*

Higher estimates of EFA for large males than large females in autumn (research traps) has been reported in previous studies (Table 4). One possible mechanism for the higher autumn catchability of males is an earlier recovery from molting for males (Tremblay & Eagles 1997). Data were not sufficient to evaluate any gender differences in molt timing in Lobster Bay. The low percentage of soft lobsters on the bottom and in traps in September, coupled with the observation that most lobsters had new shells, indicates the major molt was several weeks earlier.

The EFA estimates for Lobster Bay are at the low end of those from other areas but are most similar to those for Port L'Hebert, the area closest to Lobster Bay in terms of size structure and fishery management (same fishing season and minimum legal size). The differences between the EFA estimates for Lobster Bay and those for the other areas could arise from differences in population size structure and the behavioral interaction between sizes, sites, conditions while trapping, or in molt timing.

Table 4. Estimates of autumn EFA for large males versus large females.

Sex	CL (mm)	EFA (m^2/trap)	Area	Source
Males	70-80	169	Port L'Hebert 1987	Miller 1995
	80-90	380	'	'
	70-80	210	Port L'Hebert 1988	'
	80-90	185	'	'
	70-80	257	Sydney	'
	80-90	500	'	'
	66-75	576	Baie des Chaleurs	Tremblay et al. 1998b
	76-86	860	'	'
	70-80	118	Lobster Bay	Present study
	80-90	272	'	'
Females	70-80	120	Port L'Hebert 1987	Miller 1995
	80-90	281	'	'
	70-80	139	Port L'Hebert 1988	'
	80-90	89	'	'
	70-80	84	Sydney	'
	80-90	138	'	'
	66-75	584	Baie des Chaleurs	Tremblay et al. 1998b
	76-86	453	'	'
	70-80	100	Lobster Bay	Present study
	80-90	74	'	'

4.3.3 *June EFA estimates*

The finding that large females had higher EFA's in June than September is novel. Higher female catchability in spring supports the conclusions of tagging studies in northern Nova Scotia, where females 80-85 mm CL were found to have higher return rates in spring than either smaller females or males of similar sizes (Tremblay et al. 1998a). It also provides an explanation for why many fishers report more large females in the catch in late spring. In other areas elevated catch rates of females relative to males in late spring have been associated with an earlier male molt, which makes them less catchable than females (Tremblay & Eagles 1997). Ideally the EFA estimates would have confidence intervals but this would require replication within a season. The fact that the trends with sex and size for the Lobster Bay EFAs in September are similar to previous autumn studies, while those for June are markedly different, indicates the seasonal effect is real. Future studies should establish whether the June catchability trends are generally applicable.

The June estimates of EFA were less than 3 weeks after the commercial fishery ended, and in a relative sense, probably are representative of the late spring fishing season in Lobster Bay. In fact, spring 1997 was colder than normal, and to the extent that seasonal changes in catchability are affected by temperature the current EFA estimates are likely typical of the latter part of the fishing season in a 'normal' year. While the trends with size and gender are likely to hold in the fishing season, it is not known how applicable the absolute value of the EFAs are to the commercial fishing season. If they were applicable, commercial catch rates could be converted to densities. However, fishermen in the area use a variety of trap types and baits which will affect catchability. In addition they often fish traps at distances close enough that they must compete for catch (e.g. less than 10 m apart). Given these variables, applying the EFAs generated here to commercial catch rates would be inappropriate.

There are several implications of higher catchability of large females in spring. First if length composition data are used to estimate mortality (e.g. by comparing numbers in adjacent molt classes – see Cobb & Caddy, 1989) then consideration should be given to when the length data are collected. If collected in spring, then large females would be over represented in the catch and mortality would be under-estimated. If collected in the fall, large females would be under-represented and mortality overestimated. How sensitive the mortality estimates are to higher catch-ability of females should be examined. A second implication is that abundance indices for females based on spring catch rate will be inflated relative to other times of year. Lastly, it appears that large females are exploited to a greater degree by trap fishing in late spring relative to other times of the year. Thus, conservation of large reproductive females would be increased by reducing fishing effort in spring.

This study further supports the contention of Miller (1995a) that fall EFA estimates have limited application to other times of the year, even in a relative sense. Other seasons when it would be of interest to estimate EFA would be late fall and early spring. Such periods would be more challenging because of weather and the activities of the commercial fishery.

Estimates of effective fishing area have two potential applications: generation of density estimates from commercial catch rates, and application to fishery-independent trap surveys. At this point the more controlled application of a fishery-independent survey is closer to reality. September is a good time to conduct a fishery independent trap survey (after the molt, when catch rates are high) if catchability variation can be accounted for. Replicated September trap surveys elsewhere indicate there can be substantial variation in catch rate even over 2 weeks, probably because of weather related changes in catchability (Tremblay, unpublished data). It might be possible to correct for such variation in catchability by using EFA estimates generated at the time of the trap survey.

ACKNOWLEDGEMENTS

I thank the divers and dive tenders from Moncton, St. Andrews and Halifax for making this project possible: M. Comeau, C. Curtis, D. Duggan, A. MacIntyre, P. Lawton, S. Nolan, R. Rangely, D. Robichaud, G. Robichaud, F. Savoie, M. Strong, R. Vienneau. I also thank B. Porter for providing his vessel for the dive and trap surveys, and M. Lanteigne for his support. R. Miller and an anonymous referee are thanked for their comments on the manuscript. This project is a contribution from CLAWS (Canadian Lobster Atlantic Wide Studies), a Fisheries and Oceans Canada initiative.

REFERENCES

Caddy, J.F. 1979. Some considerations underlying definitions of catchability and fishing effort in shellfish fisheries, and their relevance for stock assessment purposes. *Can. Dept. of Fisheries and Oceans. Fish. Mar. Serv. MS Rep. 1489.*

Cobb, J.S. & Caddy, J.F. 1989. The population biology of decapods. In: J.F. Caddy (ed.) *Marine Invertebrate Fisheries: their assessment and management*: 327-374. New York: John Wiley.

Fogarty, M.J. 1995. Populations, fisheries, and management. In: J.R. Factor (ed.) *Biology of the lobster* (Homarus americanus): 111-137. New York: Academic Press.

Hudon, C. 1994. Large-scale analysis of Atlantic Nova Scotia American lobster (*Homarus americanus*) landings with respect to habitat, temperature, and wind conditions. *Can. J. Fish. Aquat. Sci.* 51: 1308-1321.

Jernakoff, P. & Phillips, B.F. 1988. Effect of a baited trap on the foraging movements of juvenile western rock lobsters *Panulirus cygnus* George. *Aust. J. Mar. Freshwater Res.* 39: 185-192.

Krouse, J.S. 1989. Performance and selectivity of trap fisheries for crustaceans. In: J.F. Caddy (ed.), *Marine Invertebrate Fisheries: their assessment and management*: 307-325. New York: John Wiley.

McQuinn, I.H., Gendron, L. & Himmelman, J.H. 1988. Area of attraction and effective area fished by a whelk (*Buccinum undatum*) trap under variable conditions. *Can. J. Fish. Aquat. Sci.* 45: 2054-2060.

Miller R.J. 1975. Density of the commercial spider crab, *Chionoecetes opilio*, and calibration of effective area fished per trap using bottom photography. *J. Fish. Res. Board Can.* 32: 761-768.

Miller R.J. 1990. Effectiveness of crab and lobster traps. *Can. J. Fish. Aquat. Sci.* 47: 1228-1251.

Miller R.J. 1995a. Catchability coefficients for American lobster (*Homarus americanus*). *ICES Mar. Sci. Symp.* 199: 349-356.

Miller R.J. 1995b. Fishery regulations and methods. In: J.R. Factor (ed.) *Biology of the lobster (Homarus americanus)*: 89-109. New York: Academic Press.

Morgan, G.R. 1974. Aspects of the population dynamics of the Western rock lobster, *Panulirus cygnus* George. II Seasonal changes in the catchability coefficient. *Aust. J. mar. Freshwat. Res.* 25: 249-259.

Paloheimo, J.E. 1963. Estimation of catchabilities and population sizes of lobsters. *J. Fish. Res. Board Can.* 20: 59-88.

Ricker, W.E. 1975. Computation and interpretation of biological statistics of fish populations. *Bull. Fish. Res. Board Can.* 191: 1-382.

Roddick, D.L. & Miller, R.J. 1992. Spatial and temporal overlap of the American lobster *(Homarus americanus)* and sea scallop (*Placopecten magellanicus*) as related to the impact of inshore scallop dragging. *Can. J. Fish. Aquat. Sci.* 49: 1486-1492.

Skud, B.E. 1979. Soak time and the catch per pot in an offshore fishery for lobsters (*Homarus americanus*). *Rapp. P.-v. Reun. Cons. Int. Explor. Mer.* 175: 190-196.

Tremblay, M.J. & Eagles, M.D. 1997. Molt timing and growth of the lobster, *Homarus americanus*, off northeastern Cape Breton Island, Nova Scotia. *J. Shellfish Res.* 16: 383-394.

Tremblay, M.J., Eagles, M.D. & Black, G.A.P. 1998a. Movements of the lobster, *Homarus americanus*, off northeastern Cape Breton Island, with notes on lobster catchability. *Can. Tech. Rep. Fish. Aquat. Sci.* 2220: IV + 32 p.

Tremblay, M.J., Lanteigne, M. & Mallet, M. 1998b. Size-specific estimates of lobster catchability in the Baie des Chaleurs based on traps with different entrance ring sizes. *Can. Tech. Rep. Fish. Aquat. Sci.* 2222: IV + 16 p.

The effects of varying dietary protein levels on growth and survival of juvenile and pre-adult redclaw (*Cherax quadricarinatus*)

JESUS T. PONCE-PALAFOX, JOSE LUIS ARREDONDO-FIGUEROA &
MIGUEL ANGEL MORENO-RODRIGUEZ
*Planta Experimental de Produccion Acuicola. Departamento de Hidrobiologia. Division de Cien-
cias Biologicas y de la Salud, Mexico. Universidad Autonoma Metropolitana Iztapalapa, Mexico
and Centro de Investigaciones Biologicas. Universidad Autonoma del Estado de Morelos, Cuernavaca, Morelos, Mexico*

ABSTRACT

A 12 week feeding trial was conducted in tanks with juveniles (2.0 g) and pre-adults
(8.0 g) redclaw *Cherax quadricarinatus* to examine the effects of dietary protein levels on growth and survival. Three practical diets were formulated to contain 35, 40,
and 45% protein. At the end of the experiment, weight gain and SGR of pre-adults
redclaw fed with diet containing 35% protein were significantly ($p < 0.05$) higher
than redclaw fed with 40 and 45% protein diets. These results indicate that the diets
formulated for the redclaw appear to be suitable and that the diet containing 35% may
be the most adequate.

1 INTRODUCTION

In recent years, the culture of *Cherax spp.* has been increasing in several countries of
the world. Thus, the determination of the protein level required in the diet is essential
for any redclaw culture initiative. A major factor influencing crayfish growth and
survival is nutrition (Aiken & Waddy 1992). Feed is the most expensive component
in most commercial fish, shellfish, and crustacean farms. The protein fraction is typically the most expensive component in crayfish diets (Huner & Meyers 1979, Acrefors et al. 1992). Under natural circumstances zooplankton has been shown to be a
major component of the juvenile crayfish food (Tcherkashina 1977). Alternatively,
under artificial conditions, reasonable survival and growth of juvenile crayfish has
been achieved on formulated diets (Huner et al. 1975, Celada et al. 1989). Redclaw
has been cultivated using fresh zooplankton and a formulated flake diet (Jones 1995).
Research conducted with other freshwater crayfish suggests that the optimum protein
requirements range from 25 to 35% (D'Abramo & Robinson 1989, Reigh et al. 1993)
and at least 15-20% of the diet should be of animal origin (Huner & Meyer, 1979,
d'Abramo & Robinson 1989). Dietary protein quantity and quality have profound effects on growth, feed utilization and composition of crayfish. Dietary protein requirements for *Cherax* have been determined for different species and stages of
growth, and under a variety of conditions for example, *Cherax tenuimanus*, 25 to
40% (Morrissy 1984) and 16 to 33% (Morrissy 1989); *Cherax destructor*, 30% (Jones

et al. 1996), and *Cherax quadricarinatus*, 23 to 52% (Webster et al. 1994). However, investigations concerning the evaluation of various dietary proteins in redclaw have been limited. Moreover, a dietary protein requirement of the juveniles and pre-adults redclaw *Cherax quadricarinatus* is not known.

The aim of this study was determine the effects of varying dietary protein levels on growth and survival of juveniles and pre-adults redclaw *Cherax quadricarinatus* when cultured under controlled conditions in a recirculating water system.

2 MATERIAL AND METHODS

2.1 *Experimental diets*

One feeding trial was conducted using a block-randomized design. Diets were formulated to contain 35%, 40%, and 45% of protein and equal proportions of lipids, ash, and available sources of energy (Table 1). Mineral and vitamin premixes were added to equal or exceed recommended levels of these nutrients for crayfish (NRC 1983). Proximate composition of the diets was confirmed by chemical analysis.

2.2 *Fish and feeding trials*

Juveniles and pre-adults of redclaw crayfish were reared in the Planta Experimental de Produccion Acuicola for approximately 12 weeks in a 1000 liter tank, using a flow-through culture system. A diet of live zooplankton (*Moina* spp.) and commercial pellet (Purina 40% protein) was fed twice daily before making a transition to feeding with the experimental diets. Fifteen, 19 liter polyethylene tanks were incorporated into a recirculating system with a submerged bio-filter.

Table 1. Ingredients and proximate composition (% dry weight) of test diets. [1]Commercial prepared and contained (as g/100 g dry weight): crude protein 50%; crude lipid, 14.23 and ash, 18.25. [2]Webster et al. (1994). [3]Sigma Chemical Company, St. Louis, Mo.

Ingredients	Dietary protein (%)		
	35	40	45
Fish meal[1]	31.29	36.00	44.00
Soybean meal	16.32	27.00	15.00
Sorghum meal	30.88	20.00	51.00
Corn gluten	10.00	6.00	11.00
Cod liver oil	4.00	4.00	4.00
Corn oil	2.00	2.00	2.00
Mineral and vitamin premix[2]	5.00	5.00	5.00
Ethoxyquin[3]	0.03	0.03	0.03
Composition			
Available energy (kJ/g dry weight)	19.4	20.5	19.9
Crude protein	35.4	39.9	45.2
Lipid	12.4	13.1	12.5
Fiber	1.4	1.2	1.1
Crude ash	6.4	6.8	7.0

Table 2. Water quality parameters in the experimental system. Min. = minimum value; Max. = maximum value.

Parameters	Min.	Max.	Mean
Temperature (°C)	26.0	31.0	28.42
Dissolved oxygen (mg/l)	4.65	6.65	5.43
pH	8.30	8.8	8.65
Total ammonia nitrogen (mg/l)	0.00	0.0336	0.008
Un-ionized ammonia (mg/l)	0.00	0.0138	0.002

Water returning to the bio-filter was tested each week for Total Ammonia Nitrogen (TAN) and converted to de-ionized ammonia using the tables provided by Piper et al. (1982). The pH was measured with an Orion pH probe. Flow rates were determined weekly in 15 randomly selected tanks along the experimental period. Approximately 30% of the water in each tank were replaced each day. Dissolved oxygen was recorded using YSI model 57 DO meter. Water temperature was measured using a mercury thermometer. Water quality parameters are provided in Table 2.

Animals were fed twice daily at 0800 h and 1800 h. Feces, exuviae, and the remaining food were siphoned from each tank at 0700 h daily. Four juveniles and four pre-adults redclaw with mean weight 2.0 g and 8.0 g respectively were placed in each of three replicate tanks for every dietary treatment. Equivalent stocking density was 25 animals/m^2. Tanks were randomly allocated to treatments.

Weight gain, feed efficiency, specific growth rate (SGR) and protein efficiency ratios (PER) were measured as response criteria. Weight gain was expressed as a percent of initial weight, and the feed conversion ratio (FCR) was calculated based on the total wet body weight gained by the redclaw and the total amount of dry food given to each group. PER was calculated as total weight gain divided by total protein fed.

2.3 *Statistical analysis*

Performance data of redclaw fed by various diets were subjected to analysis of variance using the STATISTICA Package (Release 5, '97 Edition). If the analysis of variance indicated that significant differences ($p < 0.05$) were present, comparison of treatment means was made by a Tukey test.

3 RESULTS

Final weight gain and specific growth rate (SGR) of juvenile redclaw fed a diet containing 35 and 40% protein were not significantly different ($p < 0.05$). However, pre-adult redclaw fed experimental diets containing 35% protein had significantly higher ($p < 0.05$) weight gain and SGR than those fed experimental diets containing 40 and 45% protein (Table 3). Pre-adult redclaw fed a 35% protein had a numerically higher percentage survival (100%) than redclaw fed the experimental diets 40 and 45%. However, differences in percentage survival were not statistically significant ($p < 0.05$), possibly due to variation within each treatment. Protein efficiency ratio (PER)

Table 3. Effect of increasing levels of dietary protein on weight gain, SGR, PER, FCR and survival by *C. quadricarinatus*. SGR = Specific growth rate; PER = Protein efficiency ratio; FCR = Feed conversion ratio. Different superscript letters in the same column, represent significant differences (p < 0.05).

Dietary Protein levels (%)	Weight gain (%)	SGR	PER	FCR	Survival (%)
Juveniles					
35	$362^a \pm 1.5$	$2.2^a \pm 0.04$	$1.03^a \pm 0.01$	$1.4^a \pm 0.09$	$60^b \pm 0.7$
40	$395^a \pm 1.1$	$2.4^a \pm 0.06$	$1.12^a \pm 0.04$	$1.1^a \pm 0.03$	$95^a \pm 0.4$
Pre-adults					
35	$520^b \pm 1.9$	$4.2^b \pm 0.08$	$1.48^b \pm 0.09$	$1.6^a \pm 0.04$	$100^a \pm 0.9$
40	$410^a \pm 1.4$	$3.1^a \pm 0.04$	$1.17^a \pm 0.02$	$2.1^b \pm 0.04$	$95^a \pm 0.6$
45	$370^a \pm 1.2$	$2.2^a \pm 0.03$	$1.05^a \pm 0.01$	$2.3^b \pm 0.03$	$80^a \pm 0.8$

values of pre-adult redclaw fed diets containing 35% protein were greater (p < 0.05) than those with 40 and 45% protein.

4 DISCUSSION

Growth of redclaw fed a 35% protein diet in the present study was similar to growth rates observed in other studies in experimental tanks (Morrissy 1989, Webster et al. 1994). Survival values were comparable or greater than values reported by Webster et al. (1994) and Jones (1995) in *Cherax quadricarinatus*. Other studies indicate that a diet containing 33 to 52% protein appears suitable for use in rearing small juvenile redclaw for the first 5 weeks after release from the female. Our results indicate that a formulated diet with a protein level of 35% appears to be adequate for juvenile and pre-adult redclaw. Generally PER decreased with increasing dietary protein level in pre-adults. Diets used in this study seem to be a starting point for further studies in determining nutritional requirements of redclaw.

REFERENCES

Acrefors, H., Castell, J.D., Boston, L.D. Raty, P. & Svensson, M. 1992. Standard experimental diets for crustacean nutrition research. II. Growth and survival of juvenile crayfish *Astacus astacus* (Linné) fed diets containing various amounts of protein, carbohydrate, and lipid. *Aquaculture* 104: 341-356.

Aiken, D.E. & Waddy, S.L. 1992. The growth process in crayfish. *Rev. Aquat. Sci.* 6: 335-381.

Celada, J.D., Carral, J.M., Gaudioso, V.R., Temino, C. & Fernandez, R. 1989. Response of juvenile freshwater crayfish (*Pacifastacus leniusculus* Dana) to several fresh and artificially compounded diets. *Aquaculture* 76: 67-78.

d´Abramo, L.R. & Robinson, E.H. 1989. Nutrition of crayfish. CRC *Critical Reviews in Aquatic Science* 1: 711-728.

Huner, J.V., Meyers, S.P. & Avault, Jr., J.W. 1975. Response and growth of freshwater crayfish to an extruded, water-stable diet. *Freshwater Crayfish* 2: 149-158.

Huner, J.V. & Meyers, S.P. 1979. Dietary protein requirements of the red crawfish, *Procambarus clarkii* (Girard) (Decapoda, Cambaridae), grown in a close system. *Proc. World Mariculture Society* 10: 751-760.

Jones, C.M. 1995. Production of juvenile redclaw crayfish, *Cherax quadricarinatus* (von Martens) (Decapoda, Parastacidae) II. Juvenile nutrition and habitat. *Aquaculture* 138: 239-245.

Jones, P.L., De Silva, S.S. & Mitchell, B.D. 1996. Effects of replacement of animal protein by soybean meal on growth and carcass composition in juvenile Australian freshwater crayfish. *Aquaculture International* 4: 339-359.

Morrissy, N.M. 1984. Assessment of artificial feeds for battery culture of a freshwater crayfish, marron (*Cherax tenuimanus*) (Decapoda: Parastacidae). *Dept. Fish. Wildl. West. Aust. Rept.* 63: 1-43.

Morrissy, N.M. 1989. A standard reference diet for crustacean nutrition research. IV. Growth of freshwater crayfish *Cherax tenuimanus*. *J. World Aquaculture Society* 20: 114-117.

National Research Council. 1983. *Nutrient requirements of warm water fishes and shellfishes.* Washington: National Academy Press.

Piper, R.G., McElwain, I.B., Orme, L.E., McCarren, J.P., Fowler, L.G. & Leonard, J.R. 1982. *Fish hatchery management.* Washington: United States Department of the Interior, Fisheries and Wildlife.

Reigh, R.C., Braden, S.L. & Laprarie, R.J. 1993. Substitution of soybean protein in formulated diets for red swamp crawfish *Procambarus clarkii*. *J. World Aquaculture Society* 24: 329-338.

Tcherkashina, N.Y. 1977. Survival, growth and feeding dynamics of juvenile crayfish (*Astacus leptodactylus cubanicus*) in ponds and the River Don. *Freshwater Crayfish* 3: 95-100.

Webster, C.D., Goodgame-Tiu, L.S., Tidwell, J.H. & Rouse, D.B. 1994. Evaluation of practical feed formulation with different protein levels for juvenile red claw crayfish (*Cherax quadricarinatus*). *Trans. Kentucky Acad. Sci.* 55: 108-112.

A comparison between the biology and the exploitation level of two pink shrimp (*Aristeus antennatus*) stocks from two different areas in the Spanish Mediterranean

M. GARCÍA-RODRIGUEZ & A. ESTEBAN

I.E.O., Centro Oceanográfico de Murcia, San Pedro del Pinatar (Murcia), Spain

ABSTRACT

This study compares the biology and fishery of *Aristeus antennatus* in the Ibiza Channel and in the Gulf of Vera. Monthly sampling data were obtained from commercial catches landed in two representative ports (Santa Pola and Garrucha) from each area from 1992 through 1994. The sizes of the catches showed similar length distribution, with sexual dimorphism in size, although those from Ibiza were slightly smaller. The biological parameters were very similar between the two areas. The values of biomass, yields and mortality obtained through LCA, VPA, and Y/R analysis showed an exploitation scheme that was slightly skewed towards overexploitation, which was clearer in Ibiza. The results showed that the resource was more important in the case of the Gulf of Vera.

1 INTRODUCTION

The pink shrimp (*Aristeus antennatus*) is a demersal species that is found on the muddy bottoms of the slopes of the continental shelf, more specifically in zones close to submarine canyons. Its distribution area is very wide being found in the Mediterranean and in the Atlantic south of the Iberian Peninsula and reaching as far as the Portuguese coasts (Arrobas & Ribeiro-Cascalho 1987). They are found at depths that range from 200 m down to bottoms deeper than 2000 m, although they concentrate on muddy bottoms between depths of 350 and 800 m (Cartes 1991, 1994; Cartes & Sardá 1989, 1992, 1993; Sardá & Cartes 1994).

In the Spanish Mediterranean littoral, it is a species exploited fundamentally in the Balearic Islands as well as along the peninsular coast. It can be considered as a mono-specific fishery that does not suffer discards due to its high commercial value. In the case of the Ibiza Channel, vessels that have their base in various ports along the peninsular coast (Altea, Jávea, Calpe, and Denia) undertake shrimp exploitation almost exclusively, and they carry out their sales in the ports of Alicante, Villajoyosa, and Santa Pola. In the case of the Gulf of Vera, the exploitation is undertaken by the fleets of two ports: Aguilas, which trawls the fishing grounds of the northern part of the Gulf, and Garrucha, which exploits the fishing grounds of the southern part (Fig. 1).

721

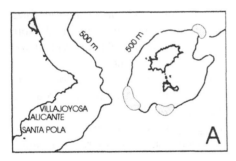

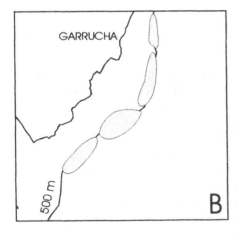

Figure 1. Location of the selected areas (A = Ibiza; B = south of Vera) and ports, pointing out the principal fishing grounds (dotted areas).

Aristeus antennatus has been the object of numerous studies that have included biological, (Massuti & Daroca 1978, Sardá & Demestre 1987, Martinez-Baños et al. 1990, Demestre & Carbonell 1994) and reproductive and fecundity aspects (Relini Orsi 1980, Orsi-Relini & Semeri, 1983, Demestre & Fortuño 1992). In addition, analyses of its exploitation have been carried out in some areas (Demestre & Lleonart 1993, Demestre & Martín 1993; Demestre et al. 1994, Martínez-Baños 1997). However, despite the existence of these studies it was considered relevant to carry out this study in order to contribute to the current knowledge on pink shrimp biology. In addition, this study was necessary due to the fact that the fishing activity is undertaken in two specific areas, which allows a high level of precision to be achieved in terms of collecting data. The study, therefore, compares the biology, the fishery, and its exploitation pattern in the two selected areas with an analysis of the exploitation state, which consequently will contribute to our knowledge of the current fishing activity on this species.

2 MATERIAL AND METHODS

2.1 Sampling

The data used in this study are based on monthly samples of the pink shrimp (*Aristeus antennatus*) carried out on landings in the port of Santa Pola. This was done from vessels that operated in the Ibiza Channel and in the port of Garrucha, and from vessels that operated in the southern part of the Gulf of Vera. The samples were taken from January 1992 to December 1994 and followed a random stratified scheme for the commercial categories (small, medium, and large) present in the catch.

Sampling was based on the catches from two different vessels each month. A sample was taken from each category, with the individuals being sexed and measured by their cephalo-thoracic length (CL) in mm. They were also weighed as a whole and each commercial category was weighed separately. The number of individuals was then extrapolated, by commercial category, to the total catch from each vessel, the total port landing of the day, and finally to the total monthly catch by commercial categories, thereby obtaining a single size composition by month and port. In total, 144 sampling operations were carried out in both areas that led to a total of 32,323 specimens being measured. A total of 16,951 individuals were individually weighed. Finally, the index of sexual maturity and the presence of spermatophores in females also was recorded.

2.2 Growth

From the compiled data, the parameters of the size-weight relationship were determined, with cephalothoracic length (CL) – total weight (W) being adjusted to a potential relationship in the form: $W = a * CL^b$, where W is the weight in grams, CL is the cephalothoracic length in millimetres, a and b are parameters that need to be estimated, *b* being the coefficient of allometry.

For the study of the size-age relationship, the Von Bertalanffy growth model (VBGF) was used with the expression: $L_t = L_\infty (1 - e^{-k(t-t0)})$, where L_∞ is the maximum theoretical size, L_t is the size at age t, K is the growth factor and t_0 is the age at which the size is 0. For the estimation of these parameters, the FISAT program (Gayanilo et al. 1994) was applied to monthly size distributions. These were grouped in size classes of 2 mm, with a running mean of three consecutive classes, taking the 'best combination' (Rn) of the parameters (FISAT programme, ELEFAN subprogramme), for the monthly size distributions of the whole period (1992-1994).

2.3 *Reproduction*

Maturity in females was determined by macroscopic observation, adopting a scale of five maturity stages, modified from Relini-Orsi & Relini (1979) that only recognized the first four of the following: A – virgin, B – developing, C – pre-spawning, D – spawning, and E – resting. Male maturity was determined macroscopically as a function of the shortening of the rostrum and the presence or not of petasma fusion (Sardá & Demestre 1989, Sardá & Cartes 1997). The sexual ratios were calculated for size class in order to evaluate the predominance of each sex in each size range. The percentages of individuals spawning, active, and at rest were determined from the monthly maturity states. These were spawning = stage D, active = stages B + C, and at rest = stages A + E. Finally, the percentages of maturity by size for each sex were calculated to determine the 50% size at first maturity adjusted to a logistic function.

2.4 *Fishery*

In order to obtain an overview of the development of the catches landed in recent decades, data were collected on the annual landings of *Aristeus antennatus* taken in the studied ports, thus compiling a historic record of the yearly shrimp catches landed from 1976 to 1994. To detect any seasonality in the landings, monthly landing data of shrimps in the port of Santa Pola during 1992, 1993, and 1994 were also compiled.

A series of analyses based on the sampling data was performed to study the exploitation state of the resource. First, the monthly size distributions by sex were grouped into 2 mm size classes based on a running mean on three consecutive classes, which gave a clear distribution of sizes by year and port. These annual frequencies were then averaged for the three years of sampling, which led to the influence of the various annual recruitments being smoothed, and thereby gave a clear distribution of sizes.

The growth and size-weight relationship parameters used were those calculated in this work. The natural mortality (M) was calculated from the formulae of Pauly (Pauly, 1980) and Djabali (Djabali et al. 1994). Carapace length was transformed to the total length in cm, according to the relationship proposed by Arrobas & Ribeiro-Cascalho (1987), since it was considered that this transformation had greater biological significance. The peaks of maturity were those calculated in this study, with the terminal mortality from fishing fixed at 1.5 after various sensibility tests with other values. Since actual catch data were used, the average annual catch for the species in the period 1992-1994 was taken from the port of Garrucha, which represents 55% of the total catch for the whole of the Gulf of Vera, together with that of the ports of Alicante, Santa Pola, and Villajoyosa, which represent 100% of the catches in the Ibiza Channel. For the single annual distribution of size frequencies mentioned above, a series of Length Cohort (LCA), Virtual Population (VPA), and yield per recruit (Y/R) analyses were carried out according to the catch equation for a pseudo-cohort. This was performed through the application of the VIT fisheries analysis program (Lleonart & Salat 1992) to obtain an understanding of the actual exploitation state of the pink shrimp stock in both areas.

3 RESULTS

3.1 *Size composition*

The contributions by size class and sex of the landings in both Santa Pola and Garrucha for 1992, 1993 and 1994, respectively, showed a sexual dimorphism in size (Fig. 2). The sizes of the females in the catches varied from a minimum of 15 mm to a maximum of 59 mm CL in the Ibiza Channel and from 15 to 62 mm CL in the Gulf of Vera. The males varied from a minimum of 15 mm to a maximum of 37 mm CL in Ibiza and from 17 mm to 38 mm CL in Vera. The mean size in each area was 31.2 mm and 32.4 mm CL for females, and 24.1 mm and 25.8 mm CL for males for Ibiza and Vera, respectively.

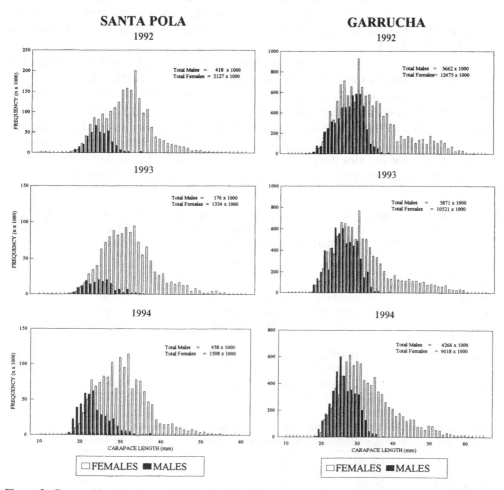

Figure 2. Composition (total contribution by size class and sex) for the shrimp landings in each selected port during 1992, 1993, and 1994.

3.2 *Growth*

The results obtained for the different size-weight relationships of *Aristeus antennatus* carried out in this study on the different groups (males, females, and the combined total) are shown in Table 1. Essentially, they give values of less than 3 for the allometric coefficients. However, they were significantly different from this value, which showed a negative allometry between the parameters considered.

The different estimations for the von Bertalanffy growth parameters obtained by the application of the FISAT statistical package (subprogram ELEFAN), shown in Table 2, gave similar results for L_∞ and growth rate values (K) in both areas.

3.3 *Reproduction*

The sexual ratio by size class showed that after an initial stage in which the males predominated significantly, the females started to dominate throughout the whole range of sizes. The mean sexual ratio for the whole period of the study was dominated by the females, with the sexual proportion being 0.80 in the Ibiza Channel and 0.67 in the Gulf of Vera.

The percentages of maturity (immature-mature) for the males showed a high ratio of mature individuals throughout the year, with the immature ratios decreasing from March to June and increasing from September to November. In the case of the females, the ratios of mated females that showed spermatophores started to increase in spring, reached the maximum during the summer (July-September), and decreased in the autumn. The spawning period was clearly shown by the percentages of spawning, active or inactive females throughout the period studied. This occurred between the months of May and October, but was especially intense in July and August.

Table 1. Results of the length-weight relationship parameters of relative growth (size-weight relationship: Weight = a × Size ^{b}) of *Aristeus antennatus* calculated for sex and area. Level of significance *** = p < 0.001, ** = p < 0.01, * = p < 0.05, and NS = p < 0.1 in the Student 't' test.

Group	a	b	err.b	signif.	r2	n
Ibiza						
Males	0.003156	2.4023	0.01727	***	0.90	2060
Females	0.002425	2.4836	0.00585	***	0.97	5955
Total	0.002526	2.4720	0.00476	***	0.97	8015
Garrucha						
Males	0.005292	2.2594	0.01435	***	0.92	2149
Females	0.002812	2.4587	0.00631	***	0.97	4312
Total	0.002868	2.4511	0.00519	***	0.97	6461

Table 2. Parameters (size-age relationship) for the VBGF model obtained using the FISAT (ELEFAN) program, for males and females of *Aristeus antennatus* belonging to the two areas.

Area/year	Group	L_∞	K	t_0
Ibiza/92-94	Males	55	0.38	-0.43
	Females	73	0.36	-0.41
Garrucha/92-94	Males	55	0.36	-0.33
	Females	75	0.40	-0.23

The percentages of maturity by size class for determination of the 50% size at first maturity showed differences in the results according from where comes the data came. For males they were 16.0 mm in Ibiza and 17.5 mm in Vera, whereas for females they were 21 mm CL in Ibiza and 22 mm in Vera.

3.4 *Catches*

A total of 50 vessels had fishing activities directed towards the pink shrimp in the Ibiza Channel fishing ground during 1993, whereas there were 43 in the southern part of the Gulf of Vera in the same year. The total annual landings for the historic series from the Ibiza Channel fluctuated around an annual average of 76 tons. The maximum volume of landings was reached in 1991 with 145 tons and the minimum in 1982 with 38 tons. For the Gulf of Vera, considering the northern and southern parts together, the annual landings fluctuated around 333 tons, with a maximum in 1991 of 517 tons and a minimum in 1989 of 220 tons. The selected port of the southern part had around 182 tons by year, with a maximum of 308 tons in 1991 and a minimum of 68 tons in 1977, which represented 55% of all the Gulf catches (Fig. 3a).

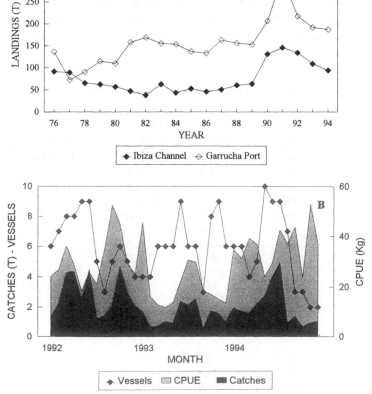

Figure 3. Historical series of annual landings for *Aristeus antennatus* from each area between 1976 and 1994 (A), and the monthly pattern of landings, vessels and CPUE in the port of Santa Pola during 1992-94 (B).

The monthly values for the development of the shrimp landings compiled during the 1992-1994 period for the Santa Pola trawling fleet showed large fluctuations (Fig. 3b), but this did not indicate any seasonality in the landings. The mean monthly catch was 2.06 tons, whereas the catch by unit of effort (CPUE), expressed in kilograms of shrimp by boat and fishing day, was 29.12 kg/boat/day based on the data compiled throughout the period considered. However, the relationships between catches and CPUE were not particularly good, mainly due to the monthly variations of the number of boats from this port that operated in the area (Fig. 3b). In the case of the port of Garrucha, the CPUE was 42.7 kg/boat/day, which was a value 1.5 times greater than in the Ibiza area.

3.5 *Exploitation scheme*

The results of the LCA analysis are shown in Table 3. The analysis was carried out on the total landings for all the ports together in the case of Ibiza and the landings from the port of Garrucha, which represented 55% of the total in the case of the Gulf of Vera. Thus the results of the analysis referred to the mean number of individuals in the sea and the biomass (mean and virgin) must be considered almost double for the Gulf of Vera.

In both areas, the mean ages were very similar to the critical age, although it was slightly higher for the males. In the case of the sizes, we found the same situation with the mean size slightly below the critical size in the females. The relationship between recruitment and growth was greater for the latter but was clearer in females. The percentage biomass caught by fishing was slightly larger in the case of Ibiza than in Vera. All the results obtained showed a difference between sexes and between ar-

Table 3. Results of the LCA for males and females of *Aristeus antennatus* for values of $M = 0.47$; $F = 1.5$, from each area.

	Ibiza males	females	Garrucha males	females
Mean no. of individuals (n × 1000)	2326	4535	14,208	1,595,500
Mean annual biomass (T)	10.5	35.4	81.7	18,489.0
SSB (gr./recruit)	1.9	2.9	3.9	7.9
Recruitment (n*1000)	272.9	289.4	1235	80,049.0
Mean age (years)	0.78	0.62	1.07	0.96
Critical age (year)	0.76	0.68	1.17	0.99
Mean size (cm)	20.1	24.4	21.5	27.3
Critical size (cm)	20.0	26.0	23.0	29.0
Virgin biomass (T)	51.3	270.5	256.7	92,220.3
Balance biomass (D)	23.9	73.7	137.3	28,700.0
Inputs				
Recruitment (%)	36.8	18.8	31.3	14.7
Growth (%)	63.2	81.2	68.7	85.3
Outputs				
Natural mortality (%)	20.5	22.6	27.2	30.3
Biomass caught (%)	79.5	77.4	72.8	69.7
Renovation rate (Turnover)	228.9	207.9	168.0	155.2

eas, with the biomass of the females being much greater than that of the males. The differences observed in the results showed that the importance of the resource in Vera was much greater than in Ibiza.

The graph of yield by recruit for each sex showed (Fig. 4) that there was a greater yield for the females, which was almost double compared to the males, and this was not so close to the maximum sustainable yield (MSY). Although the females had a value closer to MSY than males, the slope of the curve was very smooth in both cases and similar yields could be maintained with large variations of the effort. However, they showed a greater exploitation rate than the males.

The incidence of the fishing mortality (F), and initial number of each age class and of each sex from the Virtual Population Analysis (VPA) is shown in Table 4. In Ibiza, a maximum of three age classes were observed for males and four for females.

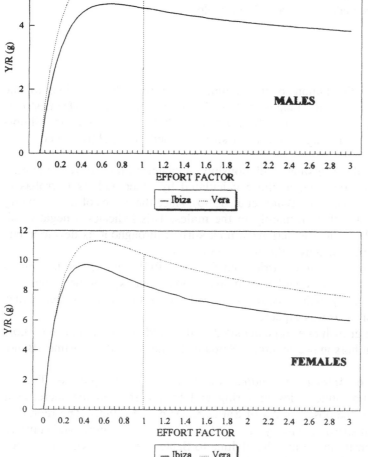

Figure 4. Comparison of the curves of Yield by Recruit (Y/R) as a function of the fishing effort from each area for males and females of *Aristeus antennatus*; actual value of F = 1.

Table 4. Results of the VPA, demonstrating the incidence of the fishing mortality (F) and the initial number, for each age class and sex of *Aristeus antennatus* in each area.

	Age	Ibiza		Garrucha	
		F	$N^o \times (1000)$	F	$N^o \times (1000)$
♂	1	0.540	4646.5	0.798	2255.1
	2	3.383	1692.8	1.815	643.4
	3	1.500	35.9	1.650	66.4
	4	-	-	1.500	8.1
♀	1	0.765	6893.3	0.289	2,147,010.0
	2	2.437	2003.6	1.261	1,005,127.0
	3	2.441	109.5	1.402	177,914.0
	4	1.500	5.9	2.908	27,342.4
	5	-	-	1.500	933.3

Meanwhile in Vera we observed four classes for males and five for females. In the case of Ibiza, the fishing pressure was centered on age class 2 in males and classes 2 and 3 in females, whereas age class 1 had low levels of exploitation that were more pronounced in the females. In Vera, the results were similar, with high pressure on age classes 2 and 3 for males, and 3 and 4 in females.

4 DISCUSSION

In both areas the size distributions of the landings, resulting in lower values for the males in all cases, show similar ranges to those found in other studies (Demestre & Martín 1993, Martinez-Baños et al. 1990), which further indicates the size dimorphism of the species. The ranges and mean size recorded are slightly lower in the case of the Ibiza Channel.

The parameters of the size-weight relationship estimated in this study are similar between the different areas and to those calculated by other authors (Arrobas & Ribeiro-Cascalho 1987, Martinez-Baños et al. 1990), since the values of the allometry coefficient b were lower than 3, mainly in the males. This indicates a negative allometry in the growth of the Mediterranean pink shrimp throughout its development, the growth in length prevailing over the growth in weight.

With respect to the estimations carried out for the VBGF parameters, the values that we obtained were almost equal between areas and do not explain the differences in size observed between populations, resulting in L_∞ values for females higher than those obtained for males, as was expected.

The values for the growth rate (K) calculated in this study result higher than those presented by other authors in nearby areas (Sardá & Demestre 1987, Martinez Baños 1997).

The reproduction of *Aristeus antennatus* is a very well known process. The decrease of the male immature ratios in spring and their corresponding increase in autumn coincide with the observations of Sardá & Demestre (1989). The reproduction period shows that spawning clearly occurred at the end of the spring and summer, with the greatest intensity in the months of June to September. There were no differences between areas, which is similar to that pointed out in other studies (Arrobas &

Ribeiro-Cascalho 1987, Martinez-Baños et al. 1990, Demestre & Martín 1993).

The sizes at first maturity (L_{50}) differ slightly, with those from the Ibiza Channel being smaller, as well as from those found in areas close to the Spanish Mediterranean, such as reported by Sardá & Demestre (1987) (L_{50} ♂ = 23-25 mm; ♀ = 27 mm), Demestre & Martín (1993) for Catalonia (L_{50} ♂ = 21 mm; ♀ = 26 mm), and Martinez-Baños et al. (1990) for Murcia (L_{50} ♂ = 21-22 mm; ♀ = 27-28 mm). This could be attributed to either a greater presence of small sized individuals, which sampling in this part of the population would intensify and thereby affecting the results, or to a response to an inadequate exploitation. In any case, the maturation stage both for males and females would have to take place within the first year of life.

The values obtained for biomass and yields showed an exploitation that was slightly towards overexploitation, with the rate of growth over fishing being greater in the Ibiza Channel than in Vera Gulf, as shows the lower mean sizes and the lower sizes of first maturity founded in the Ibiza Channel. In both cases, the situation is not so close to the optimum as has been reported in other studies from nearby areas (Demestre & Lleonart 1993, Demestre & Martín 1993, Demestre et al. 1994, Martínez Baños 1997). However, the resource is able to cope with variations of the fishing effort with the yields hardly altering, demonstrating a high biomass production rate (turnover), although the mortality vector has a greater effect on the older age classes, especially in the females. This could be due to the fact that not all the resource is equally accessible, with exploitation based on the largest sizes. The females predominate in this case leaving an important part of the population mainly constituted by males in deeper areas.

ACKNOWLEDGEMENTS

We would like to thank J.A. Martinez Madrid, Toni Romero and Diego Terrones for their collaboration in the sampling and compilation of data. Also we wish to extend our thanks to the Fishermen's Associations of Alicante, Villajoyosa, Santa Pola, and Garrucha for their collaboration in collecting data.

REFERENCES

Arrobas, I. & Ribeiro-Cascalho, A. 1987. On the biology and fishery of *Aristeus antennatus* (Risso, 1816) in the south Portuguese coast. *Inv. Pesq.* 51 (suppl. 1): 233-243.

Cartes, J.E. 1991. Análisis de las comunidades y estructura trófica de los crustáceos decápodos batiales del Mar Catalán. Ph.D. thesis, University of Barcelona.

Cartes J.E. 1994. Influence of depth and seasons on the diet of the deep aristeid *Aristeus antennatus*, along the continental slope (400-2300) in the Catalan Sea (western Mediterranean). *Mar. Biol.* 120: 639-48.

Cartes, J.E. & Sardá, F. 1989. Feeding ecology of the deep-water aristeid crustacean *Aristeus antennatus*. *Mar. Ecol. Prog. Ser.* 54: 229-238

Cartes, J.E. & Sardá, F. 1992. Abundance and diversity of decapod crustaceans in the deep Catalan Sea (western Mediterranean). *J. Nat. Hist.* 26: 1305-1323.

Cartes, J.E. & Sardá, F. 1993. Zonation of deep-sea decapod fauna in the Catalan Sea (western Mediterranean). *Mar. Ecol. Prog. Ser.* 94: 27-34.

Demestre, M. & Carbonell, A. 1994. Growth of the shrimp *Aristeus antennatus*. In: *Etùde pour l'amenagement et la gestión des peches en Mediterranee occidentale*. Rapport final. CEE FAR, 474 pp. (mimeo).

Demestre, M., Carbonell, A. & Martinez-Baños, P. 1994. Analysis of the exploited population of *Aristeus antennatus* (Risso, 1816) in three adjacent zones in the Spanish Mediterranean. In: *Etùde pour l'amenagement et la gestión des peches en Mediterranee occidentale*. Rapport final. CEE FAR, 474 pp. (mimeo).

Demestre, M. & Fortuño, J.M. 1992. Reproduction of the deep-water shrimp *Aristeus antennatus* (Decapoda: Dendrobranchiata). *Mar. Ecol. Prog. Ser.* 84: 41-51.

Demestre, M. & Lleonart, J. 1993. Population dynamics of *Aristeus antennatus* (Decapoda: Dendrobranchiata) in the north-western Mediterranean. *Sci. Mar.* 57: 183-189.

Demestre, M. & Martin, P. 1993. Optimum exploitation of a demersal resource in the western Mediterranean: the fishery of the deep-water shrimp *Aristeus antennatus* (Risso 1816). *Sci. Mar.* 57: 175-182.

Djabali, F., Mehailia, A., Koudil, M. & Brahmi, B. 1994. A reassessment of equations for predicting natural mortality in Mediterranean teleosts. *Naga, ICLARM Q. January*: 33-34.

Gayanilo Jr, F.C., Sparre, P. & Pauly, D. 1994. The FAO-ICLARM Stock Assessment Tools (FISAT) User's Guide. *FAO Computerized Information Series (Fisheries)* 6. Rome, FAO, 186 pp.

Lleonart, J. & Salat, J. 1992. VIT Programa de análisis de Pesquerías. *Inf. Téc. Sci. Mar.* 168-169.

Martinez-Baños, P. 1997. Dinámica de poblaciones de la gamba *Aristeus antennatus* (Crustacea, Decapoda) en las zonas de Murcia, Almería e Ibiza. Análisis global en el Mediterráneo Español. Ph Thesis. Universidad de Murcia.

Martinez-Baños, P., Vizuete, F. & Más, J. 1990. Aspectos biológicos de la Gamba Roja (*Aristeus antennatus* Risso, 1816) a partir de las pesquerias del S. E. de la Peninsula Ibérica. *Actas VIº Simp. Ibér. Est. Bentos Marino*, 235-243.

Massuti, M. & Daroca, E. 1978. Introducción al estudio de la biología de la gamba roja (*Aristeus antennatus*) de las pesquerias del sur de Mallorca. *Trab. Comp. Dep. Pesca Inst. Esp. Oceanogr.* 264-277.

Orsi Relini, L. & Semeria, M. 1983. Oogenesis and fecundity in bathyal penaeid prawns, *Aristeus antennatus* and *Aristaeomorpha foliacea*. In: *Rapp. Comm. Int. Mer Médit.* 28, 3.

Pauly, D. 1980. On the interrelationships between natural mortality, growth parameters and mean environmental temperature in 175 fish stocks. *J. Cons. Int. Explor. Mer.* 39: 175-192.

Relini Orsi, L. 1980. Aspetti riproduttivi in *Aristeus antennatus* (Risso, 1816) (Decapoda, Pennaeide). *Mem. Biol. Marina e Oceanogr.* Suppl. X: 285-289.

Relini Orsi, L. & Relini, G. 1979. Pesca e riproduzione del gambero rosso *Aristeus antennatus* nel Mar Ligure. *Quad. Civ. Staz. Idrobiol. Milano* 7.

Sardá, F. & Cartes, J.E. 1994. Distribution, abundance and selected biological aspects of *Aristeus antennatus* (Risso, 1816) (Decapoda, Aristeidae) in the deep water habits in the western Mediterranean. *Bios (Thessaloniki)* 1: 59-73.

Sardá, F. & Cartes, J.E. 1997. Morphological features and ecological aspects of early juvenile specimens of the aristeid shrimp *Aristeus antennatus* (Risso 1816). *Mar. Freshwater Res.* 48: 73-77.

Sardá, F. & Demestre, M. 1987. Estudio biológico de la gamba *Aristeus antennatus* (Risso, 1816) en el Mar Catalán (N.E. de España). *Inv. Pesq.* 51 (Suppl. 1): 213-232.

Sardá, F. & Demestre, M. 1989. Shortening of the rostrum and rostral variability in *Aristeus antennatus* (Risso, 1816). *J. Crust. Biol.* 9: 570-577.

Performance of the von Bertalanffy growth curve in penaeid shrimps: A critical approach

FERNANDO D'INCAO & DUANE B. FONSECA
Department of Oceanography – FURG, Rio Grande, Brazil

ABSTRACT

A literature survey was done on growth curves – based on von Bertalanffy's growth function – in shrimps of the genus *Penaeus*. A total of 108 curves were analyzed, values of parameters k and L_∞ were recorded. For standardization, the appropriateness was tested using the inverted von Bertalanffy growth function, considering previously estimated k and L_∞, and t_0 being zero. The results point out that a high number of studies estimated k values around 1/year corresponding to longevity of 3.5 years. On the other hand some results indicated longevity higher than 3.5 years with a maximum of 7.5 years. The results of this work indicate that longevity of several growth curves is higher than this value. The reason for this is linked with estimates of k. In our opinion, this parameter is usually underestimated in crustaceans resulting in higher longevity. This bias is not perceived when the growth curves are not related to the biology of the species under investigation. This work suggests that growth curves for crustaceans must be tested through longevity estimators using at least 99% of asymptotic length.

1 INTRODUCTION

The exoskeleton in crustaceans determines a discontinuous growth pattern. This pattern represents a succession of molts separated by intermolt periods. Extensive growth occurs after ecdysis, while limited growth occurs at intermolt periods because the membranes are capable of restricted extension. No such thing as bony structures are found in crustaceans, which prevent traditional age determination. Growth parameters are then estimated by modal progression analysis.

The length-frequency methods rely on cohort identification, which should be related to age groups. Cohort identification is subjective, which brings uncertainty to crustacean growth analysis based on length frequency distributions. Consequently, there are no elements to test the accuracy of the estimates using length-based methods. Age determination methods have been developed for crustaceans (Sheehy 1990, Sheehy et al. 1996). It appears necessary to evaluate crustacean growth curves. Fonseca (1998) proposed to validate the crustacean growth curves by longevity estimates, assuming that the asymptotic length of the von Bertalanffy growth curve has biological significance.

2 METHODS

A literature survey was done on growth curves based on von Bertalanffy's (1938) growth function:

$$Lt = L_\infty [1 + e^{-k(t + to)}] \qquad (1)$$

Where Lt is length at time t, k the growth coefficient and L_∞ is the asymptotic length. Here the length unit was mm and k at an annual basis.

The von Bertalanffy's equation parameters, L_∞ and k, of twenty *Penaeus*-like species were compiled. Longevity estimates were calculated for each curve by the inverse of von Bertalanffy's equation;

$$t_{max} = t_0 - (1/k) \ln [1 - (Li / L_\infty)] \qquad (2)$$

Where Lt was arbitrarily considered equal to 95% (Taylor 1962) and 99% of the asymptotic length and $t_0 = 0$. Maximum longevity (t_{max}) was considered as a function of the annual growth coefficient (k).

3 RESULTS AND DISCUSSION

Table 1 shows the growth parameters and maximum longevity estimates for 108 compiled growth curves. Maximum longevity estimates for Lt equal to 99% of the asymptotic length varied between 0.24 and 7.38 years and the annual growth coefficient between 0.62 and 18.88. Three groups of maximum longevity can be identified (Fig. 1). It is clear that very high values of longevity are associated with low k values, and low longevity is associated with high k values. The most appropriate longevity estimates have to be in accordance with the shrimp life cycle. Some have considered the penaeid maximum longevity around 30 months and Garcia & Le Reste (1981) proposed around 24 months. We consider that the most adequate longevity estimates range from 1.5 to 2.5 years.

Some authors suggest several valid k values for penaeid shrimps. For Garcia & Le Reste (1981) k values vary between 1.8 and 3.6, while Pauly et al. (1984) suggested k ranging between 0.25 and 2.5. The range proposed by Garcia & Le Reste is considered in this study more consistent with the penaeid life span, with k equal to 1.8 and the longevity reaching 20 months.

Taylor's method results in lower longevity values than the 99% method (Fig. 1). This is a consequence of the difficulty to interpret this parameter biologically, which introduces subjectivity in the analysis. Beverton & Holt (1959) demonstrated the relationships among maximum age, natural mortality, and k for different groups of fish. They emphasized that its statistical significance is limited because the accuracy of the data used is generally unknown and the values obtained are in some cases certainly over- or underestimated.

Table 1. Von Bertalanffy's growth curve parameters of the *Penaeus* literature survey and longevity estimates to 95% and 99% of asymptotic length.

Species	k	L ∞	T max 0,95 L∞	T max 0,99 L ∞	Author
P. aztecus	3.796	42.9	0.79	1.21	McCoy 1968
P. aztecus	1.92; 3.07	58.9; 41.9	1.56; 0.98	2.39; 1.50	Chávez 1973a
P. aztecus	8.89; 16.48	36.6; 30	0.34; 0.18	0.52; 0.28	McCoy 1972
P. aztecus	4.06; 4.06	47.8; 39.7	0.74; 0.74	1.14; 1.14	Parrack 1979
P. aztecus	7.85; 18.87	42.9; 30.9	0.38; 0.16	0.59; 0.24	Cohen & Fishman 1980
P. brasiliensis	1.34	196	2.24	3.44	MMA 1996
P. californiensis	2.03; 1.3	44.1; 47.3	1.48; 2.30	2.27; 3.54	Galicia 1976
P. californiensis	2.24	49.9	1.34	2.06	Lluch 1974
P. canaliculatus	2.55; 3.07	32.6; 25.3	1.18; 0.98	1.81; 1.50	Choy 1988
P. duorarum	2.55; 2.44	45.7; 46.4	1.18; 1.23	1.81; 1.88	Iversen & Jones 1961
P. duorarum	3.69; 4.42	38; 34	0.81; 0.68	1.25; 1.04	Kutkuhn 1966
P. duorarum	2.86; 2.39	52.1; 40.8	1.05; 1.25	1.61; 1.93	Berry 1967
P. duorarum	9.78; 11.28	34.5; 27	0.31; 0.27	0.47; 0.41	McCoy 1972
P. duorarum	5.2; 6.03	44.6; 31.6	0.58; 0.50	0.89; 0.76	Cohen & Fishman 1980
P. duorarum	1.2	48	2.50	3.85	Pauly et al. 1984
P. duorarum	2.6; 3.2; 3.4	203; 226; 225	1.15; 0.94; 0.88	1.77; 1.44; 1.35	Gracia 1995
P. esculentus	1.6; 2.6; 3.5; 2.9; 3.9; 3.3	49.2; 43.6; 34.7; 34.7; 39.9; 34.1	1.87; 1.15; 0.86; 1.03;0.77; 0.91	2.88;1.77;1.32; 1.59;1.18;1.4	Watson & Turnbull 1993
P. esculentus	2.6; 2.6	40.9; 32.6	1.15; 1.15	1.77; 1.77	White 1975
P. esculentus	2.13; 1.77	44.8; 37.5	1.41; 1.69	2.16; 2.6	Kirkwood & Somers 1984
P. indicus	1; 1.2; 1; 1; 1.2; 1; 1; 1.2; 1.1; 1; 1.2	210; 205; 207; 226; 220; 224; 214; 206; 213; 41.5; 40.5	3.0; 2.5; 3; 3; 2.5; 3; 3; 2.5; 2.72; 3.3; 2.5	4.6; 3.8. 4.6; 4.6; 3.8; 4.6; 4.6; 3.8; 4.2; 4.66; 3.85	Agasen & Del Mundo 1987
P. indicus	1.9; 2	39.2; 38.6	1.52; 1.48	2.33; 2.27	Kurup & Rao 1975
P. indicus	4.47; 4.84	42.4; 29.9	0.67; 0.62	1.03; 0.95	Le Reste & Marcille 1979
P. indicus	1.97; 1.82	43.5; 45.8	1.52; 1.65	2.33; 2.53	Devi 1986
P. kerathurus	0.78	41.9	3.84	5.90	Pauly et al. 1984
P. latisulcatus	0.88; 0.73	53.9; 44.1	3.39; 4.12	5.21; 6.33	Wallner 1985
P. merguiensis	4.16	38	0.72	1.11;	Lucas et al. 1979
P. merguiensis	6.03; 7.07	35.3; 29.4	0.50; 0.42	0.76; 0.65	Frusher et al. 1985
P. merguiensis	0.62; 1.19	50.9; 36.7	4.80; 2.5	7.38; 3.85	Achuthankutty & Parukelar 1986
P. merguiensis	1.61	50.3	1.86	2.86	Dwiponggo et al. 1986
P. merguiensis	1.04; 1.3	51.5; 44.5	2.88; 2.3	4.43; 3.54	Sumiono 1987
P. notialis	3.22; 3.85	42.8; 28.6	0.93; 0.78	1.43; 1.2	Lhome & Garcia 1984
P. ocidentalis	2.25	284.5	1.33	2.05	Barreto 1993
P. orientalis	0.93; 0.88	56.5; 45.8	3.2; 3.39	4.92; 5.21	Deng 1981
P. paulensis	1.27; 1.05	192; 248.3	2.36; 2.85	3.63; 4.39	D'Incao 1984
P. plebejus	5.2; 5.2	49; 40	0.58; 0.58	0.89; 0.89	Lucas 1974
P. plebejus	2.49; 3.07	59.5; 45.4	1.20; 0.98	1.85; 1.5	Glaister & MacDonall 1987
P. semisulcatus	2.08; 2.6	47.7; 48.2	1.44; 1.15	2.21; 1.77	Jones & Zalinge 1981
P. semisulcatus	1.09; 0.94	53.2; 48.2	2.74; 3.2	4.22; 4.92	Mathews et al. 1987
P. semisulcatus	3.17; 1.09	62.2; 38.1	0.94; 2.74	1.45; 4.22	Kirkwood & Somers 1984

Table 1. Continued.

Species	k	L ∞	T_{max} 0,95 L∞	T_{max} 0,99 L ∞	Author
P. setiferus	0.83; 0.78; 2.29; 2.7	59.2; 51.7; 59.2; 51.9	3.60; 3.84; 1.31; 1.11	5.54; 5.9; 2; 1.7	Lindner & Anderson 1956
P. setiferus	1.09	55.3	2.74	4.22	Klima 1974
P. setiferus	1.25; 0.83	58.5; 50.9	2.4; 3.6	3.69; 5.54	Pauly et al. 1984
P. setiferus	1.77; 2.91	54.5; 51.3	1.69; 1.03	2.60; 1.58	Gracia & Soto 1986
P. stylirostris	2.18; 2.91	49.9; 44.7	1.37; 1.03	2.11; 1.58	Galicia 1976
P. stylirostris	2.132	50.6	1.41	2.16	Lluch 1974
P. stylirostris	1.1	201	2.72	4.19	Isaac et al. 1992
P. subtilis	1.2; 1.1; 1.1; 1.1; 1; 1.2; 1.1; 1.2;1.12	177; 219; 220; 187; 225; 171; 218; 172; 216	2.5; 2.8; 2.8; 2.8; 3; 2.4; 2.8; 2.5; 2.7	3.9; 4.3; 4.3; 4.2; 4.6; 3.7; 4.3; 3.8; 4.1	Isaac et al. 1992
P. vannamei	3.172	43.4	0.94	1.45	Chavez 1973b
P. vannamei	1.248	30.7	2.40	3.69	Menz & Blake 1980

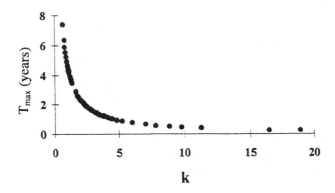

Figure 1. Relationship between k and T_{max} (L_{max} = 99% L_{∞}), related to 21 *Penaeus*-like species.

The k values in crustaceans, except for some demersal species, are usually high. However, one may discuss how high is a high value. In growth analysis, it is possible to produce more than one satisfactory mathematical adjustment. The k values around 1, 2, 3 produce different longevity estimates depending on whether the 95 or 99% of the asymptotic length is assumed. Results are shown below (L_{∞} = 42.02 mm; t_o = 0; *P. paulensis*; d'Incao 1984):

$$t_{max} = t_o - (1/k) \ln [1 - (39.92/42.02)]$$

$$t_{max} = 0 - (1/1) \ln (1 - 0.95) \cong 3.0 \text{ years}$$

$$t_{max} = 0 - (1/2) \ln (1 - 0.95) \cong 1.5 \text{ years}$$

$$t_{max} = 0 - (1/3) \ln (1 - 0.95) \cong 1.0 \text{ years}$$

$$t_{max} = t_o - (1/k) \ln [1 - (41.60/42.02)] \text{ years}$$

$$t_{max} = 0 - (1/1) \ln (1 - 0.99) \cong 4.6 \text{ years}$$

$t_{max} = 0 - (1/2) \ln (1 - 0.99) \cong 2.3$ years

$t_{max} = 0 - (1/3) \ln (1 - 0,99) \cong 1.5$ years

Observing these alternatives, longevity estimates with k values around 1 are considered high for penaeid shrimp. On the other hand, values around 3 are considered low. The k values around 2 are well adjusted to the *Penaeus* life span; in this case longevity estimates ranging from 1.5 to 2.3 are related to field observations. The best option in our opinion is then k = 2 and t_{max} = 2.3 because it permits higher values of k can be obtained without changing the adequate adjustment to longevity. If we consider the *Penaeus*-like life span, Taylor's method presents lower longevity estimates. Straight adjustment between the growth estimation and the species' life characteristics is important to decide the most appropriate estimate of k and of longevity.

We have to analyze carefully the generalizations of growth studies and to take into consideration that the relationships were proposed for fish populations. Our experience shows that the use of these relations in crustacean growth studies do not favor the detection of possible errors in the absence of good life span knowledge.

REFERENCES

Bertalanffy, L. von. 1938. A quantitative theory of organic growth (Inquities on growth laws II). *Hum. Biol.* 10: 181-213.

Beverton, R.J.H. & Holt, S.J. 1959. A review of lifespan and mortality rates of fish in nature and the relation to growth and other physiological characteristics. In G.E.W. Wolstenholme & M.O'Connor (eds), *Ciba foundation colloquium on ageing. The lifespan of animals* 5: 142-177. London: Churchill.

d'Incao, F. 1984. Estudo sobre o crescimento de *Penaeus (Farfantepenaeus) paulensis* Pérez-Farfante, 1967 da Lagoa dos Patos, RS, Brasil (Decapoda, Penaeidae). *Atlântica* 7: 73-84.

Fonseca, D.B. 1998. *Kalliapseudes schubartii* (Crustacea, Tanaidacea) comparação entre metodologias de análise de crescimento e dinâmica populacional da espécie. MSc thesis. Fundação Universidade do Rio Grande, 100p.

Garcia, S. & Le Reste, L. 1981. Life cycles, dynamics exploitation and management of coastal penaeid shrimp stocks. *Fish.Tech.Pap. FAO* 203: 1-215.

Pauly, D., Ingles, J. & Neal, R. 1984. Application to shrimp stocks of objective methods for the estimation of growth, mortality and recruitment – related parameters from length-frequency data (ELEFAN I and II), p. 220-234. In J.A. Gulland & B.J. Rothschild (eds), *Penaeid shrimps – their biology and management*. Fishing News Books.

Sheehy, M.R.J. 1990. Widespread occurrence of fluorescent morphological lipofucsin in the crustacean brain. *J. Crust. Biol.* 10: 613-622.

Sheely, M.R.J., Greenwood, J.G. & Fielder, D.R. 1996. More accurate chronological age determination of crustaceans from field situations using the physiological age marker, lipofucsin. *Mar. Biol.* 121: 237-245.

Taylor, C.C. 1962. Growth equations with metabolic parameters. *J. Cons. Perm. Int. Explor. Mer* 27: 270-286.

Relative abundance and growth of *Macrobrachium heterochirus* between 1983-1984 and 1996-1997, Huitzilapan River basin, Veracruz, Mexico

MARTHA A. HERNÁNDEZ-GUZMÁN & JAVIER CRUZ-HERNÁNDEZ
Licenciatura en Biología, División de Ciencias Biológicas y de la Salud, Mexico

L.M. MEJÍA-ORTÍZ & J.A. VICCON-PALE
ADEFRNR, Dept. El hombre y su ambiente, U.A.M. -Xochimilco, Mexico

P. ORTEGA
Departamento de Física y Química Teórica, Facultad de Química, Mexico

ABSTRACT

The aim of this study is to show the results of a study regarding the relationship between relative abundance and altitude, and the relationship between body length, body weight, and spatial distribution of the *Macrobrachium heterochirus* prawns collected from the Huitzilapan River basin in Veracruz, Mexico. The collections were made during two periods: one in 1983 to 1984, and the other in 1996 to 1997. These results indicate that the horizontal distribution of the relative abundance over the two periods was very similar and that the average body weights and lengths of the captured organisms were greater during the first period than in the second. When the analyses were controlled for gender, the results for males were equal, and the existing differences for females were not significant. During this period (1984-1996), the riparian population fished prawns in an uninterrupted and increasingly intensive way. This can be attributed to the fact that commercial prawns do not achieve the same length and body weight as those captured in the river.

1 INTRODUCTION

Decapods (crabs, crayfish, lobsters, prawns, and shrimp) are the largest known crustaceans. They are both costly and delicious, and some are harvested on farms (Vázquez & Villalobos 1980). In the Huitzilapan River basin of Veracruz, Mexico, the prawns *Macrobrachium acanthurus*, *Macrobrachium carcinus*, and *Macrobrachium heterochirus* are treasured for their delicate flavor. Fishermen sell this expensive product to tourists and in nearby towns. This accounts for the dramatic increase in prawn fishing in recent years and for the lack of regulation of this practice. Along this river, prawn fishing is still done by traditional methods that employ rather primitive casting nets and traps known as *atarrayas* and *nasas*.

Around 1980, prawn fishers in the Huitzilapan River basin complained that prawns were becoming scarce. That was when we began to monitor and characterize the abundance of the populations, as well as the body weight and size of individuals belonging to these species. We discovered that, in effect, in several of the smaller tributaries where there was a large amount of organic matter, these animals can no longer be found. However, in the main trunk of the river, they were not only abundant, some individuals were actually larger than those recorded by Holthuis (1952): *M. carcinus* = 23.3 cm and *M. heterochirus* 13.5 cm vs. González-Cervantes et al. (1981): *M. carcinus* = 24.2 cm y *M. heterochirus* 15.5 cm. Later, during the 1983-1984 collecting period, the maximum sizes were: *M. carcinus* = 24 cm and *M. heterochirus* = 19.6 cm. Because the time elapsed between the first collection and the second was so short, we decided that it would be better to allow more time to pass before collecting again, and, thus, be able to make a better comparison. During the waiting period, there were two factors that made a comparison of data even more pressing: first, the growth of the human riparian population, which led to increased prawn fishing and greater water pollution; and second, the construction of a dam. Below are some of the results.

M. heterochirus ranges from the Gulf of Mexico to Brazil (Holthuis 1952, Chace & Hobbs 1969). According to Chace & Hobbs (1969), both the adults and the young appear to be confined to ripple areas or to low waterfalls. This prawn species is a fast swimmer that can move upstream or downstream at great speeds.

González-Cervantes et al. (1981) maintained that the best model for describing the weight to size relationship for *Macrobrachium carcinus* and *M. heterochirus* in the Huitzilapan River basin was, respectively, $W = aL^b$, with $W = 0.011L^{3.23}$ and $W = 0.187L^{2.215}$. Later, Cruz-Márquez & García-Arteaga (1984) found that the scant presence of *M. heterochirus* in the basin may be attributable to the levels of salinity, temperature, and dissolved oxygen documented in the lower part of the basin. They also found that the levels of dissolved oxygen and the temperatures recorded in the upper basin limit the presence of *M. carcinus* and that pH does not affect the presence of either species. Also, in this basin, Viccon-Pale & Corona-García (1992) found that when the traditional fishing practices are followed, *M. carcinus* contributes a greater share of moist biomass to the catch and that *M. heterochirus* is more abundant in the catch. Regarding the effectiveness of fishing implements, they also observed that the *nasa* is better for capturing *M. carcinus* and the *atarraya* for capturing *M. heterochirus*.

Chavez-Alarcón & Chavez (1976) stated that the weight to size relationship in the *Macrobrachium carcinus* prawn population living in the Actopan River of Veracruz, Mexico (due north of the Huitzilapan River) can be represented by the exponential function that takes the form of $W = 0.00000899L^{3.182}$. Later, Valenti et al. (1994) found that the weight to size relationship in females of the same species living in the Ribeira de Iguape River in Sao Paulo, Brazil (24°S, 47°W), can be represented by a model that also has the form $W = 0.00873L^{3.28}$. Other prawn species that inhabit the Huitzilapan River, along with *M. carcinus* and *M. heterochirus,* are *M. acanthurus* and *M. olfersii*. The weight to size relationship in the *M. acanthurus* population that shares the Ribeira de Iguape River with *M. carcinus*, can be expressed by the model $W = 0.00635L^{3.25}$ (Valenti et al. 1987). Pereira de Barros (1995) states that in *M. olfersii* that inhabit Santa Catarina Creek in southern Brazil (28°01'S, 48°35'W), the weight to size relationship is described by the model $W = 0.000942L^{2.9285}$ for the females and $W = 0.00127L^{2.7371}$ for the males. Horne & Beisser (1977) state that dams obstruct upstream migration of some species belonging to the genus *Macrobrachium*, and they further postulate that increased dam construction along the main river ways may limit prawn habitat to the coastal regions and alter distribution patterns.

In this study, we show the results of the comparison of the abundance and relative growth of the prawn *M. heterochirus* over an initial one year period (1983-1984) with individuals collected during a second one year period twelve years later (1996-1997).

2 STUDY AREA

The Huitzilapan River basin is located between 19°10' and 19°35'N and 96°15' to 97°15'W in the state of Veracruz, Mexico (Fig. 1). It covers 2827 square kilometers, and is bordered in the north by the Actopan River, in the south by the Jamapa River, and in the west by the interior hydrological region number eighteen. The source of the main affluent of the Huitzilapan River is in the Sierra Madre Oriental at an elevation of about 3350 meters. It flows from there until it joins the Barranca Grande River, 3 km to the north of Barranca Grande peak. The Barranca Grande River´s source lies 3400 m above sea level in southwestern portion of the Cofre de Perote. It flows through deep canyons past steep slopes until joining the Resumidero River at 1350 meters above sea level. Here, its name changes to the Pescados River. At a place called Las Juntas, it receives the waters of the Chico River. Los Pescados flows for 1.25 km until the Texolo River enters from the left side. Throughout this region, the river negotiates very deep canyons whose peaks crown 500 meter walls with 80° slopes, and where white water is more abundant than pools. Further downstream the inverse is increasingly the case as the river approaches the coast. Close to the town named El Carrizal, Veracruz at 200 meters above sea level, a derivation dam was built a few years ago. The collecting channel flows past flat, arable land and meanders until it enters the Lagartos River at Paso Marino, which is where the derivation dam, La Antigua, is located. La Antigua River continues and is later joined by the Paso de Ovejas River. They flow together until La Antigua discharges into the Gulf of Mexico (Anonymous 1970).

In some parts of the high and moderately high regions of the Huitzilapan River basin, it rains all year long, and when it's not raining it's enveloped in fog especially in fall and winter. Summer is the season when the basin receives the greatest rainfall (60 to 500 mm). Spring is the dry season in the lower regions of the basin.

3 METHODS

As mentioned previously, this study was undertaken to provide local prawn fishers with some answers to their questions. Since much of the river is very inaccessible, most of the data were taken from specimens caught by the locals. They fish with *atarrayas* and *nasas*. The *atarraya* is a net whose radius measures 1.6 meters on the average, with 1.4 cm openings, and an approximate weight of 6 kg. The *nasa* is a cone made of reeds with three main parts: a funnel 8 cm. in diameter with a length of 8 cm through which the animal enters a covered structure. The height of these cones is usually about 48 cm, rising to its apex from a 24 cm base.

Specimens were captured in two periods: (May, June, September, October, November, January, February, March, and April) of 1983-1984 and (June, October, November, February, and November) of 1996-1997. Sampling stations were located in places where the local people normally go to collect prawns: 1. La Antigua (5 meters

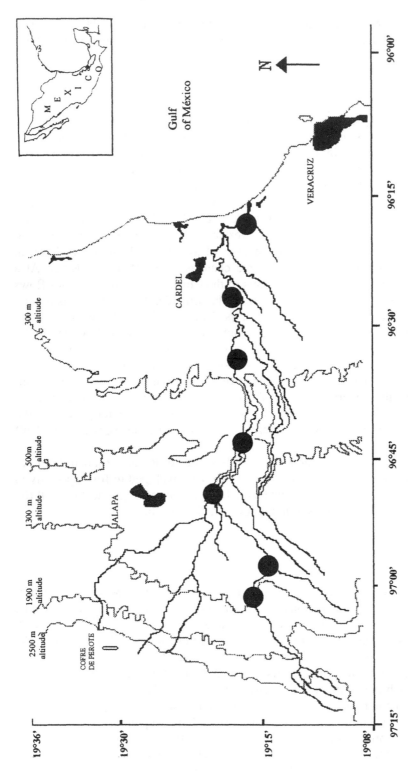

Figure 1. Study area, basin of the Huitzilapan river, Veracruz, Mexico.

above sea level), 2. Paso Mariano (105 msl), 3. Crucero-Rinconada (145 msl), 4. Apazapan-Jalcomulco (280 msl), 5. Las Juntas (535 msl), 6. Tlaltetela-Pinillo (705 msl), and 7. Barranca Grande (1045 msl) (see fig. 1). During the collection period, we accompanied the fishermen to the above mentioned collection sites at night, reserving one night per site. At about 6 P.M., the fishermen would cast their nets and lay their traps, trying to place them in the deeper parts of the river. They would collect both the nets and the traps at about 6 A.M. the next day. The fishermen who used *atarrayas* started casting their nets at about 10 P.M. and would finish close to midnight. They moved up and down the river from one bank to the other.

The captured prawns were then classified for gender and weighed with a digital scale (Ohaus ± 0.01g). The total length (from the rostrum to the telson) was also recorded. Data analyses for the relationship between length and weight and the comparison of the same in both periods were done by using Statgraphics 2.1 software.

4 RESULTS

4.1 *First period*

A total of 1632 organisms (657 females and 975 males) were collected during the first period (1983-1984). In Figure 2, one should note that the lowest abundance percentages occurred at the elevations of 1045 msl, 0.96% and 5 msl, 0.06% while the highest abundance percentages occurred at the elevations of 145 msl, 37.5% and 105 msl, 26.54%. The smallest recorded size was 4 cm, the largest was 19.6 cm, the average length was 9.19 cm, and the most common catch measured between 7.2 and 8 cm. The lowest weight recorded was 2.2 grams, the highest was 99.7 g, the average was 17.5 g, and the most common catch weighed from 8 to 10 g (Figs 3a and 3b). Table 1 shows the comparison results when using Mann-Whitney tests. The P-values that allow us to identify the similarities and differences in size and weight averages of the prawns captured during both periods should be emphasized.

For the first period, the best option for describing the relationship between body weight and length was the exponential model, which covers the whole population, ($W = 0.0165705L^{3.04355}$), and is controlled for gender (Figs 4 and 5). However, the differences between constant and exponent values vary by gender and should be noted (males: $W = 0.0214418L^{2.94144}$ and females: $W = 0.0156539L^{3.05324}$).

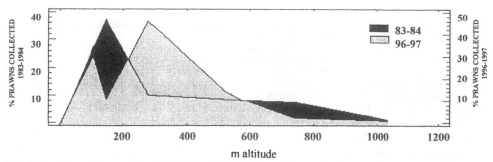

Figure 2. Comparison of the proportions of the prawn *M. heterochirus* collected in both periods (1983-1984 and 1996-1997), according to sea level meters.

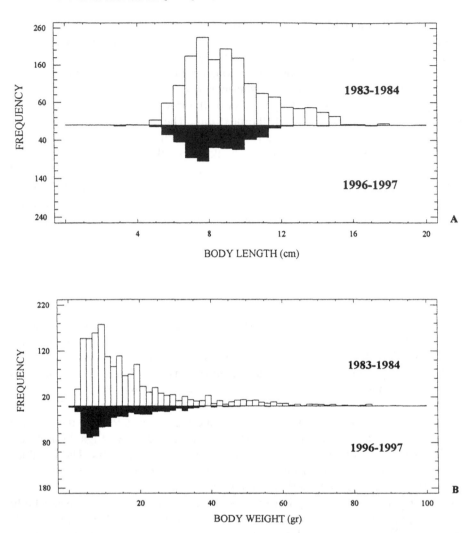

Figure 3. Comparison of frequency histograms of body length (A) and body weight (B) of *M. heterochirus*, in both periods (1983-1984 and 1996-1997).

Table 1. Comparison of the averages of the body weight (g) and the total length (cm) of *M. heterochirus* in both study periods (1983-1984 and 1996-1997).

1983-1984	P-value Mann-Whitney test	1996-1997
Av. length of male = 9.8	≥ 4.40351	Av. length of male = 8.825
Av. weight of male = 17.8	≥ –1.81866	Av. weight of male = 14.795
Av. length of female = 7.8	= 0.189996	Av. length of female = 7.5
Av. weight of female = 8.3	= 0.741794	Av. weight of female = 7.9
Av. length population = 8.8	> 0	Av. length population = 8.1
Av. weight population = 12.4	> 0	Av. weight population = 10.3

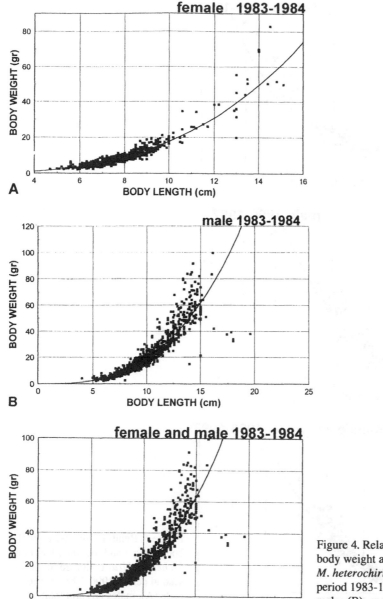

Figure 4. Relationships between the body weight and the body length of *M. heterochirus* collected in the first period 1983-1984, for females (A), males (B), and total organisms (C), respectively.

4.2 Second Period

Some 530 organisms were collected (257 females and 273 males) during the second period (1996-1997). In Figure 2, one can see that the lowest abundance percentages occurred in the collection sites with elevations of 5 meters above sea level, 1.12% ; 740 msl, 1.87%; and 1045 msl, 0.18% and that the highest abundance percentages occurred in the collection sites with elevations of 105, 28.94%, and 280 msl, 45.66%.

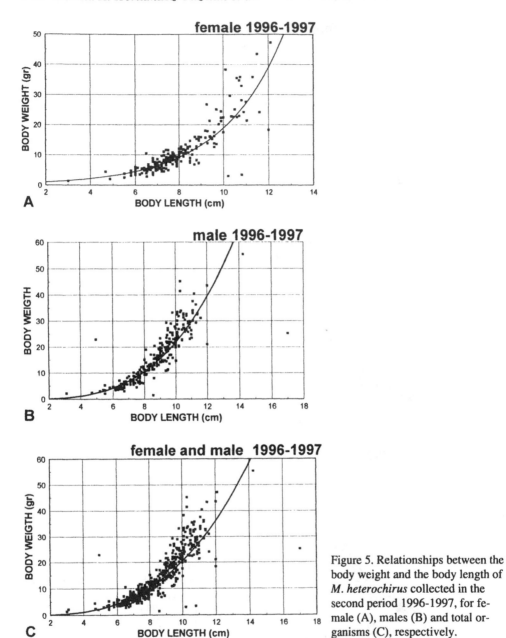

Figure 5. Relationships between the body weight and the body length of *M. heterochirus* collected in the second period 1996-1997, for female (A), males (B) and total organisms (C), respectively.

The smallest size recorded was 3 cm, the largest was 17 cm, the average was 8.2 cm, and the most common catch weighed from 5 to 6.3 g (Figs 3a and 3b).

Again, the best option for describing the relationship between body weight and length was the exponential model, which covers the whole population, $W = 0.151046L^{3.1308}$ and the male population $W = 0.0121816L^{3.2518}$. However, for the female population, this relationship was best described by the exponential model [$W = \exp(-0.67573 + 0.361373L)$].

5 DISCUSSION

Due to the difficulties involved in getting to the collection sites and to budgetary and time constraints, it was not possible to use techniques that would have allowed us to standardize the data, like capture-recapture or catch per unit effort. This would have allowed for interpretations based on information obtained through more rigorous methods, but every possible effort was made to collect reliable measurements. Since to date we know of no other study undertaken from this perspective, which gives this study the chance to serve as a platform for raising new problems and methodological questions, we are eager to present the above results and the following considerations.

First, due to the methodological restrictions, the comparisons, and particularly those regarding abundance, these considerations should be qualified with the necessary reservations. Thus, in Figure 2 it was decided not to show the abundance of the prawns that were collected in each sampling season. Rather, we present the percentages based on the sum total of prawns in each season for each period. One can see that the curve representing the first period percentages is uni-modal, and the one representing the second period percentages is bimodal with the main peak shooting toward higher elevations. However, we can conclude that the pattern of spatial distribution remained the same over the two periods.

By comparing the frequencies of body weight and length (Figs 3a and 3b), the average weights and lengths (Table 1) and the model parameters, one can postulate that, as regards the whole population, the organisms captured in the second period were smaller than those caught in the first period. By controlling for gender, the parameters of the female population went virtually unchanged. The need to perform this kind of analysis is due to sexual dimorphism that is very distinguishable in the males by the size of their second pair of pereiopods.

Changes in the quantity of specimens collected, as well as in their body weight and length, may possibly be explained by the fact that during the last decade the harvest of these prawns has increased. Although the collection methods employed in this region are relatively primitive (Viccon-Pale & Corona-García 1992) they have been modified. Traps fabricated with synthetic netting with small openings have replaced the nasas, which were originally made of rattan. As a result, fewer prawns can escape and more small prawns are captured. Furthermore, the habitat of these organisms has been drastically modified with increasing water pollution tied to human population growth, the generation of more domestic waste, and by the construction of a dam in the central part of the basin. This construction took place between the first and second collection periods. Finally, because middlemen keep the greatest share of the profit, and because real wages have consistently declined, prawn collectors must capture more prawns in order to earn a living.

6 CONCLUSIONS

Several points can be made.

First, the curve representing the altitude distribution of the relative abundance percentages of the first period is unimodal, and the one representing the second period is bimodal. The latter has a main peak that shoots toward higher altitudes. Despite this fact, one can consider that the spatial distribution patterns are very similar.

Second, the weight and size averages of *M. heterochirus* were greater in the first collection period (1983-1984) than in the second (1996-1997).

Third, the models that are best suited for describing the relative growth of the *M. heterochirus* captured during the first collection period (1983-1984) were:

Whole population: $W = 0.0165705L^{3.04355}$

Females: $W = 0.0156539L^{3.05324}$

Males: $W = 0.0214418L^{2.94144}$.

In the second collection period (1996-1997), the models best suited for describing the relative growth of the *M. heterochirus* captured during that period were:

Whole population: $W = 0.151046L^{3.1308}$

Females $W = \exp. (-0.67573+0.361373L)$

Males $W = 0.0121816L^{3.2518}$.

REFERENCES

Anonymous 1970. Región Hidrológica No. 28. Parcial: Ríos Actopan, La Antigua y Jamapa. *S.R.H. Boletín Hidrológico* 43: 86-87.

Cruz Márquez, M.G. & García Arteaga, M.C. 1984. Cuatro factores abióticos (Temperatura, Oxígeno disuelto, Salinidad y pH) y su relación con la composición de la colecta comercial de dos especies de langostinos (*Macrobrachium carcinus* y *Macrobrachium heterochirus*) en el río Huitzilapan, Ver. *Informe de Servicio Social U.A.M.–Xochimilco*, 43 pp.

Chace F.A. Jr. & Hobbs, H.H. Jr. 1969. The freshwater and terrestrial decapod crustaceans of the West Indies with special reference to Dominica. *Bull. Natl. Mus. US* 292: 1-258.

Chávez-Alarcón, Z. & Chávez, E. 1976. Introducción al conocimiento de la biología del langostino (*Macrobrachium carcinus* L.) en el estado de Veracruz. *Memorias del simposio sobre Biología y Dinámica Poblacional de Camarones, Guaymas*: 13-23.

González Cervantes, R.M., Loría Saviñon Y.M., Parés Sevilla H.M. & Ramírez Silva L.H. 1981. Evaluación del recurso langostino en los ríos Teoxolo y Chilontla en el Municipio de Teocelo Veracruz. *Informe final de Servicio Social U.A.M.–Xochimilco*, 30 pp.

Holthuis, L.B. 1952. A general revision of the Palaemonidae (Crustacea: Decapoda: Natantia) of the Americas. II Subfamily Palaemoninae. *Occ. Pap. Allan Hancock Found.* 12: 1-396.

Horne, F. & Beisser, S. 1977. Distribution of river shrimp in the Guadalupe and San Marcos rivers of Central Texas, USA (Decapoda, Caridea). *Crustaceana* 33: 56-60.

Pereira de Barros, M. 1995. Datos biológicos sobre *Macrobrachium olfersii* (Wiegmann, 1836) (Decapoda, Palaemonidae) da Praia da Vigia, Garopaba, Santa Catarina, Brasil. *Biociencias* 3: 239-252.

Valenti, W.C., Mello, J.T.C. de & Lobao, V.L. 1987. Crescimento de *Macrobrachium acanthurus* (Wiegmann, 1836) do rio Ribeira de Iguape (Crustacea, Decapoda, Palaemonidae). *Revista Brasileira de Biología* 47: 349-355.

Valenti, W.C., Mello, J.T.C. de & Lobao, V.L. 1994. Maturation and growth curves of *Macrobrachium carcinus* (Linnaeus) (Crustacea, Decapoda, Palaemonidae) from Ribeira de Iguape river, Southern Brazil. *Revista Brasileira de Zoologia* 11: 649-658.

Vázquez G.L. & Villalobos A. 1980. *Arthropoda Parte II Mandibulata*. Mexico D.F.: Universidad Nacional Autónoma de México.

Viccon-Pale J.A. & Corona García, S. 1992. Composition de la capture artisanale des crevettes *Macrobrachium carcinus* (L.) et *M. heterochirus* (Wiegmann) dans la rivière Huitzilapan, Veracruz, Mexique. *Première Conference Européenne sur les Crustacés*. Paris: Muséum National d'Histoire Naturelle, U. Pierre Marie Curie, Ecole Normale Supérieure de Paris: 170.

Assessment of the Carmel Bay spot prawn, *Pandalus platyceros*, resource and trap fishery adjacent to an ecological reserve in central California

KYRA L. SCHLINING
Moss Landing Marine Laboratories, California, USA and Monterey Bay Aquarium Research Institute, California, USA

JEROME D. SPRATT
California Department of Fish and Game, Monterey, California, USA

ABSTRACT

Due to a recent 5-fold increase in spot prawn, *Pandalus platyceros,* landings in the Monterey Bay area, data were collected at sea to establish the current status of the resource, beginning with the Carmel Canyon trap fishery. Prawn traps were set within the boundaries of the Carmel Bay Ecological Reserve, as well as in the neighboring fishing grounds in order to quantify the importance of the reserve relative to the spot prawn resource. Significant differences were found inside the reserve compared to outside for locations both close to and far away from the reserve and include higher mean catch-per-unit-effort (kg per trap), higher male to female sex ratio, and differences in mean carapace lengths. These differences suggest an underlying distribution mechanism more complex than simply a nearshore to offshore gradient of nutrients or physical factors. Spot prawn mean carapace lengths sampled during this study were found to be 2-3 millimeters smaller than for prawns sampled in 1967-1968 during the same months, December through March. The majority of prawns were found to be ovigerous from September through March providing information to support a seasonal closure if needed. However, the Carmel Canyon spot prawn resource currently appears to be in healthy condition, though it is essential that monitoring be continued in the future if catches continue to increase.

1 INTRODUCTION

Spot prawns, *Pandalus platyceros*, are a species of caridean shrimp found from Unalaska, Alaska to San Diego, California, and also off the coast of Japan (Butler 1964). As are many other shrimp in the family Pandalidae, spot prawns are protandric hermaphrodites and spend their first three years as males. After spawning as males, they transform into females and can potentially spawn for one or two more years. Spot prawns are the largest pandalid shrimp found off the Pacific coast of the United States and Canada and can live up to six years reaching a maximum carapace length (CL) of 63 mm or a maximum total length (TL) of 30 cm (Barr 1973, Butler 1964, Balsiger 1981, Sunada 1986).

1.1 *History of the fishery*

The California spot prawn fishery originated during the 1930s when fishers from the port of Monterey began landing spot prawns taken incidentally in octopus traps (Sunada & Richards 1992). Spot prawn landings for the Monterey area ranged between 400-1500 kg per year until the late 1960s when landings increased by an order of magnitude to 4000-9000 kg per year where they remained until the early 1990s (CDFG 1995). Prawn trawlers from outside regions entered the Monterey prawn fishery in 1994, and landings increased dramatically to a record of 31,746 kg. The trawlers have continued to fish the area over the past four years and combined trap and trawl landings remain high, currently between 12,000-32,000 kg per year (Fig. 1). As a result of this recent activity spot prawns have become one of the most valuable fisheries in the Monterey area, commanding an ex-vessel price of US$ 16.50 per kg and bringing in a total value of US$ 500,000 in 1997. By comparison, the California statewide spot prawn landings exceeded 340,000 kg in 1997 with a value of nearly US$ 5 million, of which the majority were landed in southern California.

Due to the recent growth in the central California spot prawn fishery and the paucity of current data on the local resource, the California Department of Fish and Game, in cooperation with Moss Landing Marine Laboratories, initiated a program to collect detailed biological and fishery information on the spot prawn, beginning with the Carmel Canyon trap fishery. The objectives of the study were: 1. To assess the current status of the spot prawn resource in the Carmel Canyon area using fishery dependent data, 2. To determine the importance of the Carmel Bay Ecological Reserve relative to the spot prawn resource in central California, and 3. To assess any changes in resource status over time.

1.2 *Study area*

Carmel Canyon is located off the coast of central California, south of Monterey Bay (Fig. 2). The head of the canyon, at a depth of 100 m, is within 0.4 km of shore and has been a prime spot prawn fishing target for many years. The California Fish and

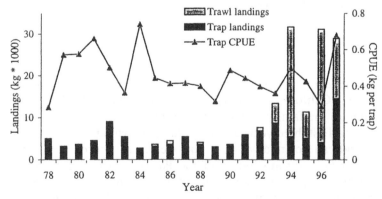

Figure 1. Monterey area spot prawn landings by gear (kilograms × 1000) and catch-per-unit-effort (kilograms per trap) for the Carmel Canyon trap fishery from 1978-1997. Data from the California Department of Fish and Game.

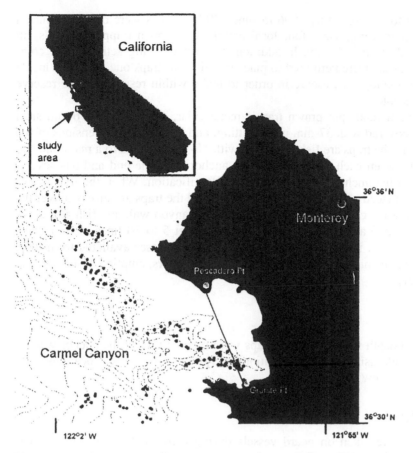

Figure 2. Location of Carmel Canyon, California. Area east of the line connecting Pescadero Point and Granite Point encompasses the Carmel Bay Ecological Reserve. Dots represent locations sampled. Depth contours are shown in 100 m intervals.

Game Commission established the Carmel Bay Ecological Reserve on February 20, 1976, beginning a very controversial closure of the head of the canyon to commercial prawn fishing. The reserve is demarcated simply by a line-of-sight from Pescadero Point (121°57′00″W 36°33′36′N) to Granite Point (121°56′24″W 36°31′12′N). The park rangers patrolling nearby Point Lobos Ecological Reserve reliably report unauthorized boats entering the reserve vicinity which along with the published regulations has kept the area virtually unfished for the past 22 years. Until the present study, the effectiveness of the reserve had never been evaluated relative to resource management.

2 METHODS

2.1 *Sampling procedure*

Data on spot prawns were collected from commercial spot prawn trapping vessels during day trips to sea from July 1996 to June 1997. Catches were observed once a week, weather permitting, using four local commercial prawn trapping vessels, all approximately 12-18 m in length. In addition to normal fishing activities the commercial prawn vessels were permitted to place two lines of traps once a month inside the Carmel Bay Ecological Reserve in order to make within reserve/outside reserve comparisons possible.

A typical commercial spot prawn trap is rectangular, constructed of 6-mm steel reinforcing bar covered with 37-mm mesh netting, and has outside dimensions of 0.4 m × 0.4 m × 1 m. The traps are fished in lines with 10-12 traps tied to a main line and roughly 20 m between each trap. Each line is anchored at one end and has several floats attached to the anchor marked with boat identification. When the trap line is set, the anchor is placed at a depth of 200 to 300 m and the traps are dropped over the edge of the submarine canyon. Traps hang down the canyon wall and fish at a depth range of between 200 and 400 m. Vessels generally set 5 to 10 lines of traps each day. Primary bait used in the trap fishery is fetid fish carcasses available from local fish processing companies. Traps were pulled every 24 hours, emptied, re-baited, and reset immediately.

2.2 *GPS*

The location and depth for each line of traps were recorded when the anchor was released over the side using a Global Positioning System (GPS). Line locations were accurate to within a few meters.

2.3 *Biological data*

Biological data were taken on board vessels during active fishing operations. The variance between lines (MS = 134) was far greater than the variance between traps (MS = 30) and traps on a line were not independent, therefore, we combined all traps on one line into one sample. After incidental species were removed, each sample of prawns (one line) was weighed to the nearest 0.1 kg. Individual prawns were not weighed due to problems with resolving small weights accurately on a moving vessel. Catch-per-unit-effort (CPUE) was calculated for each line by dividing the total catch weight per line by total number of traps per line to produce an average catch in kilograms per trap. A Mann-Whitney test was used to compare the median CPUE from the 26 lines set within the reserve to 26 samples from each of two locations outside the reserve, close to the reserve boundary and far away from the reserve boundary.

When the catch was small (< 50 prawns per line), all prawns were processed; but, when the catch grew too large a random sub-sample of 1-2 kg (approximately 30 prawns) was taken from each line for further observations on size and sex. Sizes were measured using carapace length (CL), from the base of the eyestalk to the posterior mid-dorsal edge of the carapace, to the nearest millimeter using callipers. Carapace lengths were tested for significant differences between the same 52 samples located outside the reserve as indicated above, 26 close to and 26 far from the reserve, as well as, for differences in lengths within and outside but close to the boundary of the reserve for male, female, and transitional prawns.

Sex was determined by examining the 2nd pleopod for presence of the *appendix masculina*, a sex characteristic present in the male prawns. The animals were categorized as either male, female, or transitional (in the process of changing from a male to a female). The male to female sex ratios were analyzed for significant differences inside and outside of the reserve using a Pearson χ^2-test. Gravid females were also noted.

2.4 *Historical data*

From 1967 to 1968 the California Department of Fish and Game sampled spot prawns landed at Monterey. These measurements were taken dockside after the vessels were unloaded, while our present study was done at sea. Carapace lengths were analyzed for differences between these data from December 1967 through March 1968 and our data from the same months in 1996-97 (Mann-Whitney test on medians). For comparison, we additionally reported prawn sizes from data gathered by Dahlstrom on his 1963 cruise off the California coast aboard the *M.V. Alaska*.

Access to fishing activity logs containing information on catch, number of traps, and catch location from 1978 to the present, were made available through the California Department of Fish and Game. This information allowed CPUE to be tracked over nearly 20 years. Mean monthly CPUE from these historical logs were compared to mean monthly CPUE observed inside and outside the reserve during this study, including both close and far locations.

3 RESULTS

Fifty-two lines were sampled outside of the reserve, 26 in locations close to the reserve boundary and 26 far from the reserve. The results of the Mann-Whitney test verified that the median CPUE (kg per trap) close to the reserve was significantly greater than the median CPUE for the samples further away from the reserve (U = 448.5, $p < 0.043$) and the median CPUE within the reserve was significantly greater than the median CPUE close to the reserve (U = 476.0, $p < 0.012$, Table 1). However, the mean CPUE for both of the outside reserve locations compare favorably with the mean yearly CPUE over the past 20 years (Fig. 3).

The same 52 lines mentioned above yielded 1104 prawns from locations close to the reserve boundary, 766 prawns from far away from the reserve, and 879 prawns from inside. Carapace length was not related to depth of anchor (inside reserve $R^2 = 0.0102$, outside reserve $R^2 = 0.0001$) therefore, we did not include depth as a factor when looking at differences in CL inside and outside of the reserve. For the males, CL was found to be significantly larger far from the reserve, with a mean of 37.9 mm, than close to reserve, with a mean of 35.1 mm (U = 84855.5, $p < 0.001$, Table 1) and also significantly larger close to the reserve than inside the reserve where the mean was 34.0 mm (t = −4.284, $p < 0.001$, Table 1). On the contrary, for the females, CL was significantly smaller far from the reserve, 47.8 mm, when compared to close to the reserve, 48.8 mm (U = 41853.5, $p = 0.002$), however there were no significant differences found between locations inside the reserve, 49.2 mm, and close to the reserve (t = 1.340, $p = 0.181$). The CL for transitional prawns was significantly larger far from the reserve, 44.7 mm, when compared to close, 41.0 mm, (U = 1924.5, $p <$

Table 1. Statistical results for CPUE (kilograms per trap), male to female sex ratios, and carapace lengths (mm) for male, transitional, and female spot prawns, close to and far away from, inside and outside of the Carmel Bay Ecological Reserve, California and for 1967-1968 versus 1996-1997.

	Factor 1	Mean	SE	N	Factor 2	Mean	N	SE	Test statistic	df	p-value	Significant?
CPUE	Close reserve	0.43	0.05	26	Far reserve	0.32	26	0.05	*U = 448.5	1	0.043	Yes
CPUE	Inside reserve	0.94	0.13	26	Close reserve	0.43	26	0.05	*U = 476.0	1	0.012	Yes
Male CL	Close reserve	35.1	0.16	759	Far reserve	37.9	392	0.16	*U = 84855.5	1	< 0.001	Yes
Male CL	Inside reserve	34.0	0.19	562	Close reserve	35.1	759	0.16	*t = −4284	1319	< 0.001	Yes
Trans CL	Close reserve	41.0	0.45	101	Far reserve	44.7	76	0.23	*U = 1924.5	1	< 0.001	Yes
Trans CL	Inside reserve	43.4	0.52	32	Close reserve	41.0	101	0.45	*U = 2147.0	1	0.005	Yes
Female CL	Close reserve	48.8	0.23	244	Far reserve	47.8	298	0.21	*U = 41,853.5	1	0.002	Yes
Female CL	Inside reserve	49.2	0.19	285	Close reserve	48.8	244	0.23	*t = 1340	527	0.181	No
Ratio M:F	Inside reserve	1:1.97	n/a	N/a	Outside reserve	1:1.59	n/a	n/a	$\chi^2 = 7085$	1	0.008	Yes
Male CL	1967-68	39.0	0.21	410	1996-97	37.4	1232	0.12	*t = −6586	1640	< 0.001	Yes
Trans CL	1967-68	44.1	0.15	278	1996-97	44.0	172	0.19	*t = −0.612	448	0.541	No
Female CL	1967-68	50.2	0.11	472	1996-97	47.9	447	0.18	*U = 65,854.0	1	< 0.001	Yes

*Means were compared with a t-test when assumptions of normality and homoscedasticity were met and medians were compared with a Mann-Whitney test when assumptions were violated.

0.001) but significantly larger inside the reserve, 43.4 mm, than close to the reserve (U = 2147.0, p = 0.005).

There was a significant difference in sex ratios for inside compared to outside of the reserve with a larger number of males to females located inside ($\chi^2_{,1}$ = 7.085, p = 0.008, Table 1). The ratio of males to females within the reserve was 1.97:1 compared to an outside reserve ratio of 1.59:1 (Fig. 4). Size frequency distributions yielded a smaller number of transitionals, smaller males sizes, and larger female sizes inside the reserve compared to outside (Figs 5a, b).

Male and female prawn sizes were found to be significantly smaller in the 1996-1997

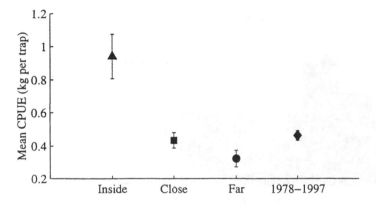

Figure 3. Mean catch-per-unit-effort (kilograms per trap) for inside the Carmel Bay Ecological Reserve and two locations outside the reserve, close to the boundary and far from the reserve, compared to the mean from 1978-1997. Bars are standard error. Historical data from the California Department of Fish and Game.

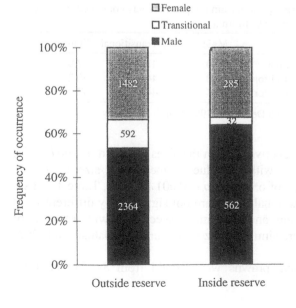

Figure 4. Sex ratios for spot prawns outside and inside of the Carmel Bay Ecological Reserve, California. Numbers indicate number of prawns sampled.

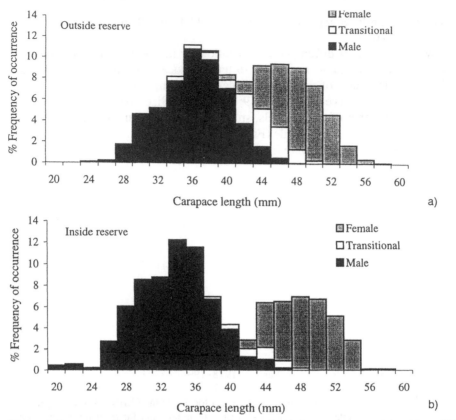

Figure 5. Frequency distribution of carapace lengths (mm) for spot prawns a) outside and b) inside of the Carmel Bay Ecological Reserve, California. (N values: Outside reserve M = 2364, T = 592, F = 1482; Inside reserve M = 562, T = 32, F = 283).

Table 2. Mean carapace lengths (mm) for spot prawns from the present study compared to past observations on spot prawn carapace lengths in the California area.

Authors	Location	Male	Transitional	Female
Dahlstrom 1963	Monterey, California	37.0	41.8	47.9
CDFG 1967-68	Carmel Canyon, California	39.0	44.1	50.2
Present study 1996-97*	Carmel Canyon, California	37.4	44.0	47.9

*Mean carapace lengths for data collected from December 1996-March 1997 only.

samples (37.4 mm and 47.9 mm, respectively) than the sizes recorded in 1967-1968, (39.0 mm and 50.2 mm, respectively) with a t-value for mean male size of –6.586, and a U-value for median female size of 65854.0 ($p < 0.001$ for both, Table 1, Figs 6 a, b). However, the 1996-1997 transitional CLs were not significantly different from their 1967-1968 counterparts (44.0 mm and 44.1 mm, respectively) with a t-value of –0.612 ($p = 0.541$). Overall CLs were similar to sizes measured by Dahlstrom (1963) for all sexes (Table 2).

The majority (60-90%) of female prawns were gravid from September 1996 through March 1997 (Fig. 7).

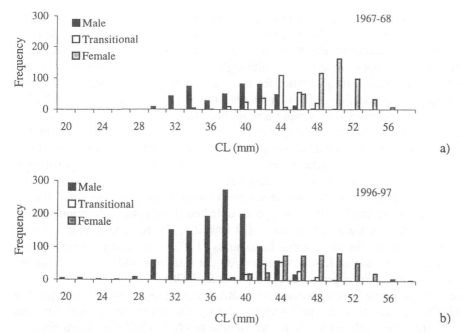

Figure 6. Carapace length frequencies (mm) for male, transitional, and female spot prawns from the Carmel Canyon trap fishery for a) December 1967-March 1968 and b) December 1996-March 1997. Historical data from the California Department of Fish and Game. (N values: 1967-1968 M = 410, T = 278, F = 472; 1996-1997 M = 1232, T = 172, F = 447).

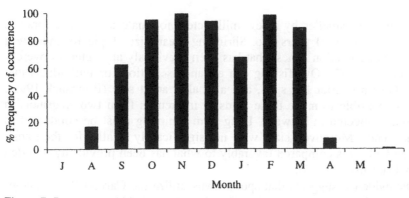

Figure 7. Percent gravid female spot prawns from Carmel Canyon trap fishery, shown by month (July 1996-June 1997). (N = 1754).

4 DISCUSSION

The significantly larger catch-per-unit-effort observed within the reserve compared to both outside locations is one indication that the reserve may indeed be functioning for the benefit of the spot prawn resource. Conversely, the gradient in CPUE from far to close to inside the reserve could simply be induced by environmental factors changing with distance offshore, and not reserve effects. If this were the case, one would expect to see similar size trends for all sexes of prawns. However, carapace lengths were significantly smaller inside of the reserve for the male prawns, while the trend was directed in the opposite direction, although not significant for female lengths, and lengths were significantly smaller for transitionals close to the reserve than for either inside or far locations. These patterns suggests an underlying distribution mechanism more complex than simply a near shore to offshore gradient of nutrients or other physical factors such as temperature and salinity. Further study is needed to resolve the factors in question and to determine how much influence the reserve has on the elevated CPUE and whether reserve spillover is responsible for supporting the high CPUE close to the boundary of the reserve.

There was a significant difference in sex ratio noted inside and outside of the reserve. According to Sunada (1986), a typical, unfished, hermaphroditic population is expected to exhibit a male to female ratio of at least 2:1. The male to female sex ratio was close to expected inside reserve boundaries (1.97:1) but much lower outside (1.60:1). The lower male to female ratio outside the reserve could potentially be related to commercial fishing pressures (Charnov et al. 1978).

Fishers in California have expressed reservations regarding the need for sequential seasonal closures for trawlers and trappers during the ovigerous season for spot prawns. There is currently a closure for trawlers in southern California from November 1-January 31 (CDFC 1997). The fishers contended that such closures are purposeless as the prawns are gravid all year long. However, our gravid peaks correspond almost exactly to what was noted by Sunada (1986) and CDFG (1995) and are similar to peaks observed by Butler (1964) off the coast of British Columbia, Canada. These findings support the effectiveness of a seasonal closure as a useful management tool.

Carapace lengths are smaller by a few millimeters for male and female prawns than for prawns measured 30 years ago. Shrinking mean sizes typically indicate a gradual decrease in population fitness that has been previously attributed to overexploitation (Bohnsack 1993). Overfishing can enhance selection for individuals that mature earlier, have a shorter life span, and a smaller adult size (Bohnsack 1993). However, it is impossible to make these kinds of inferences from two snapshots in time without data collected in between. Long term sampling must be conducted to clarify possible trends. Moreover, sizes were not significantly smaller for the transitional prawns and all sizes compared favorably to what has been previously reported for this area by Dahlstrom (1963).

Although the evidence suggests that spot prawns utilize the Carmel Bay Ecological Reserve it is obviously difficult to quantify the effect of the reserve on the spot prawn resource. The boundary for the Carmel Bay Ecological Reserve is an arbitrary line drawn between significant geographic points on land, and may in fact have little spatial relation to the distribution of the spot prawn resource. Our data implies that the reserve could be acting as a refuge for the smaller male prawns and future studies should focus on looking for evidence of juvenile prawns in the shallower areas of Carmel Canyon. If the juveniles are using the shallows as a nursery ground, as we suspect they may be, it would lend further support that the reserve is working towards the benefit of the resource. Subsequently, establishing reserves in other submarine canyon heads could be an important part of the solution to resource management.

In spite of the increase in fishing pressure to the area, Monterey prawn trap landings have remained steady at around 4340-6803 kg per year and 1997 was, in fact, a successful year for the prawn trap fishers. CPUE was also relatively high for 1997 at 0.68 kg and CPUE tracked over the past 30 years shows no evidence of recent change in trends that could possibly be attributed to the sudden rise in landings. These factors indicate that the prawn resource in Carmel Canyon is currently in healthy condition, possibly aided by the influence of the reserve. However, past examples from prawn trap and trawl logbook data in southern California have shown drastic reductions in CPUE correlated with large increases in landings for 1984-1994 (CDFG 1995). Sunada (1984) also witnessed a drop in CPUE with a sudden increase in total catch for ridgeback prawns. Continued monitoring of this fishery will show how abundance is affected over time by persistently high or further increases in landings.

ACKNOWLEDGEMENTS

The authors would like to express our sincerest gratitude to the people involved with this project. J.Field was a tremendous help during field sampling in rough seas. G. Matsumoto, R. Sherlock, C. Braby, K. Rodgers-Walz, and B. Schlining contributed invaluable input on figures and editorial advice. Thank you to S. Service, R. Sherlock, and J. Geller for providing essential advice on statistical analyses. The primary author would like to thank her advisors M. Yoklavich, J. Nybakken, and J. Connor for their constructive criticism. We are also very grateful for financial support from the California Department of Fish and Game, the Monterey Bay Aquarium Research Institute, the Earl and Ethel Myers Oceanographic Trust, and the David and Lucile Packard Foundation. A special thank you to the Aliotti family of commercial prawn trappers for their gracious cooperation.

REFERENCES

Balsiger, J.W. 1981. A review of pandalid shrimp fisheries in the northern hemisphere: Proceedings of the International Pandalid Shrimp Symposium. *Sea Grant Report* 81: 1-335

Barr, L. 1973. Studies of spot shrimp, *Pandalus platyceros*, at Little Port Walter, Alaska. *Mar. Fish. Rev.* 35: 65-66.

Bohnsack, J.A. 1993. Marine reserves: They enhance fisheries, reduce conflicts, and protect resources. *Oceanus*, Fall 1993: 63-71.

Butler, T.H. 1964. Growth, reproduction and distribution of pandalid shrimps in British Columbia. *J. Fish. Res. Bd. Canada* 21: 1403-1452.

California Department of Fish and Game. 1995. *Draft Environmental Document: Spot prawn commercial fishing regulations.* State of California: The Resources Agency, Department of Fish and Game.

California Fish and Game Commission. 1997. *California Code of Regulations, Title 14. Natural Resources.* State of California: California Fish and Game Commission.

Charnov, E.L., Gotshall, D.W. & Robinson, J.G. 1978. Sex ratio: Adaptive response to population fluctuations in pandalid shrimp. *Science* 200: 204-206.

Dahlstrom, W. 1963. Cruise report 63-A–1, prawn-shrimp. California Department of Fish and Game Marine Resource Operations 3pp.

Sunada, J.S. 1984. Spot prawn (*Pandalus platyceros*) and ridgeback prawn (*Sicyonia ingentis*) fisheries in the Santa Barbara Channel. *CalCOFI Rep.* 15: 100-104.

Sunada, J.S. 1986. Growth and reproduction of spot prawns in the Santa Barbara Channel. *Calif. Fish and Game* 72: 83-93.

Sunada, J.S., & Richards, J.B. 1992. Crustacean resources: spot prawn. In W.S. Leet, C.M.Dewees, & C.W. Haugen (eds), *California's Living Marine Resources and Their Utilization*: 10-11. California Sea Grant Publication UCSGEP–92-12.

Remarks on distribution and fishery biology of some *Plesionika* species (Decapoda, Pandalidae) in the southern Adriatic basin (Mediterranean Sea)

ROBERTA MARSAN, NICOLA UNGARO, CHIARA A. MARANO &
M. CRISTINA MARZANO
Laboratorio di Biologia Marina, Molo Pizzoli (Porto), Bari, Italy

ABSTRACT

Two trawl surveys were carried out during the summers of 1996 and 1997 in the South Adriatic area (MEDITS Project – Mediterranean Trawl Survey, sponsored by European Union – DG XIV) in order to assess demersal resources. For the first time the whole southern Adriatic Sea (Italian waters and Albanian waters) was investigated using the same vessel and sampling gear. Five different species of *Plesionika* were found among the crustaceans: *P. heterocarpus*, *P. martia*, *P. gigliolii*, *P. acanthonotus*, and *P. edwardsii*. *P. heterocarpus* and *P. martia* represented more than 95% of the total species catch from the surveys; *P. acanthonotus* and *P. edwardsii* were rarely found. *P. heterocarpus* was caught from 113 to 535 m depth; *P. martia*, *P. gigliolii*, *P. edwardsii* and *P. acanthonotus* below 300 m depth. The most abundant species, *P. heterocarpus* and *P. martia*, had length (carapace length, CL in mm) frequency distribution ranges (whole sampling area) of 10-17 mm and 11-26 mm, respectively. Ovigerous females of *P. heterocarpus* and *P. martia* were found to exceed 11 mm CL and 13.7 mm CL, respectively. Highest yields (kg/km^2) for the most abundant species were found on southeastern bottoms of the investigated area.

1 INTRODUCTION

The genus *Plesionika* represents an important fraction (because of its fairly good commercial value) of South Adriatic trawl fishery resources (Petruzzi et al. 1988, Pastorelli et al. 1996). Different species of *Plesionika* are usually included, without identification, within the marketable group 'shrimps'. During the summers of 1996 and 1997 two trawl surveys were carried out in the southern Adriatic Sea (Italian waters and Albanian waters) (MEDITS Project – Mediterranean Trawl Survey, sponsored by European Union – DG XIV) in order to assess demersal resources. For the first time the whole area was investigated using the same vessel, sampling gear and methodology (Bertrand 1996).

Survey data improved the knowledge of species distribution and relative abundance on trawlable bottoms of the said mentioned basin. In this paper, the authors focus on the analysis and added information on spatial and bathymetric distribution of various species of *Plesionika* in relation to their biology.

2 MATERIALS AND METHODS

Biological data about *Plesionika* were collected during June-July surveys using a 10 mm cod-end trawl net. Some 112 stations were sampled from 10 m to 800 m depth (Fig. 1). The sampling design of the first survey was randomly stratified (five bathymetric strata: 10-50 m, 51-100 m, 101-200 m, 201-500 m, 501-800 m depth) (Bertrand 1996), and selected points were re-sampled the following year.

The species of *Plesionika* sampled were identified, counted, and weighed. Representative sub-samples of each species were measured (carapace length, CL) to the nearest 0.1 mm, according to sex. Samples of ovigerous female were taken to estimate the relation between egg number (total number of newly laid eggs) and carapace length. All eggs carried by sampled females were counted using a binocular microscope. To calculate egg number for all ovigerous specimens, volumetric extrapolation was used. The *Plesionika* catch data were analyzed to estimate biomass yields (kg/km^2) (Bertrand 1996). Carapace length data produced length frequency distribution per species, according to sex. Correlation (Pearson parametric coefficient and Spearman non-parametric coefficient) and regression parameters were calculated between mean carapace length (for each sample) and depth, egg number, and carapace length.

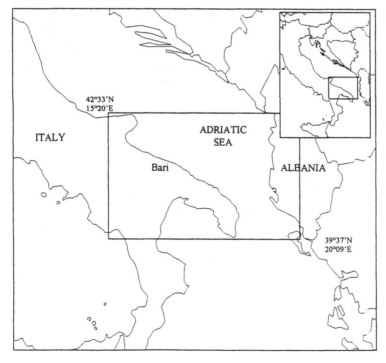

Figure 1. Investigated area.

3 RESULTS AND DISCUSSION

Five species of *Plesionika* were collected and identified during the two surveys, between 113 m and 565 m depth, in the southern Adriatic Sea: *Plesionika acanthonotus, P. edwardsii, P. gigliolii, P. heterocarpus,* and *P. martia.* Seven species of this genus have been reported for the Mediterranean Sea (also *Plesionika antigai* and *Plesionika narval*) (Falciai & Minervini 1992). The species identified were the same as found in similar surveys carried out in other Mediterranean basins (Relini et al. 1986, Mura 1987, Abellò et al. 1988, Righini & Auteri 1989, Company & Sardà 1997, Rinelli et al. 1998, Carbonell & Abellò in press, Politou et al. in press). Relative depth ranges for each species caught in the southern Adriatic Sea (Table 1) were similar to previously published observations. A small difference was found with *P. heterocarpus* distribution: During our surveys the species was found only below 100 m depth, while the upper limit in Spanish Mediterranean waters was shallower (Abellò et al. 1988).

The following were observed with respect to abundance and biomass of each species (Fig. 2). 1. *Plesionika acanthonotus* was found in two samples off Albania only (one in MEDITS '96 and the other in MEDITS '97); relative abundance of this species was low, with 0.02% and 0.03% of all *Plesionika* in 1996 and 1997, respectively. 2. *Plesionika edwardsii* was found in one sample (two specimens) off Italy during MEDITS '96 only, with a relative abundance of 0.2%. 3. *Plesionika gigliolii* was mostly found off Albanian with abundance of 1.98% and 0.12% in '96 and '97, respectively. 4. *Plesionika heterocarpus* was one of the most abundant and frequent species, distributed over the entire area, at around 200-500 m in depth and in the southeastern surveyed zones. Relative abundance was of 37.64% and 13.58% in MEDITS '96 and '97, respectively. 5. *Plesionika martia* was the most abundant species caught in the investigated area and was distributed mostly below 500 m depth and in the southeastern surveyed zones. Relative abundance was 60.26% and 86.26% in MEDITS '96 and '97, respectively.

The abundance and vertical distribution of *P. heterocarpus* and *P. martia* agree with the observations made by Carbonell & Abello (in press) and Company & Sardà (1997) in Spanish Mediterranean waters. *P. martia* and *P. heterocarpus* together represented more than 95% of total catch of *Plesionika* while other species had relatively low abundance.

It was possible to establish length-frequency distribution by sex, carapace length – eggs number relationship and mean size – depth correlation for these two main spe-

Table 1. Species of *Plesionika* – depth range and number of specimens caught.

Species	Italian waters depth (m)		n°	Albanian waters depth (m)		n°	Total area depth (m)		n°
	min	max		min	max		min	max	
Plesionika martia	364	565	2421	510	560	1990	364	565	4411
Plesionika heterocarpus	113	535	808	292	521	1775	113	535	2583
Plesionika edwardsii	366	366	2	–	–	–	366	366	2
Plesionika gigliolii	399	399	1	296	560	88	296	560	89
Plesionika acanthonotus			–	298	310	8	298	310	8

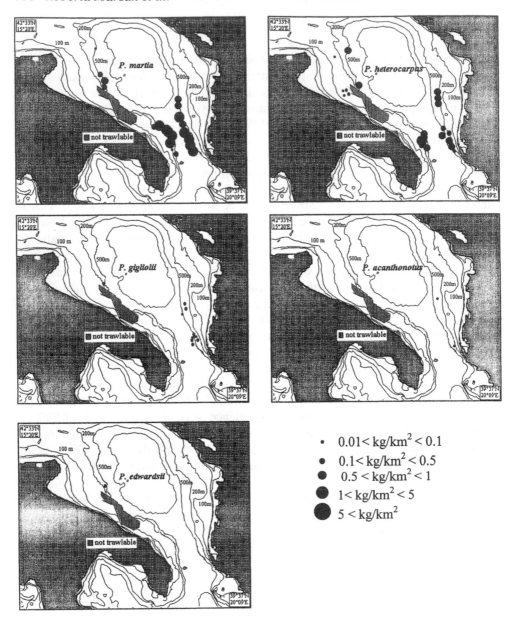

Figure 2. Catch yield (kg/km²) distribution of species of *Plesionika* from the MEDITS surveys (mean values between the surveys).

cies only. *P. martia* carapace length ranged from 11-26 mm (mostly between 18-20 mm), with the largest shrimp being collected off Italy (Fig. 3). *P. heterocarpus* carapace length ranged from 10-17 mm on the southern Adriatic trawlable bottoms (survey 1997); the length distribution referring to Italian and Albanian waters showed overlap (modal lengths at 12-14 mm) (Fig. 4). The largest *P. martia* tended to be females.

Southern Adriatic Sea (whole area)

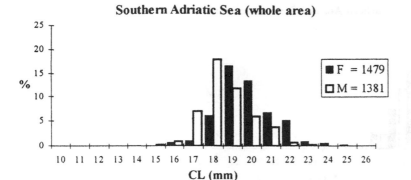

South-western Adriatic Sea (Italian waters)

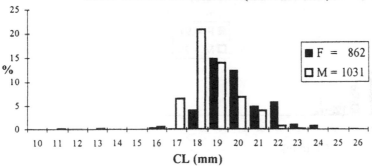

South-eastern Adriatic Sea (Albanian waters)

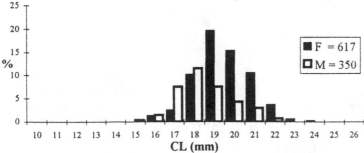

Figure 3. *Plesionika martia* length frequency distribution in southern Adriatic Sea (sub-samples from 1997 survey).

The length range of *P. martia* and *P. heterocarpus* seems to be smaller with respect to other authors' observations (Carbonell & Abello in press, Company & Sardà 1997), but this last result could be affected by our sample size. Mean size of *P. heterocarpus* was significantly correlated ($p < 0.05$) with depth both by using the Pearson ($r = 0.67$) and the Spearman coefficient ($r = 0.58$), as already found by other authors (Company & Sardà 1997). *P. martia* didn't show the same significant correlation, probably because of the restricted bathymetric range that was sampled (maximum sampling depth in our surveys $\cong$ 600 m). In fact, a significant correlation was ob-

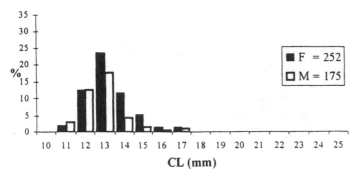

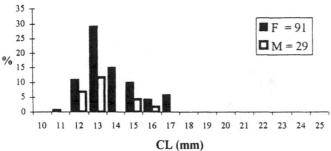

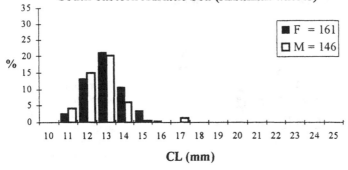

Figure 4. *Plesionika heterocarpus* length frequency distribution in Southern Adriatic Sea (sub-samples from 1997 survey).

served between mean size and depth for the same species by Company & Sardà (1997) over a wider depth range (150-1100 m depth).

The sampling occurred during the wide reproductive season of *P. martia* and *P. heterocarpus* (Falciai & Minervini 1992, Company & Sardà 1997), and ovigerous females were found above 13.7 mm CL and 11 mm CL for *P. martia* and *P. heterocarpus,* respectively. The former size is lower in comparison with other Mediterranean areas (Carbonell & Abelló in press, Company & Sardà 1997).

The following data on potential fecundity were used to determine size (CL) to egg number relationships for each of the two species. Linear regressions are presented because of their highest determination coefficients (0.69 for *P. heterocarpus*; 0.70 for *P. martia*) with respect to other models:

P. *martia* – eggs number = –8008 + 719.33 × CL(mm).
P. *heterocarpus* – eggs number = –3958 + 426.40 × CL(mm).

4 CONCLUSIONS

In this paper, additional information about the demersal crustacean fauna of the southern Adriatic basin is reported (Petruzzi et al. 1988, Vaso & Gjiknuri 1993, Pastorelli et al. 1996), particularly for the genus *Plesionika* for the first time.

Preliminary results about biomass index (kg/km^2) distribution patterns, potential fecundity, and stock length structure (referring to the most abundant species) could be useful for fishery management purposes, considering the commercial value of the species and the local fishery effort. However, the depth distribution of the most abundant species (most of specimens live below 200 m depth) and the ovigerous female minimum size (11-14 mm CL) probably prevent stock overexploitation, since the local demersal fishery uses trawl-net and the minimum authorized mesh size for the net is 40 mm.

REFERENCES

Abellò, P., Valladares, F.J. & Castellón, A. 1988. Analysis of the structure of decapod crustacean assemblages off the Catalan coast (North-West Mediterranean). *Mar. Biol.* 98: 39-49.

Bertrand, J. 1996. Campagne internationale de chalutage démersal en Méditerranée (MEDITS). Campagne 1996. Rapport final. *Rapport de contrat CEE-IFREMER-IEO-SIBM-NCMR (MED/95/19/54/65/27)*, Vol. 1-2: 1-71 + annexes, 1-88 + annexes.

Carbonell, A. & Abelló, P. in press. Distribution characteristics of pandalid shrimps (Decapoda: Caridea: Pandalidae) along the western Mediterranean. *6th CCDM-Florence 1996.*

Company, J.B. & Sardà, F. 1997. Reproductive patterns and populations characteristics of five deep-water pandalid shrimps in the western Mediterranean along a depth gradient (150-1100 m). *Mar. Ecol. P.S.* 148: 49-58.

Falciai, L. & Minervini, R. 1992. *Guida dei crostacei decapodi d'Europa.* Padua: Franco Muzzio Editore.

Mura, M. 1987. Crostacei Decapodi batiali della Sardegna meridionale. *Rend. Sem. Fac. Sci. Univ. Cagliari,* 57: 189-199.

Pastorelli, A.M., Vaccarella, R., Marano, G. & Ungaro, N. 1996. I crostacei dei fondi strascicabili del Basso Adriatico. *Nova Thalassia* 12: 27-35.

Petruzzi, T., Pastorelli, A.M. & Marano, G. 1988. Notes on the distribution of commercial crustaceans in the southern Adriatic, Trawl-Survey 1985-86. *FAO Fish. Rep.* 394: 213-221.

Politou, C.Y., Karkani, M. & Dokos, J. in press. Distribution of decapods caught during MEDITS surveys in Greek waters. In *Assessment of demersal resources by direct methods in the Mediterranean and the adjacent Seas, Pisa (Italy) 18-21 March 1998.*

Relini, G., Peirano, A. & Tunesi, L. 1986. Osservazioni sulle comunità strascicabili del Mar Ligure Centro-Orientale. *Boll. Mus. Ist. Biol. Univ. Genova,* 52 suppl., 139-161.

Righini, P. & Auteri, R. 1989. Distribuzione batimetrica dei crostacei decapodi raccolti durante le campagne di pesca nel Tirreno meridionale. *Oebalia* XV-2 (n.s.): 763-767.

Rinelli, P., Spanó, N. & Greco, S. 1998. Distribuzione dei crostacei decapodi ed echinodermi dei fondi strascicabili del Mar Tirreno meridionale. *Biol. Mar. Medit,* 5: 211-217.

Vaso, A. & Gjiknuri, L. 1993. Decapod crustaceans of the Albanian Coast. *Crustaceana* 65: 390-407.

Early ontogeny of proventriculus and mandible in the mud crab, *Scylla serrata* (Decapoda: Brachyura), and its implication on larviculture

GILDA N. LOYA-JAVELLANA
Southeast Asian Fisheries Development Center, Aquaculture Department, The Philippines

PHILIPPA J.R. UWINS
University of Queensland, Center for Microscopy and Microanalysis, St. Lucia, Australia

ABSTRACT

The morphological transformation of the proventriculus and the mandible of *Scylla serrata*, from the newly emerged zoea to the megalopa stage, was examined by light and scanning electron microscopy. The early zoea larva, compared with the later zoea and megalopa larvae, exhibited delayed and poor development of the gastric mill and gland filter. The mandibular gnathal lobe is generally complete in the zoea, but the molar ridge is lacking in that of the megalopa. In addition to the timing of appearance and development of the gastric mill and the gland filter, the degree of sclerotization of the teeth, setation, and size of the cardiac chamber are amongst the major ontogenetic changes that occurred in the proventriculus during larval development of *S. serrata*. Such changes have implications on stage-specific feeding abilities, behavior, and culture of the larvae.

1 INTRODUCTION

In the mud crab, *Scylla serrata*, as in other brachyurans, the larva that emerges from the egg is called the zoea. Five zoeal stages occur for 12-14 days, after which the larva metamorphoses into a megalopa that persists for another 8-10 days. The megalopa metamorphoses then into a first crab instar, which undergoes juvenile development prior to becoming an adult.

In some decapod crustaceans, changes in habitats and/or feeding habits during postembryonic development accompany morphological transformations in the mouth parts (e.g., Lavalli & Factor 1995, Loya-Javellana & Fielder 1997) and the foreguts (e.g., Factor 1989, 1995, Icely & Nott 1992, Loya-Javellana et al. 1994). The foreguts at the early postembryonic stages of development have been examined in few brachyurans (e.g. *Portunus pelagicus*: Shinkarenko 1979, *Menippe mercenaria*: Factor 1982, *Ranina ranina*: Minagawa & Takashima 1994), and data were rarely discussed in the context of larviculture.

Scylla serrata is economically important in the western Indo-Pacific region, but expansion of its culture is hindered by the unreliable supply of seed from the wild. To date, hatchery production of seed is yet far from becoming commercially viable.

Several factors that may affect survival rates have been identified, and feeding is one that needs further attention (Latiff & Musa 1995). The effects of various diets on larval growth, survival, and development have been studied in several crab species over the past decades (e.g. Brick 1974, Sulkin 1975). However, optimization of feeding is yet difficult to achieve based on these indices alone. Biological parameters directly related to feeding should be investigated in order to understand the specific feeding requirements of the mud crab larvae. This understanding is essential in developing a reliable feeding regime for their culture.

The present study describes the morphological transformation of the foregut, with notes on that of the mandible, of *S. serrata* from the newly emerged zoea 1 to megalopa larva. The implication of such developmental changes on larviculture is discussed.

2 MATERIALS AND METHODS

All larvae were obtained from the experimental mass rearing runs in 1996-1997 at the Southeast Asian Fisheries Development Center – Aquaculture Department (SEAF-DEC-AQD) (Iloilo, Philippines). The females that produced the larvae were taken either from the wild or the broodstock held in captivity. The samples were starved for 1-6 h, depending on size, to clear the gut prior to histological and scanning electron microscopy preparation.

2.1 *Light microscopy*

Whole animals at all larval stages were fixed in buffered (pH 7.2) formalin overnight, then rinsed with distilled water for approximately 10 minutes. Specimens were dehydrated through a graded series of alcohol (70-100%) and toluene. They were embedded in paraffin (Paraplast) at 58°C, serially sectioned at 6 μm (transversal, longitudinal and frontal), and routinely stained with Mayer's haematoxylin and eosin. Photomicrography was accomplished using a compound microscope Olympus CH–2, equipped with an automatic exposure photomicrographic system.

2.2 *Scanning electron microscopy*

Freshly dissected larvae that were frozen previously and then thawed, were examined and photographed with an ElectroScan Environmental Scanning Electron Microscope (c/o Center for Microscopy and Microanalysis, University of Queensland) operated at 10-15 kV.

3 RESULTS

The foregut of the larvae consists of an esophagus and a proventriculus that is further divided into an anterior cardiac and a posterior pyloric chamber. The cardiac and the pyloric chamber house the gastric mill (except in newly emerged zoea 1 stage) and the gland filter, respectively, which are the main digestive organs in the proventriculus. The gastric mill is a complex of calcified teeth, which triturate the food that enters the cardiac chamber. The gland filter screens the chyme that enters the ventral pyloric region prior to reaching the hepatopancreas. Other notable structures include the cardiopyloric valve at the junction of the cardiac and pyloric chambers, and the setal screen on the ventral floor of the cardiac chamber. The mandible, which is the mouth part consisting of a pair of calcified gnathal lobes and a pair of mandibular palps, guards the opening of the esophagus into the buccal cavity.

The proventriculus of the megalopa larvae is more complete compared with that of the zoea larvae. Hence, descriptions of that (and the mandible) of the megalopa are presented first, followed by those of the zoea stages.

3.1 *Megalopa*

The proventriculus of the megalopa is larger and more complex, compared with that of the zoea. It is already equipped with intricate patterns of ossicles, setae, valves, plates, and folds. The gastric mill, which is fully developed and heavily sclerotized, consists of a conspicuous median tooth and a pair of lateral teeth. The median tooth has a bicuspid anterior portion that becomes a rectangular molar-like plate posteriorly and ends with tooth-like setae. Cuticular folds and short conical setae are already present at both sides of the median tooth. The lateral teeth (Plate 1a) are conspicuously differentiated into some robust denticles ventrally and some conical teeth (Plate 1b) dorsally. There are many protuberances on the surface of the denticles.

The gland filter (Plate 1c), which is a pair of hemispherical structures characterized by an intricate pattern of filtering setae along several rows of longitudinal channels that are supported by an interampullary ridge, is already developed. The convex filter press, which complements the concavity of the gland filter, appears with a thick layer of muscles and considerable setation. The pyloric folds that guard the opening of the route from the dorsal pyloric channel to the gland filter are developed. The pair of ventral channels that connects the gland filter to the ventral floor of the cardiac chamber is already equipped with a rigid setal screen.

The cardiopyloric valve is prominent, setose, truncated, and in a conspicuous dorso-ventral position juxtaposed with the median tooth. Its medial portion is somewhat extended posteriorly in the pyloric region, bearing strong conical setae that are directed dorsally.

The mandible (Plate 1d, e) of the megalopa has a pair of gnathal lobes with no molar ridge. Each lobe has a prominent oral incisor ridge formed as a concave and round plate possessing a median, a dorsal, and a ventral tooth. The inner surface of the gnathal lobe has many spinules. The mandibular palp (Plate 1e) is segmented and well developed.

3.2 *Zoea*

The proventriculus of the early zoea is markedly simpler and smaller compared with that of the later zoea and megalopa stages. The gastric mill is lacking in the newly emerged zoea 1 larva. At late zoea 1, the gastric mill appears as vestigial cuticular thickenings that become slightly sclerotized at zoea 2 stage. The cardiac proventriculus is conspicuously setose already in these early zoea stages. At zoea 3, the median and lateral teeth of the gastric mill appear differentiated and sclerotized considerably (Plate 2a). Its median tooth is bicuspid anteriorly and extends as a somewhat rectangular plate posteriorly. The robust ventral and the conical dorsal teeth of the lateral teeth are already apparent. The gastric mill teeth increase in size and sclerotization as the larva progresses from zoea 3. Setation in the walls and the floor of the cardiac proventriculus increases as the larva progresses.

The gland filter is poorly developed in the zoea 1 larva. It appears to have no setal armature along the few longitudinal canals. However, small pyloric folds and very few short, spine-like setae borne upon the surface of the filter press that complements the gland filter are already present. A simple setal armature starts to appear and the longitudinal canals start to increase in the gland filter of the zoea 2 larva. Setation increases in number, and the pyloric folds and the muscular filter press increase in size as the larva progresses from zoea 2 stage.

A small and dorsally setose cardiopyloric valve is present in zoea 1 larva. Thenceforth, it increases in size and setation with each stage. Its dorsal setae become teeth-like as the larva progresses.

The mandible (Plate 2b) of the zoea has a pair of gnathal lobes with a molar and an incisor ridge. The molar ridge is equipped with protuberances and denticles, whereas the incisor ridge is lined orally with pointed teeth. These ridges increase in size and number of denticles and teeth with each zoea stage. The segmented mandibular palp is lacking.

4 DISCUSSION

The lack of knowledge of the feeding biology of mud crab larvae is probably the major hindrance in establishing a reliable feeding regime during larviculture. Much information on the relative food value of various laboratory diets for different crab larvae is available (e.g., Johns et al. 1981, Sulkin 1975). However, knowledge of the functional implication of the structure of the mouth parts and foreguts during early ontogeny is essential in understanding the stage-specific feeding abilities and behavior of the mud crab larvae.

The general structure of the mandible and the foregut components of the zoea and megalopa of *S. serrata* resembles those of the other crab larvae that have been examined (see Factor 1989). The zoeal mandible consists of a pair of gnathal lobes with an incisor ridge equipped with pointed teeth on the oral edges, and an inner molar ridge with protuberances. The megalopal gnathal lobes are markedly plate-like and concave, have no molar part, but have an incisor ridge equipped with a blade-like edge and one acute medial tooth. These indicate that food capture, piercing and crushing or mastication of food can be accomplished by the zoea using its mandibles. The megalopa mandible loses its masticatory ability but retains its capture and piercing abilities, while additionally possessing some cutting function. The pointed, incisor-like and molar-shaped gastric mill teeth, cardiopyloric valve, and/or intricate setal armature and folds of the cardiac and pyloric chambers constitute effective trituration and filtration systems for the food that enters through the esophagus.

The timing of appearance and development of the gastric mill and gland filter in the foregut is variable among the crab species that have been examined. For example,

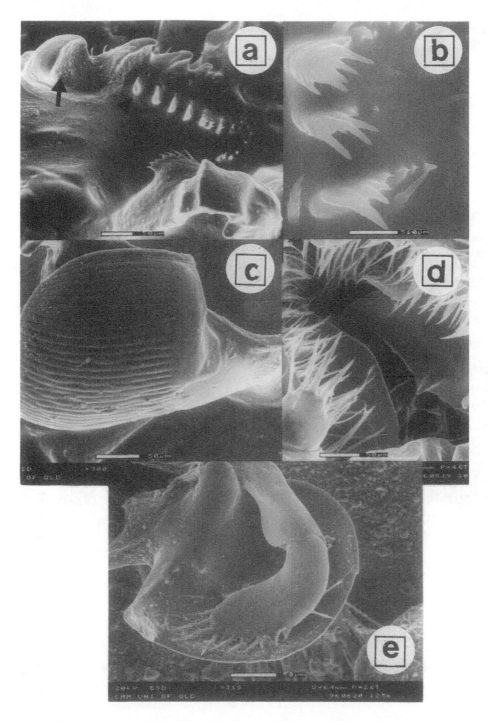

Plate 1. Proventriculus and mandible of *Scylla serrata* megalopa larvae. a. Lateral teeth showing the robust denticles (arrow) and conical setae (arrowhead). b. Closer view of the conical setae of the lateral teeth. c. Gland filter. d. Mandibular gnathal lobe external surface surrounded by other mouth-part appendages. e. Mandibular gnathal lobe inner surface showing the well-developed mandibular palp.

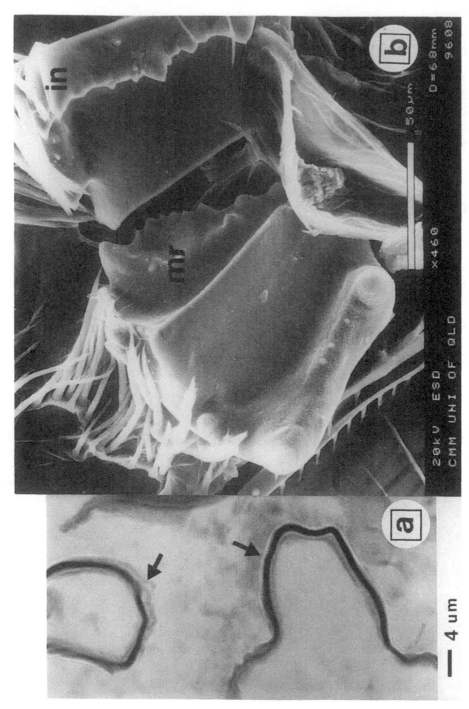

Plate 2. Gastric mill teeth and mandible of *Scylla serrata* zoea larvae. a. Sclerotized teeth of zoea–3 larvae. b. Mandibular gnathal lobe showing the teeth-edged incisor ridge (in) and developed molar ridge (mr).

in *M. mercenaria* (Factor 1982) and *P. pelagicus* (Shinkarenko 1979), no gastric mill was observed in all the zoea stages; in *R. ranina* (Minagawa & Takashima 1994), gastric mill dentition is well developed in all the zoea stages (except for the vestigial median tooth in the zoea 1 stage). In *S. serrata* (present study), gastric mill teeth are vestigial at the late zoea 1 stage, and become slightly sclerotized at zoea 2 stage. Increase in sclerotization, size and differentiation of the teeth starts to be conspicuous only at zoea 3 stage, and continues with successive stages. Such delay seems to be accompanied by the late development of the setal armature of the gland filter, which starts to appear at zoea 2 stage. In *R. ranina* (Minagawa & Takashima 1994) and in other decapod postembryonic stages that have been examined (e.g., lobster, Factor 1981; penaeid, Lovett & Felder 1989; crayfish, Loya-Javellana et al. 1994), early appearance of the setal armature of the gland filter was noted. However, the gland filter of the caridean shrimps *Crangon septemspinosa* and *Palaemonetes varians* (Le Roux 1971, Regnault 1972: cited from Lovett & Felder 1989) appears at the late larval and second larval stage, respectively.

In the zoea 1 larvae of *S. serrata*, the presence of complete mandibular gnathal lobes and a poorly developed foregut indicates that physical breakdown of food is accomplished solely by the mandible. This somewhat limited capacity to process food mechanically may require small-sized food that does not need to be much triturated. Hence, suspended organic particles, soft or very minute zooplankton, and highly digestible artificial food are likely to be the most suitable for the zoea 1 larvae. The need for a readily available food, in which a quick processing by the mandible may be sufficient, is indicated. A rudimentary or less robustly built gastric mill is often associated with microphagous diet (Woods 1994).

During larval rearing attempts using *Artemia* nauplii and/or the rotifer *Brachionus plicatilis* as food, heavy mortality generally occurs during the first intermolt period or at first molting. Deficient feeding for the zoea 1 larvae and deterioration of water quality affect early larval survival. Nonetheless, relatively better survival of the early zoea larvae of *S. serrata* fed the rotifer alone, compared with those fed *Artemia* nauplii alone, was commonly observed during rearing attempts at SEAFDEC-AQD. This is probably due to better acceptability of the rotifer in terms of size, mobility and physical structure, compared with *Artemia* nauplii. Stages following zoea 2 may require larger and more complicated food types that can be mechanically processed efficiently by their well-developed mandible, gastric mill teeth and filtration system.

The delay in the development of the foregut during zoea 1 stage also implies that feeding is not as critical as that in the later zoea stages in which the mandible and the gastric mill are both fully developed. Interestingly, this presumption is consistent with the finding that *S. serrata* larvae are able to tolerate relatively long periods of starvation within the first intermoult period (Mann & Parlato 1995). It was noted that starvation within the first 24 h following emergence does not result in significant mortality at first molting, and does not influence the time to commence molting to zoea 2 stage. At 28°C, zoea 1 larvae that were starved for 24 h following emergence and then were fed had better survival rate compared with those that were fed continuously. Such finding is in contrast with those in the king crab, *Paralithodes camtschaticus* (Paul & Paul 1980), *Menippe mercenaria, Panopeus herbstii, Neopanope sayi, Sesarma cinereum, Libinia emarginata* (Anger et al. 1981), the shore crab, *Carcinus maenas* (Dawirs 1986), and the red frog crab, *R. ranina* (Minagawa 1988) in which a short initial starvation was detrimental to survival and development. In *R. ranina*, relatively developed gastric mill teeth (i.e., lateral teeth) and gland filter are already present as early as zoea 1 (instar–1) stage (Minagawa & Takashima 1994), indicating that early feeding might indeed be critical in this species.

In the present study, the gastric mill and mandibular teeth of the later zoeal larvae are predominantly pointed which make them essentially zooplanktivorous, as has been generally established in the zoea larvae. In the megalopa larvae, the apparent cutting and piercing abilities of the oral edges of their gnathal lobes, and the presence of variously-shaped and -sized, well-sclerotized, robust gastric mill teeth make highly diverse benthic food resources suitable. With their assumption of a semi-benthic larval life, megalopa larvae are capable of accepting a wider array and larger sizes of solid food, compared with the planktonic zoea larvae.

In megalopae, the masticatory function of the mandible is shifted to the well developed, robust teeth of the gastric mill (see also Factor 1989). The concave plate-like shape of the gnathal lobes, absence of the inner molar ridge, appearance of the large-sized foregut and a developed mandibular palp in their mandible indicate a high rate of food intake or foregut filling required, suggesting the occurrence of peaks of feeding activity. Pulse-feeding behavior, as also speculated in that of *R. ranina* (Minagawa & Takashima 1994), may be a physio-ecological adaptation of the megalopa in response to their developing aggressive tendencies. Feeding may be regulated as a compromise between risk of predation and feeding requirements, which depend on the load or filling of the foregut and assimilated nutrients. An intermittent feeding schedule consistent with a potentially occurring feeding rhythm may be considered in rearing the megalopa larvae. Such strategy will minimize degradation of water quality and maximize food utilization. Qualitative analyses of the gastric fluid production during larval development and descriptions of the feeding rhythms in the megalopa larvae are potential areas of research that will enhance the understanding of the feeding biology of the mud crab larvae, and improve corresponding larviculture techniques.

ACKNOWLEDGMENTS

We thank the Southeast Asian Fisheries Development Center-Aquaculture Department for funding the project, the Center for Microscopy and Microanalysis (University of Queensland) for the use of their electron microscopes, and the Dutch Ministry of Foreign Affairs for funding the attendance of the senior author to the ICC–4. We appreciate the active participation of Danilo S. Javellana in rearing the larvae with success.

REFERENCES

Anger, K., Dawirs, R.R., Anger,V. & Costlow, J.D. 1981. Effects of early starvation periods on zoeal development of brachyuran crabs. *Biol. Bull. Mar. Biol. Lab. WoodsHole* 161: 199-212.
Brick, R.W. 1974. Effects of water quality, antibiotics, phytoplankton and food on survival and development of larvae of *Scylla serrata* (Crustacea: Portunidae). *Aquaculture* 3: 231-244.

Dawirs, R.R. 1986. Influence of limited starvation periods on growth and elemental composition (C,N,H) of *Carcinus maenas* (Decapoda:Portunidae) larvae reared in the laboratory. *Mar. Ecol. Prog. Ser.* 12: 301-308.

Factor, J.R. 1981. Development and metamorphosis of the digestive system of larval lobsters, *Homarus americanus* (Decapoda:Nephropidae). *J. Morphol.* 169: 225-242.

Factor, J.R. 1982. Development and metamorphosis of the feeding apparatus of the stone crab, *Menippe mercenaria* (Brachyura, Xanthidae). *J. Morphol.* 172: 299-312.

Factor, J.R. 1989. Development of the feeding apparatus in decapod crustaceans. *Crust. Issues* 6: 185-203.

Factor, J.R. 1995. The digestive system. In: J.R. Factor (ed.), *Biology of the Lobster Homarus americanus*: 395-440. San Diego: Academic Press.

Icely, J.D. & Nott, J.A. 1992. Digestion and absorption: digestive system and associated organs. In: F.W. Harrison (ed.), *Microscopic Anatomy of Invertebrates. Vol. 10, Decapod Crustcea*: 147-201. New York: Wiley-Liss Inc.

Johns, D.M., Berry, W.J. & McLean, S. 1981. International study on *Artemia* XXI. Investigations into why some strains of *Artemia* are better food sources than others. Further nutritional work with larvae of the mud crab, *Rithropanopeus harrissi. J. World Maricult. Soc.* 12: 303-314.

Latiff, F.A. & Musa, C.U. 1995. *The Biology of Mudcrab and Its Hatchery Production*. Department of Fisheries Malaysia: Ministry of Agriculture.

Lavalli, K.L. & Factor, J.R. 1995. The feeding appendages. In: J.R. Factor (ed.), *Biology of the Lobster Homarus americanus*: 349-393. San Diego: Academic Press.

Le Roux, A.A. 1971. Etude anatomique et fonctionelle du proventricule des larves et des juveniles du premier stade de *Palaemonetes varians* (Leach) (Decapode, Natantia). I. Données anatomiques. *Bull. Soc. Zool. Fr.* 96: 127-140.

Lovett, D.L. & Felder, D.L. 1989. Ontogeny of gut morphology in the white shrimp *Penaeus setiferus* (Decapoda, Penaeidae). *J. Morph.* 201: 253-272.

Loya-Javellana, G.N., Fielder, D.R. & Thorne, M.J. 1994. Ontogeny of foregut in the tropical freshwater crayfish, *Cherax quadricarinatus* von Martens, 1868 (Parastacidae: Decapoda). *Inv. Repr. Dev.* 25: 49-58.

Loya-Javellana, G.N. & Fielder, D.R. 1997. Developmental trends in the mouthparts during growth of juvenile to adult of the tropical freshwater crayfish, *Cherax quadricarinatus* von Martens, 1868 (Decapoda: Parastacidae). *Inv. Repr. Dev.* 32: 167-175.

Mann, D. & Parlato, D. 1995. Influence of starvation on larval growth and survival of the mud crab, *Scylla serrata* (Forskål). *Proc. Sci. Res. Workshop on Mud crabs*. Broome, October, 1995 AUTHOR-PAGES.

Minagawa, M. 1988. Influence of starvation on survival, growth, feeding success and external morphology of the larvae of *Ranina ranina* (Crustacea, Decapoda, Ranidae). *Suisanzoshoku* 36: 227-230.

Minagawa, M., & Takashima, F. 1994. Developmental changes in larval mouthparts and foregut in the red frog crab, *Ranina ranina* (Decapoda: Raninidae). *Aquaculture* 126: 61-71.

Paul, A.J. & Paul, J.M. 1980. The effect of early starvation on later feeding success of king crab zoeae. *J. Exp. Mar. Biol. Ecol.* 44: 247-251.

Regnault, M. 1972. Developpement de l'estomac chez les larvaes de *Crangon septemspinosa* Say (Crustacea, Decapoda, Crangonidae); son influence sure le mode de nutrition. *Bull. Mus. Natl. Hist. Nat. Zool.* 67: 841-856.

Shinkarenko, L. 1979. Development of the larval stages of the blue swimming crab, *Portunus pelagicus* L. (Portunidae: Decapoda:Crustacea). *Aust. J .Mar. Freshwat. Res.* 30: 485-503.

Sulkin, S.D. 1975. The significance of diet in the growth and development of larvae of the blue crab, *Callinectes sapidus* Rathbun, under laboratory conditions. *J. Exp. Mar. Biol. Ecol.* 20: 119-135.

Woods, C.M.C. 1995. Functional morphology of the foregut of the spider crab *Notomithrax ursus* (Brachyura: Majidae). *J. Crust. Biol.* 15: 220-227.

Observations on decapod crustaceans from trawlable bottoms in the southern Tyrrhenian Sea (western Mediterranean)

PAOLA RINELLI, DANIELA GIORDANO, FRANCESCO PERDICHIZZI & SILVESTRO GRECO
Istituto Sperimentale Talassografico CNR, Messina, Italy

ABSTRACT

We present preliminary data for the decapod fauna on trawlable bottoms of the southern Tyrrhenian Sea (western Mediterranean). The data come from four trawl surveys carried out, from 10 to 800 m of depth, in the ambit of the programme MEDITS, funded by EU. The bathymetrical distribution of each decapod and the abundance values of species that have a considerable commmercial value are reported. Moreover, the results concerning distribution and biology of *Aristeus antennatus, Aristaeomorpha foliacea, Parapenaeus longirostris,* and *Nephrops norvegicus* are presented. The spatio-temporal distribution of these species inside the area studied did not change during the four years of observation. On the contrary, the yields showed significant differences.

1 INTRODUCTION

This paper deals with the decapod fauna of trawlable bottoms of the southern Tyrrhenian Sea in the western Mediterranean. The taxonomic composition is reported for samples collected during four experimental trawl surveys carried out in the ambit of the program MEDITS: 'Mediterranean International Bottom Trawl Survey' funded by the European Union. The bathymetric distribution of each decapod and the abundance values of species that have a considerable commercial value are recorded. In this connection, a further analysis (size structure, sex-ratio, maturity stage) was carried out on four species that represent the most important commercial decapods for the demersal trawl fisheries operating in the southern Tyrrhenian Sea: the meso-bathyal aristeids *Aristeus antennatus* and *Aristaeomorpha foliacea,* the penaeid *Parapenaeus longirostris,* and the nephropid *Nephrops norvegicus.* There is still a lack of organic information about the distribution and biological characteristics of these valuable species in the investigated area (Arena & Bombace 1970, Arculeo et. al. 1992 a,b, Greco et al. 1994, Spedicato et al. 1995).

2 MATERIALS AND METHODS

A total of 111 hauls were sampled from 10 to 800 meters depth during four MEDITS

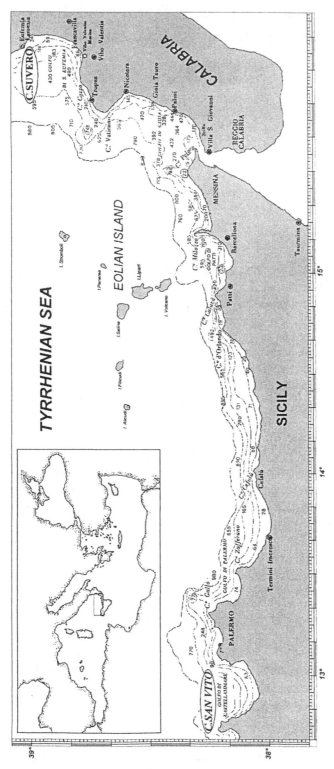

Figure 1. Sampling area.

trawl surveys carried out each June from 1994 to 1997. We applied a stratified random sampling frame (Bertrand et al. 1997), using a commercial vessel equipped with a net, type GOC 73, especially designed by IFREMER (Sète) for this project. According to the scientists involved, the selectivity of the trawl net should be as low as possible so as to have a complete picture of the population sampled, thus the mesh opening of the codend was limited to 20 mm (stretched mesh). The vertical opening of the net was 2.9 m and the horizontal one was 16 m. However, the complete design of the net is reported in the MEDITS report (Bertrand et al. 1996).

The area surveyed (Fig. 1) is located between Capo Suvero (Calabria) and Capo San Vito (Sicily) and has a total trawlable surface of 7256 km^2. The whole area was divided into five strata having the following depth range and surfaces: A: 10-50 m, 622 km^2; B: 50-100 m, 1003 km^2; C: 100-200 m, 1224 km^2; D: 200-500 m, 1966 km^2; E: 500-800 m, 2441 km^2.

The shrimp were weighed and measured on board. On fresh specimens, the 'oblique' carapace length (CL), from the posterior margin of the orbit to the posterior hind edge of the carapace, was recorded with a dial calipers, to the nearest millimeter. Sex was determined macroscopically looking for the presence of the thelycum or the petasma, respectively, for females and males. The degree of female sexual maturity was evaluated using an empirical 4-stage scale using different methods applicable to various species: *A. foliacea* (Levi & Vacchi 1988); *A. antennatus* (Relini Orsi & Relini 1979); *P. longirostris*, (de Ranieri et al. 1986); *N. norvegicus*. The sex ratio was expressed as females to males (F:M).

The yields were expressed as abundance index in weight (kg/km^2) and number (N/km^2). Length frequency distributions (LFD) by sex were analyzed by pooling together different years to average out the variability due to the sampling noise. The LFD of *N. norvegicus* are not shown because of the scarcity of the collected specimens.

3 RESULTS

Altogether 45 species of decapod crustaceans were found, the complete faunistic list with the bathymetric range of each species is reported in Table 1. Among the species collected, most were found in a wide bathymetrical range, and only 10 showed an infra-circalittoral distribution (0-120 m) and 17 were strictly bathyal (over 200 m depth).

The following results were obtained regarding the four most important commercial species, considered as 'target species' in the MEDITS programme.

3.1 *Aristeus antennatus*

The blue and red shrimp *A. antennatus* was caught between 500 and 800 meters depth. The total catch of the four surveys was 12.38 kg corresponding to 1318 individuals, coming from 28 hauls of which 14 were positive. The mean catch rate values in weight as well as in number of each survey (Table 2) ranged from 0.57 kg/km^2 with 30 N/km^2 (1995) to 10.17 kg/km^2 with 1276 N/km^2 (1997). The distribution of the species in the whole area was scattered; the areas located in the Gulf of Patti and in the waters in front of Capo Zafferano yielded the most. The latter area showed lower values than the former, in which the abundance was generally higher (maximum value: 78.8 kg/km^2 in 1997).

Table 1. List of species collected, with relative bathymetric distribution.

Species	Depth (m) min	max
Solenoceridae		
Solenocera membranacea	380	570
Aristeidae		
Aristeomorpha foliacea	500	800
Aristeus antennatus	500	800
Penaeidae		
Parapenaeus longirostris	100	800
Pasiphaeidae		
Pasiphaea multidentata	550	560
Pasiphaea sivado	550	800
Alpheidae		
Alpheus glaber	100	110
Pandalidae		
Plesionika heterocarpus	380	400
Plesionika antigai	440	460
Plesionika giglioli	350	350
Plesionika martia	550	560
Plesionika edwardsii	200	800
Crangonidae		
Pontocaris cataphracta	20	30
Pontocaris lacazei	200	350
Nephropidae		
Nephrops norvegicus	200	800
Polychelidae		
Polycheles typhlops	440	650
Diogenidae		
Paguristes eremita	60	70
Dardanus arrosor	30	60
Paguridae		
Anapagurus laevis	30	600
Pagurus alatus	540	550
Pagurus excavatus	20	550
Pagurus excavatus var. *meticolosus*	20	360
Pagurus prideaux	110	350
Galatheidae		
Munida intermedia	350	380
Dorippidae		
Medorippe lanata	30	350
Calappidae		
Calappa granulata	50	200
Latreillidae		
Latreillia elegans	80	260
Homolidae		
Homola barbata	110	125
Portunidae		
Bathynectes longipes	110	120
Bathynectes maravigna	550	560
Liocarcinus maculatus	90	110
Liocarcinus bolivari	60	70

Table 1. Continued.

Species	Depth (m)	
	min	max
Liocarcinus depurator	30	280
Macropipus tuberculatus	140	540
Geryonidae		
Geryon longipes	530	630
Xanthidae		
Pilumnus inermis	75	80
Pilumnus spinifer	60	70
Goneplacidae		
Goneplax rhomboides	70	540
Parthenopidae		
Parthenope macrochelos	120	300
Majidae		
Maja goltziana	180	300
Maja crispata	80	90
Pisa nodipes	80	650
Anamathia rissoana	540	550
Inachus dorsettensis	90	100
Macropodia longipes	40	540

Table 2. Catch frequency and mean catch rate values with relative grand mean and standard deviation (s.d.) of *Aristeus antennatus* by survey, in stratus E (500-800 m). 28 hauls, 14 of them positive, catch frequncy = 50%.

Year	N/km^2	Grand mean	s.d.	Kg/km^2	Grand mean	s.d.
1994	334			6.41		
1995	30	432	15.82	0.57	4.60	3.93
1996	88			1.23		
1997	1276			10.17		

The sex ratio (Table 3) was always strongly in favor of females. The most significance value (8.7; $\chi^2 = 116.8$ with $p < 0.05$) was recorded in 1994. The size structure observed during the four years of surveys was very similar; the carapace length ranged from 20 mm to 58 mm for females and from 18 mm to 32 mm for males. The length frequency distributions showed one modal size for males (Fig. 2a) corresponding to 26 mm. However, females (Fig. 2b) showed a polymodal pattern. Out of 1318 sexed animals, 861 were females of which the 45% were at the third stage of maturity (Table 4). Mature females were found all over the studied area.

3.2 *Aristaeomorpha foliacea*

The giant red shrimp *A. foliacea* showed an irregular distribution along the studied area and was caught predominantly in the Gulf of S. Eufemia and in the Gulf of Patti from 500 to 800 meters depth. The total catch was 49.67 kg with 3792 animals. Out of 28 hauls carried out in the deepest stratum, 26 were positive. The mean catch rate (Table 5) in the four years ranges from 527 to 2223 N/km^2 (grand mean = 1424; stan-

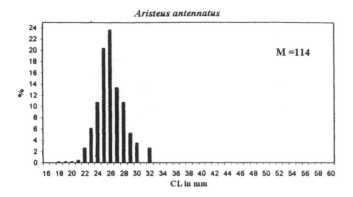

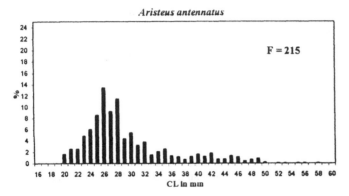

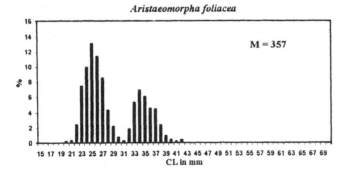

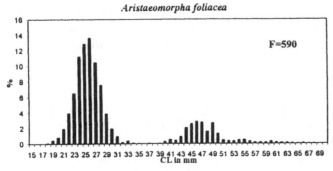

Figure 2. Length frequency distributions by sex (M: males; F: females), calculated as mean of the four years, of *Aristeus antennatus* and *Aristaeomorpha foliacea*.

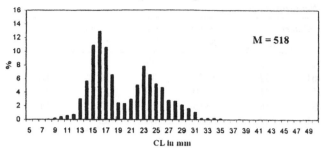

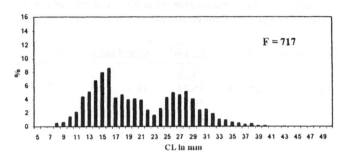

Figure 2. Continued.
Parapenaeus longirostris.

Table 3. Sex ratio of *A. antennatus*, *A. foliacea*, *P. longirostris* and *N. norvegicus*. In the second and third columns, the total number of females (F) and males (M) by survey is reported.

	M	F	RS (F/M)	$\chi 2$
	Aristeus antennatus			
1994	19	166	8.7	116.8
1995	0	21	–	–
1996	3	58	19.3	49.6
1997	435	616	1.4	123.8
	Aristaeomorpha foliacea			
1994	257	435	1.7	0.9
1995	422	761	1.8	97.1
1996	175	210	1.2	3.2
1997	577	955	1.7	93.3
	Parapenaeus longirostris			
1994	351	403	1.1	5.6
1995	551	769	1.4	36.0
1996	350	545	1.6	42.5
1997	822	1151	1.4	54.9
	Nephrops norvegicus			
1994	0	25	–	1.7
1995	10	14	1.4	0.7
1996	14	4	0.3	5.6
1997	18	12	0.7	1.2

Table 4. Total catches of females and maturity stages of *Aristeus antennatus, Aristaeomorpha folia-cea, Parapenaeus longirostris,* and *Nephrops norvegicus.*

Stage	A. antennatus		A. foliacea		P. longirostris		N. norvegicus	
	N°	%	N°	%	N°	%	N°	%
I	106	12	210	9	1119	39	7	13
II	283	33	761	32	1004	35	10	18
III	383	45	955	41	430	15	21	38
IV	89	10	435	18	315	11	17	31
Total	861	100	2361	100	2868	100	55	100

Table 5. Catch frequency and mean catch rate values with relative grand mean and standard deviation (s.d.) of *Aristaeomorpha foliacea* by survey, in stratum E: 500-800 m. 28 hauls, 26 with positive catch, catch frequency = 92%.

Year	N/km^2	Grand mean	s.d.	kg/km^2	Grand mean	s.d.
1994	1269			25.28		
1995	1677	1424	618.68	16.23	18.74	9.37
1996	527			10.29		
1997	2223			23.14		

dard deviation = 618.68) and from 10.29 to 25.28 kg/km^2 (grand mean = 18.74; standard deviation = 9.37). The highest value was recorded in 1997 in the Gulf of San Eufemia (70 kg/km^2). The sex ratio (Table 3) always tended to favor females with variable values comprised between 1.2 (1996) and 1.8 (1995). Carapace length showed homogeneous values in the four years: 18-68 mm for females and 19-42 for males. The size frequency distributions showed two modal classes corresponding to 26 and 35 mm for males (Fig. 2c) and 26 and 47 mm for females (Fig. 2d). 2361 females were recorded of which the 41% were at the third stage of maturity (Table 4).

3.3 *Parapenaeus longirostris*

The pink shrimp *P. longirostris* was caught from 100 to 800 m in depth. This species was widespread in the whole sampling area (catch frequency 75%). The total catch in four surveys was 23.30 kg corresponding to 4942 individuals coming from 80 hauls carried out between 100 and 800 m of which 61 were positive. The mean catch rate (Table 6) ranged between 9 and 2205 N/km^2 and from 0.12 to 8.3 kg/km^2. The highest yields were generally recorded from 100 to 200 meters of depth and the areas in which the species was more abundant are located in the Gulf of Patti, beside Capo Zafferano, and in the Gulf of Castellammare. The sex ratio was always over 1 and the values ranged from 1.1 (1994) to 1.6 (1996). With regard to the size structure of the populations, the females ranged from 8 to 43 mm and the males from 8 to 38 mm. The length frequency distributions (Fig. 2) for both sexes appeared to be polymodal. Fifty-eight percent of the collected animals were females of which 39% were at the first stage of maturity (Table 3).

3.4 *Nephrops norvegicus*

The Norway lobster *N. norvegicus* was captured from 200 to 800 meters of depth. This species showed very low yields in the whole area studied; a total catch of 4.93 kg, corresponding to 97 animals, was collected in 25 positive hauls out of 56 carried out in the two deepest strata. The mean catch rate values for each survey are reported in Table 7. The zones in which the species was more abundant were the Gulf of S. Eufemia and the Gulf of Patti. With regard to sex ratio the values were over 1 in 1995; in the last two years the males predominated (Table 3). As regards the sexual maturity of the 55 females specimens collected, 38% were at the third stage (Table 4).

Table 6. Catch frequency and mean catch rate values with relative grand mean and standard deviation (s.d.) of *Parapenaeus longirostris* by survey. Strata C, D, E. 80 hauls, 61 of them positive, catch frequency = 75%.

Year	N/km^2	Grand mean	s.d.	kg/km^2	Grand mean	s.d.
			100-200 m			
1994	437			3.15		
1995	1674	1392	646.55	5.52	5.36	2.34
1996	1252			4.45		
1997	2205			8.30		
			200-500 m			
1994	550			5.33		
1995	552	532	216.5	5.91	5.35	1.89
1996	208			2.44		
1997	818			7.70		
			500-800 m			
1994	181			2.08		
1995	77	79.5	63.43	0.85	0.93	0.72
1996	9			0.12		
1997	51			0.68		

Table 7. Catch frequency and mean catch rate values with relative grand mean and standard deviation (s.d.) of *Nephrops norvegicus* by survey. Strata D & E. 56 hauls, 25 of them positive, catch frequency = 45%.

Year	N/km^2	Grand mean	s.d.	kg/km^2	Grand mean	s.d.
			200-500 m			
1994	16			0.97		
1995	27	21.25	4.44	1.41	1.20	0.31
1996	18			1.6		
1997	24			0.83		
			500-800 m			
1994	34			1.75		
1995	7	12.75	12.32	0.23	0.71	0.61
1996	4			0.51		
1997	6			0.33		

4 DISCUSSION AND CONCLUSION

The spatio-temporal distribution of these species inside the entire area studied did not change during the four years of observation. However, the yields showed significant differences. On the whole, the abundance values for *A. antennatus* are not constant over the four year period, with two peaks in the samplings of 1994 and 1997; the abundance value, in the latter year, was four times that of 1994, and a good 43 times that of 1995. As to *A. foliacea*, the abundance values recorded were about twice those of *A. antennatus*. In addition, the maximum values for this species were those observed in 1994 and 1997. A remarkable yearly fluctuation was therefore observed in both aristeids. This phenomenon, which has not yet been fully explained, had already been reported in other areas of the Mediterranean, for example in the Ligurian Sea (Relini & Orsi Relini 1987) and the Gulf of Lion, where the red shrimp disappeared completely after 1984 (Campillo 1994). Furthermore, from the comparison with the data reported by nearby Italian working units, related to the same common program, an analogous fluctuation can be observed, with a decrease in the 1995 values (Bertrand et al. 1996). The size structure of *A. foliacea* indicated marked differences in growth between the sexes. The distributions showed that both females and males are present in the young classes, in the range from 34 to 40 mm, a mode composed prevalently of males appeared, and over 42 mm the distribution is composed only of females.

Regarding the pink shrimp, *P. longirostris,* the yields increased from 1994 to 1997 in the lowest strata (C, D) and decreased in the deepest one (E). It is worth noting that from 100 and 200 meters of depth the animals are smaller than in the deeper strata. Bathymetric segregation related to size was also recorded by other authors (Ribeiro-Cascalho & Arrobas 1987). The structure of the sampled populations showed a wide range of sizes related to the several size/age classes.

The Norway lobster did not show, during the four years, significant variations in abundance values from 200 to 500 meters depth, on the contrary in the stratum E from 1994, it decreased. On the whole, however, the remarkable scarcity of *N. norvegicus* in the sea bottom under observation is evident. However, in the nearby areas of the central-southern Tyrrhenian Sea and the Sicilian Channel (Bertrand et al. 1996) this species showed higher yields. It is, however, worth pointing out that as to the Norway lobster, a notorious hole-dweller, low values recorded can be attributed to the experimental net used in this project (Fiorentini, pers. comm.).

In conclusion, the preliminary data presented in this work, both on the specific composition of the decapod fauna and on the biology of demersal species with a high commercial value, should be considered as a first step bridging the information gap still existing in this area.

ACKNOWLEDGEMENTS

We wish to particularly thank Dr N. Spanò, from the University of Messina, for her help in classification of the decapod species.

REFERENCES

Arculeo, M., Baino, R., Abella, A. & Riggio, S. 1992a. Accrescimento e riproduzione di *Aristeus antennatus* (Crustacea, Decapoda) nel Golfo di Castellammare (Sicilia N/W). *Oebalia*, suppl. XVII: 117-118.

Arculeo, M., Galioto, G. & Cuttitta, A. 1992b. Aspetti riproduttivi in *Parapenaeus longirostris* (Crustacea, Decapoda) nel golfo di Castellammare (Sicilia N/W). *Biol Mar., suppl. Notiziario SIBM* (1): 307-308.

Arena, P. & Bombace, G. 1970. Bionomie benthique et faune ichthyologique des fonds de l'étage circalittoral et bathyal des golfes de Castellammare (Sicilie nord-occidentale) et de Patti (Sicilie nord-orientale). *Journées ichthyol., XXII CIESM ROMA*: 145-156.

Bertrand, J., Gil de Sola, L., Papacostantinou, C., Relini, G. & Souplet, A. 1996. Campagne internationale de chalutage démersal en Méditerranée (MEDITS 95). Campagne 1995. Rapport general (Vol.I), manuel des protocoles (VOL.II) et indeces de biomasse et distributions en tailles (Vol.III). Raport de contract CEE-IFREMER-IEO-SIBM-NCMR (etude 94/047, 011, 057, 051: 96 pp., 27 pp., 98 pp.).

Bertrand, J., Gil de Sola, L., Papacostantinou, C., Relini, G. & Souplet, A. 1997. An international bottom trawl survey in the Mediterranean: the MEDITS programme. *ICES CM 1997/Y*: 3-16.

Campillo, A. 1994. Bio-ecology of *Aristeus antennatus* in the French mediterranean. *N.T.R.-I.T.P.P. Special Pubblications* 3: 25-26.

De Ranieri, S., Biagi, F. & Mori, M. 1986. Note sulla biologia riproduttiva di *Parapenaeus longirostris* (Lucas) nel Tirreno settentrionale. *Nova Thalassia*, suppl. 3, 8: 627-628.

Greco, S., Perdichizzi, F., Spalletta, B., Capecchi, D. & Giordano, D. 1994. *Aristeomorpha foliacea* and *Aristeus antennatus* in the Southeastern Tyrrhenian Sea. *N.T.R.-I.T.P.P. Special Pubblications* 3: 37-38.

Levi, D. & Vacchi, M. 1988. Macroscopic scale for simple and rapid determination of sexual maturity in *Aristeomorpha foliacea* (Risso, 1826) (Decapoda, Penaeidae). *J. Crust. Biol.* 8: 532-538.

Relini Orsi, L. & Relini, G. 1979. Pesca e riproduzione del gambero rosso *Aristeus antennatus* (Decapoda, Penaeidae) nel Mar Ligure. *Quad. Civ. Staz. Idrobiol.* 7: 39-62.

Relini, G. & Orsi Relini, L. 1987. The decline of red shrimps stocks in the Gulf of Genova. *Inv. Pesq.* 51 (suppl. 1): 245-260.

Ribeiro-Cascalho, A. & Arrobas, I. 1987. Observation on the biology of *Parapenaeus longirostris* (Lucas, 1846) from the south coast of Portugal. *Inv. Pesq.* 51: 201-212.

Spedicato, M.T., Greco, S., Lembo, G., Perdichizzi, F. & Carbonara, P. 1995. Prime valutazioni sulla struttura dello stock di *Aristeus antennatus* (Risso, 1816) nel Tirreno centro meridionale. *Biol. Mar. Medit.* 2: 239-244.

Non-interactive coexistence of two parasitic copepods of *Caranx hippos* (Carangidae) in eastern Venezuela

ABUL K. BASHIRULLAH

Instituto Oceanográfico de Venezuela,Universidad de Oriente, Venezuela

ABSTRACT

One hundred and fifty common jack, *Caranx hippos* (Carangidae) belonging to three age groups were examined for parasitic copepods on the gills of each fish. Of the 1327 copepods recovered, 56.1% were *Lernanthropus giganteus* and 43.9% were *Caligus constrictus*. Prevalence, abundance, and distribution of each species on the gill arches are described. No significant difference in infestation was found between the two sides of the branchial chamber, but copepods preferred the second and third gill arches. Single occurrences of *L. giganteus* and *C.constrictus* were recorded in 22.55% and 17.65% of *C. hippos*, respectively, while joint occurrences were found in 77.45% of infected hosts. The association between the age of fish and the abundance of copepods was significant (P < 0.01). Fager index indicated positive affinity between *L. giganteus* and *C. constrictus*. Measure of niche breadth of *L. giganteus* was larger than for *C. constrictus,* and the niche overlap was markedly asymmetrical.

1 INTRODUCTION

Fish parasites tend to be poorer in terms of species diversity than the parasite community in birds (Kennedy et al. 1986), which is not universally factual (e.g., see Koskivaara et al. 1992). Niche preferences, microhabitat distribution, and interspecific relationships of gill parasites have been studied mainly from fish with only a few congeneric species occurring in considerable numbers on a host (Suydam 1971, Wootten 1974, Hanek & Fernando 1978, Buchmann 1988, and Bates & Kennedy 1990). Fish parasite communities are likely to be isolationist in nature, mainly because of the low diversity and abundance that make vacant niches available and competition over resources unnecessary (Kennedy 1985, Dzika & Szymanski 1989), while competition and interactive site segregation have also been observed (Buchmann 1988).

The jack mackerel, *Caranx hippos*, is a migratory fish, the adult reaching the eastern coast of Venezuela annually, mostly between May to August, while the juveniles are found throughout the year in the shallow water of the Gulf of Cariaco, Venezuela. The sub-adults are caught occasionally in the deeper waters of this region. *C. hippos* harbors a wide range of ectoparasites on their gills (Rodriguez 1983) and the interrelationships of four species of monogeneic trematodes on this host have been reported (Bashirullah & Rodriguez 1992).

Thus the present study was instigated in order to: 1) elucidate the distribution and the degree (prevalence and mean intensity) of infection of parasitic copepods present on the gills of *C. hippos*, 2) define the microhabitat distribution of copepods on the gills, and 3) examine the relationships of these copepods within their host.

2 MATERIALS AND METHODS

Details of the methods of catching and examining jack mackerel are given elsewhere (Bashirullah & Rodriguez 1992). A total of 150 fish were examined for parasitic copepods. The fish were subdivided into three groups according to their size and age. A total of 744 *Lernanthropus giganteus* (Lernanthropidae) and 583 *Caligus constrictus* (Caligidae) were collected for the present study. These copepods are always found on the gills of marine fishes. The terms 'prevalence,' 'intensity,' and 'mean intensity' are used according to the definitions of Margolis et al. (1982).

2.1 *Statistical treatments*

Fish were grouped into three age groups based on size and a preliminary otolith examination. These are: Age group 1, fish measuring 140-200 mm; group II, fish of size 201-270 mm and group III, adults of 550-820 mm. Interaction based on prevalence, between each of the two copepods were analyzed by the Fager index of affinity (Fager 1957, Southwood 1975) and the coefficient of association. The coefficient is expressed:

$$I_{ai} = 2[(j_i/ A + B) - 0.5] \qquad (1)$$

Where j_i = the total number of individuals of species A and B in samples of joint occurrences in a species pair; A = number of individuals of species A in all samples; B= total number of individuals of species B in the samples. The coefficient has a normal range of -1(no association) to $+ 1$ (complete association).

Niche breadth and overlap of two copepods were determined considering all gill arches as a resource set, in which individual gills represented resource states. Niche measures in this set indicated the degree of resource use and the potential of interactions between co-occurring copepods along the length of the gill. This was calculated using the equations of Choe & Kim (1989)

$$B_i = \frac{1}{n}\sum_{h=1}^{n} P_{ih}^2 \quad \text{and} \quad O_{ij} = \frac{\sum_{h=1}^{n} P_{ih}P_{jh}}{\sum_{h=1}^{n} P_{ih}^2}, \qquad (2)$$

where B_i is the niche breadth of species i. O_{ij} is the niche overlap of the species i with species j. P_{ih} is the proportion of species i in the hth resource state, n is the number of states in each resource set.

3 RESULTS

The overall composition of the samples is detailed in Table 1. The fish all belonged to three year classes and the sizes varied between 140-820 mm. A total of 150 fish were autopsied of which 102 (68%) were infected from which 1327 copepods were collected. Of these 56.1% were *L. giganteus* and 43.9% were *C. constrictus*. The adult fish had higher infections with both copepod species compared to the other two size groups (Table 3). No significant difference was found in respect to the parasite preference for the left or the right side of the fish in the adults. Both of the species favored the 3rd gill arches (57% of *L. giganteus* as compared to 51% for *C.constrictus*), followed by the 2nd gill. The least preferred was the 1st gill arch (Table 1). One-way analysis of variance on the abundance of *L. giganteus* and a posteriori analysis (95% SNK, Newman-Keul test) for fishes invaded first by *L. giganteus* and thereafter by *C. constrictus* indicated this was very significant ($P < 0.01$). Similar results were given for *C. constrictus* and thereafter by *L. giganteus*.

Only 22.55% of infected *C. hippos* harbored 60.22% of *L .giganteus* and 17.65% of fish carried 55.75% of *C. constrictus* in their respective single occurrence (Table 2). Both species appeared rarely on the same side or region of the hemibranchs. Both species were never found on the same gill filament or very close to each other. Most of the *C. constrictus* were found on the internal face of the hemibranch. There seems to have microhabitat segregation.

Table 1. Occurrence of *Lernanthropus giganteus* and *Caligus constrictus* on different gill arches of *Caranx hippos* (± standard error).

Gill Arches	*L. giganteus*	%	*C. constrictus*	%	Total
1st Gill	10	1.34	14	2.4	24
2nd Gill	284	38.17	247	42.37	531
3rd Gill	425	57.12	300	51.46	725
4th Gill	25	3.36	22	3.77	47
Total	744		583		1327
Mean Intensity	7.29		6.0		
Abundance	4.96 ± 0.45		3.88 ± 0.35		
Minimum	0		0		
Maximum	32		17		
Prevalence	68		64.7		

Table 2. Single and concurrent infection of two copepods, *Lernanthropus giganteus* and *Caligus constrictus,* on 102 infected *Caranx hippos.* (Number in column is number of copepods collected and their percentage and % of joint occurrence was calculated on the basis of total parasites collected.)

Gill Arches	Single Occur.		Single Occur.		Joint Occur.	
	L. giganteus	%	*C. constrictus*	%		%
1st Gill	5	0.67	11	1.89	8	0.6
2nd Gill	193	25.94	138	23.67	200	15.07
3rd Gill	234	31.45	161	27.62	330	24.87
4th Gill	16	2.15	15	2.57	16	1.2
Total	448	60.22	325	55.75	554	41.75

Table 3. Association between age of the fish and the abundance of *Lernanthropus giganteus* and *Caligus constrictus* on *Caranx hippos* (both species present at the same time; Chi-square value $X^2_{(2;0.01)} = 9.210$ very significant).

Size of Fish	Age of Fish	L. giganteus		C. constrictus		Total by rows	
		no.	%	no.	%	no.	%
140-200 mm	1+	249	33.47	191	32.76	440	33.16
201-280	2+	154	20.70	120	20.58	274	20.5
580-820	5+	341	45.3	272	46.65	613	46.19
Total by Columns		744	56.1	583	43.9	1327	100

Table 4. Measure of Niche breadth (B) and Overlap (0) *of Lernanthropus giganteus* (L) and *Caligus constrictus* (C) on gill arches of *Caranx hippos*.

	Individual gill arches			
	Gill 1	Gill 2	Gill 3	Gill 4
B_L	0.39	0.15	0.14	0.12
B_C	0.03	0.07	0.73	0.07
O_{LC}	0.11	0.52	0.56	0.56
O_{CL}	1.03	1.18	1.05	0.87

The total Chi-square analyses were used to observe the association between the age of the fish and the abundance of *L. giganteus* and *C. constrictus*. This indicates a very significant association (Table 3). Fager (1957) suggested an index of affinity, which relates the possibilities of the joint occurrence of two species to the sum of their total occurrences. The Fager index of affinity is regarded basically as a sounder technique and as a preliminary step for delimitation of communities (Custer & Pence 1981). Fager's index indicated positive affinity between *L. giganteus* and *C. constrictus* on the gills of their host ($I_{AB} = 0.79$, $t = 8.13**$) and the coefficient of association was given at 0.44. These two species form a mixed population having a strong affinity between them.

The measures of overall niche breadth of *L. giganteus* were larger than those of *C. constrictus* (Table 4). The great majority of both species were concentrated on the 2nd and 3rd gill arches. Both species occupied niches of similar sizes in the same resource states and coexisted without interaction.

4 DISCUSSION

Extensive literature on the spatial distribution of gill parasites, especially copepods exists (Shields & Tidd 1974, Hanek & Fernando 1978, Van den Broek 1979, Davey 1980). Most of the parasites studied generally prefer to attach on the 2nd and 3rd gill arches of the host. Hughes & Morgan (1973) stated that the degree of infection of the gills is probably related to the ventilation volume and the particular pattern of current flow across the gills. Paling (1968) showed experimentally that most of the ventilation current flows over the second and third pair of gills, less over the first pair on each side. Davey (1980) observed that both sexes of *Lernanthropus kroyeri* preferred the medial section of the second gill. Van den Broek (1979) stated that a parasite's location cannot always be explained in terms of respiratory currents and concluded that the combination of factors which may be responsible for their final distribution is unknown.

The increased prevalence and mean intensity of *L. giganteus* and *C. constrictus* with increasing age of *C. hippos* appear to result from the increased exposure, by random chance, along their migration route.

Most parasite communities exhibit habitat segregation in the host, which Brooks (1980) termed as non-interactive population if the site specificity is independent of the presence or absence of other parasites. *L. giganteus* and *C. constrictus* co-occurred in 77.45% of the hosts examined that harbored only 41.75% of parasites, but the regions of attachment were segregated either on the internal, or the external face of the hemibranch. Site segregation is predominantly an active process on the part of the parasite. For some, parasite site selection is a continuing process, which may result in parasites alternating between different microhabitats (Holmes 1973). Inglis (1971) considers that spatial segregation is undoubtedly the most prevalent means of niche specialization. Competition appears to be rare in natural communities largely due to low densities of interacting species. A similar explanation can be applied to the lack of competition in many ectoparasite communities (Marshall 1981). Species co-occurrence is common in fish hosts. Apparently, no major harmful effects by copepods in *C. hippos* have been observed and there seems to be no predation by other fish to regulate the parasite numbers.

The Fager index demonstrated a positive affinity between *L. giganteus* and *C. constrictus* occurring on the gills of *C. hippos*. Thus, the positive associations indicating a clumped distribution of these copepods noted herein appear to relate to common, similar requirements for certain intrinsic or extrinsic factors. Pence & Meinzer (1979) and Custer & Pence (1981) proposed a similar interpretation.

The niche breadth of *L. giganteus* is quite large compared to *C. constrictus* on their host. Niche-overlap measures of *L. giganteus* and *C. constrictus* were markedly asymmetrical. The tendency of *C. constrictus* to exploit the niche of *L. giganteus* was consistently greater. It was so great, that a rare case of niche overlap measure exceeding one was found (Table 4) when each gill arch was treated as a resource state. This probably happened because *C. constrictus* is a broader niched species than *L. giganteus,* and the attachment process of the former species may facilitate easy shifting of location on gill arches. This requires further evidence through experimental manipulation and behavioral observation.

REFERENCES

Bates, R.M. & Kennedy, C.R. 1990. Interactions between the Acanthocephalans *Pomphorhynchus laevis* and *Acanthocephalus angulae* in rainbow trout: testing an exclusion hypothesis. *Parasitol.* 100: 435-444.

Bashirullah, A.K. & Rodriguez, J.C. 1992. Spatial distribution and interrelationship of four Mongenoidea of jack mackerel, *Caranx hippos* (Carangidae) in the North-East of Venezuela. *Acta Cient. Venez.* 43: 125-128.

Brooks, D.R. 1980. Allopatric speciation and non-interactive parasite community structure. *Syst. Zool.* 29: 192-203.

Buchmann, K. 1988. Spatial distribution of *Pseudodactylogyrus angullae* and *P.bini* (Monogenea) on the gills of the European eel, *Anguilla anguilla. J. Fish. Biol.* 32: 801-802.

Choe, J.C & Kim K.C. 1989. Microhabitat selection and coexistence in feather mites (Acari: Analgoidea) on Alaskan seabirds. *Oecologia* 79: 10-14.

Custer, J.W. & Pence, D.B. 1981. Ecological analyses of helminth populations of wild canids from the Gulf coastal prairies of Texas and Louisiana. *J. Parasitol.* 67: 289-307.

Davey, J.T. 1980. Spatial distribution of the copepod parasite *Lernanthropus kroyeri* on the gills of bass *Dicentrarchus labrax. J. Mar. Biol. Assc. UK* 60: 1061-1067.

Dzika, E. & Szymanski, S. 1989. Co-occurrence and distribution of Monogenea of the genus *Dactylogyrus* on gills of bream, *Abramis brama. Acta Parasit. Polonica* 34: 1-14.

Fager, E. 1957. Determination and analysis of recurrent groups. *Ecology* 38: 586-595.

Hanek, G. & Fernando, C.H. 1978. Spatial distribution of gill parasites of *Lepomis gibbosus* (L) and *Ambloplites rupestris. Can. J. Zool.* 56: 1235-1240.

Holmes, J.C. 1973. Site selection by parasitic helminths: interspecific interactions, site segregation, and their importance to the development of helminth communities. *Can. J. Zool.* 51: 333-347.

Hughes, G.M. & Morgan, M. 1973. The structure of fish gills in relation to their respiratory function. *Biol. Rev.* 48: 419-475.

Inglis, W.G. 1971. Speciation in parasitic nematodes. *Adv. Parasitol.* 9: 185-223.

Kennedy, C.R., 1985. Site segregation by species of Acanthocephala in fish, with special reference to eels, *Anguilla anguilla. Parasitol.* 90: 375-390.

Kennedy, C.R., Bush, A.O. & Aho, J.M. 1986. Patterns in helminth communities: why are birds and fish different? *Parasitol.* 93: 205-215.

Koskivaara, M., Valtonen, E.T. & Vouri, K.M. 1992. Microhabitat distribution and coexistence of *Dactylogyrus* species (Monogenea) on the gills of roach. *Parasitol.* 104: 273-281.

Margolis, L., Esch, G.W, Holmes, J.C, Kuris, A.M., & Schad, G.A. 1982. The use of ecological terms in Parasitology (reports of an ad hoc committee of the American Society of Parasitologists). *J. Parasitol.* 68: 131-133.

Marshall, A.G. 1981. *The Ecology of Ectoparasitic Insects.* London: Academic Press.

Pence, D.B. & Meinzer, W.P. 1979. Helminths of the coyote, *Canis latrans*, from the Rolling Plains of Texas. *Int. J. Parsitol.* 9: 339-344.

Paling, J.E. 1968. A method of estimating the relative volume of water flowing over the different gills of freshwater fish. *J. Exp. Biol.* 48: 533-544.

Rodriguez, J.C. 1983. Taxonomia, distribución y ecología de monogeneos y copepodos en branquias de jurel, *Caranx hippos* (Fam: Carangidae) en la zona Oriental de Venezuela. Thesis Licenciatura, Facultad de Ciencia, Universidad de Oriente.

Southwood, T.R. 1975. *Ecological Methods.* London: Chapman & Hall.

Suydam, E.L. 1971. The mico-ecology of three species of monogentic trematodes of fishes from the Beaufort-Cape Hatteras area. *Proc. Helminth. Soc. Wash.* 38: 240-246.

Shields, R.J. & Tidd, W.M. 1974. Site selection on hosts by copepodids of *Lernaea cyprinacea* (Copepoda). *Crustaceana* 27: 226-230.

Van den Broek, W.L.F. 1979. Copepod ectoparasites of *Merlangius merlangus* and *Platichthys flesus. J. Fish. Biol.* 14: 371-380.

Wootten, R. 1974. The spatial distribution of *Dactylogyrus amphibothrium* on the gills of ruffe *Gymnocephalus cernua* and its relation to the relative amounts of water passing over the parts of the gills. *J. Helminth.* 48: 167-174.

7 Biodiversity

Conservation implications arising from a systematic review of the Tasmanian freshwater crayfish genus *Parastacoides* (Decapoda: Parastacidae)

BRITA HANSEN & ALASTAIR M.M. RICHARDSON
School of Zoology, University of Tasmania, Tasmania, Australia

ABSTRACT

Species richness and endemism are significant features of biodiversity, so taxonomic reviews can change the basis of biodiversity conservation issues. A recent review of the Tasmanian endemic freshwater crayfish genus *Parastacoides* has identified 14 species, rather than the single species (consisting of three sub-species) previously recognized.

Most of the range of *Parastacoides* lies within the WHA (Western Tasmania World Heritage Area), and since the previous sub-species had widespread distributions, they were thought to be well-protected. However, several of the newly recognized species lie outside the protected area, and some have very restricted distributions. This raises new conservation issues, as these species may be vulnerable to man made threats such as forestry, mining and hydro-electric power development. The issues extend beyond the conservation of *Parastacoides* species themselves, since some of them are keystone species in their effects on soils and the provision of habitat for other invertebrates.

1 INTRODUCTION

Assessing and understanding biodiversity is fundamental to conservation, however, there are a range of definitions. Biodiversity may be used simply to mean the number of species, or may be expanded to include genetic diversity and even ecosystem diversity (Crozier 1992, Franklin 1993). The dilemma facing many biologists when confronted with the issue of the 'biodiversity crisis' is where to place the emphasis for research; should the emphasis be on the variability of life-forms (the number of species existing), or on life-processes (the ecological, phylogenetic and historical importance of species). The assessment of biodiversity is rendered more difficult by new methods in molecular studies; the discovery of cryptic species and component evolutionary lineages are rendering suspect assumptions regarding protection of underlying genetic and evolutionary diversity via the conservation of single species (Avise 1989, Brooks et al. 1992).

We would argue that, regardless of the definition used, species richness and endemism are still significant features. However, to concentrate on the importance of species numbers alone precludes the importance of some epiphenomena; scientifically important information relating to historical and spatial evidence relating taxa may be lost if species number is the only criterion used for conservation purposes. Sound taxonomy must therefore be fundamental to conservation and management issues; taxonomic revision can alter the conservation status and management issues surrounding specific taxa. A recent review of the Tasmanian endemic freshwater crayfish genus *Parastacoides* (Hansen & Richardson in prep.) is a case in point; details of the review, and the implications for the conservation of *Parastacoides* species this review has brought to the fore, are discussed below.

2 SYSTEMATICS

The most recent review had been by Sumner (1978), and that allowed only a single species, albeit divided into three sub-species (Fig. 1). However, since that time, ex-

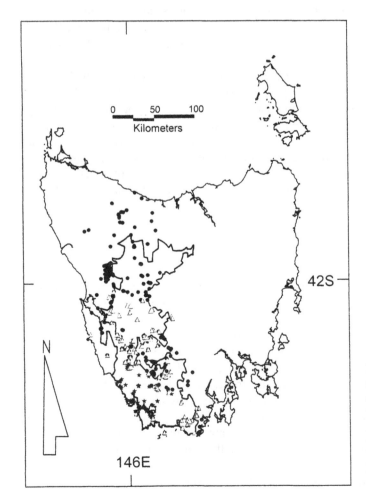

Figure 1. Map of Tasmania showing the distribution of current *Parastacoides* sub-species with the World Heritage Area indicated in outline. '•' = *Parastacoides tasmanicus tasmanicus*, Δ = *P.t. inermis* and ★ = *P.t. insignis).*

tensive collection had been undertaken in very remote regions of Tasmania. The diversity revealed in collections from the Lower Gordon Scientific Survey in the late 1970s and the Wilderness Ecosystem Baseline Studies in the late 1980s, in particular, suggested the need for a review.

Unpublished electrophoretic examination (M. Adams & P. Baverstock, IMVS, Adelaide, in 1983, T. Krasniki, Zoology Department, University of Tasmania, in 1989) suggested that as many as fourteen species should be recognized. A study, undertaken with the aim of uncovering morphological characters which were capable of diagnosing the electromorphs, proved successful (Hansen 1996); and a revised taxonomy suggesting fourteen species is consequently in preparation (Hansen & Richardson in prep). Codes, based on locality, are used throughout this paper to distinguish each new species; the full names cannot be used till the species are described. The resultant distributions of the new species (Fig. 2) proved intriguing; some species appear to have large, exclusive distributions, whilst others have restricted, sometimes overlapping distributions, i.e., centers of endemism.

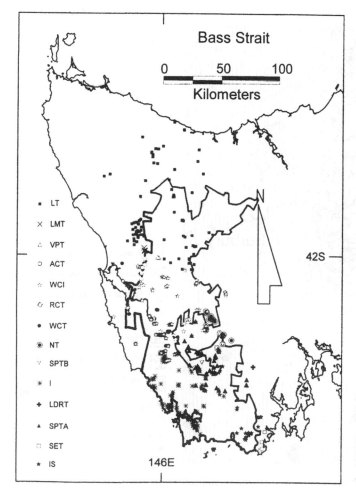

Figure 2. Map of Tasmania showing the distribution of the putative *Parastacoides* species, with the World Heritage Area indicated in outlines. Species codes are used to indicate new species until publication of description.

3 SIGNIFICANCE OF *PARASTACOIDES*

Much of the range occupied by *Parastacoides* lies within the area listed as a World Heritage site in 1986: the Western Tasmanian World Heritage Area. The taxonomic status of *Parastacoides* at that time suggested that all three sub-species were well protected (Fig. 1). However, with the revised taxonomy, and the subsequent re-examination of distributions, the conservation status of some species must be reassessed (Fig. 2).

Figure 3 indicates one of the centers of endemism. Six of the newly recognized species are to be found in this area; three of them are confined to this area alone. One of the species, LMT, has been collected only from areas which now lie under the waters of Lake Burbury, a hydro-electric lake created by a dam commenced in 1983 and commissioned in 1992. As indicated by Figure 3, the land tenure position of the area is very complex, and the conservation status of several of the new species is therefore uncertain. The conservation status of other species, for example SET and LDRT, is also unresolved (Fig. 2). LDRT is presently known from only one small

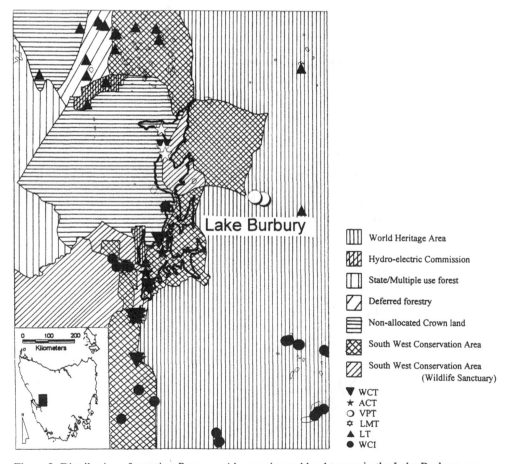

Figure 3. Distribution of putative *Parastacoides* species and land tenure in the Lake Burbury area.

river drainage system in an area currently allocated for forestry operations. The species designated LT appears to have a large and exclusive range (Fig. 2). However, a molecular study is currently underway to determine whether it is indeed one species. Evidence of glaciation (Colhoun et al. 1996), as well as the slow dispersal rates of *Parastacoides* (Richardson & Swain 1991) suggest that populations from the main drainage systems may have been isolated from each other for very long periods of time. Speciation may have occurred during this period, creating a number of cryptic species. If this proves to be the case, some of these species are likely to be highly vulnerable to forestry operations and agricultural land clearing.

Rawlinson (1974) and Davis (1986), have discussed Bass Strait as a barrier to the dispersal of organisms in either a northward or a southward direction. *Parastacoides* species have a long history of occupancy of Tasmania with pre-Pleistocene (Knott 1976) or Gondwanan (Brown 1990) origins in Tasmania. They should be considered scientifically important taxa, both because of their relatively undisturbed distribution patterns and for their relictual status.

Parastacoides species commonly burrow extensively in acid peat land. This appears to be a unique phenomenon since burrowing crayfish do not extensively colonize acidic peat elsewhere (Growns & Richardson 1988). Densities of over one per square meter have been noted (Lake & Newcombe 1975, Growns & Richardson 1988). They represent an important avenue of gas exchange in peat soils; the metabolism of the peat immediately surrounding the burrow is enhanced (Richardson 1983), and burrows are associated with an increase in plant production (Richardson & Wong 1995). Crayfish burrows also provide habitat for a discrete fauna, the pholeteros (Lake 1977), elements of which are not found in any other microhabitat (Horwitz 1989). *Parastacoides* species have scientific value as keystone species; in summer burrows represent the only available water for aquatic species, whilst in winter they act as a refuge from surface freezing (Horwitz 1989). *Parastacoides* species represent one of the major species groups delineating the faunal break known as Tyler's Line (Shiel et al. 1989, Mesibov 1994). This faunal break corresponds with a series of ecotones that may act as isolation barriers by preventing the dispersal of organisms across them. However, Tyler's Line represents a relatively broad band, and whilst no *Parastacoides* species are found to the east of this faunal break, the distribution boundary of the species does not fully correspond to the eastern boundary of the break. It is important to conserve these boundaries, as they represent potentially valuable regions for the reconstruction of historical population processes (Mesibov 1994) and serve as benchmarks against which to monitor the effect of climatic change.

4 DISCUSSION AND CONCLUSION

The review of the genus *Parastacoides* has substantially increased both the number of species believed to exist, but probably more importantly, it has defined the distribution, and as a consequence, the conservation status of species within the genus. What then of conservation and management issues? We suggest that one of the more important implications to arise from the study is one of conservation management – where to place the emphasis on conservation.

From a scientific viewpoint, the study of biological diversity confronts us with some of the most fundamental questions in evolutionary biology. Questions such as what defines a species, how they arise and how variation in diversity occurs across time and space, immediately arise (Cracraft 1994). Geological change and the ensuing vicariant speciation are highly significant processes in their effect on biodiversity (Brooks et al. 1992, Cracraft 1994, Riddle 1996). Vicariance biogeography is therefore potentially the most significant basis for elucidating the history of biotas (Cracraft 1994). It has been noted that biodiversity has a tendency to be clumped in 'hot spots', or areas of endemism. These areas are often concurrent to areas which exhibit historically high rates of vicariant events. Information from phylogenetics and vicariance biogeography can establish areas of endemism the conservation of which will ensure the preservation of evolutionary potential. This has the potential to answer questions about the susceptibility of species and ecosystems to environmental stresses, such as climate change, and raising hypotheses regarding the history of these (Brookes et al. 1992, Cracraft 1994).

Most phylogeographic study has been done on vertebrates (Lee et al. 1996, Wenink et al. 1996). It is not clear whether the patterns and processes involved in vicariance speciation and dispersal are the same or fundamentally different for invertebrates. Does scale matter? Differences in scale, mobility and generation turnover suggest that patterns derived from the study of vertebrates may not predict invertebrate biodiversity well. Mesibov (1994) points out that the recognition of faunal breaks and fine scale mapping of the distributions of some Tasmanian invertebrates suggests that it would be unwise to suggest that these species would be found more than even a few kilometers from known locations.

The clear geographical isolation imposed by Bass Strait makes Tasmania an ideal biogeographic laboratory in which to examine the principles of vicariant or dispersal biogeography, and to study Pleistocene recovery by invertebrate fauna in the Southern Hemisphere. *Parastacoides* provides a unique opportunity to study a relatively speciose taxon in relatively untouched habitat; however, the problem of conservation and management issues raised by the systematic review must be solved. The majority of *Parastacoides* species are well protected within the WHA, however, the most important centers of endemism lie outside its boundaries. If the current management paradigm suggests that the most important issue in the conservation of biodiversity is to conserve the greatest number of species, we may miss the opportunity presented by centers of endemism, and the epiphenomena they represent, to study the historical biodiversity they may elucidate. These centers of endemism would provide a valuable tool for the study of vicariance biogeography, making a clear case for their conservation.

REFERENCES

Avise, J.C. 1989. A role for molecular geneticists in the recognition and conservation of endangered species. *TREE* 4: 279-281.

Brooks, D.R., Mayden, R.L. & McLennan, D.A. 1992. Phylogeny and biodiversity: conserving our evolutionary legacy. *TREE* 7: 55-58.

Brown, M.J. 1990. Buttongrass moorland ecosystems. In S.J. Smith & M.R. Banks (eds), *Tasmanian Wilderness – World Heritage Values*: 109-113. Hobart: Royal Society of Tasmania.

Colhoun, E.A., Hannan, D. & Kiernan, K.K. 1996. Late Wisconsin Glaciation of Tasmania. *Pap. Proc. R. Soc. Tasmania* 130(2): 33-45.

Cracraft, J. 1994. Species diversity, biogeography, and the evolution of biotas. *Amer. Zool.* 34: 33-47.

Crozier, R.H. 1992. Genetic diversity and the agony of choice. *Biol. Conserv.* 61: 11-15.

Davis, J.L. 1986. Revision of the Australian Psephenidae (Coleoptera): systematics, phylogeny and historical biogeography. *Aust. J. Zool. Supplementary Series* 119: 1-97.

Franklin, J.F. 1993. Preserving biodiversity: species, ecosystems, or landscapes? *Ecol. Appl.* 3: 202-205.

Growns, I.O. & Richardson, A.M.M. 1988. Diet and burrowing habits of the freshwater crayfish, *Parastacoides tasmanicus tasmanicus* Clark (Decapoda:Parastacidae). *Aust. J. Mar. Freshwat. Res.* 39: 525-534.

Hansen, B., 1996. Preliminary systematic study of the freshwater crayfish genus *Parastacoides* Clark (Decapoda: Parastacidae). Unpublished Honours thesis, University of Tasmania.

Horwitz, P. 1989. The faunal assemblage (or Pholeteros) of some freshwater crayfish burrows in southwest Tasmania. *Bull. Aust. Soc. Limnol.* 12: 29-36.

Knott, B. 1976. Systematic studies on the Phreatoicoidea (Order Isopoda) with a discussion on the phylogeny and zoogeography of other freshwater malacostracan crustaceans from Australia and Tasmania. Unpubl. Ph.D. Thesis. University of Tasmania.

Lake, P.S. 1977. Pholeteros – the faunal assemblage found in crayfish burrows. *Aust. Soc. Limnol. Newsletter* 15: 57-60.

Lake, P.S. & Newcombe, K.J. 1975. Observations on the ecology of the crayfish *Parastacoides tasmanicus* (Decapoda; Parastacidae) from south-western Tasmania. *Austral. Zool.* 18: 197-213.

Lee, T.E., Riddle, B.R. & Lee, P.L. 1996. Speciation in the desert pocket mouse (*Chaetopus penicillatus* Woodhouse). *J. Mamm.* 77: 58-68.

Mesibov, R. 1994. Faunal breaks in Tasmania and their significance for invertebrate conservation. *Mem. Queensl. Mus.* 36: 133-136.

Rawlinson, P.A., 1974. Biogeography and ecology of the reptiles of Tasmania and the Bass Strait region. In Biogeography and Zoology in Tasmania, Williams, W.D. (ed.). W. Junk Publishers, The Hague. pp 291-334.

Riddle, B. 1996. The molecular phylogeographic bridge between deep and shallow history in continental biotas. *TREE.* 11: 207-211

Richardson, A.M.M. 1983. The effect of the burrows of a crayfish on the respiration of the surrounding soil. *Soil Biol. and Bioch.* 18: 197-213.

Richardson, A.M.M. & Swain, R.R. 1991. Pattern and persistence in the burrows of two species of the freshwater crayfish, *Parastacoides* (Decapoda: Parastacidae), in southwest Tasmania. *Mem. Queensl. Mus.* pp 283.

Richardson, A.M.M. & Wong, V. 1995. The effect of a burrowing crayfish, *Parastacoides sp.*, on the vegetation of Tasmanian wet heathlands *Freshwat. Crayfish* 10: 174-182.

Sheil, R.J., Koste, W.W. & Tan, L.W. 1989. Tasmania revisited: rotifer communities and habitat heterogeneity. *Hydrobiol.* 186/187: 239-245.

Sumner, C.E. 1978. A revision of the genus *Parastacoides* Clark (Crustacea: Decapoda: Parastacidae). *Aust. J. Zool.* 26: 809-821.

Wenink, P.W., Baker, A.J., Rosner, H.L. & Tilanus, M.G.J. 1996. Global mitochondrial DNA phylogeography of holarctic breeding dunlins (*Calidris alpina*). *Evolution* 50: 318-330.

Terrestrial evolution in Crustacea: The talitrid amphipod model

A.M.M. RICHARDSON & R. SWAIN
School of Zoology, University of Tasmania, Australia

ABSTRACT

The Tasmanian talitrid fauna includes examples of almost all the proposed phylogenetic and ecological groups in the family and provides a unique opportunity to make deductions about the route by which talitrids colonized land. Four potential routes are considered, and the distributions of Tasmanian talitrids are described for each. Sandy shores are eliminated as potential routes because of the sharp discontinuity between the difficult sand habitat and adjacent terrestrial vegetation; rocky shores are eliminated for similar reasons and for the lack of angiosperm food in the intertidal zone. Saltmarshes and estuaries remain as likely routes that still retain extensive overlaps in the distributions of intertidal and terrestrial species. Some difficulties are associated with both of these routes: the hypersaline barrier at the landward edge of saltmarshes and the lack of any physiological evidence that terrestrial talitrids have passed through a freshwater stage in their evolution. We conclude that more than one route may have been involved.

1 INTRODUCTION

Very few families of animals have distributions that span the full range of habitats from the sea, through all types of intertidal zones, to dry land and also fresh water. The talitrid amphipods span this range and can be described as a translittoral family (Chester 1992). The Talitridae have an almost worldwide distribution in the intertidal zone, though they are absent from ice-scoured shores at high latitudes and they are less abundant in the tropics than in the temperate zones. Terrestrial and freshwater species are found in both hemispheres, but the most extensive radiations onto land have been in the southern hemisphere, particularly in Australia, New Zealand and South Africa. In these regions ancient continental faunas have developed, whereas many of the other terrestrial and freshwater species elsewhere seem to be the result of local evolution from supralittoral ancestors (Bousfield 1984).

Phylogenetic hypotheses for the Talitridae have been developed only in very broad terms. There seems little doubt that the group arose from a hyalid-like ancestor living in the intertidal zone. Bousfield (1984) used new characters, particularly the presence or absence of cusps on the peraeopod dactyls and the dentition of the left mandibular lacinia mobilis, to suggest some major subdivisions. The earliest talitrid probably resembled the extant genus *Protorchestia* (South America, New Zealand, Australia), with a 5-dentate lacinia mobilis and without dactylar cusps. A line of simplidactylate, 4-dentate animals diverged early, giving rise to the palustral group (Group 1: northern and southern hemispheres, tropical and temperate) and the simplidactylate landhoppers (Group 4i: temperate Australia and South Africa). The cuspidactylate line gave rise to both 5- and 4-dentate beachfleas (Group 2) and sandhoppers (Group 3), and from within the 4-dentate beachflea line, the cuspidactylate landhoppers (Group 4ii). No detailed phylogenetic analysis has yet been carried out to elucidate the sequence of these events, but Conceição et al. (1998) have recently suggested from allozyme data that a Cretaceous date for the initial radiation of the Talitroidea may be too early.

In Australasia, the most diverse radiation of talitrids appears today in southeast Australia, and particularly Tasmania (Friend 1987, Richardson 1993, Richardson 1996). The Tasmanian fauna includes representatives of all but one group from Bousfield's (1984) informal classification (Table 1) and most groups are represented by several species.

Tasmania's climate is characterized by steep gradients in rainfall, and to a lesser extent temperature, from west to east, and the relatively low levels of human disturbance mean that many sea-land transitions exist more or less intact, with the natural vegetation cover remaining. This combination of phylogenetic and ecological diversity provides a valuable opportunity to reconstruct the evolution of the terrestrial habit in this family, one of the two groups of crustaceans that have become fully euterrestrial, thus no longer requiring access to standing water to reproduce (Powers & Bliss 1983).

Table 1. Ecomorphological groups of Talitridae (Bousfield 1984) with the genera and numbers of species found in Tasmania. 4- or 5-dentate refers to the number of teeth on the left lacinia mobilis; simpli- and cuspidactylate refer to the absence or presence of cusps on the peraeopod dactyls.

Group Number	Group Name	Genera	Spp. #
1	*Palustral talitrids, or marsh hoppers*	*Eorchestia*	2
2	*Beachfleas*		
	Group i (5-dentate, simplidactylate)	*Protorchestia*	1
	Group ii (5-dentate, cuspidactylate)	Undescribed	3
	Group iii (4-dentate, cuspidactylate)	Several undescribed	6
3	*Sandhoppers*		
	Group i (4-dentate)	Several undescribed	3
	Group ii (5-dentate)	Several undescribed	5
4	*Landhoppers*		
	Group i (Simplidactylate)	*Protaustrotroides*	1
		Austrotroides	4
		Neorchestia	1
		Mysticotalitrus	2
		Arcitalitrus	3
		Keratroides	6
	Group ii (Cuspidactylate)	*Orchestiella*	2
		Tasmanorchestia	3

In this paper, we examine distributional data collected in Tasmania over the four main types of sea-land transition: sandy beaches, rocky shores, saltmarshes and estuaries, and interpret them in terms of the route(s) which talitrids may have taken onto land.

2 SEA-LAND TRANSITIONS

2.1 *Sandy beaches*

Sandy beaches comprise a substantial proportion of the Tasmanian coastline on the north, west and upper east coasts, where they may be backed by a range of native vegetation types from dry grassland to wet sclerophyll forest. In a recent survey of 222 beaches around the Tasmanian coast (Richardson et al. 1996, 1998), at least 17 species of talitrid were recorded. Most of these (10) were undescribed sandhoppers, but an undescribed beachflea genus was also recorded, which has developed a highly convergent morphology to those sandhopper species that forage over the mid shore at low tide scavenging on diatoms and other particulate material. Five species of landhoppers were also recorded in pitfall traps set between the strandline and the terrestrial vegetation.

On many beaches, three or more species of sandhoppers can be found. Shepherd (1994) described the distributions of three species at Fortescue Bay in southeast Tasmania. An undescribed species of sandhopper with a gracile body, long walking legs, and large protuberant eyes (convergent with the characteristics of *Pseudorchestoidea* spp. in Bousfield 1982) forages at the mid-tide level during low tides. At the strandline, *Talorchestia marmorata* (Haswell) colonizes piles of cast kelp, and at the very highest level of the beach an undescribed 4-dentate sandhopper emerges from deep burrows to forage at night on the pioneer grasses and other high strandline plants. Immediately inland, in woody coastal scrub, several species of landhoppers are found, including one coastal species (*sensu* Friend 1987), *Austrotroides* sp., which is confined to the immediate coastal zone within 100 m of the high water mark.

At South Cape Rivulet beach on the far southern part of the west coast, Richardson et al. (1991) recorded the distributions of talitrid species along transects from the mid-tide level across the strandline and up to 80 m into the adjacent wet sclerophyll forest. Eleven species of talitrid were present, three sandhoppers with similar distributions (though only one species is common to both sites) to those at Fortescue Bay described above, and 8 species of landhoppers, 3 of which show coastal distributions. At this site, and many others in the wetter parts of the island, the transition from beach to coastal scrub is very sharp, with a band of pioneering grasses less than 5 m wide followed by a closed canopy of woody vegetation. The transition between the distributions of sandhoppers and landhoppers was found to be correspondingly sharp: sandhoppers and landhoppers replaced each other completely within a range of 2 m.

2.2 *Rocky shores*

Rock platforms and various grades of boulder shores are extensive on all Tasmanian coasts. The rock platforms of the east coast of Tasmania were surveyed by Richardson et al. (1997a), but this survey yielded few talitrids since the sites surveyed were rock platforms which allowed little or no accumulation of kelp, and provided no shelter for beachfleas. The beachfleas and palustral amphipods found on these shores are effectively confined to kelp accumulations or boulder/pebble/gravel shores where they can find shelter.

The Tasmanian beachflea fauna is diverse (ca 10 species) but undescribed, apart from *Orchestia australis* Fearn-Wannan. Some of the more robust species are confined to boulder shores, as is the palustral species *Eorchestia rupestris* Richardson.

No transects across boulder shores and above the high water mark have been examined in detail, but casual observations suggest that the rocky shore species do not extend any distance above high water, except in situations where maritime turf or saltmarsh has developed in the spray zone of very exposed shores (Wong et al. 1993).

2.3 *Salt marshes*

Mangroves do not extend as far south as Tasmania, but salt marshes, largely dominated by glassworts (*Sarcocornia* spp.) and the shrubby salt bush (*Arthrocnemum arbusculum*), are common on sheltered shores. These are found largely on the east and north coasts, but saltmarshes are also present in the large inlets of the west coast (Macquarie Harbour and Port Davey/Bathurst Harbour), or occasionally as a narrow band above exposed rocky shores where there is intense deposition of spray.

Richardson et al. (1998) surveyed the crustacean and mollusc fauna on 65 salt marshes around the Tasmanian coast. They recorded 17 species of talitrids: 2 marshhoppers, 5 beachfleas and 10 landhoppers, of which only the two marshhoppers (*Eorchestia palustris* and an undescribed *Eorchestia* sp.) were confined to salt marsh. Richardson et al. compared the distributions of talitrids with vegetation patterns and soil characteristics. The widespread beachflea *Orchestia australis* was associated with almost all vegetation types, but was only weakly represented in drier sites characterized by the rush *Schoenus nitens*. Marsh hoppers frequently occurred in quadrants characterized by *Samolus repens, Selliera radicans* and *Juncus kraussii*, plants typical of poorly drained sites of relatively low salinity (Kirkpatrick & Glasby 1981). Coastal landhoppers, *Austrotroides maritimus* and *Tasmanorchestia* spp., were only common in quadrats dominated by *Leptocarpus brownii*, a plant typical of low salinity sites.

Richardson & Mulcahy (1996) examined the distributions of talitrids on Lutregala saltmarsh on Bruny Island, off the southeast coast of Tasmania. Eight talitrid species were present: one marsh hopper, one beachflea, three coastal landhoppers and three forest landhoppers. The distributions of several of these species overlapped substantially. The beachflea, *Orchestia australis,* had the widest range, from the wettest, most saline sites to relatively dry tussock grassland at the extreme high water mark. The marsh hopper *Eorchestia palustris* overlapped substantially with *O. australis,* but was found within a narrower band of salinities, excluding the most saline sites, and in wetter situations than the beachflea. Coastal landhoppers were found in dry saline grassland at and above the extreme high water mark where they overlapped with the beachflea in the seaward direction and with forest landhoppers in the landward direction.

2.4 *Estuaries*

The salt marshes examined by Richardson et al. (in press) included estuarine, reedy marshes in the lower reaches of the larger rivers. These marshes support dense populations of *Eorchestia palustris* and in some places *Orchestia australis*. In estuarine reedy marshes regularly submerged by the tides, *E. palustris* was the only species present. Many of these estuarine marshes were backed by non-natural vegetation, so the transition from supralittoral to terrestrial species could not be recorded, but in a few smaller and more remote estuaries, *E. palustris* coexisted with coastal landhoppers at the landward edges of the marshes.

Chester (1992) examined the range and dynamics of a population of the marsh-hopper *Eorchestia palustris* on the estuary of North West Bay River, a small river in southern Tasmania. He showed that this species lived in near-fresh water for extended periods during the winter wet season. Tolerance experiments showed that it could survive for 35 days in seawater diluted to 9 ppt, and 20 days at 2 ppt (at laboratory temperatures between 14 and 24°C). Mulcahy (1990) recorded an ED_{50} of 25 days for the survival of *E. palustris* in air on a salt-free substrate. In South Africa, the palustral species *Eorchestia rectipalma* can be found in estuaries and in freshwater streams several kilometers inland (Griffiths 1976, A. Richardson, unpublished).

Though not strictly in an estuarine habitat, it is worth noting that a species of the primitive beachflea genus *Protorchestia, P. lakei,* has been collected in Tasmania only from fresh water runnels draining seabird rookeries on an offshore island (Richardson 1996). The salinity of the runnels is unknown, but the associated fauna of hydrobiid snails, janiriid, and phreatoicid isopods suggests that the salt content was low.

3 DISCUSSION

Little (1989) discusses the pitfalls in deducing evolutionary processes from physiological evidence, and the same problems of local adaptation and convergence apply to making deductions from ecological evidence. However, if the general conditions on present day shores reflect those of the shores of the Tethys Sea, and if the ecological responses of present day species can be taken as representative of the original colonizing species, then some deductions about the route(s) taken onto land by talitrid amphipods can be made.

Firstly, and following the conclusions already made by Bousfield (1984), it seems that sandy shores can be eliminated. Conditions in the mid to high intertidal of a sandy beach are particularly harsh climatically, osmotically and in terms of the abrasive, mobile substrate. Sandhoppers are the most morphologically modified of the talitrids, and those modifications are related to the problems of burrowing and surviving in sand. These adaptations are apparently not useful on the landward side of the sharp ecological transition at the top of the beach, perhaps because of hugely increased organic matter and root density in the soil, and no burrowing talitrids are known in coastal forest. Indeed, truly burrowing landhoppers are rare and burrow in clayey, rather than sandy, soils (Friend 1987). Sandhoppers also appear to be phylogenetically remote from either group of landhoppers, and for all these reasons sandy beaches appear to be the least likely route to land.

Rocky shores present different problems. Burrowing is impossible and the talitrid fauna consists of beachfleas and palustrals which find shelter under kelp, or in cavities between pebbles and boulders. In contrast to the sand-burrowing habit of the sandhoppers, this 'non-substrate-modifying' (Macintyre 1963), or 'nestling' (Griffiths 1976) way of life is compatible with life landward of the high tide mark. On exposed shores where spray is deposited, and there is suitable vegetation and leaf litter inland, beachfleas may extend tens or even hundreds of meters inland (Bagenal 1957). A potential difference between the rocky intertidal zone and terrestrial habitats is the change in the nature of plant litter, from algal tissues to the harder cuticles, waxes, lignin and cellulose of angiosperms. However, Moore & Francis (1985) noted that *Orchestia gammarellus* fed on algal and angiosperm litter on the strandline in proportion to their availability, and cellulases have been recorded in the gut of *O. gammarellus* (Agrawal 1964, Wildish & Poole 1970). On sheltered shores, where forest vegetation occurs immediately inland, intertidal species and landhoppers replace each other quite sharply, although not as abruptly as the sandhopper/landhopper transition on sandy shores (Richardson 1980).

Salt marshes appear to provide a more suitable route to land. The sea-land transition is gradual (at least where forest is found inland), and observations of the distributions of talitrids on salt marshes show substantial overlaps in the distributions of intertidal and terrestrial species. Angiosperm litter is present in the intertidal zone and has been incorporated in the diet of the intertidal species. As on the rocky shores, the talitrid groups present are those from which both lines of the terrestrial fauna are believed to have arisen. On the negative side, salt marshes in warmer drier regions are characterized by a hypersaline barrier at the landward edge, where evaporation produces hypersalinity in the substratum (Little 1990). This may have been a barrier to landward colonization, but its effect would be substantially reduced in high rainfall areas where closed forest vegetation backs the salt marsh. In temperate areas the hypersaline barrier would be a transient, seasonal phenomenon perhaps leading to seasonal migrations among the talitrid fauna which would increase their terrestrialization.

Estuaries also provide a gradual transition from marine to freshwater conditions, and a long shoreline that is an interface between aquatic and terrestrial habitats. Several lines of evidence suggest that the estuarine transition may have been an important route. The tolerance of palustral species to very low salinities was noted above, along with their distribution on salt marshes in low salinity areas. In Western Europe the beachflea *Orchestia cavimana* extends far inland along water courses, living in wet litter at the margins (Curry et al. 1972), with ample opportunity to colonize adjacent forest litter. One group 4ii landhopper, *Orchestiella neambulans,* is found in saturated leaf litter at the edges of rainforest streams in Tasmania, but it is as yet unknown whether its distribution extends seaward to the headwaters of estuaries. Lindeman (1991) suggests that the Central American landhopper taxa *Cerrorchestia* and *Caribitroides* arose from palustral ancestors similar to *Chelorchestia* spp. and colonized wet forests via estuaries and river margins. Bowman (1977) described *Chelorchestia vaggala* from an island in the Galapagos, where it lives in inland montane cloud forest. *Chelorchestia vaggala* has apparently evolved recently from group 1 coastal ancestors (Peck 1994). *C. costaricana* is found in mangroves on tropical American shores in the Carribean and Pacific, but in the Galapagos it appears to have a wider range including sandy and rocky supralittoral habitats and litter accumulations several kilometers from the coast (Peck 1994).

Another possibly relevant phenomenon is the presence of the family Ceinidae (or possibly Hyalellidae), also in the superfamily Talitroidea, in the fresh waters of South Africa (*Afrochiltonia*), Australia (*Austrochiltonia* and *Phreatochiltonia*) and New Zealand (*Chiltonia*). Bousfield (1984) suggests that the presence of hyalellids may have restricted the upriver penetration of talitrids in South America, but it may also be possible that these groups are the modern representatives of a talitroidean radiation in estuaries which began in the Cretaceous.

Wildish (1982) discusses the role of the driftwood niche in the colonization of land by talitrids. He has identified a small suite of beachfleas from northern hemisphere estuaries and saltmarshes which are found under decomposing driftwood at the very highest tidal levels; these species (e.g. *Orchestia aestuarensis*) are characterized by small body size, reduced activity, and reduced egg and brood size. Wildish proposes that land colonization proceeded via this route directly into the litter of tropical forests. To date, no beachfleas have been found occupying the driftwood niche on Tasmanian salt marshes and estuaries, but coastal landhoppers are quite commonly found in this habitat (Richardson & Mulcahy 1996, Richardson et al. 1997b). Driftwood species may have been replaced by landhoppers where continental landhopper faunas have developed.

Physiological evidence is ambivalent about these potential routes to land. Little (1990) discounted the freshwater route on the grounds that group 4i landhoppers, *Arcitalitrus dorrieni* (Morritt 1988) (and *Mysticotalitrus cryptus*: Morritt & Richardson 1998) are unable to produce hyposmotic urine, whereas the freshwater littoral beachflea *Orchestia cavimana* can. He also noted the relatively high overall concentration and the high concentration of some ions in landhopper haemolymph relative to terrestrial and freshwater invertebrates. Little argued that these characteristics imply the absence of a freshwater phase in the evolution of landhoppers. However, the concentration of urine may not be a reliable indicator of freshwater ancestry. Brachyuran crabs produce isosmotic urine whether they are marine, estuarine, freshwater, or terrestrial (Greenaway 1988). While dilute urine may be good evidence for freshwater ancestry, isosmotic urine does not necessarily indicate direct colonization from the sea.

Little (1990) also suggests that the very poor ability of the group 4i landhoppers *Arcitalitrus dorrieni* (Morritt 1987) and *A. sylvaticus* (Laso-Wazem 1984) (and indeed talitrids in general: Morritt 1987) to resist desiccatory water loss is an argument against their evolutionary passage through the desiccating conditions at the top of most saltmarshes. *Arcitalitrus dorrieni* is also unable to hypo-osmoregulate at concentrations above those of full seawater, but is able to tolerate a wide range of salinities (Morritt 1988). The estuarine route provides a continuous moist microhabitat from the sea to inland forest litter without a hypersaline barrier.

More recent work on landhopper osmoregulation (Morritt & Richardson unpublished) has shown a consistent pattern of hyper-iso-osmotic regulation in three species of Group 4i landhoppers over a concentration range of 40-1045 $mOsm.kg^{-1}$, while a Group 4ii species, *Tasmanorchestia annulata,* showed a hyper-hypo-osmoregulatory pattern over the same range. This suggests different evolutionary histories in the two landhopper groups, and could be taken as evidence that they took different routes to land. It may be that Group 4i ancestors colonized via the estuarine-freshwater route, while group 4ii species crossed the salt marshes. However, the apparent absence of any group 4i landhoppers from the freshwater littoral and the presence of at least one group 4ii species from that habitat (*Orchestiella neambulans*, see above) does not support this hypothesis.

An alternative hypothesis is that the group 4i and 4ii ancestors crossed the salt-marsh interface in different climatic zones, group 4ii under warmer, drier conditions where a hypersaline barrier at the top of the marshes was well developed, and group 4i in cooler wetter conditions where high rainfall and low evapotranspiration eliminated the development of hypersaline conditions, at least seasonally, as seen at the landward edge of salt marshes in southern and western Tasmania today (Wong et al. 1993, Richardson & Mulcahy 1993).

Further information is required to resolve a clearer picture of the routes by which talitrids colonized land. Data on the osmoregulatory mechanisms of species in the palustral group 1 are critical, as are strongly supported hypotheses about phylogenetic relationships, both within the Talitridae and among the families that make up the Talitroidea. These gaps can be filled, and the talitrids will be a remarkable model group for the study of the evolution of life on land.

We are already at a stage where some simple and testable hypotheses can be presented. For example, if the groups 4i and 4ii landhoppers colonized land by a fresh water and high rainfall saltmarsh route, respectively, then we would expect differences in their surface permeability, because the 4ii species show higher permeability than the descendants of the group that moved from the freshwater littoral into moist forest litter. By the same token, the microhabitats occupied by the species in the two groups should also differ, with a more eurytopic range in the 4ii's, as suggested by Bousfield (1984). The role of seasonal migration in the colonization of land could be examined by comparing the seasonal distributions of talitrids on salt marshes in summer-dry and summer-wet climatic zones. The Tasmanian fauna provides the species and habitats necessary to make these tests.

ACKNOWLEDGMENTS

We are grateful to Dr David Morritt for his helpful comments on the physiology of talitrid amphipods.

REFERENCES

Agrawal, V.P. 1964. Studies on the physiology of digestion in *Orchestia gammarella*. *Proc. Zool. Soc. Lond.* 143: 133-141.

Bagenal, T.B. 1957. The vertical range of some littoral animals on St. Kilda. *Scott. Natur.* 69: 50-51.

Bousfield, E.L. 1982. The amphipod superfamily Talitroidea in the northeastern Pacific region. 1. Family Talitridae: systematics and distributional ecology. *Natnl. Mus. Nat. Sci. Ottawa, Pub. Biol. Oceanogr.* 11: 1-73.

Bousfield, E.L. 1984. Recent advances in the systematics and biogeography of landhoppers (Amphipoda: Talitridae) of the Indo-Pacific region. In F.J. Radovsky, P.H. Raven & S.H. Sohmer (eds), *Biogeography of the Tropical Pacific*: 171-210. Honolulu: Association of Systematic Collections & Bernice P. Bishop Museum.

Bowman, T.E. 1977. *Orchestia vaggala* n.sp., a new land-hopper from the Galapagos Islands (Crustacea: Amphipoda: Talitridae). *Proc. Biol. Soc. Wash.* 90: 658-668.

Chester, E.T. 1992. The life history of the palustral talitrid *Eorchestia palustris* (Crustacea: Amphipoda: Talitridae). Unpublished Honours thesis, Hobart: University of Tasmania.

Conceição, M.B., Bishop, J.D. D. & Thorpe, J.P. 1998. Genetic relationships between ecologically divergent species of talitrid amphipod (Crustacea). *Mar. Ecol. Progr. Ser.* 165: 225-233.

Curry, A., Grayson, R.F. & Milligan, T.D. 1972. New British records of the semi-terrestrial amphipod *Orchestia cavimana. Freshwat. Biol.* 2: 55-56.

Friend, J.A. 1987. The terrestrial Amphipods (Amphipoda: Talitridae) of Tasmania: systematics and zoogeography. *Rec. Aust. Mus. Suppl.* 7: 1-85.

Greenaway, P. 1988. Ion and water balance. In W.W. Burggren & B.R. McMahon (eds), *Biology of the land crabs*: 211-248. Cambridge: Cambridge University.

Griffiths, C.L. 1976. *Guide to the benthic marine amphipods of southern Africa.* Cape Town: South African Museum.

Kirkpatrick, J.B. & Glasby, J. 1981. *Saltmarshes in Tasmania.* Occasional Paper 8. Hobart: Department of Geography, University of Tasmania.

Lazo-Wasem, E.A. 1984. Physiological and behavioural ecology of the terrestrial amphipod *Arcitalitrus sylvaticus* (Haswell, 1880). *J. Crust. Biol.* 4: 343-355.

Lindeman, D.H. 1991. Phylogeny and zoogeography of the New World terrestrial amphipods (landhoppers) (Crustacea: Amphipoda; Talitridae). *Can. J. Zool.* 69: 1104-1116.

Little, C. 1989. Comparative physiology as a tool for investigating the evolutionary routes of animals onto land. *Trans. Roy. Soc. Edinburgh: Earth Sciences* 80: 201-208.

Little, C. 1990. *The Terrestrial Invasion.* Cambridge: Cambridge University Press.

Macintyre, R.J. 1963. The supralittoral fringe of New Zealand sand beaches. *Trans. Roy. Soc. N. Z. (General)* 1: 89-103.

Moore, P.G. & Francis, C.H. 1985. Some observations of food and feeding of the supra-littoral beach-hopper *Orchestia gammarellus* (Pallas) (Crustacea: Amphipoda). *Ophelia* 24: 183-197.

Morritt, D. 1987. Evaporative water loss under desiccation stress in semiterrestrial and terrestrial amphipods (Crustacea: Amphipoda: Talitridae). *J. Exp. Mar. Biol. Ecol.* 111: 145-157.

Morritt, D. 1988. Osmoregulation in littoral and terrestrial talitroidean amphipods (Crustacea) from Britain. *J. Exp. Mar. Biol. Ecol.* 123: 77-94.

Morritt, D. & Richardson, A.M.M. 1998. Female control of the embryonic environment in a terrestrial amphipod, *Myusticotalitrus cryptus* (Crustacea). *Funct. Ecol.* 12: 351-358.

Mulcahy, M.E. 1990. Saltmarshes: a route to colonization of land for amphipod Crustacea? Unpublished Honours thesis. Hobart: University of Tasmania.

Peck, S.B. 1994. Diversity and zoogeography of the non-oceanic Crustacea of the Galapagos Islands, Ecuador (excluding terrestrial Isopoda). *Can. J. Zool.* 72: 54-69.

Powers, L.W. & Bliss, D.E. 1983. Terrestrial adaptations. In F.J. Vernberg & W. B. Vernberg (eds), *Biology of Crustacea*: 271-334. New York: Academic Press.

Richardson, A.M.M. 1980. Notes on the occurrence of *Talitrus dorrieni* Hunt (Crustacea: Amphipoda: Talitridae) in south west England. *J. Nat. Hist.* 14: 751-757.

Richardson, A.M.M. 1993. Tasmanian intertidal Talitridae (Crustacea: Amphipoda). Palustral talitrids: two new species of *Eorchestia* Bousfield 1984. *J. Nat. Hist.* 27: 267-284.

Richardson, A.M.M. 1996. *Protorchestia lakei,* new species (Amphipoda: Talitridae) from Maatsuyker Island, Tasmania, with a key to the genus and notes on the diversity of Tasmanian Talitridae. *J. Crust. Biol.* 16: 574-583.

Richardson, A.M.M., Swain, R. & Smith, S.J. 1991. Local distributions of sandhoppers and landhoppers (Crustacea: Amphipoda: Talitridae) in the coastal zone of western Tasmania. *Hydrobiologia* 223: 127-140.

Richardson, A.M.M. & Mulcahy, M.E. 1996. The distribution of talitrid amphipods (Crustacea) on a saltmarsh in southern Tasmania, in relation to vegetation and substrate. *Estuar. Coast. Shelf Sci.* 43: 801-817.

Richardson, A.M.M., Swain, R. & Shepherd, C.J. 1996. *The strandline invertebrates of beaches on the east coast of Tasmania.* Canberra: National Estate Grants Program.

Richardson, A.M.M., Swain, R. & Shepherd, C.J. 1997a. *The fauna of rock platforms on the east coast of Tasmania and Flinders Island.* Canberra: National Estate Grants Program.

Richardson, A.M.M., Swain, R. & Shepherd, C.J. 1998. *The strandline fauna of beaches on the north & west coasts of Tasmania, Flinders and King Islands.* Canberra: National Estate Grants Program.

Richardson, A.M.M., Swain, R. & Wong, V. 1997b. The crustacean and molluscan fauna of Tasmanian saltmarshes. *Pap. Proc. Roy. Soc. Tasm.* 131: 21-30.

Richardson, A.M.M., Swain, R. & Wong, V. (1998). The relationship between the crustacean and molluscan assemblages of Tasmanian saltmarshes, the vegetation and soil conditions. *Mar. Freshw. Res.* 49.

Shepherd, C.J. 1994. Behaviour and distribution of sandhoppers at Fortescue Bay. Unpublished Honours thesis. Hobart: University of Tasmania.

Wildish, D.J. 1982. Talitroidea (Crustacea, Amphipoda) and the driftwood ecological niche. *Can. J. Zool.* 60: 3071-3074.

Wildish, D.J. & Poole, N.J. 1970. Cellulase activity in *Orchestia gammarella* (Pallas). *Comp. Biochem. Physiol.* 33: 713-716.

Wong, V., Richardson, A.M.M. & Swain, R. 1993. *The Crustaceans and Molluscs of Tasmanian Saltmarshes.* Canberra: National Estate Grants Program.

Use of the mitochondrial 16S rRNA gene for phylogenetic and population studies of Crustacea

CHRISTOPH D. SCHUBART, JOSEPH E. NEIGEL & DARRYL L. FELDER
Department of Biology and Laboratory for Crustacean Research, University of Louisiana, Lafayette, USA

ABSTRACT

The mitochondrial 16S DNA is a structural, non-coding gene. Its transcript is the large subunit ribosomal RNA (16S rRNA) that in a conserved secondary structure, and in association with proteins, forms the large subunit of mitochondrial ribosomes. This gene is used frequently for molecular systematics of several animal taxa, including Crustacea. The complete 16S mtDNA sequence is so far known for only one crustacean species, the anostracan *Artemia franciscana*. However, a single region of approximately 520 base pairs has become one of the most commonly used sequences for molecular studies of crustaceans during this decade. Thereby, evolutionary questions of very different temporal and spatial scales have been addressed. To demonstrate this wide range of applications we review previous studies based on the 16S rRNA gene and present new results that resolve some of the phylogenetic relationships between brachyuran families but also allow detection of geographic variation in an intertidal crab species.

1 INTRODUCTION

At the end of the last decade, sequence data for portions of animal (and human) mitochondrial genomes (mtDNA) and their subsequent use for reconstructing phylogenetic relationships revolutionized evolutionary biology (e.g. Avise et al. 1987, Cann et al. 1987, Moritz et al. 1987). MtDNA is maternally inherited, haploid, and present in multiple copies in most cells. These features facilitate its amplification and the interpretation of sequencing results. DNA from crustacean mitochondria was first obtained by Batuecas et al. (1988) in a study of the genome organization of this organelle in *Artemia*. Four years later it was used for the first time in a crustacean molecular phylogeny (Cunningham et al. 1992). So far, the anostracan *Artemia franciscana* is the only crustacean for which the complete mitochondrial genome has been sequenced (Valverde et al. 1994). In this species, it consists of 15,822 basepairs and the typical 37 genes of which 13 are protein coding, 22 are tRNAs and 2 rRNAs.

Most of the presently published molecular phylogenies of crustaceans rely on single genes to reconstruct species relationships. Gene trees are thereby assumed to reflect species trees. Besides 16S rRNA, other commonly used mitochondrial genes are the cytochrome oxidase I (COI), cytochrome b (cyt-b), and the small subunit ribosomal RNA (12S rRNA). While COI and cyt-b are mitochondrial coding genes, the 12S and 16S mtDNAs are structural, non-coding genes. Their rRNA transcripts form part of the mitochondrial ribosomes where they play important catalytic roles during translation of mRNAs into proteins. The rRNAs have a secondary structure with stems and loops, and some regions of the molecules are active in biochemical processes and thus more conserved in their sequence. The combination of variable and conserved regions within the same gene is probably one of the reasons why 16S rRNA has become one of the most popular genes for reconstructing animal phylogenies. If properly applied, it allows the study of old phylogenetic relationships as well as fairly recent separation events. On the other hand, the popularity of this gene is probably also due to its successful use by previous workers and the limited number of alternative primers available for invertebrate molecular comparisons.

Here we summarize all previous studies that used 16S rRNA for crustacean systematics or population comparisons, present new sequences for brachyuran crustaceans, and discuss alignment procedures as well as the use of this gene as a molecular clock. To demonstrate the broad range of applications for the 16S rRNA gene, we present two extremes: a phylogeny of the Brachyura based on conserved regions of this gene and an intraspecific analysis of the grapsid marsh crab species *Sesarma reticulatum*.

2 MATERIALS & METHODS

All published information on 16S mtDNA of Crustacea was gathered by means of a literature and genetic database research. We generated additional 16S mtDNA sequences of several brachyuran taxa for phylogenetic and biogeographical comparisons. American specimens were identified by DLF and CDS, material from New Caledonia by P. Castro, and from Spain by I. López de la Rosa. All specimens from which sequences were obtained are deposited in biological collections as vouchers (see Table 1). New sequences have been deposited in the EMBL database (AJ130799-AJ130817).

All specimens used for DNA sequencing were preserved in 75-90% ethanol. Genomic DNA was isolated from the muscle tissue of walking legs or claws using a phenol-chloroform extraction (Kocher et al. 1989). Selective amplification of a ~565 basepair product from the mitochondrial 16S rRNA gene was carried out by a polymerase-chain-reaction (PCR) (35-40 cycles; 1 min 94°/1 min 50-55°/2 min 72° denaturing/annealing/extension temperatures) using primers 16Sar (5'–CGCCTGTTTA-TCAAAAACAT–3') and 16Sbr (5'–CCGGTCTGAACTCAGATCACGT–3') from Palumbi et al. (1991). PCR products were purified with 100,000–MW Millipore or Microcon 100 filters and sequenced with the ABI BigDye terminator mix in an ABI Prism 310 Genetic Analyzer.

Brachyuran sequences were aligned manually with the multisequence editing program ESEE (Cabot & Beckenbach 1989), with special consideration of the secondary structure of the gene (Machado et al. 1993, Schneider-Broussard & Neigel 1997). For our phylogenetic analysis, we used only sequences that included the complete length

Table 1. List of species for which 16S mtDNA was sequenced and used for phylogenetic analyses of brachyuran families. MNHN-B: Muséum National d'History Naturelle, Paris; R: Colllection Rudolf Diesel, Starnberg; SMF: Senckenberg Museum, Frankfurt a.M.; USLZ: University of Southwestern Louisiana Zoological Collection, Lafayette.

Species	Family	Locality of collection	Collection number
Petrolisthes armatus	Porcellanidae	Louisiana: Grande Isle	ULLZ 3779
Xantho poressa	Xanthidae	Spain: Cádiz	ULLZ 3808
Panopeus herbstii	Panopeidae	South Carolina: Charleston	ULLZ 3778
Menippe nodifrons	Menippidae	Mexico: Veracruz	ULLZ 3720
Trapezia cymodoce	Trapeziidae	New Caledonia	MNHN-B 24961
Cancer irroratus	Cancridae	Maine: Portland	ULLZ 3843
Carcinus maenas	Portunidae	New Hampshire: Hampton B.	ULLZ 3840
Callinectes sapidus	Portunidae	Louisiana: Isles Dernieres	ULLZ 3895
Epilobocera sinuatifrons	Pseudothelphusidae	Puerto Rico: Guajataca	R 199
Eudaniela garmani	Pseudothelphusidae	Trinidad	R 269
Palicus obesus	Palicidae	Gulf of Mexico	ULLZ 3852
Crossotonotus compressipes	Palicidae	New Caledonia	MNHN-B 6215
Pinnixa retineus	Pinnotheridae	Texas: Corpus Christi	ULLZ 3870
Pachygrapsus transverus	Grapsidae	Louisiana: Grande Isle	ULLZ 3782
Percnon gebbesi	Grapsidae	Puerto Rico: North coast	R 153
Sesarma reticulatum	Grapsidae	Delaware: Woodland Beach	ULLZ 3835
Cyrograpsus altimanus	Grapsidae	Argentina: Santa Clara	SMF 24545
Cardisoma crassum	Gecarcinidae	Costa Rica: Rincón	SMF 24543
Gecarcinus lateralis	Gecarcinidae	Mexico: Veracruz	ULLZ 3722

between the 16Sar and 16Sbr primers (~ 520 basepairs). The following sequences from molecular databases were included: *Scopimera globosa* (accession number AB002124), *Leipocten trigranulum* (AB002129), *Uca lactea* (AB002130), *Macrophthalmus banzai* (AB002132) (all Ocypodidae), *Mictyris brevidactylus* (AB002133) (Mictyridae), *Menippe mercenaria* (U20737), and *M. mercenaria* nuclear copy (U20738) (Menippidae). Distance matrices of sequence divergence were analyzed using Kimura 2-parameter distances and neighbor-joining (NJ) (Saitou & Nei 1987) with the program MEGA (Kumar et al. 1993). Statistical significance of groups within inferred trees was evaluated by the interior branch method (Rzhetsky & Nei 1992) and by bootstrapping the maximum parsimony (MP) analysis with 1000 replicates using the program PAUP (Swofford 1993). For the MP analysis, transversions were weighted three times greater than transitions and gaps were treated as missings.

3 RESULTS

Table 2 summarizes all presently published studies that use 16S mtDNA for either phylogenetic or population studies of crustaceans. As can be seen from this table and Table 3, most primer combinations have been used to amplify a DNA region of approximately 520 basepair length (here termed '16Sar-br region' after the corresponding universal primers designed by T.D. Kocher and S.R. Palumbi). Palumbi et al. (1991) showed variations of these primers for some animal taxa, including the arthropod *Drosophila* (Table 3). Since then, the 16Sar-br region has been clearly the most commonly sequenced region of 16S mtDNA. In few cases (not including the *Artemia* complete mtDNA sequence), crustacean workers have sequenced beyond this region, which would determine the actual sequence in the primer regions. These cases are the recent study by Kitaura et al. (1998) that revealed the sequence (5'-YGCCTGTTTATYAAAAACAT-3') for the 16Sar region in ocypodid crabs and a study by Crandall & Fitzpatrick (1996) showing the sequence (5'-CCGGTCTGA-ACTCAAATCATGT-3') for the 16Sbr region in cambarid crayfish.

Table 2. List of studies that use 16S mtDNA sequence for crustacean phylogenies or population genetics, with the number of individuals (ind.), species (sp.), genera (gen.), and families (fam.) included, as well as the sequence or alignment lengths in basepairs (bp) and primers used.

Taxonomic groups	Length	Primer combination	References
Cladocera			
Daphniidae: *Daphnia* (4 sp./13ind.)	491 bp	ar & br	Taylor et al. 1998
Copepoda			
Calanidae, Metridinidae: *Calanus, Metridia* (10sp/14ind.)	387 bp	ar-Dr & br-Dy	Bucklin et al. 1995
Calanidae: *Pseudocalanus* (2 sp./19 ind.)	250 bp	167 & br-Dy	Bucklin et al. 1998a
Calanidae: *Nannocalamus minor* (2 sp.?/155 ind)	440 bp	ar-Dr & br-Dy	Bucklin et al. 1996a
Calanidae: *Calanus* (3 sp./20 ind.)	430 bp	ar-Dr & br-Dy	Bucklin et al. 1992
Calanidae: *Calanus pacificus* (27 ind.)	449 bp	ar-Dr & br-Dy	Bucklin & Lajeunesse 1994
Calanidae: *Calanus finmarchicus* (182 ind.)	350 bp	ar-Dr & br-Dy	Bucklin & Kocher 1996, Bucklin et al. 1996b
Calanidae: review of above data on Calanidae		ar-Dr & br-Dy	Bucklin et al. 1998b
Isopoda			
Serolidae: (11 gen./15 sp.)	503 bp	ar & br	Held, in press
Amphipoda			
Lysianassidae: (4 gen./7sp./32 ind.)	179 bp	Amph1 & Amph2	France & Kocher 1996a
Lysianassidae: *Eurythenes gryllus* (95 ind.)	437 bp	ar & br-Dr, Amph1 & Amph2	France & Kocher 1996b
Euphausiacea			
Euphausiidae: (3gen./sp./6 ind.)	524 bp	ar & br	Patarnello et al. 1996
Decapoda			
Penaeidae: *Penaeus* (2 sp./7 ind.)	438 bp	ar & br	Machado et al. 1993
Penaeidae: *Penaeus* (11 sp./=40 ind.)	472 bp	ar & br	Chu et al. 1998
Penaeidae: *Metapenaeopsis* (7 sp./16 ind.)	475 bp	ar & br	Tong et al., in press

Table 2. Continued.

Taxonomic groups	Length	Primer combination	References
Palinuridae: *Palinurus* (4 sp./13 ind.)	491 bp	ar & br	Sarver et al. 1998
Nephropidae: (5 gen./7 sp.)	350 of 474 bp	ar & SB	Tam & Kornfield 1998
Cambaridae: (3 gen./38 sp./72 ind.)	554 bp	1471 & 1472	Crandall & Fitzpatrick 1996; Crandall 1998
Cambaridae & Parastacidae: (9 gen./12 sp.)	535 bp	1471 & 1472	Crandall et al. 1995
Parastacidae: (3 gen./9 sp)	~550 bp	1471 & 1472	Lawler & Crandall 1998
Parastacidae: *Euastacus* (10 sp.)	461 bp	ar & br	Ponniah & Hughes 1998
Paguridae, Lithodidae et al.: (4 fam./7 gen./10 sp./12 ind.)	420 bp	ar & br	Cunningham et al. 1992
Hippidae: *Emerita* (6 sp./9 ind.)	~400 bp	ar & 16SB	Tam et al. 1996
Menippidae: *Menippe* (2 sp./ 9 ind.)	525 bp	ar & br, ar-M & br-I	Schneider-Broussard & Neigel 1997, Schneider-Broussard et al. 1998
Portunidae: *Carcinus, Callinectes* (5 sp./47 ind.)	395 bp	Dar &Dbr	Geller et al. 1997
Grapsidae & Gecarcinidae: (26 gen.)	531 of 589 bp	ar & br, 1472	Schubart et al., in press
Grapsidae: Sesarminae (4 gen./21 sp.)	522 bp	ar & br, L12 & H16	Schubart et al. 1998
Grapsidae: *Sesarma* (2 sp./16 ind.)	526 bp	ar & br, L12 & H16	Schubart et al., 1998b
Grapsidae: *Pachygrapsus* (2 sp./ 5 ind.)	510 bp	ar & br, L12 & H16	Cuesta & Schubart, 1998
Ocypodidae et al.: (7 gen./30 sp. 24 ind.)	491	ar & br	Levinton et al. 1996
Ocypodidae: *Uca* (24 sp./27 ind.)	491	ar & br	Sturmbauer et al. 1996
Ocypodidae: (10 gen./ 20 sp.)	~1170	br, L2482, L2510, H2492, H2716, H3058, H3062	Kitaura et al. 1998

Alignment of 16S mtDNA corresponding to the 16Sar-br region from several brachyuran families confirmed a marked heterogeneity in sequence conservation among portions of this region. Even after applying secondary structure models to this alignment (Machado et al. 1993, Schneider-Broussard & Neigel 1997), the most variable regions could not be aligned with enough certainty to assume that homologous positions were being compared. We therefore excluded these regions and positions with compensatory mutations (see Discussion) from the phylogenetic analyses. Our initial alignment of 572 positions was thereby reduced to 426 basepairs, of which 215 were variable and 170 parsimony-informative. A phylogenetic analysis of brachyuran families based on these conserved regions of the 16S mtDNA was carried out with NJ and MP. The resulting tree is presented in Figure 1. Only confidence / bootstrap values above 50% of the interior-branch method (NJ) and of the bootstrap analysis (MP) are shown.

In another case study, the 16Sar-br region was sequenced for several specimens of

Table 3. List of primers used for amplifying 16S mtDNA in crustaceans.

Name	Primer sequence (5' -> 3')	Designation	References
Forward primers (L-strand)			
1. 16Sar primer and modifications			
ar:	CGCCTGTTTATCAAAAACAT	universal	Palumbi et al. 1991
ar-Dr:	CGCCTGTTTAACAAAAACAT	*Drosophila*	Palumbi et al. 1991
1471:	CCTGTTTANCAAAAACAT	Evertebrata	Crandall & Fitzpatrick 1996
Dar:	CGCCTGTTTAHYAAAAACAT	universal	Geller et al. 1997
L2510:	CGCCTGTTTAACAAAGACAT	Evertebrata	Kitaura et al. 1998
2. Internal primers to 16Sar-br region			
167:	GACGAGAAGACCCTATGAAG	Calanoida	Bucklin et al 1998a
Amph1:	GACGACAAGACCCTAAAAGG	Amphipoda	France & Kocher 1996
ar-M:	ATAAGACCCTATAAAGC	*Menippe*	Schneider-Broussard & Neigel 1997
L12:	TGACCGTGCAAAGGTAGCATAA	Grapsoidea	Schubart et al. 1998a
3. External primers to 16Sar-br region			
L2482:	GAAGGAACTCGGCAA	universal?	Kitaura et al. 1998
Reverse primers (H-strand)			
1. 16Sbr primer and modificications			
br:	CCGGTCTGAACTCAGATCACGT	universal	Palumbi et al. 1991
br-Dr:	CCGGTTTGAACTCAGATCATG	*Drosophila*	Palumbi et al. 1991
br-Dy	CCGGTTTGAACTCAGATCACGT	*Drosophila yakuba*	Bucklin et al. 1995
SB:	CTCCGGTTTGAACTCAGATC	Arthropoda	Xiong & Kocher 1991
Dbr:	CCGGTCTGAACTCAGMTCAYGT	universal	Geller et al. 1997
H3062:	CCGGTCTGAACTCAGATCA	universal	Kitaura et al. 1998
H3058	TCCGGTCTGAACTCAGATCACGTA	universal	Kitaura et al. 1998
H2492:	CAGACATGTTTTTAATAAACAGGC	~reverse of ar	Kitaura et al. 1998
2. Internal primers to 16Sar-br region			
Amph1	CGCTGTTATCCCTAAAGTA	Amphipoda	France & Kocher 1996
br-I:	CCGCCCCAGCAAAATAAA	*Menippe*	Schneider-Broussard & Neigel 1997
H16:	TTATCRCCCCAATAAAATA	Grapsoidea	Schubart et al. 1998a
H2716i	AAGTTTTATAGGGTCTTATCGTC	Ocypodoidea	Kitaura et al. 1998
3. External primers to 16Sar-br region			
1472:	AGATAGAAACCAACCTGG	Evertebrata	Crandall & Fitzpatrick 1996

the grapsid marsh crab *Sesarma reticulatum* to determine intraspecific variability and to test the hypothesis that populations from the Gulf of Mexico are distinct from populations of the American Atlantic between Massachusetts and Florida (here called Atlantic) and might represent a distinct species (Zimmerman & Felder 1991, Staton & Felder 1992, Felder & Staton 1994). The comparison of 532 basepairs of mtDNA from eight crab specimens collected at six localities between Delaware and Texas revealed the existence of five different haplotypes. From their genetic divergence, these haplotypes can be clearly separated into two distinct groups, corresponding to the Gulf of Mexico and the Atlantic (Table 4). Haplotypes from the Gulf of Mexico differed by at most three nucleotide substitutions, with the most divergent haplotype found in western Florida. Likewise, there was also little divergence among haplotypes from the Atlantic; the same haplotype was found in three individuals and the second haplotype differed by only one transition. In contrast, eight diagnostic sequence differences (2-3 transversions, 4-5 transitions and 1 indel) separate Gulf of Mexico from Atlantic haplotypes. The extent of this sequence divergence suggests a relatively long period of isolation.

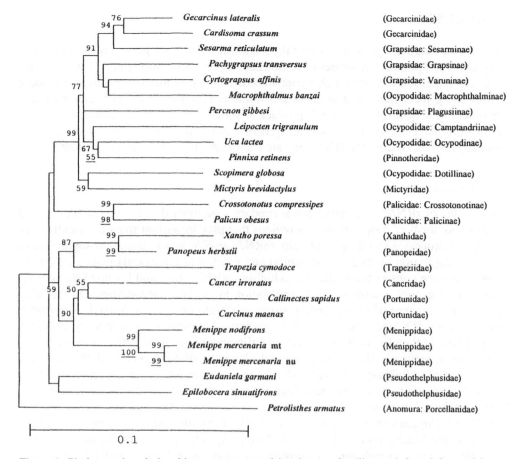

Figure 1. Phylogenetic relationships among several brachyuran families as inferred from 426 co-served basepairs of the 16S rRNA gene. The tree topology is based on a neighbor-joining analysis using Kimura 2-parameter distances. Confidence values are from an internal node test and a bootstrap maximum parsimony analysis (underlined) (only values > 50 shown); mt: mitochondrial gene, nu: nuclear copy.

Table 4. Percent genetic divergence and number of differences between 532-basepair of the 16S mt-DNA of 8 specimens of *Sesarma rerticulatum* from the Gulf of Mexico (GM) and the northwestern Atlantic (ATL); (s: Transition, v: Tranverion, i: Indel).

	GM-1	GM-2	GM-3	GM-4	ATL-1	ATL-2, 3, 4
GM-1 (Sabine Pass, TX)	–	0.2	0	0.4	1.5	1.7
GM-2 (Calcasieu Pass, LA)	1s	–	0.2	0.6	1.5	1.7
GM-3 (Cocodrie, LA)	0	1s	–	0.4	1.5	1.7
GM-4 (Alligator Point, FL)	1v, 1s	1v, 2s	1v, 1s	–	1.5	1.9
ATL-1 (Brunswick, GA)	3v, 4s, 1i	3v, 4s, 1i	3v, 4s, 1i	2v, 6s, 1i	–	0.2
ATL-2 (Brunswick, GA)	3v, 5s, 1i	3v, 5s, 1i	3v, 5s, 1i	2v, 7s, 1i	1s	–
ATL-3, 4 (Woodland Beach, DE)	3v, 5s, 1i	3v, 5s, 1i	3v, 5s, 1i	2v, 7s, 1i	1s	–

4 DISCUSSION

The broad range of applications for mitochondrial 16S rRNA gene sequences in phylogenetic and population studies of crustaceans has been demonstrated (Table 2, present results). However, certain characteristics of this gene should be taken into account when using it for molecular systematics. While not subject to the constraints of protein-coding sequences, the rRNA product is subject to specific secondary structure constraints, and its sequence is heterogeneous in the degree of conservation. Occasionally, copies of the 16S rRNA gene are translocated into the nuclear genome.

4.1 *Secondary structure*

Molecules of rRNA have conserved secondary structures that consist of base-paired stems and unpaired loops, as well as regions that play important roles in essential biochemical processes (binding of protein, mRNA, tRNA, and termination suppression). This configuration of localized structural and functional constraints places regions of high sequence conservation adjacent to highly variable ones, and is one of the reasons this gene has such a wide range of application in evolutionary studies. On the other hand, the relevance of the secondary structure should not be overlooked. Alignments of divergent rRNA sequences should begin with the underlying secondary structure, which should then be used to 'anchor' the alignment (see Kjer 1995). The secondary structure is certainly the most conserved character of this gene and therefore the alignment of individual nucleotides should be subordinated to the conservation of the overall structure to determine homologous positions. For four crustacean species, hypothetical models of secondary structures are available: *Artemia* sp. (see Palmero et al. 1988), *Daphnia* sp. (see Taylor et al. 1998), *Farfantepenaeus* (formerly *Penaeus*) *notialis* (see Machado et al. 1993) and *Menippe* sp. (see Schneider-Broussard & Neigel 1997). Even though these reconstructions are not entirely consistent with each other (probably none of them reflects the final truth, for which one would need the entire rRNA molecule to consider all potential stems), they represent the most important stem and loop regions within the 16Sar-br region. The addition of gaps to alignments (to compensate for insertions or deletions) should preferably occur in loop regions (see Kjer 1995).

Another characteristic of the secondary structure is that mutations in stem regions are normally followed by compensatory mutations on the other branch of the stem to re-establish C-G, A-U, or G-U base pairs. Such compensatory mutations help to verify stem regions and after being recognized should be regarded as single mutation events or character states (Kjer 1995).

4.2 *Different forms of mutations*

Three forms of single-base mutations must be taken into account when aligning and analyzing DNA sequences: transitions, transversions, and indels (insertions and deletions). Transitions are fairly common in comparison to transversions. They represent the change from one pyrimidine to the other, or one purine to the other and thus do not involve much structural change. In some phylogenetic analyses, transversions may therefore be weighted more than transitions.

Unlike in protein-coding genes, there is no selective pressure against shifts in the reading frame throughout an rRNA sequence. A deletion or insertion of a single base makes a coding gene non-functional, while it might not have any effect on an rRNA molecule. As a consequence, indels accumulate in rRNAs over time and result in considerable length variation of the same gene across taxa. The difficulty for molecular systematists consists in determining where to place gaps when aligning homologous sequences to ensure comparison of homologous bases. As discussed above, the alignment should be based first on secondary structure. Once the secondary structure is determined and ambiguous positions are confined to variable loop regions, gaps should be placed in a way that transitions become more common than transversions, especially if transversions are given stronger weight in subsequent phylogenetic analyses. If uncertainty about the exact alignment of variable regions remains, it is advisable not to include these regions in phylogenetic analyses.

4.3 *Molecular dating*

In three studies, the rate of molecular change has been estimated independently for the 16Sar-br region of the 16S rRNA gene in crustaceans. Cunningham et al. (1992) related the divergence of left and right-handed hermit crabs (73-78 Myr), the vicariance of Atlantic and Pacific cold-water marine faunas (35-40 Myr), and the vicariance of the amphi-Atlantic boreal marine province (2.5-3.1 Myr) to their modified Kimura distances in order to calculate the separation of king and hermit crabs. On the basis of their data, we calculated a divergence rate of 0.38% per Myr (not 2.2% per Myr as erroneously cited in Schubart et al. 1998a) for anomuran crabs. The other two available calibrations are for brachyuran crabs and are based on genetic distances of trans-isthmian species pairs across the Panama Isthmus (closure approximately 3.1 Myr ago) resulting in a divergence rate of 0.9% per Myr (Sturmbauer et al. 1996) for *Uca* and 0.65-0.88% per Myr (Schubart et al. 1998a) for *Sesarma*. The anomuran rate is strikingly lower, but more recent results from this taxon indicate that this may be artifactual (Cunningham, pers. comm.). The brachyuran rates, in contrast, are fairly similar and their use is therefore recommended for molecular dating in brachyuran crabs, especially when looking at relatively recent time periods (e.g. Schneider-Broussard et al. 1998).

Several crustacean species are characterized by a large number of different haplotypes for the 16S rRNA gene. This is especially true for planktonic species: 68 haplotypes in *Nannocalanus minor* Type I (see Bucklin et al. 1996a), 30 in *Calanus finmarchicus* (see Bucklin et al. 1996b) 13 in *Pseudocalanus moultoni* (see Bucklin et al. 1998a). In contrast, single haplotypes have been documented for two species of the green crab genus *Carcinus* (see Geller et al. 1997) and along entire coastlines for the littoral grapsid crab species *Pachygrapsus transversus* (unpubl. data). Marked intraspecific variation can possibly blur calculations of molecular clocks when single representatives are used for each species. To increase the accuracy of the clock, several individuals per population/species should be sequenced and only invariable positions used for the molecular calibration.

Caution also needs to be taken when comparing rates of sequence divergence, if regions of different length and location within the 16S rRNA gene have been used for calibration. As stated above, this gene contains highly conserved and more variable regions. Often the areas adjacent to the primers are rather conserved. Studies that do not include the entire region, most often missing the regions closest to the primers, might therefore result in an upwardly biased estimate of divergence. In some analyses of sequence data, the most variable regions are excluded, because of the difficulty of a precise alignment (e.g. this study). In this case, molecular dating cannot be performed using calibrations from other studies, unless the restricted data set is recalibrated. Molecular dating of older events often suffers from convergences and multiple substitutions that accumulate in the sequences, while dating of very recent events must take into account sequence polymorphism (see above). In all cases, molecular clocks should ideally be calibrated with geological events from a similar time period and using estimates that allow for multiple substitutions.

4.4 *Pseudogenes*

A potential risk in the use of the 16S rRNA gene for phylogenetic analyses are translocated copies of this gene in the nuclear genome. In the *Menippe mercenaria – M. adina* complex, the presence of such a pseudogene has been recently shown and discussed by Schneider-Broussard & Neigel (1997) and Schneider-Broussard et al. (1998). Similar findings with 16S mtDNA of ghost shrimp (Bilodeau, pers. comm.) and another mitochondrial gene in crabs (unpubl. observ.) suggest that the occurrence of pseudogenes is not an unusual phenomenon and is a potential source of artifacts. In the case of the *Menippe mercenaria – M. adina* complex, the closest relative, *M. nodifrons*, constitutes an outgroup to both the mitochondrial gene and the nuclear copy (Fig. 1). The pseudogene thus evolved after separation of these species. Accidental sequencing of the pseudogene in this case would not alter the phylogeny. Nevertheless, it can severely confuse results in studies of lower systematic levels or biogeographic analyses. The fact that pseudogenes in the nuclear genome are nonfunctional makes it likely that mutations occur also within normally conserved regions, and thus primers should not amplify these translocated copies after prolonged evolutionary times.

4.5 *Present results and outlook*

The two studies here presented exemplify scenarios of very different evolutionary time scales and consist of preliminary results. The geological record of brachyuran crabs suggests that most of the divergence into what we consider modern families took place 30-40 Myr ago (Glaessner 1969, Schram 1986). Our tree of a few selected brachyuran families suggests one very early split into two main groups. On one branch we find the Grapsidae, Gecarcinidae, Ocypodidae, Mictyridae and Pinnotheridae (all Thoracotremata sensu Guinot 1978), on the other we have the Xanthoidea, Portunidae, Cancridae (all Heterotremata) thus rendering support to the classification of Guinot (1978). The position of the Palicidae (monophyletic) and Pseudothelphusidae (early split between South and Central American representatives) is unresolved, which is also the case in morphological taxonomy. More ancestral Brachyura (Podotremata) were not included (see Spears et al. 1992). The phylogenetic tree furthermore suggests a paraphyly of the Grapsidae (inclusion of the Gecarcinidae and Macrophthalminae), a polyphyly of the Ocypodidae (different subfamilies paired with several other families like Pinnotheridae and Grapsidae) and a polyphyly of the Xanthoidea, justifying the splitting into different families (sensu Guinot 1978). The internal node method (NJ) is based on a standard error test and confidence values reflect statistical significance levels. Excessive weight should therefore not be placed on groupings with levels under 95%. This is corroborated by the MP bootstrap analysis that results in a large consensus for most of the tree with exception of some of the more recent separations (Xanthidae-Panopeidae, subfamilies of the Palicidae, *Menippe* sp.) (see Fig. 1). The lack of resolution with the parsimony method is possibly due to genetic saturation in the variable regions (increased numbers of homoplasies) and an insufficient substitution rate in the conserved ones. The time of 30 to 40 Myr thus seems to represent a limit for the use of the 16S rRNA gene in phylogenies. Sequencing of additional genes and representatives from other families, will potentially help to better resolve some of the suggested branching patterns in brachyuran phylogeny.

In the molecular biogeographic study of *Sesarma reticulatum*, sequences from additional specimens need to be obtained in order to determine the approximate number of haplotypes present. If the pattern of molecular divergence between the Gulf of Mexico and the Atlantic continues to hold, recognition of a new species will be supported by 16S mtDNA, confirming previous distinctions based on allozymes, color, and physiology (see Zimmerman & Felder 1991, Staton & Felder 1992, Felder & Staton 1994, Mangum & McKenney 1996). Also in other studies, 16S mtDNA clearly separates populations of crustacean species (Bucklin & Lajeunesse 1994, Bucklin et al. 1996a, France & Kocher 1996b, Sarver et al. 1998, Cuesta & Schubart 1998, Schubart et al. 1998b), the last study including the description of a new species.

Repeated new discoveries of cryptic species among plant and animal groups by molecular methods, raises the question of whether genetically distinct populations that are morphologically indistinguishable should be described as separate species. This would involve some assumptions and methodological disadvantages (identification only possible after DNA extraction). On the other hand, molecular evolution is a continuous process and isolated populations will always diverge over time. However, morphology does not necessarily change over extended time periods if ecological factors remain constant, as shown by comparison of several trans-isthmian marine invertebrate species. Along with the molecular changes there may be physiological, ecological or behavioral changes that are not readily evident in morphology, as for *Sesarma reticulatum* in studies to date (Zimmerman & Felder 1991, Staton & Felder 1992). Molecular findings can thus 'alert' systematists to potentially distinct populations and instigate studies that might reveal divergence in additional non-molecular characters. However, even without those additional findings, molecular systematics should be viewed as a highly effective tool for documenting and classifying the biological diversity of this world.

ACKNOWLEDGEMENTS

This paper was presented as an oral contribution at the Fourth International Crustacean Congress in Amsterdam in July 1998 and we wish to thank the organizers of this meeting. The present study was funded by the U.S. Department of Energy (grant no. DE-FG02-97ER12220). We extend our gratitude to Lanie Bilodeau, Pedro Castro, Keith Crandall, José Antonio Cuesta, Rudolf Diesel, Danièle Guinot, Michelle Harrison, Carol Lee, Inmaculada López de la Rosa, Harriet Perry, Eduardo Spivak, and Kit Tam for providing information, material, or sequences, assisting during collections, or identifying specimens. Jun Kitaura, Mutsumi Nishida, and Keiji Wada kindly made several complete 16Sar-br sequences available for the brachyuran phylogeny. Some sequences were generated in the laboratory of Blair Hedges (Pennsylvania State University), when CDS was funded by the Deutsche Forschungsgemeinschaft (grant Di 479-2). Danièle Guinot, Christoph Held, and one anonymous referee assisted with valuable comments on the manuscript. This is contribution no. 65 from the USL Laboratory for Crustacean Research.

REFERENCES

Avise, J.C., Arnold, J., Ball, M., Bermingham, E., Lamb, T., Neigel, J.E., Reeb, C.A. & Saunders, N.C. 1987. Intraspecific phylogeography: the mitochondrial DNA bridge between population genetics and systematics. *Ann. Rev. Ecol. Syst.* 18: 489-522.

Batuecas, B., Garesse, R., Calleja, M., Valverde, J.R. & Marco, R. 1988. Genome organization of *Artemia* mitochondrial DNA. *Nucleic Acids Res.* 16(14): 6515-6529.

Bucklin, A. & Kocher, T.D. 1996. Source regions for recruitment of *Calanus finmarchicus* to Georges Bank: evidence from molecular population genetic analysis of mtDNA. *Deep-Sea Res. II* 43: 1665-1681.

Bucklin, A. & Lajeunesse, T.C. 1994. Molecular genetic variation of *Calanus pacificus* (Copepoda: Calanoida): preliminary evaluation of genetic structure and subspecific differentiation based on mtDNA sequences. *Calif. Coop. Ocean. Fish. Invest. Rep.* 35: 45-51.

Bucklin, A., Bentley, A.M. & Franzen, S.P. 1998a. Distribution and relative abundance of *Pseudocalanus moultoni* and *P. newmani* (Copepoda: Calanoida) on Georges Bank using molecular identification of sibling species. *Mar. Biol.* 132: 97-106.

Bucklin, A., Caudill, C.C. & Guarnieri, M. 1998b. Population genetics and phylogeny of marine planktonic copepods. In K.C. Cooksey (ed.), *Molecular approaches to the study of the ocean*: 303-318. London: Chapman and Hall.

Bucklin, A., Frost, B.W. & Kocher, T.D. 1992. DNA sequence variation of the mitochondrial 16S rRNA in *Calanus* (Copepoda; Calanoida): intraspecific and interspecific patterns. *Molec. Mar. Biol. Biotechnol.* 1: 397-407.

Bucklin, A., Frost, B.W. & Kocher, T.D. 1995. Molecular systematics of six *Calanus* and three *Metridia* species (Calanoida: Copepoda). *Mar. Biol.* 121: 655-664.

Bucklin, A., Lajeunesse, T.C., Curry, E., Wallinga, J. & Garrison, K. 1996a. Molecular genetic diversity of the copepod, *Nannocalanus minor*: genetic evidence of species and population structure in the N. Atlantic Ocean. *J. Mar. Res.* 54: 285-310.

Bucklin, A., Sundt, R. & Dahle, G. 1996b. Population genetics of *Calanus finmarchicus* in the North Atlantic. *Ophelia* 44: 29-45.

Cabot, E.L. & Beckenbach, A.T. 1989. Simultaneous editing of multiple nucleic acid and protein sequences with ESEE. *Comput. Appl. Biosci.* 5: 233-234.

Cann, R.L., Stoneking, M. & Wilson, A.C. 1987. Mitochondrial DNA and human evolution. *Nature* 325: 31-36.

Chu, K.H., Chan, T.-Y. & Lavery, S. 1998. Mitochondrial DNA phylogenies for the genus *Penaeus* (Decapoda: Penaeidae). *Proceedings and Abstracts of the 4th International Crustacean Congress, Amsterdam*: 91.

Crandall, K.A. 1998. Conservation phylogenetics of Ozark crayfishes: assigning priorities for aquatic habitat protection. *Biological Conservation* 84: 107-117.

Crandall, K.A. & Fitzpatrick Jr, J.E. 1996. Crayfish molecular systematics: using a combination of procedures to estimate phylogeny. *Syst. Biol.* 45(1): 1-26.

Crandall, K.A., Lawler, S.H. & Austin, C.M. 1995. A preliminary examination of the molecular phylogenetic relationships of some crayfish genera from Australia (Decapoda: Parastacidae). *Freshwater Crayfish* 10: 18-30.

Cuesta, J.A. & Schubart, C.D. 1998. Morphological and molecular differentiation between three allopatric populations of the littoral crab *Pachygrapsus transversus* (Gibbes, 1850) (Brachyura: Grapsidae). *J. Nat. Hist.* 32: 1499-1508.

Cunningham, C.W., Blackstone, N.W. & Buss, L.W. 1992. Evolution of king crabs from hermit crab ancestors. *Nature* 355: 539-542.

Felder, D.L. & Staton, J.L. 1994. Genetic differentiation in trans-Floridian species complexes of *Sesarma* and *Uca* (Decapoda: Brachyura). *J. Crust. Biol.* 14: 191-209.

France, S.C. & Kocher, T.D. 1996a. DNA sequencing of formalin-fixed crustaceans from archival research collections. *Mol. Mar. Biol. Biotechnol.* 5(4): 304-313.

France, S.C. & Kocher, T.D. 1996b. Geographic and bathymetric patterns of mitochondrial 16S rRNA sequence divergence among deep-sea amphipods, *Eurythenes gryllus*. *Mar. Biol.* 126(4): 633-643.

Geller, J.B., Walton, E.D., Grosholz, E.D. & Ruiz, G.M. 1997. Cryptic invasions of the crab *Carcinus* detected by molecular phylogeography. *Mol. Ecol.* 6: 901-906.

Glaessner, M.F. 1969. Decapoda. In R.C. Moore (ed.), *Treatise on invertebrate paleontology. Part R, Arthropoda 4*: 399-533. Lawrence: University of Kansas Press.

Guinot, D. 1978. Principes d'une classification évolutive des Crustacés Décapodes Brachyoures. *Bull. biol. Fr. Belgique* 112(3): 211-292.

Held, C. (in press). On the phylogeny of serolid isopods (Crustacea, Isopoda, Serolidae) and the use of ribosomal expansion segments in molecular systematics. *Mol. Phyl. Evol.*

Kitaura, J., Wada, K. & Nishida, M. 1998. Molecular phylogeny and evolution of unique mud-using territorial behavior in ocypodid crabs (Crustacea: Brachyura: Ocypodidae). *Mol. Biol. Evol.* 15(6): 626-637.

Kjer, K.M. 1995. Use of rRNA secondary structure in phylogenetic studies to identify homologous positions: an example of alignment and data presentation from the frogs. *Mol. Phylogen. Evol.* 4(3): 395-407.

Kocher, T.D., Thomas, W.K., Meyer, A., Edwards, S.V., Pääbo, S., Villablanca, F.X. & Wilson, A.C. 1989. Dynamics of mitochondrial DNA evolution in animals: amplification and sequencing with conserved primers. *Proc. Natl. Acad. Sci. USA* 86: 6196-6200.

Kumar, S., Tamura, K. & Nei, M. 1993. *MEGA: Molecular evolutionary genetics analysis, version 1.01*. University Park: The Pennsylvania State University.

Lawler, S.H. & Crandall, K.A. 1998. The relationship of the Australian freshwater crayfish genera *Euastacus* and *Astacopsis*. *Proc. Linn. Soc. N.S.W.* 119: 1-8.

Levinton, J., Sturmbauer, C. & Christy, J. 1996. Molecular data and biogeography: Resolution of a controversy over evolutionary history of a pan-tropical group of invertebrates. *J. Exp. Mar. Biol. Ecol.* 203: 117-131.

Machado, E.G., Dennebouy, N., Suarez, M.O., Mounolou, J.-C. & Monnerot, M. 1993. Mitochondrial 16S-rRNA gene of two species of shrimps: sequence variability and secondary structure. *Crustaceana* 65: 279-286.

Mangum, C.P. & McKenney, A.L. 1996. Subunit composition of the crustacean hemocyanins: divergence in incipient speciation. *Biol. Bull.* 191: 33-41.

Moritz, C., Dowling, T.E. & Brown, W.M. 1987. Evolution of animal mitochondrial DNA: Relevance for population biology and systematics. *Ann. Rev. Ecol. Syst.* 18: 269-292.

Palmero, I., Renart, J. & Sastre, L. 1988. Isolation of cDNA clones coding for mitochondrial 16S ribosomal RNA from the crustacean *Artemia*. *Gene* 68: 239-248.

Palumbi, S.R., Martin, A., Romano, S., Mcmillan, W.O., Stice, L. & Grabowski, G. 1991. *The simple fool's guide to PCR. A collection of PCR protocols, version 2*. Honolulu: University of Hawaii.

Patarnello, T., Bargelloni, L., Varotto, V. & Battaglia, B. 1996. Krill evolution and the Antarctic ocean currents: evidence of vicariant speciation as inferred by molecular data. *Mar. Biol.* 126: 603-608.

Ponniah, M. & Hughes, J.M. 1998. Evolution of Queensland spiny mountain crayfish of the genus *Euastacus* Clark (Decapoda: Parastacidae): preliminary 16s mtDNA phylogeny. *Proc. Linn. Soc. N.S.W.* 119: 9-19.

Rzhetsky, A. & Nei, M. 1992. A simple method for estimating and testing minimum-evolution trees. *Mol. Biol. Evol.* 9: 945-967.

Saitou, N. & Nei, M. 1987. The neighbor-joining method: a new method for reconstructing phylogenetic trees. *Mol. Biol. Evol.* 4: 406-425.

Sarver, S.K., Silberman, J.D. & Walsh, P.J. 1998. Mitochondrial DNA sequence evidence supporting the recognition of two subspecies or species of the Florida spiny lobster *Panulirus argus. J. Crust. Biol.* 18(1): 177-186.

Schneider-Broussard, R. & Neigel, J.E. 1997. A large subunit mitochondrial ribosomal DNA sequence translocated to the nuclear genome of two stone crabs (*Menippe*). *Mol. Biol. Evol.* 14(2): 156-165.

Schneider-Broussard, R., Felder, D.L., Chlan, C.A. & Neigel, J.E. 1998. Tests of phylogeographic models with nuclear and mitochondrial DNA sequence variation in the stone crabs, *Menippe adina* and *M. mercenaria. Evolution* 52: 1671-1678.

Schram, F.R. 1986. *Crustacea.* Oxford: Oxford University Press.

Schubart, C.D., Diesel, R. & Cuesta, J.A. (in press). Molecular phylogeny, taxonomy, and evolution to terrestrial life within the American Grapsoidea (Crustacea: Brachyura). *Mol. Phyl. Evol.*

Schubart, C.D., Diesel, R. & Hedges, S.B. 1998a. Rapid evolution to terrestrial life in Jamaican crabs. *Nature* 393: 363-365.

Schubart, C.D., Reimer, J. & Diesel, R. 1998b. Morphological and molecular evidence for a new endemic freshwater crab, *Sesarma ayatum* sp. n., (Grapsidae, Sesarminae) from eastern Jamaica. *Zool. Scr.* 27: 373-380.

Spears, T., Abele, L.G. & Kim, W. 1992. The monophyly of brachyuran crabs: a phylogenetic study based on 18S rRNA. *Syst. Biol.* 41(4): 446-461.

Staton, J.L. & Felder, D.L. 1992. Osmoregulatory capacities in disjunct western Atlantic populations of the *Sesarma reticulatum* complex (Decapoda: Grapsidae). *J. Crust. Biol.* 12: 335-341.

Sturmbauer, C., Levinton, J.S. & Christy, J. 1996. Molecular phylogeny analysis of fiddler crabs: Test of the hypothesis of increasing behavioral complexity in evolution. *Proc. Natl. Acad. Sci. USA* 93: 10855-10857.

Swofford, D.L. 1993. *Phylogenetic analysis using parsimony (PAUP), version 3.1.1.* Champaign: University of Illinois.

Tam, Y.K. & Kornfield, I. 1998. Phylogenetic relationships of clawed lobster genera (Decapoda: Nephropidae) based on mitochondrial 16S rRNA gene sequences. *J. Crust. Biol.* 18(1): 138-146.

Tam, Y.K., Kornfield, I. & Ojeda, F.P. 1996. Divergence and zoogeography of mole crabs, *Emerita* spp. (Decapoda: Hippidae), in the Americas. *Mar. Biol.* 125: 489-497.

Taylor, D.J., Finston, T.L. & Hebert, P.D.N. 1998. Biogeography of a widespread freshwater crustacean: pseudocongruence and cryptic endemism in the North American *Daphnia laevis* complex. *Evolution* 52: 1648-1670.

Tong, J.G., Chan, T.-Y. & Chu, K.H. (in press). Phylogenetic analysis of some *Metapenaeopsis* species (Crustacea: Decapoda: Penaeidae) from the Indo-West Pacific based on mitochondrial DNA sequences. *J. Crust. Biol.*

Valverde, J.R., Batuecas, B., Moratilla, C., Marco, R. & Garesse, R. 1994. The complete mitochondrial DNA sequence of the crustacean *Artemia franciscana.* J. Mol. Evol. 39: 400-408.

Xiong, B. & Kocher, T.D. 1991. Comparison of mitochondrial DNA sequences of seven morphospecies of black flies (Diptera: Simuliidae). *Genome* 34: 306-311.

Zimmerman, T.L. & Felder, D.L. 1991. Reproductive ecology of an intertidal brachyuran crab, *Sesarma* sp. (nr. *reticulatum*), from the Gulf of Mexico. *Biol. Bull.* 181: 387-401.

List of contributors

Almelkar 'Ashram' Saraswatpur, G.B.: Dharwad-580 004, Karnataka State, India.

Araújo Flávio, M.P.: Departamento de Zoologia, IB, Universidade Estadual de Campinas, CxP 6109, CEP 13083-970 Campinas, SP, Brasil.

Arredondo-Figueroa, Jose Luis: Division de Ciencias Biologicas y de la Salud. Universidad Autonoma Metropolitana Iztapalapa PO Box. 55-535, Iztapalapa, Mexico DF 09340, Mexico.

Aubyn António St.: Dep. Matemática, Instituto Superior de Agronomia, P-1399 Lisboa Codex, Portugal.

Bailey, G.W.: Sea Fisheries Research Institute, Private Bag X2, Rogge Bay 8012, Cape Town, South Africa.

Barbaresi, Silvia: Dipartimento di Biologia Animale e Genetica, Via Romana 17, I-50125 Florence, Italy.

Bashirullah, Abul K.: Instituto Oceanográfico de Venezuela, Universidad de Oriente, Apartado 245, Cumana 6101, Venezuela.

Bertini, Giovana: NEBECC, Departamento de Zoologia, Instituto de Biociências, Universidade Estadual Paulista, CP 510, 18618-000 Botucatu (São Paulo), Brazil.

Borg, Joseph A.: Department of Biology, University of Malta, Msida, MSD06, Malta.

Brailovsky, P.: Departamento El Hombre y su Ambiente, Universidad Autónoma Metropolitana-Xochimilco, Calz. Hueso 1100, Col. Villa Quietud, Mexico DF, CP 04960, Mexico.

Brendonck, Luc: Laboratory of Aquatic Ecology, KU Leuven, Naamsestraat 59, B-3000 Leuven, Belgium.

Bueno, Sérgio L.S.: Departamento de Zoologia, Instituto de Biociências, Universidade de São Paulo, Rua do Matão, travessa 14 no. 101, CEP 05508-900 São Paulo, Brazil.

Calazans, Danilo: Departamento de Oceanografia, FURG, Cx. Postal 474, CEP 96201-900, Rio Grande, RS, Brasil.

Cano, Esperanza: Laboratorio de Zoología Aplicada, Dpto. Fisiología y Biología Animal, Fac. Biología, Univ. Sevilla, Avda. Reina Mercedes 6, E-41012 Seville, Spain.

Carmona-Suárez, C.A.: Centro de Ecología. Instituto Venezolano de Investigaciones Científicas, Apartado 21827, Caracas 1020-A, Venezuela.

Castro, Margarida: Centro de Ciências do Mar (CCMAR), Universidade do Algarve, Campus de Gambelas P-8000 Faro, Portugal.

Castro, Peter: Biological Sciences Department, California State Polytechnic University, Pomona, CA 91768, USA.

Cau, Angelo: Dipartimento di Biologia Animale ed Ecologia, Università agli Studi di Cagliari, Viale Poetto, I-09126 Cagliari, Italy.

Cerrano, Carlo: Dipartimento di Biologia Animale e dell'Uomo, Università di Torino, Via Accademia Albertina 17, I-10123 Torino, Italy.

Chen, Q.C.: South China Sea Institute of Oceanology, Academia Sinica, Guangzhou, China.

Clark, P.F.: Department of Zoology, The Natural History Museum, Cromwell Road, London SW7 5BD, UK.

Cobo, Valter José: Dept. de Biologia, Universidade de Taubaté – UNITAU, PCA Marcelino Monteiro 63, CEP 12020-270, Taubate, Brazil.

Cockcroft, A.C.: Marine and Coastal Management, Private Bag X2, Rogge Bay 8012, Cape Town, South Africa.

Correa, Michele O.D.: NEBECC, Departamento de Zoologia, Instituto de Biociências – Universidade Estadual Paulista, UNESP, 18618-000 Botucatu (São Paulo), Brazil.

Cristo, M.: Universidade do Algarve, UCTRA, Campus de Gambelas, P-8000 Faro, Portugal.

Cruz-Hernández, Javier: Licenciatura en Biología, División de Ciencias Biológicas y de la Salud, UAM Xochimilco, Mexico.

Cuccu, D.: Dipartimento di Biologia Animale ed Ecologia, Università agli Studi di Cagliari, Viale Poetto, I-09126 Cagliari, Italy.

Cunha, Marina R. da: Departamento de Biologia, Universidade de Aveiro, P-3810, Aveiro, Portugal.

Cvetkovic–Milicic, Dragana: Faculty of Biology, Institute of Zoology, Studentski trg 16, Belgrade, Yugoslavia.

Damiano, Maria Teresa: Dipartimento di Biologia Animale e dell'Uomo, Università di Torino, Via Accademia Albertina 17, I-10123 Italy.

Dappiano, Marco: Istituto di Zoologia, Università di Genova, Via Balbi 5, I-16125 Genova, Italy.

Donaldson, William E.: Alaska Dept. of Fish and Game, 211 Mission Rd., Kodiak AK 99615, USA.

Emmerson, W.D.: Department of Zoology, University of Transkei, P. Bag X1, Umtata, East Cape, South Africa.

Encarnação, Pedro: Centro de Ciências do Mar (CCMAR), Universidade do Algarve, Campus de Gambelas P-8000 Faro, Portugal.

Esteban, A.: IEO, Centro Oceanográfico de Murcia, San Pedro del Pinatar (Murcia), Spain.

Felder, Darryl L.: Department of Biology and Laboratory for Crustacean Research, University of Louisiana, Lafayette, LA 70504-2451, USA.

Ferrero, Enrico A.: BRAIN Center for Neuroscience, Departement of Biology, University of Trieste. Via Giorgieri 7, Trieste, I-34127 Italy.

Flores, E.: Phytoplankton Laboratory, ICMYL-UNAM, AP 173, Mexico DF, 04510, Mexico.

Follesa, M.C.: Dipartimento di Biologia Animale ed Ecologia, Università agli Studi di Cagliari, Viale Poetto, I-09126 Cagliari, Italy.

Fonseca, Duane B.: Departamento de Oceanografia, FURG, CP 474, 96201-900, Rio Grande, RS, Brazil.

Fransozo, Adilson: NEBECC, Departamento de Zoologia, Instituto de Biociências, Universidade Estadual Paulista, CP 510, 18618-000 Botucatu (São Paulo), Brazil.

Fu, Y.Y.: Department of Biology, The Chinese University of Hong Kong, Shatin, Hong Kong, China.

Galil, Bella S.: Israel Oceanographic and Limnological Research, National Institute of Oceanography, PO Box 8030, Haifa 31080, Israel.

Galvão, Renata: Departamento de Zoologia, Instituto de Biociências, Universidade de São Paulo, Rua do Matão, travessa 14 no 101. CEP: 05508-900 São Paulo, Brazil

García-Rodriguez, M.: IEO, Centro Oceanográfico de Murcia, Varadero 1, E-30740, San Pedro del Pinatar (Murcia), Spain.

Gaten, Edward: Department of Biology, University of Leicester, University Road, Leicester, LE1 7RH, UK.

Geurden, Inge: Laboratory of Aquaculture and *Artemia* Reference Center, University of Gent, Rozier 44, B-9000 Gent, Belgium.

Gherardi, Francesca: Dipartimento di Biologia Animale e Genetica, Via Romana 17, I-50125 Florence, Italy.

Giller, Paul S.: Department of Zoology & Animal Ecology, Lee Maltings, University College, Cork, Ireland.

Giordano, Daniela: Istituto Sperimentale Talassografico CNR, Spianata San Raineri 86, I-68127 Messina, Italy.

Gonçalves, F.: Departamento de Biologia da Universidade de Aveiro, Campus Universitário de Santiago, P-3810 Aveiro, Portugal.

Greco, Silvestro: Istituto Sperimentale Talassografico CNR, Spianata San Raineri 86, I-68127 Messina, Italy.

Gubanov, Vladimir G.: Institute of Biophysics, Academtown, Krasnoyarsk, 660036, Russia.

Haaga, Jan A.: National Marine Fisheries Service, Kodiak Laboratory, PO Box 1638, Kodiak, AK 99615, USA.

Han Kyung-Min: Laboratory of Aquaculture and Artemia Reference Center, University of Gent, Rozier 44, B-9000 Gent, Belgium.

Haney, James F.: Dept. of Zoology, University of New Hampshire, Durham, New Hampshire 03824, USA.

Hansen, Brita: School of Zoology, University of Tasmania, GPO Box 252-05, Hobart, Tasmania 7001, Australia.

Hartnoll, Richard G.: Port Erin Marine Laboratory, University of Liverpool, Isle of Man IM9 6JA, British Isles.

Henderson, Suzanne M.: FPRC, Buckinghamshire College, Queen Alexandra Road, High Wycombe, Buckinghamshire, HP11 2JZ, UK.

Hereu, Clara María: Departamento de Oceanografia, FURG, Cx. Postal 474, CEP 96201-900, Rio Grande, RS, Brasil.

Hernández-Guzmán, Martha A.: Licenciatura en Biología, División de Ciencias Biológicas y de la Salud, UAM Xochimilco, Mexico.

Herring, Peter J.: Southampton Oceanography Centre, Empress Dock, Southampton SO14 3ZH, UK.

Holdich, David M.: School of Biological Sciences, The University of Nottingham, Nottingham NG7 2RD, UK.

Huntingford, Felicity A.: Division of Environmental & Evolutionary Biology, Graham Kerr Building, University of Glasgow, G12 8QQ UK.

d'Incao, Fernando: Departamento de Oceanografia, FURG, CP 474, 96201-900, Rio Grande, RS, Brazil.

Jacobucci, Giuliano B.: Departamento de Zoologia, IB, Universidade Estadual de Campinas, Cx Postal 6109, CEP 13083-970 Campinas, SP, Brazil.

Jagadisha, K.: Department of Zoology, Vijaya College, Mulki-574 154, Karnataka State, India.

Jalihal, D.R.: College of Fisheries, Shirgoan, Ratnagiri-415 629, Maharashtra State, India.

Jamieson, Glen S.: Science Branch, Department of Fisheries and Oceans, Pacific Biological Station, Nanaimo, BC, V9R 5K6, Canada.

Jazdzewski, Krzysztof: Department of Invertebrate Zoology & Hydrobiology, University of Lódz, 12/16 Banacha Street, PL-90-237 Lódz, Poland.

Jeng, Ming-Shiou: Laboratory of Marine Biology, Institute of Zoology, Academia Sinica, Taipei, Taiwan.

Jerí, Teresa: Alfred Wegener Institute for Polar and Marine Research, Columbusstrasse, D-27568, Bremerhaven, Germany.

Jiménez, Agustina: Laboratorio de Zoología Aplicada, Dpto. Fisiología y Biología Animal, Fac. Biología, Univ. Sevilla, Avda. Reina Mercedes 6, E-41012 Seville, Spain.

Johnson, Magnus L.: Department of Biology, University of Leicester, University Road, Leicester, LE1 7RH, UK.

Kapiris, K.: National Centre for Marine Research, Institute of Marine Biological Resources, Agios Kosmas, GR-166 04 Hellinikon, Athens, Greece.

Kelleher, Barry: Laboratory of Aquatic Ecology, University of Nijmegen, Post Box 9010, NL-6500 GL Nijmegen, the Netherlands.

Konopacka, Alicja: Department of Invertebrate Zoology & Hydrobiology, University of Lódz, 12/16 Banacha Street, PL-90-237 Lódz, Poland.

Leite, Fosca P.P.: Departamento de Zoologia, IB, Universidade Estadual de Campinas, Cx Postal 6109, CEP 13083-970 Campinas, SP, Brasil.

Li Li: The Swire Institute of Marine Science, Department of Ecology and Biodiversity, The University of Hong Kong, Cape d'Aguilar, Shek O, Hong Kong.

Licea, S.: Phytoplankton Laboratory, ICMYL-UNAM, AP 173, Mexico DF, 04510, Mexico.

Lorenzon, Simonetta: BRAIN Center for Neuroscience, Departement of Biology, University of Trieste. Via Giorgieri 7, I-34127 Trieste, Italy.

Loya-Javellana, Gilda N.: Southeast Asian Fisheries Development Center, Aquaculture Department, Tigbauan, Iloilo 5021, The Philippines.

Maiorana, Giuseppe: Dipartimento di Biologia Animale e dell'Uomo, Università di Torino, Via Accademia Albertina 17, I-10123 Turin, Italy.

Manickchand-Heileman, S.: Benthic Ecology Laboratory, ICMYL-UNAM, A.P. 173, Mexico DF, 04510, Mexico.

Manning, Raymond B.: Department of Invertebrate Zoology, National Museum of Natural History, Smithsonian Institution, Washington, DC 20560-0163, USA.

Mantelatto, Fernando Luis Medina: NEBECC, Departamento de Biologia, Faculdade de Filosofia, Ciências e Letras de Ribeirão Preto, Universidade de São Paulo (USP), Av. Bandeirantes 3900, 14040-901, Ribeirão Preto (SP), Brazil.

Marano, Chiara Alessandra: Laboratorio di Biologia Marina, Molo Pizzoli (Porto), I-70123 Bari, Italy.

Marano, Giovanni: Laboratorio Provinciale di Biologia Marina, Molo Pizzoli (Porto) I-70123 Bari, Italy.

Marsan, Roberta: Laboratorio di Biologia Marina, Molo Pizzoli (Porto), I-70123 Bari, Italy.

Marzano, M. Cristina: Laboratorio di Biologia Marina, Molo Pizzoli (Porto), I-70123 Bari, Italy.

McKenzie, Kenneth G.: Science & Technology, Charles Sturt University, Riverina, Wagga Wagga, New South Wales, Australia 2678.

Mejía-Ortíz, L.M.: ADEFRNR, Dept. El hombre y su ambiente, U.A.M.–Xochimilco. Apdo.Post. 70-458, C. Universitaria, Mexico DF, 04510, Mexico.

Moraitou-Apostolopoulou, M.: Dept. of Zoology-Marine Biology, School of Biology, University of Athens, Panepistimioupolis, GR-157 84 Athens, Greece.

Moreira, Maria Helena: Departamento de Biologia, Universidade de Aveiro, P-3810 Aveiro, Portugal.

Moreno-Rodriguez, Miguel Angel: Planta Experimental de Produccion Acuicola, Departamento de Hidrobiologia, Division de Ciencias Biologicas y de la Salud, Universidad Autonoma Metropolitana Iztapalapa, PO Box 55-535, Mexico DF, 09340, Mexico.

Mura Marco: Department of Animal Biology and Ecology, University of Cagliari, Viale Poetto 1, I-09126 Cagliari, Italy.

Murenu, Matteo: Department of Animal Biology and Ecology, University of Cagliari, Viale Poetto 1, I-09126 Cagliari, Italy.

Ilona B. Muskó: Balaton Limnological Research Institute, Hungarian Academy of Sciences, H-8237 Tihany, Hungary.

Nakagaki, J.M.: NEBECC, Departamento de Zoologia, Instituto de Biociências, Universidade Estadual Paulista, CP 510, 18618-000 Botucatu (São Paulo), Brazil.

Negreiros-Fransozo, M.L.: NEBECC, Departamento de Zoologia, Instituto de Biociências, Universidade Estadual Paulista, CP 510, 18618-000 Botucatu (São Paulo), Brazil.

Neigel, Joseph E.: Department of Biology and Laboratory for Crustacean Research University of Southwestern Louisiana, Lafayette, LA 70504-2451, USA.

Nilssen, E.M.: Norwegian College of Fisheries, University of Tromsø, N-9037 Tromsø, Norway.

Ocete, Ma Elvira: Laboratorio de Zoología Aplicada, Dpto. Fisiología y Biología Animal, Fac. Biología, Univ. Sevilla, Avda. Reina Mercedes 6, E-41012 Seville, Spain.

Ortega, P.: Departamento de Física y Química Teórica, Facultad de Química, UNAM, Mexico DF, Mexico.

Ortega Mendoza, Ana Laura: Departamento El Hombre y su Ambiente, Universidad Autónoma Metropolitana-Xochimilco, Calz. Hueso 1100, Col. Villa Quietud, Mexico DF, 04960, Mexico.

Papaconstantinou, C.: National Centre for Marine Research, Institute of Marine Biological Resources, Agios Kosmas, GR-166 04 Hellinikon, Athens, Greece.

Pasqual, Paola: BRAIN Center for Neuroscience, Departement of Biology, University of Trieste, Via Giorgieri 7, I-34127 Trieste, Italy.

Passarella, Giuseppe: CNR, Istituto Ricerca Sulle Acque, Viale De Blasio, I-70123 Bari, Italy.

Pastorelli, Anna Maria: Laboratorio di Biologia Marina, Molo Pizzoli (Porto), I-70123 Bari, Italy.

Perdichizzi, Francesco: Istituto Sperimentale Talassografico CNR, Spianata San Raineri 86, I-68127 Messina, Italy.

Pessani, Daniela: Dipartimento di Biologia Animale e dell'Uomo, Università di Torino, Via Accademia Albertina 17, I-10123 Torino, Italy.

Petrakis, G.: National Centre for Marine Research, Institute of Marine Biological Resources, Agios Kosmas, GR-166 04 Hellinikon, Athens, Greece.

Petrov, Brigita: Faculty of Biology, Institute of Zoology, Studentski trg 16, Belgrade, Yugoslavia.

Pitcher, G.C.: Sea Fisheries Research Institute, Private Bag X2, Rogge Bay 8012, Cape Town, South Africa.

Ponce-Palafox, Jesus T.: Division de Ciencias Biologicas y de la Salud, Universidad Autónoma Metropolitana, Iztapalapa, AP 55-535, Mexico DF, 09340, Mexico.

Raddi, Andrea: Dipartimento di Biologia Animale e Genetica, Via Romana 17, I-50125 Florence, Italy.

Rae, Gordon H.: Scottish Salmon Growers Association, Drummond House, Scott Street, Perth, PH1 5EJ, Scotland.

Rafter, E.E.: Polaria, N-9005 Tromsø, Norway.

Rainbow, P.S.: Department of Zoology, The Natural History Museum, Cromwell Road, London SW7 5BD, UK.

Rajagopal, Sanjeevi: Laboratory of Aquatic Ecology, University of Nijmegen, Post Box 9010, NL-6500 GL Nijmegen, the Netherlands.

Raposo, Pedro: Instituto do Ambiente e Vida, Departamento de Zoologia, Universidade de Coimbra, P-3000 Coimbra, Portugal.

Reed, Karen J.: Department of Invertebrate Zoology, National Museum of Natural History, Smithsonian Institution, Washington, DC 20560-0163, USA.

Reigada, A.L.D.: NEBECC, Departamento de Zoologia, Instituto de Biociências – Universidade Estadual Paulista, CP 510, 18618-000 Botucatu (São Paulo), Brazil.

Reis, João: Centro de Ciências do Mar (CCMAR), Universidade do Algarve, Campus de Gambelas P-8000 Faro, Portugal.

Ribeiro, R.: Instituto do Ambiente e Vida, Departamento de Zoologia, Universidade de Coimbra, P-3000 Coimbra, Portugal.

Richardson, Alastair M.M.: School of Zoology, University of Tasmania, GPO Box 252-05, Hobart, Tasmania 7001, Australia.

Riddoch, Bruce J.: University of Botswana, Department of Biological Sciences, Private Bag 0022, Gaborone, Botswana.

Rinelli, Paola: Istituto Sperimentale Talassografico CNR, Spianata San Raineri 86, I-68127 Messina, Italia.

Ringelberg, J.: NIOO, Centre for Limnology, Rijksstraatweg 6, NL-3631 AC Nieuwersluis, the Netherlands.

Robbins, R.S.: Department of Zoology, The Natural History Museum, Cromwell Road, London SW7 5BD, UK.

Rodrigues, Sérgio de A.: Departamento de Ecologia Geral, Universidade de São Paulo, Rua do Matao, Travessa 14 no 321, 05508-900 São Paulo, SP, Brazil.

Salvi, Gabriele: Dipartimento di Biologia Animale e Genetica, Via Romana 17, I-50125 Florence, Italy.

Sankolli, K.N.: College of Fisheries, Shirgoan, Ratnagiri-415 629, Maharashtra State, India.

Santhanam, R.: Fisheries College and Research Institute, Tamilnadu Veterinary and Animal Sciences University, Tuticorin-628 008, India.

Schallreuter, Roger E.L.: Geological-Palaeontological Institute & Museum, University of Hamburg, Bundesstrasse 55, D-20146 Hamburg, Germany.

Schembri, Patrick J.: Department of Biology, University of Malta, Msida, MSD06, Malta.

Schlining, Kyra L.: Moss Landing Marine Laboratories, PO Box 450, Moss Landing, CA 95039, USA; Monterey Bay Aquarium Research Institute, PO Box 628, Moss Landing, CA 95039 0628, USA.

Schoeman, D.S.: Sea Fisheries Research Institute, Private Bag X2, Rogge Bay 8012, Cape Town, South Africa.

Schubart, Christoph D.: Department of Biology and Laboratory for Crustacean Research University of Southwestern Louisiana, Lafayette, LA 70504-2451, USA.

Secci, E.: Dipartimento di Biologia Animale ed Ecologia, Università agli Studi di Cagliari, Viale Poetto 1, I-09126 Cagliari, Italy.

Serafim, Paula: Dep. Matemática, Instituto Superior de Agronomia, P-1399 Lisboa Codex, Portugal.

Shelton, Peter M.J.: Department of Biology, University of Leicester, University Road, Leicester, LE1 7RH, UK.

Shenoy, Shakuntala: College of Fisheries, Konkan Agricultural University, Shirgoan, Ratnagiri-415 629, Maharashtra State, India.

Shimizu, Roberto M.: Departamento de Ecologia Geral, Universidade de São Paulo, Rua do Matao, Travessa 14 no 321, 05508-900 São Paulo, SP, Brazil.

Signoret, Gisèle: Departamento El Hombre y su Ambiente, Universidad Autónoma Metropolitana-Xochimilco, Calz. Hueso 1100, Col. Villa Quietud, Mexico DF, CP 04960, Mexico.

Smith, B.D.: Department of Zoology, The Natural History Museum, Cromwell Road, London SW7 5BD, UK.

Sneddon, Lynne U.: Division of Environmental & Evolutionary Biology, Graham Kerr Building, University of Glasgow, G12 8QQ, Scotland.

Soares, A.M.V.M.: Instituto do Ambiente e Vida, Departamento de Zoologia, Universidade de Coimbra, P-3000 Coimbra, Portugal.

Sorbe, Jean Claude: Laboratoire d'Océanographie Biologique, CNRS URA 197, 2 Rue Jolyet, F-33120 Arcachon, France.

Sorgeloos, Patrick: Laboratory of Aquaculture and Artemia Reference Center, University of Gent, Rozier 44, B-9000 Gent, Belgium.

Soto, L.A.: Benthic Ecology Laboratory, ICMYL-UNAM, AP 173, Mexico DF 04510, Mexico.

Spratt, Jerome D.: California Department of Fish and Game, Monterey Unit, 20 Lower Ragsdale Drive, Monterey, CA 93940, USA.

Srinivasan, A.: Fisheries College and Research Institute, Tamilnadu Veterinary and Animal Sciences University, Tuticorin-628 008, South India, India.

Stevens, Bradley G.: National Marine Fisheries Service, Kodiak Fisheries Research Centre, 301 Research Ct., Kodiak, AK 99615, USA.

Sundet, J.H.: Norwegian Institute of Fisheries and Aquaculture Ltd., N-9005 Tromsø, Norway.

Swain, Roy.: School of Zoology, University of Tasmania, GPO Box 252-05, Hobart 7001, Australia.

Swanson, Kerry M.: Geology Department, University of Canterbury, Private bag 4800, Christchurch, New Zealand.

Tam, P.F.: Department of Biology, The Chinese University of Hong Kong, Shatin, Hong Kong, China.

Taylor, Alan C.: Division of Environmental & Evolutionary Biology, Graham Kerr Building, University of Glasgow, G12 8QQ U.K.

Thessalou-Legaki, M.: Dept. of Zoology-Marine Biology, School of Biology, University of Athens, Panepistimioupolis, GR-157 84, Athens, Greece.

Thiel, Martin: Facultad de Ciencias del Mar, Universidad Católica del Norte, Larrondo 1281, Campus Guayacan, Coquimbo, Chile.

Tirelli, Tina: Dipartimento di Biologia Animale e dell'Uomo, Università di Torino, Via Accademia Albertina 17, I-10123 Torino, Italy.

Tremblay, M. John: Invertebrate Fisheries Division, Science Branch, Fisheries and Oceans Canada, Bedford Institute of Oceanography, PO Box 1006, Dartmouth, Nova Scotia B2Y 4A2, Canada.

Trubetskova, Irina L.: Estonian Marine Institute, 18 b Viljandi Rd., 11216 Tallinn, Estonia.

Turra, Alexander: Departamento de Zoologia, IB, Universidade Estadual de Campinas, CxP 6109, CEP 13083-970 Campinas, SP, Brasil.

Ungaro, Nicola: Laboratorio Provinciale di Biologia Marina, Molo Pizzoli (Porto), I-70123 Bari, Italy.

Uwins, Philippa J.R.: University of Queensland, Center for Microscopy and Microanalysis, St. Lucia, Queensland 4072, Australia.

Vaate, Abraham bij de: Institute for Inland Water Management and Waste Water Treatment (RIZA), Post Box 17, NL-8200 AA Lelystad, the Netherlands.

Velde, Gerard van der: Laboratory of Aquatic Ecology, University of Nijmegen, Post Box 9010, NL-6500 GL Nijmegen, the Netherlands.

Vernon-Carter, E. Jaime: Div. De Ciencias Biologicas y de la Salud, Universidad Autónoma Metropolitana-Iztapalapa, AP 55-535, Mexico DF, 09340, Mexico.

Viccon-Pale, J.A.: ADEFRNR, Dept. El hombre y su ambiente, UAM-Xochimilco. Apdo.Post. 70-458, C Universitaria, Mexico DF, 04510, Mexico.

Vlora, Alessandro: Laboratorio Provinciale di Biologia Marina, Molo Pizzoli (Porto), I-70123 Bari, Italy.

Wong, C.K.: Department of Biology, The Chinese University of Hong Kong, Shatin, Hong Kong, China.

Zadereev, Yegor S.: Institute of Biophysics, Academtown, Krasnoyarsk, 660036, Russia.

Zyl, D.L van: Sea Fisheries Research Institute, Private Bag X2, Rogge Bay 8012, Cape Town, South Africa.

Appendix

The Table of Contents of the sister volume to this one issued by Koninklijke Brill NV, Leiden, the Netherlands in 1999 as: *Crustaceans and the Biodiversity Crisis: Proceedings of the Fourth International Crustacean Congress, volume I* – F.R. Schram & J.C. von Vaupel Klein, editors.

Subtheme 2, Invasive Crustaceans; Biogeography

Subtheme 3, Ecology and Behaviour; Larvae

Index